Digital Terrain Analysis

Digital Terrain Analysis

Third edition, revised and enlarged

Igor V. Florinsky
Institute of Mathematical Problems of Biology, Keldysh Institute of Applied Mathematics, Russian Academy of Sciences, Pushchino, Russia

ELSEVIER

ACADEMIC PRESS
An imprint of Elsevier

Academic Press is an imprint of Elsevier
125 London Wall, London EC2Y 5AS, United Kingdom
525 B Street, Suite 1650, San Diego, CA 92101, United States
50 Hampshire Street, 5th Floor, Cambridge, MA 02139, United States

ISBN 978-0-443-24798-9

For information on all Academic Press publications
visit our website at https://www.elsevier.com/books-and-journals

Publisher: Candice Janco
Acquisitions Editor: Peter Llewellyn
Editorial Project Manager: Helena Beauchamp
Production Project Manager: Paul Prasad Chandramohan
Cover Designer: Mark Rogers

Typeset by STRAIVE, India

Contents

Preface to the third edition ix
Acknowledgments xv
Abbreviations and acronyms xvii

1. Digital terrain analysis: Past and present

1.1 Why topography? 1
1.2 A short historical overview 1
1.3 Current advances in digital terrain analysis 3
1.3.1 Factors in the development of geomorphometry 3
1.3.2 Applications 6

Part I
Principles and methods of digital terrain modeling

2. Topographic surface and its characterization

2.1 Topographic surface 17
2.1.1 Definition and limitations 17
2.1.2 Contour line and special points 18
2.1.3 Morphometric variables and their types 20
2.2 Local morphometric variables 24
2.2.1 Overview 24
2.2.2 Gradient 26
2.2.3 Aspect 30
2.2.4 Curvatures 31
2.2.5 Deflections 46
2.3 Nonlocal morphometric variables 49
2.3.1 Catchment and dispersive areas 50
2.3.2 Topographic position index 52
2.4 Structural morphometric variables 52
2.5 Two-field-specific morphometric variables 54
2.5.1 Solar morphometric variables 54
2.5.2 Wind morphometric variables 57
2.5.3 View morphometric variables 59
2.6 Combined morphometric variables 61
2.6.1 Local-nonlocal morphometric variables 61
2.6.2 Local-local morphometric variables 62
2.7 Landform classifications 62
2.7.1 Gaussian landform classification 63
2.7.2 The Efremov-Krcho landform classification 64
2.7.3 The Shary landform classification 67
2.7.4 Other geomorphometric landform classifications 67

3. Digital elevation models

3.1 DEM generation 73
3.1.1 Conventional topographic surveys 73
3.1.2 Kinematic GNSS surveys 75
3.1.3 Stereophotogrammetry of optical remote sensing images 75
3.1.4 Structure-from-motion technique for unmanned survey imagery 78
3.1.5 Laser altimetry 80
3.1.6 Synthetic aperture radar techniques 81
3.1.7 Echo sounding 82
3.1.8 Satellite radar altimetry 82
3.1.9 Optical sensing of bathymetry 85
3.1.10 Soil and geological core drilling 85
3.1.11 Three-dimensional seismic survey 85
3.1.12 Ice and ground penetrating surveys 85
3.1.13 Digitizing of contours 87
3.1.14 Fusion of data from different sources 87

3.2 DEM grids 89
3.2.1 Plane grids 89
3.2.2 Spheroidal grids 91
3.2.3 Dense point clouds 92
3.2.4 Datums and coordinate systems 93
3.3 DEM resolution 94
3.4 DEM interpolation 95
3.4.1 General 95
3.4.2 Selected methods 96

4. Calculation methods

4.1 The Evans-Young method 99
4.2 Calculation of local morphometric variables on a plane square grid 100
4.2.1 Motivation 100
4.2.2 Formulas 102
4.2.3 Method validation 106
4.3 Calculation of local morphometric variables on a spheroidal equal angular grid on a sphere and an ellipsoid of revolution 112
4.3.1 Motivation 112
4.3.2 Formulas 112
4.3.3 Linear sizes of a spheroidal equal angular window 117
4.3.4 Discussion 118
4.4 Calculation of nonlocal morphometric variables 118
4.4.1 Plane square grid 118
4.4.2 Spheroidal equal angular grid on a sphere and an ellipsoid of revolution 120
4.5 Calculation of structural morphometric variables 122
4.5.1 Conventional algorithms 122
4.5.2 Horizontal deflection 123
4.6 Calculation of two-field-specific morphometric variables 125
4.6.1 Calculation of solar morphometric variables 125
4.6.2 Calculation of wind morphometric variables 126
4.6.3 Calculation of view morphometric variables 128
4.7 Calculation of combined morphometric variables 129
4.8 Calculation of landform classifications 131
4.9 Calculations on a triaxial ellipsoid 131
4.9.1 Motivation 131
4.9.2 Coordinate systems 133
4.9.3 Solution for the inverse geodetic problem 134
4.9.4 Length of meridian and parallel arcs 135
4.9.5 Cell area 136
4.9.6 General algorithms for geomorphometric calculations 136

5. Errors and accuracy

5.1 Sources of DEM errors 139
5.1.1 DEMs from conventional topographic surveys 139
5.1.2 DEMs from kinematic GNSS topographic surveys 139
5.1.3 DEMs from stereophotogrammetry of optical remote sensing images 139
5.1.4 UAV- and SfM-based DEMs 140
5.1.5 Lidar DEMs 142
5.1.6 Interferometric DEMs 142
5.1.7 Bathymetric DEMs 143
5.1.8 Subsurface DEMs 144
5.1.9 Contour-based DEMs 144
5.1.10 Fused DEMs 145
5.2 Estimation of DEM accuracy 145
5.3 Calculation accuracy of local morphometric variables 146
5.3.1 Motivation 146
5.3.2 RMSE formulas for local morphometric variables 147
5.3.3 RMSE formulas for the partial derivatives 149
5.3.4 RMSE mapping 151
5.4 Ignoring the sampling theorem 151
5.4.1 Motivation 151
5.4.2 Materials and data processing 154
5.4.3 Results and discussion 154
5.5 The Gibbs phenomenon 157
5.5.1 Motivation 157
5.5.2 Materials and data processing 158
5.5.3 Results and discussion 158
5.6 Grid displacement 160
5.6.1 Motivation 160
5.6.2 Materials and data processing 160
5.6.3 Results and discussion 161
5.7 Linear artifacts 163
5.7.1 Motivation 163
5.7.2 Isotropy of local morphometric variables 164

6. Filtering

6.1 Tasks of DTM filtering 167
6.1.1 Decomposition of the topographic surface 167
6.1.2 Denoising 167
6.1.3 Generalization 168
6.2 Methods of DTM filtering 170
6.2.1 Trend-surface analysis 170
6.2.2 The Filosofov method 171
6.2.3 Two-dimensional discrete Fourier transform 172
6.2.4 Two-dimensional discrete wavelet transform 173
6.2.5 Smoothing 175
6.2.6 Data elimination 178
6.2.7 Cutting method 179
6.3 Two-dimensional singular spectrum analysis 181
6.3.1 Formulas 182
6.3.2 Materials and data processing 184
6.3.3 Results and discussion 186

7. Universal spectral analytical modeling

7.1 Motivation 191
7.2 Method 191
7.2.1 Calculation of expansion coefficients 192
7.2.2 The Fejér summation 192
7.2.3 Reconstruction of the approximated function 193
7.2.4 Calculation of derivatives 193
7.3 Algorithm 194
7.4 Materials and data processing 195
7.5 Results and discussion 206

8. Mapping and visualization

8.1 Peculiarities of morphometric mapping 209
8.2 Combined visualization of morphometric variables 211
8.3 Combining hill-shaded maps with soil and geological data 212
8.4 Cross sections 213
8.5 Three-dimensional modeling 214
8.5.1 Three-dimensional terrain modeling 214
8.5.2 Three-dimensional subsurface modeling 222
8.6 Virtual globes 228
8.6.1 Desktop globes 228
8.6.2 Web globes 229

Part II
Digital terrain modeling in soil science

9. Influence of topography on soil properties

9.1 Introduction 241
9.2 Local morphometric variables and soil 241
9.3 Nonlocal morphometric variables and soil 242
9.4 Discussion 242

10. Adequate resolution of models

10.1 Motivation 245
10.2 Theory 246
10.3 Field study 248
10.3.1 Study site 248
10.3.2 Materials and methods 250
10.3.3 Results and discussion 252

11. Predictive soil mapping

11.1 The Dokuchaev hypothesis as a central idea of soil predictions 255
11.2 Early models 256
11.3 Current predictive methods 258
11.3.1 Classification of methods 258
11.3.2 Mathematical approaches 259
11.3.3 Small-scale predictive models and upscaling 262
11.3.4 Prediction accuracy 267
11.4 Topographic multivariable approach 268

12. Analyzing relationships in the topography-soil system

12.1 Motivation 271
12.2 Study sites 271
12.3 Materials and methods 272
12.3.1 Field work 272
12.3.2 Laboratory analyses 275
12.3.3 Data processing 276

12.4 **Results and discussion** 281
12.4.1 Variability in relationships between soil and morphometric variables 281
12.4.2 Topography and denitrification 289

Part III Digital terrain modeling in geology

13. Folds and folding

13.1 **Introduction** 295
13.2 **Fold geometry and fold classification** 295
13.3 **Predicting the degree of fold deformation and fracturing** 296
13.4 **Folding models and the *Theorema Egregium*** 297

14. Lineaments and faults

14.1 **Motivation** 301
14.2 **Theory** 303
14.3 **Method validation** 305
14.3.1 Materials and data processing 305
14.3.2 Results and discussion 305
14.3.3 Strike, dip, and displacement estimation 307
14.4 **Two case studies** 307
14.4.1 The Crimean peninsula 308
14.4.2 The Kurchatov city area 313

15. Accumulation zones and fault intersections

15.1 **Motivation** 319
15.2 **Study area** 320
15.3 **Materials and methods** 320
15.4 **Results and discussion** 322

16. Global topography and tectonic structures

16.1 **Motivation** 325
16.2 **Materials and data processing** 327
16.3 **Results and discussion** 337
16.3.1 General interpretation 337
16.3.2 Global helical structures 338

Part IV Digital terrain modeling in glaciology and polar research

17. Glacier motion and evolution

17.1 **Introduction** 345
17.2 **Glacier mass balance** 345
17.3 **Ice flow velocity** 347
17.4 **Glacier force balance** 347

18. Crevasses

18.1 **Motivation** 349
18.2 **Study area** 350
18.3 **Materials and methods** 351
18.3.1 UAS characteristics 351
18.3.2 UAS surveying 352
18.3.3 Data processing 352
18.4 **Results and discussion** 354

19. Catastrophic glacier events

19.1 **Motivation** 357
19.2 **Study area** 357
19.3 **Materials and methods** 358
19.3.1 UAS surveying 358
19.3.2 Data processing 359
19.4 **Results and interpretation** 363

20. Antarctic oases

20.1 **Motivation** 365
20.2 **Study area** 365
20.3 **Materials and methods** 366
20.3.1 Geomorphometric modeling 366
20.3.2 Field geomorphometric interpretation 367
20.4 **Results** 368
20.5 **Discussion** 374

21. Concluding remarks and pending problems

21.1 **Geomorphometry today** 379
21.2 **Theory** 379
21.3 **Data processing** 380
21.4 **Applications** 381

References 383
Index 451

Preface to the third edition

Topography, one of the most important components of the environment, is quantitatively studied by the science of geomorphometry, which is also known as digital terrain modeling or digital terrain analysis. The first and second editions of this book have originally focused on theory and methods of digital terrain analysis as well as its applications in the context of multiscale problems of soil science and geology. Since the preparation of the first and second editions in 2010–11 and 2015–16, respectively, considerable progress has been made in geomorphometry. By the end of the first quarter of the 21st century, it had evolved into a science with a well-developed physical and mathematical theory and a powerful set of computational methods and algorithms. More importantly, digital terrain modeling has become an integral part of almost all spatially distributed geoscientific studies. This has necessitated the preparation of the third, significantly revised and expanded edition of the book.

In particular, Chapter 1 (Digital terrain analysis: Past and present) was enlarged several times to include a detailed state-of-the-art of geomorphometry. Chapter 2 (Topographic surface and its characterization) was supplemented with information on wind and view morphometric variables as well as geomorphometric landform classifications; most of the figures were replaced with new ones. Chapter 3 (Digital elevation models) now includes a description of structure-from-motion techniques used to process imagery from unmanned aerial surveys as well as information on dense point clouds. Chapter 4 (Calculation methods) was supplemented with formulas and algorithms for the calculation of wind and view morphometric variables as well as details of morphometric computations on the surface of a triaxial ellipsoid. In Chapter 5 (Errors and accuracy), a section describing sources of DEM errors was structured and expanded. Chapter 8 (Mapping and visualization) was supplemented with information on three-dimensional terrain modeling with multiple-source illumination as well as web virtual globes. A new Part IV (Digital terrain modeling in glaciology and polar research) has been prepared. It includes four new chapters: Chapter 17 (Glacier motion and evolution), Chapter 18 (Crevasses), Chapter 19 (Catastrophic glacier events), and Chapter 20 (Antarctic oases). Chapter 21 (Concluding remarks and pending problems) was updated. In this edition, the total number of illustrations has increased by about 50% (216 figures vs. 155 in the second edition). The list of references is substantially updated and amended (2267 references vs. 1285 ones in the second edition).

The book is divided into four parts. Part I represents the main concepts, principles, and methods of digital terrain modeling. Part II discusses various aspects of the use of digital terrain analysis in soil science. Part III looks at applications of digital terrain modeling in geology. Part IV considers key aspects of geomorphometric modeling in glaciology and polar research.

Chapter 1 overviews the development of digital terrain analysis and geomorphometry. First, we briefly point out the role of topography as one of the most important components of the environment. Second, we present a short historical overview of the progress of digital terrain modeling from the mid-1950s to the mid-2010s. Second, we present a review of the development of theory, methods, and applications of geomorphometry in 2016–24. In particular, we analyze factors that have caused progress in geomorphometry in recent years. These include widespread use of unmanned aerial survey and photogrammetry, development of tools and methods for survey of submarine topography, emergence of new publicly available global, quasi-global, and regional digital elevation models, and implementation of machine learning techniques. Finally, we discuss the recent application of geomorphometry for solving multiscale problems of geomorphology, hydrology, soil science, geology, glaciology, speleology, geobotany, forestry, zoogeography, oceanology, planetology, landslide science, avalanche and fire hazards, remote sensing, urban studies, and archaeology.

Part I, discussing theoretical and methodical issues of digital terrain analysis, comprises seven chapters, Chapters 2–8.

Chapter 2 discusses the basic physical and mathematical concepts of digital terrain analysis. First, we introduce the notion of the topographic surface and its limitations. Second, we discuss contour lines, special and nonspecial points, slope lines, and flow lines. Third, we present definitions, formulas, and physical meanings for five main groups of morphometric variables, such as local, nonlocal, structural, two-field-specific, and combined morphometric attributes. In particular, we present the derivation of formulas for slope gradient, slope aspect, twelve curvatures, rotor, as well as horizontal and vertical deflections. Among nonlocal morphometric variables, we are concerned with catchment and dispersive areas as well as the topographic position index. Among two-field-specific morphometric variables, we discuss solar attributes (reflectance, insolation, and sky view factor), wind variables (wind effect index and wind exposition index), as well as view attributes (terrain visibility, total viewshed, and topographic openness). The list of combined morphometric variables includes the topographic wetness index, stream power index, shape index, and curvedness. We discuss three fundamental types of landform classifications: the Gaussian classification, the Efremov-Krcho classification, and the Shary classification. Finally, we review some other landform classifications, such as the Pennock classification, the MacMillan classification, the Schmidt-Hewitt classification, the topographic position index-based classification, the Iwahashi-Pike classification, and the geomorphon classification.

In Chapter 3, we first present a review of techniques used to produce digital elevation models (DEMs). Specifically, we mention conventional topographic surveys, kinematic global navigation satellite system techniques, stereophotogrammetry of optical remote sensing images, structure-from-motion technique for imagery from unmanned aerial surveys, laser altimetry, synthetic aperture radar techniques, echo sounding, satellite radar altimetry, airborne optical sensing of bathymetry, soil and geological core drilling, three-dimensional seismic surveys, ice and ground penetrating surveys, digitizing of contours, as well as fusion of elevation data from different sources. Second, we describe the main types of DEM grids, viz., plane irregular and regular grids, spheroidal equal angular grids, as well as dense point clouds. DEM datums and coordinate systems are briefly mentioned. Third, we discuss issues of DEM resolution, in particular, the fundamental sampling theorem and its three consequences. Finally, there is a brief review of interpolation methods used in digital terrain analysis.

Chapter 4 deals with calculation methods of digital terrain modeling. First, we describe the Evans-Young method, a classic finite-difference technique for computing local topographic attributes. Second, we present two methods to derive models of local morphometric variables from DEMs based on plane square grids as well as spheroidal equal angular grids at a sphere and an ellipsoid of revolution. The method for plane square grids uses the approximation of the third-order polynomial to the 5×5 moving window by the least squares. The method for spheroidal equal angular grids is based on the approximation of the second-order polynomial to the 3×3 moving window by the least squares. We briefly review techniques to compute nonlocal and combined morphometric variables on plane square grids and spheroidal equal angular grids, approaches to detect structural lines including calculation of horizontal deflection as well as algorithms for derivation of two-field-specific morphometric variables and landform classifications. Finally, we present analytical and computational solutions, which provide the basis for geomorphometric modeling on a triaxial ellipsoid.

Chapter 5 investigates the problem of errors and accuracy in digital terrain models (DTMs). First, we look at the sources and types of DEM errors as well as the methods of their detection and analysis. Second, we describe a method to estimate the calculation accuracy of digital models of local morphometric attributes using the criterion of root mean square error of a function of measured variables. Third, we discuss artifacts (false landforms), which arise during DEM processing. Such artifacts are caused by: (1) the ignoring of the sampling theorem; (2) the Gibbs phenomenon; and (3) discretization errors due to grid displacement. Finally, we explore the intrinsic properties of local morphometric variables as a possible cause of orthogonal and diagonal linear artifacts.

Chapter 6 considers various aspects of geomorphometric filtering. There are three key tasks of geomorphometric filtering: (1) DTM decomposition, that is, separation of high- and low-frequency components of the topographic surface in order to study its structure and elements of different scales; (2) DTM denoising; and (3) DTM generalization. We briefly describe methods of DTM filtering: trend-surface analysis, the Filosofov method, two-dimensional (2D) discrete Fourier transform, 2D discrete wavelet transform; mean, median, Gaussian, and feature-preserving smoothing; data elimination from DTMs, as well as the cutting method. The algorithm for the 2D singular spectrum analysis is described in detail.

There are three key mathematical problems in digital terrain analysis: (1) DEM interpolation; (2) DEM generalization and denoising; and (3) computation of morphometric variables through calculating partial derivatives of elevation. Traditionally, these three problems are solved separately by means of procedures implemented using different methods and algorithms. In Chapter 7, we present a universal spectral analytical method based on high-order orthogonal expansions using the Chebyshev polynomials of the first kind with the subsequent Fejér summation. The method is intended for the processing of regularly spaced DEMs within a single, end-to-end framework including DEM global approximation, denoising, smoothing, generalization, calculation of the partial derivatives of elevation, and derivation of local morphometric variables. The method is characterized by good performance (i.e., a distinct monotonic convergence of the approximation) and high speed of data processing. This method is an effective alternative to the common techniques of DEM processing.

Chapter 8 examines the peculiarities of geomorphometric visualization. For the correct perception of morphometric maps, it is reasonable (1) to apply layer tinting, (2) to subdivide values of morphometric variables into intervals relative to the zero value, (3) to select contrast color schemes, and (4) to employ a logarithmic transform. We describe key approaches to geomorphometric visualization: combined displaying of several morphometric variables on a single map, integration of geological or soil information with hill-shaded maps, cross-section construction, three-dimensional (3D) terrain modeling with single- and multiple-source illumination, as well as 3D geological and soil modeling. Finally, we discuss virtual morphometric globes.

Part II, concerning soil application of digital terrain analysis, consists of four chapters, Chapters 9–12.

It is well known that topography is one of the soil-forming factors. Topography influences soil properties through two key "tools": (1) the gravity-driven lateral migration and accumulation of water and (2) spatial differentiation of the temperature regime of slopes. Chapter 9 discusses the main regularities of the influence of topography on spatial distribution of soil properties exemplified by soil moisture content. In particular, we explain the role of slope gradient and aspect, horizontal, vertical, and mean curvatures, as well as catchment area and topographic wetness index, which are responsible for the spatial distribution and redistribution of water in the landscape. Also, we briefly discuss the influence of morphometric characteristics of subsurface topography and tillage practices on soil properties.

The central problem of the combined analysis of morphometric and soil data is the selection of the correct value for the DTM grid spacing. Chapter 10 describes a method for determining adequate grid spacing based on the concept of representative elementary volume. The method includes the following steps: (1) deriving a set of DTMs using different values for the grid spacing; (2) analyzing correlations between a soil property and a morphometric variable, calculated with different values of the grid spacing; (3) plotting a graph of correlation coefficients between the soil property and the morphometric variable versus the grid spacing values; and (4) finding smooth portions of the graph indicating intervals of the adequate grid spacing. The method is exemplified by studying the relations between topography and soil moisture.

Chapter 11 looks at predictive soil mapping, a branch of soil science. It is demonstrated that Vasily Dokuchaev explicitly defined its central idea and made a statement of the problem in 1886. In predictive soil mapping, it is common to use morphometric variables as key predictors of soil properties. We briefly review predictive soil-mapping methods developed in the precomputer era. Then we present a classification and a brief review of predictive soil-mapping methods based on digital terrain modeling and various mathematical approaches (e.g., multiple regression analysis, hybrid geostatistical approaches, fuzzy logic, discriminant analysis, artificial neural networks, and decision trees). We also discuss problems of small-scale soil prediction and the upscaling of relationships in the topography-soil system. Finally, we describe our approach toward mapping the spatial distribution of the physical, chemical, and biological properties of soil based on digital terrain modeling and multiple regression analysis.

Chapter 12 presents two case studies on geomorphometry-based analysis of relationships between topography and soil. Study sites were located in the Canadian prairies. We studied: (1) temporal variability in the influence of topography on soil properties; (2) changes in the influence of topography on soil dynamic properties depending on the soil layer depth; and (3) the effect of topography on the activity of denitrifiers under different soil moisture conditions. We demonstrate that: (1) spatial distribution of soil dynamic properties depends on topographic variables only if soil moisture content is higher than some threshold value; (2) dependence of soil dynamic properties on morphometric attributes may both decrease and increase as the soil layer depth increases; and (3) there exists temporal variability in the relationships between the spatial distribution of soil dynamic properties and morphometric attributes.

Part III, discussing the geological application of geomorphometry, includes four chapters, Chapters 13–16.

Folds of various scales are the most abundant and studied geological features. Chapter 13 probes into the applications of data on curvatures of the land and stratigraphic surfaces in research on folds and folding processes. In particular, such data are used: (1) to describe the geometry of folds and to classify folds; (2) to predict the degree of fold deformation and strain as well as the fracture orientation and fracture density of folded strata; and (3) to estimate the degree of plasticity of folds as well as to reconstruct paleotopography using Gauss's *Theorema Egregium.*

Lineaments are usually associated with faults and linear zones of fracturing, bending deformation, and increased permeability of the crust. Chapter 14 describes a method to reveal and classify topographically expressed lineaments. This method is based on the derivation of horizontal and vertical curvatures from DEMs of the land or stratigraphic surfaces. Lineaments revealed on horizontal curvature maps are associated with faults formed mostly by horizontal tectonic movements (i.e., strike-slip faults). Lineaments recognized through the mapping of vertical curvature relate to faults formed mainly by vertical motions (i.e., dip-slip and reverse faults) and thrusting. Lineaments recorded on maps of both horizontal and vertical curvatures indicate, as a rule, oblique-slip and gaping faults. The method is validated using artificial DEMs with modeled faults. Finally, we present the results of the method's implementation for seismically active, mountainous terrain as well as a platform plain region.

In many ways, topography controls overland and intrasoil gravity-driven transport of water. At the same time, valley networks, determining principal routes of overland flows, can often be connected with fault networks, which serve as pathways for the upward transport of groundwater. For tectonic terrains, it has been established that topographically expressed zones of flow accumulation, as a rule, coincide with sites of fault intersections due to increased rock fracturing. Chapter 15 looks at relationships between zones of flow accumulation and natural phenomena a priori associated with fault intersections (i.e., sites of intensive rock fracturing and sites of springs/boreholes with abnormally high discharges). It was found that the phenomena under study are spatially correlated with the location of accumulation zones. This testifies that in accumulation zones, soil moisture depends on both the upward transport of deep groundwater and the accumulation of overland lateral water flows. In other words, topographically expressed accumulation zones/fault intersections are areas of contact and interaction between overland and deep substance flows.

In the past few decades, there have been proposals suggesting that there are tectonically and topographically expressed, global linear (helical) structures. Chapter 16 examines this hypothesis using spheroidal digital terrain modeling and global DEMs of the Earth, Mars, Venus, and the Moon. Local and nonlocal topographic variables were calculated and mapped for the entire surface of the planetary bodies. Digital terrain analysis provided support for the existence of global lineaments: on catchment area maps of the Earth, it was possible to detect five mutually symmetrical pairs of helical structures encircling the planet from pole to pole. On catchment area maps of Mars and Venus, it was also possible to detect several helical structures. All these structures are apparently associated with traces of the planetary rotational stresses.

Part IV, concerning the application of digital terrain analysis in glaciology and polar research, includes four chapters, Chapters 17–20.

A glacier is a natural object with a constantly and dynamically changing surface. General characteristics of glacier changes, motion, and evolution are usually obtained by comparing multitemporal cartographic and geophysical data, aerial and satellite imagery as well as satellite altimetry measurements. Velocity and direction of movement of a glacier surface are traditionally estimated by geodetic methods using multitemporal, high-precision measurements of marked ground points. In Chapter 17, we discuss geomorphometry-based studies of glacier motion and evolution, which are generally focused on three interrelated issues characterizing the glacier state and behavior, namely: glacier mass balance, ice flow velocity, and glacier force balance. Such studies, as a rule, include comparative analysis and processing of multitemporal digital elevation and morphometric models of the glacier surface.

For polar research, one of the important support tasks is the monitoring of sledge routes and other visited areas for detection of open crevasses and revealing of hidden, snow-covered ones. In Chapter 18, we discuss the possibility of revealing hidden crevasses with geomorphometric treatment of high-resolution glacier DEMs produced using imagery from unmanned aerial surveys. The study was conducted on a sledge route from the Progress to Vostok Stations, East Antarctica. The unmanned aerial survey resulted in a set of aerial images with a resolution of 6 cm. From these images, we produced glacier DEMs with resolutions of 25 cm, 50 cm, and 1 m for a typical area with snow-covered crevasses. From these DEMs, we derived models of a series of morphometric variables. For the 1-m gridded DEM, mapping some morphometric attributes allowed us to detect hidden crevasses. The approach is functional, at least, for revealing crevasses with a width of 1.5 m and wider, covered by the snow layer with a thickness of up to 1.2 m.

In Chapter 19, we describe and interpret a catastrophic subsidence in the Dålk Glacier, Princess Elizabeth Land, East Antarctica, which occurred on the morning of January 30, 2017. We analyzed and geomorphometrically processed a data set of high-resolution aerial images and video records from unmanned aerial surveys captured ten days before, one hour after, and ten days after the event. We reconstructed a subsidence mechanism as follows: An englacial cavern measuring about 885,000 m^3 with a thin roof existed before the event. There was a channel of the surface and near-subsurface meltwater flow from the Boulder Lake to the Prydz Bay passing near the cavern. An intensive meltwater flow began from the Boulder Lake about a day before the event. The meltwater accumulated on the cavern rooftop led to its initial collapse. In several days, the meltwater completed the process of subsidence and collapse of the cavern. The event led to the formation of an ice depression about 40,000 m^2 in size and up to 43 m deep. Geomorphometric analysis demonstrated that ice-surface hydrological processes could cause rapid, deep subsidences of large areas in glaciers, without involving subglacial geothermal or volcanic activity as well as englacial water outbursts from subglacial lakes.

Antarctic oases are glacier-free areas of the Antarctic coastal zone characterized by the presence of three essential features: (1) a local climate largely determined by the adjacent or surrounding ice sheet; (2) unfrozen water in the form of seasonal streams and unfrozen lakes; and (3) primitive soils, plants, and biota. Chapter 20 presents the results of geomorphometric modeling and mapping of the Larsemann Hills (Princess Elizabeth Land, East Antarctica), which is one of the most popular Antarctic oases. As input data, we used a fragment of the Reference Elevation Model of Antarctica. From the extracted and edited DEM, we derived and mapped a set of 17 morphometric variables, including slope, aspect, horizontal, vertical, minimal, and maximal curvatures, catchment area, as well as topographic wetness index. We also conducted a field geomorphometric interpretation to provide correct physical geographic, geological, and geomorphological descriptions of morphometric maps. In the fieldwork, we carried out 54 ft routes with a total length of about 422 km. The derived morphometric maps can be helpful for revealing topographically manifested geological lineaments (i.e., fractures, faults, and folds), predicting floodwater discharge pathways along the lake cascades, differentiating slopes by insolation and wind impact levels, as well as identifying accumulation zones and fracture/fault intersection sites.

Chapter 21 concludes the book by summarizing its main topics and describing important pending problems in digital terrain modeling. First, we discuss some aspects of the geomorphometric theory, such as the role of a classification of morphometric variables based on their physical and mathematical meaning as well as the interpretation ambiguity of roughness indices and Haralick textural characteristics in the context of quantitative analysis of topography. Second, we look at pending problems in geomorphometric computation and modeling, for example, calculation of nonlocal and structural morphometric variables, digital terrain modeling of small celestial bodies with complex irregular shapes, and choice of DEM grid spacing. Finally, we explore some promising application issues, in particular, Antarctic research studies, whose sharp increase becomes possible due to unprecedented number of new Antarctic DEMs created over the past several years.

The book is addressed to geoscientists—geomorphometrists, geomorphologists, soil scientists, geologists, glaciologists, geographers, and GIS scientists—at scholar, lecturer, and postgraduate student levels, with mathematical skills. The book is also intended for GIS practitioners in industry and research laboratories focusing on geoscientific research. The book can also serve as a textbook for advanced undergraduate or graduate-level courses.

I would appreciate receiving readers' suggestions and pointing out of errors.

Igor V. Florinsky

Progress Station, Princess Elizabeth Land, Antarctica, April 2023 – Pushchino, Moscow Region, Russia, July 2024

Acknowledgments

This book is based on my 35-year research experience I gained from my studies in digital terrain analysis at the Laboratory of Soil Mapping, Institute of Soil Science and Photosynthesis, Russian Academy of Sciences (1989–92), Institute of Mathematical Problems of Biology, Keldysh Institute of Applied Mathematics, Russian Academy of Sciences (from 1993 till now), Manitoba Land Resource Unit, Agriculture and Agri-Food Canada (1998–2000), Department of Soil Science, University of Manitoba (2000–02), as well as Russian Antarctic Expedition, Arctic and Antarctic Research Institute (2021–23).

Section 6.3 is based on the collaborative work with N.E. Golyandina and K.D. Usevich (Department of Statistical Modeling, St. Petersburg University). Chapter 7 includes results obtained together with A.N. Pankratov (Institute of Mathematical Problems of Biology, Keldysh Institute of Applied Mathematics, Russian Academy of Sciences). Sections 8.5.1.1 and 8.6.1 are based on my collaboration with S.V. Filippov (Institute of Mathematical Problems of Biology, Keldysh Institute of Applied Mathematics, Russian Academy of Sciences). Section 8.6.2 uses results from my collaborative project with A.S. Garov and I.P. Karachevtseva (Extraterrestrial Laboratory, Moscow State University of Geodesy and Cartography). Soil data analyzed in Section 10.3 were collected together with G.A. Kuryakova and P.A. Shary (Institute of Soil Science and Photosynthesis, Russian Academy of Sciences). Chapter 12 includes selected results from collaborative studies with R.G. Eilers and G.R. Manning (Manitoba Land Resource Unit, Agriculture and Agri-Food Canada), as well as D.L. Burton and S.K. McMahon (Department of Soil Science, University of Manitoba). Chapters 18 and 19 present some results from my projects performed in collaboration with D.P. Bliakharskii (Department of Cartography and Geoinformatics, St. Petersburg University), who performed unmanned aerial survey and field works during the 62nd Russian Antarctic Expedition.

I am grateful to all my colleagues, friends, and beloved women for their assistance, criticism, enthusiasm, and moral support.

Abbreviations and acronyms

API	application programming interface
ALOS	Advanced Land Observing Satellite
ANOVA	analysis of variance
ASTER	Advanced Spaceborne Thermal Emission and Reflection Radiometer
AVHRR	Advanced Very High Resolution Radiometer
AUV	autonomous underwater vehicle
BGE	Blender game engine
CPU	central processing unit
DEM	digital elevation model
DoD	DEM of difference
DSM	digital soil mapping
DSM	digital surface model
DTED	Digital Terrain Elevation Data
DTM	digital terrain model
D8	eight-node single-flow direction
EU-DEM	European DEM
FPDEMS	feature-preserving DEM smoothing
EGM2008	Earth Gravitational Model 2008
ERS	European Remote-Sensing Satellite
ET	eigentriple
GCP	ground control point
GCS	ground control station
GEBCO	General Bathymetric Chart of the Oceans
GIS	geographical information system
GLONASS	Global Navigation Satellite System
GLSL	OpenGL shading language
GNSS	global navigation satellite system
GMTED	Global Multi-Resolution Terrain Elevation Data
GPR	ground penetrating radar
GPS	global positioning system
GPU	graphics processing unit
GSD	ground sample distance
HD	high definition
HTTP	hypertext transfer protocol
IBCAO	International Bathymetric Chart of the Arctic Ocean
IBCSO	International Bathymetric Chart of the Southern Ocean
ICESat	Ice, Cloud, and Land Elevation Satellite
IDW	inverse distance weighted
InSAR	interferometric synthetic aperture radar
KML	Keyhole Markup Language
K–S	Kolmogorov-Smirnov test
Lidar	light detection and ranging
LOD	level of detail
LOLA	Lunar Orbiter Laser Altimeter
LoS	line of sight
LRO	Lunar Reconnaissance Orbiter

MFD	multiple-flow direction
MOLA	Mars Orbiter Laser Altimeter
MSS	MultiSpectral Scanner
NIR	near infrared
OpenGL	Open Graphics Library
PDS	planetary data system
PPK	post-processed kinematic
PRISM	Panchromatic Remote-Sensing Instrument for Stereo Mapping
QMC	quasi-Monte Carlo
RAE	Russian Antarctic Expedition
RES	radio-echo sounding
ROV	remotely operated vehicle
RTK	real-time kinematic
REMA	Reference elevation model of Antarctica
RMSE	root mean square error
SAR	synthetic aperture radar
SD	standard deviation
SfM	structure-from-motion
SPOT	Satellite Pour l'Observation de la Terre
SRTM	Shuttle Radar Topography Mission
SSA	singular spectrum analysis
SQL	structured query language
TanDEM-X	TerraSAR-X add-on for Digital Elevation Measurement
TIN	triangulated irregular network
UAS	unmanned aerial system
UAV	unmanned aerial vehicle
UTM	Universal Transverse Mercator
UV	axes of the 2D texture in 3D graphics
WASM	WebAssembly
WebGL	Web Graphics Library
WGS84	World Geodetic System 1984
1D	one-dimensional
2D	two-dimensional
2D-SSA	two-dimensional singular spectrum analysis
3D	three-dimensional
4D	four-dimensional

A	slope aspect
A_E	eastness
A_N	northness
A_r	relative terrain aspect
CA	catchment area
DA	dispersive area
D_{kh}	horizontal deflection
D_{kv}	vertical deflection
E	difference curvature
g, h, k, m	the third-order partial derivatives of elevation
$\dot{g}$	gravitational acceleration constant
$\mathbf{g}$	gravitational acceleration vector
$\mathbf{g_t}$	tangential component of the gravitational acceleration vector
$\mathbf{g_n}$	normal component of the gravitational acceleration vector
G	slope gradient
H	mean curvature
i, j, and k	unit vectors in the directions of x-, y-, and z-axes
I	insolation
IS	shape index
K	Gaussian curvature
K_a	accumulation curvature

k_h	horizontal (tangential) curvature
k_{he}	horizontal excess curvature
k_{max}	maximal curvature
k_{min}	minimal curvature
k_p	plan curvature
K_r	ring curvature
k_v	vertical (profile) curvature
k_{ve}	vertical excess curvature
M	unsphericity
$Moist$	soil moisture
$\mathbf{n}$	a unit external normal vector
N	a set of nonspecial points
p, q	the first-order partial derivatives of elevation
r, s, t	the second-order partial derivatives of elevation
R	reflectance
R^2	coefficient of determination
rot	rotor
SCA	specific catchment area
SDA	specific dispersive area
SPI	stream power index
SVF	sky view factor
TO	topographic openness
TPI	topographic position index
TWI	topographic wetness index
V	terrain visibility (singular viewshed)
V_Σ	total viewshed
w	square grid spacing
w_α	angular step
WEf	wind effect index
WEx	wind exposition index
z	elevation
∇^2	Laplacian
β	ellipsoidal latitude
δ	wind direction angle
Θ	solar azimuth angle
λ	longitude
ξ	contour line length
σ	horizon angle
σ_w	horizon angle in the wind direction
φ	latitude
ψ	solar elevation angle
Ψ_s	sky view factor
υ	morphometric variable
ω	ellipsoidal longitude

Chapter 1

Digital terrain analysis: Past and present

1.1 Why topography?

Topography is one of the most important components of the environment, one of the main elements of geosystems, and the framework of a landscape.

Being a result of the interaction of endogenous and exogenous processes of different scales, topography can reflect geological structures and govern drainage systems (Penck, 1924; Gerasimov, 1959; Meshcheryakov, 1965; Ollier, 1981; Ufimtsev, 1988; Keller and Pinter, 2002; Scheidegger, 2004; Lopatin, 2008; Brocklehurst, 2010; Burbank and Anderson, 2012). At the same time, topography is one of the main factors determining processes taking place in the near-surface layer of the planet (Huggett and Cheesman, 2002).

First, topography determines key prerequisites for gravity-driven overland and intrasoil lateral transport of water and other substances (Kirkby and Chorley, 1967; Speight, 1980). Second, topography influences microclimatic and meteorological characteristics (temperature, moisture, as well as exposure to wind and sun), controlling hydrological and thermal regimes of a terrain (Geiger, 1927; Romanova, 1977; Kondratyev et al., 1978; Raupach and Finnigan, 1997; Huggett and Cheesman, 2002, chap. 3; Oke, 2002).

In this connection, topography affects the spatial distribution of vegetation cover (Yaroshenko, 1961; Franklin, 1995; Huggett and Cheesman, 2002, chap. 6) and is one of the soil-forming factors (Dokuchaev, 1883; Zakharov, 1913; Neustruev, 1927; Jenny, 1941; Huggett, 1975, 2023; Fridland, 1976; Gerrard, 1981; Huggett and Cheesman, 2002, chap. 5; Schaetzl and Anderson, 2005). In addition, topography predetermines slope processes (Young, 1972), glacier development (Menzies, 2002), and a wide range of human activities, such as urban and road construction, land use, and agricultural practices (Huggett and Cheesman, 2002, chap. 7).

Due to these facts, qualitative and quantitative topographic information and the methods of its processing and interpretation are widely used in geosciences. The scientific discipline whose subject is the quantitative study of topography is called *geomorphometry*, *digital terrain modeling*, or *digital terrain analysis* (Pike, 2000). In this book, we use these three synonymous terms interchangeably.

1.2 A short historical overview[a]

Before the 1990s, topographic maps were the main source of quantitative information on topography. They were analyzed using geomorphometric techniques, which originated in geomorphology, to calculate manually morphometric variables (e.g., slope gradient, drainage density, horizontal curvature, catchment area) and produce morphometric maps (Vakhtin, 1930; Weinberg, 1934a; Chentsov, 1940; Volkov, 1950; Clarke, 1966; Devdariani, 1967; Pannekoek, 1967; Mark, 1975b; Gardiner and Park, 1978; Stepanov et al., 1984; Lastochkin, 1987).

Conventional manual geomorphometric techniques have received wide acceptance in geological studies (Berlyant, 1960). They have been utilized to represent graphically the shape of deposits (Sobolevsky, 1932), to explore oil-/gas-bearing and ore-controlling structures (Levorsen, 1927; Filosofov, 1960; Murray Jr., 1968; Volchanskaya, 1981; Guberman et al., 1997), to analyze the block structure of the Earth's crust (Orlova, 1975; Glasko and Rantsman, 1996), to study seismicity (Gelfand et al., 1972), and so on.

In soil science, conventional geomorphometric techniques have been used to investigate relationships between soil cover and topography (Dokuchaev, 1891; Ototzky, 1901), to predict soil properties (Aandahl, 1948; Troeh, 1964; Walker et al., 1968; Romanova, 1963, 1970, 1971), to produce soil maps (Anisimov et al., 1977; Stepanov et al., 1987,

a. A detailed historical overview of digital terrain modeling can be found elsewhere (Pike et al., 2009).

Digital Terrain Analysis. https://doi.org/10.1016/B978-0-443-24798-9.00024-8

1998; Stepanov, 1989; Stepanov and Loshakova, 1998), and to study regularities in the soil cover structure and its relations with geological features (Filatov, 1927; Stepanov and Sabitova, 1983; Kuryakova and Florinsky, 1991; Stepanov, 1996).

In hydrology, conventional geomorphometric techniques have been applied to describe and study the morphology of drainage basins and stream channels (Horton, 1945; Strahler, 1957), as well as to predict quick response flow from a storage/contributing area relationship (Beven and Kirkby, 1979).

In the mid-1950s, a new research field—digital terrain modeling—emerged in photogrammetry (Rosenberg, 1955). Within its framework, two-dimensional discrete functions of elevation called digital elevation models (DEMs) became the main source of information on topography. DEMs were used to calculate digital terrain models (DTMs), two-dimensional discrete functions of morphometric variables. Initially, digital terrain modeling was mainly utilized in engineering fields, for example, to produce raised-relief maps or physical three-dimensional (3D) models using computer-controlled milling machines (Spooner Jr. et al., 1957; Lyubkov and Martynenko, 1963) as well as to design highways (Miller and Leflamme, 1958) and railways (Konovalov, 1960).

In the late 1960s, methods of digital terrain modeling began to be used in geosciences. In particular, a research avenue aroused in geology focusing on studies of geological structures using DEMs of both the land surface and stratigraphic surfaces (Muñoz-Espinoza, 1968; Abelsky and Lastochkin, 1969; Robinson et al., 1969). One of the pioneering works was conducted by Belonin and Zhukov (1968), who analyzed the uplift evolution of an elevated block using digital models of the Gaussian, mean, and principal curvatures of geological surfaces.

The first effective methods to calculate morphometric variables were developed in the 1970s and 1980s (Young, 1978; Evans, 1979; Zevenbergen and Thorne, 1987; Jenson and Domingue, 1988; Martz and de Jong, 1988). However digital terrain modeling was still relatively uncommon in geoscientific studies of this period. Only in the 1980s were the first attempts made to analyze the influence of topography on soil properties as well as to model soil properties with digital topographic data (Sinai et al., 1981; Burt and Butcher, 1985; Kachanoski et al., 1985a,b; Pennock et al., 1987). In geology, researchers began to use DEMs for revealing weakly manifested geological structures, such as lineaments and faults (Moore and Simpson, 1983; Schowengerdt and Glass, 1983; Onorati et al., 1987; Zeilik et al., 1989).

In the 1970s–90s, advances in computer technology were responsible for the gradual transition from conventional manual geomorphometry to digital terrain modeling (Evans, 1972; Koshkarev, 1982; Burrough, 1986; Serbenyuk, 1990). This was supported by the development of the physical and mathematical theory of the topographic surface (Krcho, 1973; Evans, 1979; Shary, 1991, 1995; Koenderink and van Doorn, 1994; Rudy, 1999) (for its current state, see Chapter 2.).

In the late 1980s and 1990s, digital geomorphometry began to be used in geomorphological mapping (Dikau, 1988, 1989, 1990; Dikau et al., 1995; Dikau and Schmidt, 1999), studies of erosion (Martz and de Jong, 1987, 1990 1991; Desmet and Govers, 1995, 1996, 1997) and landslide hazards (Carrara et al., 1991; Carrara and Guzzetti, 1995), as well as to predict meteorological characteristics such as heat stress (Menz, 1988), radiation from sky and terrain (Dozier and Frew, 1990), as well as wind sheltering and snow accumulation (Lapen and Martz, 1993, 1996).

During the same period, intensive studies were carried out in hydrology, where researchers began to apply systematically digital terrain analysis to predict the spatial distribution of hydrological characteristics within the TOPMODEL (Section 11.2) framework (Quinn et al., 1991; Beven and Kirkby, 1993; Beven and Moore, 1993; Quinn and Beven, 1993; Beven et al., 1995; Beven, 1997; Saulnier et al., 1997).

In the 1990s, digital terrain analysis began to be used in glaciology. Using cartographic materials, Suetova and Chistov (1993) created two small-scale DEMs of glacial and subglacial surfaces for a fragment of Antarctica and calculated its approximate ice volume as the difference between the two DEMs. Introducing the concept of glacier geomorphometry, Etzelmüller and Sollid (1997) analyzed long-term changes in the glacier surface using a series of multi-temporal DEMs of glaciers.

In the first 15 years of the 21st century, further advances in aerial, space, and geophysical technologies opened new horizons for digital terrain modeling. High-resolution DEMs of the land surface became available (Tarolli, 2014) due to the progress in kinematic global navigation satellite system (GNSS) survey (Section 3.1.2), lidar aerial survey (Section 3.1.5), and structure-from-motion (SfM) techniques (Section 3.1.4). Quasi-global DEMs with a grid spacing of 1″—SRTM1 and ASTER GDEM—were produced using satellite surveys (Farr et al., 2007; Hato et al., 2009). Free public access to these materials via the Internet (USGS, 2015; Japan Space Systems, 2009–2019) extended the capabilities for conducting DTM-based soil, geological, hydrological, and other studies at various scales. The release of three new bathymetric DEMs—IBCAO version 3 (Jakobsson et al., 2012), IBCSO version 1 (Arndt et al., 2013), and GEBCO_2014 (Weatherall et al., 2015)—offered new opportunities for carrying out regional marine geomorphic, geological, and geophysical research. Two global DEMs—SRTM30_PLUS (Becker et al., 2009) and SRTM15_PLUS (Sandwell et al.,

2014) with the grid spacings of 30″ and 15″, respectively—provided a framework for tectonic investigations at regional, continental, and global scales. The advances in 3D seismic survey (Chopra and Marfurt, 2005, 2007a) (Section 3.1.11) enhanced the production of DEMs of geological (stratigraphic) surfaces.

By the end of the first quarter of the 21st century, digital terrain modeling had evolved into the science of quantitative modeling and analysis of the topographic surface and the relationships between topography and other natural and artificial components of geosystems. Currently, geomorphometry is widely used to solve various multiscale problems of geomorphology, soil science, geology, hydrology, geobotany, glaciology, oceanology, planetology, and other sciences.

At different stages of its development, theoretical and applied aspects of digital terrain analysis have been described in a series of analytical reviews (McCullagh, 1988; Moore et al., 1991; Shary et al., 1991; Weibel and Heller, 1991; Band, 1993; Franklin, 1995; Florinsky, 1998b, 2017a, 2021; Pike, 1995, 2000, 2001; Deng, 2007; Brocklehurst, 2010; Minár et al., 2016, 2024; Lv et al., 2017; Franklin, 2020; Kharchenko, 2020; Guth et al., 2021; Xiong et al., 2022; Trevisani and Florinsky, 2023) and books (Felicísimo, 1994a; Wilson and Gallant, 2000; El-Sheimy et al., 2005; Li et al., 2005; Hengl and Reuter, 2009; Wilson, 2018).

1.3 Current advances in digital terrain analysis

In this section, we discuss progress in geomorphometry that has occurred since the publication of the previous, second edition of this book in 2016. Articles in international journals belonging to the 1st and 2nd quartiles (Q1 and Q2) of the JCR Web of Science Core Collection were predominantly used for this survey. If a journal had several different quartiles, we used its highest quartile in 2016–24. We believe that such an approach allowed us to select the strongest and most original articles from several thousand geomorphometry-related papers published in that period.

1.3.1 Factors in the development of geomorphometry

In the last several years, significant progress in digital terrain analysis has been caused by a series of factors related to further development of remote sensing technologies, publicly available DEMs, and methods for data processing.

1.3.1.1 Development of remote sensing technologies

Three groups of technologies have had a critical influence on progress in digital terrain modeling in the last decade, such as unmanned aerial survey, airborne lidar survey, and bathymetric survey techniques.

The wide spread of two linked technologies—unmanned aerial systems (UASs) and SfM techniques (Section 3.1.4)—had the strongest impact on the development and dissemination of large-scale, DTM-based studies of virtually all elements of landscapes. UASs are used for performing aerial surveys of relatively small areas (Toth and Jóźków, 2016; Bhardwaj et al., 2016b; Singh and Frazier, 2018; Xiang T. et al., 2019; Yao et al., 2019; Gaffey and Bhardwaj, 2020), whereas SfM techniques are applied for processing UAS-derived imagery (Eltner et al., 2016; Smith et al., 2016; Anderson et al., 2019; Gomez, 2022). The use of these two technologies allows reproducible generation of DEMs (Clapuyt et al., 2016; Hendrickx et al., 2019) with an ultra-high, centimeter-range resolution and accuracy (Girdraux et al., 2017; James et al., 2017a; Florinsky et al., 2018b; Guisado-Pintado et al., 2019; Kurkov and Kiseleva, 2020). To ensure the required DEM accuracy, ground control points (GCPs) are usually utilized during aerial surveys. If it is difficult or impossible to set or use GCPs, the direct georeferencing approach can be applied. In this case, reference data are high-precision coordinates of the image projection centers determined with on-board GNSS receivers (Gabrlik et al., 2018; Bliakharskii et al., 2019a; Chudley et al., 2019).

Another important factor in the progress of geomorphometry was the spread of a long-developed technology—airborne laser scanning (Section 3.1.5)—for producing high-resolution DEMs with regional coverage (Toth and Jóźków, 2016; Gomez, 2022). This made it possible to appear free, publicly available DEMs with a high, meter-range resolution for many areas of Europe and North America. Application of such DEMs can be found in many publications (e.g., Florinsky et al., 2019; Guth and Geoffroy, 2021). Moreover, airborne laser scanning experiences now a renaissance due to the development of UAS-based lidars, which are used for ultra-detailed surveys of relatively small areas (Kurkov et al., 2021; Curcio et al., 2022; Li et al., 2024).

In recent years, there has been further development of tools and methods for the acquisition of bathymetric DEMs marked by high resolution and accuracy. The studies were focused on three objects: (1) deep and shelf zones of the seas and oceans; (2) coastal shallow waters; and (3) river beds. In the first case, one can use multibeam echo sounding (Section 3.1.7) from surface and underwater guided and autonomous platforms (Wölfl et al., 2019; Zwolak et al., 2020;

Zhao D. et al., 2021). In the second case, photogrammetric processing of aerial and space images, as well as dense point clouds from lidar aerial survey, can be applied (Collin et al., 2018; Hodúl et al., 2018; Agrafiotis et al., 2019; Ashphaq et al., 2021). In the third case, researchers utilize UAS-based lidars as well as sonars operated by UASs (Alvarez et al., 2018; Kasvi et al., 2019; Legleiter and Harrison, 2019).

1.3.1.2 Development of free publicly available DEMs

The emergence of new, original and compiled, quasi-global, global, and regional, publicly available DEMs of medium and high resolution (Mudd, 2020) was extremely important for the development and dissemination of geomorphometric methods as a recognized means for analyzing spatial data. The accuracy and quality of such DEMs still leave much to be desired. For the scientific community, however, the importance of free, publicly available DEMs rather than commercial ones can hardly be overestimated (Schumann and Bates, 2018). Such DEMs allow carrying out research studies on different problems at regional, continental, and global scales without additional costs.

Among the new original, publicly available DEMs created in 2016–24, we can distinguish two quasi-global and three regional ones:

- Two quasi-global terrestrial DEMs with 1″ grid spacing: (1) Advanced Land Observing Satellite World 3D digital surface model (ALOS AW3D30 DSM) versions 1–3.2 (Tadono et al., 2016) and (2) Advanced Spaceborne Thermal Emission and Reflection Radiometer Global DEM (ASTER GDEM) version 3 (Abrams et al., 2020). These DEMs were photogrammetrically created from ALOS and Terra satellite images, respectively. Generally, ALOS AW3D30 DSM is the most accurate original quasi-global DEM among its analogs (Boulton and Stokes, 2018; Caglar et al., 2018; Florinsky et al., 2018c, 2019; Jain et al., 2018; Li and Zhao, 2018; Liu et al., 2019; González-Moradas and Viveen, 2020; Kumar et al., 2020; Uuemaa et al., 2020; Carrera-Hernández, 2021; Kramm and Hoffmeister, 2022). ASTER GDEM has low accuracy and is crowded with artifacts (Purinton and Bookhagen, 2017; Boulton and Stokes, 2018; Florinsky et al., 2018c, 2019; Gesch, 2018; Graf et al., 2018; Grohmann, 2018; Liu et al., 2019; González-Moradas and Viveen, 2020; Kumar et al., 2020; Liu et al., 2020; Carrera-Hernández, 2021; Karlson et al., 2021; Kramm and Hoffmeister, 2022). The international expert meeting "Global DEM Benchmarking" (Strobl et al., 2021) recommended abandoning the use of ASTER GDEM.
- Three regional terrestrial DEMs of the Arctic, Antarctica, and high mountain Asia were derived photogrammetrically from WorldView-1, WorldView-2, and WorldView-3 satellite stereo images. These DEMs include (1) 2-m gridded ArcticDEM version 3 with a pan-Arctic coverage (Porter et al., 2018), (2) Reference Elevation Model of Antarctica (REMA), whose version 1.1 had grid spacings of 2 and 8 m, whereas version 2 has grid spacings of 2 and 10 m (Howat et al., 2019), and (3) the High Mountain Asia DEM with a grid spacing of 8 m (Shean, 2017). These three DEMs describe mixed topography consisting of bare earth and ice/snow surfaces.

The goals of creating compiled DEMs are: (1) to produce global DEMs describing the entire topography of the planet, including land, submarine, and subglacial topography and (2) to improve existing DEMs by filtering noise, errors, and artifacts as well as filling voids. These goals are achieved using heterogeneous complementary data sources (Section 3.1.14). In 2016–24, there were two new compiled, publicly available global DEMs:

- SRTM15+ DEM version 2 with a grid spacing of 15″ (Tozer et al., 2019) in which SRTM-3 DEM, ASTER GDEM, ArcticDEM, and REMA were used to describe the land and glacier topography, while the submarine topography was described with bathymetric data and predictive submarine topographic models based on satellite altimetry data (Section 3.1.8).
- General Bathymetric Chart of the Oceans (GEBCO) DEMs versions _2019, _2020, _2021, _2022, and _2023 with a grid spacing of 15″ (GEBCO, 2023a, b) based on the annually refining SRTM15+ DEM.

Newly improved, compiled, and publicly available quasi-global terrestrial DEMs created in 2016–24 include the following models:

- NASADEM with a grid spacing of 1″ created by re-processing the SRTM radar data using improved synthetic aperture radar (SAR) methods and newly available quasi-global terrestrial DEMs for filling voids (Crippen et al., 2016).
- Multi-Error-Removed Improved-Terrain (MERIT) DEM (Yamazaki et al., 2017) with a grid spacing of 3″ created by merging and editing ALOS AW3D30 DSM and SRTM-3 DEM.
- A 1″-gridded DEM created using SRTM-1 DEM and ASTER GDEM version 2 (Yue et al., 2017).

- Copernicus GLO-30 and GLO-90 DEMs (Strobl, 2020) with a grid spacing of 1″ and 3″ based on the edited and hydrologically corrected, commercial quasi-global WorldDEM[b] (Rizzoli et al., 2017). It was recently found that Copernicus GLO-30 outperforms other 1″-gridded quasi-global DEMs (Guth and Geoffroy, 2021; Karlson et al., 2021; Bielski et al., 2024), or, at least, Copernicus GLO-30 and ALOS AW3D30 have comparable high overall quality (Li H. et al., 2022; Trevisani et al., 2023a). However, their relative performances depend on local landscape properties. Copernicus GLO-30 quality is better within urbanized and vegetation-free areas with gentle slopes and relatively low short-range roughness. ALOS AW3D30 quality is better in rougher terrain and on steeper slopes. Both models have poor quality on steep slopes, where they cannot describe fine-scale morphology (Trevisani et al., 2023a).
- Forest and Buildings Removed DEM (FABDEM) with a grid spacing of 1″ derived from the Copernicus GLO-30 DEM (Hawker et al. 2022), which was corrected by multimodal deep learning techniques (Hong et al., 2021).

Data on publicly available, global and quasi-global DEMs are summarized in Table 3.1.

For the years 2016–24, a list of newly compiled, publicly available regional DEMs includes eight models, six of which describe the Antarctic topography:

- A bathymetric DEM, the International Bathymetric Chart of the Arctic Ocean (IBCAO) version 4 (Jakobsson et al., 2020) with a grid spacing of 200 m based on multi- and single-beam echo sounding data and digitized isobaths from hydrographic maps.
- A bathymetric DEM—the International Bathymetric Chart of the Southern Ocean (IBCSO) version 2 (Dorschel et al., 2022) with a grid spacing of 500 m based on multi- and single-beam echo-sounding data. The DEM covers the area south of 50°S and is distributed in two versions: (1) a version with glacier topography on the Antarctic continent, which was mainly derived from REMA, while the land topography of South America and some sub-Antarctic islands was derived from ALOS AW3D30 DSM and (2) a version with subglacial topography derived from the BedMachine Antarctica DEM (see later). Bathymetric gaps were mostly filled with the satellite altimetry-predicted submarine topography (Section 3.1.8) from SRTM15+ DEM.
- BedMachine version 3 DEM of the subglacial and submarine topography of Greenland (Morlighem et al., 2017) with a grid spacing of 150 m created by the mass conservation method (Morlighem et al., 2011) using ice-penetrating radar measurements (Section 3.1.12) and multibeam echo-sounding data (Section 3.1.7).
- BedMachine Antarctica DEM of the subglacial topography of Antarctica (Morlighem et al., 2020) with a grid spacing of 500 m created by the mass conservation method (Morlighem et al., 2011) using ice thickness data from airborne ice-penetrating radio-echo sounding (Section 3.1.12) and ice flow velocity data from satellite interferometry.
- CryoSat-2 DEM of the Antarctic glacier topography (Slater et al., 2018) with a grid spacing of 1 km generated from observations recorded by the CryoSat-2 radar altimeter (Section 3.1.8).
- TanDEM-X PolarDEM 90 m of the Antarctic glacier and ice-free topography (Wessel et al., 2021) with a grid spacing of 90 m and an absolute elevation accuracy of 3.74 m on blue-ice regions. The DEM was generated from the commercial quasi-global WorldDEM and calibrated with data from satellite laser altimeter Ice, Cloud, and Land Elevation Satellite-2 (ICESat-2) (Section 3.1.5).
- ICESat-2 DEM of the Antarctic glacier topography (Shen et al., 2022) with a grid spacing of 500 m created on the basis of ICESat-2 data (Section 3.1.5).
- Gapless-REMA100 DEM of the Antarctic glacier and ice-free topography (Dong et al., 2022) with a grid spacing of 100 m created on the basis of REMA, which voids were filled using multisource data from ICESat-2 DEM, TanDEM-X PolarDEM 90 m, and CryoSat-2 DEM.

1.3.1.3 Development of data processing methods

In 2016–24, progress in four areas of data processing played an important role in the development of digital terrain analysis:

(1) Development of new methods for processing remote sensing data to create global, quasi-global, and regional DEMs (Shean et al., 2016; Tadono et al., 2016; Rizzoli et al., 2017; Novaczek et al., 2019; Abrams et al., 2020; Leong and Horgan, 2020; Zhou et al., 2021; Dorschel et al., 2022). In particular, Harper and Sandwell (2024) proposed a new

b. The commercial WorldDEM with a grid spacing of 0.4″ (Zink et al., 2014; Rizzoli et al., 2017) was obtained by a radargrammetric method (Section 3.1.6) using the TanDEM-X satellite data. Currently, WorldDEM is the most accurate quasi-global DEM (Rizzoli et al., 2017; Gesch, 2018; Gonzalez and Rizzoli, 2018; Grohmann, 2018; Gdulová et al., 2020; González-Moradas and Viveen, 2020; Vassilaki and Stamos, 2020). In terms of root mean square error, its vertical accuracy is 1.4 m (Wessel et al., 2018). However, WorldDEM includes minor errors and artifacts because synthetic aperture radar interferometry (Section 3.1.6) is challenging over steep and irregular mountain slopes due to foreshortening and shadowing effects.

approach for predicting global bathymetry from a large collection of echo-sounding depth measurements (Section 3.1.7) and satellite-derived gravity anomaly data (Section 3.1.8) using neural network regression; the obtained predicted depth model improved on the root mean square error of the previous model by 16%, from 165 to 138 m.

(2) Development of new methods for DEM preprocessing with the following functions:
- Merging complementary DEMs and filling voids in DEMs (Fuss et al., 2016; Yue et al., 2017; Tozer et al., 2019; Guan et al., 2020; Jakobsson et al., 2020; Li S. et al., 2022).
- Correcting DEMs for vegetation cover and human-made objects (Lindsay, 2016c; O'Loughlin et al., 2016; Yamazaki et al., 2017; Gevaert et al., 2018; Meddens et al., 2018; Pijl et al., 2020; Hong et al., 2021; Magruder et al., 2021; Van Nieuwenhuizen et al., 2021a).
- Filtering, smoothing, and removing high-frequency noise, errors, and artifacts in DEMs (Yamazaki et al., 2017; Yue et al., 2017; Ma et al., 2018; Tian et al., 2018; Tozer et al., 2019; Jakobsson et al., 2020; Cai et al., 2023).
- Accurate interpolating DEMs to minimize interpolation errors and artifacts (Lindsay et al., 2019; Chen et al., 2022; Rodriguez-Perez and Sanchez-Carnero, 2022).
- Matching DEMs and hydrographic networks (Barnes, 2016; Lindsay, 2016a, b; Zhou G. et al., 2016, 2017).
- Evaluating DEM accuracy (Mesa-Mingorance and Ariza-López, 2020; Polidori and El Hage, 2020; Ariza-López and Reinoso-Gordo, 2021; Bielski et al., 2024).

In particular, these methods are important because noise, errors, and artifacts existing in any DEM are significantly amplified in morphometric models calculated from DEMs (Lecours et al., 2017a, b; Graf et al., 2018; Moudrý et al., 2018; O'Neil et al., 2019; Riihimäki et al., 2021) (Section 5.4.1).

(3) Development of methods for two-dimensional (2D) and 3D visualization of DEMs. The methods of the 2D group are focused on improving the digital implementation of classical cartographic approaches to the representation of topography, such as contouring (Arundel et al., 2018; Samsonov et al., 2019), hachuring (Geisthövel and Hurni, 2018), hill shading (Jenny et al., 2020) (Sections 2.5.1.1 and 8.3), as well as generalization of topography (Duan et al., 2017; Behrens et al., 2018b; Wu et al., 2019; Jenny, 2020; Feciskanin and Minár, 2021; Samsonov, 2022) (Section 6.1.3) and simultaneous mapping of several morphometric characteristics (Chu et al., 2019) (Section 8.2). The methods of the 3D group allow 3D, high-quality visualization of large DEMs in real time (Müller et al., 2016; Zhai et al., 2016; Bladin et al., 2018; Chen et al., 2018; Graciano et al., 2018; Thornton et al., 2018; Florinsky and Filippov, 2019c, 2021) (Sections 8.5.1.1 and 8.6).

Achievements of computational methods (Chapter 4) and 3D visualization were implemented in the hardware-software complex "Tangible Landscape" (Petrasova et al., 2018). In this complex, two models of a landscape—its digital and physical 3D models—are connected by a cycle of 3D scanning of the physical 3D model, digital modeling, and projecting the results on the physical 3D model. The user can manually change the physical 3D model by sculpting new landforms, placing vegetation patterns, etc. These changes are entered into the digital model by the scanner, after which simulated processes (e.g., slope erosion) are recalculated, and then the results are projected onto the physical 3D model.

(4) Development of machine learning methods (Guo et al., 2016), in particular, random forest algorithms (Belgiu and Drăguţ, 2016; Hengl et al., 2018), and their adaptation to spatial data (Valentine and Kalnins, 2016; Zhang L. et al., 2016; Ma L. et al., 2019). These methods are used to create predictive mapping models of soils, vegetation, etc., in which one of the predictor groups (covariates of training samples) are digital models of morphometric variables derived from DEMs (Chapters 11 and 12).

1.3.2 Applications

Geomorphometric modeling is systematically used in many sciences, such as geomorphology, hydrology, geology, soil science, glaciology, speleology, geobotany, forestry, zoogeography, oceanology, planetology, landslide science, avalanche and fire hazard studies, urban studies, archeology, and others. Let us consider these research fields in detail.

1.3.2.1 Geomorphology

Methods of geomorphometry are so frequently utilized in geomorphology (Sofia, 2020) that sometimes geomorphometry is mistakenly considered a quantitative geomorphology. Geomorphometric modeling, often integrated with machine learning methods, is most widely used in the following research directions:

(1) Automated classifying of landforms and topographic elements based on the sign or value intervals of morphometric variables (Mashimbye and Loggenberg, 2023). Landform classification, or DEM segmentation methods (Section 2.7)

(Libohova et al., 2016; Gruber et al., 2017; Camiz and Poscolieri, 2018; Arundel and Sinha, 2019; Du et al., 2019; Li S. et al., 2020; Na et al., 2021; Nowosad and Stepinski, 2021; Kharchenko, 2023) are used in geomorphological classification, modeling, and mapping of vast territories (Argyriou et al., 2016a; Gorini and Mota, 2016; Luo et al., 2016; Dekavalla and Argialas, 2017a; Amatulli et al., 2018; Bugnicourt et al., 2018; Doumit, 2018; Iwahashi et al., 2018; Mithan et al., 2019; Alvioli et al., 2020; Dyba and Jasiewicz, 2022; Siqueira et al., 2022; Gawrysiak and Kociuba, 2023) as well as in soil science and some other fields (Sections 1.3.2.3 and 1.3.2.9). Notice, however, that a human remains necessary to correct geomorphological models and maps based on the formal quantitative landform classification (Van der Meij et al., 2022).

(2) Automated identifying modern and relict-specific landforms, such as terraces (Clubb et al., 2017; Dai W. et al., 2019a, 2020; Zhao F. et al., 2021), drumlins (Sărăşan et al., 2019), periglacial patterned grounds (Mather et al., 2019), glacial landforms, in particular, till and outwash plains, end and hummocky moraines (Janowski et al., 2022), glaciovolcanic structures (Pedersen, 2016), alluvial fans (Pipaud and Lehmkuhl, 2017), glacial boulders (Nevalainen et al., 2016), linear dunes (Shumack et al., 2020), tidal creeks (Zhao et al., 2019), as well as paleomoraines and traces of paleoseismic soil liquefaction (Middleton et al., 2020).

(3) Studying fluvial processes and landforms. This research direction includes, in particular, analyzing the evolution of valleys and river channels (Bakker and Lane, 2017; Kasprak et al., 2017), spatial patterns and intensity of erosional dissection of mountains (Różycka and Migoń, 2022), debris flow basins (Nikolova et al., 2021), as well as predicting and classifying the river channel types according to their dominant shape and substrate (Lane et al., 2017; Guillon et al., 2020).

(4) Modeling connectivity of water and sediment flows within watersheds (Heckmann et al., 2018). Such works are important for understanding the links between different parts of a watershed and the influence of its configuration, landforms, and human-made objects on the course and continuity of gravity-driven migration and accumulation of substances (Bigelow et al., 2016; Nicoll and Brierley, 2017; Cucchiaro et al., 2019; James et al., 2019; Marchi et al., 2019; Martini et al., 2019; Bhadra et al., 2021; Bollati and Cavalli, 2021; Jing et al., 2022).

(5) Modeling and forecasting gully erosion (Bergonse and Reis, 2016; Mitusov et al., 2017; Pourghasemi et al., 2017; Shary et al., 2017a; Garosi et al., 2018, 2019; Liu K. et al., 2018; Zabihi et al., 2018; Dai et al., 2019b; Gayen et al., 2019; Orti et al., 2019; Tien Bui et al., 2019b; Walker et al., 2020). Studies are often carried out using multi-temporal DEMs and their differences, which are analyzed to estimate the removed material volume and the erosion rate (Cavalli et al., 2017; Cucchiaro et al., 2020). This research direction also includes studies of the material transport by debris flows (Loye et al., 2016) as well as soil erosion (Quijano et al., 2016; Li et al., 2018a, b; Tarolli et al., 2019b; Jenčo et al., 2020; Lu S. et al., 2020), including those depending on the soil tillage technology. DEMs with a grid spacing of 0.5–5 cm are often used in such works.

(6) Modeling and studying aeolian processes and landforms. Some of them are analysis of shifting dunes migration and changes in sand volumes (Brunier et al., 2016; Danchenkov et al., 2019; Grohmann et al., 2020; Rotnicka et al., 2020; Carvalho et al., 2021), evolution of sand islands (Houser et al., 2017; Patton et al., 2019), as well as classification of fixed dunes and determining the direction of paleowaves that formed such dunes (Bernhardson and Alexanderson, 2017).

(7) Studying karst landforms and processes (Silva et al., 2017; Yang et al., 2019), including identification and mapping of karst sinkholes (Wu et al., 2016; Zumpano et al., 2019; Kim et al., 2019; Moreno-Gómez et al., 2019; Vajedian and Motagh, 2019), including those hidden by forest (Hofierka et al., 2018) and agricultural activities (Čeru et al., 2017).

(8) Studying cryolithogenic, glacial, and periglacial processes and landforms, such as colifluction surges and terraces, patterned grounds, polygonal-block microtopography, and frost bumps (Dąbski et al., 2017; Van der Sluijs et al., 2018; Queen et al., 2021), colluvial and alluvial fans (Tomczyk and Ewertowski, 2017), moraine evolution (Tonkin et al., 2016; Westoby et al., 2016; Ely et al., 2017; Midgley et al., 2018), rock glacier dynamics (Colucci et al., 2016; Piermattei et al., 2016; Dall'Asta et al., 2017; Micheletti et al., 2017), as well as meso- and microtopography around gas emission craters (Kizyakov et al., 2017, 2018).

(9) Studying topography of volcanic areas (Gomez, 2018) and its changes after eruptions, in particular, the analysis of lava flows and channels (Favalli and Fornaciai, 2017; Dirscherl and Rossi, 2018; Dai and Howat, 2017), scoria cones (Vörös et al., 2021, 2022), collapsed volcano calderas (Dirscherl and Rossi, 2018), and areas and volumes of teff deposits and their erosion (Dai C. et al., 2020). Most of such works involve the analysis of multi-temporal DEMs. In addition, digital terrain modeling allows for studying the typology and evolution of glaciovolcanic landforms (Pedersen et al., 2020) as well as reconstructing the topography of eroded paleovolcanoes (Karátson et al., 2016).

(10) Quantifying the impact of human activity on the landscape (Tarolli and Sofia, 2016; Tarolli et al., 2019a). Geomorphometric analysis allows identifying ancient and modern anthropogenic landscapes because their morphometric characteristics are statistically different from those of natural landscapes (Brecheisen et al., 2019; Xiang J. et al., 2019; Cao et al., 2020).

(11) Modeling landscape evolution as an interaction result of tectonic and erosion processes (Codilean et al., 2006; Brocklehurst, 2010; Burbank and Anderson, 2012, chap. 11). Most of the works are devoted to real territories (Andreani and Gloaguen, 2016; Elez et al., 2016; Grieve et al., 2016; Hergarten et al., 2016; Owono et al., 2016; Xiong et al., 2016, 2017; Zhang et al., 2017b; Blomdin et al., 2018; Szymanowski et al., 2019; Noriega-Londoño et al., 2020; Delchiaro et al., 2021; Różycka et al., 2021; Byzov and Sankov, 2024). In addition to DEMs, geological, geophysical, and climatic data are used to analyze and study landscape evolution. Also, mathematical models of landscape evolution are being developed (Clubb et al., 2016; Moon et al., 2017; Theodoratos et al., 2018; Galin et al., 2019; Keylock et al., 2021; Byzov and Sankov, 2024). In this case, the object of modeling is a DEM of an abstract territory, which gradually changes in the course of calculations. This approach is convenient and illustrative: visualization of changes in the DEM and morphometric models makes it possible to assess the adequacy of the mathematical model, to see how changes in its parameters work, and, if necessary, to correct them.

Landscapes with a concentration of sandstone rock outcrops and clay erosion pillars (e.g., the Monument Valley and Bryce Canyon in Utah) are difficult to survey, construct a DEM, visualize, and analyze. Typical elements of such sites are steep walls, negative slopes, and overhangs. Such landscapes include rock chaoses and many closed micro-depressions and are marked by poor hydrological connectivity (Jancewicz and Poręba, 2022; Jancewicz et al., 2019). For these terrains, it is possible to obtain DEMs by combined airborne and ground-based laser scanning, as in the case of kurums (Šašak et al., 2019); 3D imaging is optimal; and in digital terrain modeling (Migoń and Kasprzak, 2016; Duszyński et al., 2017) some limitations may arise like in the case of caves (Section 1.3.2.6).

On the Earth's surface, small impact craters should be widespread. However, they are weakly manifested in topography due to the post-impact erosion and sedimentation processes. A promising research direction is revealing and studying such craters, including the influences of the pre-impact topography, impact processes, and post-impact modifications on the final morphology of the craters. Application of digital terrain analysis allows to delimit boundaries of the craters and estimate their diameters, ellipticities, slope gradients, depths, and volumes (Włodarski et al., 2017).

1.3.2.2 Hydrology

The systematic use of geomorphometry in hydrology began in the 1980s to model the hydrological regime of watersheds. There is a well-known, spatially distributed hydrological model TOPMODEL (Beven et al., 2021) (Section 11.2), in which a combined local–nonlocal morphometric variable, topographic wetness index (Section 2.6.1), is used to take into account the influence of topography (Meles et al., 2020). The connection between geomorphometry and hydrology has been cemented in the names of some morphometric variables (e.g., flow attributes, catchment area—Sections 2.2.1 and 2.3.1), although they can be used to analyze any surface regardless of the presence of water (cf. Sections 1.3.2.9 and 1.3.2.10).

Mathematical modeling of the hydrological regime involves the analysis of DEMs and morphometric models, soil, lithological, and meteorological data, and allows predicting precipitation, surface runoff, evaporation, infiltration, etc. (Guida et al., 2016; Wiekenkamp et al., 2016; Peckham et al., 2017). Machine learning algorithms, such as random forest and maximum entropy ones, which use morphometric variables in the training samples, give good results for groundwater prediction in arid regions (Rahmati et al., 2016).

DEM analysis is common for identifying watersheds (Lindsay, 2016c; Liu et al., 2020a), classifying them by hydrological characteristics (Gaucherel et al., 2017; Gao et al., 2018) and response to precipitation (Papageorgaki and Nalbantis, 2016), identifying wetlands (O'Neil et al., 2018), and determining reservoir volumes (Zhang S. et al., 2016; Liu et al., 2020b; Zhu et al., 2021). Typical tasks performed with geomorphometric analysis are predicting floods due to water level rise in rivers, lakes, and reservoirs (Koriche and Rientjes, 2016; Abdel-Fattah et al., 2017; Cunha et al., 2017; Wing et al., 2019; Medvedev et al., 2020; Chekina and Khromykh, 2023) as well as modeling flash flood susceptibility (Khosravi et al., 2018; Tien Bui et al., 2019a; Nhu et al., 2020; Ngo et al., 2021). An ultra-high-resolution DEM analysis allows for investigating seasonal brine microflows and associated changes in the surface of salt plateau (Bhardwaj et al., 2019b).

The high-resolution ArcticDEM (Section 1.3.1.2) has proven useful for mapping the drainage network of the Arctic (Lu X. et al., 2020) as well as for studying seasonal fluctuations in water levels in small, remote Arctic rivers, which are logistically difficult to monitor (Dai et al., 2018).

1.3.2.3 Soil science

Topography is one of the main factors in soil formation (Chapter 9 and Section 11.1). For about 40 years, geomorphometric modeling has been widely used to study the spatial distribution of physical, chemical, and biological properties of soils at different scales (Chapter 11). Digital terrain analysis is one of the cornerstones of pedometrics—the modern quantitative soil science (Burrough et al., 1994; Webster, 1994; McBratney et al., 2018, 2019; Wadoux et al., 2021, 2023)—because geomorphometry is one of the key bases for predictive digital soil mapping, the primary focus of pedometrics for about three decades (McBratney et al., 2000, 2003; Scull et al., 2003; Lagacherie and McBratney, 2007; Walter et al., 2007; Lagacherie, 2008; MacMillan, 2008; Grunwald, 2009; Heung et al., 2016; Minasny and McBratney, 2016; Malone et al., 2018; Ma Y. et al., 2019; Shary, 2023).

The current progress in predictive digital soil mapping is discussed in Section 11.3. Our analysis of such studies performed in 2016–24 (Tables 11.1 and 11.2) showed that their distinctive feature is the transition from statistical and geostatistical models used for prediction and mapping of soil properties to machine learning ones (Lamichhane et al., 2019). It should be emphasized, however, that in order to obtain an adequate predictive soil model, the correct selection of predictors, rather than the mathematical apparatus, plays a decisive role.

A promising research direction is the concept of teleconnection in the spatial modeling of soil cover (Behrens et al., 2019). The concept considers interregional connections between geosystems. This concept is close to the modeling of stream connectivity in geomorphology (Section 1.3.2.1). In such studies, the use of geomorphometric methods is natural because nonlocal and combined morphometric variables (e.g., catchment area and topographic wetness index—Sections 2.3 and 2.6) can serve as metrics of mass transfer over relatively long distances.

1.3.2.4 Geology

In geology, geomorphometry is used to analyze DEMs of both the land surface and surfaces of stratigraphic horizons and geological bodies (Chapters 13 and 14). In the latter case, a 3D seismic survey (Section 3.1.11) can be performed to create DEMs (Wu et al., 2017; Zhou L. et al., 2016; Tesch et al., 2018). The key tasks are the analysis of geological composition as well as detection of active fault zones (Ren et al., 2018) and geological structures weakly expressed in the topography, such as lineaments (Yeomans et al., 2019; Elmahdy et al., 2021, 2022; Kelka et al., 2022; El Koundi et al., 2024), fracture zones, folds, and dikes (Booysen et al., 2019). The latter task (Chapter 14) can be solved due to the high sensitivity of local morphometric variables (Section 2.2) to minor changes in elevation (Section 5.4.1).

A classical approach to prospecting geology—the Filosofov method (Section 6.2.2), identifying anomalous patterns of the drainage and thalweg networks, which can indicate hydrocarbon traps, was successfully implemented using geomorphometric methods (Ibanez et al., 2016).

Morphometric models are used to refine geological maps of sedimentary basins in remote territories with forest cover (Valeriano and Rossetti, 2017) as well as to identify and map zones of neotectonic deformations for seismic risk assessment (Argyriou et al., 2016b, 2017b). The analysis of a multiscale data set—ALOS AW3D30 DSM, archived Corona images, unmanned aerial images, and UAS-based morphometric data—allowed to clarify the conditions of mud volcanism development in the Taman Peninsula and to detect inactive mud volcanic formations hidden by agricultural activities (Skrypitsyna et al., 2020).

1.3.2.5 Glaciology

Glaciological processes are usually reflected in the glacier topography and can depend on the characteristics of ice–bedrock contact surface. Therefore, geomorphometric methods are used to analyze both the glacier and subglacial topographies (Chapters 17–19).

DEMs of glacier surfaces can be photogrammetrically created using archived aerial photographs (Mertes et al., 2017; Mölg and Bolch, 2017), satellite scenes (Dehecq et al., 2016; Ragettli et al., 2016; Podgórski et al., 2019), ground images (Piermattei et al., 2016; Watson et al., 2017), as well as laser scanning data (Bhardwaj et al., 2016a). However, the development of glacial geomorphometry (Bhardwaj et al., 2016b; Gaffey and Bhardwaj, 2020) became possible primarily due to the progress in the technology of unmanned aerial survey (Sections 1.3.1.1 and 3.1.4) adapted to the conditions of the high mountains and polar regions (Bliakharskii, 2019; Bliakharskii et al., 2019a; Ewertowski et al., 2019; Taylor et al., 2021; Li et al., 2023).

The interrelated problems of glaciology, which are solved with digital terrain analysis, are studies of the dynamics of glacier motion and evolution (Pogorelov et al., 2017; Seier et al., 2017; Rossini et al., 2018; Chudley et al., 2019; Xue et al., 2021; Qiao et al., 2023) (Chapter 17), including determination of the horizontal and vertical speed of glacier movement (Kraaijenbrink et al., 2016a; Vincent et al., 2016; Gibson et al., 2017; Wigmore and Mark, 2017; Skrypitsyna et al., 2023), estimation of the glacier melting rate (Brun et al., 2016, 2018; Ragettli et al., 2016; Thompson et al., 2016; Mallalieu et al., 2017; Watson et al., 2017; Zhou Y. et al., 2017; Bash and Moorman, 2020), and determination of the glacier mass balance (Brun et al., 2017) as well as its surface temperature (Kraaijenbrink et al., 2018). The use of UAS-derived DEMs, characterized by ultra-high resolution and high accuracy, made it possible to detect high spatial variability in glacier movement (Wigmore and Mark, 2017), glacier melting (Buri et al., 2016), and even minor changes in glacier surface geometry occurring over short (from 24 h) time intervals (Bash et al., 2018). When solving these research tasks, differences between multi-temporal DEMs are usually calculated and analyzed (Chapter 17). A finer, more sophisticated geomorphometric analysis allows studying the dependence of glacier dynamics on slope geometry (Li Y. et al., 2017; Sam et al., 2018).

Models and maps of morphometric variables derived from high-resolution DEMs of the glacier surface make it possible to identify weakly visible ice crevasses hidden by snow bridges, which pose a danger to glaciologists and polar explorers (Florinsky and Bliakharskii, 2019a; Foroutan et al., 2019; Ishalina et al., 2021) (Chapter 18). Rapid unmanned aerial survey and geomorphometric modeling provide an opportunity to study the risk of glaciological disasters (Fugazza et al., 2018; Florinsky and Bliakharskii, 2019b; Li T. et al., 2020) (Chapter 19).

Studying the Bellingshausen Ice Dome (King George Island, Antarctica), Suleymanov et al. (2024) applied elements of geomorphometric modeling and predictive digital soil mapping (Section 1.3.2.3) to evaluate and predict the influence of glacier topography on the spatial distribution of physical and chemical properties of sediments on glacier surface.

In contrast to other applications of geomorphometry, DEMs and morphometric models are rarely used in machine learning for glaciological research. In particular, machine learning was applied to identify and classify glaciers covered by dense surface moraine, which are indistinguishable from adjacent ice-free valleys on the images (Alifu et al., 2016, 2020), as well as to identify ice cliffs and ice ponds on glaciers of this type (Kraaijenbrink et al., 2016b).

The popular task of determining snow cover thickness (Bühler et al., 2016, 2017; De Michele et al., 2016; Cimoli et al., 2017; Avanzi et al., 2018), which is solved by analyzing DEMs obtained in different seasons (e.g., autumn and spring), can also conditionally be referred to glacial geomorphometry.

1.3.2.6 Speleology

Caves have a number of properties and features—e.g., space confinement and the presence of speleosculptures—that make it difficult to model their topography. To create cave DEMs, ground laser scanning technology (Section 3.1.5) is predominantly used (Idrees and Pradhan, 2016). A photogrammetric system was recently developed to create DEMs of underground cavities using a photo camera (Shcherbakov et al., 2023).

Geomorphometric modeling of caves is used to identify the underground hydrographic network, phreatic channels, mineral crusts on the walls, etc. (Gallay et al., 2016; Fabbri et al., 2017; De Waele et al., 2018). However, the theory of geomorphometry was developed in the topographic surface approximation, one of the limitations of which is the single valuedness of the elevation function (see Limitation 2.1 in Section 2.1.1). This means that one elevation value z corresponds to one planimetric coordinate (x, y). In the case of a cave, the function is multivalued: one plan coordinate (x, y) has at least two elevation values z_1 and z_2. Therefore, not all morphometric variables can be mathematically correctly used to analyze the cave topography. Form attributes, which are gravity field invariants (Section 2.2.1), are suitable for these purposes without limitations (Gallay et al., 2016). Flow attributes, which are gravity field-specific variables (Section 2.2.1), can only be used to analyze the topography of the cave floor and the lower part of its walls.

1.3.2.7 Geobotany and forestry

Topography has direct and indirect impacts on the vegetation cover as a whole and individual plant communities. This is because topography controls the hydrological and thermal regimes of slopes as well as soil properties. The functioning of forest ecosystems depends on soil fertility indicators, which are significantly influenced by topography (Scholten et al., 2017). In geobotany, digital terrain analysis is predominantly used for predictive mapping of vegetation cover properties on the basis of statistical and machine learning models, in which predictors or covariates are morphometric variables (Florinsky and Kuryakova, 1996; Sharaya and Shary, 2011).

Morphometric indices are good predictors for characteristics of tropical rainforest structure (e.g., basal area of trees) even on plain territories (Bispo et al., 2016). Morphometric variables together with spectral and textural characteristics of satellite images were used to predict forest volume and productivity of *Eucalyptus* plantations (Dos Reis et al., 2019, 2022) as well as to map dominant plant communities in mixed forests (Liu Y. et al., 2018) and savannas (Hurskainen et al., 2019). Geodiversity as a function of morphometric attributes is a controlling factor for the richness of species diversity of vascular plants (Bailey et al., 2017, 2018). A model of the aboveground biomass in the Norwegian boreal ecozone was constructed using the forest canopy height model (calculated as the difference between ArcticDEM and Norway DEM), Sentinel-2 mosaic, and forest inventory data (Puliti et al., 2020). An analysis of the relationship between the distribution of arid sparse forest shrubs and fracture zones identified from hill-shaded maps (Section 2.5.1.1) allowed Guirado et al. (2018) to identify groundwater-dependent arid ecosystems.

Note that the use of morphometric data to explain and predict the spatial variability of forest characteristics gives good results only for a tree community as a whole and some common species. For rare species, spatial prediction can be difficult (Bispo et al., 2017). However, geomorphometric modeling can correctly describe the spatial variability of herbaceous communities in steppe and forest-steppe ecozones (Alexander et al., 2016; Shary et al., 2017b).

1.3.2.8 Zoogeography

There are few known examples of the use of digital terrain modeling in zoogeography to study the influence of land topography on the spatial distribution or behavior of animals. The rarity of such works is associated with the difficulty of obtaining statistically representative zoological samples, the mobility of objects of research, and the high level of animal adaptability to environmental conditions: the higher the organization of a species, the less its representatives are affected and controlled by topography (with the obvious exception of altitude zonality).

There were several studies of relationships between topographic characteristics and winter survival of spotted-winged *Drosophila* populations on the northwestern coast of North America (Thistlewood et al., 2018) as well as between topography and habitats of South American frugivorous ungulates, rodents, birds, and small cats (Nagy-Reis et al., 2017a, b). For *Drosophila,* some dependence on elevation and topographic wetness index (a measure of soil moisture accumulation—Section 2.6.1) was found. For frugivores and felines, however, the impact of topography turned out to be negligibly small.

1.3.2.9 Oceanology

The development of marine geomorphometry (Lecours et al., 2016b) was facilitated by two key factors. First, the Nippon Foundation and the General Bathymetry Chart of the Ocean (GEBCO) group of experts initiated an international project, Seabed 2030, to create a global, publicly available bathymetric DEM with high resolution and accuracy (Wölfl et al., 2019). Most of this DEM will be created from multibeam echo-sounding data (Section 3.1.7). Second, autonomous underwater vehicle technology is currently undergoing rapid development (González-García et al., 2020). The analysis of bathymetric DEMs is in demand in marine geomorphological and biological studies.

Methods based on principles of DEM classification (Section 2.7) were developed to classify submarine landforms (Dekavalla and Argialas, 2017b) and to map hierarchically continental margin topography (Sowers et al., 2020). Applying the boosted regression tree technique with local morphometric variables (Section 2.2) as covariates, made it possible to construct prediction maps of the bottom sediment distribution with subdivision into mud, sand, and gravel fractions (Misiuk et al., 2018). A landform classification technique based on the bathymetric position index (an analog of the topographic position index—Sections 2.3.2 and 2.7.4) provided an opportunity for predictive mapping of benthic habitats in the Amazon Shelf (Lavagnino et al., 2020) and the Mediterranean algal reefs (Marchese et al., 2020). The use of models of local morphometric variables (Section 2.2) made it possible to predict the distribution of cold-water coral colonies (Tong et al., 2016; Bargain et al., 2017).

For the Arctic Ocean, digital terrain modeling helped clarify the role of the submarine topography in controlling seawater properties and local deep currents, which are crucial for heat distribution in this region. It was found that even minor elements of the submarine topography affect the water exchange between neighboring basins with different densities and salinities (Björk et al., 2018).

Geomorphometric analysis allowed Copping et al. (2018) to determine the reasons for whale shark aggregations in some coastal areas. These areas are much shallower than the neighboring ones, but they are at the same time located much closer to the deep-water zones, and their slopes are much steeper than in other places. Such features of the submarine topography cause the rise of nutrient-enriched cold waters (upwelling) and plankton development and thus attract species with a filtration type of feeding, which include whale sharks.

Digital terrain analysis of multi-temporal bathymetric DEMs of the Lido Inlet, which connects the Adriatic Sea with the Venetian Lagoon, allowed to study changes in the submarine topography, in particular, the evolution of submarine dunes caused by the construction and operation of hydraulic engineering structures protecting Venice from floods (Toso et al., 2019).

1.3.2.10 Planetology

Methods of digital terrain analysis can be applied to DEMs of solid planets, moons, asteroids, and comets. In this case, researchers use two types of DEMs. First, there are free, publicly available global DEMs with medium resolution and accuracy for Mars (Smith et al., 2003), Venus (Ford et al., 1992), the Moon (Neumann, 2008), and some other celestial bodies. Second, one can specially create DEMs of relatively small areas of celestial bodies with higher resolution and accuracy. DEMs of this sort can be generated from data captured by particular satellite sensors.

The following multiscale tasks are typically the focus of DTM-based planetological research:

- Studying the geological composition of territories as well as identifying and analyzing particular geological structures, such as sinuous rilles in lunar seas—traces of small lava channels or collapsed subsurface lava tubes (Podda et al., 2020), Martian scoria cones (Vörös and Székely, 2022), etc.
- Analyzing topographic roughness as an indirect sign of the genesis and relative age of the terrain and underlying rocks (Fa et al., 2016; Pardo-Igúzquiza and Dowd, 2021).
- Analyzing impact crater morphometry, which provides information on the age, rheological properties of rocks, and erosion rate (Robbins et al., 2017; Wang et al., 2020; Villaça et al., 2021).
- Analyzing the structure of paleodeltas and fans (Goudge et al., 2017), gully networks (Conway and Balme, 2016), dune migration (Cardinale et al., 2016; Runyon et al., 2017; Vaz et al., 2017), and erosion patterns (Brusnikin et al., 2016) in studies of fluvial, aeolian, and slope processes on Mars.
- Analyzing and mapping past and planned spacecraft landing sites (Karachevtseva et al., 2017; Bhardwaj et al., 2019a).
- Geomorphological and geomorphometric multiscale mapping (Wang J. et al., 2017; Florinsky, 2018b).

One of the most intriguing research studies of recent years was the study of Ontario Lacus—the hydrocarbon lake on Saturn's satellite Titan (Dhingra et al., 2018). It was not clear why there is only one large lake in the southern hemisphere of Titan, whereas the neighboring deeper depressions remain dry. Analyzing a catchment area (a nonlocal morphometric variable—Section 2.3.1) derived from the Titanian DEM, it was found that (1) Ontario Lacus catchment area is 5.5% of the Titan's total area, and (2) the ratio of the catchment area to the lake area equals 83. After hydrocarbon rains, this catchment could provide a surface runoff of such a volume of methane and ethane, which would be enough to fill Ontario Lakus in less than half a year of Titan (~15 Earth years). Dry depressions could have been formed by karst-like processes of rock dissolution (Hayes et al., 2017).

1.3.2.11 Landslide science

Geomorphometric methods are systematically used for detection of landslides and predictive mapping of landslide hazards (Zhong et al., 2020) as well as for inventory and classification of landslides (Niculiță, 2016; Migoń et al., 2017; Pánek et al., 2019). This is because the slope shape usually plays a crucial role in the formation of landslides, all other factors being equal (i.e., soil and rock characteristics, hydrological regime, precipitation amount, and human-made load). Even if a rock property is a key factor in landslide development (e.g., the presence of monoclinal flysch rocks), DEM analysis remains a necessary element in the study of such territories (Břežný and Pánek, 2017).

In 2016–24, a distinctive feature of DTM-based landslide studies was the extensive use of machine learning approaches (Pourghasemi and Rahmati, 2018). For predictive mapping of landslide hazard, researchers applied logistic regression (Chen et al., 2017b; Chang et al., 2019; Tanyas et al., 2019; Jacobs et al., 2020), artificial neural networks (Vasu and Lee, 2016; Chen et al., 2017a; Ghorbanzadeh et al., 2019), algorithms of random forest (Chen et al., 2017b; Chang et al. 2019; Ghorbanzadeh et al., 2019; Sîrbu et al., 2019; Maxwell et al., 2020) and decision trees (Chen et al., 2017b), methods of support vector machine (Chen et al., 2017a; Chang et al., 2019; Ghorbanzadeh et al., 2019; Hu Q. et al., 2020), fuzzy logics (Feizizadeh et al., 2017; Hong et al., 2017), and analytic hierarchy process (Nicu and Asăndulesei, 2018; Nachappa et al., 2020) as well as object-based image analysis (Feizizadeh et al., 2017; Nachappa et al., 2020). The local G_i^* statistic (Shi W. et al., 2018) and the geomorphon classification approach (Luo and Liu, 2018) (Section 2.7.4) were also utilized.

As one of the predictors/covariates groups in the above-cited studies, landslide researchers used models of morphometric variables (e.g., slope and aspect, horizontal and vertical curvature, topographic wetness index—Chapter 2) derived from high-resolution DEMs obtained primarily by airborne laser surveys (Section 3.1.5). In addition to morphometry, data on other factors of landslide formation (e.g., lithology, hydrology, precipitation, and vegetation cover) were also used in those models.

A combined analysis of two data sets—hill-shaded maps (Section 2.5.1.1) used for identifying inactive Holocene landslides as well as data on archaeological monuments located on these landslides—allowed Niculiță et al. (2019) to determine the chronology and evolution of landslide activity in the Moldavian Plateau. If several lidar surveys of a site were performed before and after the displacement of rock masses and, as a result, a series of multi-temporal DEMs is available, it is possible to conduct a detailed analysis of the topographic changes that occurred and to map the distribution of displaced masses along the slope (Gatter et al., 2018).

1.3.2.12 Avalanche hazard

The problem of avalanche hazard prediction (Eckerstorfer et al., 2016) is relatively close to the problem of landslide detection (Section 1.3.2.11). This is because of two factors. First, the shape of slopes largely controls the distribution, accumulation, and redistribution of snow masses. Second, the avalanche descent can be considered as a gravitational slope process. Thus, the DTM-based methods for identifying and predictive mapping of the avalanche hazard are close to the approaches to landslide detection and predictive mapping, as discussed earlier (Kumar et al., 2016, 2017, 2019; Veitinger et al., 2016; Bühler et al., 2018; Choubin et al., 2019; Rahmati et al., 2019).

1.3.2.13 Fire hazard

Topography is an important factor influencing fire hazard and fire behavior in natural landscapes (Pettinari and Chuvieco, 2020). This is because topography controls (a) the distribution of vegetation cover (Section 1.3.2.7), which is the fuel for fire and (b) microclimatic characteristics influencing the course of fire, such as wind rose, temperature, relative humidity, and precipitation, which in turn depends on the altitude, slope gradient, aspect, and shape. In particular, the slope gradient (Section 2.2.2) can influence fire propagation, as the fire preheats the vegetative cover in front of the fire front when it moves upward, thus drying it out and accelerating its ignition.

Comparing three methods of statistics and machine learning (i.e., boosted regression tree, generalized additive model, and random forest), Pourtaghi et al. (2016) demonstrated that it is necessary to use morphometric variables controlling the spatial distribution of moisture in the landscape in predictive mapping of forest fire hazard. For a detailed discussion of the topographic control of soil moisture, see Chapter 9.

1.3.2.14 Urban studies

The use of geomorphometry in urban studies is fraught with difficulties. In a city, the natural topography is changed and supplemented by artificial elements with different covers (Łajczak et al., 2021). Buildings, road networks, and other human-made objects play the role of "topography." All this has an impact on the thermal and hydrological regime of the territory. Therefore, digital surface models (DSMs), including envelope surface of buildings, should be used as input data rather than DEMs (cf. Definitions 3.1 and 3.3 in Chapter 3).

Although obtaining DSMs and 3D models of the city has become a routine procedure (Liang et al., 2017), their geomorphometric analysis is nontrivial. This is due in particular to the fact that such DSMs are full of elevation function discontinuities related to vertical walls of houses that inevitably lead to artifacts in models due to the Gibbs phenomenon (Section 5.5).

The key directions of urban geomorphometry are analyzing topography for construction of new buildings and roads to minimize an impact on the landscape (Loro et al., 2017), modeling urban surface water flows (Leitão et al., 2016; Qin et al., 2018) and flooding (Darabi et al., 2019; Huang et al., 2019; Rong et al., 2020; Muthusamy et al., 2021), analyzing traffic noise propagation (Kharchenko et al., 2023), wind effects (Kharchenko, 2016; Shi et al., 2017; García-Romero et al., 2019), and temperature regime (Shi Y. et al., 2018; Hofierka et al., 2020; Badaro-Saliba et al., 2021; Chang et al., 2021; Doumit and Sakr, 2021) as well as their impact on the comfort of living. Studies have begun on adapting methods of predictive digital soil mapping (Section 11.3) to urban conditions (Mohamed, 2020).

As part of the assessment of the aesthetic and recreational properties of the urban landscape, Tabrizian et al. (2020) proposed a method for mapping the rehabilitation potential of park landscapes. A high-resolution textured DEM is used to analyze viewsheds (Section 2.5.3) and to calculate morphometric indicators of landscape composition. A predictive map is constructed on the basis of a statistical model calculated by comparing the indicators for a sample of landscapes and subjective assessments of their rehabilitation potential obtained through a survey using virtual reality technology.

1.3.2.15 Archeology

In archaeology, methods of geomorphometry are applied, first of all, to search for hidden traces of human activity (buried ruins of ancient settlements and fortifications, abandoned roads, etc.), which, as a rule, are poorly expressed in the microtopography (Luo L. et al., 2019; Gillings et al., 2020). The solution to this problem is possible due to the sensitivity of partial derivatives of elevation (Section 5.4.1), which are used to derive local morphometric variables from DEMs (Chapters 2 and 4). Even a usual hill shading (Section 2.5.1.1) allows revealing ruins of settlements (Canuto et al., 2018; Masini et al., 2018) and temples (Evans, 2016), abandoned mines (Gallwey et al., 2019), burial mounds (Freeland et al., 2016; Wang S. et al., 2017), military traces (e.g., trenches, explosive craters, dugouts) (Gheyle et al., 2018), and so on.

Archaeologists commonly use DEMs obtained by aerial lidar survey (Štular et al., 2021) (Section 3.1.5), which makes it possible to study areas with dense forest cover (Evans, 2016; Freeland et al., 2016; Wang S. et al., 2017; Canuto et al., 2018; Guyot et al., 2018; Inomata et al., 2018; Masini et al., 2018; Gallwey et al., 2019; Luo L. et al., 2019). To analyze morphometric models derived from DEMs, researchers apply traditional methods of image processing and pattern recognition (Wang S. et al., 2017; Masini et al., 2018) as well as machine learning techniques (Guyot et al., 2018). For example, a random forest algorithm was used for automated detection of burial mounds of Bronze Age cultures in northeastern

Romania (Niculiţă, 2020). A convolutional neural network previously trained to detect lunar craters was utilized for mapping test pits and shallow excavations of abandoned tin and copper mines in England (Gallwey et al., 2019).

DEMs obtained by photogrammetric methods (Section 3.1.3) from archival aerial photographs can be useful for archaeological reconnaissance of territories under strong anthropogenic pressure (Papworth et al., 2016; Sevara et al., 2018). To document the ongoing excavations and their supplementary exploration, it is easier to work with DEMs generated from unmanned aerial survey images (Waagen, 2019) (Section 3.1.4).

In planning underwater archaeological studies, it is optimal to use high-resolution bathymetric DEMs derived from multibeam echo-sounding data (Menna et al., 2018) (Section 3.1.7). To record underwater excavations and to 3D visualize studied objects (e.g., sunken structures and ships), one can use DEMs obtained by the photogrammetric processing of underwater images (Yamafune et al., 2017; Pacheco-Ruiz et al., 2018) (Sections 3.1.3 and 3.1.4).

Modeling and visualization of the topography of archaeological sites and adjacent territories can help to reconstruct paleoenvironmental conditions (Popa and Knitter, 2016; Tapete et al., 2017) and social and economic relations in ancient societies (Evans, 2016; Chase and Chase, 2017; O'Reilly et al., 2017; Richards-Rissetto, 2017; Inomata et al., 2018; Garrison et al., 2019), as well as paleolandscape perception by the ancient population (Knitter et al., 2019). A particular task is to analyze the influence of topography on the choice of settlement sites in antiquity. For the Minoan civilization (Crete Island), Argyriou et al. (2017a) demonstrated that there were no relationships between the settlement locations and the topography in the Early Minoan period, but a significant relationship emerged later when the level of social organization increased.

Geomorphometric modeling and 3D visualization are also used in documentation, museification, and analysis of the shape and texture of archaeological artifacts, such as ceramics (Garstki, 2017) and petroglyphs (Horn et al., 2022). This research direction is close to studies using digital terrain modeling for analysis of microtopography of geological samples (Pollard et al., 2004), metrological control of micro- and nanotopography of various items (Pike, 2001; Dedkova and Florinsky, 2023), analysis and control of road pavement conditions (Barbarella et al., 2022), as well as forensic analysis of footwear and wheel marks (Larsen et al., 2021). In those cases, researchers produce and analyze DEMs with micro- and even nanometer-range resolution and accuracy. However, such applications of digital terrain analysis are beyond the scope of this review and book.

Part I

Principles and methods of digital terrain modeling

Chapter 2

Topographic surface and its characterization

2.1 Topographic surface

2.1.1 Definition and limitations

In this book, we consider the following six types of real surfaces studied in geosciences (Fig. 2.1):

1. *Land topography*: the surface between the lithosphere and the atmosphere (the biosphere and the anthroposphere are considered as removed).
2. *Submarine topography*: the surface between the lithosphere and the hydrosphere, seabed, or ocean floor.
3. *Glacier topography*: the surface between ice sheets/glaciers (a part of the cryosphere) and the atmosphere.[a]
4. *Subglacial topography*: the surface between ice sheets/glaciers and the lithosphere, or the ice-bedrock contact surface.
5. *Subterranean*, or *subsurface topography*: contact surfaces between soil horizons, stratigraphic horizons, and geological bodies.
6. *Extraterrestrial topography*: the surface between the lithosphere of a solid planetary body and space or the atmosphere, if it exists.

These real surfaces are too complex for rigorous mathematical treatment because they are not smooth and regular (Shary, 2008). However, in practice, it is sufficient to approximate them by the topographic surface.

Definition 2.1 (Topographic surface). A topographic surface is a closed, oriented, continuously differentiable, two-dimensional manifold S in the three-dimensional Euclidean space E^3.

Definition 2.2 (Elevation). Elevation is the distance from a reference surface to a given point of the topographic surface measured along the direction of the gravitational acceleration vector (**g**) at the given point.

Elevation can be positive and negative, upward and downward, respectively, from the reference surface.

The five key limitations are valid for the topographic surface (Evans, 1979; Mark, 1979; Shary, 1991, 1995):

Limitation 2.1 (Single valuedness). The topographic surface is uniquely defined by a continuous, single-valued bivariate function

$$z = f(x, y), \tag{2.1}$$

where z is elevation, x and y are planimetric coordinates.

In particular, this condition means that subterranean cavities, grottos, and similar landforms are ignored.

Limitation 2.2 (Smoothness). The function z (Eq. 2.1) is smooth; that is, the topographic surface has derivatives of all orders.

In this book, we systematically use the first- (p and q), second- (r, t, and s), and third-order (g, h, k, and m) partial derivatives of elevation:

$$p = \frac{\partial z}{\partial x}, \quad q = \frac{\partial z}{\partial y}, \quad r = \frac{\partial^2 z}{\partial x^2}, \quad t = \frac{\partial^2 z}{\partial y^2}, \quad s = \frac{\partial^2 z}{\partial x \partial y}, \quad g = \frac{\partial^3 z}{\partial x^3}, \quad h = \frac{\partial^3 z}{\partial y^3}, \quad k = \frac{\partial^3 z}{\partial x^2 \partial y}, \quad m = \frac{\partial^3 z}{\partial x \partial y^2}. \tag{2.2}$$

a. Do not confuse with glacial topography, which is a type of land topography created by glacier impact, such as glacier movement, melting, and so on.

Digital Terrain Analysis. https://doi.org/10.1016/B978-0-443-24798-9.00018-2

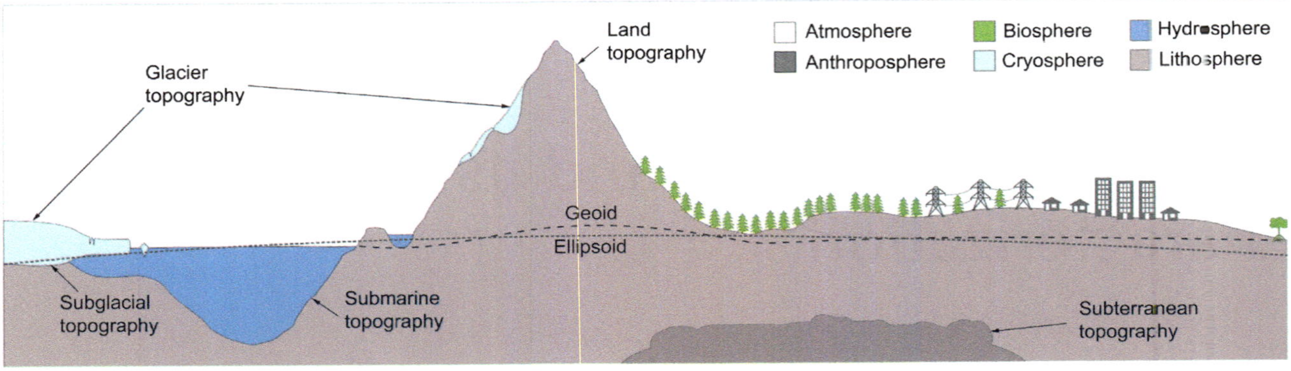

FIG. 2.1 Types of real surfaces studied in geosciences. *(Modified from Guth, P.L., Van Niekerk, A., Grohmann, C.H., Muller, J.-P., Hawker, L., Florinsky, I.V. et al., 2021. Digital elevation models: terminology and definitions. Remote Sens. (Basel) 13 (18), 3581, Fig. 1, © Authors, distributed under the terms and conditions of the Creative Commons Attribution (CC BY) license.)*

Limitation 2.3 (Uniform gravity). The topographic surface is situated in a uniform gravitational field with the gravitational acceleration vector **g** parallel to the *Z*-axis and directed downwards:

$$\mathbf{g} = -\dot{g}\mathbf{k}, \tag{2.3}$$

where $\dot{g}$ is the gravitational acceleration constant and **k** is a unit vector in the directions of the *Z*-axis (Fig. 2.2A).

This condition is important because the topographic surface is situated in the gravity field. Description of the topographic surface by measuring elevations (Section 3.1) implicitly considers the direction of the gravitational acceleration vector. Notice that Limitation 2.3 can be realistic for sufficiently small portions of the geoid for which the equipotential surface can be considered a plane. Thus, we obtain the next limitation.

Limitation 2.4 (Planar size limitedness). The planar size of the topographic surface is essentially less than the planet's radius.

It is generally assumed that the planet's curvature may be ignored if the size of the surface portion is less than at most 0.1 of the average radius of the planet.

Limitation 2.5 (Scale dependentness). The topographic surface is a scale-dependent surface (Clarke, 1988).

This condition means that a fractal component of topography can be considered a high-frequency noise. In this book, we do not discuss fractal topographic models (Clarke, 1988; Xu et al., 1993; Ivanov, 1994; McClean and Evans, 2000).

2.1.2 Contour line and special points[b]

Definition 2.3 (Contour line). In a uniform gravitational field (Eq. 2.3), a contour line is a set of intersection points of the topographic surface *S* and a horizontal plane (Shary, 1991).

A contour line is a plane curve (Fig. 2.2B and C) described by the equation:

$$z(x, y) = \text{const.} \tag{2.4}$$

The sufficient condition for a contour line (Eq. 2.4) to be a differentiable curve may be obtained from the theorem of implicit functions (Nikolsky, 1977a, § 7.16). This condition is as follows:

$$p^2 + q^2 > 0, \tag{2.5}$$

where p and q are the first-order partial derivatives of elevation (Eq. 2.2). Here, a curve is a segment of a contour line for which a one-to-one and unique solution of Eq. (2.4) exists in the form of $y = y(x)$ or $x = x(y)$ in the vicinity of each point. The inequality (Eq. 2.5) is valid for a set N of nonspecial points. A contour line is a curve at nonspecial points.

b. The mathematics of Section 2.1.2 is based on the materials from the appendix prepared by P.A. Shary for the first edition of this book. *(Modified from Shary, P.A., 2012. The mathematical basis of local morphometric variables. In: Florinsky, I.V., Digital Terrain Analysis in Soil Science and Geology, first ed. Academic Press, Amsterdam, The Netherlands, pp. 289–313, Section A.1.2, reprinted with permission from Elsevier.)*

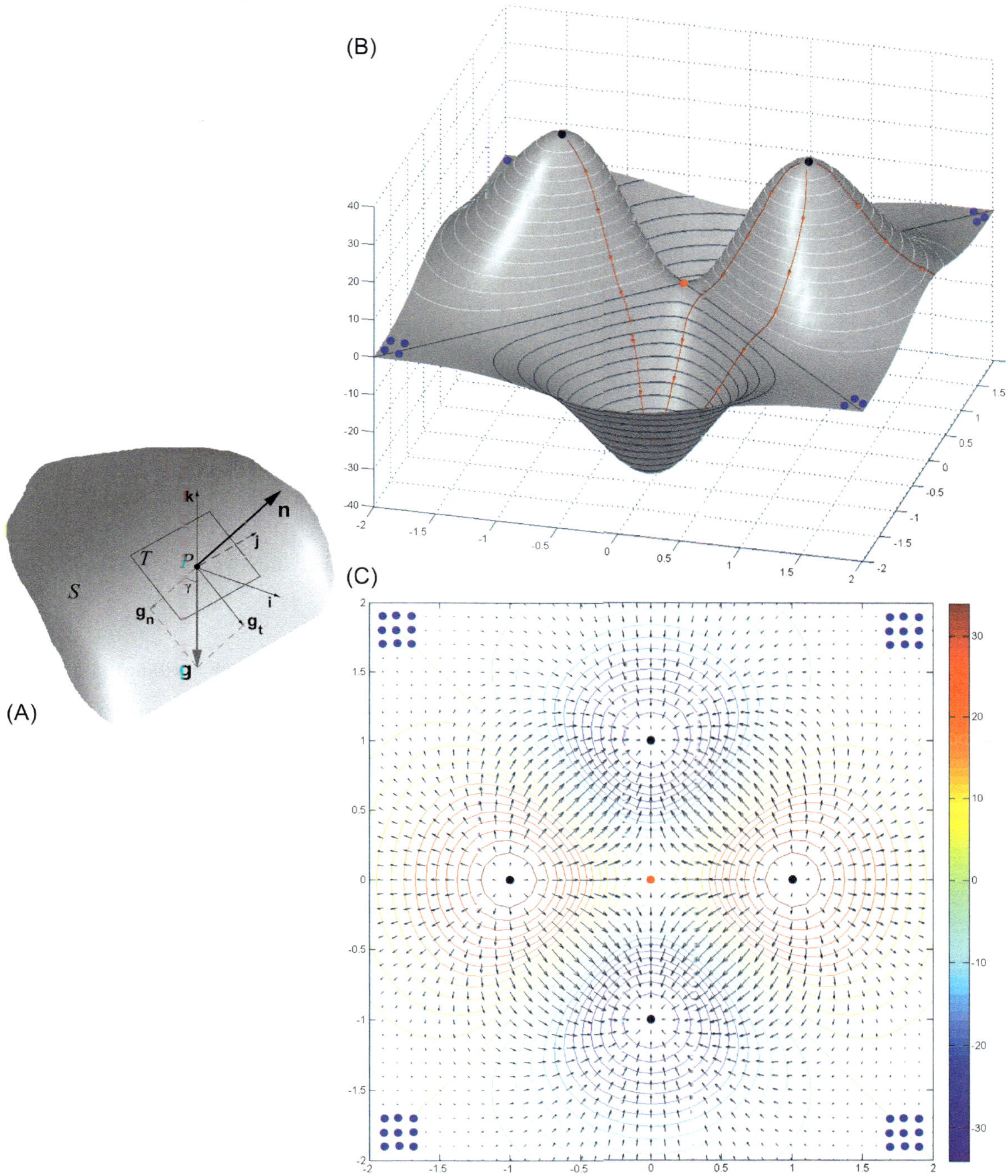

FIG. 2.2 Key vectors and curve families of the topographic surface: (A) Vectors at a given point P of the topographic surface S: **g**, gravitational acceleration vector; $\mathbf{g_t}$ and $\mathbf{g_n}$, tangential and normal components of the gravitational acceleration vector, respectively; **i**, **j**, and **k**, unit vectors in the directions of x-, y-, and z-axes (east, north, and upward), respectively; **n**, external normal; T, tangent plane; γ, gradient angle. (B) Contour and slope lines at a surface: *black and white lines*, contour lines; *red lines with arrows*, slope lines. (C) Contour and slope lines as well as special points at a surface projection to the horizontal plane: *color lines*, contour lines, *black lines with arrows*, flow lines. *Black, red,* and *blue dots*, special points: local extrema, saddle, and horizontal planes, respectively.

Definition 2.4 (Special points). Special points are a set O of points (x, y) for which

$$p^2 + q^2 = 0 \tag{2.6}$$

(Nikolsky, 1977a, §§ 7.23 and 7.24; Shary, 1991).

At special points, a contour line may not be a curve in the usual sense defined previously. For example, this may be an isolated point (local extrema, such as peaks and pits), a saddle, or a horizontal plane (Mark, 1979) (Fig. 2.2B and C). In Sections 2.2.2–2.2.5, it is shown that not all morphometric variables can be defined at special points.

Let P be an opened rectangle with edges parallel to the X- and Y-axes, and $\overline{P}$ is its closure. Let also Q be an opened rectangular parallelepiped with faces parallel to coordinate axes such that Q contains $\overline{P}$. Let N_Q denote a set of all nonspecial points of Q, and let $\overline{O} = \overline{P} \backslash N_Q$ be the set of all special points of $\overline{P}$. The following theorem is valid:

Theorem 2.1 (Theorem on special points). The set N of nonspecial points is open, while the set $\overline{O}$ of special points is closed (Shary, 1991).

Proof The proof of the first part follows from continuity of the function p^2+q^2. If a point $(x, y) \in Q$ is nonspecial (i.e., $p^2+q^2>0$), then all points in its vicinity belong to N_Q. Thus, N_Q consists of internal points; that is, N_Q is an open set. The intersection of two open sets, $N_Q \cap P$ is, on the one hand, an open set. On the other hand, this intersection is a subset of the nonspecial points of P, that is, N. The proof of the second part follows from the definition of $\overline{O}$ as a difference between the closed set $\overline{P}$ and the open set N_Q. Thus, $\overline{O}$ is a closed set. ▪

For the topographic surface, Theorem 2.1 means that special points are usually rare because $\overline{O}$ is the closed set.

2.1.3 Morphometric variables and their types

Definition 2.5 (Morphometric variable). A morphometric (or topographic) variable (or attribute) is a single-valued bivariate function $v = u(x, y)$ describing the properties of the topographic surface.

In this book, we mainly discuss fundamental topographic variables associated with concepts of differential geometry and the theory of the topographic surface (Shary, 1991, 1995; Shary et al., 2002b). Such an approach is connected with the concept of general geomorphometry, which is defined as

> *"the measurement and analysis of those characteristics of landform which are applicable to any continuous rough surface. ... General geomorphometry as a whole provides a basis for the quantitative comparison even of qualitatively different landscapes, and it can adapt methods of surface analysis used outside geomorphology"*
>
> (Evans, 1972, p. 18).

There are five main types of morphometric variables:

1. Local variables (Section 2.2)
2. Nonlocal variables (Section 2.3)
3. Structural variables (Section 2.4)
4. Two-field-specific variables (Section 2.5)
5. Combined variables (Section 2.6)

Notice that for morphometric variables, terms *local* and *nonlocal* are used regardless of the study scale or data resolution. These terms are associated with the mathematical sense of a particular variable. A local morphometric attribute describes the surface geometry in the vicinity of a given point of the surface. A nonlocal topographic variable characterizes a relative position of a given point of the surface.

Being a morphometric variable, elevation does not belong to any of the types specified above. All topographic attributes are derived from elevation data.

Alternative classifications of morphometric variables can be found elsewhere (Shary et al., 2002b; Evans and Minár, 2011; Minár et al., 2013).

Morphometric variables of all types are considered in detail in Sections 2.2–2.6. Table 2.1 summarizes definitions and interpretations of discussed morphometric attributes; their formulas are consolidated in Table 2.2.

A series of three-dimensional (3D) morphometric maps of Mount Ararat illustrate mathematical concepts for topographic variables discussed in this chapter. Fig. 2.3 shows a 3D map of the elevation of Mount Ararat.

TABLE 2.1 Definitions and interpretations of morphometric variables.

Variable, notation, and unit	Definition and interpretation
Local morphometric variables	
Form attributes	
Minimal curvature, k_{min}, m^{-1}	A curvature of a principal section with the lowest value of curvature at a given point of the topographic surface (Gauss, 1828). $k_{min}>0$ corresponds to local convex landforms, while $k_{min}<0$ relates to elongated concave landforms (e.g., hills and valleys, respectively) (Section 2.2.4.5)
Maximal curvature, k_{max}, m^{-1}	A curvature of a principal section with the highest value of curvature at a given point of the topographic surface (Gauss, 1828). $k_{max}>0$ corresponds to elongated convex landforms, while $k_{max}<0$ relates to local concave landforms (e.g., crests and holes, respectively) (Section 2.2.4.5)
Mean curvature, H, m^{-1}	A half-sum of curvatures of any two orthogonal normal sections at a given point of the topographic surface (Young, 1805). H represents two accumulation mechanisms of gravity-driven substances—convergence and relative deceleration of flows—with equal weights (Section 2.2.4.4)
Gaussian curvature, K, m^{-2}	A product of maximal and minimal curvatures. According to *Theorema egregium*, K retains values in each point of the topographic surface after its bending without breaking, stretching, or compressing (Gauss, 1828) (Section 2.2.4.6)
Unsphericity curvature, M, m^{-1}	A half-difference of maximal and minimal curvatures (Shary, 1995). M shows the extent to which the shape of the topographic surface is nonspherical at a given point; $M=0$ on a sphere (Section 2.2.4.7)
Flow attributes	
Slope gradient, G, °	An angle between the tangential and horizontal planes at a given point of the topographic surface (Lehmann, 1799). Relates to the velocity of gravity-driven flows (Section 2.2.2)
Slope aspect, A, °	An angle between the north direction and the horizontal projection of the two-dimensional vector of gradient counted clockwise, from 0° to 360°, at a given point of the topographic surface. Relates to the direction of gravity-driven flows (Section 2.2.3)
Horizontal curvature, k_h, m^{-1}	A curvature of a normal section tangential to a contour line at a given point of the topographic surface (Krcho, 1983; Shary, 1991). A measure of flow convergence and divergence. Gravity-driven overland and intrasoil lateral flows converge and diverge where $k_h<0$ and $k_h>0$, respectively (Section 2.2.4.2)
Vertical curvature, k_v, m^{-1}	A curvature of a normal section having a common tangent line with a slope line at a given point of the topographic surface (Aandahl, 1948; Speight, 1974; Shary, 1991). A measure of relative deceleration and acceleration of gravity-driven flows. Overland and intrasoil lateral flows are decelerated and accelerated where $k_v<0$ and $k_v>0$, respectively (Section 2.2.4.3)
Difference curvature, E, m^{-1}	A half-difference of vertical and horizontal curvatures (Shary, 1995). Comparing two accumulation mechanisms of gravity-driven substances, E shows to what extent the relative deceleration of flows is higher than flow convergence at a given point of the topographic surface (Section 2.2.4.9)
Horizontal excess curvature, k_{he}, m^{-1}	A difference of horizontal and minimal curvatures (Shary, 1995). k_{he} shows to what extent the bending of a normal section tangential to a contour line is larger than the minimal bending at a given point of the topographic surface (Section 2.2.4.10)
Vertical excess curvature, k_{ve}, m^{-1}	A difference of vertical and minimal curvatures (Shary, 1995). k_{ve} shows to what extent the bending of a normal section having a common tangent line with a slope line is larger than the minimal bending at a given point of the topographic surface (Section 2.2.4.10)
Accumulation curvature, K_a, m^{-2}	A product of vertical and horizontal curvatures (Shary, 1995). A measure of the extent of flow accumulation at a given point of the topographic surface (Section 2.2.4.12)
Ring curvature, K_r, m^{-2}	A product of horizontal excess and vertical excess curvatures (Shary, 1995). K_r describes flow line twisting (Section 2.2.4.11)
Plan curvature, k_p, m^{-1}	A curvature of a contour line at a given point of the topographic surface (Sobolevsky, 1932; Krcho, 1973) (Section 2.2.4.1)
Rotor, rot, m^{-1}	A curvature of a flow line (Shary, 1991). Describes flow line twisting (Section 2.2.4.8)

Continued

TABLE 2.1 Definitions and interpretations of morphometric variables—cont'd

Variable, notation, and unit	Definition and interpretation
Deflections	
Horizontal deflection, D_{kh}, m^{-2}	A derivative of horizontal curvature by the contour line length. A measure for the deviation of k_h from loci of the extreme curvature of the topographic surface (Shary and Stepanov, 1991) (Section 2.2.5.1)
Vertical deflection, D_{kv}, m^{-2}	A derivative of vertical curvature by the contour line length. A measure for the deviation of k_v from loci of the extreme curvature of the topographic surface (Jenčo et al., 2009) (Section 2.2.5.2)
Nonlocal morphometric variables	
Catchment area, *CA*, m^2	An area of a closed figure formed by a contour segment at a given point of the topographic surface and two flow lines coming from upslope to the contour segment ends (Speight, 1974). A measure of the contributing area (Section 2.3.1)
Dispersive area, *DA*, m^2	An area of a closed figure formed by a contour segment at a given point of the topographic surface and two flow lines going downslope from the contour segment ends (Speight, 1974). A measure of a downslope area potentially exposed by flows passing through a given point (Section 2.3.1)
Specific catchment area, *SCA*, m^2/m	A ratio of an area *CA* to the length of a contour segment (Speight, 1974) (Section 2.3.1)
Specific dispersive area, *SDA*, m^2/m	A ratio of an area *DA* to the length of a contour segment (Speight, 1974) (Section 2.3.1)
Topographic position index, *TPI*	A difference between an elevation at a given point of the topographic surface and a mean elevation of a specified neighborhood area of that point (Weiss, 2001). A measure of the relative position of a given point of the topographic surface within a neighborhood area (Section 2.3.2)
Two-field-specific morphometric variables	
Solar morphometric variables	
Reflectance, *R*	A measure of the brightness of an illuminated surface (Horn, 1981) (Section 2.5.1.1)
Insolation, *I*, %	A measure of the topographic surface illumination by solar light flux (Shary et al., 2005) (Section 2.5.1.2)
Horizon angle, σ, °	The maximum angle from a given point of the topographic surface toward any other point in a given azimuth within a selected distance (Section 2.5.1.3)
Sky view factor, Ψ_s	A measure of the ratio of diffuse irradiance at a given point of the topographic surface to that on an unobstructed horizontal surface (Dozier and Frew, 1990) (Section 2.5.1.3)
Wind morphometric variables	
Relative terrain aspect, A_r, °	An absolute value of the difference between slope aspect and wind direction angle at a given point of the topographic surface (Böhner and Antonić, 2009) (Section 2.5.2). A simple measure of topographic exposure to wind
Horizon angle in the wind direction, σ_w, °	The maximum angle from a given point of the topographic surface toward any other point in a wind direction azimuth within a selected distance (Section 2.5.2). A simple measure of topographic exposure to wind
Wind effect index, *WEf*	A measure of an exposition of slopes to the wind flow of a particular direction at a given point of the topographic surface (Böhner, 2004) (Section 2.5.2)
Wind exposition index, *WEx*	A measure of an average exposition of slopes to wind flows of all possible directions at a given point of the topographic surface (Böhner, 2004) (Section 2.5.2)
View morphometric variables	
Terrain visibility, *V*	A set of points of the topographic surface, which are visible from the view point (Yoeli, 1985) (Section 2.5.3)
Total viewshed, V_Σ, m^2	A measure of a topographic surface area visible from a given point of the topographic surface (Kim et al., 2004) (Section 2.5.3)
Topographic openness, *TO*, °	A measure of dominance or enclosure of a given point of the topographic surface (Yokoyama et al., 2002) (Section 2.5.3)

TABLE 2.1 Definitions and interpretations of morphometric variables—cont'd

Variable, notation, and unit	Definition and interpretation
Combined morphometric variables	
Local-nonlocal morphometric variables	
Topographic wetness index, *TWI*	The logarithm of a ratio of *CA* to *G* at a given point of the topographic surface. A measure of the extent of flow accumulation (Beven and Kirkby, 1979) (Section 2.6.1)
Stream power index, *SPI*	The logarithm of a product of *CA* and *G* at a given point of the topographic surface. A measure of the potential flow erosion (Moore et al., 1991) (Section 2.6.1)
Local-local morphometric variables	
Shape index, *IS*	A continual form of the discrete Gaussian landform classification. $IS>0$ relates to convex landforms, while $IS<0$ corresponds to concave landforms. $\|IS\|=0.5, \ldots, 1$ relates to elliptic surfaces, while $\|IS\|=0, \ldots, 0.5$ relates to hyperbolic ones (Koenderink and van Doorn, 1992) (Section 2.6.2)
Curvedness, *C*, m^{-1}	The root mean square of maximal and minimal curvatures. A measure of the magnitude of surface bending regardless of its shape (Koenderink and van Doorn, 1992) (Section 2.6.2)

TABLE 2.2 Formulas of morphometric variables.

Variable	Formula
Local morphometric variables	
Form attributes	
Minimal curvature	$k_{min} = H - M = H - \sqrt{H^2 - K}$ (Shary, 1995)
Maximal curvature	$k_{max} = H + M = H + \sqrt{H^2 - K}$ (Shary, 1995)
Mean curvature	$H = \frac{1}{2}(k_{min} + k_{max}) = \frac{1}{2}(k_h + k_v) = -\frac{(1+q^2)r - 2pqs + (1+p^2)t}{2\sqrt{(1+p^2+q^2)^3}}$ (Gauss, 1828, Shary, 1991)
Gaussian curvature	$K = k_{min}k_{max} = H^2 - M^2 = \frac{rt - s^2}{(1+p^2+q^2)^2}$ (Gauss, 1828; Shary, 1995)
Unsphericity curvature	$M = \frac{1}{2}(k_{max} - k_{min}) = \sqrt{H^2 - K}$ (Shary, 1995)
Flow attributes	
Slope gradient	$G = \arctan\sqrt{p^2+q^2}$ (Shary et al., 2002b)
Slope aspect	$A = -90[1 - \mathrm{sign}(q)](1 - \|\mathrm{sign}(p)\|) + 180[1 + \mathrm{sign}(p)] - \frac{180}{\pi}\mathrm{sign}(p)\arccos\left(\frac{-q}{\sqrt{p^2+q^2}}\right)$ (Shary et al., 2002b)
Horizontal curvature	$k_h = H - E = -\frac{q^2r - 2pqs + p^2t}{(p^2+q^2)\sqrt{1+p^2+q^2}}$ (Krcho, 1983; Shary, 1991)
Vertical curvature	$k_v = H + E = -\frac{p^2r + 2pqs - q^2t}{(p^2+q^2)\sqrt{(1+p^2+q^2)^3}}$ (Krcho, 1973; Young, 1978; Shary, 1991)
Difference curvature	$E = \frac{1}{2}(k_v - k_h) = \frac{q^2r - 2pqs + p^2t}{(p^2+q^2)\sqrt{1+p^2+q^2}} - \frac{(1+q^2)r - 2pqs + (1+p^2)t}{2\sqrt{(1+p^2+q^2)^3}}$ (Shary, 1995)
Horizontal excess curvature	$k_{he} = k_h - k_{min} = M - E$ (Shary, 1995)
Vertical excess curvature	$k_{ve} = k_v - k_{min} = M + E$ (Shary, 1995)
Accumulation curvature	$K_a = k_hk_v = H^2 - E^2 = \frac{(q^2r - 2pqs + p^2t)(p^2r + 2pqs + q^2t)}{[(p^2+q^2)(1+p^2+q^2)]^2}$ (Shary, 1995)

Continued

TABLE 2.2 Formulas of morphometric variables—cont'd

Variable	Formula
Ring curvature	$K_r = k_{he}k_{ve} = M^2 - E^2 = \left[\frac{(p^2-q^2)s - pq(r-t)}{(p^2+q^2)(1+p^2+q^2)}\right]^2$ (Shary, 1995)
Plan curvature	$k_p = -\frac{q^2r - 2pqs + p^2t}{\sqrt{(p^2+q^2)^3}}$ (Krcho, 1973; Young, 1978)
Rotor	$rot = \frac{(p^2-q^2)s - pq(r-t)}{\sqrt{(p^2+q^2)^3}}$ (Shary, 1991)
Deflections	
Horizontal deflection	$D_{kh} = \frac{q^3g - 3pq^2k + 3p^2qm - p^3h}{\sqrt{(p^2+q^2)^3(1+p^2+q^2)}} - k_h rot\frac{2+3(p^2+q^2)}{1+p^2+q^2}$ (Shary and Stepanov, 1991; Florinsky, 2009b)
Vertical deflection	$D_{kv} = \frac{q^3m - p^3k + pq[q(2k-h) - p(2m-g)]}{\sqrt{(p^2+q^2)^3(1+p^2+q^2)^3}} - rot\left[\frac{2(r+t)}{\sqrt{(1+p^2+q^2)^3}} + k_v\frac{2+5(p^2+q^2)}{(1+p^2+q^2)}\right]$ (Jenčo et al., 2009; Florinsky, 2017a)
Nonlocal morphometric variables	
Specific catchment area	$\frac{\partial SCA}{\partial \sigma} = 1 - SCA \cdot k_p$ (Gallant and Hutchinson, 2011)
Two-field-specific morphometric variables	
Reflectance (Lambertian model)	$R_L = \frac{1 - p\sin\theta\cot\psi - q\cos\theta\cot\psi}{\sqrt{1+p^2+q^2}\sqrt{1+(\sin\theta\cot\psi)^2+(\cos\theta\cot\psi)^2}}$ (Horn, 1981)
Insolation	$I = 50\frac{\{1+\mathrm{sign}[\sin\psi - \cos\psi(p\sin\theta + q\cos\theta)]\}[\sin\psi - \cos\psi(p\sin\theta + q\cos\theta)]}{\sqrt{1+p^2+q^2}}$ (Shary et al., 2005)
Combined morphometric variables	
Local-nonlocal morphometric variables	
Topographic wetness index	$TWI = \ln\left(1 + \frac{CA}{10^{-3} + \tan G}\right)$ (Beven and Kirkby, 1979)
Stream power index	$SPI = \ln(1 + CA \cdot \tan G)$ (Moore et al., 1991)
Local-local morphometric variables	
Shape index	$IS = \frac{2}{\pi}\arctan\frac{H}{\sqrt{H^2 - K}}$ (Koenderink and van Doorn, 1992)
Curvedness	$C = \sqrt{\frac{k_{max}^2 + k_{min}^2}{2}}$ (Koenderink and van Doorn, 1992)

g, h, k, m, r, t, s, p, and *q*, the partial derivatives of elevation (Eq. 2.2); θ and ψ, solar azimuth and solar elevation angles, respectively.

2.2 Local morphometric variables

2.2.1 Overview

Definition 2.6 (Local morphometric variable). A local morphometric variable is a single-valued bivariate function $v = u(x, y)$ describing the geometry of the topographic surface in the vicinity of a given point of the surface (Speight, 1974) along directions determined by one of the two pairs of mutually perpendicular normal sections.

A normal section is a curve formed by the intersection of a surface with a plane containing the normal vector to the surface at a given point. An infinite number of normal sections can be constructed through each point of the topographic surface, but only two pairs of them have importance for digital terrain analysis.

The first pair of mutually perpendicular normal sections is the principal sections APA' and BPB' (Fig. 2.4A), which are well known from differential geometry (Pogorelov, 1957, § 7.3; Nikolsky, 1977a, § 7.24). These are normal sections with extreme (maximal and minimal) bending at a given point P of the surface.

The second pair of mutually perpendicular normal sections CPC' and DPD' (Fig. 2.4B) is allocated by gravity (Shary, 1991). The normal section DPD' includes the gravitational acceleration vector **g** and has a common tangent line with a slope

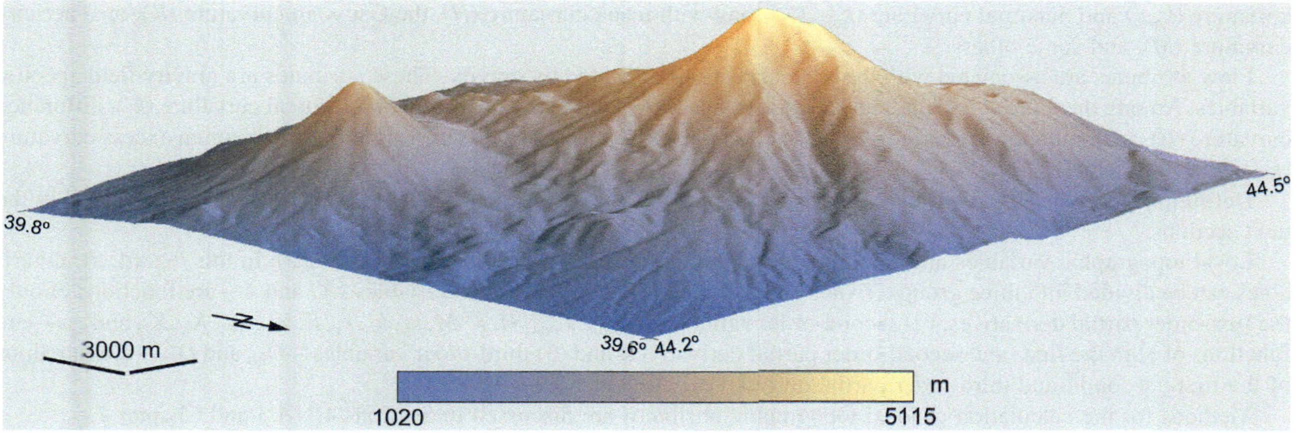

FIG. 2.3 Mount Ararat: elevation. The area is located between 44.2°E and 44.5°E and 39.6°N and 39.8°N (the area size is $18' \times 12'$, that is, 25.725 km × 22.204 km). The spheroidal equal angular digital elevation model (DEM) was derived from the SRTM1 DEM (USGS, 2015). The DEM includes 779,401 points (the elevation matrix 1081 × 721). The grid spacing is 1″, that is, the linear sizes of the 1″ × 1″ cell are 23.82 m × 30.84 m at the mean latitude, 39.7° N. Elevation approximation was carried out by the universal spectral analytical method (Chapter 7) using 300 expansion coefficients. Data were processed and visualized with the author's script for the MATLAB software (Florinsky and Pankratov, 2016). Vertical exaggeration is 2×. *(Modified from Florinsky, I.V., 2017a. An illustrated introduction to general geomorphometry. Progr. Phys. Geogr. 41 (6), 723–752, Fig. 2, reproduced with permission from Sage.)*

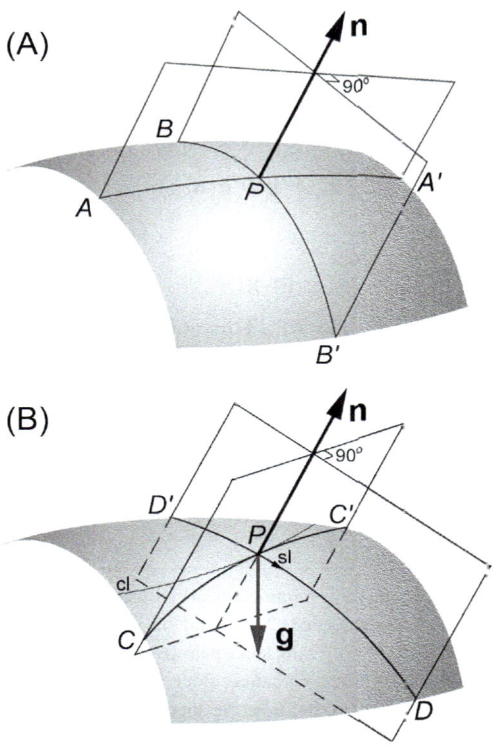

FIG. 2.4 Two pairs of normal sections at a given point P of the topographic surface: (A) Principal sections APA' and BPB'. (B) Normal sections CPC' and DPD' allocated by gravity. *cl*, contour line; **g**, gravitational acceleration vector; **n**, external normal vector to the surface; *sl*, slope line. *((A) Modified from Lisle, R.J., Toimil, N., Aller, J., Bobillo-Ares, N., Bastida, F., 2010. The hinge lines of non-cylindrical folds. J. Struct. Geol. 32 (2), 166–171, Fig. 2, reproduced with permission from Elsevier.)*

line (Section 2.2.2.1) at a given point P of the topographic surface. The normal section CPC' is perpendicular to the section DPD' and tangential to a contour line at a given point P of the topographic surface.

Two classes of local morphometric variables—form attributes and flow attributes—are related to the two pairs of normal sections (Shary, 1995; Shary et al., 2002b).

Form attributes are associated with the principal sections. These attributes are gravity field invariants; that is, they do not depend on the direction of the gravitational acceleration vector **g**. Among these are two principal curvatures—minimal curvature (k_{min}) and maximal curvature (k_{max})—along with mean curvature (H), the Gaussian curvature (K), unsphericity curvature (M), and some others.

Flow attributes are associated with the normal sections allocated by gravity. These attributes are gravity field-specific variables. Among these are slope gradient (G), slope aspect (A), horizontal curvature (k_h), vertical curvature (k_v), difference curvature (E), accumulation curvature (K_a), ring curvature (K_r), vertical excess curvature (k_{ve}), horizontal excess curvature (k_{he}), and some others.[c]

Definitions, derivation of formulas, and interpretations of local morphometric variables are discussed in detail in the next sections.

Local topographic variables are functions of the partial derivatives of elevation (Table 2.2). In this regard, local variables can be divided into three groups (Evans and Minár, 2011): (1) first-order variables—G and A—are functions of only the first-order partial derivatives; (2) second-order variables—k_{min}, k_{max}, H, K, M, k_p, k_h, k_v, E, k_{he}, k_{ve}, K_a, K_r, and rot—are functions of both the first- and second-order partial derivatives; and (3) third-order variables—D_{kh} and D_{kv}—are functions of the first-, second-, and third-order partial derivatives.

Methods for the calculation of local topographic attributes are discussed in Sections 4.1–4.3 and Chapter 7.

2.2.2 Gradient[d]

There are two components of the gravitational acceleration vector (**g**) (Eq. 2.3): tangential ($\mathbf{g_t}$) and normal ($\mathbf{g_n}$) (Fig. 2.2A):

$$\mathbf{g} = \mathbf{g_t} + \mathbf{g_n}. \tag{2.7}$$

On the surface S, gravity-driven flows are governed by $\mathbf{g_t}$. Let us find an expression for $\mathbf{g_t}$ (Shary, 1991).

n is a unit external normal vector (Fig. 2.2A) to S at a point $\mathbf{P}_0 \in S$. An equation of a tangent plane T to S is as follows:

$$(\mathbf{P} - \mathbf{P}_0, \mathbf{n}) = 0, \tag{2.8}$$

where round brackets denote the scalar product. The end of **g**, which begins at the point $\mathbf{P}_0$, is $\mathbf{P}_0+\mathbf{g}$. The distance from the point $\mathbf{P}_0+\mathbf{g}$ to the tangent plane is $|(\mathbf{P}_0+\mathbf{g}-\mathbf{P}_0,\mathbf{n})| = |(\mathbf{n},\mathbf{g})|$. It follows from Eq. (2.7) that $\mathbf{g_t}=\mathbf{g}-\mathbf{g_n}$, but $|\mathbf{g_n}|$ is the distance $|(\mathbf{n},\mathbf{g})| = -(\mathbf{n},\mathbf{g})$. The direction of $\mathbf{g_n}$ coincides with the direction of $-\mathbf{n}$, so $\mathbf{g_n}=\mathbf{n}(\mathbf{n},\mathbf{g})$. Finally,

$$\mathbf{g_t} = \mathbf{g} - \mathbf{n}(\mathbf{n}, \mathbf{g}). \tag{2.9}$$

The inequality

$$(\mathbf{n}, \mathbf{g}) < 0 \tag{2.10}$$

means that the angle between **n** and **g** is greater than 90°. This corresponds to Limitation 2.1 (Section 2.1.1), excluding caves, infinitely steep slopes, and so on.

The expression for **n** to S is as follows (e.g., Zevenbergen and Thorne, 1987):

$$\mathbf{n} = \frac{-p\mathbf{i} - q\mathbf{j} + \mathbf{k}}{\sqrt{1 + p^2 + q^2}}, \tag{2.11}$$

---

c. To avoid further misunderstanding, we should briefly note some terminological confusion surrounding local morphometric variables. First, horizontal curvature is often confused with plan curvature (for details, see Sections 2.2.4.1 and 2.2.4.2). Second, some authors call horizontal and vertical curvatures tangential and profile curvatures, respectively (e.g., Shary, 2006). Third, some petroleum geologists and geophysicists use the terms strike curvature, dip curvature, dip angle, and azimuth for horizontal curvature, vertical curvature, slope gradient, and slope aspect, respectively (e.g., Roberts, 2001).

d. The mathematics of Sections 2.2.2–2.2.5.1 is based on the materials from the appendix prepared by P.A. Shary for the first edition of this book. *(Modified from Shary, P.A., 2012. The mathematical basis of local morphometric variables. In: Florinsky, I.V., Digital Terrain Analysis in Soil Science and Geology, first ed. Academic Press, Amsterdam, The Netherlands, pp. 289–313, Sections A.1.1, A.1.3, A.2.1, A.3, and A4, reprinted with permission from Elsevier.)*

where p and q are the first-order partial derivatives of elevation (Eq. 2.2); **i**, **j**, and **k** are unit vectors in the directions of X-, Y-, and Z-axes, respectively (Fig. 2.2A). For the uniform gravitational field (Eq. 2.3), the inequality (Eq. 2.10) is always valid for $f(x, y)$:

$$(\mathbf{n}, \mathbf{g}) = -\frac{\dot{g}}{\sqrt{1 + p^2 + q^2}} < 0, \tag{2.12}$$

where g is the gravitational acceleration constant. It follows from Eqs. (2.3), (2.9), and (2.11):

$$\mathbf{g_t} = -\dot{g}\frac{p\mathbf{i} + q\mathbf{j} + (p^2 + q^2)\mathbf{k}}{1 + p^2 + q^2}. \tag{2.13}$$

For the uniform gravitational field (Eq. 2.3), the following equations are valid:

$$\cos\gamma = \left(-\mathbf{n}, \frac{\mathbf{g}}{\dot{g}}\right) = \frac{1}{\sqrt{1 + p^2 + q^2}}, \tag{2.14}$$

$$\sin\gamma = \sqrt{\frac{p^2 + q^2}{1 + p^2 + q^2}}, \tag{2.15}$$

$$\tan\gamma = \sqrt{p^2 + q^2}, \tag{2.16}$$

where γ is the angle between **k** and **n** (Fig. 2.2A) (Shary, 1991). This angle is slope steepness or gradient. Dimensionless gradient ($\overline{G}$) is

$$\overline{G} = \tan\gamma = \sqrt{p^2 + q^2}. \tag{2.17}$$

Gradient in degrees (G) is

$$G = \arctan\sqrt{p^2 + q^2}. \tag{2.18}$$

Definition 2.7 (Slope gradient). Slope gradient is the angle between the normal vector **n** and the Z-axis unit vector **k** (Fig. 2.2A) (or, equivalently, an angle between the tangential and horizontal planes) at a given point of the topographic surface.

Slope gradient is a nonnegative variable. The unit of G is degree. G values range from 0° to 90°.

The velocity of gravity-driven flows is associated with G. Soil-related interpretations of G can be found in Section 9.2.

It follows from Eqs. (2.13) and (2.15):

$$|\mathbf{g_t}| = \dot{g}\sin\gamma. \tag{2.19}$$

This means that the tangential component of the gravitational force is proportional to a gradient factor (GF) (Strahler, 1952):

$$GF = \frac{|\mathbf{g_t}|}{\dot{g}} = \sin\gamma = \frac{\sqrt{p^2 + q^2}}{\sqrt{1 + p^2 + q^2}}. \tag{2.20}$$

One obtains from Eqs. (2.17) and (2.20):

$$GF = \frac{\overline{G}}{\sqrt{1 + \overline{G}^2}}. \tag{2.21}$$

Fig. 2.5A represents a 3D map of G for Mount Ararat.

2.2.2.1 Slope line and flow line

Definition 2.8 (Slope line). A slope line (or gradient line) is a space curve defined for the set of nonspecial points N on the surface S. At each point of the curve, the direction of the tangent to the curve coincides with the direction of the tangential component of the gravitational force $\mathbf{g_t}$ (Cayley, 1859) (Fig. 2.2B).

Slope lines are directed downslope because the z-component of $\mathbf{g_t}$ is negative on the set N. At special points, slope lines are not defined because $p = q = 0$, so $\mathbf{g_t} = 0$. Thus, the direction of $\mathbf{g_t}$ is not defined.

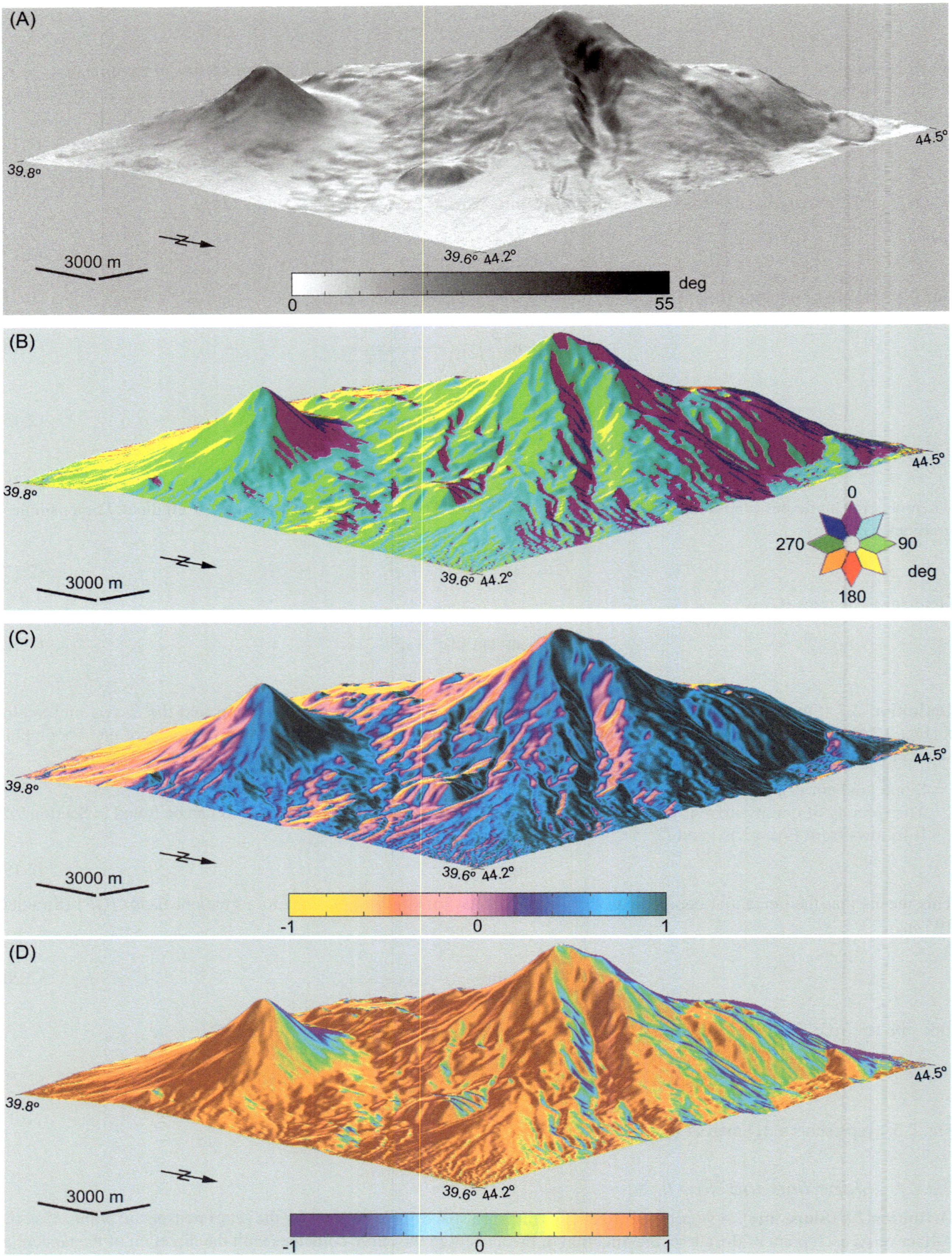

FIG. 2.5 Mount Ararat, slope gradient and aspect: (A) Slope gradient. (B) Slope aspect. (C) Northness. (D) Eastness. Calculation was carried out by the universal spectral analytical method (Chapter 7). Data were processed and visualized with the author's script for the MATLAB software (Florinsky and Pankratov, 2016). Vertical exaggeration is 2×. For the elevation data description, see Fig. 2.3. *((A) and (B) Modified from Florinsky, I.V. 2017a. An illustrated introduction to general geomorphometry. Progr. Phys. Geogr. 41 (6), 723–752, Fig. 3A and B, reproduced with permission from Sage.)*

Definition 2.9 (Flow line). A flow line is a plane curve, which is a projection of a slope line to a horizontal plane (Fig. 2.2C).

A unit tangent vector to the slope line is

$$\mathbf{r}'_{\eta} = \frac{dx}{d\eta}\mathbf{i} + \frac{dy}{d\eta}\mathbf{j} + \frac{dz}{d\eta}\mathbf{k}, \tag{2.22}$$

where η is a slope line length;

$$\left(\frac{dx}{d\eta}\right)^2 + \left(\frac{dy}{d\eta}\right)^2 + \left(\frac{dz}{d\eta}\right)^2 = 1. \tag{2.23}$$

From Eqs. (2.13) and (2.19), one obtains that for the set N

$$\mathbf{r}'_{\eta} = -\frac{p\mathbf{i} + q\mathbf{j} + (p^2 + q^2)\mathbf{k}}{\sqrt{(p^2 + q^2)(1 + p^2 + q^2)}}. \tag{2.24}$$

Theorem 2.2 (Theorem on orthogonality). On the set of nonspecial points N, any contour line is orthogonal to any slope line at the point of their intersection. On the set N, contour lines and flow lines are also orthogonal (Shary, 1991).

Proof For the first part of the theorem, it is sufficient to prove that on the set N

$$\left(\mathbf{r}'_{\eta}, \mathbf{r}'_{\xi}\right) = 0, \tag{2.25}$$

where ξ is a contour line length and $\mathbf{r}'_{\xi}$ is a unit tangent vector to the contour line. Let us take the derivative of the contour line (Eq. 2.4) with respect to ξ. For a plane curve, Eq. (2.23) is

$$\left(\frac{dx}{d\xi}\right)^2 + \left(\frac{dy}{d\xi}\right)^2 = 1. \tag{2.26}$$

One obtains

$$p\frac{dx}{d\xi} + q\frac{dy}{d\xi} = 0. \tag{2.27}$$

Thus, on the set N

$$\mathbf{r}'_{\xi} = \pm\frac{-q\mathbf{i} + p\mathbf{j}}{\sqrt{p^2 + q^2}}. \tag{2.28}$$

The choice of sign is determined by the contour line orientation. Therefore, Eq. (2.25) is proved as the first part of the theorem.

Using Eq. (2.24), one obtains an expression for a tangent vector $\mathbf{r}'_{\sigma}$ to a flow line on the set N:

$$\mathbf{r}'_{\sigma} = -\frac{p\mathbf{i} + q\mathbf{j}}{\sqrt{p^2 + q^2}}, \tag{2.29}$$

where σ is a flow line length. It follows from Eqs. (2.28) and (2.29) that $(\mathbf{r}_{\sigma}', \mathbf{r}_{\xi}') = 0$ for the set N. Therefore, flow lines are orthogonal to contour lines. ▪

Let us demonstrate that the direction of the maximal decrease of elevation is the direction of a flow line for each nonspecial point. Indeed, on the set N, the directional derivative along a unit vector $\mathbf{m}$ is

$$\left|\frac{\partial z}{\partial \mathbf{m}}\right| = |(\nabla \mathbf{z}, \mathbf{m})| \le |\nabla \mathbf{z}| = \sqrt{p^2 + q^2}, \tag{2.30}$$

where $\nabla\mathbf{z}$ is a two-dimensional vector of gradient:

$$\nabla \mathbf{z} = p\mathbf{i} + q\mathbf{j}. \tag{2.31}$$

Therefore, the direction of $-\nabla\mathbf{z}$ gives the maximal decrease of elevation among all possible directions. However, this is the direction of $\mathbf{r}'_{\sigma}$ along a flow line.

2.2.3 Aspect

Definition 2.10 (Slope aspect). Slope aspect (A) is the angle between the north direction $\mathbf{j}$ (Fig. 2.2A) and the horizontal projection of the direction of a two-dimensional vector of gradient $-\nabla \mathbf{z}$ counted clockwise from 0° to 360°.

A unit tangent vector to a flow line $\frac{-\nabla \mathbf{z}}{|\nabla \mathbf{z}|}$ is $\mathbf{r}'_\sigma$. On the set N, the scalar product $(\mathbf{r}'_\sigma, \mathbf{j})$ is the cosine of the angle A between $\mathbf{r}'_\sigma$ and $\mathbf{j}$:

$$\cos A = \left(\mathbf{r}'_\sigma, \mathbf{j}\right) = \frac{-q}{\sqrt{p^2+q^2}}. \tag{2.32}$$

Considering that the inequality $q>0$ refers to southern slopes ($90° < A \le 180°$ if $p \le 0$, $180° < A < 270°$ if $p>0$), $A = 90°$ ($p<0$) or $A = 270°$ ($p>0$) if $q=0$, and the inequality $q<0$ refers to northern slopes ($0° < A < 90°$ if $p<0$, $270° < A < 360°$ if $p>0$), for the set N one obtains:

$$\begin{cases} A = 0 + \dfrac{180}{\pi}\arccos\left(\dfrac{-q}{\sqrt{p^2+q^2}}\right) & \text{for} \quad p \le 0,\ q<0 \\ A = 90 + \dfrac{180}{\pi}\arccos\left(\dfrac{-q}{\sqrt{p^2+q^2}}\right) & \text{for} \quad p \le 0,\ q \ge 0 \\ A = 180 + \dfrac{180}{\pi}\arccos\left(\dfrac{-q}{\sqrt{p^2+q^2}}\right) & \text{for} \quad p > 0,\ q \ge 0 \\ A = 270 + \dfrac{180}{\pi}\arccos\left(\dfrac{-q}{\sqrt{p^2+q^2}}\right) & \text{for} \quad p > 0,\ q<0 \end{cases}, \tag{2.33}$$

where arccosine is in radians.

One may use the function

$$\operatorname{sign}(x) = \begin{cases} 1 & \text{for} \quad x>0 \\ 0 & \text{for} \quad x=0, \\ -1 & \text{for} \quad x<0 \end{cases} \tag{2.34}$$

and transform Eq. (2.33) into a single formula valid on the set N:

$$A = -90[1-\operatorname{sign}(q)](1-|\operatorname{sign}(p)|) + 180[1+\operatorname{sign}(p)] - \frac{180}{\pi}\operatorname{sign}(p)\arccos\left(\frac{-q}{\sqrt{p^2+q^2}}\right). \tag{2.35}$$

On the set N, a sine of A may be deduced as a z-component of the vector product $[\mathbf{r}'_\sigma, \mathbf{j}]$

$$\sin A = \left[\mathbf{r}'_\sigma, \mathbf{j}\right]_z = \begin{vmatrix} \mathbf{i} & \mathbf{j} & \mathbf{k} \\ \dfrac{-p}{\sqrt{p^2+q^2}} & \dfrac{-q}{\sqrt{p^2+q^2}} & 0 \\ 0 & 1 & 0 \end{vmatrix}_z = \frac{-p}{\sqrt{p^2+q^2}}. \tag{2.36}$$

Slope aspect is a nonnegative circular variable. The unit of A is degree. A values range from 0° to 360°, and both of these values correspond to the north direction.

A relates to the direction of gravity-driven flows. For soil-related interpretations of A, see Section 9.2.

Fig. 2.5B shows a 3D map of A for Mount Ararat.

2.2.3.1 Northness and eastness

Since the slope aspect is a circular variable, it cannot be used in linear statistical analysis. To avoid this problem, two auxiliary local indices can be applied: northness (A_N) and eastness (A_E) (King et al., 1999):

$$A_N = \cos A, \tag{2.37}$$

$$A_E = \sin A. \tag{2.38}$$

Northness and eastness are dimensionless variables. A_N is equal to 1, −1, and 0 on the northern, southern, and eastern/western slopes, respectively. A_E is equal to 1, −1, and 0 on the eastern, western, and northern/southern slopes, respectively. Thus, these variables accentuate northern/southern and eastern/western trends in the spatial distribution of slopes.

Fig. 2.5C and D represents 3D maps of A_N and A_E, respectively, for Mount Ararat.

2.2.4 Curvatures

In this book, we systematically use the notion of curvature. Intuitively, it is clear that the curvature of a plane curve is the amount by which the curve deviates from the straight line, while the curvature of a surface is the amount by which the surface deviates from the flat plane. A curvature can be usually determined for any point of a curve or surface. In particular, the curvature of a surface can be defined as the rate of change of the tangent vector **t** moved along the curve *ds* on the surface (Fig. 2.6). Details can be found in textbooks on differential geometry (e.g., Pogorelov, 1957).

To have a flexible mathematical tool for deducing formulas of curvatures, let us use a general formula of curvature of a smooth curve (Nikolsky, 1977a, § 6.9):

$$|\mathbf{r}''_\lambda| = \sqrt{\left(\frac{d^2x}{d\lambda^2}\right)^2 + \left(\frac{d^2y}{d\lambda^2}\right)^2 + \left(\frac{d^2z}{d\lambda^2}\right)^2}, \tag{2.39}$$

where λ is the length of a curve, and $\mathbf{r}''_\lambda$ is the second derivative of the vector **r** with respect to λ.

In this book, we assume (unless otherwise specified) that curvature is positive for convex landforms (e.g., hills and ridges), and negative for concave ones (e.g., depressions and valleys).

2.2.4.1 Plan curvature

Definition 2.11 (Plan curvature). Plan curvature (k_p) is the curvature of a contour line at a given point of the topographic surface (Krcho, 1973; Young, 1978; Evans, 1979).

This variable can be positive, negative, or zero. The unit of k_p is m^{-1}.

To derive a formula for k_p, let us substitute a contour line length ξ to Eq. (2.39) and then develop the derivatives included in this formula. It follows from Eq. (2.28) that on the set N

$$\frac{dx}{d\xi} = \pm\frac{-q}{\sqrt{p^2 + q^2}}, \tag{2.40}$$

$$\frac{dy}{d\xi} = \pm\frac{p}{\sqrt{p^2 + q^2}}, \tag{2.41}$$

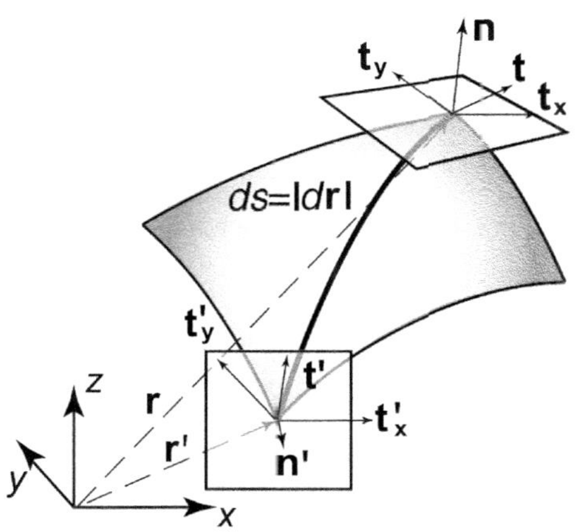

FIG. 2.6 The curvature of a surface can be defined as the rate of change of the tangent vector **t** moved along the curve *ds* on the surface (thick line). **n**, a unit normal to the surface at the initial point defined by the vector **r**; $\mathbf{t}_x$ and $\mathbf{t}_y$, tangent vectors defining the tangent plane at the initial point; *primed symbols*, corresponding vectors at the end point. *(From Bergbauer, S., Pollard, D.D., 2003. How to calculate normal curvatures of sampled geological surfaces. J. Struct. Geol. 25 (2), 277–289, Fig. 1, reproduced with permission from Elsevier.)*

$$\frac{dp}{d\xi} = r\frac{dx}{d\xi} + s\frac{dy}{d\xi}, \tag{2.42}$$

$$\frac{dq}{d\xi} = s\frac{dx}{d\xi} + t\frac{dy}{d\xi}, \tag{2.43}$$

where r, s, and t are the second-order partial derivatives of elevation (Eq. 2.2). Thus, on the set N

$$\frac{d^2x}{d\xi^2} = -\frac{p}{(p^2+q^2)^2}\left(q^2r - 2pqs + p^2t\right), \tag{2.44}$$

$$\frac{d^2y}{d\xi^2} = -\frac{q}{(p^2+q^2)^2}\left(q^2r - 2pqs + p^2t\right). \tag{2.45}$$

On the set N, the required vector

$$\mathbf{r}''_{\xi} = \frac{d^2x}{d\xi^2}\mathbf{i} + \frac{d^2y}{d\xi^2}\mathbf{j}, \tag{2.46}$$

$$\mathbf{r}''_{\xi} = -k_p\mathbf{r}'_{\sigma}, \tag{2.47}$$

where $\mathbf{r}'_{\sigma}$ is a unit tangent vector to a flow line (Eq. 2.29) and

$$k_p = -\frac{q^2r - 2pqs + p^2t}{\sqrt{(p^2+q^2)^3}}. \tag{2.48}$$

Let us prove that k_p is a quantitative measure of flow line divergence and convergence. To clarify the meaning of divergence div($\mathbf{r}'_{\sigma}$) of the unit tangent vector $\mathbf{r}'_{\sigma}$ to a flow line, let us apply the Gauss-Ostrogradsky theorem (Nikolsky, 1977b, § 13.10). Let an arbitrary fixed point (x, y) belongs to the set N and be surrounded by a small circle T_{ε} of radius $\varepsilon > 0$, which also belongs to N, with an external normal $\mathbf{n}'$ to a positively oriented boundary Γ_{ε} of the circle T_{ε}. According to the two-dimensional case of the Gauss-Ostrogradsky theorem,

$$\iint_{T_\varepsilon} \operatorname{div}(\mathbf{r}'_{\sigma})\,dx\,dy = \int_{\Gamma_\varepsilon} (\mathbf{r}'_{\sigma}, \mathbf{n}')\,d\xi, \tag{2.49}$$

where $d\xi$ is the differential of Γ_{ε}. Using the theorem on averages (Nikolsky, 1977a, § 5.8) and taking a limit, one obtains

$$\operatorname{div}(\mathbf{r}'_{\sigma}) = \lim_{\varepsilon\to 0,\, \varepsilon>0}\left(\frac{1}{\pi\varepsilon^2}\right)\int_{\Gamma_\varepsilon} (\mathbf{r}'_{\sigma}, \mathbf{n}')\,d\xi, \tag{2.50}$$

where the integral is the flow of a unit vector $\mathbf{r}'_{\sigma}$ through Γ_{ε}. This means that div($\mathbf{r}'_{\sigma}$) is a quantitative measure of flow line divergence.

Theorem 2.3 (Theorem on flow divergence). The divergence of the unit tangent vector $\mathbf{r}'_{\sigma}$ to a flow line is equal to k_p (Shary, 1995).

Proof Calculating div($\mathbf{r}'_{\sigma}$) with Eq. (2.29), one obtains for the set N

$$\operatorname{div}(\mathbf{r}'_{\sigma}) = \frac{\partial\left(\frac{-p}{\sqrt{p^2+q^2}}\right)}{\partial x} + \frac{\partial\left(\frac{-q}{\sqrt{p^2+q^2}}\right)}{\partial y} = k_p. \tag{2.51}$$

Therefore, plan curvature is a quantitative measure of flow line divergence. ▪

Flow lines converge where $k_p < 0$ and diverge where $k_p > 0$; $k_p = 0$ refers to parallel flow lines.

2.2.4.2 Horizontal curvature

Definition 2.12 (Horizontal curvature). Horizontal (or tangential) curvature (k_h) is the curvature of a normal section tangential to a contour line at a given point of the topographic surface (Krcho, 1983; Shary, 1991; Mitášová and Hofierka, 1993).

This normal section is one of the pair of mutually perpendicular normal sections allocated by gravity at a given point of the topographic surface (Fig. 2.4B) (Section 2.2.1). Horizontal curvature can be positive, negative, or zero. The unit of k_h is m^{-1}.

To deduce the formula of k_h, one may use the Meusnier theorem (Nikolsky, 1977a, § 7.24):

$$k_0 = k\cos\theta, \tag{2.52}$$

where k_0 is a curvature of a normal section passing through a point **P** on the surface S and tangential to a curve Γ; k is a curvature $k=|\mathbf{r}''_\lambda|$ of Γ belonging to S and passing through **P**; and θ is an angle between vectors $\boldsymbol{\nu} = \frac{\mathbf{r}''_\lambda}{|\mathbf{r}''_\lambda|}$ and **n** to S at **P**, $\cos\theta=(\mathbf{n},\boldsymbol{\nu})$. Thus, the Meusnier theorem may be rewritten as

$$k_0 = \left(\mathbf{n}, \mathbf{r}''_\lambda\right). \tag{2.53}$$

Considering a contour line as the curve Γ (that is, $\lambda=\xi$) and using Eqs. (2.11) and (2.15) for **n** and $\sin\gamma$, one obtains for the set N

$$k_h = k_p \sin\gamma, \tag{2.54}$$

where $0 < \gamma < \frac{\pi}{2}$. Thus,

$$k_h = -\frac{q^2 r - 2pqs + p^2 t}{(p^2+q^2)\sqrt{1+p^2+q^2}} \tag{2.55}$$

k_p and k_h have the same signs at nonspecial points, which follows from Eqs. (2.43) and (2.55). Therefore, flow lines converge where $k_h<0$ and diverge where $k_h>0$; $k_h=0$ refers to parallel flow lines (Fig. 2.7A).

This can be interpreted as follows: gravity-driven overland and intrasoil lateral flows converge where $k_h<0$, and they diverge where $k_h>0$. Geomorphologically, k_h mapping allows the revealing of ridges and valley spurs. For soil-related interpretations of k_h, see Section 9.2.

Fig. 2.8A represents a 3D map of k_h for Mount Ararat.

2.2.4.3 Vertical curvature

Definition 2.13 (Vertical curvature). Vertical (or profile) curvature (k_v) is the curvature of a normal section having a common tangent line with a slope line at a given point of the topographic surface (Aandahl, 1948; Krcho, 1973; Young, 1978; Evans, 1979).

This normal section is one of the pair of mutually perpendicular normal sections allocated by gravity at a given point of the topographic surface (Fig. 2.4B) (Section 2.2.1). Vertical curvature can be positive, negative, or zero. The unit of k_v is m^{-1}.

It is known that contour lines and slope lines are mutually orthogonal spatial curves (Theorem 2.2). Let us apply the Euler theorem (Nikolsky, 1977a, § 7.24) to mean curvature (H) (Section 2.2.4.4). For the set N, we obtain

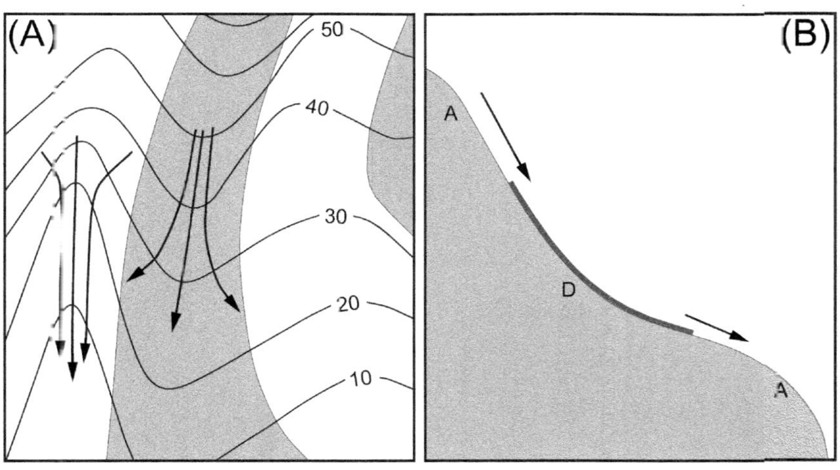

FIG. 2.7 Two accumulation mechanisms of gravity-driven flows: (A) Planimetric view. Flow convergence/divergence: flow lines converge where $k_h<0$ (convergence areas, *white*) and diverge where $k_h>0$ (divergence areas, *gray*). (B) Profile view. Flow relative deceleration/acceleration: flow decelerates where $k_v<0$ (relative deceleration areas, *D*) and accelerates where $k_v>0$ (relative acceleration areas, *A*). *(Modified from Shary, P.A., Sharaya, L.S., Mitusov, A.V., 2002b. Fundamental quantitative methods of land surface analysis. Geoderma. 107 (1–2), 1–32, Fig. 4, reproduced with permission from Elsevier.)*

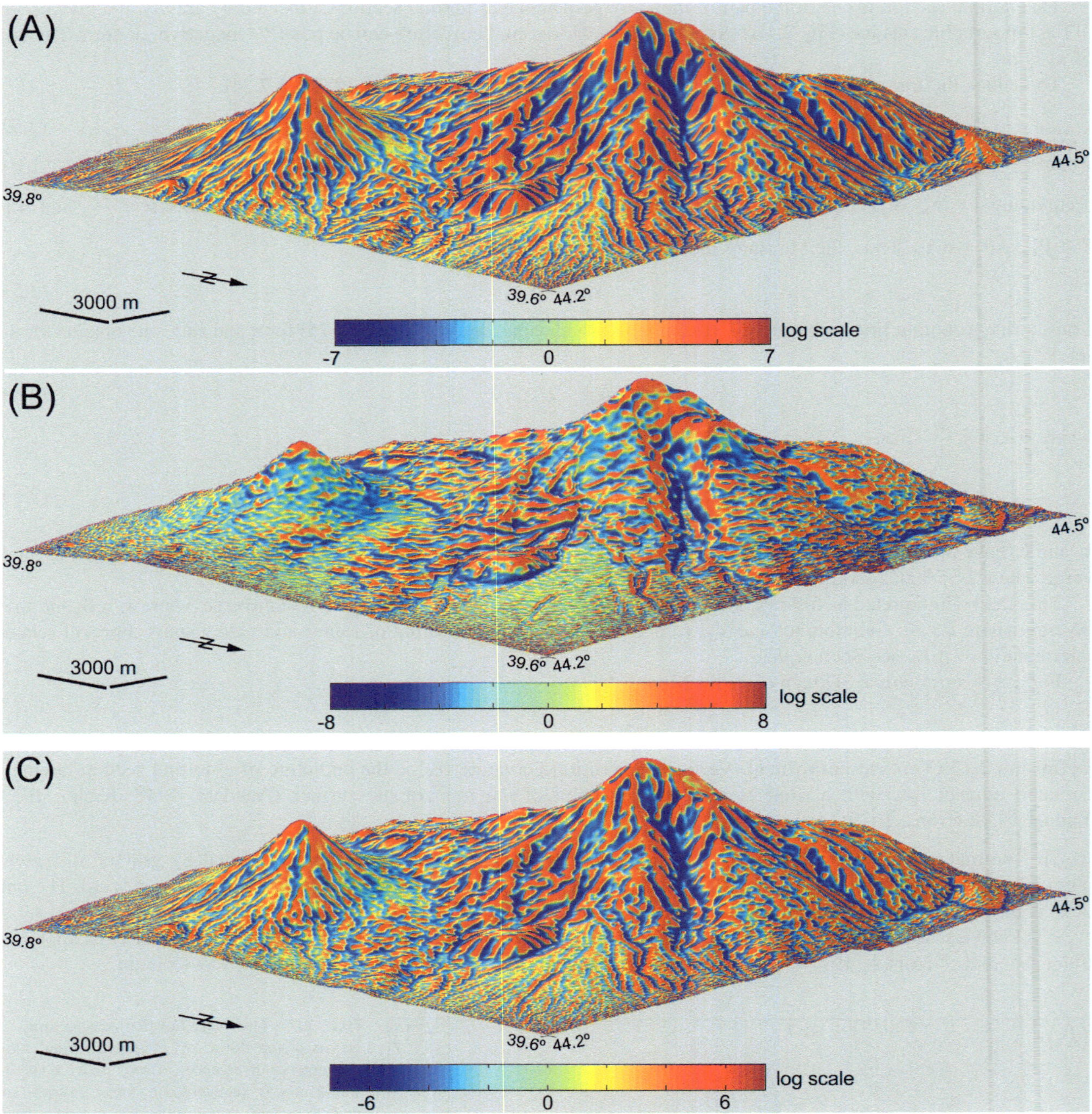

FIG. 2.8 Mount Ararat, curvatures: (A) Horizontal curvature. (B) Vertical curvature. (C) Mean curvature.

$$H = \frac{1}{2}(k_h + k_v). \tag{2.56}$$

Substituting Eqs. (2.55) and (2.64) to Eq. (2.56), one obtains for the set N

$$k_v = -\frac{p^2 r + 2pqs + q^2 t}{(p^2 + q^2)\sqrt{(1 + p^2 + q^2)^3}}. \tag{2.57}$$

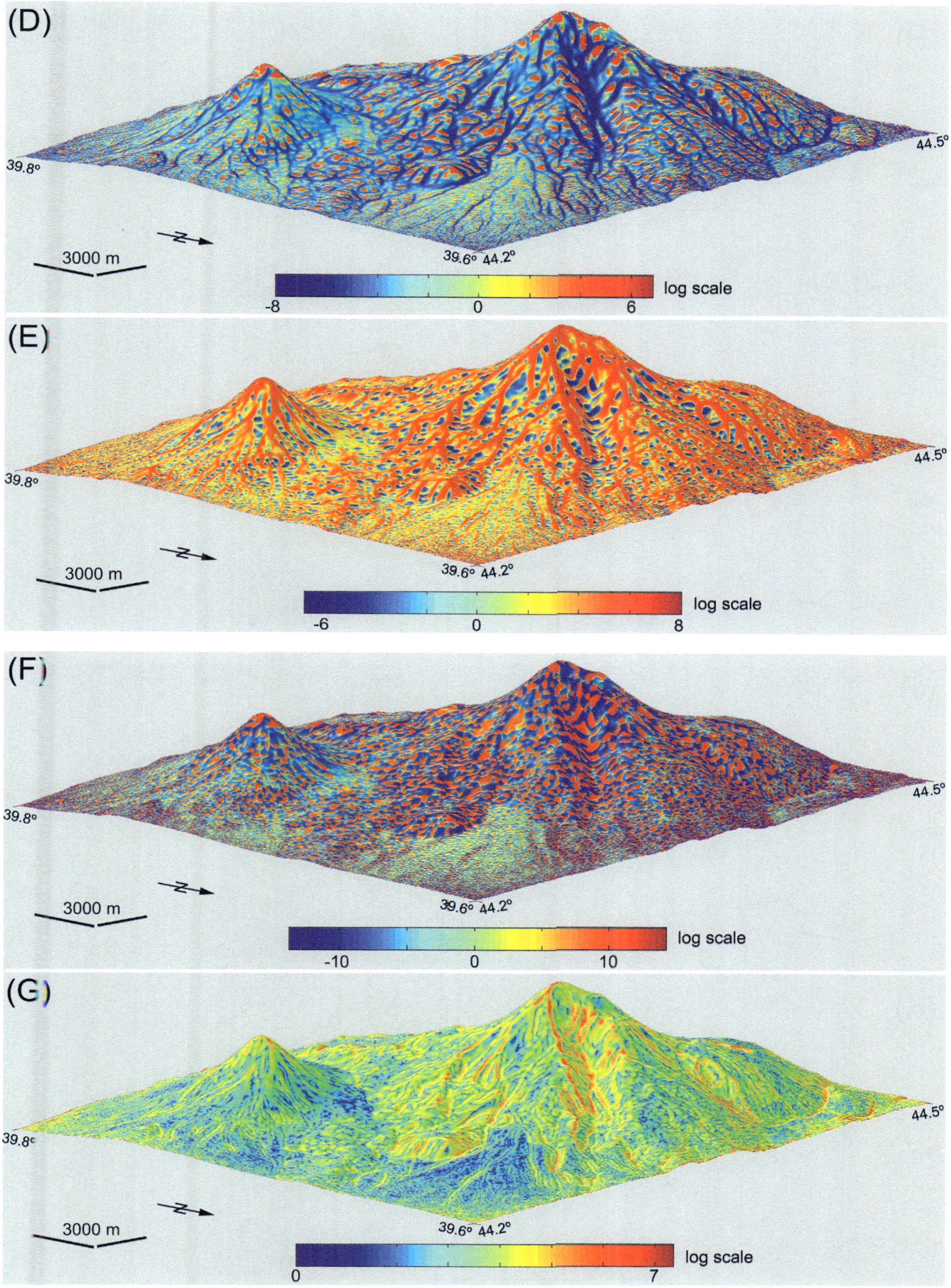

FIG. 2.8, CONT'D (D) Minimal curvature. (E) Maximal curvature. (F) Gaussian curvature. (G) Unsphericity curvature.

(Continued)

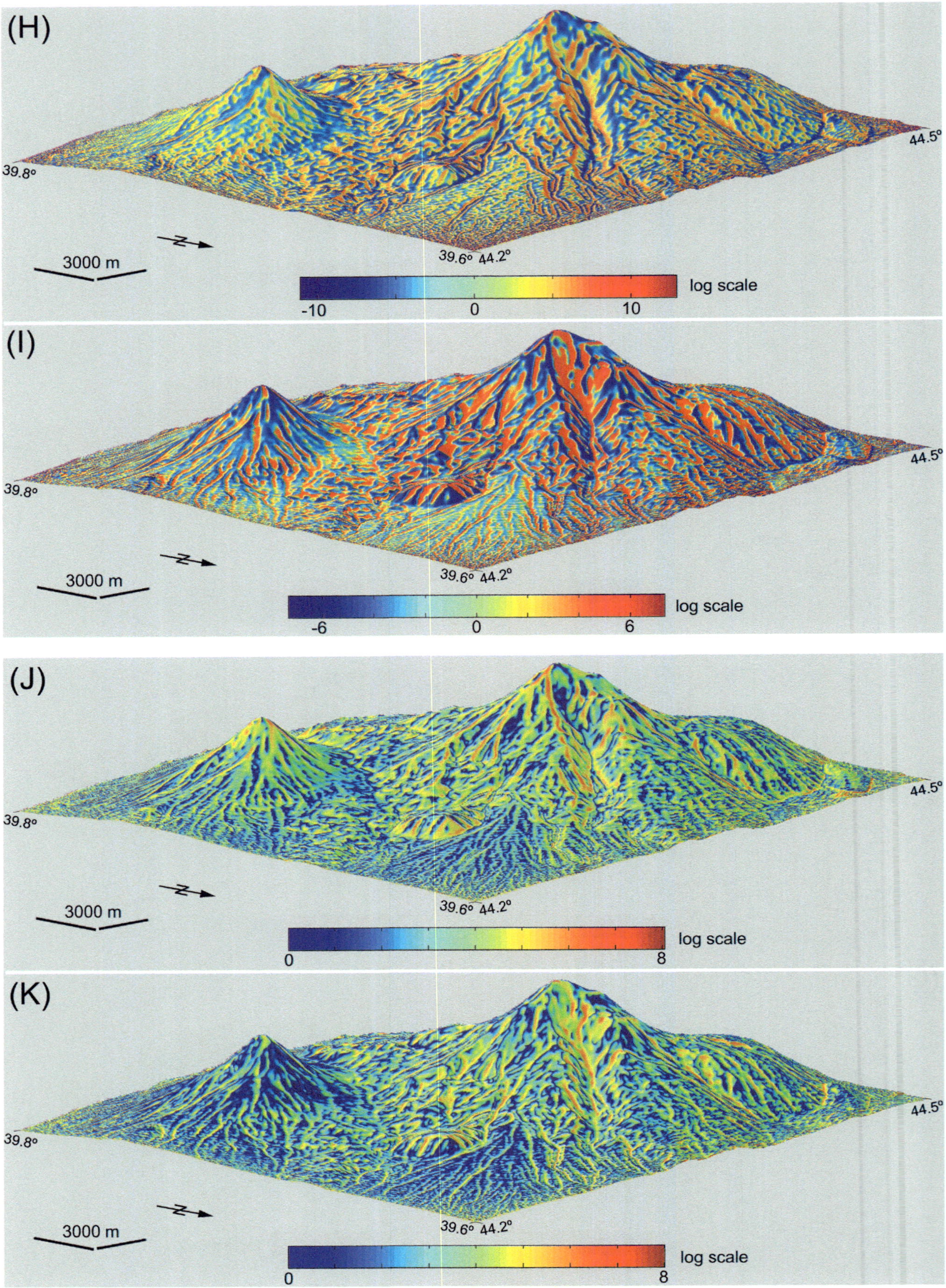

FIG. 2.8, CONT'D (H) Rotor. (I) Difference curvature. (J) Horizontal excess curvature. (K) Vertical excess curvature.

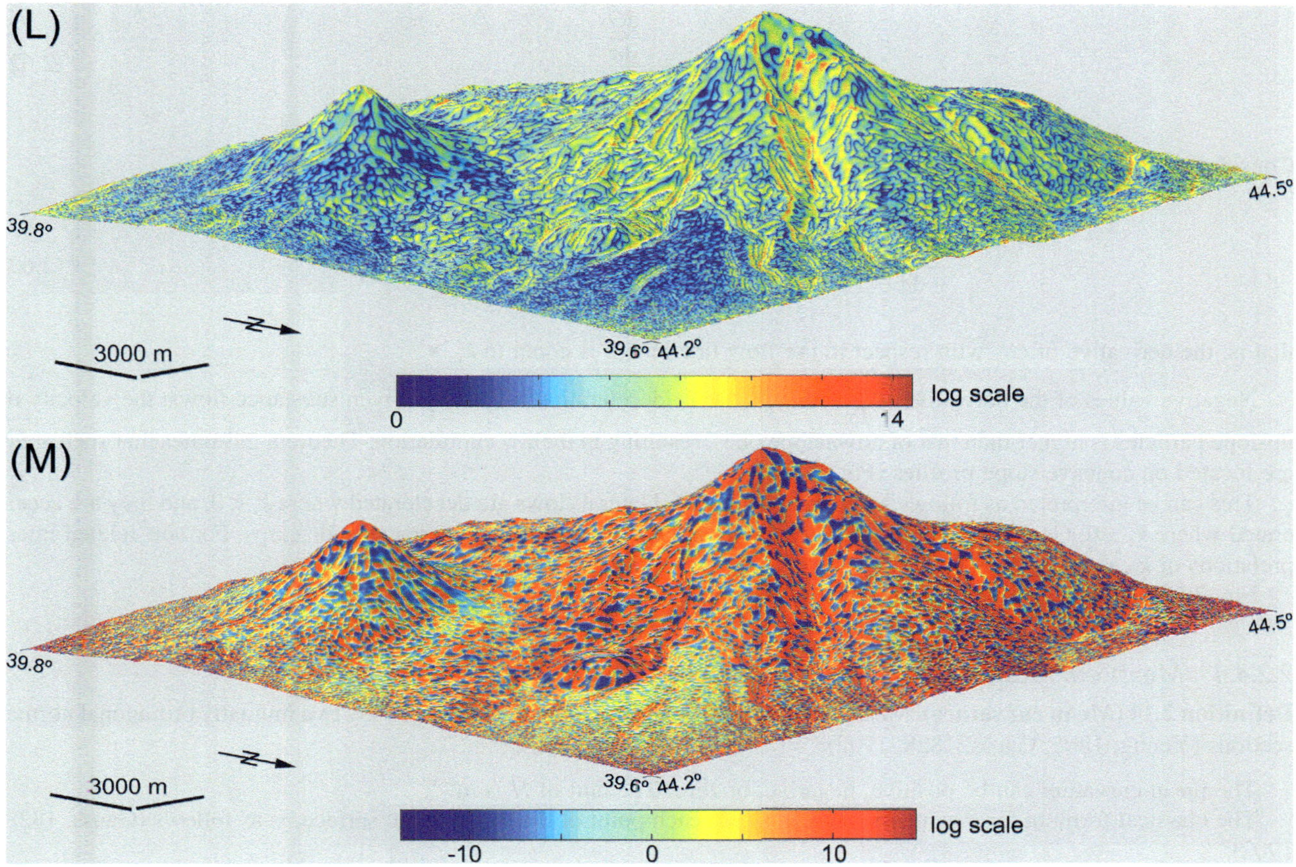

FIG. 2.8, CONT'D (L) Ring curvature. (M) Accumulation curvature. To deal with the large dynamic range of these variables, their values were logarithmically transformed by Eq. (8.1) with $m=2$ for the total curvatures, $m=1$ for other variables, $n=5$. Calculation was carried out by the universal spectral analytical method (Chapter 7). Data were processed and visualized with the author's script for the MATLAB software (Florinsky and Pankratov, 2016). Vertical exaggeration is $2^{\times}$. For the elevation data description, see Fig. 2.3. *(Modified from Florinsky, I.V., 2017a. An illustrated introduction to general geomorphometry. Progr. Phys. Geogr. 41 (6), 723–752, Figs. 3F–M and 4A–F, reproduced with permission from Sage.)*

Theorem 2.4 (Theorem on flow deceleration). On the set N, k_v is a derivative of GF by a flow line length σ (Shary, 1995; Shary et al., 2002b).

Proof On the set N, the first derivative of elevation z with respect to a flow line length σ is

$$\frac{\mathrm{d}z}{\mathrm{d}\sigma} = p\frac{\mathrm{d}x}{\mathrm{d}\sigma} + q\frac{\mathrm{d}y}{\mathrm{d}\sigma} = -\sqrt{p^2+q^2} = -\overline{G}. \tag{2.58}$$

In this case, expressions for $\frac{\mathrm{d}x}{\mathrm{d}\sigma}$ and $\frac{\mathrm{d}y}{\mathrm{d}\sigma}$ were taken from Eq. (2.29).

On the set N, the second derivative of z with respect to σ is:

$$\frac{\mathrm{d}^2 z}{\mathrm{d}^2\sigma} = -\frac{\mathrm{d}\overline{G}}{\mathrm{d}\sigma} = -\frac{\mathrm{d}\sqrt{p^2+q^2}}{\mathrm{d}\sigma} = -k_v\sqrt{(1+p^2+q^2)^3}. \tag{2.59}$$

In this case, expressions for $\frac{\mathrm{d}p}{\mathrm{d}\sigma}$ and $\frac{\mathrm{d}q}{\mathrm{d}\sigma}$ are as follows:

$$\frac{\mathrm{d}p}{\mathrm{d}\sigma} = \frac{\mathrm{d}p}{\mathrm{d}x}\frac{\mathrm{d}x}{\mathrm{d}\sigma} + \frac{\mathrm{d}p}{\mathrm{d}y}\frac{\mathrm{d}y}{\mathrm{d}\sigma} = r\frac{\mathrm{d}x}{\mathrm{d}\sigma} + s\frac{\mathrm{d}y}{\mathrm{d}\sigma}, \tag{2.60}$$

$$\frac{\mathrm{d}q}{\mathrm{d}\sigma} = \frac{\mathrm{d}q}{\mathrm{d}x}\frac{\mathrm{d}x}{\mathrm{d}\sigma} + \frac{\mathrm{d}q}{\mathrm{d}y}\frac{\mathrm{d}y}{\mathrm{d}\sigma} = s\frac{\mathrm{d}x}{\mathrm{d}\sigma} + t\frac{\mathrm{d}y}{\mathrm{d}\sigma}. \tag{2.61}$$

It follows from Eq. (2.59) that on the set N

$$k_v = \frac{\frac{d\overline{G}}{d\sigma}}{\sqrt{\left(1+\overline{G}^2\right)^3}}. \tag{2.62}$$

Considering Eq. (2.21), one obtains for the set N

$$\frac{d(GF)}{d\sigma} = \frac{\frac{d\overline{G}}{d\sigma}\left(\sqrt{1+\overline{G}^2} - \frac{\overline{G}^2}{\sqrt{1+\overline{G}^2}}\right)}{1+\overline{G}^2} = \frac{\frac{d\overline{G}}{d\sigma}}{\sqrt{\left(1+\overline{G}^2\right)^3}} = k_v, \tag{2.63}$$

that is, the derivative of GF with respect to the flow line length is equal to k_v. ▪

Negative values of the derivative $\frac{d(GF)}{d\sigma}$ mean relative deceleration of gravity-driven substance flows: the velocity of upslope particles is higher than that of downslope ones, resulting in their accumulation. Theorem 2.4 states that such areas are located on concave slope profiles (Fig. 2.7B).

This can be interpreted as follows: overland and intrasoil lateral flows are decelerated where $k_v < 0$, and they are accelerated where $k_v > 0$. Geomorphologically, k_v mapping allows the revealing of terraces and scarps. For soil-related interpretations of k_v, see Section 9.2.

Fig. 2.8B shows a 3D map of k_v for Mount Ararat.

2.2.4.4 Mean curvature

Definition 2.14 (Mean curvature). Mean curvature (H) is a half-sum of curvatures of any two mutually orthogonal normal sections (Young, 1805; Gauss, 1828, 1902).

The mean curvature can be positive, negative, or zero. The unit of H is m^{-1}.

The classical formula for mean curvature, valid for each point of the topographic surface, is as follows (Gauss, 1828, 1902):

$$H = -\frac{(1+q^2)r - 2pqs + (1+p^2)t}{2\sqrt{(1+p^2+q^2)^3}}. \tag{2.64}$$

It follows from Eq. (2.56) that mean curvature represents two accumulation mechanisms of gravity-driven substances (Fig. 2.7)—convergence and relative deceleration of flows—with equal weights at a given point of the topographic surface. For soil-related interpretations of H, see Section 9.2.

Fig. 2.8C represents a 3D map of H for Mount Ararat.

2.2.4.5 Principal (minimal and maximal) curvatures

Principal curvatures are curvatures of the two principal sections (Fig. 2.4A) (Section 2.2.1).

Definition 2.15 (Minimal curvature). Minimal curvature (k_{min}) is the curvature of a principal section with the lowest value of curvature at a given point of the topographic surface (Gauss, 1828, 1902).

Definition 2.16 (Maximal curvature). Maximal curvature (k_{max}) is the curvature of a principal section with the highest value of curvature at a given point of the topographic surface (Gauss, 1828, 1902).

Minimal and maximal curvatures can be positive, negative, or zero. The unit of k_{min} and k_{max} is m^{-1}.

It follows from the Euler theorem (Nikolsky, 1977a, § 7.24) that mean curvature is

$$H = \frac{1}{2}(k_{\min} + k_{\max}). \tag{2.65}$$

From Eqs. (2.65) and (2.67), one obtains that

$$k_{\max}, k_{\min} = H \pm \sqrt{H^2 - K}, \tag{2.66}$$

where K is the Gaussian curvature (Section 2.2.4.6); signs "+" and "–" refer to k_{max} and k_{min}, respectively.

Geomorphologically, $k_{min} > 0$ corresponds to local convex landforms, while $k_{min} < 0$ relates to elongated concave landforms (e.g., hills and valleys, respectively). $k_{max} > 0$ corresponds to elongated convex landforms, while $k_{max} < 0$ relates to local concave landforms (e.g., crests and holes, respectively).

Fig. 2.8D and E shows 3D maps of k_{min} and k_{max} for Mount Ararat.

2.2.4.6 Gaussian curvature

Definition 2.17 (Gaussian curvature). The Gaussian (or total) curvature (K) is the product of maximal and minimal curvatures of normal sections at a given point of the topographic surface (Gauss, 1828, 1902):

$$K = k_{\min} k_{\max}. \tag{2.67}$$

The Gaussian curvature can be positive, negative, or zero. The unit of K is m^{-2}.

The classical formula of the Gaussian curvature is as follows (Gauss, 1828, 1902):

$$K = \frac{rt - s^2}{\left(1 + p^2 + q^2\right)^2}. \tag{2.68}$$

According to *Theorema egregium*, the Gaussian curvature retains values at each point of the topographic surface after its bending without breaking, stretching, or compressing (Gauss, 1828). For details and examples, see Section 13.4.

Fig. 2.8F represents a 3D map of K for Mount Ararat.

2.2.4.7 Unsphericity curvature

Definition 2.18 (Unsphericity curvature). Unsphericity curvature (M) is a half-difference of maximal and minimal curvatures (Shary, 1995):

$$M = \frac{1}{2}\left(k_{\max} - k_{\min}\right). \tag{2.69}$$

Unsphericity curvature is a nonnegative variable. The unit of M is m^{-1}.

It follows from Eqs. (2.66) and (2.69):

$$M = \sqrt{H^2 - K}. \tag{2.70}$$

To obtain an explicit formula for M, one should transform the radicand to a sum of squares. This may be done using Eqs. (2.68) and (2.64) for K and H:

$$\begin{aligned} M &= \sqrt{\frac{\left[(1 + q^2)r - 2pqs + (1 + p^2)t\right]^2}{4(1 + p^2 + q^2)^3} - \frac{rt - s^2}{(1 + p^2 + q^2)^2}} \\ &= \sqrt{\frac{\left[(1 + q^2)r - 2pqs + (1 + p^2)t\right]^2 - 4(1 + p^2 + q^2)(rt - s^2)}{4(1 + p^2 + q^2)^3}}. \end{aligned} \tag{2.71}$$

Let α denote the following expression:

$$\alpha = \left[\left(1 + q^2\right)r - 2pqs + \left(1 + p^2\right)t\right]^2 - 4\left(1 + p^2 + q^2\right)\left(rt - s^2\right). \tag{2.72}$$

Now Eq. (2.71) may be rewritten:

$$M = \sqrt{\frac{\alpha}{4(1 + p^2 + q^2)^3}}. \tag{2.73}$$

Let us transform α:

$$\begin{aligned} \alpha = {} & \left(1 + q^2\right)^2 r^2 + 4\left(1 + p^2\right)\left(1 + q^2\right)s^2 + \left(1 + p^2\right)^2 t^2 - 4pq\left(1 + q^2\right)sr \\ & -4pq\left(1 + p^2\right)st + 2\left(p^2q^2 - 1 - p^2 - q^2\right)rt. \end{aligned} \tag{2.74}$$

This is the second-order polynomial

$$\alpha = a^2r^2 + b^2s^2 + c^2t^2 + 2drs + 2est + 2frt, \tag{2.75}$$

where a^2, b^2, c^2, d, e, and f are

$$\begin{cases} a^2 = (1+q^2)^2, \\ b^2 = 4(1+p^2)(1+q^2), \\ c^2 = (1+p^2)^2, \\ d = -2pq(1+q^2), \\ e = -2pq(1+p^2), \\ f = p^2q^2 - 1 - p^2 - q^2. \end{cases} \tag{2.76}$$

α may be represented as a sum of squares

$$\begin{aligned} \alpha &= (Ar+Bt)^2 + (Cr+Ds+Et)^2 = \\ &= \left(A^2+C^2\right)r^2 + D^2s^2 + \left(B^2+E^2\right)t^2 + 2CDrs + 2DEst + 2(AB+CE)rt \end{aligned} \tag{2.77}$$

with unknown coefficients A, B, C, D, and E. Equations for their determination may be found from the comparison of Eqs. (2.74) and (2.77):

$$\begin{cases} A^2 + C^2 = a^2, \\ CD = d, \\ D^2 = b^2, \\ DE = e, \\ B^2 + E^2 = c^2, \\ AB + CE = f. \end{cases} \tag{2.78}$$

One may express A, B, C, D, and E through a^2, b^2, c^2, d, e, and f using Eq. (2.75):

$$\begin{cases} A = \sqrt{a^2 - \dfrac{d^2}{b^2}} = \sqrt{\dfrac{a^2b^2 - d^2}{b^2}}, \\ B = \sqrt{c^2 - \dfrac{e^2}{b^2}} = \sqrt{\dfrac{b^2c^2 - e^2}{b^2}}, \\ C = \dfrac{d}{b}, \\ D = d, \\ E = \dfrac{e}{b}, \end{cases} \tag{2.79}$$

and substitute expressions for a^2, b^2, c^2, d, e, and f from Eq. (2.76). The choice of sign for b is not unique; choosing $b = -2\sqrt{(1+p^2)(1+q^2)}$ one obtains

$$D = b = -2\sqrt{(1+p^2)(1+q^2)}, \tag{2.80}$$

$$C = \frac{d}{b} = \frac{-2pq(1+q^2)}{-2\sqrt{(1+p^2)(1+q^2)}} = \frac{pq\sqrt{1+q^2}}{\sqrt{1+p^2}}, \tag{2.81}$$

$$E = \frac{e}{b} = \frac{-2pq(1+p^2)}{-2\sqrt{(1+p^2)(1+q^2)}} = \frac{pq\sqrt{1+p^2}}{\sqrt{1+q^2}}, \tag{2.82}$$

$$A = \sqrt{\frac{4(1+q^2)^3(1+p^2) - 4p^2q^2(1+q^2)^2}{4(1+p^2)^3(1+q^2)}} = \sqrt{\frac{(1+q^2)(1+p^2+q^2)}{(1+p^2)}}, \tag{2.83}$$

$$B = -\sqrt{\frac{4(1+p^2)^3(1+q^2) - 4p^2q^2(1+p^2)^2}{4(1+p^2)(1+q^2)}} = \sqrt{\frac{(1+p^2)(1+p^2+q^2)}{(1+q^2)}}. \tag{2.84}$$

Now calculate:

$$AB + CE = -(1+p^2+q^2) + p^2q^2 = p^2q^2 - 1 - p^2 - q^2 = f, \tag{2.85}$$

that is, the last equation in Eq. (2.78) is automatically satisfied, and therefore the polynomial α may be represented as a sum of squares (Eq. 2.77).

Substituting Eqs. (2.80)–(2.84) into Eq. (2.77) for α, and then substituting α into Eq. (2.73), one can finally obtain an explicit formula for M:

$$M = \sqrt{\frac{1}{4(1+p^2+q^2)^3}\left\{\left(r\sqrt{\frac{1+q^2}{1+p^2}} - t\sqrt{\frac{1+p^2}{1+q^2}}\right)^2(1+p^2+q^2) + \left(pqr\sqrt{\frac{1+q^2}{1+p^2}} - 2s\sqrt{(1+q^2)(1+p^2)} + pqt\sqrt{\frac{1+p^2}{1+q^2}}\right)^2\right\}}. \tag{2.86}$$

Because $M=0$ on a sphere, M values show the extent to which the shape of the topographic surface is nonspherical at a given point.

Fig. 2.8G demonstrates a 3D map of M for Mount Ararat.

2.2.4.8 Rotor

Definition 2.19 (Rotor). Rotor (*rot*) is a curvature of a flow line (Shary, 1991).

The rotor can be positive, negative, or zero. The unit of *rot* is m^{-1}.

To deduce a formula for the rotor, let us determine a flow line curvature $|\mathbf{r}''_\sigma|$ and compare it with $|\mathrm{rot}(\mathbf{r}'_\sigma)|$, where $\mathbf{r}'_\sigma$ is a unit tangent vector to a flow line. It follows from Eq. (2.29) that on the set N

$$\frac{dx}{d\sigma} = -\frac{p}{\sqrt{p^2+q^2}}, \tag{2.87}$$

$$\frac{dy}{d\sigma} = -\frac{q}{\sqrt{p^2+q^2}}. \tag{2.88}$$

Therefore,

$$\frac{dp}{d\sigma} = r\frac{dx}{d\sigma} + s\frac{dy}{d\sigma} = -\frac{pr+qs}{\sqrt{p^2+q^2}}, \tag{2.89}$$

$$\frac{dq}{d\sigma} = s\frac{dx}{d\sigma} + t\frac{dy}{d\sigma} = -\frac{ps+qt}{\sqrt{p^2+q^2}}, \tag{2.90}$$

$$\begin{aligned}\frac{d^2x}{d\sigma^2} &= -\frac{-\left[-(pr+qs) - \dfrac{p\left(p\dfrac{dp}{d\sigma} + q\dfrac{dq}{d\sigma}\right)}{\sqrt{p^2+q^2}}\right]}{p^2+q^2} \\ &= -\frac{-\{-(pr+qs)(p^2+q^2) + p[p(pr-qs)+q(ps+qt)]\}}{(p^2+q^2)^2} \\ &= \frac{p^3r + p^2qs + pq^2r + q^3s - p^3r - p^2qs - p^2qs - pq^2t}{(p^2+q^2)^2} \\ &= \frac{pq^2(r-t) - qs(p^2+q^2)}{(p^2+q^2)^2} = \left(\frac{-q}{\sqrt{p^2+q^2}}\right)\frac{(p^2-q^2)s - pq(r-t)}{\sqrt{(p^2+q^2)^3}},\end{aligned} \tag{2.91}$$

$$
\begin{aligned}
\frac{d^2y}{d\sigma^2} &= -\frac{-\left[-(ps+qt)-\dfrac{q\left(p\dfrac{dp}{d\sigma}+q\dfrac{dq}{d\sigma}\right)}{\sqrt{p^2+q^2}}\right]}{p^2+q^2} \\
&= -\frac{-\{-(ps+qt)(p^2+q^2)+q[p(pr+qs)+q(ps+qt)]\}}{(p^2+q^2)^2} \\
&= \frac{p^3s+p^2qt+pq^2s+q^3t-p^2qr-pq^2s-pq^2s-q^3t}{(p^2+q^2)^2} \\
&= \frac{ps(p^2-q^2)-p^2q(r-t)}{(p^2+q^2)^2} = \left(\frac{p}{\sqrt{p^2+q^2}}\right)\frac{(p^2-q^2)s-pq(r-t)}{\sqrt{(p^2+q^2)^3}}.
\end{aligned}
\tag{2.92}
$$

Comparing Eqs. (2.91), (2.92), and (2.28), one finds that on the set N

$$
\mathbf{r}''_{\sigma} = \pm\frac{(p^2-q^2)s-pq(r-t)}{\sqrt{(p^2+q^2)^3}}\mathbf{r}''_{\xi}. \tag{2.93}
$$

The choice of sign is determined by the contour line orientation. Thus, on the set N, the flow line curvature is

$$
|\mathbf{r}''_{\sigma}| = \left|\frac{(p^2-q^2)s-pq(r-t)}{\sqrt{(p^2+q^2)^3}}\right|, \tag{2.94}
$$

where the sign of the rotor is not taken into account.

The rotor is also a vector with components

$$
\mathrm{rot}(\mathbf{a}) = \begin{vmatrix} \mathbf{i} & \mathbf{j} & \mathbf{k} \\ \dfrac{\partial}{\partial x} & \dfrac{\partial}{\partial y} & \dfrac{\partial}{\partial z} \\ a_x & a_y & a_z \end{vmatrix}. \tag{2.95}
$$

Substituting here components of $\mathbf{r}'_{\sigma}$ (Eq. 2.29) instead of a_x, a_y, and a_z and considering their independence of z, one obtains for the set N

$$
\begin{aligned}
\mathrm{rot}(\mathbf{r}'_{\sigma}) &= -\left\{\frac{\partial}{\partial x}\left(\frac{q}{\sqrt{p^2+q^2}}\right) - \frac{\partial}{\partial y}\left(\frac{p}{\sqrt{p^2+q^2}}\right)\right\}\mathbf{k} \\
&= -\frac{s(p^2+q^2)-q(pr+qs)-s(p^2+q^2)+p(ps+qt)}{\sqrt{(p^2+q^2)^3}}\mathbf{k} \\
&= -\frac{(p^2-q^2)s-pq(r-t)}{\sqrt{(p^2+q^2)^3}}\mathbf{k}.
\end{aligned}
\tag{2.96}
$$

Comparing Eqs. (2.96) and (2.94), one can see that on the set N

$$
\mathrm{rot}(\mathbf{r}'_{\sigma}) = \pm|\mathbf{r}''_{\sigma}|\mathbf{k}. \tag{2.97}
$$

According to the rotor meaning, the flow line turns clockwise, where the factor by which $\mathbf{k}$ is multiplied is negative. For this reason, the choice of sign for scalar rotor *rot* is opposite to the z-component of rot($\mathbf{r}_{\sigma}'$). Therefore, one obtains that the flow line turns clockwise if *rot* is positive, while it turns counterclockwise in the opposite case. Finally, on the set N

$$
rot = \frac{(p^2-q^2)s-pq(r-t)}{\sqrt{(p^2+q^2)^3}}. \tag{2.98}
$$

Fig. 2.8H represents a 3D map of *rot* for Mount Ararat.

2.2.4.9 Difference curvature

Definition 2.20 (Difference curvature). On the set N, the difference curvature (E) is a half-difference of vertical and horizontal curvatures (Shary, 1995):

$$E = \frac{1}{2}(k_v - k_h). \tag{2.99}$$

The difference curvature can be positive, negative, or zero. The unit of E is m^{-1}.

Using Eq. (2.56), one obtains that $k_v = 2H - k_h$ and $E = H - k_h$. From here and Eqs. (2.64) and (2.55), it follows that on the set N

$$E = \frac{q^2 r - 2pqs + p^2 t}{(p^2 + q^2)\sqrt{1 + p^2 + q^2}} - \frac{(1 + q^2)r - 2pqs + (1 + p^2)t}{2\sqrt{(1 + p^2 + q^2)^3}}. \tag{2.100}$$

It follows from Eqs. (2.99) and (2.56) that on the set N

$$k_v = H + E, \tag{2.101}$$

$$k_h = H - E. \tag{2.102}$$

It follows from Eqs. (2.65) and (2.69):

$$k_{\max} = H + M, \tag{2.103}$$

$$k_{\min} = H - M. \tag{2.104}$$

It follows from Eqs. (2.67), (2.103), and (2.104):

$$K = H^2 - M^2, \tag{2.105}$$

Let us consider an interrelationship between E and M. According to the Euler theorem (Nikolsky, 1977a, § 7.24), the curvature k of a normal section, turned to angle φ from the principal section with the curvature k_{max}, is a function of k_{max}, k_{min}, and φ:

$$k(\varphi) = k_{\max} \cos^2\varphi + k_{\min} \sin^2\varphi. \tag{2.106}$$

Let θ be the smallest angle between a plane of the normal section with the curvature k_v and a plane of the principal section with the curvature k_{max}. As follows from Theorem 2.2, the plane of the normal section with the curvature k_h is orthogonal to the plane of the normal section with the curvature k_v; that is, $k_h = k\left(\theta + \frac{\pi}{2}\right)$. From here and the Euler theorem (Nikolsky, 1977a, § 7.24), it follows:

$$k_v = k_{\max} \cos^2\theta + k_{\min} \sin^2\theta, \tag{2.107}$$

$$k_h = k_{\max} \cos^2\left(\theta - \frac{\pi}{2}\right) + k_{\min} \sin^2\left(\theta + \frac{\pi}{2}\right). \tag{2.108}$$

Eq. (2.108) can be transformed:

$$k_h = k_{\max} \sin^2\theta + k_{\min} \cos^2\theta. \tag{2.109}$$

Subtracting Eq. (2.109) from Eq. (2.107), one obtains:

$$\begin{aligned} k_v - k_h &= k_{\max}\left(\cos^2\theta - \sin^2\theta\right) + k_{\min}\left(\sin^2\theta - \cos^2\theta\right) \\ &= (k_{\max} - k_{\min})\left(\cos^2\theta - \sin^2\theta\right) = (k_{\max} - k_{\min})\cos(2\theta). \end{aligned} \tag{2.110}$$

It follows from Eqs. (2.110), (2.69), and (2.99) that on the set N

$$E = M\cos(2\theta), \tag{2.111}$$

$$|E| \le M. \tag{2.112}$$

Comparing two accumulation mechanisms of gravity-driven substances (Fig. 2.7), E shows to what extent the relative deceleration of flows is higher than flow convergence at a given point of the topographic surface.

Fig. 2.8I demonstrates a 3D map of E for Mount Ararat.

2.2.4.10 Horizontal excess and vertical excess curvatures

Definition 2.21 (Horizontal excess curvature). On the set N, horizontal excess curvature (k_{he}) is the difference of horizontal and minimal curvatures (Shary, 1995):

$$k_{he} = k_h - k_{\min}. \tag{2.113}$$

Definition 2.22 (Vertical excess curvature). On the set N, vertical excess curvature (k_{ve}) is the difference of vertical and minimal curvatures (Shary, 1995):

$$k_{ve} = k_v - k_{\min}. \tag{2.114}$$

These curvatures are nonnegative variables because k_{min}, by its definition (Section 2.2.4.5), is smaller than a curvature of any other normal section at the same point of the surface S. The unit of k_{he} and k_{ve} is m^{-1}.

Using these definitions and Eqs. (2.101)–(2.104), one obtains that on the set N

$$k_{ve} = M + E, \tag{2.115}$$

$$k_{he} = M - E. \tag{2.116}$$

k_{he} shows to what extent the bending of a normal section tangential to a contour line is larger than the minimal bending at a given point of the topographic surface. In its turn, k_{ve} shows to what extent the bending of a normal section having a common tangent line with a slope line is larger than the minimal bending at a given point of the topographic surface.

Fig. 2.8J and K represents 3D maps of k_{he} and k_{ve}, respectively, for Mount Ararat.

2.2.4.11 Ring curvature

Definition 2.23 (Ring curvature). Ring curvature (K_r) is a product of horizontal excess and vertical excess curvatures (Shary, 1995).

Ring curvature is a nonnegative variable. The unit of K_r is m^{-2}.

It follows from Eqs. (2.115) and (2.116) that on the set N

$$K_r = k_{ve}k_{he} = (M + E)(M - E) = M^2 - E^2. \tag{2.117}$$

Using Eqs. (2.117), (2.70), and (2.102), one finds that on the set N

$$K_r = 2Hk_h - k_h^2 - K. \tag{2.118}$$

Substituting Eqs. (2.55), (2.64), and (2.68) to Eq. (2.118), one obtains for the set N

$$\begin{aligned} K_r = {} & \frac{[(1+q^2)r - 2pqs + (1+p^2)t](q^2r - 2pqs + p^2t)}{(p^2+q^2)(1+p^2+q^2)^2} \\ & - \frac{(q^2r - 2pqs + p^2t)^2}{(p^2+q^2)^2(1+p^2+q^2)} - \frac{rt - s^2}{(1+p^2+q^2)^2}. \end{aligned} \tag{2.119}$$

Let $k_h' = q^2r - 2pqs + p^2t$, then the numerator of K_r is

$$\begin{aligned} & (p^2+q^2)(r+t)k_h' + (p^2+q^2)(k_h')^2 - (k_h')^2 - (p^2+q^2)(k_h')^2 - (p^2+q^2)^2(rt - s^2) \\ & = (p^2+q^2)(r+t)(q^2r - 2pqs + p^2t) - (q^2r - 2pqs + p^2t)^2 - (p^2+q^2)^2(rt - s^2) \\ & = (q^2r - 2pqs + p^2t)[(p^2+q^2)(r+t) - q^2r + 2pqs - p^2t] - (p^2+q^2)^2(rt - s^2) \\ & = (q^2r - 2pqs + p^2t)(p^2r + 2pqs + q^2t) - (p^2+q^2)^2(rt - s^2) \\ & = p^2q^2r^2 - 2p^3qsr + p^4rt + 2pq^3sr - 4p^2q^2s^2 + 2p^3qst + q^4rt - 2pq^3st \\ & \quad + p^2q^2t^2 - (p^4 + 2p^2q^2 + q^4)(rt - s^2) \\ & = p^2q^2r^2 - 2pq(p^2 - q^2)sr + (p^4 + q^4)rt + 2pq(p^2 - q^2)st - 4p^2q^2s^2 + p^2q^2t^2 \\ & \quad - (p^4 + 2p^2q^2 + q^4)rt + (p^4 + 2p^2q^2 + q^4)s^2 \\ & = p^2q^2r^2 - 2pq(p^2 - q^2)sr - 2p^2q^2rt + 2pq(p^2 - q^2)st + (p^2 - q^2)^2s^2 + p^2q^2t^2 \\ & = p^2q^2(r - t)^2 - 2pq(p^2 - q^2)s(r - t) + (p^2 - q^2)^2s^2 \\ & = [(p^2 - q^2)s - pq(r - t)]^2. \end{aligned} \tag{2.120}$$

Therefore, on the set N

$$K_r = \left[\frac{(p^2 - q^2)s - pq(r - t)}{(p^2 + q^2)(1 + p^2 + q^2)}\right]^2. \tag{2.121}$$

Comparing Eqs. (2.121) and (2.98), one finds that on the set N

$$K_r = \frac{\overline{G}^2}{\left(1 + \overline{G}^2\right)^2} rot^2. \tag{2.122}$$

This means that K_r describes flow line twisting, but unlike rotor, K_r does not consider the direction of a flow line turn (Shary et al., 2002b).

Theorem 2.5 (Theorem on ring curvature). For any subset of N, where elevation z may be expressed as a function $\sqrt{(x - x_0)^2 + (y - y_0)^2}$, K_r is equal to zero (Shary, 1995).

Proof Let ρ, ϕ be plane polar coordinates with the pole at a given point (x_0, y_0). Then elevation $z(x,y) = z(\rho)$. For this function

$$E = -\frac{R - \frac{P}{\rho}(1 + P^2)}{2\sqrt{(1 + P^2)^3}}, \tag{2.123}$$

$$H = -\frac{\left|R - \frac{P}{\rho}(1 + P^2)\right|}{2\sqrt{(1 + P^2)^3}}, \tag{2.124}$$

where $R = \frac{d^2z}{d\rho^2}$, $P = \frac{dz}{d\rho}$. Therefore, $M = |E|$. According to Eq. (2.117), on the set N

$$K_r = (|E| + E)(|E| - E) = 0. \tag{2.125}$$

■

Fig. 2.8L shows a 3D map of K_r for Mount Ararat.

2.2.4.12 Accumulation curvature

Definition 2.24 (Accumulation curvature). Accumulation curvature (K_a) is a product of vertical and horizontal curvatures (Shary, 1995):

$$K_a = k_h k_v. \tag{2.126}$$

Accumulation curvature is a nonnegative variable. The unit of K_a is m^{-2}.

Using Eqs. (2.55) and (2.57), one obtains for the set N

$$K_a = k_h k_v = \frac{(q^2 r - 2pqs + p^2 t)(p^2 r - 2pqs + q^2 t)}{[(p^2 + q^2)(1 + p^2 + q^2)]^2}. \tag{2.127}$$

Also, using Eqs. (2.101) and (2.102), one obtains for the set N

$$K_a = H^2 - E^2. \tag{2.128}$$

Theorem 2.6 (Theorem on accumulation curvature). For any subset of N, where elevation z may be expressed as a function of polar angle ϕ with a pole at a point (x_0, y_0), accumulation curvature is equal to zero (Shary, 1995).

Proof In this case, $z(x,y) = z(\phi)$. For this function

$$H = E = -\tau \Big/ \left(2\rho^2 \sqrt{\left(\frac{1 + \omega^2}{\rho^2}\right)^3}\right). \tag{2.129}$$

where $\tau = \frac{d^2z}{d\phi^2}$ and $\omega = \frac{dz}{d\phi}$. It follows from Eqs. (2.128) and (2.129) that for the set N

$$K_a = (H + E)(H - E) = 0. \tag{2.130}$$

■

Accumulation curvature can be interpreted as a measure of the extent of flow accumulation at a given point of the topographic surface (Section 2.7.2).

Fig. 2.8M represents a 3D map of K_a for Mount Ararat.

K, K_r, and K_a are called total curvatures (the unit is m^{-2}); k_{min}, k_{max}, k_h, k_v, k_{he}, and k_{ve} are called simple curvatures (the unit is m^{-1}); and H, E, and M are called independent curvatures (the unit is m^{-1}). Total and simple curvatures can be expressed by elementary formulas using the independent curvatures (Shary, 1995, 2012) (Table 2.2).

Theorem 2.7 (Theorem of total curvatures). $K_a = K + K_r$ for the set N (Shary, 1995).

Proof It was proved earlier that $K_a = H^2 - E^2$, $K = H^2 - M^2$, and $K_r = M^2 - E^2$ for the set N (Eqs. (2.128), (2.105), and (2.117)). The proof follows from these formulas. ■

Twelve simple, total, and independent curvatures mentioned previously constitute a complete system of curvatures (Shary, 1995). Mathematical completeness of the curvature system means the possibility of expressing all curvatures of this system in terms of other curvatures of the same system; using any other variables in such operations is redundant.

2.2.4.13 Laplacian

Definition 2.25 (Laplacian). Laplacian of elevation ($\nabla^2 z$) is

$$\nabla^2 z = \frac{\partial^2 z}{\partial x^2} + \frac{\partial^2 z}{\partial y^2} = r + t \tag{2.131}$$

(Nikolsky, 1977a, § 7.26).

Laplacian can be positive, negative, or zero. The unit of $\nabla^2 z$ is m^{-1}.

Theorem 2.8 (Theorem on Laplacian). $\nabla^2 z \approx -2H$ at $p, q \to 0$ (Shary et al., 2002b).

Proof The limit value of H at $p, q \to 0$ can be found from Eq. (2.64): it is equal to $-\frac{1}{2}(r+t)$. Therefore, $\nabla^2 z = r + t \approx -2H$ at $p, q \to 0$. ■

Gradient $\overline{G}$ (Eq. 2.17) also tends to zero at $p, q \to 0$. Thus, it follows from this theorem that $\nabla^2 z$ is close to $-2H$ in gently sloping terrains.

2.2.5 Deflections

Deflections are third-order local morphometric variables. They are functions of the first-, second-, and third-order partial derivatives of elevation. This group of variables includes, in particular, horizontal and vertical deflections.

2.2.5.1 Horizontal deflection

Definition 2.26 (Horizontal deflection). Horizontal deflection (or generating function) (D_{kh}) is a derivative of k_h by the contour line length (Shary and Stepanov, 1991):

$$D_{kh} = \frac{dk_h}{d\xi}, \tag{2.132}$$

where ξ is a contour line length.

This variable can be positive, negative, or zero. The unit of D_{kh} is m^{-2}.

It follows from Eqs. (2.132) and (2.55):

$$\frac{dk_h}{d\xi} = \frac{\partial k_h}{\partial p}\frac{dp}{d\xi} + \frac{\partial k_h}{\partial q}\frac{dq}{d\xi} + \frac{\partial k_h}{\partial r}\frac{dr}{d\xi} + \frac{\partial k_h}{\partial s}\frac{ds}{d\xi} + \frac{\partial k_h}{\partial t}\frac{dt}{d\xi}. \tag{2.133}$$

For $\frac{dp}{d\xi}$ and $\frac{dq}{d\xi}$, see Eqs. (2.42) and (2.43). Similarly, for the set N

$$\frac{\mathrm{d}r}{\mathrm{d}\xi} = g\frac{\mathrm{d}x}{\mathrm{d}\xi} + k\frac{\mathrm{d}y}{\mathrm{d}\xi}, \tag{2.134}$$

$$\frac{\mathrm{d}s}{\mathrm{d}\xi} = k\frac{\mathrm{d}x}{\mathrm{d}\xi} + m\frac{\mathrm{d}y}{\mathrm{d}\xi}, \tag{2.135}$$

$$\frac{\mathrm{d}t}{\mathrm{d}\xi} = m\frac{\mathrm{d}x}{\mathrm{d}\xi} + h\frac{\mathrm{d}y}{\mathrm{d}\xi}, \tag{2.136}$$

where g, h, k, and m are the third-order partial derivatives of elevation (Eq. 2.2). For $\frac{\mathrm{d}x}{\mathrm{d}\xi}$ and $\frac{\mathrm{d}y}{\mathrm{d}\xi}$, see Eqs. (2.40) and (2.41).

After differentiation, the terms of Eq. (2.133) are as follows:

$$\frac{\partial k_h}{\partial p}\frac{\mathrm{d}p}{\mathrm{d}\xi} = \pm\frac{2(pt-qs)(ps-qr)}{\sqrt{(p^2+q^2)^3}\sqrt{1+p^2+q^2}} \pm \frac{(ps-qr)p[2+3(p^2+q^2)](q^2r-2pqs+p^2t)}{\sqrt{(p^2+q^2)^5}\sqrt{(1+p^2+q^2)^3}}, \tag{2.137}$$

$$\frac{\partial k_h}{\partial q}\frac{\mathrm{d}q}{\mathrm{d}\xi} = \pm\frac{-2(pt-qs)(ps-qr)}{\sqrt{(p^2+q^2)^3}\sqrt{1+p^2+q^2}} \pm \frac{(pt-qs)q[2+3(p^2+q^2)](q^2r-2pqs+p^2t)}{\sqrt{(p^2+q^2)^5}\sqrt{(1+p^2+q^2)^3}}, \tag{2.138}$$

$$\frac{\partial k_h}{\partial r}\frac{\mathrm{d}r}{\mathrm{d}\xi} = \pm\frac{q^2(qg-pk)}{\sqrt{(p^2+q^2)^3}\sqrt{1+p^2+q^2}}, \tag{2.139}$$

$$\frac{\partial k_h}{\partial s}\frac{\mathrm{d}s}{\mathrm{d}\xi} = \pm\frac{2pq(pm-qk)}{\sqrt{(p^2+q^2)^3}\sqrt{1+p^2+q^2}}, \tag{2.140}$$

$$\frac{\partial k_h}{\partial t}\frac{\mathrm{d}t}{\mathrm{d}\xi} = \pm\frac{-p^2(ph-qm)}{\sqrt{(p^2+q^2)^3}\sqrt{1+p^2+q^2}}. \tag{2.141}$$

The sum of these terms is:

$$\begin{aligned} D_{kh} &= \frac{\mathrm{d}k_h}{\mathrm{d}\xi} = \frac{\partial k_h}{\partial p}\frac{\mathrm{d}p}{\mathrm{d}\xi} + \frac{\partial k_h}{\partial q}\frac{\mathrm{d}q}{\mathrm{d}\xi} + \frac{\partial k_h}{\partial r}\frac{\mathrm{d}r}{\mathrm{d}\xi} + \frac{\partial k_h}{\partial s}\frac{\mathrm{d}s}{\mathrm{d}\xi} + \frac{\partial k_h}{\partial t}\frac{\mathrm{d}t}{\mathrm{d}\xi} \\ &= \pm\frac{(ps-qr)p[2+3(p^2+q^2)](q^2r-2pqs+p^2t)}{\sqrt{(p^2+q^2)^5(1+p^2+q^2)^3}} \\ &\quad \pm\frac{(pt-qs)q[2+3(p^2+q^2)](q^2r-2pqs+p^2t)}{\sqrt{(p^2+q^2)^5(1+p^2+q^2)^3}} \\ &\quad \pm\frac{q^2(qg-pk)+2pq(pm-qk)-p^2(ph-qm)}{\sqrt{(p^2+q^2)^3(1+p^2+q^2)}}. \end{aligned} \tag{2.142}$$

After simple algebraic operations, one obtains:

$$\begin{aligned} D_{kh} &= \pm\frac{q^3g-3pq^2k+3p^2qm-p^3h}{\sqrt{(p^2+q^2)^3(1+p^2+q^2)}} \\ &\pm\frac{[2+3(p^2+q^2)](q^2r-2pqs+p^2t)}{\sqrt{(p^2+q^2)^5(1+p^2+q^2)^3}}\left[-pqr+(p^2-q^2)s+pqt\right]. \end{aligned} \tag{2.143}$$

The choice of sign is determined by the contour line orientation. The sign of D_{kh} is not important because D_{kh} is zero within concave and convex spurs, where $|k_h|$ is maximal. Thus, D_{kh} (or its zero values) might be considered as the generating function of structural lines (Section 2.4). For the sign "–", Eq. (2.143) may be rewritten as follows (Florinsky, 2009b):

$$D_{kh}=\frac{q^3g-3pq^2k+3p^2qm-p^3h}{\sqrt{(p^2+q^2)^3(1+p^2+q^2)}}+\frac{[2+3(p^2+q^2)](q^2r-2pqs+p^2t)}{\sqrt{(p^2+q^2)^5(1+p^2+q^2)^3}}\left[(p^2-q^2)s-pq(r-t)\right]. \tag{2.144}$$

Considering Eqs. (2.55) and (2.98), one finally obtains for the set N

$$D_{kh}=\frac{q^3g-3pq^2k+3p^2qm-p^3h}{\sqrt{(p^2+q^2)^3(1+p^2+q^2)}}-k_h rot\frac{2+3(p^2+q^2)}{1+p^2+q^2}. \tag{2.145}$$

Horizontal deflection measures the deviation of k_h from the loci of the extreme curvature of the topographic surface.

Fig. 2.9A shows a 3D map of D_{kh} for Mount Ararat.

2.2.5.2 Vertical deflection

Definition 2.27 (Vertical deflection). Vertical deflection (or tangent change of profile curvature) (D_{kv}) is a derivative of k_v by the contour line length ξ (Jenčo et al., 2009):

$$D_{kv}=\frac{\mathrm{d}k_v}{\mathrm{d}\xi}. \tag{2.146}$$

This variable can be positive, negative, or zero. The unit of D_{kv} is m^{-2}.

It follows from Eqs. (2.146) and (2.57):

$$\frac{\mathrm{d}k_v}{\mathrm{d}\xi}=\frac{\partial k_v}{\partial p}\frac{\mathrm{d}p}{\mathrm{d}\xi}+\frac{\partial k_v}{\partial q}\frac{\mathrm{d}q}{\mathrm{d}\xi}+\frac{\partial k_v}{\partial r}\frac{\mathrm{d}r}{\mathrm{d}\xi}+\frac{\partial k_v}{\partial s}\frac{\mathrm{d}s}{\mathrm{d}\xi}+\frac{\partial k_v}{\partial t}\frac{\mathrm{d}t}{\mathrm{d}\xi}. \tag{2.147}$$

For $\frac{\mathrm{d}p}{\mathrm{d}\xi}$, $\frac{\mathrm{d}q}{\mathrm{d}\xi}$, $\frac{\mathrm{d}r}{\mathrm{d}\xi}$, $\frac{\mathrm{d}s}{\mathrm{d}\xi}$, and $\frac{\mathrm{d}t}{\mathrm{d}\xi}$, see Eqs. (2.42), (2.43), and (2.134)–(2.136). For $\frac{\mathrm{d}x}{\mathrm{d}\xi}$ and $\frac{\mathrm{d}y}{\mathrm{d}\xi}$, see Eqs. (2.40) and (2.41).

After differentiation, the terms of Eq. (2.147) are as follows:

$$\frac{\partial k_v}{\partial p}\frac{\mathrm{d}p}{\mathrm{d}\xi}=\pm\frac{(ps-qr)}{\sqrt{(p^2+q^2)^3(1+p^2+q^2)^3}}\left[-2(pr+qs)\pm p(p^2r+2pqs+q^2t)\left(\frac{2}{p^2+q^2}\pm\frac{3}{1+p^2+q^2}\right)\right], \tag{2.148}$$

$$\frac{\partial k_v}{\partial q}\frac{\mathrm{d}q}{\mathrm{d}\xi}=\pm\frac{(pt-qs)}{\sqrt{(p^2+q^2)^3(1+p^2+q^2)^3}}\left[-2(ps+qt)\pm q(p^2r+2pqs+q^2t)\left(\frac{2}{p^2+q^2}\pm\frac{3}{1+p^2+q^2}\right)\right], \tag{2.149}$$

$$\frac{\partial k_v}{\partial r}\frac{\mathrm{d}r}{\mathrm{d}\xi}=\pm\frac{p^2(qg-pk)}{\sqrt{(p^2+q^2)^3(1+p^2+q^2)^3}}, \tag{2.150}$$

$$\frac{\partial k_v}{\partial s}\frac{\mathrm{d}s}{\mathrm{d}\xi}=\pm\frac{2pq(qk-pm)}{\sqrt{(p^2+q^2)^3(1+p^2+q^2)^3}}, \tag{2.151}$$

$$\frac{\partial k_v}{\partial t}\frac{\mathrm{d}t}{\mathrm{d}\xi}=\pm\frac{-q^2(ph-qm)}{\sqrt{(p^2+q^2)^3(1+p^2+q^2)^3}}. \tag{2.152}$$

After summing these five terms (Eqs. 2.148–2.152) within Eq. (2.147), performing simple algebraic operations, and choosing the sign "+" (the choice of the sign is determined by the contour line orientation), one obtains:

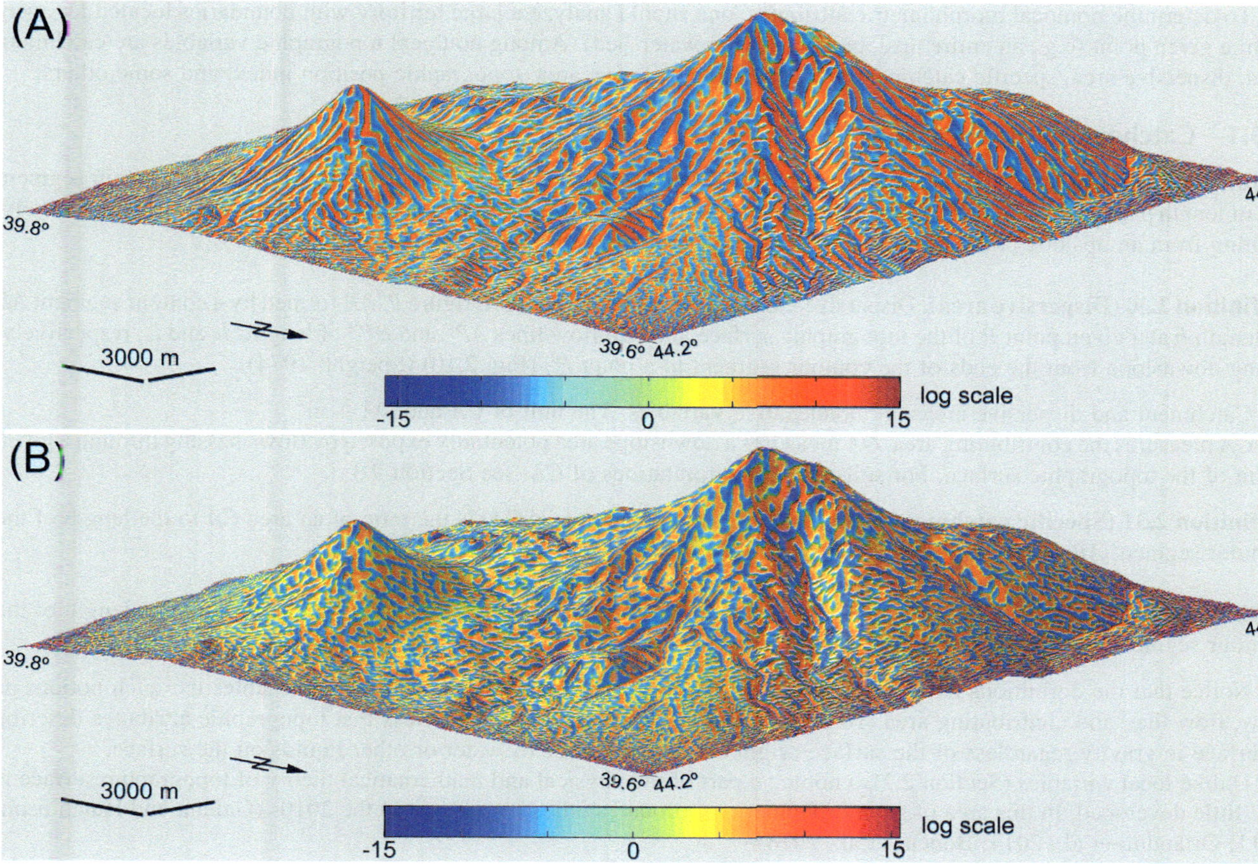

FIG. 2.9 Mount Ararat, deflections: (A) Horizontal deflection. (B) Vertical deflection. To deal with the large dynamic range of these variables, their values were logarithmically transformed by Eq. (8.1) with $m=2$ and $n=4$. Calculation was carried out by the universal spectral analytical method (Chapter 7). Data were processed and visualized with the author's script for the MATLAB software (Florinsky and Pankratov, 2016). Vertical exaggeration is $2^{\times}$. For the elevation data description, see Fig. 2.3. *(Modified from Florinsky, I.V., 2017a. An illustrated introduction to general geomorphometry. Progr. Phys. Geogr. 41 (6), 723–752, Fig. 3N–O, reproduced with permission from Sage.)*

$$D_{kv} = \frac{q^3 m - p^3 k + pq[q(2k-h) - p(2m-g)]}{\sqrt{(p^2+q^2)^3(1+p^2+q^2)^3}} + \frac{2(r+t)(pq(r-t) - s(p^2-q^2))}{\sqrt{(p^2+q^2)^3(1+p^2+q^2)^3}} + \frac{(p^2 r + 2pqs + q^2 t)((p^2-q^2)s - pq(r-t))(2 - 5p^2 + 5q^2)}{(p^2+q^2)\sqrt{(1+p^2+q^2)^3}\sqrt{(p^2+q^2)^3(1+p^2+q^2)}}. \quad (2.153)$$

Considering Eqs. (2.57) and (2.98), one finally obtains for the set N:

$$D_{kv} = \frac{q^3 m - p^3 k + pq[q(2k-h) - p(2m-g)]}{\sqrt{(p^2+q^2)^3(1+p^2+q^2)^3}} - rot\left[\frac{2(r+t)}{\sqrt{(1+p^2+q^2)^3}} + k_v \frac{2+5(p^2+q^2)}{(1+p^2+q^2)}\right]. \quad (2.154)$$

Vertical deflection measures the deviation of k_v from the loci of the extreme curvature of the topographic surface.

Fig. 2.9B shows a 3D map of D_{kv} for Mount Ararat.

2.3 Nonlocal morphometric variables

Definition 2.28 (Nonlocal morphometric variable). A nonlocal morphometric variable is a single-valued bivariate function $v=u(x,y)$ describing the relative position of a given point of the topographic surface (Speight, 1974).

To determine nonlocal morphometric attributes, one should analyze a large territory with boundaries located far away from a given point (e.g., an entire upslope portion of a watershed). Among nonlocal topographic variables are catchment area, dispersive area, specific catchment area, specific dispersive area, topographic position index, and some others.

2.3.1 Catchment and dispersive areas

Definition 2.29 (Catchment area). Catchment area (*CA*) is the area of a closed figure $P'AB$ formed by a contour segment AB of length b at a given point P of the topographic surface and two flow lines $P'A$ and $P'B$ of lengths l_1 and l_2, respectively, coming from an upslope point P' to the ends of the contour segment (Fig. 2.10) (Speight, 1974).

Definition 2.30 (Dispersive area). Dispersive area (*DA*) is an area of a closed figure $P''AB$ formed by a contour segment AB of length b at a given point P of the topographic surface and two flow lines AP'' and BP'' of lengths l_3 and l_4, respectively, going downslope from the ends of the contour segment to a point P'' (Fig. 2.10) (Speight, 1974).

Catchment and dispersive areas are nonnegative variables. The unit of *CA* and *DA* is m^2.

CA measures the contributing area. *DA* measures a downslope area potentially exposed by flows passing through a given point of the topographic surface. For soil-related interpretations of *CA*, see Section 9.3.

Definition 2.31 (Specific catchment area). Specific catchment area (*SCA*) is the ratio of an area *CA* to the length of the contour segment AB of length b (Fig. 2.10) (Speight, 1974).

Definition 2.32 (Specific dispersive area). Specific dispersive area (*SDA*) is the ratio of an area *DA* to the length of the contour segment AB of length b (Fig. 2.10) (Speight, 1974).

Notice that the definitions and interpretations of the discussed nonlocal morphometric variables use such notions as flow, flow line, and contributing area. To avoid misunderstanding, we should stress that topographic attributes describe a surface in gravity regardless of the surface origin and the existence of water or other liquids on the surface.

Unlike local variables (Section 2.2), a nonlocal part of the physical and mathematical theory of topographic surface is still little developed. In this area of geomorphometry, profound studies began only in the 2010s (Gallant and Hutchinson, 2011; Orlandini et al., 2014; Bonetti et al., 2018).

In particular, Gallant and Hutchinson (2011) developed a differential equation for specific catchment area:

$$\frac{\partial SCA}{\partial \sigma} = 1 - SCA \cdot k_p, \tag{2.155}$$

where σ is the length of a central flow line within the figure $P'AB$; k_p is plan curvature. This expression can be applied to calculate *SCA* at any point of the topographic surface by constructing a slope line upslope from a given point to a hilltop and integrating Eq. (2.155) down the same slope line; an initial condition is $SCA = 0$ at the hilltop.

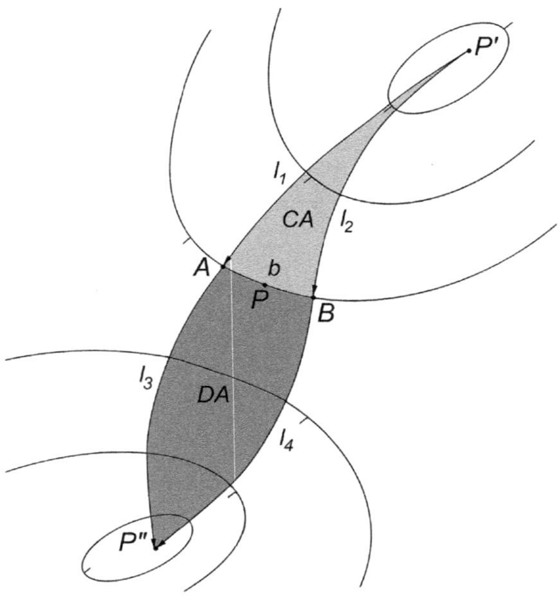

FIG. 2.10 Scheme for the definitions of catchment and dispersive areas. b, the length of a contour line segment AB; *CA* and *DA*, areas of figures $P'AB$ (*light gray*) and $P''AB$ (*dark gray*), respectively; l_1, l_2, l_3, and l_4, the lengths of slope lines $P'A$, $P'B$, AP'', and BP'', respectively.

A part of the theory of topographic surface concerning catchment and dispersive areas will hopefully be completed in the future. The diversity of methods for the calculation of nonlocal variables (Section 4.4) is the indirect result of the under-development of a mathematical theory.

Fig. 2.11A and B represents 3D maps of *CA* and *DA*, respectively, for Mount Ararat.

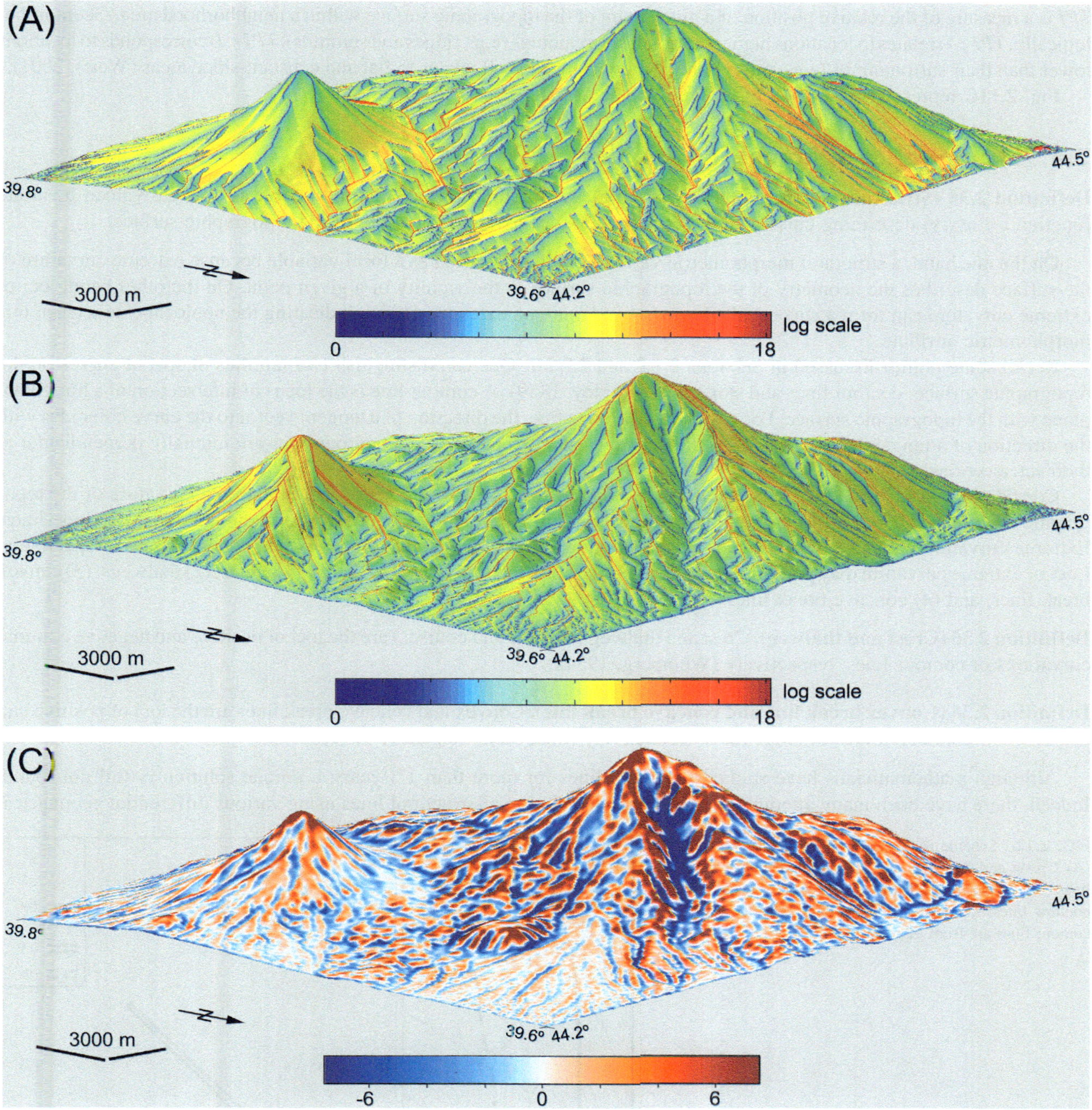

FIG. 2.11 Mount Ararat, nonlocal morphometric variables: (A) Catchment area. (B) Dispersive area. Calculation was carried out by the Martz-de Jong method (Martz and de Jong E., 1988) adapted to spheroidal equal angular grids (Section 4.4.2). To deal with the large dynamic range of these variables, their values were logarithmically transformed by Eq. (8.1). Data were processed by the LandLord software (Florinsky, 2016) and visualized with the author's script for the MATLAB software. (C) Topographic position index. Calculation was carried out with a window radius of 300 m. Data were processed by the SAGA software (Conrad et al., 2015) and visualized with the author's script for the MATLAB software. Vertical exaggeration is 2×. For the elevation data description, see Fig. 2.3. *((A) and (B) Modified from Florinsky, I.V., 2017a. An illustrated introduction to general geomorphometry. Progr. Phys. Geogr. 41 (6), 723–752, Fig. 5, reproduced with permission from Sage.)*

2.3.2 Topographic position index

Definition 2.33 (Topographic position index). Topographic position index (*TPI*) is the difference between an elevation z_i at a given point of the topographic surface and a mean elevation $\bar{z}$ of a specified neighborhood area of that point (Guisan et al., 1999; Wilson and Gallant, 2000, p. 74; Weiss, 2001):

$$TPI = z_i - \bar{z}. \tag{2.156}$$

TPI is a measure of the relative position of a given point of the topographic surface within a neighborhood area. Geomorphologically, $TPI > 0$ relates to locations higher than their surroundings (e.g., ridges and summits). $TPI < 0$ corresponds to locations lower than their surroundings (e.g., valleys and depressions). $TPI \approx 0$ relates to flat and constant-slope areas (Weiss, 2001).

Fig. 2.11C represents a 3D map of *TPI* for Mount Ararat.

2.4 Structural morphometric variables

Definition 2.34 (Structural morphometric variable). A structural morphometric variable is a single-valued bivariate function $\upsilon = u(x, y)$ describing curvilinearly distributed loci of extreme curvature of the topographic surface.

On the one hand, a structural morphometric variable can be classified as a local variable because extreme curvature of the surface describes the geometry of the topographic surface in the vicinity of a given point. On the other hand loci of extreme curvature can form a curve (called *a structural line*), an extended object, indicating the nonlocal character of this morphometric attribute.

Recall some points discussed in Sections 2.1.2 and 2.2.2.1. One can distinguish two families of spatial curves on the topographic surface: contour lines and slope lines (Cayley, 1859). A contour line is the locus of intersection of a horizontal plane with the topographic surface. For any point of a slope line, the direction of a tangent vector to the curve coincides with the direction of a tangential component of the gravity vector. Slope lines and contour lines are mutually perpendicular at their intersections.

Considering these families of spatial curves, one can distinguish two groups of loci of extreme curvature of the topographic surface: (1) a locus of extreme curvature of contour lines and (2) a locus of extreme curvature of slope lines. Extreme curvature varies in sign: one can set the positive and negative signs to convex and concave areas, respectively. Loci of extreme curvature may partially describe four types of structural lines (Fig. 2.12): (1) crests, (2) thalwegs, (3) convex break lines, and (4) concave break lines.

Definition 2.35 (Crest and thalweg). Crest and thalweg (or ridge and courses) are the loci of positive and negative extreme curvatures of contour lines, respectively (Weinberg, 1934a).

Definition 2.36 (Convex break line and concave break line). Convex and concave break lines are the loci of positive and negative extreme curvatures of slope lines, respectively (Shary and Stepanov, 1991).

Although mathematicians have studied structural lines for more than 170 years, a unique solution is still not found. Indeed, there have been many attempts for analytical description of structural lines using various differential geometrical

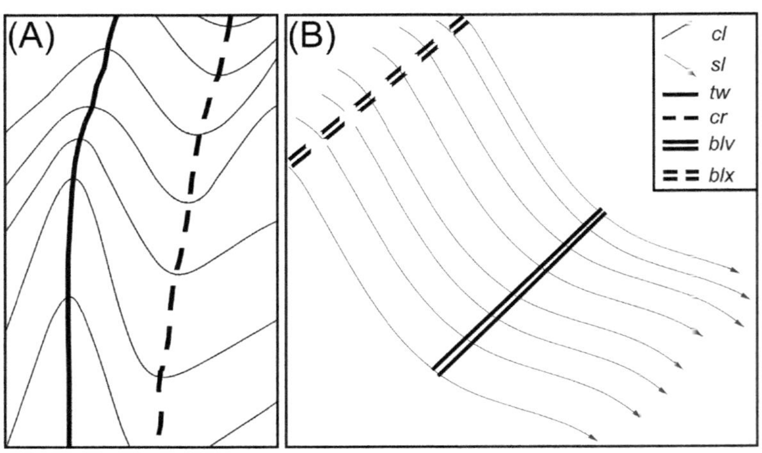

FIG. 2.12 Scheme for the definitions of structural lines: (A) Crests and thalwegs. (B) Convex and concave break lines. *blv*, concave break line; *blx*, convex break line; *cl*, contour lines; *cr*, crest; *sl*, slope lines; *tw*, thalweg. The contour lines are from Fig. 2.7.

local criteria (De Saint-Venant, 1852; Boussinesq, 1872; Jordan, 1872; Breton de Champ, 1877)—see historical reviews by Koenderink and van Doorn (1993, 1994). Shary and Stepanov (1991) argued that structural lines may be analytically described using the horizontal deflection (Section 2.2.5.1), that is, the derivative of k_h. Criteria of those authors are as follows: ridges and convex break lines relate to the locus of zero values of the horizontal deflection ($D_{kh} = 0$) within divergence areas ($k_h > 0$), while thalwegs and concave break lines correspond to the locus of $D_{kh} = 0$ within areas with $k_h < 0$ (see details in Section 4.5.2). Minár et al. (2013) suppose that it is possible to describe structural lines analytically using loci of extrema of both horizontal and vertical curvatures.

Alternatively, one can consider ridge and valley lines as two topologically connected tree-like hierarchical structures. Maxwell (1870) defined a ridge as a slope line connecting a sequence of local maximal and saddle points, and a thalweg as a slope line connecting a sequence of local minimal and saddle points. Rothe (1915) argued that crests and thalwegs are singular solutions of the differential equation of the slope lines. Following this idea, Koenderink and van Doorn (1993) believed that crests and thalwegs are the special types of slope lines where other slope lines converge. In this context, Koenderink and van Doorn (1993, p. 12) have stressed:

> *"… the ever ongoing quest for local criteria for courses and ridges ('ridge finders') is bound to fail: There is no such animal. … Examples … unequivocally demonstrate that several of the criteria for courses and ridges in common use today (such as the loci of vertices, of steepest descent along an isohypse, of maximum height along the direction of extremal 2nd-order directional derivative) are plainly erroneous in the strict sense. Of course such loci may turn out to have their uses in certain settings, but they are definitely not courses and ridges in the original sense of these words (which we take to apply to the idealized physics of drainage in the terrestrial landscape)."*

We hope that this scientific dispute will be resolved in the future. As for catchment and dispersive areas (Section 2.3.1), the diversity of methods for revealing structural lines (Section 4.5) is the consequence of the underdevelopment of a correct mathematical theory.

Fig. 2.13A and B shows 3D maps of thalwegs and crests, respectively, for Mount Ararat.

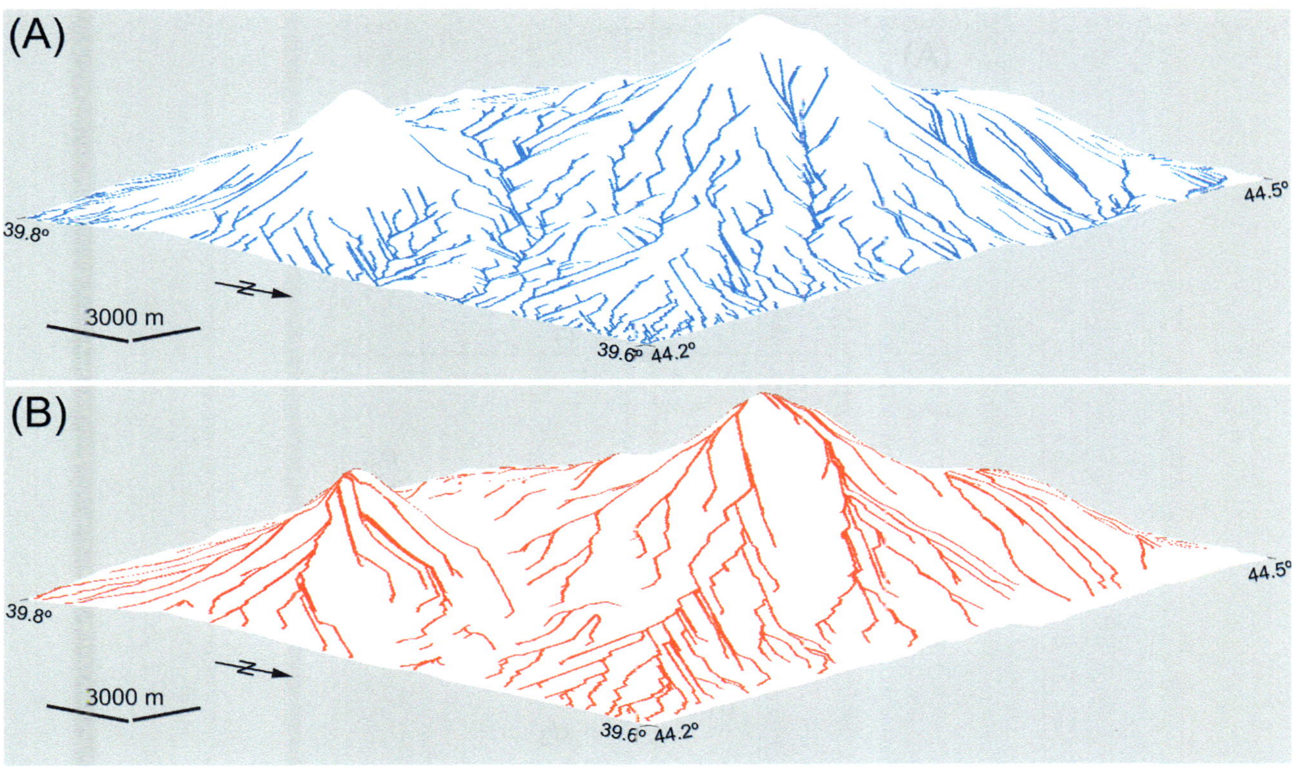

FIG. 2.13 Mount Ararat, structural lines: (A) Thalwegs. (B) Crests. Calculation was carried out by a D8 cell-to-cell flow routing algorithm (Sections 4.4 and 4.5.1). Four orders of thalwegs and crests were revealed. Data were processed by the SAGA software (Conrad et al., 2015) and visualized with the author's script for the MATLAB software. Vertical exaggeration is 2×. For the elevation data description, see Fig. 2.3.

2.5 Two-field-specific morphometric variables

Definition 2.37 (Two-field-specific morphometric variable). A two-field-specific morphometric variable is a single-valued bivariate function $\upsilon = u(x, y)$ describing relations between the topographic surface located in gravity field and other fields, such as solar irradiation, wind flow, and view field (Evans and Minár, 2011).

2.5.1 Solar morphometric variables

Definition 2.38 (Solar morphometric variable). A solar morphometric variable is a single-valued bivariate function $\upsilon = u(x, y)$ describing relations between the topographic surface and solar irradiation.

Among the solar morphometric variables are reflectance, insolation, sky view factor, and some other related functions.

2.5.1.1 Reflectance

Definition 2.39 (Reflectance). Reflectance (R) is a measure of the brightness of an illuminated topographic surface at a given point (Horn, 1981).

Reflectance is a nonnegative, dimensionless variable. R values are usually normalized as $\min(R) = 0$ and $\max(R) = 1$.

For a surface element, the apparent brightness depends on its orientation relative to the viewer's position and the light source. Reflectance can be estimated with various models of light reflectance from a surface, such as the Lambertian model, Lommel-Seeliger law, Minnaert function, and others (Horn, 1981).

The Lambertian surface (or ideal diffuser) reflects all light and appears equally bright from all viewing directions. The Lambertian reflectance (R_L) has the following expression (Horn, 1981, § 27):

$$R_L = \frac{1 - p \sin\theta \cot\psi - q \cos\theta \cot\psi}{\sqrt{1 + p^2 + q^2}\sqrt{1 + (\sin\theta \cot\psi)^2 + (\cos\theta \cot\psi)^2}}, \tag{2.157}$$

where Θ and ψ are the solar azimuth and elevation angles (Fig. 2.14A), respectively.

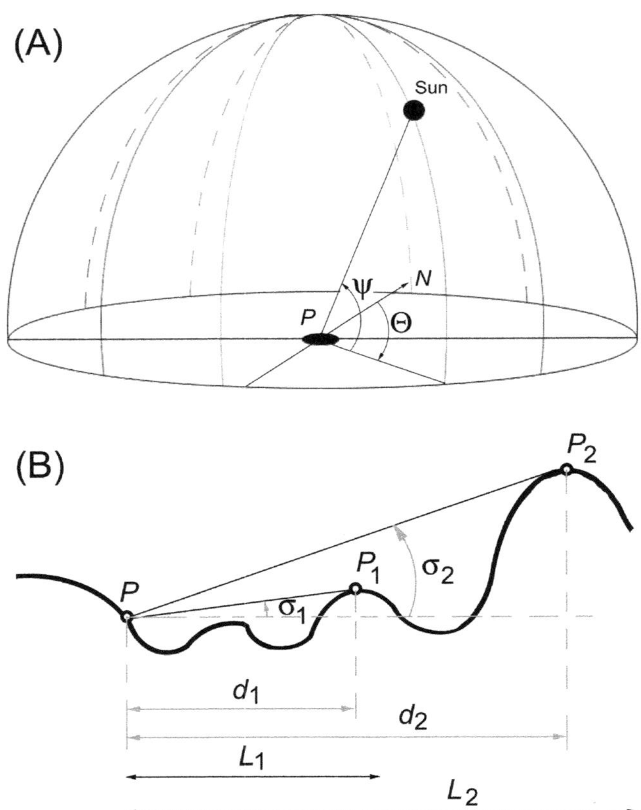

FIG. 2.14 Schemes for the definitions of solar morphometric variables: (A) Position of the sun in the sky relative to a given point P of the topographic surface. N, north direction; θ, solar azimuth angle; ψ, solar elevation angle. (B) Horizon angle for a given point P of the topographic surface and a certain azimuth depending on the search distance. *Bold curve*, profile of the topographic surface; d_1 and d_2, distances from the point P to points with the highest elevations P_1 and P_2 within search distances L_1 and L_2, respectively; σ_1 and σ_2, horizon angles for search distances L_1 and L_2, respectively.

The Lommel-Seeliger model includes a random distribution of similar particles suspended in a transparent medium. The Lommel-Seeliger reflectance (R_{LS}) can be estimated using the following formula (Horn, 1981, § 35):

$$R_{LS} = \frac{1 + 1/\sqrt{1 + (\sin\theta\cot\psi)^2 + (\cos\theta\cot\psi)^2}}{1 + \sqrt{1 + (\sin\theta\cot\psi)^2 + (\cos\theta\cot\psi)^2}/(1 - p\sin\theta\cot\psi - q\cos\theta\cot\psi)}. \tag{2.158}$$

Reflectance maps are very popular in cartography. Traditionally, they are called hill-shaded maps; procedures for their drawing are called hill shading. These procedures have long been performed by manual methods (Imhof, 1965, 1982, chap. 9; Horn, 1981, §§ 2–4). Currently, their automated versions are known as analytical hill shading (Yoeli, 1965; Brassel, 1974; Jenny, 2001; Jenny et al., 2020). Hill-shaded maps are widely used in geosciences because they can perfectly represent the shape of the topographic surface (Sections 8.2 and 8.5). An advantage of hill-shaded maps is the possibility of modeling such positions of the sun, which are impossible in nature. This offers the potential for revealing some topographically expressed geological structures (Section 14.1).

Fig. 2.15A demonstrates a 3D hill-shaded map for Mount Ararat.

In hill shading, the usual single-source illumination may be deficient for topographically complex terrains. To avoid this problem, algorithms of multiple-source illumination are periodically applied (Mark, 1992; Hobbs, 1999; Smith and Clark, 2005; Huffman, 2014, 2017; Kennelly and Stewart, 2014; Veronesi and Hurni, 2014; Marston and Jenny, 2015; Tzvetkov, 2018).

In remote sensing, reflectance maps are applied to correct the topographic effect in remotely sensed images. The topographic effect is the variation in radiance from an inclined surface compared to the radiance from a horizontal surface as a function of the surface orientation relative to the light source and sensor position. Correction of the topographic effect allows the reduction of mistakes in the classification of remotely sensed images, as this reduces the brightness variation in a scene. A review of this issue can be found elsewhere (Franklin, 2020).

2.5.1.2 Insolation[e]

Definition 2.40 (Insolation). Insolation (I) is a proportion of maximal direct solar irradiation at a given point of the topographic surface, with the sun's position determined by solar azimuth and solar elevation (Fig. 2.14A) (Shary et al., 2005).

Insolation is a nonnegative variable measured in percent.

Solar rays may be considered parallel, so a unit vector to the sun from a given point of the surface S is

$$\mathbf{m} = \mathbf{i}\,\sin\theta\cos\psi + \mathbf{j}\,\cos\theta\cos\psi + \mathbf{k}\,\sin\psi. \tag{2.159}$$

Solar irradiation intensity is proportional to the cosine of the angle between $\mathbf{m}$ and the unit normal $\mathbf{n}$ to S: it is $(\mathbf{m}, \mathbf{n})$ if this angle is smaller than 90°, and it is null in the opposite case (the shadow side of a hill). The inequality $(\mathbf{m}, \mathbf{n}) > 0$ corresponds to angles smaller than 90°, so this proportion is

$$\frac{1}{2}(\mathbf{m}, \mathbf{n})[1 + \operatorname{sign}(\mathbf{m}, \mathbf{n})]. \tag{2.160}$$

Therefore,

$$I = 50(\mathbf{m}, \mathbf{n})[1 + \operatorname{sign}(\mathbf{m}, \mathbf{n})]. \tag{2.161}$$

From Eqs. (2.159) and (2.11), one can obtain:

$$(\mathbf{m}, \mathbf{n}) = \frac{\sin\psi - \cos\psi(p\sin\theta + q\cos\theta)}{\sqrt{1 + p^2 + q^2}}. \tag{2.162}$$

Substituting Eq. (2.162) into Eq. (2.161), one obtains the final expression:

$$I = 50\frac{\{1 + \operatorname{sign}[\sin\psi - \cos\psi(p\sin\theta + q\cos\theta)]\}[\sin\psi - \cos\psi(p\sin\theta + q\cos\theta)]}{\sqrt{1 + p^2 + q^2}}. \tag{2.163}$$

I can be defined at each point of the topographic surface.

e. The mathematics of Section 2.5.1.2 is based on the materials from the appendix prepared by P.A. Shary for the first edition of this book. *(Modified from Shary, P.A., 2012. The mathematical basis of local morphometric variables. In: Florinsky, I.V., Digital Terrain Analysis in Soil Science and Geology, first ed. Academic Press, Amsterdam, The Netherlands, pp. 289–313, Section A.2.2, reprinted with permission from Elsevier.)*

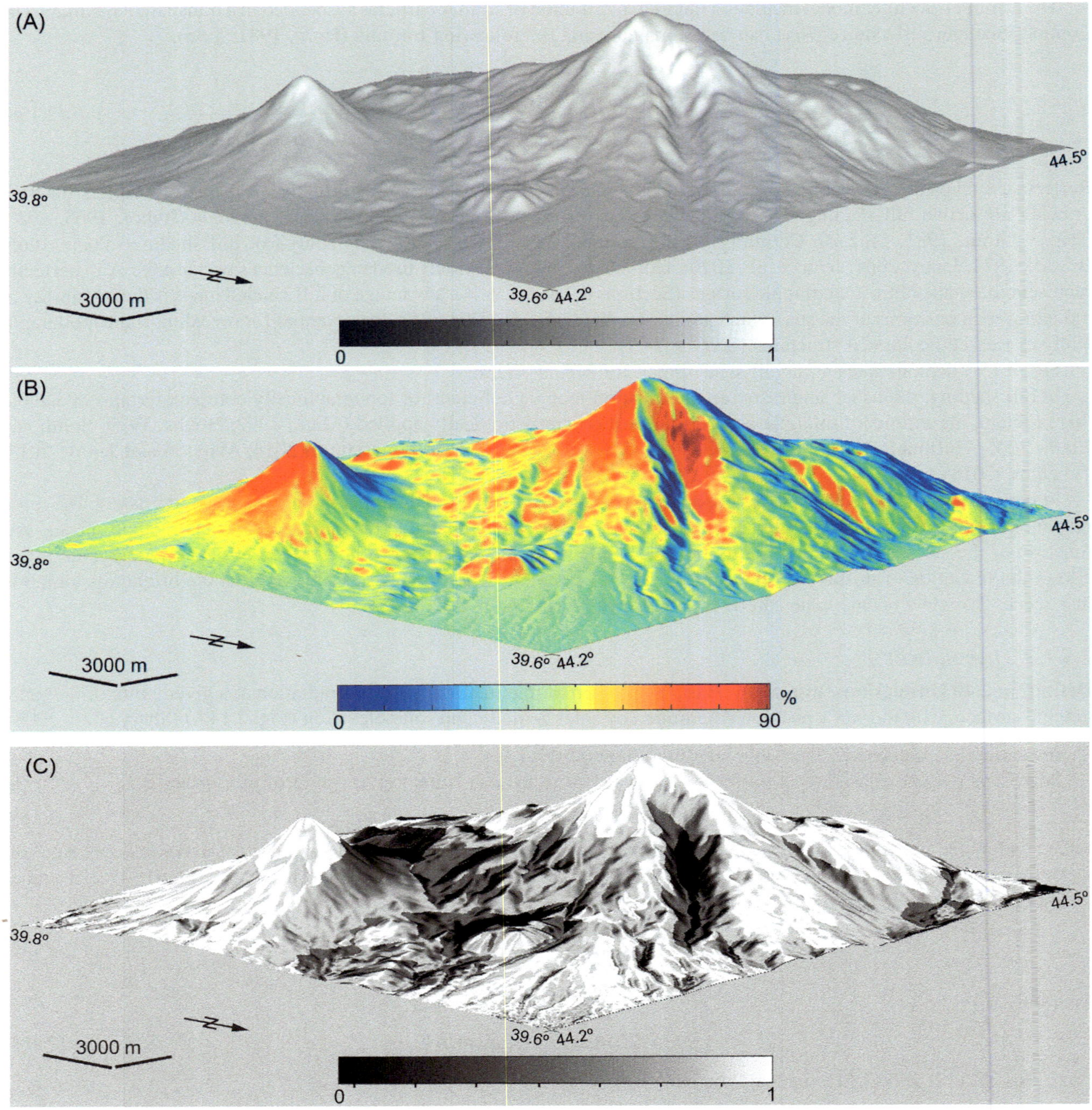

FIG. 2.15 Mount Ararat, solar morphometric variables: (A) Reflectance (the Lambertian model; $\theta = 45°$; $\psi = 35°$). (B) Insolation ($\theta = 90°$; $\psi = 20°$). Calculations were carried out by the universal spectral analytical method (Chapter 7). Data were processed and visualized with the author's script for the MATLAB software (Florinsky and Pankratov, 2016). (C) Sky view factor ($L = 10$ km). Data were processed by the SAGA software (Conrad et al., 2015) and visualized with the author's script for the MATLAB software. Vertical exaggeration is $2^{\times}$. For the elevation data description, see Fig. 2.3. *((A) and (B) Modified from Florinsky, I.V., 2017a. An illustrated introduction to general geomorphometry. Progr. Phys. Geogr 41 (6), 723–752, Fig. 6, reproduced with permission from Sage.)*

The thermal regime of slopes depends in part on the incidence of solar rays on the land surface, so it depends on both G and A (Section 9.2). Insolation directly describes this incidence and therefore better considers the thermal regime. Reviews of insolation-based solar radiation models can be found elsewhere (Dubayah and Rich, 1995; Böhner and Antonić, 2009).

It is obvious that insolation and Lambertian reflectance (Section 2.5.1.1) are, in the physical sense, some sort of "opposite" variables.

Fig. 2.15B shows a 3D map of I for Mount Ararat.

2.5.1.3 Sky view factor

Definition 2.41 (Horizon angle). Horizon angle (σ) is the maximum angle from a given point of the topographic surface toward any other point in a given azimuth within a distance L (Böhner and Antonić, 2009).

The horizon angle (Fig. 2.14B) can be estimated by the following expression:

$$\sigma = \arctan\left(\frac{\Delta z}{d}\right)_{\max}, \tag{2.164}$$

where d is the distance to the point with higher elevation $(z+\Delta z)$; $d \leq L$.

Horizon angle is a nonnegative variable ranging from 0° to 90°. The unit of σ is degree. Horizon angle can be defined at each point of the topographic surface. The concept of horizon angle is used in determining some solar (see later), wind (Section 2.5.2), and visibility (Section 2.5.3) topographic attributes.

Definition 2.42 (Sky view factor). Sky view factor (Ψ_s) is a measure of the ratio of diffuse irradiance at a given point of the topographic surface to that on an unobstructed horizontal surface (Dozier and Frew, 1990; Böhner and Antonić, 2009).

For the full unit circle, Ψ_s has the following expression:

$$\psi_s = \frac{1}{2\pi}\int_0^{2\pi}\left[\cos G \cos^2\sigma + \sin G \cos(\Phi - A)(90 - \sigma - \sin\sigma\cos\sigma)\right]\mathrm{d}\Phi, \tag{2.165}$$

where Φ is the azimuth direction; G is the slope gradient; A is the slope aspect; and σ is the horizon angle for a given point of the topographic surface (Dozier and Frew, 1990; Böhner and Antonić, 2009).

Sky view factor is a nonnegative dimensionless variable ranging from 0 for a completely obstructed topographic surface to 1 for a completely unobstructed topographic surface (peaks, ridges, and horizontal surfaces). Ψ_s can be defined at each point of the topographic surface.

Fig. 2.15C shows a 3D map of Ψ_s for Mount Ararat.

Methods for the calculation of solar topographic variables are discussed in Section 4.6.1.

2.5.2 Wind morphometric variables

Definition 2.43 (Wind morphometric variable). A wind morphometric variable is a single-valued bivariate function $\upsilon = u(x, y)$ describing the relationship between the topographic surface and the wind field.

A group of wind morphometric variables includes relative terrain aspect, horizon angle in the wind direction, wind shelter index, wind effect index, wind exposition index, etc. (Lapen and Martz, 1993; Antonić and Legović, 1999; Winstral et al., 2002; Böhner, 2004; Plattner et al., 2004; Böhner and Antonić, 2009; Gerlitz et al., 2015).

Definition 2.44 (Relative terrain aspect). Relative terrain aspect (A_r) is an absolute value of the difference between slope aspect and wind direction at a given point of the topographic surface (Böhner and Antonić, 2009).

A_r has the following expression:

$$A_r = |A - \delta|, \tag{2.166}$$

where A is slope aspect (Eq. 2.35) and δ is wind direction (Fig. 2.16).

Relative terrain aspect is a nonnegative variable. The unit of A_r is degree. Its values vary between 0° for points with a strict windward positions and 180° for points with a strict leeward positions.

After transformation of A_r into $\cos A_r$, its values vary between 1 and −1 (windward and leeward positions, respectively). Therefore, northness (Section 2.2.3.1) can be interpreted as A_r with respect to the north wind (Böhner and Antonić, 2009).

A_r is a simple measure of topographic exposure to wind taking into account the orientation of the surface but ignoring the influence of G and remote landforms on the wind field.

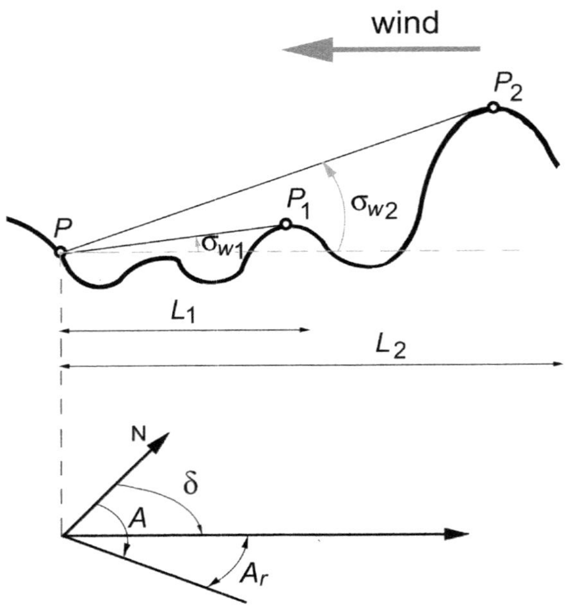

FIG. 2.16 Scheme for the definitions of wind morphometric variables. *A*, slope aspect at a given point *P* on the topographic surface; A_r, relative terrain aspect; *bold curve*, profile of the topographic surface; *N*, north direction; *δ*, wind direction; σ_{w1} and σ_{w2}, horizon angles in the wind direction for a given point *P* and search distances L_1 and L_2, respectively.

Definition 2.45 (Horizon angle in the wind direction). Horizon angle in the wind direction (σ_w) is the maximum angle from a given point of the topographic surface toward any other point in the wind direction azimuth within a distance *L* (Fig. 2.16).

Horizon angle in the wind direction is a nonnegative variable ranging from 0° to 90°. The unit of σ_w is degree. σ_w can be defined at each point of the topographic surface.

σ_w considers the influence of remote landforms on the wind field because higher values of σ_w may provide greater shelter from wind (Böhner and Antonić, 2009). However, the influence of *G* and *A* is ignored in this case.

There are two more complex, related wind morphometric variables—wind effect index and wind exposition index—which are empiric functions originally developed for numerical meteorological modeling (Böhner, 2004, 2005).

Definition 2.46 (Wind effect index). Wind effect index (*WEf*) is a measure of the exposition of slopes to the wind flow of a certain wind direction at a given point of the topographic surface.

In general, *WEf* is a product of two normalized dimensionless functions of mean weighted horizon angles for windward and leeward directions (f_{σ_W} and f_{σ_L}, respectively):

$$WEf = f_{\sigma_W} \cdot f_{\sigma_L}. \tag{2.167}$$

Definition 2.47 (Wind exposition index). Wind exposition index (*WEx*) is a measure of the average exposition of slopes to wind flows in all possible wind directions at a given point of the topographic surface.

WEx is determined for the full unit circle. *WEx* is the average of *WEf* (Eq. 2.167) over all possible wind directions, $\delta = [0°, 360°($.

WEf and *WEx* are positive, dimensionless variables. Their values below 1 relate to wind-shadowed areas, while their values above 1 relate to wind-exposed areas.

WEf can be used as a proxy for snow accumulation on the lee sides of landforms if this index is estimated for a prevailing wind direction. *WEx* can describe an integral picture of the topographic prerequisites of wind shelter/expose level for all possible wind directions. Both indices can be utilized in modeling of soil and vegetation cover distribution in mountainous regions and other areas with severe weather conditions.

Wind morphometric variables can be defined at each point of the topographic surface.

Fig. 2.17A and B shows 3D maps of *WEf* and *WEx*, respectively, for Mount Ararat.

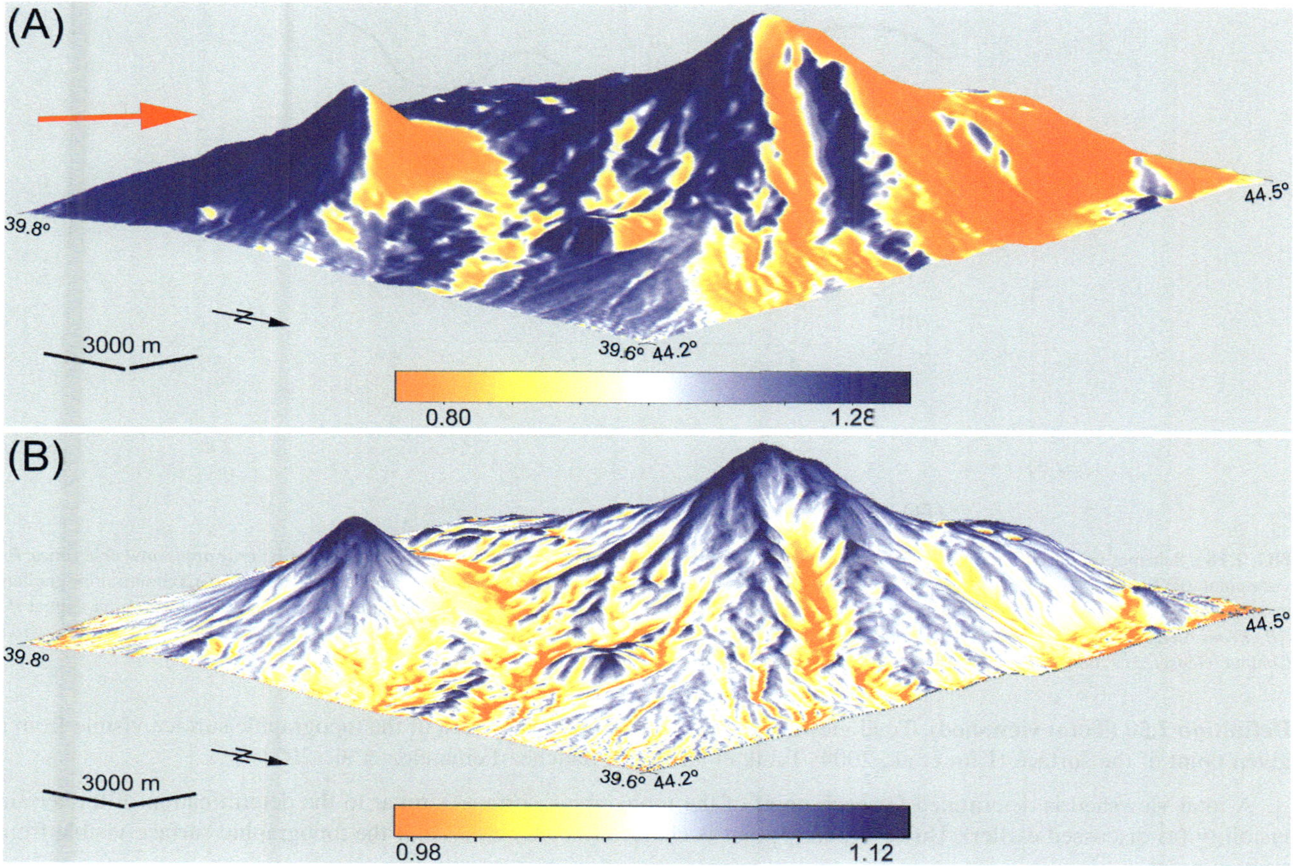

FIG. 2.17 Mount Ararat, wind morphometric variables: (A) Wind effect index calculated with a search distance of 1 km; *red arrow*, wind direction of 135°. (B) Wind exposition index calculated with a search distance of 30 km and an angular step of 15°. Data were processed by the SAGA software (Conrad et al., 2015) and visualized with the author's script for the MATLAB software. Vertical exaggeration is 2×. For the elevation data description, see Fig. 2.3.

Methods for the calculation of wind topographic variables are discussed in Section 4.6.2.

The wind morphometric variables are proxies of interaction between the wind field and topographic surface. For comprehensive prediction of wind characteristics and wind-related processes, topographic data are used in dynamic models of wind fields near the ground (Troen and Lundtang Petersen, 1989, chap. 8; Emeis and Knoche, 2009; Mann et al., 2017; Badger et al., 2021).

2.5.3 View morphometric variables

Definition 2.48 (View morphometric variable). A view morphometric variable is a single-valued bivariate function $v = u(x, y)$ describing relations between the topographic surface and the field of view.

A family of view morphometric variables includes several related functions, such as terrain visibility, total viewshed, topographic openness, etc.

Definition 2.49 (Terrain visibility). Terrain visibility (or singular viewshed) is a set V of points P_i of the topographic surface, which are visible from the view point P_v (Yoeli, 1985; Lee, 1991; De Floriani et al., 1994; Nagy, 1994; Fisher, 1995).

Terrain visibility is determined for one certain P_v. A point P_i of the topographic surface is visible from P_v if these points can be connected by a straight-line segment without intersecting the topographic surface (Fig. 2.18A).

Terrain visibility is a Boolean function; it takes 1 or 0 when a point P_i is visible or nonvisible, respectively, from P_v.

A result of the terrain visibility estimation is a binary viewshed map: visible and nonvisible sections of the topographic surface can, for instance, be presented in two different colors.

Fig. 2.19A shows a 3D map of the terrain visibility for a certain view point at Mount Ararat.

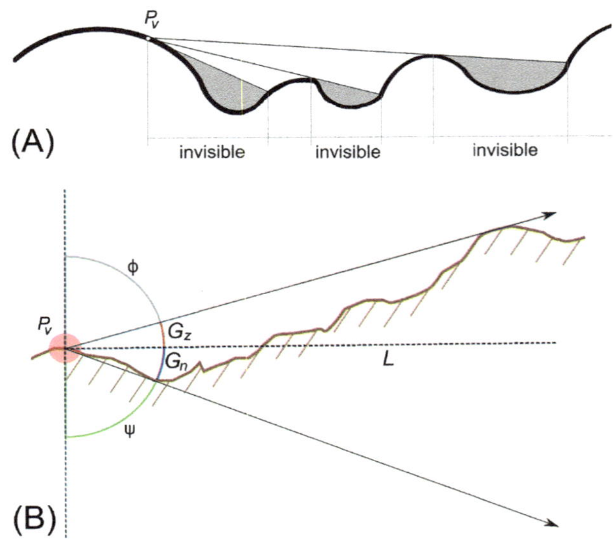

FIG. 2.18 Schemes for the definitions of view morphometric variables: (A) Terrain visibility. *Bold curve*, terrain profile; *gray areas*, invisible zones; P_v, view point. (B) Topographic openness. ϕ and ψ, zenith and nadir angles, respectively, at a view point P_v; L, a radial distance; G_z, a maximum slope gradient to a neighboring point with a greater elevation than at P_v; G_n, a maximum slope gradient from a neighboring point with a lower elevation than at P_v. *((B) From Alonso-Sarría, F., Gomariz-Castillo, F., Cánovas-García, F., 2018. A new approach to the openness index for landform characterisation. Comput. Geosci. 119, 68–79, Fig. 2A, reproduced with permission from Elsevier.)*

Definition 2.50 (Total viewshed). Total viewshed (V_Σ) is a measure of the area of the topographic surface visible from a given point of the surface (Kim et al., 2004; Tabik et al., 2013; Sanchez-Fernandez et al., 2021).

A total viewshed is determined for each point of the topographic surface, similar to the determination of the terrain visibility (as discussed earlier). However, each point is characterized by an area of the topographic surface visible from a given point.

Total viewshed is a nonnegative variable. The unit of V_Σ is m^2.

Definition 2.51 (Topographic openness). Topographic openness (*TO*) is a measure of dominance or enclosure of a given point of the topographic surface (Yokoyama et al., 2002; Alonso-Sarría et al., 2018).

Topographic openness, incorporating the terrain visibility concept (as discussed earlier), relates to how wide a terrain can be viewed from any position. There are two types of topographic openness: positive and negative *TO*. Positive *TO* expresses the degree of dominance of a point's location on the topographic surface, while negative *TO* describes the degree of closure of a point's location by the topographic surface.

Positive and negative *TO* are averages of the zenith (ϕ) and nadir (ψ) angles, respectively, measured along all possible directions from a view point P_v using a radial distance (L) (Fig. 2.18B) as follows:

$$\phi = 90^\circ - G_z, \tag{2.168}$$

$$\psi = 90^\circ - G_n, \tag{2.169}$$

where G_z and G_n are the greatest slope gradients between P_v and all the points located within L in a given direction, whose elevations are higher or lower, respectively, than that of P_v (Yokoyama et al., 2002; Alonso-Sarría et al., 2018).

Topographic openness is a nonnegative variable. The unit of *TO* is degree.

Fig. 2.19B and C shows 3D maps of the positive and negative *TO*, respectively, for Mount Ararat.

Terrain visibility, total viewshed, and topographic openness can be defined at each point of the topographic surface.

Models of view morphometric variables can be useful, for instance, in the location of forest fire towers, cell towers, radio and television transmitters, pilot training simulation, navigation, scenic landscape assessment, etc. (De Floriani et al., 1994; Nagy, 1994; Fisher, 1995; Kim et al., 2004; Sanchez-Fernandez et al., 2021). Topographic openness can be used as a geometric signature to classify landforms (Yokoyama et al., 2002; Prima et al., 2006; Alonso-Sarría et al., 2018).

Methods for the calculation of topographic variables are discussed in Section 4.6.3.

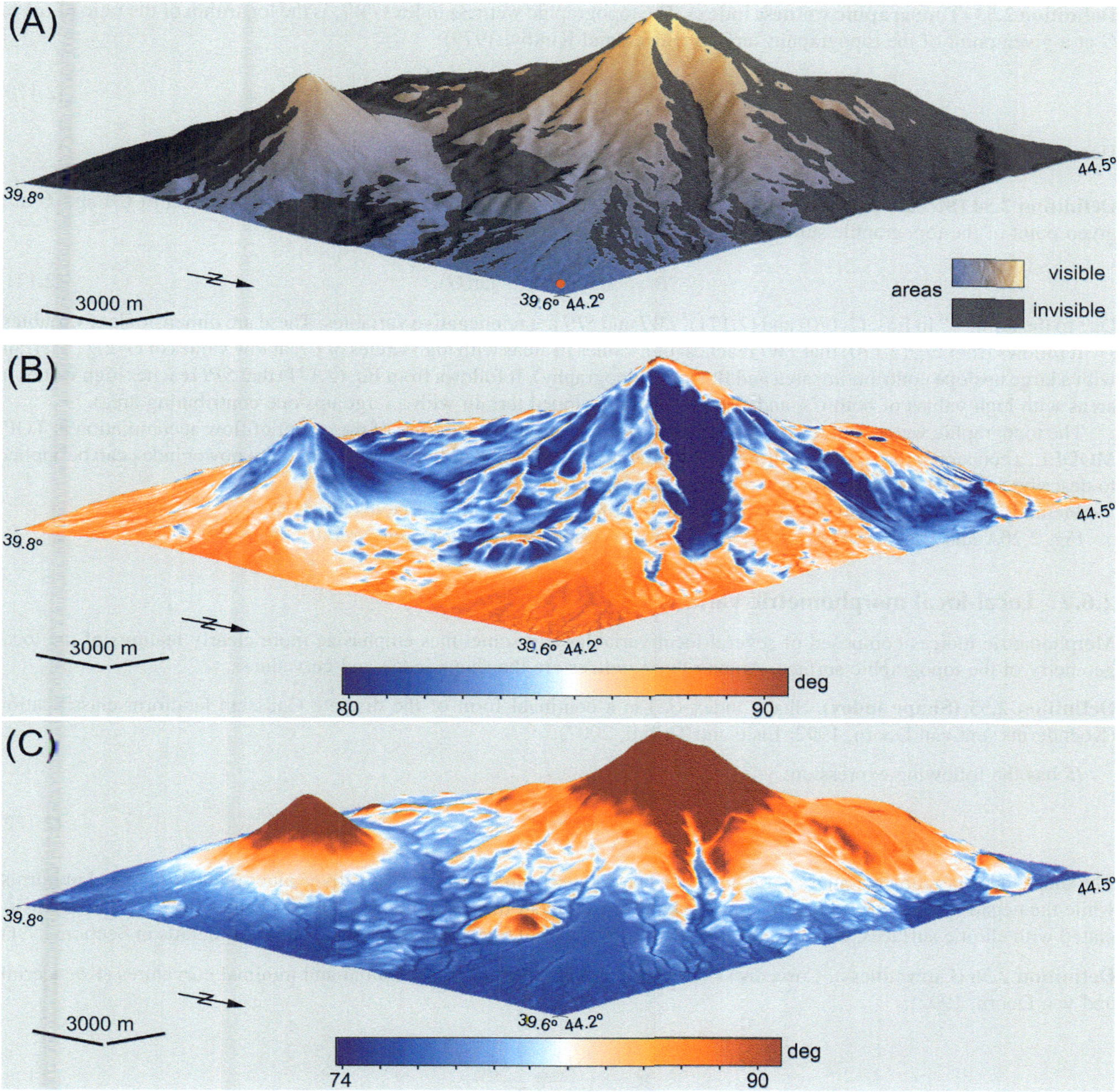

FIG. 2.19 Mount Ararat, view morphometric variables: (A) Terrain visibility. *Red dot*, a viewpoint with a relative height of 100 m. The binary viewshed map is draped over the elevation map (Fig. 2.3). Invisible areas are darkened. (B) Positive openness and (C) negative openness calculated with an 8-direction line tracing and a radial distance of 10 km. Data were processed by the SAGA software (Conrad et al., 2015) and visualized with the author's script for the MATLAB software. Vertical exaggeration is $2^{\times}$. For the elevation data description, see Fig. 2.3.

2.6 Combined morphometric variables

Definition 2.52 (Combined morphometric variable). A combined morphometric variable is a single-valued bivariate function $v = u(x, y)$ composed of several local and/or nonlocal morphometric variables.

2.6.1 Local-nonlocal morphometric variables

Morphometric indices composed of local and nonlocal variables consider both the local geometry of the topographic surface and the relative position of a point of the surface. Among these variables are the topographic wetness index and the stream power index.

Definition 2.53 (Topographic wetness index). The topographic wetness index (*TWI*) is the logarithm of the ratio of *CA* to *G* at a given point of the topographic surface (Beven and Kirkby, 1979):

$$TWI = \ln\left(1 + \frac{CA}{10^{-3} + \tan G}\right). \tag{2.170}$$

Here, the term 10^{-3} is used to avoid division by zero for the case of horizontal planes.

Definition 2.54 (Stream power index). The stream power index (*SPI*) is the logarithm of the product of *CA* and *G* at a given point of the topographic surface (Moore et al., 1991):

$$SPI = \ln(1 + CA \cdot \tan G). \tag{2.171}$$

Due to the term "1" in Eqs. (2.170) and (2.171), *TWI* and *SPI* are nonnegative variables. These are dimensionless variables.

It follows from Eq. (2.170) that *TWI* reaches high values in areas with high values of *CA* at low values of *G* (e.g., a terrain with a large upslope contributing area and flat local topography). It follows from Eq. (2.171) that *SPI* reaches high values in areas with high values of both *CA* and *G* (e.g., a highly sloped terrain with a large upslope contributing area).

The topographic wetness index (or topographic index) is used as a measure of the extent of flow accumulation in TOPMODEL, a concept of distributed hydrological modeling (Beven and Kirkby, 1979). The stream power index can be applied to describe potential flow erosion and related landscape processes (Moore et al., 1991).

Methods for the calculation of combined local-nonlocal topographic variables are discussed in Section 4.7.

Fig. 2.20A and B shows 3D maps of *TWI* and *SPI*, respectively, for Mount Ararat.

2.6.2 Local-local morphometric variables

Morphometric indices composed of several local variables can sometimes emphasize more clearly features of the local geometry of the topographic surface. Among these indices are the shape index and curvedness.

Definition 2.55 (Shape index). Shape index (*IS*) is a continual form of the discrete Gaussian landform classification (Koenderink and van Doorn, 1992; Lisle and Toimil, 2007).

IS has the following expression:

$$IS = \frac{2}{\pi} \arctan \frac{H}{\sqrt{H^2 - K}}. \tag{2.172}$$

The shape index is a dimensionless variable. It can take values from −1 to 1. The positive values relate to convex landforms, while the negative ones correspond to concave landforms. Under these conditions, absolute values from 0.5 to 1 are associated with elliptic surfaces, whereas absolute values from 0 to 0.5 relate to hyperbolic ones (see details in Section 2.7.1).

Definition 2.56 (Curvedness). Curvedness (*C*) is the root mean square of maximal and minimal curvatures (Koenderink and van Doorn, 1992):

$$C = \sqrt{\frac{k_{max}^2 + k_{min}^2}{2}}. \tag{2.173}$$

This is nonnegative variable. The unit of *C* is m^{-1}. Curvedness measures the magnitude of surface bending regardless of its shape. Flat areas have low values of *C*, while areas with sharp bending are marked by high values of *C*.

Fig. 2.21A and B shows 3D maps of *IS* and *C*, respectively, for Mount Ararat.

2.7 Landform classifications

Signs and value intervals of local and nonlocal morphometric variables (Sections 2.2 and 2.3) are commonly used in diverse combinations according to various formal rules. These procedures are carried out in order to move from a quantitative description of portions of the topographic surface to a qualitative categorical description, that is, to perform terrain segmentation or classification of landforms or landform elements.

Definition 2.57 (Landform classification). A landform classification is a categorical description of a limited set of landforms and/or landform elements/units, which is based on formal rules applied to signs and/or value intervals of morphometric variables.

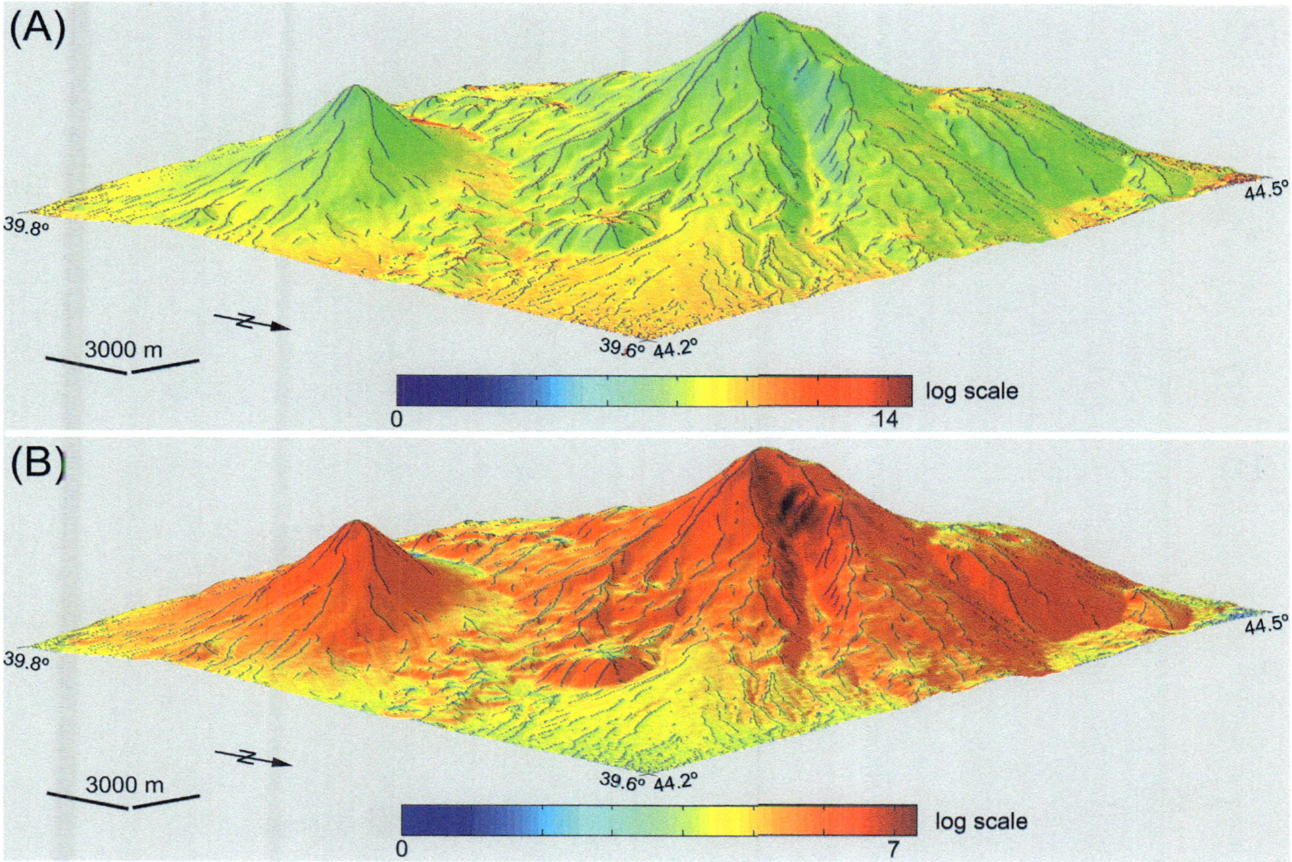

FIG. 2.20 Mount Ararat, combined local-nonlocal morphometric variables: (A) Topographic wetness index. (B) Stream power index. Calculation was carried out by the Martz-de Jong method (Martz and de Jong E., 1988) adapted to spheroidal equal angular grids (Section 4.4.2). Data were processed by the LandLord software (Florinsky, 2016) and visualized with the author's script for the MATLAB software. Vertical exaggeration is 2×. For the elevation data description, see Fig. 2.3. *(Modified from Florinsky, I.V., 2017a. An illustrated introduction to general geomorphometry. Progr. Phys. Geogr. 41 (6), 723–752, Fig. 7, reproduced with permission from Sage.)*

On one hand, landform classifications based on morphometric attributes are of theoretical importance to geomorphometry. On the other hand, landform classifications are applied to solve various problems of geosciences. There are three fundamental quantitative approaches to classifying landforms using information on the local geometry of the surface:

1. The Gaussian classification based on the signs of the Gaussian and mean curvatures.
2. The Efremov-Krcho classification using the signs of horizontal and vertical curvatures.
3. The Shary classification based on the signs of the Gaussian, mean, difference, horizontal, and vertical curvatures.

2.7.1 Gaussian landform classification

The Gaussian landform classification is based on signs of two morphometric variables—K and H—belonging to form attributes. In other words, this classification considers a pair of principal sections whose position is independent of gravity (Section 2.2.1) (Fig. 2.4A).

According to Gauss (1828, 1902), positive values of K describe an elliptic surface, negative values of K relate to a hyperbolic surface, while zero values of K define a parabolic surface. As applied to the topographic surface, its elliptic portions can correspond to hills and closed depressions, its hyperbolic portions can relate to saddles, and its parabolic portions can correlate with ridges, valleys, and plane areas.

To distinguish between hills and closed depressions, convex and concave saddles, as well as ridges and valleys, one should combine data on the signs of the Gaussian and mean curvatures (Sections 2.2.4.6 and 2.2.4.4) as follows (Shary, 1991, 1995; Koenderink and van Doorn, 1992):

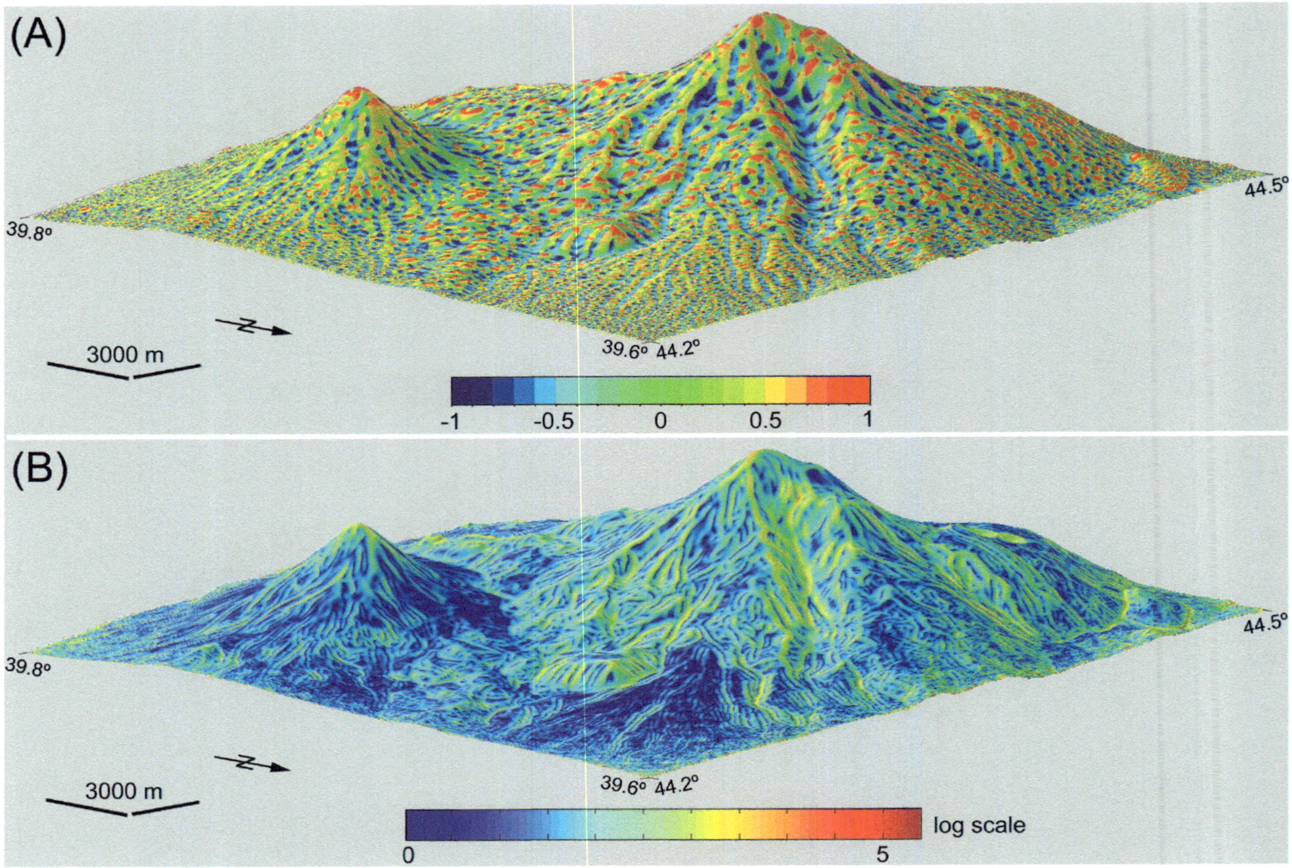

FIG. 2.21 Mount Ararat, combined local-local morphometric variables: (A) Curvedness. (B) Shape index. Calculation was carried out by the universal spectral analytical method (Chapter 7). Data were processed and visualized with the author's script for the MATLAB software (Florinsky and Pankratov, 2016). Vertical exaggeration is 2×. For the elevation data description, see Fig. 2.3. *(Modified from Florinsky, I.V., 2017a. An illustrated introduction to general geomorphometry. Progr. Phys. Geogr. 41 (6), 723–752, Fig. 4G and H, reproduced with permission from Sage.)*

- $K > 0$ with $H > 0$ describes hills, while $K > 0$ with $H < 0$ defines closed depressions.
- $K < 0$ with $H > 0$ relates to convex saddles, while $K < 0$ with $H < 0$ describes concave saddles.
- $K = 0$ with $H > 0$ corresponds to ridges, while $K = 0$ with $H < 0$ defines valleys.

Fig. 2.22A illustrates the scheme of this classification using terms of structural geology because the Gaussian classification is systematically used in geological studies (Roberts, 2001; Bergbauer and Pollard, 2003; Bergbauer et al., 2003; Pollard and Fletcher, 2005, chap. 3; Lisle and Toimil, 2007; Mynatt et al., 2007a). In nature, however, portions of the topographic surface with zero values of the Gaussian or mean curvatures are rarely observed (Section 2.7.3). Thus, the number of landform types in the Gaussian classification may be reduced from nine to four (i.e., antiformal saddle, synformal saddle, dome, and basin—Fig. 2.22A). The use of this classification in structural geology is discussed in Sections 6.2.7 and 13.2.

Fig. 2.23A represents a 3D map of the reduced Gaussian classification for Mount Ararat.

2.7.2 The Efremov-Krcho landform classification

The Efremov-Krcho classification is based on signs of two topographic variables—k_v and k_h—belonging to flow attributes. In other words, this classification considers a pair of normal sections whose position is allocated by gravity (Section 2.2.1) (Fig. 2.4B).

This classification was first described qualitatively by Efremov (1949, pp. 119–120) and then quantitatively by Krcho (1983). According to the Efremov-Krcho classification, there are nine landform elements distinguished by various combinations of k_v and k_h signs: $k_h > 0$ with $k_v < 0$, $k_h > 0$ with $k_v = 0$, $k_h > 0$ with $k_v > 0$, $k_h = 0$ with $k_v < 0$, $k_h = 0$ with $k_v = 0$, $k_h = 0$ with $k_v > 0$, $k_h < 0$ with $k_v < 0$, $k_h < 0$ with $k_v = 0$, and $k_h < 0$ with $k_v > 0$ (Fig. 2.22B).

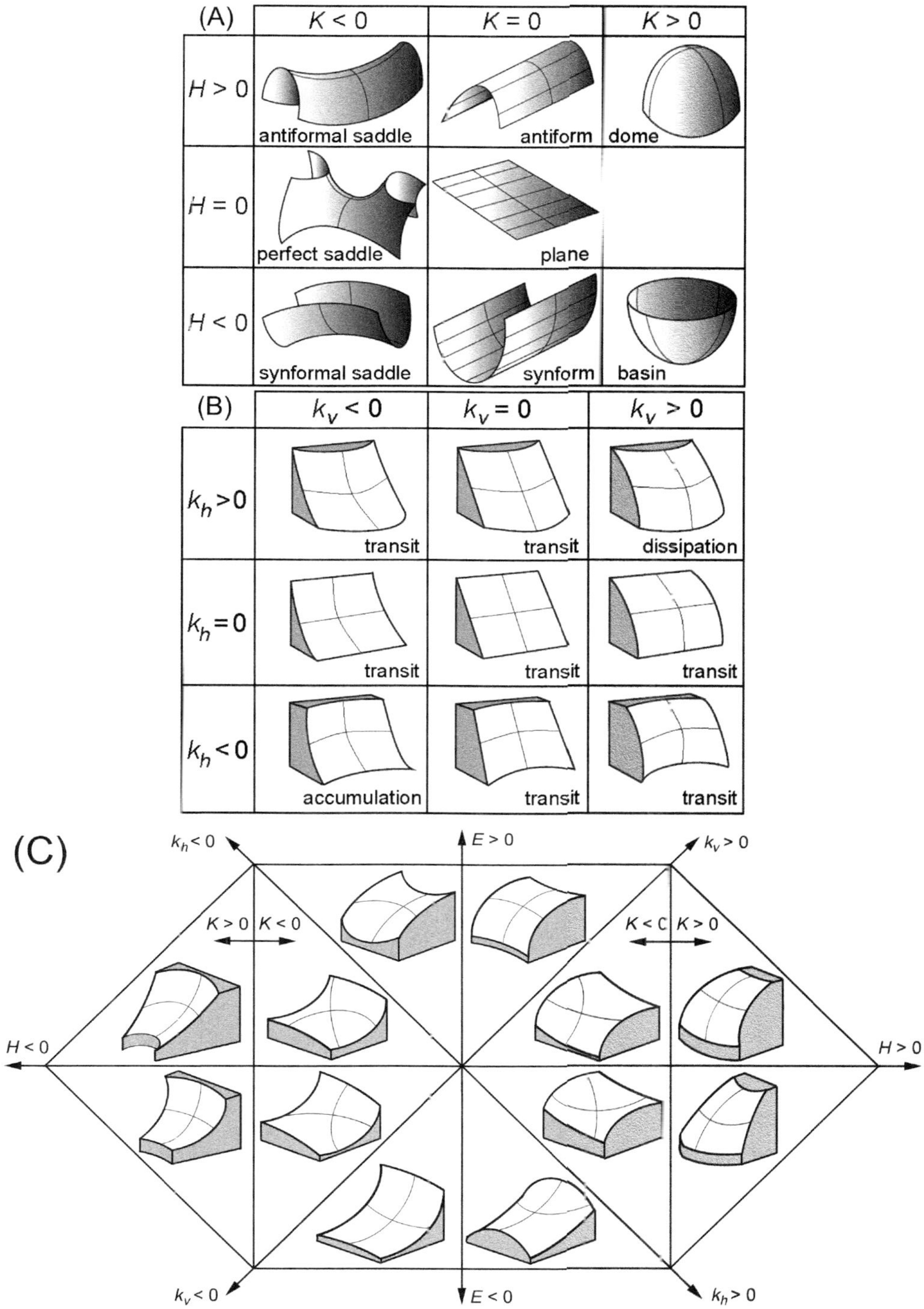

FIG. 2.22 Schemes for the definitions of fundamental landform classifications: (A) The Gaussian classification. Shape terms from structural geology are indicated. (B) The Efremov-Krcho classification. Zones of relative accumulation, transit, and dissipation are indicated. (C) The Shary classification. Twelve main types of landforms are displayed. *((A) Modified from Mynatt, I., Bergbauer, S., Pollard, D.D., 2007a. Using differential geometry to describe 3-D folds. J. Struct. Geol. 29 (7), 1256–1266, Fig. 2, reproduced with permission from Elsevier.)*

Recall that local accumulation of a gravity-driven flow is controlled by two mechanisms: convergence and relative deceleration of flows (Shary, 1991) (Fig. 2.7). Flow convergence is controlled by k_h: a flow diverges where $k_h > 0$ and converges where $k_h < 0$ (Kirkby and Chorley, 1967) (Sections 2.2.4.2). Flow relative deceleration is determined by k_v: a flow tends to accelerate where $k_v > 0$, and to decelerate where $k_v < 0$ (Speight, 1974) (Section 2.2.4.3).

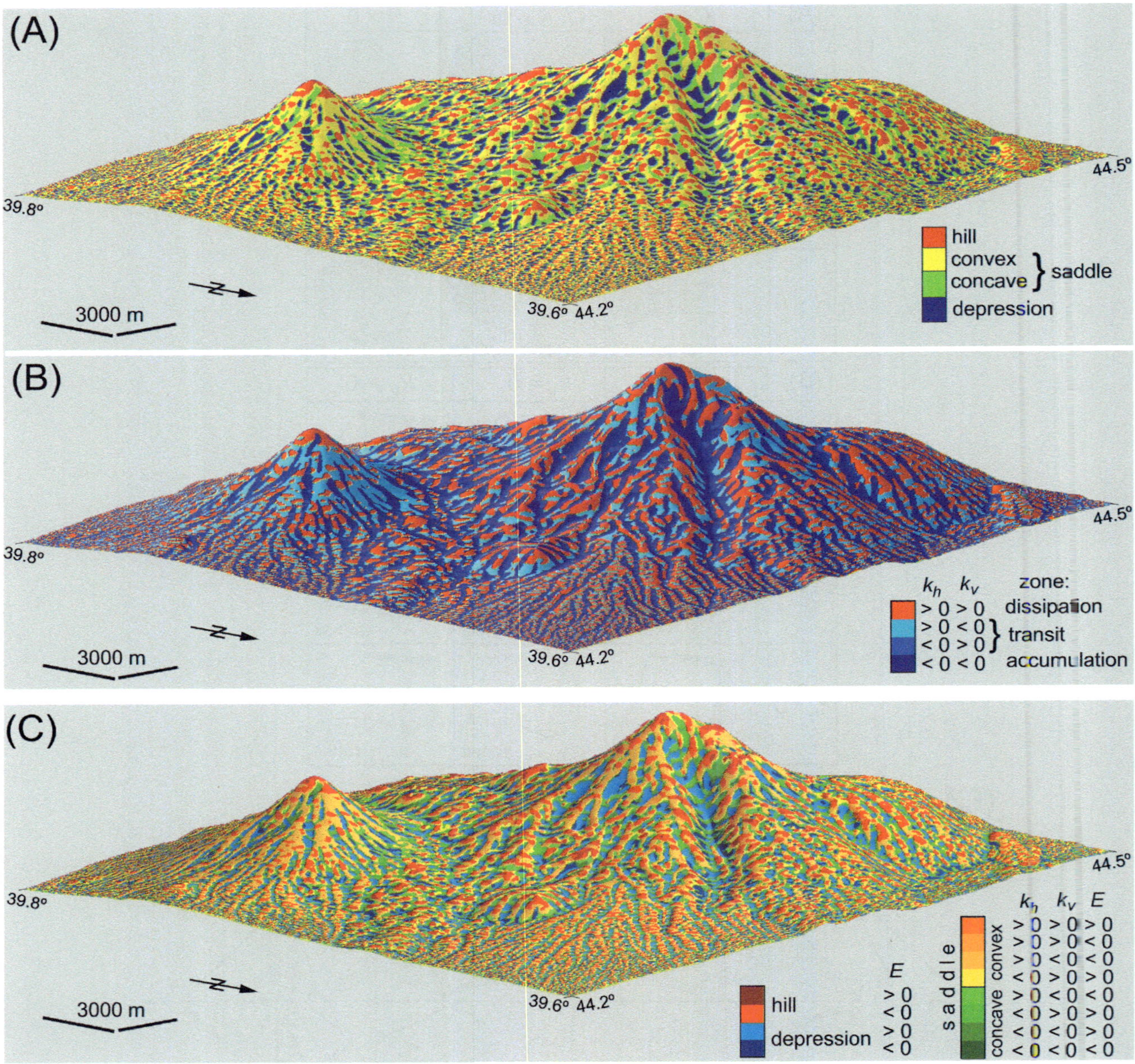

FIG. 2.23 Mount Ararat, fundamental landform classifications: (A) The Gaussian classification. (B) The Efremov-Krcho classification with zones of relative accumulation, transit, and dissipation. (C) The Shary classification. Data were processed by the author's method (Section 4.3) with the LandLord software (Florinsky, 2016) and visualized with the author's script for the MATLAB software. Vertical exaggeration is $2^{\times}$. For the elevation data description, see Fig. 2.3.

There is a concurrent action of convergence and relative deceleration of flows within areas characterized by both $k_h < 0$ and $k_v < 0$. These areas are called relative accumulation zones (Shary et al., 1991). If divergence and relative acceleration of flows act simultaneously ($k_h > 0$ and $k_v > 0$), such areas are called relative dissipation zones. Areas with other combinations of the k_h and k_v signs (no concurrent action of "unidirectional" processes) are called transit zones. It should be noted that we consider zones of relative accumulation rather than dead-end depressions. A flow may pass through many relative accumulation zones before entering into a dead-end depression.

According to the landform classifications by k_v and k_h sings, accumulation zones can be defined as concave-concave landforms, while dissipation zones can be described as convex-convex landforms. Another seven types of landforms can be assigned to transit zones (Fig. 2.22B). As for the Gaussian classification (Section 2.7.1), portions of the topographic surface with zero values of horizontal or vertical curvatures are rarely observed (see details in Section 2.7.3). Therefore, the number of landform types in the Efremov-Krcho classification can also be reduced from nine to four.

Fig. 2.23B represents a 3D map of the reduced Efremov-Krcho classification for Mount Ararat.

If the mapping of accumulation, transit, and dissipation zones is carried out by a simple combination of k_h and k_v maps, one can visualize only the spatial distribution of these zones without quantitative estimation of the probable degree of flow accumulation. To solve this problem, Shary (1995) proposed the use of data on K_a (Section 2.2.4.12) and H (Section 2.2.4.4). Negative values of K_a correspond to transit zones, while positive values of K_a correspond to both accumulation and dissipation zones. They can be distinguished using data on H. Positive values of K_a with negative values of H correspond to accumulation zones, while positive values of K_a with positive values of H correspond to dissipation zones.

Maps of relative accumulation zones represent areas where geometrical peculiarities of the topography provide conditions for the local accumulation of gravity-driven substances, such as water (meteoric water, soil moisture), dissolved and suspended substances (salts, mineral and organic particles, etc.), and other liquids (e.g., petroleum products). In this connection, maps of relative accumulation zones are used in geosciences: to predict slope instability (Lanyon and Hall, 1983; Florinsky, 2007b); to study and model soil properties at field, regional, and continental scales (Pennock et al., 1987; Florinsky et al., 1999, 2002; Florinsky and Eilers, 2002; Shary et al., 2002a; Shary, 2005); to estimate the risk for soil secondary salinization (Florinsky et al., 2000); to predict soil degradation and contamination along pipelines; to explore placer deposits (Florinsky, 2007b); to search meteoritic matter (Florinsky, 2008f); and so on. Relative accumulation zones may coincide with intersections of lineaments, faults, and fracture zones. Thus, they constitute areas of contact and interaction between overland and intrasoil lateral water flows and upward flows of deep groundwater (Florinsky, 2000) (Chapter 15). The Efremov-Krcho landform classification was used in geomorphological mapping (Dikau, 1989).

2.7.3 The Shary landform classification

Shary (1995) proved that neither the Gaussian classification nor the Efremov-Krcho classification are complete. He proposed a new landform classification, which incorporates two previous classifications as special cases: each landform type of the Gaussian and Efremov-Krcho classifications is a combination of several landform types of the Shary classification.

This classification is based on the signs of five curvatures: the Gaussian, mean, difference, horizontal, and vertical ones (Section 2.2.4). According to the Shary classification, the topographic surface is described by 46 landform types. Twelve of them are main types (Fig. 2.22C) (Shary et al., 2005) including:

- Two types of hills defined by (1) $K>0$, $H>0$ with $E>0$ and (2) $K>0$, $H>0$ with $E<0$ (see block diagrams in two rightmost triangles—Fig. 2.22C).
- Two types of depressions defined by (1) $K>0$, $H<0$ with $E>0$ and (2) $K>0$, $H<0$ with $E<0$ (see block diagrams in two leftmost triangles—Fig. 2.22C).
- Four types of convex saddles defined by (1) $K<0$, $H>0$ with $k_h>0$, $k_v>0$, $E>0$; (2) $K<0$, $H>0$ with $k_h>0$, $k_v>0$, $E<0$; (3) $K<0$, $H>0$ with $k_h>0$, $k_v<0$, $E<0$; and (4) $K<0$, $H>0$ with $k_h<0$, $k_v>0$, $E>0$ (see block diagrams in four triangles just to the right of the E axis—Fig. 2.22C).
- Four types of concave saddles defined by (1) $K<0$, $H<0$ with $k_h>0$, $k_v<0$, $E<0$; (2) $K<0$, $H<0$ with $k_h<0$, $k_v>0$, $E>0$; (3) $K<0$, $H<0$ with $k_h<0$, $k_v<0$, $E>0$; and (4) $K<0$, $H<0$ with $k_h<0$, $k_v<0$, $E<0$ (see block diagrams in four triangles just to the left of the E axis—Fig. 2.22C).

In nature, these 12 landform types can be found with a frequency of 1/12. 34 landform types are rare: they are observed with a frequency of less than 0.0001. These are landforms for which one of the five curvatures has zero value. On maps, rare landforms are represented as lines and points usually located between landforms belonging to the main types (Shary, 1995; Shary et al., 2005).

Compared to the Gaussian and Efremov-Krcho classifications, the higher number of landform types in the Shary classification leads to its higher flexibility. For instance, Buivydaite and Mozgeris (2007) found that soil units and classes correlate strongly with the landform types derived from the Shary classification rather than the Gaussian and Efremov-Krcho ones.

Fig. 2.23C demonstrates a 3D map of the Shary classification for Mount Ararat.

2.7.4 Other geomorphometric landform classifications

Besides the three fundamental landform classification schemes (Sections 2.7.1–2.7.3), there are other geomorphometric approaches for landform classification. They are as a rule based on elements of the theory of the topographic surface using signs and/or interval values of curvatures, slope gradients, and some other variables as classification criteria. Some of those classifications were originally developed and adopted into frameworks of geomorphological and/or pedological concepts. Let us take a brief look at some of such classifications.

Pennock et al. (1987) developed a landform classification based on ranges of three topographic variables: vertical and horizontal curvatures as well as slope gradients. k_v sign was the primary classification criterion; it was used to recognize convex shoulders, linear backslopes, and concave footslopes. These three classes were subdivided into convergent and divergent ones using k_h sign. G value of 3° was used to distinguish between backslopes and level landforms. Thus, seven landform classes were recognized, such as divergent and convergent shoulders, backslopes, and footslopes as well as level elements (summits and toeslopes) (Fig. 2.24A).

It is quite obvious that this classification is a modification of the Efremov-Krcho landform classification (cf. Figs. 2.22B and 2.24A). The Pennock classification was used for studies of the relationship between topography and soil properties in hummocky terrain in southern Saskatchewan, Canada (Pennock et al., 1987).

Providing geomorphometric support to precision farming research in the neighboring Canadian province of Alberta, MacMillan et al. (2000) expanded the seven-unit Pennock landform classification (as discussed earlier) into a classification based on 15 landform entities:

- *Upper slope*: level crest, divergent shoulder, and upper depression;
- *Mid-slope*: backslope, divergent backslope, convergent backslope, terrace, saddle, and midslope depression;
- *Lower slope*: footslope, toeslope, fan, lower slope mound, level lower slope, and lower slope depression.

Each of these landform classes was characterized by interval values of slope gradient, vertical and horizontal curvatures, topographic wetness index, as well as relative relief. The classification procedure included the application of a fuzzy rule base to digital models of the above morphometric variables. The use of the fuzzy logic approach made it possible to reduce the number of recognized landform classes depending on the complexity of the terrain and the scale of the study (MacMillan et al., 2000).

The MacMillan landform classes are closely associated with the movement and accumulation of water and dissolved substances in the landscape and are different in terms of soil characteristics and crop yields, especially in the conditions of the Canadian prairies. This classification approach was used in predictive soil mapping (Section 11.3).

Schmidt and Hewitt (2004) proposed a three-step, fuzzy rule-based landform classification. First, landforms are classified at some scale using a combination of two schemes: (1) the Efremov-Krcho classification based on signs of horizontal and vertical curvatures (Section 2.7.2) and (2) selected elements from the Wood (1996) classification based on signs of maximal and minimal curvatures (Section 2.2.4.5). Values of all four curvatures are divided into three categories (Fig. 2.24B): positive (convex forms), zero (straight forms), and negative (concave forms) values. k_h and k_v are used to describe nine landform units for sloping areas, whereas k_{max} and k_{min} are used to define six landform units for low-gradient (flat) areas. Thus, the first step of the Schmidt-Hewitt classification results in 15 landform units described by geomorphic terms as follows (Fig. 2.24B): shoulder spur ($k_h > 0$ and $k_v > 0$), shoulder slope ($k_h = 0$ and $k_v > 0$), shoulder hollow ($k_h < 0$ and $k_v > 0$), spur ($k_h > 0$ and $k_v = 0$), back slope ($k_h = 0$ and $k_v = 0$), hollow ($k_h < 0$ and $k_v = 0$), foot spur ($k_h > 0$ and $k_v < 0$), foot slope ($k_h = 0$ and $k_v < 0$), foot hollow ($k_h < 0$ and $k_v < 0$), peak ($k_{max} > 0$ and $k_{min} > 0$), ridge ($k_{max} > 0$ and $k_{min} = 0$), plain ($k_{max} = 0$ and $k_{min} = 0$), saddle ($k_{max} > 0$ and $k_{min} < 0$), channel ($k_{max} = 0$ and $k_{min} < 0$), and pit ($k_{max} < 0$ and $k_{min} < 0$).

Second, the terrain is classified at a higher scale according to landscape position using three categories, such as hill, hillslope, and valley. This step is performed by the Black Top Hat function, a gray-level image analysis approach (Rodriguez et al., 2002). Third, two obtained classifications are combined, that is, landforms of 15 classes are subdivided according to the landscape position of three classes (as a result, up to 45 landform classes may be obtained). Such an approach to landform classification takes into account both the local geometry of the surface and the position of the landform in the landscape (similar to combined local-nonlocal morphometric variables—Section 2.6).

The Schmidt-Hewitt landform classification was used in soil predictive mapping (Schmidt et al., 2005; Barringer et al., 2008). Notice that when analyzing relatively small areas with relatively uniform topography, the first step of this classification may be used individually, without subsequent second and third steps.

Fig. 2.25A shows a 3D map of the first, 15-landform units step of the Schmidt-Hewitt classification for Mount Ararat.

Weiss (2001) proposed to use the topographic position index (Section 2.3.2) for landform classification by thresholding TPI values. Applying a threshold of a standard deviation (SD) of TPI to two TPI models derived for small and large scales, TPI_S and TPI_L, respectively (that is, using small and large neighborhood areas in TPI derivation), one can classify the landscape into 10 landform categories (Fig. 2.24C). These include: (1) deep narrow canyon or deeply incised V-shape valley; (2) midslope drainage or shallow valley; (3) upland incised drainage or depression; (4) U-shape valley; (5) plain or broad flat area; (6) broad open slope; (7) upper slope, or flat ridge top, or flat hilltop; (8) local ridge or hill in valley;

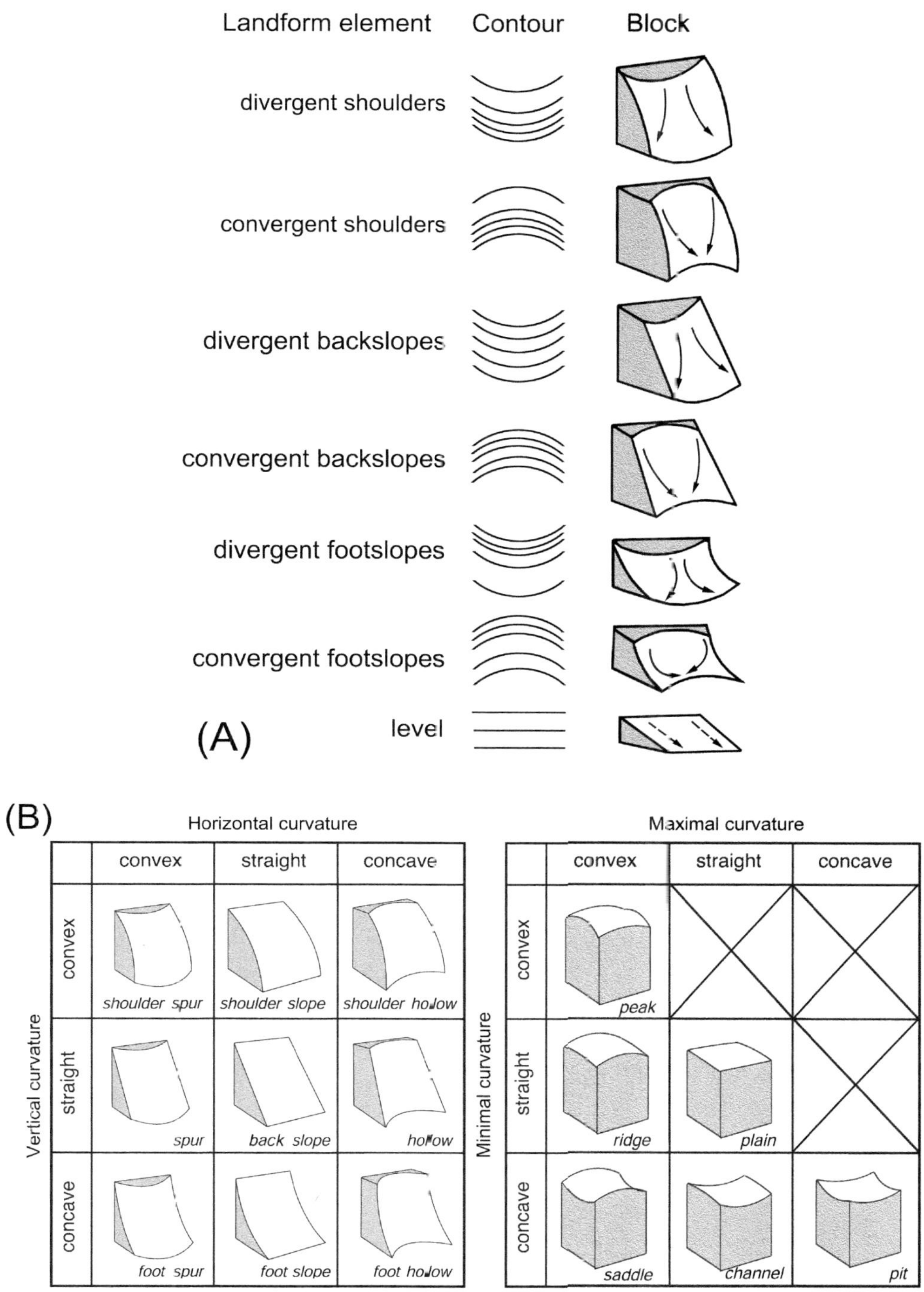

FIG. 2.24 Schemes for the definitions of selected landform classification: (A) The Pennock classification: contour and block diagram depictions of the seven landform classes. (B) The Schmidt-Hewitt classification. *Left*, sloping landforms. *Right*, low-gradient landforms. (C) The *TPI*-based classification. *Upper*, classification threshold criteria: TPI_S and TPI_L, topographic position indices derived for small and large neighborhood areas, respectively; *G*, slope gradient; SD, standard deviation. *Lower*, examples of the threshold criteria application (see text for details): *blue* and *red double arrows*, small and large neighborhood areas, respectively; *brown curve*, terrain profile. (D) The Iwahashi-Pike classification: nested-means half/half partitioning of three input variables (i.e., slope gradient, local convexity, and surface texture) into 8, 12, and 16 output classes. *Shaded* and *open rectangles*, points with variable values below and above the mean value of the variable, respectively; *rectangles within two dot-dashed boxes*, same as those immediately above. Rectangle sizes are symbolic, they do not represent number of points. (E) The geomorphon classification: block diagrams for the 10 most common landform elements and their corresponding ternary representation of relative elevations; *red*, *green*, and *blue dots*, higher, same, and lower relative elevations, respectively,

(Continued)

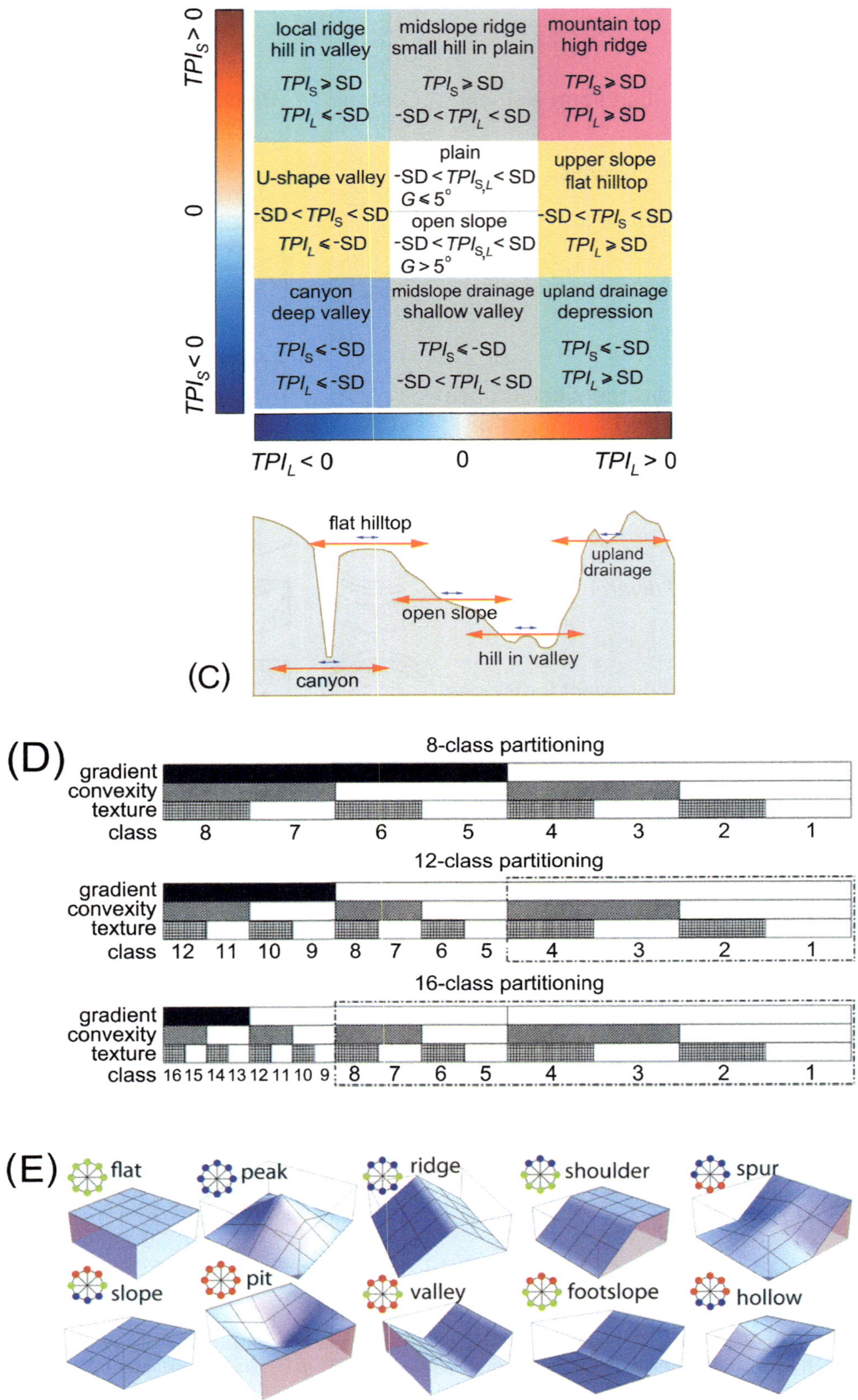

FIG. 2.24, CONT'D between a given point of the topographic surface and its surroundings along the eight compass directions. *((A) From Pennock, D.J., Zebarth, B.J., de Jong, E., 1987. Landform classification and soil distribution in hummocky terrain, Saskatchewan, Canada. Geoderma 40 (3–4), 297–315, Fig. 1, reproduced with permission from Elsevier. (B) Modified from Schmidt, J., Hewitt, A., 2004. Fuzzy land element classification from DTMs based on geometry and terrain position. Geoderma 121 (3–4), 243–256, Fig. 3, reproduced with permission from Elsevier. (D) From Iwahashi, J., Pike, R.J., 2007. Automated classifications of topography from DEMs by an unsupervised nested-means algorithm and a three-part geometric signature. Geomorphology 86 (3–4), 409–440, Fig. 3, reproduced with permission from Elsevier. (E) From Jasiewicz, J., Stepinski, T.F., 2013. Geomorphons—a pattern recognition approach to classification and mapping of landforms. Geomorphology 182, 147–156, Fig. 3, reproduced with permission from Elsevier.)*

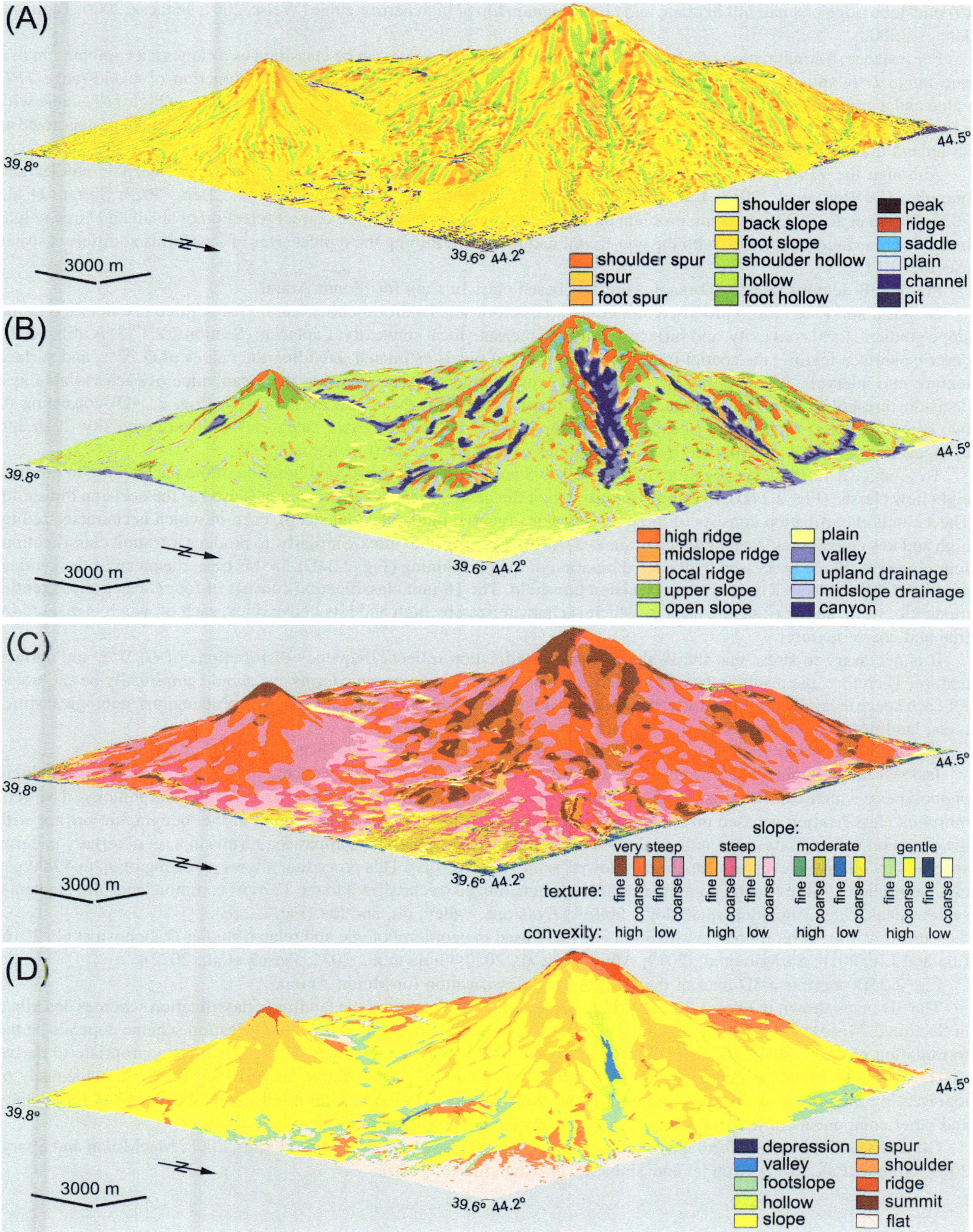

FIG. 2.25 Mount Ararat, selected landform classifications: (A) The Schmidt-Hewitt classification (the first, 15-landform units step). (B) The *TPI*-based classification. Small and large neighborhood areas are 100 and 1000 m, respectively. (C) The uncalibrated Iwahashi-Pike classification. (D) The geomorphon classification. The search distance is 10 km. Data were processed by the SAGA software (Conrad et al., 2015) and visualized with the author's script for the MATLAB software. Vertical exaggeration is 2×. For the elevation data description, see Fig. 2.3.

(9) midslope ridge or small hill in plain; and (10) mountain top or high narrow ridge (Weiss, 2001; Jenness, 2006; Tagil and Jenness, 2008).

For instance, a combination of a low TPI_S value and a low TPI_L value can be classified as a canyon; a combination of a mid-range TPI_S value and a high TPI_L value can be categorized as a flat hilltop; a combination of a mid-range TPI_S value and a mid-range TPI_L value with a relatively high G value can be classified as an open slope; a high TPI_S value with a low TPI_L value can be attributed to a hill in a valley; whereas a low TPI_S value with a high TPI_L value can be categorized as an upland drainage (Fig. 2.24C).

Although the *TPI*-based landform classification and its modifications are widely used in geomorphic studies and mapping (De Reu et al., 2013; Lavagnino et al., 2020; Marchese et al., 2020; Ghosh and Bera, 2023; Shirani et al., 2023; Siervo et al., 2023), the main problem of this approach is the difficulty in correct selection of small and large neighborhood area values associated with the non-trivial nature of determining the typical sizes of landforms at different scales (Section 10.1).

Fig. 2.25B demonstrates a 3D map of the *TPI*-based classification for Mount Ararat.

Iwahashi and Pike (2007) proposed an iterative, nested-means landform classification using three taxonomic criteria: slope gradient, local convexity, and surface texture. To measure local convexity, Laplacian (Section 2.2.4.13) is utilized. To describe surface texture, the spatial frequency of peaks and pits is estimated. Dividing the values of G, $\nabla^2 z$, and surface texture into intervals, it is possible to produce 8, 12, or 16 landform classes using the mean value of each variable as a decision threshold. A single iteration of the partitioning results in the 8-unit classification (Fig. 2.24D) consisting of two categories of G (gentle and steep), each of which is characterized by two categories of local convexity (high and low), each of which is also marked by two categories of surface texture (fine and course).

To produce a 12-unit classification, four classes 5–8 are involved in the second iteration of the partitioning to produce eight new classes (Fig. 2.24D). In this case, the mean of the gentler half of the whole area is used as the decision threshold. The 12-unit classification consists of three G categories (gentle, moderate, and steep), each of which is characterized by high and low convexities, each of which is marked by fine and coarse textures. Similarly, to produce a 16-unit classification, four classes 9–12 are processed by the third iteration of the partitioning (Fig. 2.24D). In this case, the mean of the gentlest quarter of the whole area is used as the decision threshold. The 16-unit classification consists of four G categories (gentle, moderate, steep, and very steep), each of which is characterized by high and low convexities, each of which is marked by fine and coarse textures.

It is necessary to stress that the Iwahashi-Pike classification reflects frequency distributions of G, $\nabla^2 z$, and surface texture. Therefore, the resulting landform classes are undefined in geomorphic terms and should empirically be calibrated for each specific landscape. Generally, the Iwahashi-Pike classification maps reflect geomorphic regions, geological structures, landforms, and slope processes (Iwahashi and Pike, 2007).

Fig. 2.25C demonstrates a 3D map of the uncalibrated Iwahashi-Pike classification for Mount Ararat.

Jasiewicz and Stepinski (2013) proposed a landform classification using the concept of geomorphon (*geomor*phologic *phono*types), which can be considered a simple ternary pattern, an archetype of certain terrain morphology. The geomorphon classification is based on the comparison of the elevation value at a given point of the topographic surface with the elevation values of its surroundings along eight compass directions. The comparison results in a set of ternary patterns specified by relative elevations (i.e., higher, lower, same) (Fig. 2.24E). This procedure allows revealing of up to 498 types of geomorphons. All revealed geomorphons are then generalized (reclassified) to the 10 most common landform elements, such as peak, ridge, shoulder, spur, slope, hollow, footslope, valley, pit, and flat (Fig. 2.24E).

Currently, the geomorphon classification is widely used in geomorphologic and related studies (Libohova et al. 2016; Luo and Liu, 2018; Sărăşan et al., 2019; Atkinson et al., 2020; Gioia et al., 2021; Petsch et al., 2022).

Fig. 2.25D presents a 3D map of the geomorphon classification for Mount Ararat.

The 3D maps shown in Figs. 2.23 and 2.25 testify that the application of the landform classification schemes described in Section 2.7 leads to very different results. The correctness of choosing a specific classification scheme depends on the formulation of the problem, the type of terrain, and the user's experience. It seems that it is most appropriate to derive several variants of landform classification and then to carry out their comparative analysis considering the adequacy of representing properties of the real topography and the correctness of describing relationships between the topography and other components of the landscape (if the user has the goal of studying these relationships).

Other discussions on landform classifications can be found elsewhere (Minár and Evans, 2008; MacMillan and Shary, 2009; Gruber et al., 2017; Arundel and Sinha, 2019; Kharchenko, 2020).

Chapter 3

Digital elevation models

Definition 3.1 (DEM). A digital elevation model (DEM) is a two-dimensional (or bivariate) discrete function of elevation of the topographic surface.

Definition 3.2 (DTM). A digital terrain model (DTM) is a two-dimensional discrete function of a morphometric variable describing a topographic surface.

Among DTMs are DEMs, digital models of slope gradient, horizontal curvature, catchment area, and other morphometric variables (Table 2.1). For any portion of the topographic surface, digital models of all morphometric variables are derived from a DEM of this portion.

Definition 3.3 (DSM). A digital surface model (DSM)[a] is a two-dimensional discrete function of elevation of the top of objects situated on the terrain.

DSM does not describe the topographic surface but rather a surface consisting of portions of the land surface, buildings, vegetation cover, and other natural and human-made objects (Fig. 3.1).

3.1 DEM generation

Data for DEMs can be collected by various ground-based, remote sensing, and laboratory techniques, which can be allocated to the following 14 main groups: (1) conventional topographic surveys, (2) kinematic global navigation satellite system (GNSS) surveys, (3) stereophotogrammetry of optical remote sensing images, (4) structure-from-motion (SfM) processing of unmanned survey imagery, (5) synthetic aperture radar (SAR) techniques, (6) laser altimetry, (7) echo sounding, (8) satellite radar altimetry, (9) optical sensing of bathymetry, (10) soil and geological drilling, (11) three-dimensional (3D) seismic survey, (12) ice and ground penetrating surveys, (13) digitizing of contours, and (14) fusion of data from different sources.

The techniques of the first six groups are commonly used to produce data for the DEMs of land topography. The approaches of the groups 7–9 are applied to describe submarine topography. The methods of the group 8 are utilized to generate data for DEMs of glacier topography. The techniques of the groups 10–12 are commonly used to quantify subterranean and subglacial topography. The procedures of the group 13 are universal; they may be used to produce DEMs of the land, submarine, glacier, subglacial, and subterranean topography. The techniques of the group 14 are usually applied to generate DEMs on the global or continental scales, including portions of the land, submarine, glacier, and subglacial topography.

It is important to remember that most of the raw data gained from the application of these techniques are DSMs rather than DEMs. Additional steps of data processing (i.e., filtering and/or editing) should be done to derive DEMs from DSMs.

Let us briefly consider the general principles of DEM production for each group of techniques.

3.1.1 Conventional topographic surveys

Conventional ground-based topographic surveys are the most traditional techniques for producing high-resolution DEMs of relatively small agricultural, forested, and urban areas. These techniques are based on ground observations and measurements of horizontal and vertical angles between points on the land surface, distances between the points, and their relative elevations. Optical and optoelectronic instruments are utilized in these surveys.

a. Do not confuse this abbreviation with DSM as digital soil mapping (Chapter 11).

Digital Terrain Analysis. https://doi.org/10.1016/B978-0-443-24798-9.00016-9

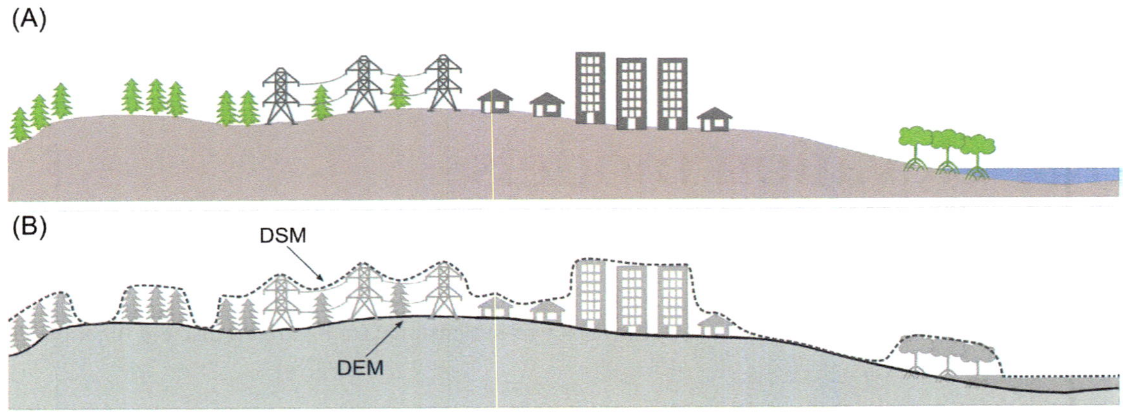

FIG. 3.1 A scheme of a terrain (A) described by two types of digital models (B): a digital surface model and a digital elevation model. *Dashed line*, DSM; *solid line*, DEM. *(Modified from Guth, P.L., Van Niekerk, A., Grohmann, C.H., Muller, J.-P., Hawker, L., Florinsky, I.V. et al., 2021. Digital elevation models: terminology and definitions. Remote Sens. (Basel) 13 (18), 3581, Fig. 4A and B, © Authors, distributed under the terms and conditions of the Creative Commons Attribution (CC BY) license.)*

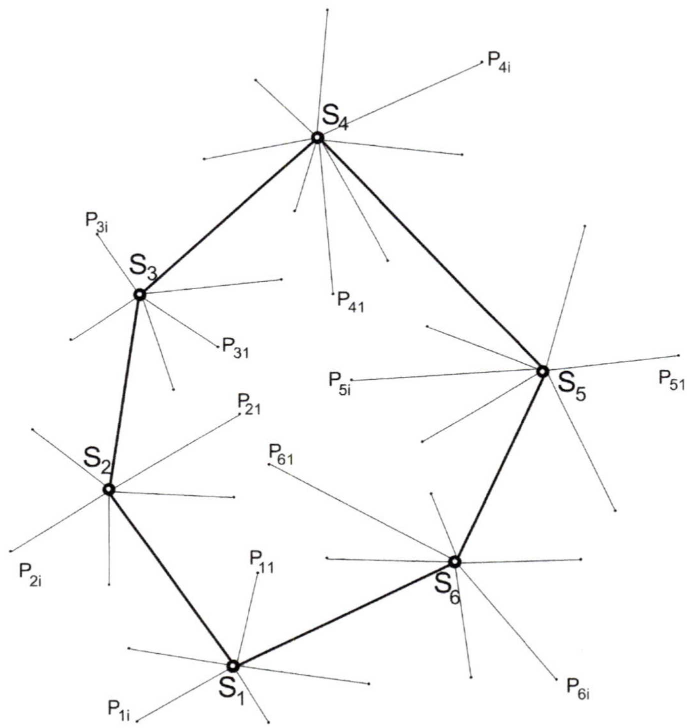

FIG. 3.2 A scheme for a total station traversing. P_{11}, … P_{1i}, …, P_{61} …, P_{6i}, points observed from traverse stations; S_1, …, S_6, traverse stations.

Two types of surveys are commonly used for DEM generation: (1) differential and trigonometric leveling (Yurov, 1976; Bakanova, 1983, § 22; Ghilani, 2022, chaps. 4 and 5), and (2) tacheometric surveying/traversing with optical tacheometers (Malyavsky and Zharnovsky, 1974; Boyko and Limontov, 1975; Bakanova, 1983, § 21) or laser total station systems (Ghilani, 2022, chaps. 8 and 9). In particular, DEM construction with total station traversing is based on closed polygon traverses starting and finishing at the same point of the land surface (Fig. 3.2). A traverse is a set of consecutive lines with intervisible ends (traverse stations) whose lengths and directions are determined from observations. From each traverse station, a surveyor chooses a set of points located at characteristic points of the terrain (e.g., along structural lines). Then angle and distance measurements of selected points are conducted from a particular traverse station. DEM generation is completed after a series of final adjustments and computations (Bolshakov et al., 1989, pt. 2; Ghilani, 2022, chap. 10) to determine the coordinates (x, y, z) of the points.

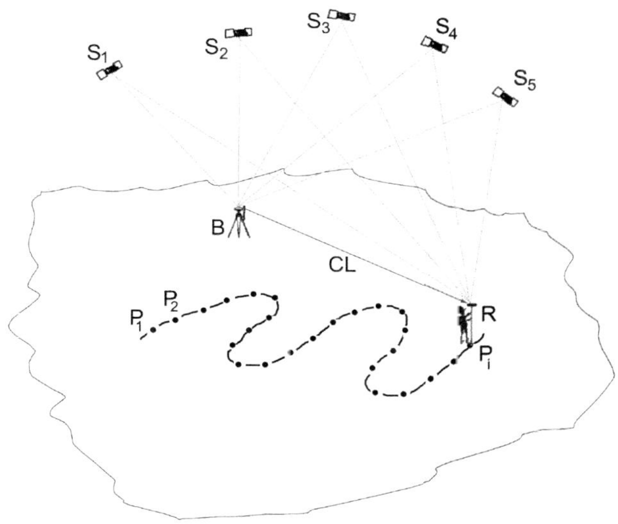

FIG. 3.3 A scheme for an RTK GNSS survey. *CL*, communication link between the base station and rover; *B*, base station; P_1, P_2, ... P_i, points with measured coordinates; *R*, rover; S_1, ..., S_5, satellites.

High-resolution DEMs of relatively small areas obtained by conventional topographic surveys are used in large-scale soil, geological, and other studies of landscapes (e.g., Pennock et al., 1987; King et al., 1999; Florinsky and Kuryakova, 2000).

3.1.2 Kinematic GNSS surveys

GNSS are satellite-based positioning systems—the US's Global Positioning System (GPS), Russia's Global Navigation Satellite System (GLONASS), China's BeiDou Navigation Satellite System, and the European Union's Galileo—providing geodetic coordinates of ground points (Ghilani, 2022, chap. 13). A GNSS device receives radio signals from a satellite constellation. Measurements from at least four satellites are necessary to determine the coordinates (x, y, z) of a ground point.

Real-time kinematic (RTK) GNSS survey is a technique based on the measurements of the phase of the signal's carrier wave. Two GNSS receivers are used: a base stationary receiver and a rover receiver mounted on an all-terrain vehicle or moved by an operator (Fig. 3.3). The base station receiver samples data from satellites and transmits it, together with its known position, via radio to the roving receiver. The rover receives the transmitted data and processes the GNSS measurements obtained at both the base and rover receivers to determine the position of the rover (Awange, 2012, pt. 1; Madry, 2015, chap. 4; Ghilani, 2022, chap. 15). As a result, a user can generate high-resolution DEMs of relatively small non-forested areas (Panin and Gelman, 1997; Schmidt J.P. et al., 2003; Nico et al., 2005b; Kumar et al., 2008; Awange, 2012, chap. 8).

Kinematic GNSS surveys offer fast and cost-effective DTM-based solutions for many detailed and large-scale problems of soil science (Clark and Lee, 1998; Florinsky et al., 2002), geology (Baldi et al., 2000; Pearce et al., 2006; Fiore Allwardt et al., 2007; Nazari et al., 2009), and other geosciences.

3.1.3 Stereophotogrammetry of optical remote sensing images

Stereophotogrammetry is the classical way to produce DEMs from stereo pairs obtained by optical remote sensing, which is a group of passive remote sensing approaches collecting radiation reflected and emitted from the observed surfaces (Lobanov, 1984; Kraus, 2007; Linder, 2016; Chibunichev, 2022). Optical remote sensing deals with a part of electromagnetic spectrum from the visible to the near-infrared (NIR) bands. DEMs can be derived from stereo pairs of aerial, satellite, and terrestrial images.

The basic idea of stereophotogrammetry is as follows: Suppose that there are two images of the same object taken from two different positions (Fig. 3.4). If it is possible to reconstruct geometric parameters of the photographing condition (i.e., camera parameters, position, and orientation), a stereo model of the object can be constructed. Then the coordinates (x, y, z) of any point represented in both images can be computed. Stereo correlation, or image matching algorithms, are used for this purpose (Potuckova, 2004).

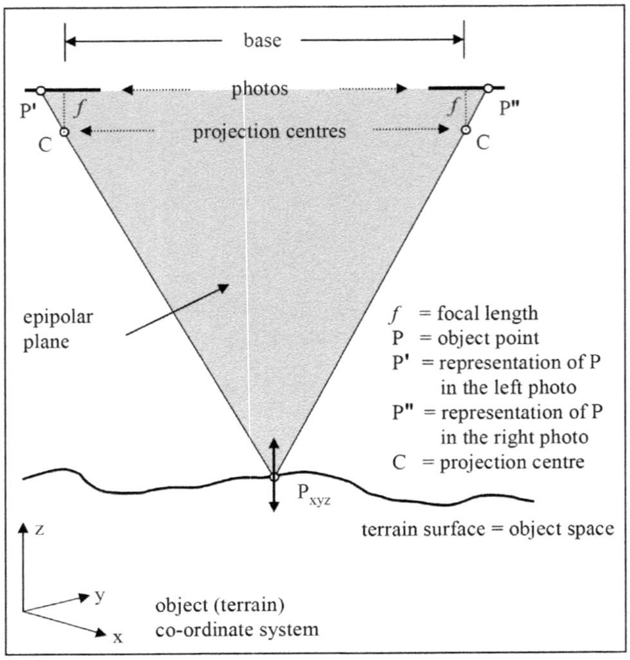

FIG. 3.4 The geometry of an oriented stereo model. Changing the height of point *P* on the land surface leads to a linear motion (left-right) of the points *P′* and *P″* within the photos. *(From Linder, W., 2016. Digital Photogrammetry: A Practical Course, fourth ed. Springer, Berlin, Germany, Fig. I.1, reproduced with permission from Springer Nature.)*

Automated photogrammetric generation of DEMs can be done either by processing digitized stereo pairs from an analog camera (Boyko and Limontov, 1975; Claus, 1984; Makarovic, 1984a,b; Jensen, 1995) or by handling digital stereo images (Lascelles et al., 2002; Egels and Kasser, 2002, chap. 3; Linder, 2016, § 4.6).

To generate DEMs, satellite stereo images from various platforms can be utilized (Toutin, 2000), such as SPOT (Gugan and Dowman, 1988; Chen and Rau, 1993; Al-Rousan et al., 1997; Toutin, 2006; Korona et al., 2009), Ikonos (Toutin, 2004b), QuickBird (Toutin, 2004a), WorldView-1, -2, and -3 (Shean, 2017; Porter et al., 2018; Howat et al., 2019), and the like. It is common practice to use stereo pairs from the same platform. For instance, a quasi-global, 1″-gridded model of the land surface—ASTER GDEM (Table 3.1)—was produced by photogrammetric processing of 15-m resolution NIR stereo images taken by the Advanced Spaceborne Thermal Emission and Reflection Radiometer (ASTER) sensor of the Terra spacecraft from 2000 to 2008 (Welch R. et al., 1998; Hirano et al., 2003; Toutin, 2008; Abrams et al. 2020)

TABLE 3.1 Global and quasi-global spheroidal equal angular DEMs of the Earth.[a]

Title	Coverage	Grid spacing	References
TerrainBase	Global	5′	Row et al. (1995) and NOAA (1995)
ETOPO5	Global	5′	NOAA (1988, 1993)
ETOPO2	Global	2′	Smith and Sandwell (1997) and NOAA (2001, 2006)
ETOPO1	Global	1′	Amante and Eakins (2009) and NOAA (2009)
ETOPO 2022	Global	15″	MacFerrin et al. (2021, 2022) and NOAA (2022)
GTOPO30	Quasi-global (land)	30″	Gesch et al. (1999) and USGS (1996)
GLOBE	Quasi-global (land)	30″	Hastings and Dunbar (1998, 1999) and GLOBE Task Team and Others (1999)
ACE	Quasi-global (land)	30″	Berry et al. (2002) and EAPRS Lab (2002)
ACE2	Global: Quasi-global (land):	5′ and 30″ 3″	Berry et al. (2010) and ACE2 Team (2008)

TABLE 3.1 Global and quasi-global spheroidal equal angular DEMs of the Earth—cont'd

Title	Coverage	Grid spacing	References
SRTM3	Quasi-global (land) 56°S–60°N 180°W–180°E	3″	Rabus et al. (2003) and Farr et al. (2007)
SRTM1	Quasi-global (land) 56°S–60°N 180°W–180°E	0°–50° N, S: 1″ >50° N, S: 2″ × 1″	Rabus et al. (2003), Farr et al. (2007), and USGS (2015)
SRTM30	Quasi-global (land)	30″	Farr et al. (2007) and NASA (2010)
SRTM30_PLUS	Global	30″	Becker et al. (2009) and Sandwell et al. (2008)
SRTM15_PLUS, versions 1 and 2	Global	15″	Olson et al. (2014), Sandwell et al. (2014), and Tozer et al. (2019)
GEBCO_2014	Global	30″	Weatherall et al. (2015) and GEBCO (2014)
GEBCO_2019—GEBCO_2023	Global	15″	GEBCO (2023a,b)
GMTED2010	Quasi-global (land) 90°S–84°N: 56°S–84°N: 180°W–180°E	 30″ 15″ and 7.5″	Danielson and Gesch (2011) and USGS (2011)
EarthEnv-DEM90	Quasi-global (land) 60°S–83°N 180°W–180°E	3″	Robinson et al. (2014) and Robinson and NCEAS ENO Working Group (2014)
ASTER GDEM, versions 1–3	Quasi-global (land) 83°S–83°N, 180°W–180°E	1″	Hato et al. (2009), Fujisada et al. (2012), Abrams et al. (2020), JPL (2009), and Japan Space Systems (2009–2019)
ALOS World 3D	Quasi-global (land)	0.15″	Tadono et al. (2014)
ALOS AW3D30 DSM	Quasi-global (land)	1"	Tadono et al. (2016) and JAXA EORC (2015–2023)
NASADEM	Quasi-global (land) 56°S–60°N 180°W–180°E	1″	Crippen et al. (2016) and NASA JPL (2021)
WorldDEM	Quasi-global (land)	0.4″	Zink et al. (2014) and Rizzoli et al. (2017)
WorldDEM Neo	Quasi-global (land)	0.15″	Airbus DS (2021)
MERIT DEM	Quasi-global (land)	3″	Yamazaki et al. (2017) and Yamazaki (2018)
Copernicus GLO-30	Quasi-global (land)	1″	Strobl (2020) and ESA (2022)
Copernicus GLO-90	Quasi-global (land)	3″	Strobl (2020) and ESA (2022)
FABDEM	Quasi-global (land)	1″	Hawker et al. (2022) and Neal and Hawker (2023)

[a] *The table presents only civil, free, and commercial DEMs. Information on military DEMs, such as DTED, including declassified ones, can be found elsewhere (e.g., NIMA, 2000).*

(Fig. 3.5). Other quasi-global model—AW3D30 DSM (Table 3.1)—was photogrammetrically derived from 2.5-m resolution, visible-band stereo images acquired by the Panchromatic Remote Sensing Instrument for Stereo Mapping (PRISM) onboard the Advanced Land Observing Satellite (ALOS) from 2006 to 2011 (Tadono et al. 2014, 2016).

A DEM may also be created using two photos obtained from different satellites, for example, taking one image from Landsat MSS and another one from NOAA AVHRR (Akeno, 1996).

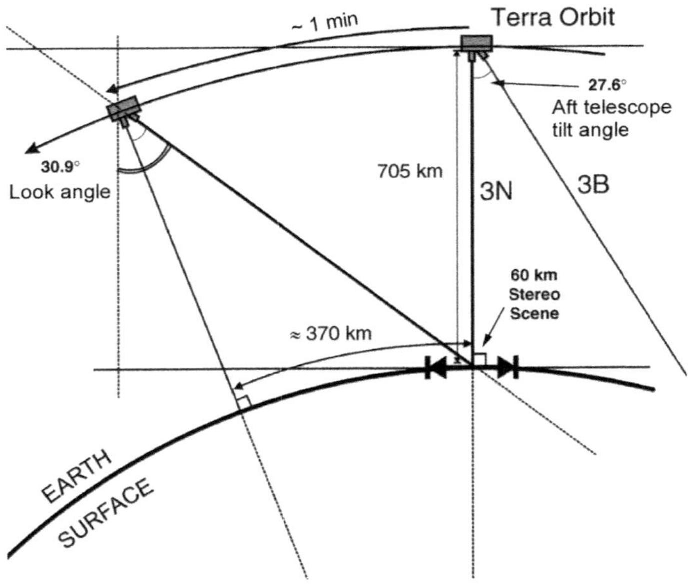

FIG. 3.5 A simplified scheme of the along-track stereo geometry of the ASTER system. Two near-infrared spectral bands, 3N and 3B (nadir and aft-looking telescopes), generate a stereo pair with a base-to-height ratio of about 0.6. *(Modified from Hirano, A., Welch, R., Lang, H., 2003. Mapping from ASTER stereo image data: DEM validation and accuracy assessment. ISPRS J. Photogrammet. Remote Sens. 57 (5–6), 356–370, Fig. 1, reproduced with permission from Elsevier.)*

Airborne and satellite stereo images can be used to obtain DEMs not only of land and glacier topography but also of shallow stream beds (Dietrich, 2016) and submarine topography of clear shallow waters (Collin et al., 2018; Hodúl et al., 2018). Notice that in through-water photogrammetry, the effect of light refraction at the air-water boundary, when features appear shallower than they actually are, should be corrected (Fryer and Kniest, 1985; Butler et al., 2002; Murase et al. 2008).

Terrestrial, or close-range stereo photogrammetry (Remondino and El-Hakim, 2006; Luhmann et al., 2020), is implemented to create very high-resolution DEMs (up to several millimeters) for super-detailed field and laboratory studies of soil erosion (Warner, 1995; Hancock and Willgoose, 2001). Super-detailed stereo photography is also applied to produce DEMs in laboratory physical modeling of tectonic processes such as folding and faulting (Fischer and Keating, 2005; Keating and Fischer, 2008).

Photogrammetrically derived DEMs are used in soil science (Odeh et al., 1991; Ziadat et al., 2003), geology (Chorowicz et al., 1991; Babiker and Gudmundsson, 2004), glaciology (Etzelmüller and Sollid, 1997; Etzelmüller, 2000; Fox and Gooch, 2001; Fox and Cziferszky, 2008), and other geosciences in a wide range of scales.

3.1.4 Structure-from-motion technique for unmanned survey imagery

Although the SfM technique has been originated and developed in computer vision (Civera et al., 2012; Özyeşil et al., 2017; Szeliski, 2022), SfM in fact is an extension of the conventional stereophotogrammetry (Westoby et al., 2012; Fonstad et al., 2013; Carrivick et al., 2016; Eltner et al., 2016; Smith et al., 2016; Anderson et al., 2019; Chibunichev, 2022; Gomez, 2022). To make a three-dimensional reconstruction of the structure or shape of an object, an SfM technique processes a set of overlapping photos from multiple viewpoints (Fig. 3.6A). Camera positions and scene geometry are simultaneously reconstructed by the automatic identification of matching image features, which are tracked from image to image. Initial estimates of camera positions and object coordinates are iteratively refined by nonlinear least-squares minimization. SfM images can be collected by a moving operator or vehicle.

Unmanned aerial systems (UASs), unmanned aerial vehicles (UAVs), or drones are increasingly used in sciences and industry (Watts et al., 2012; Nex and Remondino, 2013; Colomina and Molina, 2014; Shahbazi et al., 2014; Whitehead and Hugenholtz, 2014; Whitehead et al., 2014; Pajares, 2015; Bhardwaj et al., 2016b; Toth and Jóźków, 2016; Singh and Frazier, 2018; Aber et al., 2019; Xiang T. et al., 2019; Yao et al., 2019; Gaffey and Bhardwaj, 2020; Li et al., 2023). UASs include various types of aerial systems (e.g., blimps, kites, gliders, airships, and copters) equipped with various types of sensors, such as visible band, multispectral, and thermal cameras, light detection and ranging (lidar) scanners, SAR, and so on. In the context of DEM production, UASs equipped with visible-band cameras are of main interest.

UASs and SfM-based photogrammetry introduced a low-cost alternative to manned aerial photogrammetry (Eisenbeiß, 2009; Aber et al., 2019). UAS-derived imagery is utilized for generation of DEMs with an ultra-high, centimeter-range

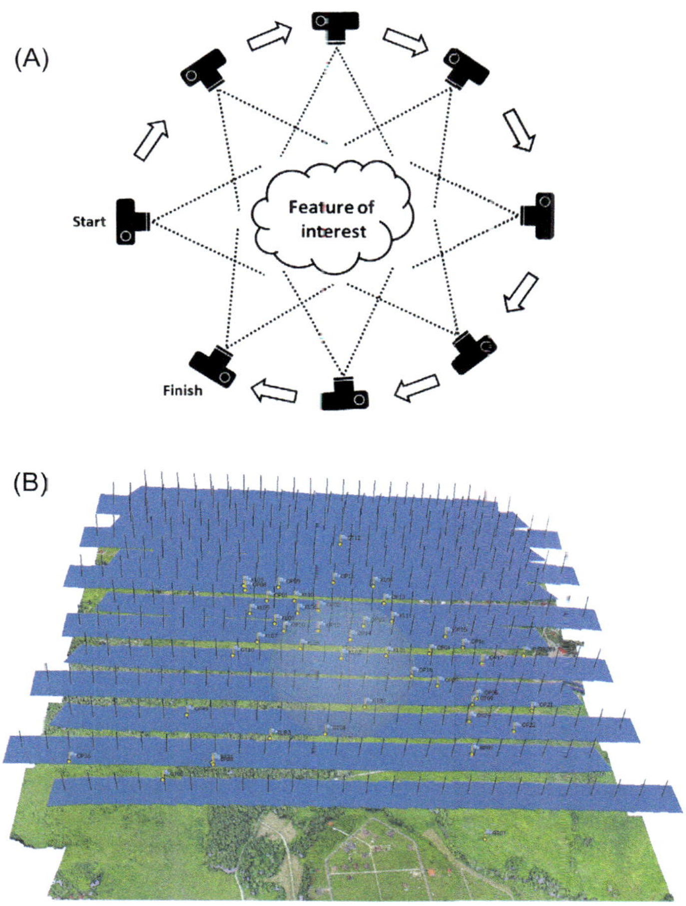

FIG. 3.6 Structure-from-motion and unmanned aerial surveys: (A) SfM technique requires multiple overlapping photos as input data. (B) Typical image overlapping in unmanned aerial surveys. *Blue stripes*, overlapping images along flight stripes; *black vertical lines*, image projection centers. *((A) From Westoby, M.J., Brasington, J., Glasser, N.F., Hambrey, M.J., Reynolds, J.M., 2012. 'Structure-from-Motion' photogrammetry: a low-cost, effective tool for geoscience applications. Geomorphology 179, 300–314, Fig. 1, reproduced with permission from Elsevier. (B) Modified from Kurkov, V., Bliakharskii, D., Florinsky, I., 2016. Application of unmanned aerial surveying for geomorphometric modelling. Izvestia Vuzov. Geodesy Aerophotosurveying 60 (6), 69–77 (in Russian), Fig. 2.)*

resolution and accuracy (Westoby et al., 2012; Carrivick et al., 2016; Smith et al., 2016; Gindraux et al., 2017; James et al., 2017a; Florinsky et al., 2018b; Guisado-Pintado et al., 2019; Kurkov and Kiseleva, 2020). Compared to the conventional manned aerial survey, one of the features of the unmanned aerial survey is the excessive overlap of images: forward and side overlaps should be not less than 70% and 50%, respectively (Fig. 3.6B). This makes it possible to avoid photogrammetric gaps in DEMs caused by the instability of too light UAVs during flight due to wind gusts.

To ensure the required DEM accuracy, ground control points (GCPs) are typically used in unmanned aerial surveys. However, if it is difficult or impossible to set or use GCPs (e.g., in severe conditions of mountain glaciers and polar regions), one can apply a direct georeferencing approach (Chiang et al., 2012; Pfeifer et al., 2012; Turner et al., 2012, 2014; Hugenholtz et al., 2016; Gabrlik et al., 2018; Bliakharskii et al., 2019a; Chudley et al., 2019). In this case, the reference data are the high-precision coordinates of the image projection centers determined with onboard GNSS receivers. However, to achieve centimeter-range accuracy of UAS-derived data, one should: (1) use geodetic-grade, onboard GNSS receivers; (2) utilize cameras with the central leaf shutter; and (3) provide a precision of 1 ms and higher for synchronization between imaging and navigation sensors (Bliakharskii et al., 2019a).

If an unmanned aerial survey is performed without GCPs, data processing includes the following key steps:

1. Determining the high-precision coordinates of image projection centers. These data are then utilized for the post-processed kinematic (PPK) direct georeferencing (Turner et al., 2014).
2. Aligning aerial images by the least-squares bundle adjustment with data from the camera field calibration (the aerial triangulation).

3. Building a dense point cloud (Section 3.2.3) by algorithms belonging to the group of semi-global matching methods (Hirschmüller, 2008; Chibunichev et al., 2015; Chibunichev, 2022, § 6.4 and 6.5).
4. Ground filtering of a dense point cloud (Section 3.2.3), that is, removing the points belonging to above-ground objects (vegetation, buildings, etc.) to retrieve bare earth points. Various types of methods are used for this purpose, such as the adaptive triangulated irregular network algorithm, the multiscale curvature classification, the cloth simulation filtering, and the 2D progressive morphological algorithms (Tan et al., 2018; Yilmaz et al., 2018; Yilmaz, 2021; Qin et al., 2023).
5. Generating a DEM by interpolation (Section 3.4) of the ground-filtered dense point cloud.
6. Producing orthomosaics.

The SfM technique can similarly be used to create bathymetric DEMs of submarine topography when photo or video recording is performed by remotely operated vehicles (ROVs) or autonomous underwater vehicles (AUVs) (Fallati et al., 2022; Lucas et al., 2022).

One can use DEMs obtained from UAS-based data by the SfM technique in detailed and large-scale geoscientific studies of relatively small areas, for example, to study and model erosion (D'Oleire-Oltmanns et al., 2012), coastal (Mancini et al., 2013; Brunier et al., 2016), and fluvial (Jeziorska et al., 2016) processes, to reveal geological features (Bemis et al., 2014; Johnson et al., 2014), to monitor and study glaciers (Bliakharskii et al., 2019a,b), etc.

3.1.5 Laser altimetry

Laser altimetry is a group of active remote sensing techniques using the following general scheme (Fig. 3.7). A laser altimeter instrument emits laser pulses toward the surface under survey. A part of the radiation reflected from the surface returns and is detected by the instrument receiver. Then the distance (D) from the instrument to the surface can be calculated considering the speed of light (v): $D = vt/2$, where t is the time it takes for the laser beam to travel from the laser instrument to the surface and back. Knowing the angle of the mirror rotation φ, one can determine the coordinates of a point on the surface in the coordinate system of the instrument. Knowing the external orientation elements of the laser instrument at this moment, one can calculate the coordinates of the point in the coordinate system of the object. The external orientation elements of the laser instrument during survey are determined using a position and orientation determination onboard unit, which includes an inertial measurement device and a GNSS receiver. The result of the lidar scanning is a dense point cloud (Section 3.2.3) with coordinates (x, y, z) and the intensity value of the reflected signal measured for each point (Baltsavias, 1999a,b; Wehr and Lohr, 1999; Kraus, 2007, chap. 8; Shan and Toth, 2008; Mallet and Bretar, 2009; Yu et al., 2010; Chibunichev, 2022, chap. 9).

Laser altimetry exists in three forms: satellite laser altimetry, airborne lidar scanning, and terrestrial lidar survey.

Satellite laser altimetry is widely used for DEM generation. First, two global DEMs of the Moon were produced: the low-resolution DEM based on the Clementine mission data (Smith et al., 1997; Zuber et al., 1996) and the medium-resolution DEM based on the Lunar Reconnaissance Orbiter LOLA data (Smith D.E. et al., 2010; Neumann, 2008). Second, the global medium-resolution DEM of Mars was created using the Mars Global Surveyor MOLA data (Smith et al., 1999, 2003). Third, DEMs for the ice sheets of Antarctica and Greenland were generated from ICESat and ICESat-2 data (Schutz et al., 2005; Dimarzio et al., 2005; Shen et al., 2022). Such DEMs are used in medium- and small-scale geological and planetological studies (e.g., Zuber et al., 1994, 2000; Byrne, 2007).

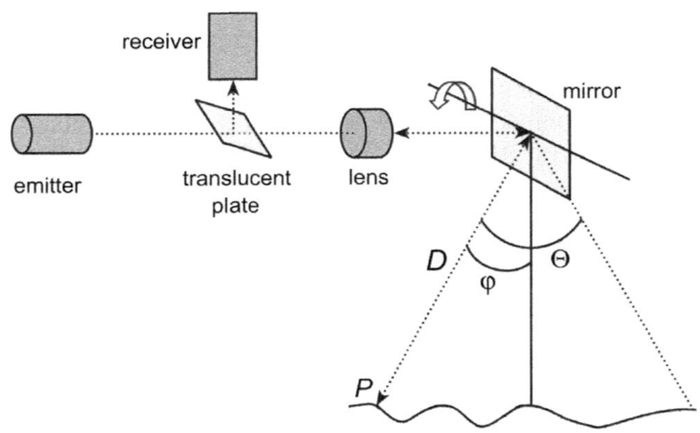

FIG. 3.7 A principle scheme of laser altimetry scanning. D, distance from the instrument to a given point P of the surface; Θ, laser scanner field of view; φ, mirror rotation angle.

Airborne lidar scanning is commonly applied to create high-resolution DEMs of both the land surface (Baltsavias, 1999a; Wehr and Lohr, 1999; French, 2003; Brovelli et al., 2004; Lloyd and Atkinson, 2006; Liu, 2008; Bretar and Chehata, 2010; Selezneva, 2013) and the seafloor, down to −70 m in clear waters (Finkl et al., 2005; Gao, 2009; Agrafiotis et al., 2019; Chibunichev, 2022, § 9.3). Although the problem of effective ground filtering of lidar point clouds (i.e., removing points related to vegetation cover and human-made structures) is still not fully resolved (Axelsson, 1999; Aguilar and Mills, 2008; Leigh et al., 2009; Mongus and Žalik, 2012; Yilmaz, 2021; Qin et al., 2023), lidar aerial surveys offer fast (but relatively expensive) DTM-based solutions for large-scale problems of soil science (Ballabio and Comolli, 2010; Greve et al., 2010), geology (Haugerud et al., 2003; Mynatt et al., 2007b), geomorphology (Gomez, 2022), and other geosciences. Currently, airborne laser scanning is experiencing a renaissance due to the active use of UAS-based lidars (UAV-lidar technology) for producing ultra-high-resolution DEMs of relatively small areas (Kurkov et al., 2021; Curcio et al., 2022; Li et al., 2024).

Terrestrial lidars are applied to create high-resolution DEMs of outcrops (Bellian et al., 2005; Buckley et al., 2008; Hodgetts, 2013), which can be included in large-scale, 3D geological models (Section 8.4).

3.1.6 Synthetic aperture radar techniques

SAR is a group of active remote sensing techniques. The SAR antenna transmits electromagnetic radiation at microwave frequencies that is reflected from the area of interest. Then the radar echoes are received by the instrument's receiver. SARs are usually mounted on moving platforms, such as aircraft and spacecraft (Leberl, 1990; Polidori, 1997; Kampes, 2006). To produce DEMs, four SAR techniques can be applied (Toutin and Gray, 2000): clinometry, polarimetry (Schuler et al., 1996, 1998), stereoscopy using stereo pairs of radar images (Gelautz et al., 2003; Toutin, 2010), and interferometry (Zebker and Goldstein, 1986; Zebker et al., 1994a, b; Bürgmann et al., 2000).

Interferometric SAR (InSAR) techniques are in the most common use. InSAR utilizes phase-difference measurements derived from two radar images, which can be obtained by one-pass or two-pass collection modes (Fig. 3.8). In the one-pass mode, a platform has two SARs, but only one of them, the primary antenna, transmits pulses. The radar echoes are received by both antennas, primary and secondary. In the two-pass mode, a platform with a single SAR makes two flights over the target area.

The global low-resolution DEM of Venus was created using SAR data from the Magellan mission (Pettengill et al., 1991; Ford et al., 1992). The InSAR technique was applied in the Shuttle Radar Topography Mission (SRTM) to prepare

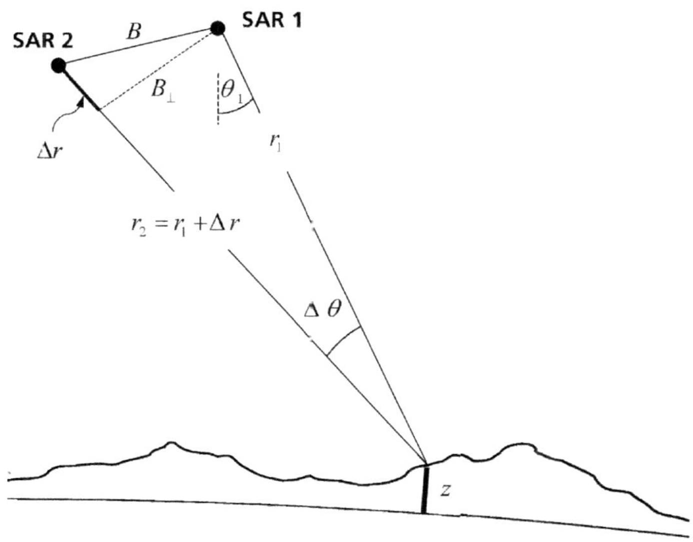

FIG. 3.8 Across-track InSAR configuration. Two SARs, SAR1 and SAR2, fly on parallel tracks (the flight paths of antennas are perpendicular to the plane of the drawing). The SARs view the land surface from slightly different directions. The separation of the flight paths is baseline B; its component perpendicular to the look direction is the effective baseline $B_\perp$. Given the sensor locations and the two ranges, r_1 and r_2, every point imaged on the land surface can be mapped from the images. Interferometry uses the phase information of every pixel to measure the parallaxes $\Delta r = r_2 - r_1$. Θ_1, SAR1 look angle. *(From Rabus, B., Eineder, M., Roth, A., Bamler, R., 2003. The Shuttle Radar Topography Mission—a new class of digital elevation models acquired by spaceborne radar. ISPRS J. Photogrammet. Remote Sens. 57 (4), 241–262, Fig. 1, reproduced with permission from Elsevier.)*

the SRTM1, SRTM3, and SRTM30 DEMs (Table 3.1), the quasi-global medium-resolution DEMs of the Earth (Rabus et al., 2003; Farr et al., 2007). The SRTM30 DEM was later used to describe most of the land topography in the global medium-resolution SRTM30_PLUS and SRTM15_PLUS DEMs (Table 3.1) (Section 3.1.14). The large one-pass spaceborne SAR interferometer based on two similar radar satellites, TerraSAR-X and TanDEM-X, was intended for production of the WorldDEM (Table 3.1), the recent quasi-global high-resolution DEM (Krieger et al., 2007; Rossi et al., 2012; Zink et al., 2014; Rizzoli et al., 2017).

SAR data from ERS-1 and Radarsat satellites were utilized to produce DEMs of the ice sheets of Greenland (Joughin et al., 1996) and Antarctica (Liu et al., 2001), respectively. Data from the SAR of the Cassini satellite were used to create a digital model of near-shore bathymetry for the hydrocarbon lake Ontario Lacus on Titan, the largest moon of Saturn (Hayes et al., 2010).

Sometimes, ground-based radar interferometry is used to generate high-resolution DEMs of small areas (Nico et al., 2005a).

SAR-based DEMs are, for example, utilized in medium- and small-scale soil studies (Hengl et al., 2007a; Mora-Vallejo et al., 2008) as well as geological and planetological research studies (Solomon et al., 1991; Florinsky, 2008c).

3.1.7 Echo sounding

Echo sounding is a group of sonar hydrographic, active remote sensing techniques. They are used to determine the distance between a vessel and submarine topography by transmitting sound pulses into water. The depth of water is determined by the time lag between the emission and return of a pulse as well as the sound speed in the water. Echo sounding can be performed by single-beam, multibeam, and side scan systems (De Moustier, 1988; Bourillet et al., 1996; Mayer, 2006; Blondel, 2009).

Common single-beam echo sounders transmit a single beam oriented toward the vessel's nadir. They usually utilize short (<2 ms) low-frequency (<20 kHz) signals. Single-beam systems return only one depth value per ping, providing data on the seabed below the surveying vessel (Fig. 3.9A). More modern and advanced multibeam echo sounders operate at frequencies from 12 to 455 kHz (for deep and shallow waters, respectively). They transmit several (up to 100–240) relatively narrow beams produced by transducer arrays, covering a wide swath on each side of the vessel track (120°–150°). Multibeam systems simultaneously measure the depth of each beam. Measurements are marked by rather high resolution (some percent of the water depth) and low vertical errors (<1% of the water depth) (Fig. 3.9B). Side scan sonars transmit two beams—a broad beam in the vertical plane and a narrow beam in the horizontal plane—with frequencies ranging from 6.5 kHz to 1 MHz. These systems can achieve resolutions ranging from 60 m to 1 cm. Side scan sonars (towed separately and close to the seabed) can cover a large portion of the underwater topography away from the vessel, from a few tens of meters to a few tens of kilometers (De Moustier, 1988; Bourillet et al., 1996; Mayer, 2006; Blondel, 2009). The use of AUVs as a platform for echo sounders allows for accelerating the generation of accurate, detailed DEMs of the submarine topography (Wynn et al., 2014; Zwolak et al., 2020).

Echo-sounding data are used to produce bathymetric DEMs for rivers, lakes, seas, and oceans (Hall, 1996; Sherstyankin et al., 2006; Gasperini et al., 2007; Jakobsson et al., 2008, 2012; Alvarez et al., 2018; Collin et al., 2018; Hodúl et al., 2018; Kasvi et al., 2019; Legleiter and Harrison, 2019; Wölfl et al., 2019). For example, the 200-m gridded DEM for the Drake Passage was produced by a compilation of precise multibeam bathymetric data obtained on 120 cruises between 1992 and 2015 (Bohoyo et al., 2019). Echo-sounding data are systematically used to calibrate satellite gravity predictions of bathymetry (Section 3.1.8) in the global SRTM30_PLUS and SRTM15_PLUS DEMs (Table 3.1). The Nippon Foundation–GEBCO Seabed 2030 Project has been launched to create a global, high-resolution, and accurate bathymetric DEM based on multibeam measurements (Mayer et al., 2018).

Bathymetric, multibeam-based DEMs can be utilized in marine geological (Kukowski et al., 2001, 2008), geomorphic, and biological (Section 1.2.2.9) studies of different scales.

3.1.8 Satellite radar altimetry

Satellite radar altimetry is an active remote sensing approach. Satellite radar altimeters (pulse-limited radars) are commonly used to measure the shape of the sea surface and to reveal marine gravity anomalies (Chelton et al., 2001) (Fig. 3.10).

Ship track echo-sounding data (Section 3.1.7) are too sparse to fully define the submarine topography. Sandwell et al. (2006) noted that "current bathymetric charts are inadequate for many of these applications because only a small fraction of the seafloor has been surveyed. Modern multibeam echo sounders provide the best resolution, but it would take more than 200 ship-years … to complete the job." At the same time, tracks of radar altimetry satellites cover the oceans more densely

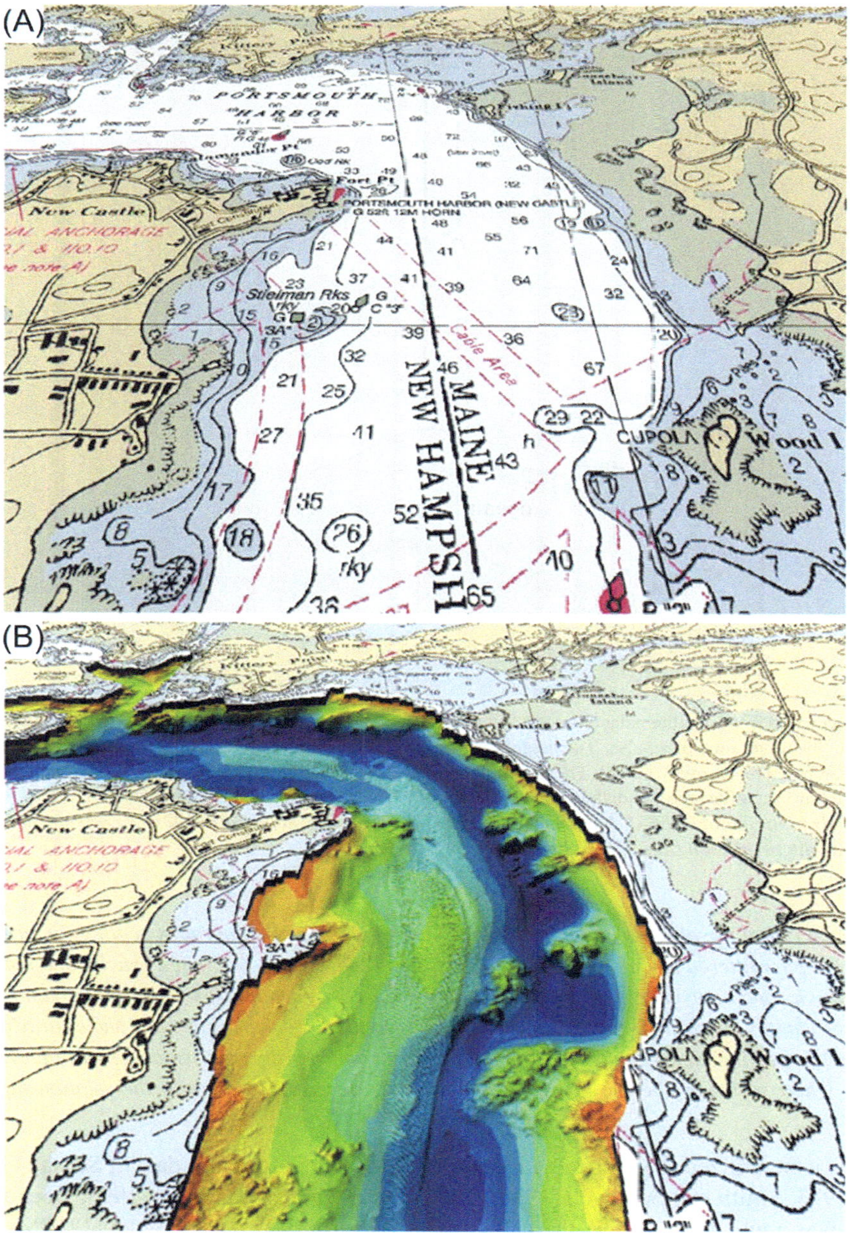

FIG. 3.9 Comparison of single- and multibeam echo sounding: (A) Single-beam sonar data allows the construction of traditional hydrographic charts, which consist of selected soundings and contours interpolated from sparse sounding. (B) Multibeam sonar data offering complete bathymetric coverage of the seafloor. *(From Mayer, L.A., 2006. Frontiers in seafloor mapping and visualization. Mar. Geophys. Res. 27 (1), 7–17, Fig. 2, reproduced with permission from Springer Nature.)*

than ship tracks. Since gravity anomalies can be associated with seafloor topography, it is possible to predict bathymetry from satellite gravity anomaly data (Smith and Sandwell, 1994, 1997; Sandwell and Smith, 1997, 2001; Smith, 1998; Sandwell et al., 2006). Later, we describe the Smith and Sandwell method of bathymetric prediction after Marks et al. (2010, p. 225):

> *"The Smith and Sandwell (1994, 1997) bathymetric model combines depths derived from satellite gravity data with in situ data such as soundings and shorelines. … Gravity anomalies are Gaussian filtered to pass full wavelengths <160 km, and then converted to bathymetry with a gravity-to-topography scaling factor assuming appropriate geologic properties locally calibrated by ship soundings. The predicted bathymetry is then combined with regional low-pass filtered (passing full wavelengths >160 km)*

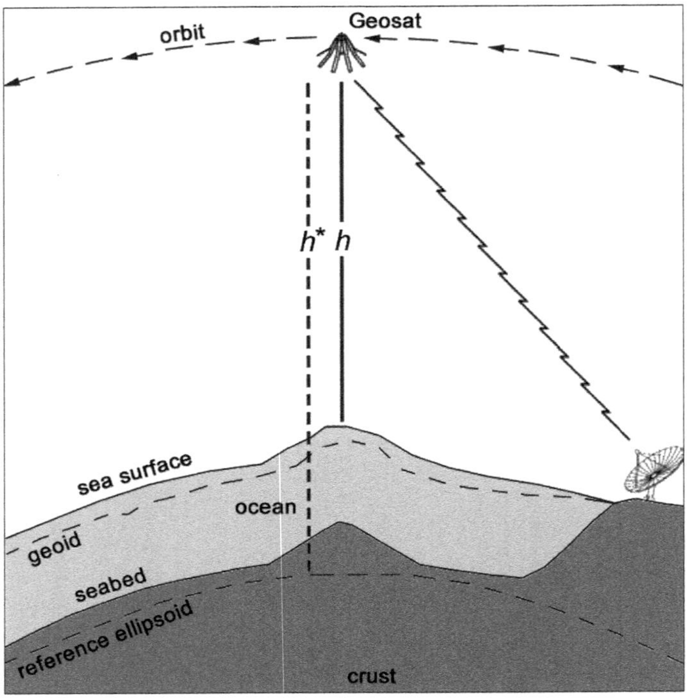

FIG. 3.10 Bathymetric prediction from satellite radar altimetry. A pulse-limited satellite radar altimeter measures the distance h between its antenna and the sea surface by recording the travel time of a pulse. The height of the altimeter above the reference ellipsoid, h^*, is estimated from a global tracking network and orbit calculation based on a gravity model. The height of the ocean surface above the ellipsoid is the difference between h^* and h. The permanent undulations of the sea surface reflect the geoid; they are associated with local and regional gravity anomalies. *(From Sandwell, D.T., Smith, W.H.F., 1997. Marine gravity anomaly from Geosat and ERS-1 satellite altimetry. J. Geophys. Res. Solid Earth 102 (B5), 10039–10054, Fig. 1, © American Geophysical Union, reproduced with permission from John Wiley & Sons.)*

> *bathymetric soundings to form the altimetric bathymetry model. The gravity-to-topography scaling factor is estimated from the empirical ratio of band-pass filtered soundings to band-pass filtered, downward-continued gravity anomalies, in the prediction (15–160 km) band. Gravity and bathymetry may be correlated in the prediction band, if sediment is thin and seafloor relief is high enough; at longer wavelengths the seafloor topography is usually isostatically compensated and shows little or no correlation with gravity. In most areas, the scaling factor implies a geologically reasonable density contrast at the seafloor. Exceptions occur over very tall seamounts, where a linear filtering process cannot be adequate to characterize the relationship between depth and gravity."*

Various algorithms were utilized to produce bathymetric DEMs from altimetric data of Seasat, Geosat, and ERS-1 satellites (Dixon et al., 1983; Smith and Sandwell, 1994, 1997; Ramillien and Cazenave, 1997; Becker et al., 2009). In particular, this approach was applied to generate the bathymetric layer for the following global DEMs: ETOPO5, ETOPO2, ETOPO1, ETOPO 2022, SRTM30_PLUS, and SRTM15_PLUS (Table 3.1).

Bathymetric models are periodically updated using (1) the latest satellite gravity field to improve bathymetry from the prediction band and (2) new echo soundings to improve local gravity-to-topography scale factors (Watts et al., 2006; Marks et al., 2010). Thus, the accuracy of the predicted bathymetric DEMs increases.

Radar altimeters can also be applied for surveying relatively flat land and ice surfaces. Some altimeters are specifically designed for ice surface surveys (Chelton et al., 2001, pp. 31–32). Altimetric data from Seasat, Geosat, ERS-1, and CryoSat-2 satellites were used to produce DEMs of the Antarctic and Greenland ice sheets (Zwally et al., 1983, 1987, 1994; Slater et al., 2018). ERS-1 radar altimetry data were included in the DEM of Antarctica (Liu et al., 1999). Such DEMs can be, for instance, utilized in small-scale geological investigations.

Berry (2000) argued the possibility of producing low-resolution DEMs of land topography from satellite radar altimetry retracked by using an expert system approach. It has been implemented in ACE and ACE2 (Table 3.1), two quasi-global DEMs (Berry et al., 2002, 2010).

3.1.9 Optical sensing of bathymetry

Airborne optical sensing of bathymetry is a group of passive remote sensing approaches. They are based on the principle that the total amount of radiative energy reflected from a water column is a function of water depth (Gao, 2009). The water depth is estimated from brightness values of the image. These approaches may be applied to clear waters only. They can be used to create DEMs of shallow seafloors, which are hazardous for navigation and thus echo-sounding (Ji et al., 1992; Mishra et al., 2004; McIntyre et al., 2006). Such DEMs may be used in large- and medium-scale marine geomorphic, biological, and geological research.

3.1.10 Soil and geological core drilling

Soil and geological core drilling are standard field procedures in soil science and geology. To obtain information on the depth of soil and stratigraphic horizons, one can use core drills, which are mechanisms allowing one to remove unbroken cores (cylinders of soil/rock materials) from the ground (Acker III, 1974; Australian Drilling Industry Training Committee, 1997). Subsequent visual and/or laboratory analysis of such cores makes it possible to determine the boundaries between subsurface horizons and their depths.

Data from soil and geological core drilling are used to generate DEMs of the contact surfaces of soil and stratigraphic horizons, as well as DEMs of subglacial topography (Lythe et al., 2001). Such DEMs can be useful for large-scale soil studies (Florinsky and Arlashina, 1998; Mendonça-Santos et al., 2000; Chaplot and Walter, 2003; Zakharchenko and Zakharchenko, 2006; Scott and Needelman, 2007) as well as medium- and small-scale geological research studies (Belonin and Zhukov, 1968; Robinson et al., 1969; Lemon and Jones, 2003; Groshong Jr., 2006; Kaufmann and Martin, 2008).

3.1.11 Three-dimensional seismic survey

A 3D seismic survey is a group of geophysical, active remote sensing techniques. They are based on the phenomenon of reflection of sound energy from layer boundaries in the subsurface. An energy source is an explosion or vibration on/near the topographic (land, ice sheet, or seabed) surface. Recorders are located along several parallel lines (Fig. 3.11A). The return time of reflected waves depends on the depth of the reflecting horizon as well as the sound velocity between the topographic surface and the horizon. A user obtains an initial set of "surfaces" of two-way travel time (the unit is millisecond), which relate to contact surfaces between adjacent geological horizons. A total set of such time-structure matrices forms a 3D seismic volume (Fig. 3.11B). Then two-way travel time can be converted to the depth of a particular geological surface (Brown, 2011; Liner, 2016).

Sometimes researchers apply methods of digital terrain modeling to time-structure matrices without converting them to DEMs (e.g., Roberts, 2001; Chopra and Marfurt, 2007a,b; Hart and Sagan, 2007). These works are not discussed in this book because such an approach is incorrect: all three coordinates of the DEM matrix should have the same units (i.e., meters) to correctly estimate the partial derivatives of elevation (Eq. 2.2).

A three-dimensional seismic survey is used to produce medium- and low-resolution DEMs of subterranean or subsurface topography, that is, contact surfaces of stratigraphic horizons and geological structures. Such DEMs are commonly applied in medium- and small-scale geological studies and offshore petroleum exploration (Bergbauer et al., 2003; Wynn and Stewart, 2003; Ustinova and Ustinov, 2004; Roberts et al., 2009). Data from seismic surveys were also used to create a DEM of the subglacial topography of Antarctica (Lythe et al., 2001).

3.1.12 Ice and ground penetrating surveys

Ice and ground penetrating surveys are geophysical, active remote sensing approaches.

Ice-penetrating surveys, or radio-echo sounding (RES), can be performed from aircraft and snow vehicles as well as by a moving operator. RES is used to measure the ice thickness and depth of subglacial topography (Robin et al., 1969; Bogorodsky et al., 1985; Siegert, 1999; Plewes and Hubbard, 2001).

Ice-penetrating RES systems usually consist of a transmitter and receiver as well as transmitting and receiving antennas moved over the snow/ice cover. The transmitter generates a pulse of radio waves in the range of 30–1000 MHz, which propagates into the ice cover and reflects off boundaries between the media with distinct dielectric properties (e.g., the

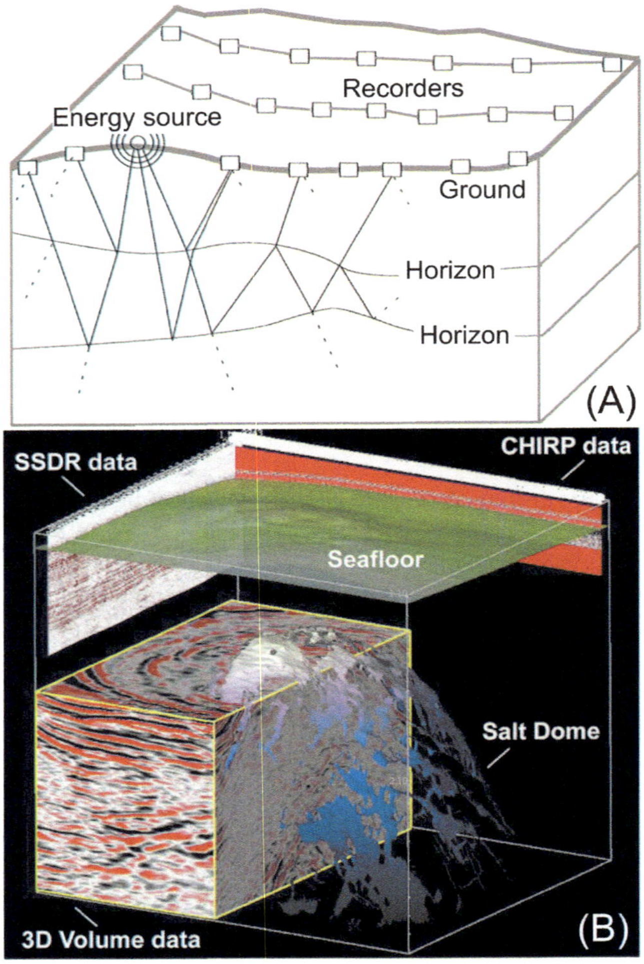

FIG. 3.11 Three-dimensional seismic survey: (A) General scheme. (B) A multiple-resolution seismic dataset for the Woolsey Mound, Gulf of Mexico: multichannel 3D volume data; *SSDR*, surface-source deep-receiver single channel data; *CHIRP*, compressed high-intensity radiated pulse AUV sub-bottom profiler data. *((A) From Ma, C., Rokne, J., 2004. 3D seismic volume visualization. In: Zhang, D.D., Kamel, M., Baciu, G. (Eds.), Integrated Image and Graphics Technologies. Kluwer, New York, NY, pp. 241–262, Fig. 13.1, reproduced with permission from Springer Nature. (B) From Macelloni, L., Simonetti, A., Knapp, J.H., Knapp, C.C., Lutken, C.B., Lapham, L.L., 2012. Multiple resolution seismic imaging of a shallow hydrocarbon plumbing system, Woolsey Mound, Northern Gulf of Mexico. Mar. Petrol. Geol. 38 (1), 128–142, Fig. 3, reproduced with permission from Elsevier.)*

ice sheet and glacier bed). The receiver records these reflections (Fig. 3.12). Since the velocities of radio waves in air and ice are known, the time delay between transmission and reception can be converted into the ice thickness. RES data are collected along flight or trail lines (Siegert, 1999; Plewes and Hubbard, 2001).

Data from ice-penetrating RES were used to produce medium- and low-resolution DEMs of subglacial topography in Greenland (Gogineni et al., 2001; Bamber et al., 2001b, 2013), Antarctica (Lythe et al., 2001; King et al., 2009; Fretwell et al., 2013), and other glaciated territories (Welch B.C. et al., 1998). Such DEMs can be utilized in large-, medium- and small-scale glaciological and geological studies.

Ground penetrating radars (GPRs) (Plewes and Hubbard, 2001; Jol, 2009) are currently widely used for RES in large-scale studies of the internal properties of ice masses. These include the generation of DEMs of ice-bedrock contacts and 3D modeling of internal ice structures, such as interglacial cavities, streams, and lakes. In the latter case, however, the main problem is a correct interpretation of internal reflection horizons that cannot be done without ice-core-based calibration.

GPR surveys (Jol, 2009; Carrick Utsi, 2017) can also be applied to create DEMs of soil-bedrock contacts (Lapen et al., 1996; Carless et al., 2021).

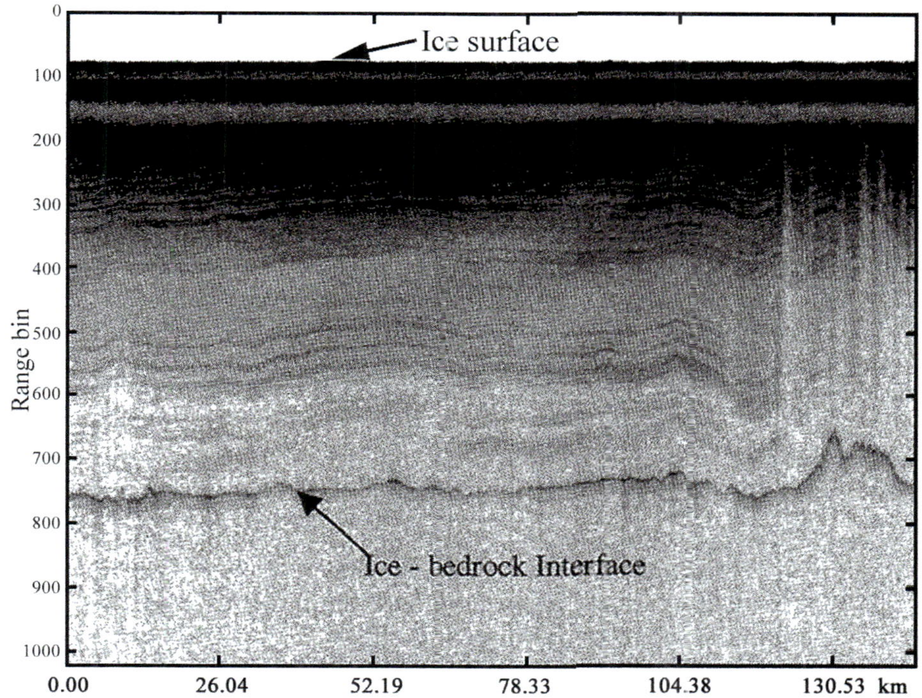

FIG. 3.12 Ice-penetrating radio-echo sounding. An echogram from data collected over an ice sheet. *(Modified from Gogineni, S., Tammana, D., Braaten, D., Leuschen, C., Akins, T., Legarsky, J. et al., 2001. Coherent radar ice thickness measurements over the Greenland ice sheet. J. Geophys. Res. Atmos. 106 (D24), 33761–33772, Fig. 2A, © American Geophysical Union, reproduced with permission from John Wiley & Sons.)*

3.1.13 Digitizing of contours

Although the digitizing of contours per se is a laboratory approach, contours on traditional topographic paper maps are, in most cases, the results of conventional topographic surveys and photogrammetric processing of aerial stereo pairs. Isobaths on hydrographic maps are usually the result of echo sounding. Contours of stratigraphic surfaces on geological maps are, as a rule, based on the results of drilling.

Digitizing a contour consists of the transformation of the contour line continuum into a discrete sequence of georeferenced points located along the contour. In digitizing, one may use ancillary cartographic information, such as elevation/depth values for summits and bottoms as well as structural lines. Digitizing can be performed manually using (1) a digitizer when a paper map is handled (Found, 1970; Boyko, 1980; Vasmut, 1983) or (2) a computer mouse and a scanned map on the monitor screen (on-screen digitizing). In the latter case, the digitizing can be semi-automated (Mor and Lamdan, 1972; Yoeli, 1975; Helvaci and Bayram, 2004).

Digitizing of contours was applied to generate national DEM databases in the United States (USGS, 1993), Soviet Union (Khalugin et al., 1992), Canada (Natural Resources Canada, 1997), and other countries, to produce early versions of military Digital Terrain Elevation Data (DTED) (US DMA, 1993), as well as to describe the global topography in the Digital Chart of the World (Danko, 1992; US DMA, 1992). Depth contours were digitized to create a substantial part of the DEM of the Arctic Ocean floor (Jakobsson et al., 2008), the DEM of the Baltic Sea floor (Seifert et al., 2001), and the like.

In terms of relief representation, old large-scale topographic maps (e.g., 1:25,000-scale Soviet topographic maps) still have a much higher resolution and accuracy than some modern high-resolution DEMs, such as SRTM1 or ASTER GDEM (Table 3.1). In this regard, digitizing contours remains relevant for DEM production.

Contour-based DEMs are used to solve medium- and small-scale problems of soil science (Bell et al., 1992; Gessler et al., 1995), geology (Morris, 1991; Onorati et al., 1992), and other geosciences.

3.1.14 Fusion of data from different sources

The fusion of various DEMs and other datasets, which were earlier prepared by different techniques, is a laboratory approach. It includes mutual linking of data of different origins and coverage as well as filling voids by interpolation.

The data fusion is commonly applied to produce global and quasi-global DEMs (Table 3.1), DEMs of large, poorly accessible areas, as well as DEMs of submarine and subglacial topography.

Let us exemplify the various options for data fusion with the following DEM sequence: GTOPO30 → SRTM30 → SRTM30_PLUS → SRTM15_PLUS version 1 (Table 3.1). The earliest model, GTOPO30, was created using several digital and paper map sources (Gesch et al., 1999). The key sources were (1) DTED Level 1 (US DMA, 1993), produced primarily by digitizing the contours of maps of different origins (Guth, 1999; Hastings and Dunbar, 1999, p. 14) and (2) Digital Chart of the World (Danko, 1992; US DMA, 1992) based on digitized contours. Then SRTM30 was produced from GTOPO30 by replacing the GTOPO30 cells with the SRTM1 30 × 30 averaged cells, except for Antarctica and territories north of 60° N (Farr et al., 2007). The SRTM30_PLUS was the result of a fusion of SRTM30 with: (1) the ICESat elevation data for the Antarctica ice sheet; (2) ocean bathymetry predicted from satellite radar altimetric data calibrated using ~300 million echo soundings; and (3) the International Bathymetric Chart of the Arctic Ocean (IBCAO) data for the ocean bathymetry north of 80° N (Becker et al., 2009). To compile SRTM15_PLUS DEM version 1, Olson et al. (2014) used (1) ASTER GDEM 30 × 30 averaged cells for replacing the GTOPO30 cells of the northern areas of Eurasia and North America not covered by the SRTM data and (2) the CryoSat-2 DEMs (Helm et al., 2014) for replacing the previous description of the Greenland and Antarctica ice sheets.

There are two bathymetric low-resolution DEMs covering huge areas of the ocean floor of the Earth's polar regions: the IBCAO DEM (Jakobsson et al., 2008, 2012, 2020; IBCAO Compilation Group, 2012, 2020) and the International Bathymetric Chart of the Southern Ocean (IBCSO) DEM (Arndt et al., 2013; Dorschel et al., 2022; IBCSO Compilation Group, 2022). These DEMs were produced using the fusion of data on single- and multibeam echo soundings collected by various ship surveys as well as digitized isobaths from old nautical charts. To describe adjacent land areas in the Arctic, portions from the Global Multi-Resolution Terrain Elevation Data (GMTED2010) (Danielson and Gesch, 2011) were included in the IBCAO version 3 (Jakobsson et al., 2012). To describe the subglacial topography of Greenland, data from the BedMachine version 3 (Morlighem et al., 2017) were incorporated into the IBCAO version 4.0 (Jakobsson et al., 2020).

In the IBCSO version 1 (Arndt et al., 2013), the glacier and subglacial topography of Antarctica were described using data from the Bedmap2—a set of DEMs of glacier topography, ice thickness, and subglacial topography (Fretwell et al., 2013). In the IBCSO version 2 (Dorschel et al., 2022), the glacier topography of Antarctica was derived from the Reference Elevation Model of Antarctica (REMA) (Howat et al., 2019), the subglacial topography was incorporated from the BedMachine Antarctica DEM (Morlighem et al., 2020), and the land topography of South America was derived from the ALOS AW3D30 DSM (JAXA EORC, 2015–2023), while bathymetric gaps were filled with SRTM15_PLUS version 2 (Tozer et al., 2019).

The Bedmap2 (like its precursor Bedmap—Lythe et al., 2001) was also the result of the fusion and processing of multiple data sets describing both the glacier topography and ice thickness (Fretwell et al., 2013). In particular, the DEM of the glacier topography was compiled with several satellite radar and laser altimetry datasets, including a DEM derived from combined ERS-1 and ICESat data (Bamber et al., 2009), while the DEM of the subglacial topography was constructed using datasets on the ice thickness measurements come from airborne and ground RES as well as seismic surveys.

The DEM of Greenland was also generated by a fusion of multiple data sets. Initially, it was based on the combination of data from ERS-1 and Geosat radar altimetry, aerial stereophotogrammetry, SAR interferometry, and digitized contours (Bamber et al., 2001a). Howat et al. (2014) enhanced this DEM using data from the ICESat laser altimetry and the SPIRIT DEMs, which were the result of photogrammetric processing of the SPOT-5 stereo pairs (Korona et al., 2009).

Note that the current versions of the continental/regional DEMs describing the subglacial topography of Antarctica and Greenland—BedMachine Antarctica (Morlighem et al., 2020) and BedMachine version 3 (Morlighem et al., 2017), respectively—were created not by a simple fusion and interpolation of data from different sources but using a specially developed, physically based mass conservation method (Morlighem et al., 2011). A usual data interpolation provides users with ice thickness maps and DEMs of subglacial topography, which are not consistent with ice flow dynamics and are impractical for ice flow simulations. The mass conservation method combines ice thickness data from airborne radar sounding profilers with high-resolution swath mapping of ice velocity from InSAR satellite data for obtaining ice thickness maps and DEMs of subglacial topography, which conserves mass and minimizes deviation from observations.

DEMs of stratigraphic surfaces are also compiled using data obtained by various techniques, such as 3D seismic surveys, geological drilling, and other available geological sources (Caumon et al., 2009). In some rare cases, DEMs of the surfaces of deep geological structures may be derived from indirect measurements. For instance, coordinates of earthquake hypocenters were used to generate DEMs of the oceanic lithosphere in subduction zones (Cahill and Isacks, 1992; Nothard et al., 1996).

3.2 DEM grids

From a mathematical point of view, two principally different approaches—continuous and discrete—can be used to describe the topographic surface:

1. Two-dimensional (piecewise) continuous functions can be applied to define relatively small portions of the topographic surface (Jancaitis and Junkins, 1973; Jancaitis, 1978; Strakhov, 2007; Florinsky and Pankratov, 2016). One may also use spherical functions to define the topography of the globe, hemispheres, or other vast territories (Schröder and Sweldens, 2000; Wieczorek, 2007).
2. Two-dimensional discrete functions, that is, DEMs can be employed. In the book, we mainly discuss this approach, which gained wide acceptance due to its relative simplicity.

In the latter case, one can use various types of DEM grids.

Definition 3.4 (DEM grid). A DEM grid is a network of points (grid nodes) at which elevation values are known.

From the standpoint of considering the curvature of a planet, there are two main types of DEM grids: plane and spheroidal ones.

3.2.1 Plane grids

The following types of plane grids are exploited for DEMs of relatively small portions of the topographic surface, when the curvature of the planet may be ignored (Leberl, 1973; Boyko and Limontov, 1975; Peucker, 1980a; Yoeli, 1983; Carter, 1988; Davis, 2002, chap. 5):

1. Irregular grids based on:
   (a) Random distribution of points (Fig. 3.13A),
   (b) Points located along contour lines (Fig. 3.13B),
   (c) Points located along cross-sections (Fig. 3.13C),
   (d) Points located along/at critical elements of the topographic surface, such as structural lines, local minima and maxima, and saddles.

Irregular random grids (Fig. 3.13A) are usually typical for DEMs obtained by conventional topographic and kinematic GNSS surveys (Sections 3.1.1 and 3.1.2) as well as soil and geological drilling (Section 3.1.10). Irregular grids with points located along contour lines (Fig. 3.13B) are common for DEMs resulting from computer-assisted visual stereo matching (Section 3.1.3) and contour digitizing (Section 3.1.13). Irregular grids based on cross-sections (Fig. 3.13C) are typical for DEMs obtained through digital photogrammetric techniques (Section 3.1.3), echo sounding (Section 3.1.7), ice-penetrating RES (Section 3.1.12), as well as contour digitizing (Section 3.1.13).

The latest type of irregular grids is commonly used to construct triangulated irregular networks (TINs) (Peucker et al., 1976; Van Kreveld, 1997; Li et al., 2005, chap. 5) (Fig. 3.13D). TIN is a vector-based data structure comprising a triangular network of vertices with associated coordinates (x, y, z) connected by edges. The mathematical basis for a TIN construction is the Delaunay triangulation (Delaunay, 1934; Skvortsov, 2002; Hjelle and Dæhlen, 2006).

Definition 3.5 (Delaunay triangulation, circle criterion). The Delaunay triangulation Δ of a set of points P on the plane is a triangulation where the interior of the circumcircle of any triangle in Δ contains no point from P (Hjelle and Dæhlen, 2006, § 3.5) (Fig. 3.14).

The circle criterion of the Delaunay triangulation defines a triangulation consisting of triangles that are close to being equiangular (Hjelle and Dæhlen, 2006, chap. 3).

2. Regular grids are based on equilateral triangles, squares, rectangles, hexagons, and so on. The square grid is the most popular one.

Direct generation of a DEM on a regular grid is possible, for instance, by square grid leveling (Section 3.1.1) and automated photogrammetric processing of stereo pairs (Section 3.1 3). Irregular DEMs are commonly converted to regular ones by interpolation (Section 3.4).

Definition 3.6 (Square grid). A square grid is a network of points (grid nodes), which are located at intersections of two mutually perpendicular sets of equally spaced, parallel straight lines (Fig. 3.13E).

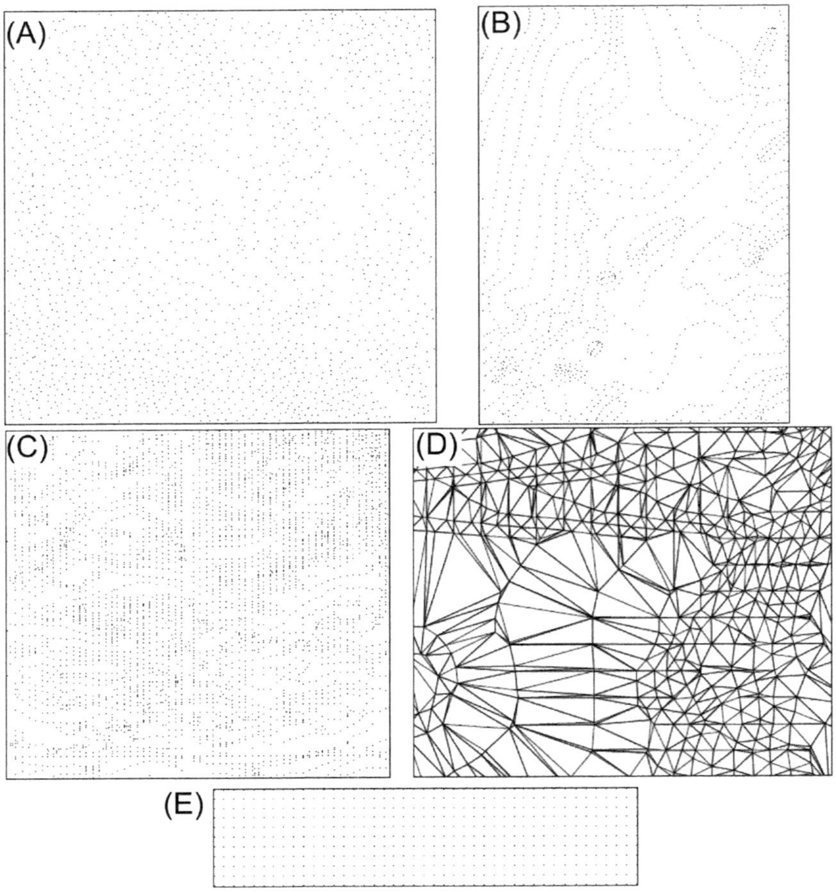

FIG. 3.13 Plane grids: (A) Plane random irregular grid. (B) Plane irregular grid with points located along contours. (C) Plane irregular grid with points located along cross-sections and structural lines. (D) Plane TIN. (E) Plane square grid.

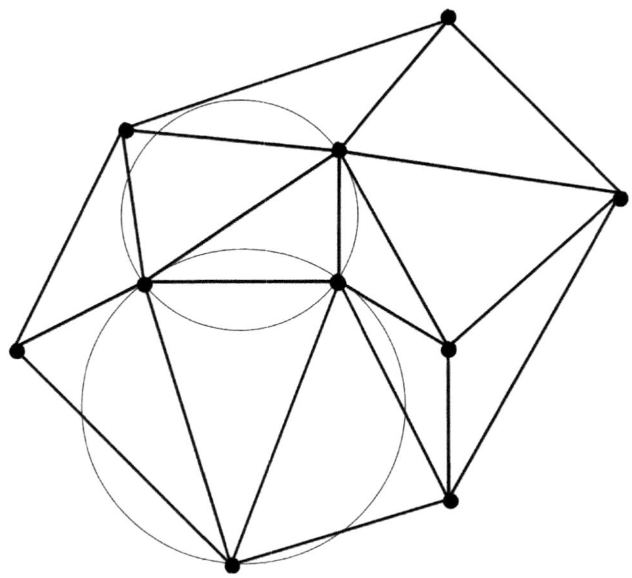

FIG. 3.14 The Delaunay triangulation.

A square grid consists of a set of grid rows or columns.

In the 1970s, there has been much debate about what type of gridding is preferable for DEMs: square grids or TINs. It has been argued that DEMs should be generated using TINs because (1) this is the informal approach considering the actual structure of topography and (2) TIN structure allows for saving machine resources (Malyavsky and Zharnovsky, 1974; Boyko and Limontov, 1975; Peucker and Chrisman, 1975; Peucker et al., 1976). However, Leberl (1973, p. 760) presciently remarked:

> *"It is expected, that sampling of DTMs in a regular point pattern is the most promising application of automated photogrammetry. Compared to other sampling methods, data procurement can be automized (electronic correlator) fully for the regular point DTM, with substantially higher reliability. In addition, the organization of data and its interpolation is greatly simplified with this type of DTM."*

Later, Kumler (1994) demonstrated that square-gridded DEMs have higher accuracy than TIN-based DEMs. Square-gridded DEMs are besides marked by simplicity of representation. The use of square grids is also reasonable because models of morphometric variables (Chapter 2) are usually derived from square-gridded DEMs (Chapter 4). Today, square-gridded DEMs dominate.

3.2.2 Spheroidal grids

Various spheroidal grids (Gold and Mostafavi, 2000; Goodchild and Kimerling, 2002; Lukatela, 2002) can be employed for describing the topography of the globe or vast territories, when the curvature of the planet may not be ignored. The most popular is the spheroidal equal angular grid for the sphere or ellipsoid of revolution.

Definition 3.7 (Spheroidal equal angular grid). A spheroidal equal angular grid is a network of points (grid nodes) that are located at intersections of two mutually perpendicular sets of parallels and meridians with equal angular spacing along latitude and longitude (Fig. 3.15).

A spheroidal equal angular grid consists of a set of grid rows or columns.

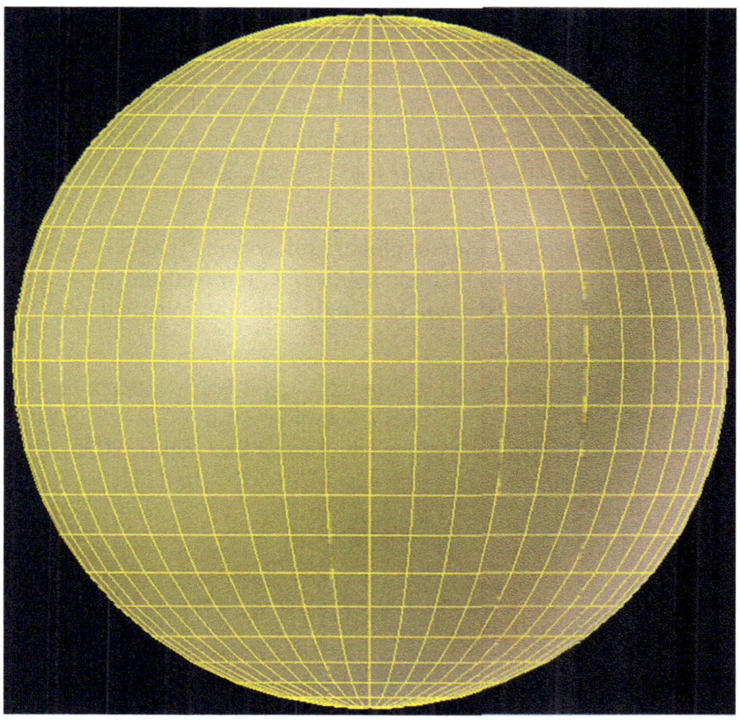

FIG. 3.15 Geometry of spheroidal equal angular grids for a sphere.

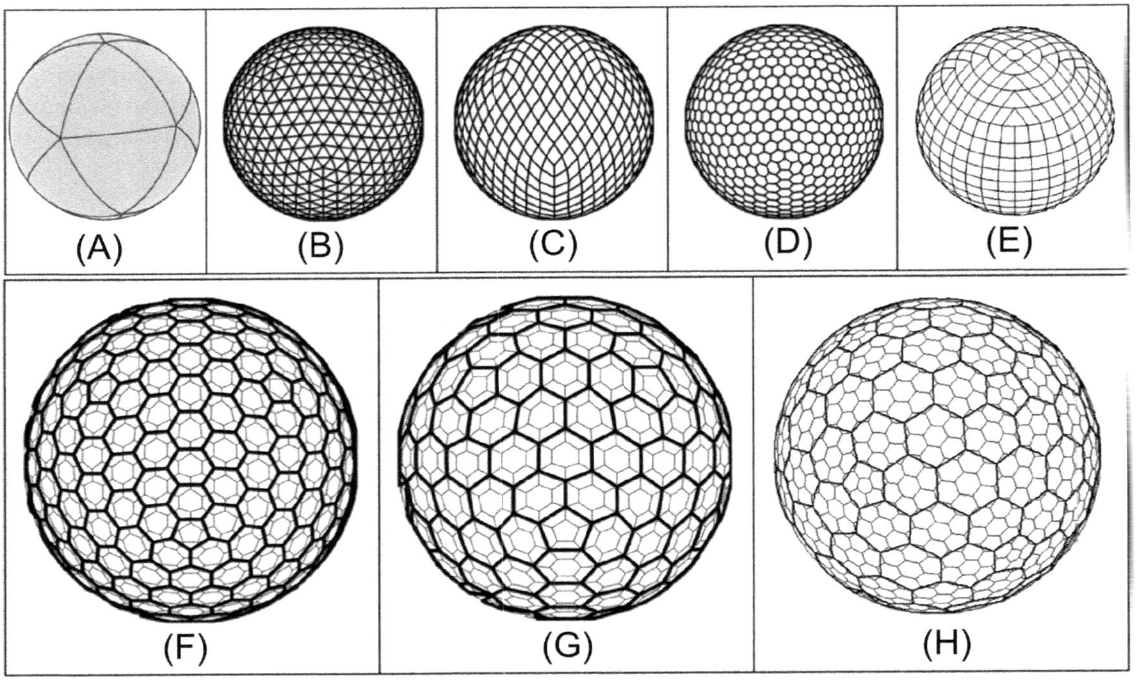

FIG. 3.16 Geometry of discrete global grid systems: (A) Icosahedron grids tessellated by spherical triangles (B), diamonds (C), hexagons (D), and squares (E). Hexagon grids of apertures 3 (F), 4 (G), and 7 (H). *(From Robertson, C., Chaudhuri, C., Hojati, M., Roberts, S.A., 2020. An integrated environmental analytics system (IDEAS) based on a DGGS. ISPRS J. Photogrammet. Remote Sens. 162, 214–228, Fig. 1A–H, © Authors, distributed under the terms and conditions of the Creative Commons Attribution (CC BY) license.)*

The spheroidal equal angular grid is the standard for global and quasi-global DEMs of the Earth (Gesch 2012) (Table 3.1), Mars (Smith et al., 2003), Venus (Ford et al., 1992; Sjogren et al., 1997), the Moon (Zuber et al., 1996; Neumann, 2008), Mercury (Becker et al., 2016), and many national DEMs (e.g., USGS, 1993; Natural Resources Canada, 1997), as well as models of various geophysical characteristics (e.g., Mooney et al., 1998).

Disadvantages of the spheroidal equal angular grid are: (1) the excess of information for poles: each pole is described not by one but by a series of points, whose number depends on the grid resolution and (2) trapezoid areas vary depending on the latitude, which may hamper further data analysis (Bjørke and Nilsen, 2004). Some discrete global grid systems are free of these disadvantages.

Definition 3.8 (Discrete global grid systems). Discrete global grid systems (DGGS) are hierarchical grids based on the tessellation of the sphere or ellipsoid by spherical Platonic solids (i.e., tetrahedron, cube, octahedron, dodecahedron, and icosahedron) and the subsequent subdivision of each face into cells by spherical triangles, diamonds, or hexagons (Fig. 3.16).

DGGS theory and algorithms are being developed within the Digital Earth concept framework (Dutton, 1999; White, 2000; Sahr et al., 2003; Zhao et al., 2008; Yuan et al., 2010; Mahdavi-Amiri et al., 2015; Robertson et al., 2020). In geomorphometry, the use of DGGS-based DEMs is still marginal (Li and Stefanakis, 2020; Li M. et al., 2022).

3.2.3 Dense point clouds

Definition 3.9 (Dense point cloud). A dense point cloud is a big set of irregularly distributed points with measured planimetric coordinates and elevations positioned in 3D space.

A dense point cloud (Buckley et al., 2008; Gomez, 2022) is a raw product of lidar scanning (Section 3.1.5) or SfM processing of visual-band images from unmanned aerial surveys (Section 3.1.4). In the case of lidar scanning, a dense point cloud is a set of measurements from returned laser pulses. In the case of SfM processing, a dense point cloud is a set of tie points determined by a matching algorithm.

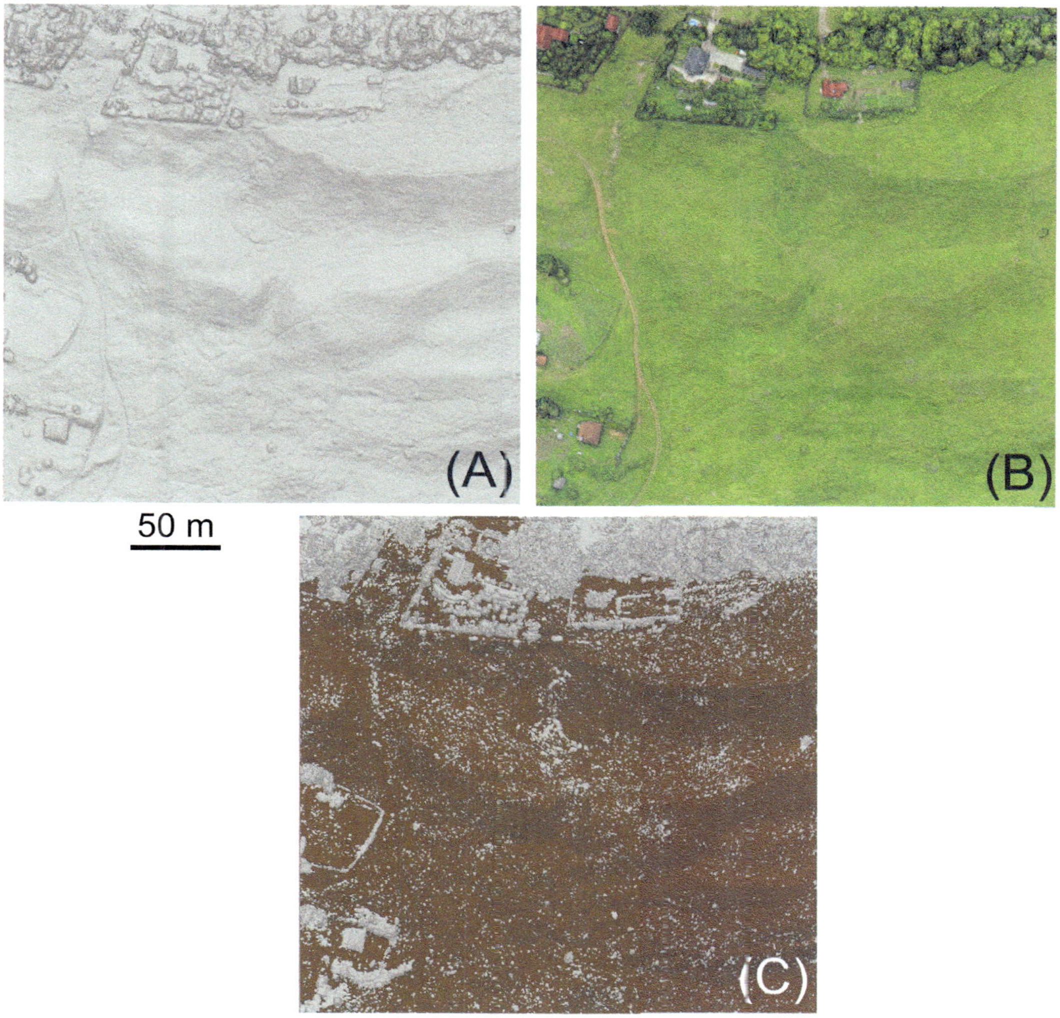

FIG. 3.17 Dense point cloud: (A) Initial point cloud visualized by hill shading. (B) Point cloud draped by a set of images. (C) Point cloud after ground filtering: *brown*, bare earth; *gray*, vegetation, and buildings. *(Modified from Kurkov, V., Bliakharskii, D., Florinsky, I., 2016. Application of unmanned aerial surveying for geomorphometric modelling. Izvestia Vuzov. Geodesy Aerophotosurveying 60 (6), 69–77 (in Russian), Fig. 4.)*

A typical dense point cloud includes several million measurements; that is, it is characterized by obvious information redundancy. However, this feature allows one to visualize dense point clouds by, for example, hill shading (Fig. 3.17A), and then to use dense point clouds for creating true-color hill-shaded images of terrains under survey (Fig. 3.17B). To drape a dense point cloud, one can apply either true-color imagery obtained during lidar scanning from a built-in or separately mounted digital camera, or digital aerial images collected by an unmanned aerial survey.

It is possible to consider dense point clouds as extensions of irregular plane grids (Section 3.2.1) in 3D space. After ground filtering (Sections 3.1.4 and 3.1.5) (Fig. 3.17C), dense point clouds are used to generate regular DEMs by interpolation (Section 3.4).

3.2.4 Datums and coordinate systems

Distinct datums and coordinate systems (Cooper, 1998) may be convenient to use at different steps of DEM generation, storage, and processing. Conversion of a DEM from one datum/coordinate system to another is a regular photogrammetric procedure (Tyuflin, 1996). To process a DEM together with other georeferenced data (e.g., remotely sensed images), all data should have a common datum and coordinate system (Franklin, 1989).

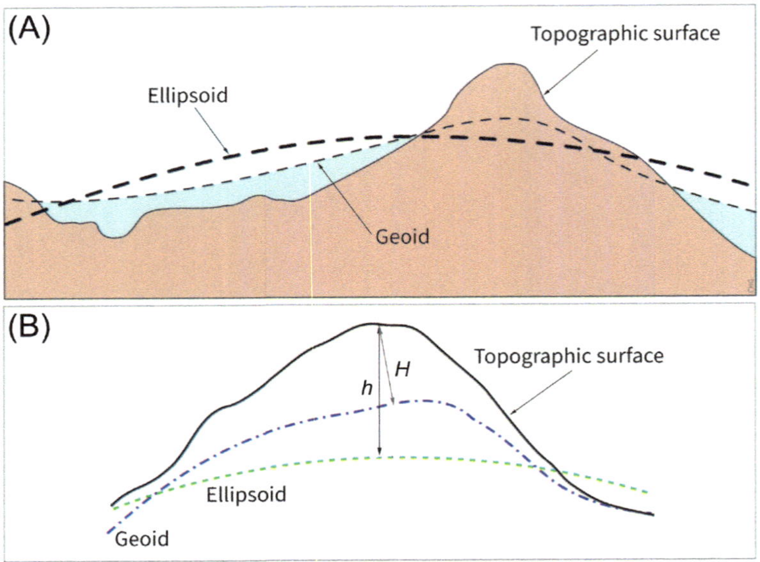

FIG. 3.18 A scheme for vertical datums (A) and elevation types (B): *H*, orthometric elevation; *h*, ellipsoidal elevation. *(Modified from Guth, P.L., Van Niekerk, A., Grohmann, C.H., Muller, J.-P., Hawker, L., Florinsky, I.V., et al., 2021. Digital elevation models: terminology and definitions. Remote Sens. (Basel) 13 (18), 3581, Fig. 2, © Authors, distributed under the terms and conditions of the Creative Commons Attribution (CC BY) license.)*

For global, quasi-global, and continental DEMs, a common horizontal datum is the World Geodetic System 1984 (WGS84, see NIMA, 1997). Defining the zero elevation, a vertical datum can be ellipsoidal or geoidal (mean sea level) (Fig. 3.18A). Geoidal datums can be global, such as the Earth Gravitational Model 2008 (EGM2008; see Pavlis et al. 2012), continental, or national. National datums may be more convenient to use for national and regional DEMs.

Depending on the datum, elevations in a DEM can be orthometric or ellipsoidal (Fig. 3.18B). Considering Definition 2.2 (Section 2.1.1), an orthometric elevation is a distance from a reference geoid or mean sea level to a given point on the topographic surface, while an ellipsoidal elevation is a distance from a reference ellipsoid to a given point on the topographic surface.

Square-gridded DEMs are usually represented in Universal Transverse Mercator rectangular coordinates. The geographical coordinate system is commonly used for spheroidal equal angular DEMs. Relative Cartesian coordinates may be convenient for DEMs of small areas.

3.3 DEM resolution

Definition 3.10 (Grid spacing). Grid spacing (w) is the distance between two neighboring grid nodes of a regular grid, which is measured along a grid row or column.

For a plane square grid, w is measured in metric units. For a spheroidal equal angular grid, w is measured in angular units (e.g., arc-seconds).

Definition 3.11 (DEM resolution). DEM resolution is the planimetric size of the smallest topographic feature detectable in a DEM.

Selection of the average density of points of an irregular DEM or the grid spacing of a regular DEM is one of the key problems of digital terrain modeling (Balce, 1987; Hengl, 2006). Influencing DEM resolution, these parameters should maintain a reasonable level of accuracy in order to describe the topographic surface using the minimum number of points.

There are several empirical approaches for selecting the w value. For example, according to the Peucker criterion, w value should be 4.3 times larger than the contour interval of a digitized map (Sasowsky et al., 1992). Ivanov and Kruzhkov (1992) presented a table to choose the w value depending on the type of topography and the root mean square error of a DEM. Let us examine this issue from a theoretical standpoint.

Theorem 3.1 (The sampling theorem, the univariate case). A one-dimensional continuous bandlimited function $z=f(x)$ with a bandwidth v can be completely determined by a set of samples $f(k\Delta x)$ if $\Delta x \leq 1/(2v)$, where Δx is the sampling interval, $-\infty \leq k \leq \infty$.

The proof can be found elsewhere (Kotelnikov, 1933, 2001; Shannon, 1949).

Theorem 3.2 (The sampling theorem, the bivariate case) A continuous bandlimited function, $z=f(x,y)$ with bandwidths v_x and v_y, can be determined by a set of samples $f(k\Delta x, l\Delta y)$ if $\Delta x \leq 1/(2v_x)$, $\Delta y \leq 1/(2v_y)$, where Δx and Δy are sampling intervals, $-\infty \leq k, l \leq \infty$.

In other words, it is possible to reconstruct a continuous function from a discrete one, if at least two samples per the shortest wavelength $\lambda_{x,y}$ were collected, $\lambda_{x,y}=1/v_{x,y}$. Practically, to avoid ambiguity in the reconstruction of a function due to limitations of the sampling theorem and interpolation effects (e.g., the Gibbs phenomenon—see Section 5.5), it is advisable to use a multiplicative factor $n=2, \ldots, 10$ to determine sampling intervals (Grigorenko, 1998): $\Delta x \leq \lambda_x/(2n)$, $\Delta y \leq \lambda_y/(2n)$.

DEM generation is a discretization of the topographic surface (Makarovič, 1973, 1977; Mark, 1975a; Stefanovic et al., 1977). The spectrum of the topographic surface is unlimited, so the condition of the sampling theorem is not met. This is a tractable problem because a researcher is usually interested in studying landforms whose typical planimetric sizes are not smaller than the threshold short wavelength $\tilde{\lambda}_{x,y}$ (Mark, 1975a). Therefore, in each given case, it is possible to consider that the topographic surface is a bandlimited function with bandwidths $\tilde{v}_{x,y}$.

For digital terrain analysis, the sampling theorem has three main consequences (Florinsky, 2002):

1. To keep the information on short-wavelength topographic features with typical planimetric sizes $\tilde{\lambda}_{x,y}$ in a DEM, one should use DEM sampling intervals $\Delta x \leq \tilde{\lambda}_x/(2n)$ and $\Delta y \leq \tilde{\lambda}_y/(2n)$, or the DEM grid spacing $w \leq \tilde{\lambda}_{x,y}/(2n)$ (Mark, 1975a). These sampling intervals or the grid spacing, set the resolution limit for all DTMs derived from the DEM.
2. DEM interpolation does not provide a way to increase the DEM resolution. In other words, it is impossible to reconstruct topographic features with typical planimetric sizes less than $\tilde{\lambda}_{x,y}$. Of course, the DEM resolution may be nominally "enhanced" by interpolation using $w \ll \Delta x$, Δy, but this procedure cannot improve the actual resolution of the DEM (Stefanovic et al., 1977; Horn, 1981; Sasowsky et al., 1992).
3. Interpolation-generated "landforms" with typical planimetric sizes less than $\tilde{\lambda}_{x,y}$ (Section 3.4) should be considered a high-frequency noise caused by the properties of an interpolator.

DEM errors caused by the ignoring of the sampling theorem are discussed in Section 5.4.

3.4 DEM interpolation

3.4.1 General

Definition 3.12 (Interpolation). Interpolation is an approach of finding intermediate values of a variable from a discrete set of its known values. For a bivariate function $z=f(x, y)$, interpolation is the process of using known values $f(x_1, y_1), f(x_1, y_2), \ldots, f(x_m, y_n)$ to find unknown values $f(x, y)$ for $x \neq x_i, y \neq y_j$, where $i=1, 2, \ldots, n; j=1, 2, \ldots, m$. A function $F(x, y)$, agreeing with $f(x, y)$ at the points $x=x_i$, $y=y_j$, is the interpolator used to compute the unknown values (Bakhvalov, 1977, chap. 2; Meijering, 2002).

Bivariate (or spatial) interpolation is usually applied to convert irregular DEMs into regular ones based on plane square grids or spheroidal equal angular girds. This procedure is, in particular, necessary because effective methods for the calculation of morphometric variables (Chapter 4) deal with DEMs based on regular grids. Interpolation is also used to refine regular DTMs.

There are dozens of interpolation methods, which can be generally divided into two main classes: global and local (Leberl, 1973; McCullagh, 1988; Hutchinson, 1989). The global methods fit a single function into the entire DEM. Such methods are closely related to methods of global approximation (Chapter 7). The local methods involve partitioning an irregular DEM into patches, fitting simple functions (e.g., splines) on each patch, and then linking the interpolated patches. Local methods are much more popular because they require less computer resources.

It is obvious that interpolation of the same irregular DEM by different methods results in distinct regular DEMs (Fig. 3.19). The selection of a particular method to interpolate a particular DEM depends on the technique of DEM generation, the type of DEM grid, the type of topography, and the purpose of interpolation. Interpolation inevitably introduces new errors and artifacts in interpolated DEMs.

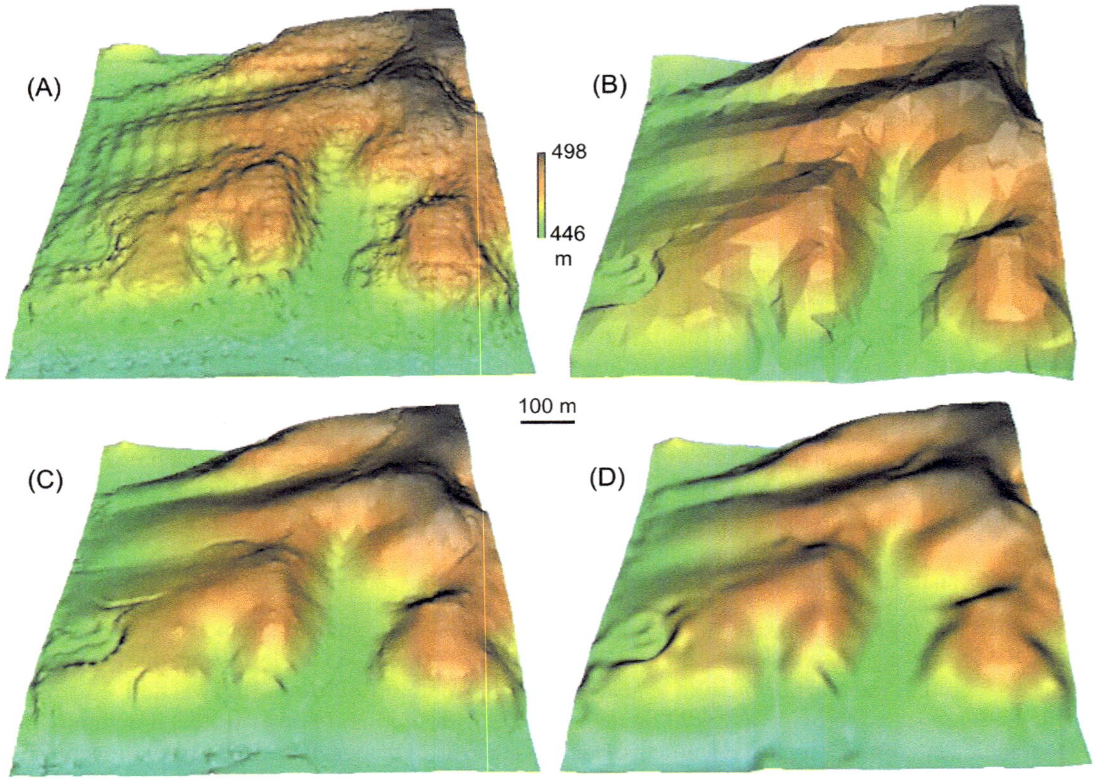

FIG. 3.19 Interpolation of an irregular DEM by several methods: (A) Inverse distance weighting. (B) Triangulation-based linear interpolation. (C). Spline with tension and stream enforcement. (D) Regularized spline with tension and smoothing. *(From Mitas, L., Mitasova, H., 2005. Spatial interpolation. In: Longley, P., Goodchild, M.F., Maguire, D.J., Rhind, D.W. (Eds.), Geographical Information Systems: Principles, Techniques, Management and Applications. second abr. ed., Wiley, Hoboken, NJ, pp. 481–492, Plate 26, reproduced with permission from John Wiley & Sons.)*

Review and comparison of interpolation methods can be found elsewhere for both DEMs based on plane grids (Leberl, 1973; Schut, 1976; Franke, 1982; Lam, 1983; McCullagh, 1988; Watson, 1992, § 2.5; Mitas and Mitasova, 2005; Li et al., 2005, chap. 6; Bagheri et al., 2014) and DEMs based on spheroidal grids (Wahba, 1981; Willmott et al., 1985; Robeson, 1997; Fasshauer and Schumaker, 1998).

3.4.2 Selected methods

Later, we briefly describe three popular methods for DEM interpolation.

1. *Inverse distance weighted (IDW) interpolation* (Watson, 1992, pp. 112, 114–116). An elevation value at a regular grid node is approximated by a weighted average of elevation values at points within a certain distance from the node. Weights are inversely proportional to a power of distance:

$$F(x,y) = \sum_{i=1}^{N} w_i f(x_i, y_i) = \frac{\sum_{i=1}^{N} f(x_i, y_i)/d_i^p}{\sum_{j=1}^{N} 1/d_j^p}, \tag{3.1}$$

   where w_i is the ith inverse distance weight; d_i is the distance from the interpolation node to the ith datum; p is the pth power of the distance (usually, $p=2$).

   Fig. 3.19A illustrates the application of the IDW method.

2. *Triangulation-based interpolation* (McCullagh, 1988; Agishtein and Migdal, 1991; Watson, 1992, pp. 130–137). This approach includes two steps. First, the Delaunay triangulation (Section 3.2.1) of irregularly distributed points is

performed. Second, for each triangle, a bivariate function is derived and then used to estimate elevation values at regular grid nodes. In linear interpolation, flat facets are fitted to each triangle In nonlinear interpolation, blending functions such as polynomials are applied to provide merging of triangles.

Fig. 3.19B illustrates the application of this approach.

3. *Thin-plate spline interpolation* (Mitas and Mitasova, 2005, p. 484). Consider the condition required to minimize the sum of the deviations from the irregular grid points and the smoothness seminorm in the spline function:

$$\sum_{j=1}^{N} \left| z_j - F\left(\mathrm{r}_j\right) \right|^2 w_j + w_0 I(F) = \text{minimum}, \tag{3.2}$$

where z_j are the values of elevation, $j=1, \ldots, N$, measured at discrete points $\mathrm{r}_j=(x_j^{[1]}, x_j^{[2]}, \ldots, x_j^{[d]})$ within a certain area of a d-dimensional space; $F(\mathrm{r})$ is a d-variate function passing through the given points; w_j and w_0 are weights; and $I(F)$ is the smoothness seminorm, a measure of roughness of the function in terms of its order of derivative relative to the function. Using a bivariate smoothness seminorm with squares of second derivatives, one obtains a thin-plate spline function. It minimizes the curvature of the topographic surface and imitates a steel sheet passing through the irregular grid points. Adding the first-order derivatives into $I(F)$, one obtains a thin-plate spline with tension (Franke, 1985) imitating an elastic membrane. A function with regular second- and higher-order derivatives can be obtained by adding higher-order derivatives into $I(F)$: the regularized spline with tension includes the sum of all derivatives with rapidly decreasing weights (Mitášová and Mitáš, 1993; Mitášová and Hofierka, 1993).
Fig. 3.19D illustrates the application of this approach.

Chapter 4

Calculation methods

4.1 The Evans-Young method

Local morphometric variables are functions of the partial derivatives of elevation (Section 2.2, Table 2.2). There is a classic technique for computing r, t, s, p, and q (Eq. 2.2) from DEMs on the plane square grid—the Evans-Young method (Young, 1978; Evans, 1979)—based on approximation of the partial derivatives by finite differences (Samarskii, 2001, § 2.2) using a 3×3 plane square-gridded moving window. The Evans-Young method is the most accurate of the related finite-difference methods (Florinsky, 1998a; Schmidt J. et al., 2003).

In the Evans-Young method, a function $z = f(x, y)$ is expressed as the second-order bivariate Taylor polynomial

$$z = \frac{rx^2}{2} + \frac{ty^2}{2} + sxy + px + qy + u, \tag{4.1}$$

where u is a residual term. The polynomial (Eq. 4.1) is fitted by least squares to the nine points of the 3×3 square-spaced window with a grid spacing of w (Fig. 4.1A). The Cartesian coordinates and elevations of the topographic surface are known for the window points $(-w, w, z_1)$, $(0, w, z_2)$, (w, w, z_3), $(-w, 0, z_4)$, $(0, 0, z_5)$, $(w, 0, z_6)$, $(-w, -w, z_7)$, $(0, -w, z_8)$, and $(w, -w, z_9)$. For the point $(0, 0, z_5)$, the polynomial coefficients, which are the partial derivatives of elevation, are estimated by the following formulas:

$$r = \frac{z_1 + z_3 + z_4 + z_6 + z_7 + z_9 - 2(z_2 + z_5 + z_8)}{3w^2}, \tag{4.2}$$

$$t = \frac{z_1 + z_2 + z_3 + z_7 + z_8 + z_9 - 2(z_4 + z_5 + z_6)}{3w^2}, \tag{4.3}$$

$$s = \frac{z_3 + z_7 - z_1 - z_9}{4w^2}, \tag{4.4}$$

$$p = \frac{z_3 + z_6 + z_9 - z_1 - z_4 - z_7}{6w}, \tag{4.5}$$

$$q = \frac{z_1 + z_2 + z_3 - z_7 - z_8 - z_9}{6w}. \tag{4.6}$$

Moving the 3×3 window along a DEM, one can calculate values of r, t, s, p, and q (and so values of local morphometric variables) for all points of the plane square-gridded DEM, except for boundary rows and columns.

The polynomial (Eq. 4.1) is *approximated* to elevation values at the window points rather than passing exactly through them. This leads to some smoothing of the elevation function within the window, that is, to the local filtering of high-frequency noise inherent for any DEM. Such filtering allows one to maintain a correct estimation of derivatives, which are sensitive to the high-frequency component of a signal (Section 5.4.1). A further merit of the Evans-Young method is the use of excess measurements: nine elevation values are utilized to estimate six coefficients of the polynomial (Eq. 4.1), which increases estimation accuracy.

Fig. 4.2 illustrates the derivation of local topographic variables by the Evans-Young method. An initial spheroidal equal angular DEM of the Molnieboi Spur (Altai Mountains, Russia) was extracted from the ASTER GDEM (Japan Space Systems, 2009–2019). The extracted DEM included 19,880 points (the elevation matrix 142×140). The grid spacing was $1'$, that is, approximately 30.90 m × 19.85 m for the latitude of the spur. Then, applying the Delaunay triangulation and a piecewise quadric polynomial interpolation (Agishtein and Migdal, 1991), we produced a square-gridded DEM (Fig. 4.2A) including 149,136 points (the elevation matrix was 312×478) with the grid spacing of 9 m. To suppress the high-frequency noise, the DEM was smoothed three times with the 3×3 moving window. Some local topographic

Digital Terrain Analysis. https://doi.org/10.1016/B978-0-443-24798-9.00023-6

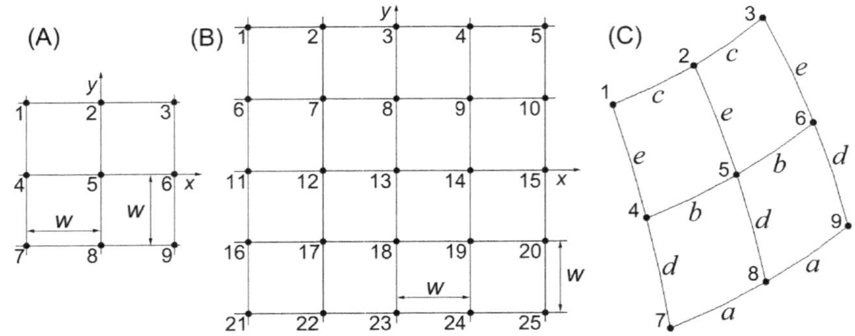

FIG. 4.1 Main types of moving windows used to calculate local morphometric variables: (A) A 3 × 3 plane square-gridded window; *w*, grid spacing (meters); 1, …, 9, numbers of the window nodes. (B) A 5 × 5 plane square-gridded window; *w*, grid spacing (meters); 1, …, 25, numbers of the window nodes. (C) A 3 × 3 spheroidal equal angular window for an ellipsoid of revolution or a sphere; *a*, *b*, *c*, *d*, *e*, linear sizes (meters) of moving window elements; 1, …, 9, numbers of the window nodes. *(Modified from Florinsky, I.V., 2010. Theory and Applications of Mathematical Cartographic Modeling of Topography (D.Sc. thesis). Institute of Mathematical Problems of Biology, Russian Academy of Sciences, Pushchino, Russia (in Russian), Fig. 2.1.)*

variables were finally calculated by the Evans-Young method (Fig. 4.2B–F). To deal with the large dynamic range of k_h, k_v, k_{min}, and k_{max}, we logarithmically transformed their values by Eq. (8.1) with $n=3$. Data were processed by the LandLord software (Florinsky, 2016).

4.2 Calculation of local morphometric variables on a plane square grid

4.2.1 Motivation

There were proposals to use some morphometric variables, which are functions of not only the first- and second-order partial derivatives of elevation but also third-order ones (Lastochkin, 1987; Shary and Stepanov, 1991; Jenčo, 1992;

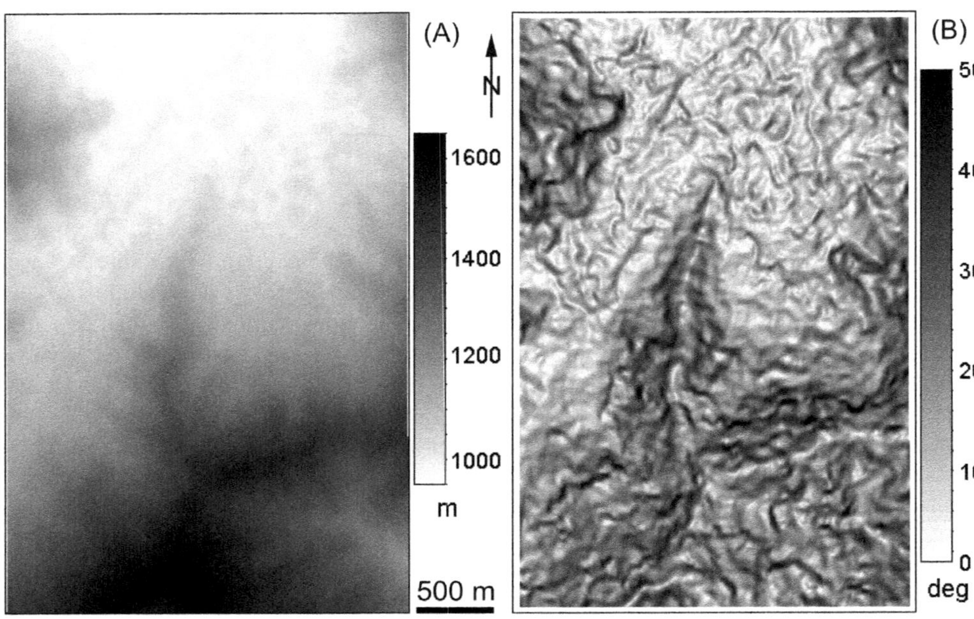

FIG. 4.2 The Molnieboi Spur, Altai (50°10′30″ N, 85°42′30″ E): (A) Elevation. (B) Slope gradient.

(Continued)

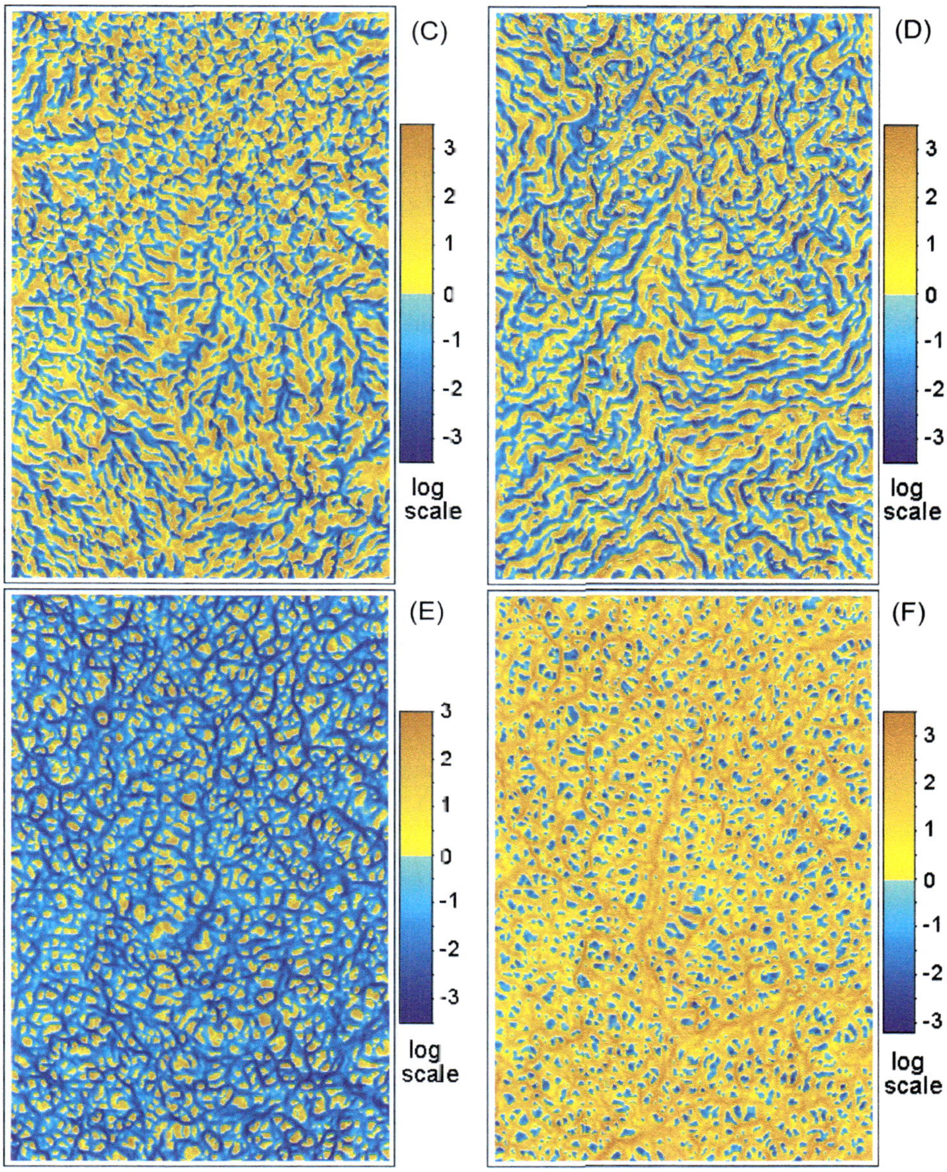

FIG. 4.2, CONT'D (C) Horizontal curvature. (D) Vertical curvature. (E) Minimal curvature. (F) Maximal curvature. Calculations were conducted on the square grid with the 3 × 3 moving window (see details in the text). *(Modified from Florinsky, I.V., Selezneva, E.V., Kulikova, A.I., 2015. GIS-based support for the complex botanical studies at the Molnieboi Spur, Altai. arXiv Preprint 1512.05143, Figs. 2E and 3.)*

Minár and Evans, 2008; Pacina, 2010; Minár et al., 2013) (Sections 2.2.5 and 4.5.2). It is obvious that the Evans-Young method and other techniques using the second-order bivariate polynomials (Section 4.1) are not suitable for estimating the third-order partial derivatives.

Next, we describe the method that can be used to derive local topographic attributes from DEMs based on plane square grids (Florinsky, 2009a, b). Like the Evans-Young algorithm, this method is simple and fast, but it has two advantages: First, it allows one to estimate the first- (p and q), second- (r, t, and s), and third-order (g, h, k, and m) partial derivatives of elevation (Eq. 2.2). Second, it has an increased accuracy and a better suppression of high-frequency noise.

4.2.2 Formulas

Using the well-known Taylor formula (Fichtenholz, 1966a, pp. 415–416; Nikolsky, 1977a, § 7.13), a function $z=f(x,y)$ can be conveniently expressed as the third-order bivariate Taylor polynomial:

$$z = \frac{1}{6}gx^3 + \frac{1}{6}hy^3 + \frac{1}{2}kx^2y + \frac{1}{2}mxy^2 + \frac{1}{2}rx^2 + \frac{1}{2}ty^2 + sxy + px + qy + u. \tag{4.7}$$

In the least-squares method (Bjerhammar, 1973), one should carry out n measurements of z to find N unknown coefficients of $z=f(x,y)$ with $n>(N+1)$. Therefore, one cannot use a 3×3 window (Fig. 4.1A) to find the 10 unknown coefficients g, h, k, m, r, t, s, p, q, and u of the third-order polynomial (Eq. 4.7) because such a window includes only nine measured values of z. To find these coefficients for the central point of a window, there is a need to use at least a 5×5 square-gridded window (Fig. 4.1B). The Cartesian coordinates and elevations of the topographic surface are known for 25 points of this window: $(-2w, 2w, z_1)$, $(-w, 2w, z_2)$, $(0, 2w, z_3)$, $(w, 2w, z_4)$, $(2w, 2w, z_5)$, $(-2w, w, z_6)$, $(-w, w, z_7)$, $(0, w, z_8)$, (w, w, z_9), $(2w, w, z_{10})$, $(-2w, 0, z_{11})$, $(-w, 0, z_{12})$, $(0, 0, z_{13})$, $(w, 0, z_{14})$, $(2w, 0, z_{15})$, $(-2w, -w, z_{16})$, $(-w, -w, z_{17})$, $(0, -w, z_{18})$, $(w, -w, z_{19})$, $(2w, -w, z_{20})$, $(-2w, -2w, z_{21})$, $(-w, -2w, z_{22})$, $(0, -2w, z_{23})$, $(w, -2w, z_{24})$, and $(2w, -2w, z_{25})$.

Let us fit the polynomial (Eq. 4.7) to the 25 points of the 5×5 window by the least squares (Bjerhammar, 1973). Writing the polynomial (Eq. 4.7) for all points of the window, we obtain a system of 25 conditional linear equations, which can be expressed as follows:

$$\boldsymbol{\alpha} = \mathbf{F}\boldsymbol{\beta}, \tag{4.8}$$

where $\boldsymbol{\alpha}$ is a 25×1 matrix of the 25 measured values of z:

$$\boldsymbol{\alpha} = \begin{pmatrix} z_1 \\ \vdots \\ z_{25} \end{pmatrix}, \tag{4.9}$$

$\boldsymbol{\beta}$ is a 10×1 matrix of the 10 unknown coefficients of the polynomial (Eq. 4.7):

$$\boldsymbol{\beta} = \begin{pmatrix} g \\ h \\ k \\ m \\ r \\ t \\ s \\ p \\ q \\ u \end{pmatrix}, \tag{4.10}$$

and $\mathbf{F}$ is a 25×10 matrix of the known coefficients of the equation system (Eq. 4.8):

$$
\mathbf{F}=\begin{pmatrix}
-\frac{4}{3}w^3 & \frac{4}{3}w^3 & 4w^3 & -4w^3 & 2w^2 & 2w^2 & -4w^2 & -2w & 2w & 1\\
-\frac{1}{6}w^3 & \frac{4}{3}w^3 & w^3 & -2w^3 & \frac{1}{2}w^2 & 2w^2 & -2w^2 & -w & 2w & 1\\
0 & \frac{4}{3}w^3 & 0 & 0 & 0 & 2w^2 & 0 & 0 & 2w & 1\\
\frac{1}{6}w^3 & \frac{4}{3}w^3 & w^3 & 2w^3 & \frac{1}{2}w^2 & 2w^2 & 2w^2 & w & 2w & 1\\
\frac{4}{3}w^3 & \frac{4}{3}w^3 & 4w^3 & 4w^3 & 2w^2 & 2w^2 & 4w^2 & 2w & 2w & 1\\
-\frac{4}{3}w^3 & \frac{1}{6}w^3 & 2w^3 & -w^3 & 2w^2 & \frac{1}{2}w^2 & -2w^2 & -2w & w & 1\\
-\frac{1}{6}w^3 & \frac{1}{6}w^3 & \frac{1}{2}w^3 & -\frac{1}{2}w^3 & \frac{1}{2}w^2 & \frac{1}{2}w^2 & -w^2 & -w & w & 1\\
0 & \frac{1}{6}w^3 & 0 & 0 & 0 & \frac{1}{2}w^2 & 0 & 0 & w & 1\\
\frac{1}{6}w^3 & \frac{1}{6}w^3 & \frac{1}{2}w^3 & \frac{1}{2}w^3 & \frac{1}{2}w^2 & \frac{1}{2}w^2 & w^2 & w & w & 1\\
\frac{4}{3}w^3 & \frac{1}{6}w^3 & 2w^3 & w^3 & 2w^2 & \frac{1}{2}w^2 & 2w^2 & 2w & w & 1\\
-\frac{4}{3}w^3 & 0 & 0 & 0 & 2w^2 & 0 & 0 & -2w & 0 & 1\\
-\frac{1}{6}w^3 & 0 & 0 & 0 & \frac{1}{2}w^2 & 0 & 0 & -w & 0 & 1\\
0 & 0 & 0 & 0 & 0 & 0 & 0 & 0 & 0 & 1\\
\frac{1}{6}w^3 & 0 & 0 & 0 & \frac{1}{2}w^2 & 0 & 0 & w & 0 & 1\\
\frac{4}{3}w^3 & 0 & 0 & 0 & 2w^2 & 0 & 0 & 2w & 0 & 1\\
-\frac{4}{3}w^3 & -\frac{1}{6}w^3 & -2w^3 & -w^3 & 2w^2 & \frac{1}{2}w^2 & 2w^2 & -2w & -w & 1\\
-\frac{1}{6}w^3 & -\frac{1}{6}w^3 & -\frac{1}{2}w^3 & -\frac{1}{2}w^3 & \frac{1}{2}w^2 & \frac{1}{2}w^2 & w^2 & -w & -w & 1\\
0 & -\frac{1}{6}w^3 & 0 & 0 & 0 & \frac{1}{2}w^2 & 0 & 0 & -w & 1\\
\frac{1}{6}w^3 & -\frac{1}{6}w^3 & -\frac{1}{2}w^3 & \frac{1}{2}w^3 & \frac{1}{2}w^2 & \frac{1}{2}w^2 & -w^2 & w & -w & 1\\
\frac{4}{3}w^3 & -\frac{1}{6}w^3 & -2w^3 & w^3 & 2w^2 & \frac{1}{2}w^2 & -2w^2 & 2w & -w & 1\\
-\frac{4}{3}w^3 & -\frac{4}{3}w^3 & -4w^3 & -4w^3 & 2w^2 & 2w^2 & 4w^2 & -2w & -2w & 1\\
-\frac{1}{6}w^3 & -\frac{4}{3}w^3 & -w^3 & -2w^3 & \frac{1}{2}w^2 & 2w^2 & 2w^2 & -w & -2w & 1\\
0 & -\frac{4}{3}w^3 & 0 & 0 & 0 & 2w^2 & 0 & 0 & -2w & 1\\
\frac{1}{6}w^3 & -\frac{4}{3}w^3 & -w^3 & 2w^3 & \frac{1}{2}w^2 & 2w^2 & -2w^2 & w & -2w & 1\\
\frac{4}{3}w^3 & -\frac{4}{3}w^3 & -4w^3 & 4w^3 & 2w^2 & 2w^2 & -4w^2 & 2w & -2w & 1
\end{pmatrix}. \tag{4.11}
$$

To find the unknown coefficients of the polynomial (Eq. 4.7), it is necessary to solve the equation

$$\boldsymbol{\beta} = \left(\mathbf{F}^{\mathrm{T}}\mathbf{F}\right)^{-1}\mathbf{F}^{\mathrm{T}}\boldsymbol{\alpha}, \tag{4.12}$$

where ^{T} denotes the matrix transpose, $^{-1}$ denotes the matrix inversion. Using matrix operations, we find a 10×25 matrix $(\mathbf{F}^{\mathrm{T}}\mathbf{F})^{-1}\mathbf{F}^{\mathrm{T}}$:

$$\left(\mathbf{F}^{\mathrm{T}}\mathbf{F}\right)^{-1}\mathbf{F}^{\mathrm{T}} = \left\| \begin{array}{cccccccc}
-\frac{1}{10w^3} & \frac{1}{5w^3} & 0 & -\frac{1}{5w^3} & \frac{1}{10w^3} & -\frac{1}{10w^3} & \frac{1}{5w^3} & 0 \\
\frac{1}{10w^3} & \frac{1}{10w^3} & \frac{1}{10w^3} & \frac{1}{10w^3} & \frac{1}{10w^3} & -\frac{1}{5w^3} & -\frac{1}{5w^3} & -\frac{1}{5w^3} \\
\frac{2}{35w^3} & -\frac{1}{35w^3} & -\frac{2}{35w^3} & -\frac{1}{35w^3} & \frac{2}{35w^3} & \frac{1}{35w^3} & -\frac{1}{70w^3} & -\frac{1}{35w^3} \\
-\frac{2}{35w^3} & -\frac{1}{35w^3} & 0 & \frac{1}{35w^3} & \frac{2}{35w^3} & \frac{1}{35w^3} & \frac{1}{70w^3} & 0 \\
\frac{2}{35w^2} & -\frac{1}{35w^2} & -\frac{2}{35w^2} & -\frac{1}{35w^2} & \frac{2}{35w^2} & \frac{2}{35w^2} & -\frac{1}{35w^2} & -\frac{2}{35w^2} \\
\frac{2}{35w^2} & \frac{2}{35w^2} & \frac{2}{35w^2} & \frac{2}{35w^2} & \frac{2}{35w^2} & -\frac{1}{35w^2} & -\frac{1}{35w^2} & -\frac{1}{35w^2} \\
-\frac{1}{25w^2} & -\frac{1}{50w^2} & 0 & \frac{1}{50w^2} & \frac{1}{25w^2} & -\frac{1}{50w^2} & -\frac{1}{100w^2} & 0 \\
\frac{31}{420w} & -\frac{11}{105w} & 0 & \frac{11}{105w} & -\frac{31}{420w} & -\frac{1}{84w} & -\frac{31}{210w} & 0 \\
-\frac{31}{420w} & \frac{1}{84w} & \frac{17}{420w} & \frac{1}{84w} & -\frac{31}{420w} & \frac{11}{105w} & \frac{31}{210w} & \frac{17}{105w} \\
-\frac{13}{175} & \frac{2}{175} & \frac{1}{25} & \frac{2}{175} & -\frac{13}{175} & \frac{2}{175} & \frac{17}{175} & \frac{22}{175}
\end{array} \right.$$

$$\begin{array}{cccccccccc}
-\frac{1}{5w^3} & \frac{1}{10w^3} & -\frac{1}{10w^3} & \frac{1}{5w^3} & 0 & -\frac{1}{5w^3} & \frac{1}{10w^3} & -\frac{1}{10w^3} & \frac{1}{5w^3} & 0 \\
-\frac{1}{5w^3} & -\frac{1}{5w^3} & 0 & 0 & 0 & 0 & 0 & \frac{1}{5w^3} & \frac{1}{5w^3} & \frac{1}{5w^3} \\
-\frac{1}{70w^3} & \frac{1}{35w^3} & 0 & 0 & 0 & 0 & 0 & -\frac{1}{35w^3} & \frac{1}{70w^3} & \frac{1}{35w^3} \\
-\frac{1}{70w^3} & -\frac{1}{35w^3} & \frac{2}{35w^3} & \frac{1}{35w^3} & 0 & -\frac{1}{35w^3} & -\frac{2}{35w^3} & \frac{1}{35w^3} & \frac{1}{70w^3} & 0 \\
-\frac{1}{35w^2} & \frac{2}{35w^2} & \frac{2}{35w^2} & -\frac{1}{35w^2} & -\frac{2}{35w^2} & -\frac{1}{35w^2} & \frac{2}{35w^2} & \frac{2}{35w^2} & -\frac{1}{35w^2} & -\frac{2}{35w^2} \\
-\frac{1}{35w^2} & -\frac{1}{35w^2} & -\frac{2}{35w^2} & -\frac{2}{35w^2} & -\frac{2}{35w^2} & -\frac{2}{35w^2} & -\frac{2}{35w^2} & -\frac{1}{35w^2} & -\frac{1}{35w^2} & -\frac{1}{35w^2} \\
\frac{1}{100w^2} & \frac{1}{50w^2} & 0 & 0 & 0 & 0 & 0 & \frac{1}{50w^2} & \frac{1}{100w^2} & 0 \\
\frac{31}{210w} & \frac{1}{84w} & -\frac{17}{420w} & -\frac{17}{105w} & 0 & \frac{17}{105w} & \frac{17}{420w} & -\frac{1}{84w} & -\frac{31}{210w} & 0 \\
\frac{31}{210w} & \frac{11}{105w} & 0 & 0 & 0 & 0 & 0 & -\frac{11}{105w} & -\frac{31}{210w} & -\frac{17}{105w} \\
\frac{17}{175} & \frac{2}{175} & \frac{1}{25} & \frac{22}{175} & \frac{27}{175} & \frac{22}{175} & \frac{1}{25} & \frac{2}{175} & \frac{17}{175} & \frac{22}{175}
\end{array}$$

$$
\left.\begin{array}{ccccccc}
-\frac{1}{5w^3} & \frac{1}{10w^3} & -\frac{1}{10w^3} & \frac{1}{5w^3} & 0 & -\frac{1}{5w^3} & \frac{1}{10w^3} \\
\frac{1}{5w^3} & \frac{1}{5w^3} & -\frac{1}{10w^3} & -\frac{1}{10w^3} & -\frac{1}{10w^3} & -\frac{1}{10w^3} & -\frac{1}{10w^3} \\
\frac{1}{70w^3} & -\frac{1}{35w^3} & -\frac{2}{35w^3} & \frac{1}{35w^3} & \frac{2}{35w^3} & \frac{1}{35w^3} & -\frac{2}{35w^3} \\
-\frac{1}{70w^3} & -\frac{1}{35w^3} & -\frac{2}{35w^3} & -\frac{1}{35w^3} & 0 & \frac{1}{35w^3} & \frac{2}{35w^3} \\
-\frac{1}{35w^2} & \frac{2}{35w^2} & \frac{2}{35w^2} & -\frac{1}{35w^2} & -\frac{2}{35w^2} & -\frac{1}{35w^2} & \frac{2}{35w^2} \\
-\frac{1}{35w^2} & -\frac{1}{35w^2} & \frac{2}{35w^2} & \frac{2}{35w^2} & \frac{2}{35w^2} & \frac{2}{35w^2} & \frac{2}{35w^2} \\
-\frac{1}{100w^2} & -\frac{1}{50w^2} & \frac{1}{25w^2} & \frac{1}{50w^2} & 0 & -\frac{1}{50w^2} & -\frac{1}{25w^2} \\
\frac{31}{210w} & \frac{1}{84w} & \frac{31}{420w} & -\frac{11}{105w} & 0 & \frac{11}{105w} & -\frac{31}{420w} \\
-\frac{31}{210w} & -\frac{11}{105w} & \frac{31}{420w} & -\frac{1}{84w} & -\frac{17}{420w} & -\frac{1}{84w} & \frac{31}{420w} \\
\frac{17}{175} & \frac{2}{175} & -\frac{13}{175} & \frac{2}{175} & \frac{1}{25} & \frac{2}{175} & -\frac{13}{175}
\end{array}\right\|. \tag{4.13}
$$

Solving Eq. (4.12), we obtain the formulas for the partial derivatives of elevation:

$$
\begin{aligned}
g = \frac{1}{10w^3} & [z_5 + z_{10} + z_{15} + z_{20} + z_{25} - z_1 - z_6 - z_{11} - z_{16} - z_{21} \\
& + 2(z_2 + z_7 + z_{12} + z_{17} + z_{22} - z_4 - z_9 - z_{14} - z_{19} - z_{24})],
\end{aligned} \tag{4.14}
$$

$$
\begin{aligned}
h = \frac{1}{10w^3} & [z_1 + z_2 + z_3 + z_4 + z_5 - z_{21} - z_{22} - z_{23} - z_{24} - z_{25} \\
& + 2(z_{16} + z_{17} + z_{18} + z_{19} + z_{20} - z_6 - z_7 - z_8 - z_9 - z_{10})],
\end{aligned} \tag{4.15}
$$

$$
\begin{aligned}
k = \frac{1}{70w^3} & [z_{17} + z_{19} - z_7 - z_9 + 4(z_1 + z_5 + z_{23} - z_3 - z_{21} - z_{25}) \\
& + 2(z_6 + z_{10} + z_{18} + z_{22} + z_{24} - z_2 - z_4 - z_8 - z_{16} - z_{20})],
\end{aligned} \tag{4.16}
$$

$$
\begin{aligned}
m = \frac{1}{70w^3} & [z_7 + z_{17} - z_9 - z_{19} + 4(z_5 + z_{11} + z_{25} - z_1 - z_{15} - z_{21}) \\
& + 2(z_4 + z_6 + z_{12} + z_{16} + z_{24} - z_2 - z_{10} - z_{14} - z_{20} - z_{22})],
\end{aligned} \tag{4.17}
$$

$$
\begin{aligned}
r = \frac{1}{35w^2} & [2(z_1 + z_5 + z_6 + z_{10} + z_{11} + z_{15} + z_{16} + z_{20} + z_{21} + z_{25}) \\
& -2(z_3 + z_8 + z_{13} + z_{18} + z_{23}) - z_2 - z_4 - z_7 - z_9 - z_{12} - z_{14} - z_{17} - z_{19} - z_{22} - z_{24}],
\end{aligned} \tag{4.18}
$$

$$
\begin{aligned}
t = \frac{1}{35w^2} & [2(z_1 + z_2 + z_3 + z_4 + z_5 + z_{21} + z_{22} + z_{23} + z_{24} + z_{25}) \\
& -2(z_{11} + z_{12} + z_{13} + z_{14} + z_{15}) - z_6 - z_7 - z_8 - z_9 - z_{10} - z_{16} - z_{17} - z_{18} - z_{19} - z_{20}],
\end{aligned} \tag{4.19}
$$

$$
\begin{aligned}
s = \frac{1}{100w^2} & [z_9 + z_{17} - z_7 - z_{19} + 4(z_5 + z_{21} - z_1 - z_{25}) + \\
& + 2(z_4 + z_{10} + z_{16} + z_{22} - z_2 - z_6 - z_{20} - z_{24})],
\end{aligned} \tag{4.20}
$$

$$p = \frac{1}{420w}\left\{44(z_4 + z_{24} - z_2 - z_{22}) + 31\left[z_1 + z_{21} - z_5 - z_{25} + 2(z_9 + z_{19} - z_7 - z_{17})\right] + 17[z_{15} - z_{11} + 4(z_{14} - z_{12})] + 5(z_{10} + z_{20} - z_6 - z_{16})\right\}, \quad (4.21)$$

$$q = \frac{1}{420w}\left\{44(z_6 + z_{10} - z_{16} - z_{20}) + 31\left[z_{21} + z_{25} - z_1 - z_5 + 2(z_7 + z_9 - z_{17} - z_{19})\right] + 17[z_3 - z_{23} + 4(z_8 - z_{18})] + 5(z_2 + z_4 - z_{22} - z_{24})\right\}. \quad (4.22)$$

Moving the 5×5 window along a DEM, one can compute values of g, h, k, m, r, t, s, p, and q (and so values of local morphometric variables) for all points of the DEM, except for two boundary rows and two boundary columns on each side of the plane square-gridded DEM.

We do not present a formula for the residual term u of the polynomial (Eq. 4.7) because it is not used in calculations of topographic variables. Eqs. (4.14)–(4.22) were developed with the Maple software (© Waterloo Maple Inc., 1981–1997).

4.2.3 Method validation

4.2.3.1 *Materials and data processing*

To illustrate the application of the described method, we used a portion of a DEM of the Stavropol Upland (North Caucasus, Russia) (Fig. 4.3). An irregular DEM (Florinsky, 2002) was produced by digitizing a topographic map (Central Board of Geodesy and Cartography, 1968). The area selected measures about 133×100 km. The irregular DEM included 2571 points.

Using the Delaunay triangulation and a piecewise quadric polynomial interpolation with matching derivatives along triangle edges (Agishtein and Migdal, 1991), we produced a square-gridded DEM with a grid spacing of 300 m. To reduce high-frequency noise in the DEM, it was smoothed three times using the 3×3 window with inverse distance weights. Applying the described method to the smoothed DEM, we derived digital models of G, A, k_h, k_v, H, K, k_{min}, k_{max}, K_a, E, K_r, k_{ve}, k_{he}, and M (Figs. 4.4 and 4.5); the grid spacing was 300 m.

We analyzed the accuracy of the described method using the criterion of the root mean square error (RMSE) of a function of measured variables (Florinsky, 2009b). It was proved that the method offers a higher accuracy for the calculation of local morphometric attributes than does the Evans-Young method, the most accurate similar technique. For details, see Section 5.3.3.1.

To compare accuracies of the Evans-Young and author's methods, we carried out the following procedures:

- For both methods, we derived digital models of RMSE of calculation of G, k_h, k_v, H, K, k_{min}, k_{max}, K_a, E, K_r, k_{ve}, k_{he}, and M. RMSE formulas (Florinsky, 1998a, 2008e, 2009b) can be found in Section 5.3.2.
- Applying the Evans-Young method to the smoothed DEM, we derived digital models of G, k_h, k_v, H, K, k_{min}, k_{max}, K_a, E, K_r, k_{ve}, k_{he}, and M.
- For each morphometric variable, we estimated a DTM difference, that is, a difference between two digital models derived by the Evans-Young and author's methods.

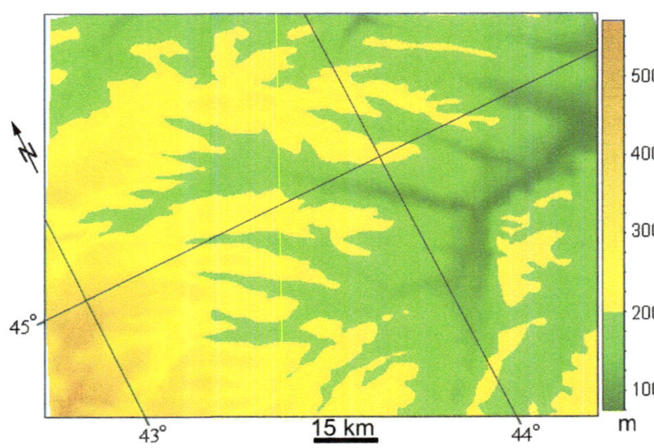

FIG. 4.3 The Stavropol Upland: elevation. *(Modified from Florinsky, I.V., 2009a. Accurate method for derivation of local topographic variables. Geodesy Cartogr. (4), 19–23 (in Russian), Fig. 1.)*

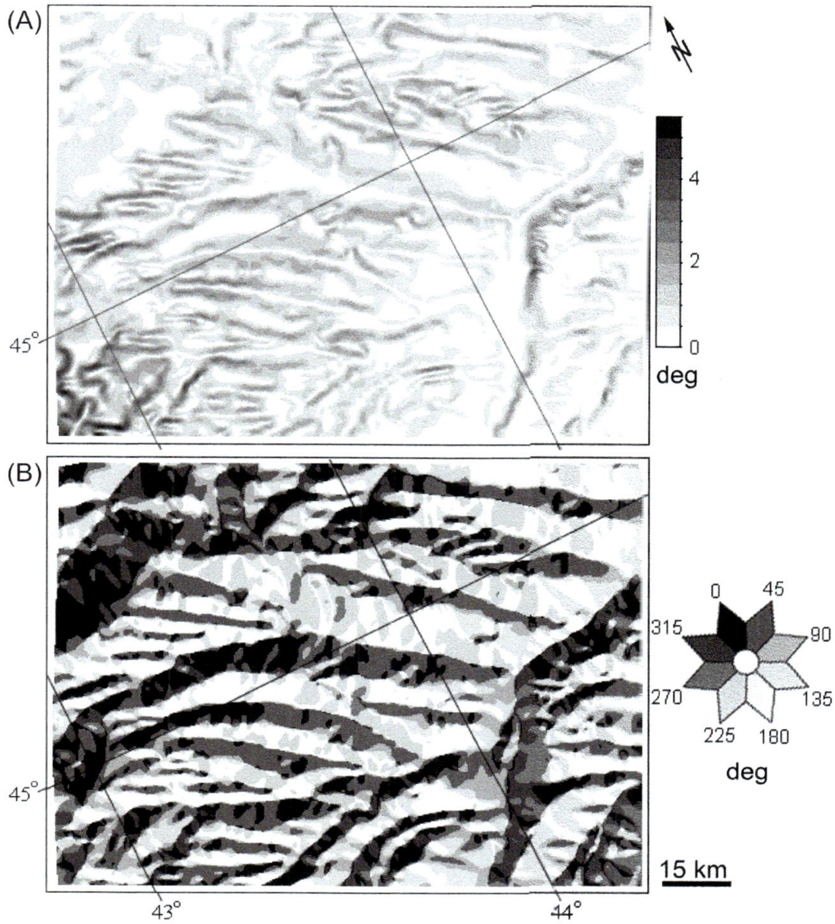

FIG. 4.4 The Stavropol Upland: (A) Slope gradient. (B) Slope aspect. Calculations were conducted on the square grid with the 5 × 5 moving window (see details in the text). For the elevation map, see Fig. 4.3. *(Modified from Florinsky, I.V., 2010. Theory and Applications of Mathematical Cartographic Modeling of Topography (D.Sc. thesis). Institute of Mathematical Problems of Biology, Russian Academy of Sciences, Pushchino, Russia (in Russian), Fig. 1.4.)*

- For each morphometric variable, we visually analyzed patterns on maps derived by both methods.
- We performed a statistical analysis of DTM differences and distributions of each morphometric attribute derived by both methods. We also carried out the Kolmogorov–Smirnov two-sample test (Daniel, 2000, chap. 8) for statistical differences between the two distributions of each topographic variable. We used samples, each including 1376 points (43×32 matrices with the grid spacing of 3000 m extracted from related DTMs).

To deal with the large dynamic range of morphometric attributes and their RMSE, we logarithmically transformed these digital models by Eq. (8.1) with $n = 5$. Terrain modeling was performed by the LandLord software (Florinsky, 2016). Statistical analysis was carried out with the Statgraphics software (© Statistical Graphics Corp., 1994–1997).

4.2.3.2 Results and discussion

There are essential differences in maps of the RMSE of morphometric variables calculated by the different methods. In particular, RMSE patterns have distinct spatial distributions (Fig. 4.6). For both methods, the RMSE dynamic ranges have approximately the same size. For example, the RMSE of k_h (m_{k_h}) has a dynamic range of about 5 units (Fig. 4.6). However, the application of the author's method lowered the boundaries of the dynamic range: it changed from 1.9–6.9 to 0.0–4.9 for m_{k_h} (Fig. 4.6A and B). The statistical distribution of the RMSE was also changed: for the author's method, the most values of m_{k_h} and RMSE of k_v (m_{k_v}) are relatively evenly distributed within a rather wide interval, from 0 to 1. For the Evans-Young method, m_{k_h} and m_{k_v} values are located within a narrow band ranging from 2.0 to 2.2 (Fig. 4.7).

The described method shares a common trait with the Evans-Young method: the polynomial (Eq. 4.7) is approximated to elevation values of the 5×5 window rather than passing exactly through them. This leads to local denoising that may enhance the calculation of derivatives and local topographic variables because they are responsive to a high-frequency component of a signal (Florinsky, 2002) (Section 5.4.1). However, Wood (1996, § 4.2) supposed that calculation of derivatives fitting the third and higher-order polynomials to $n \times n$ windows (for $n > 3$) can lead to an undue generalization of the topographic surface. This is not necessarily the case.

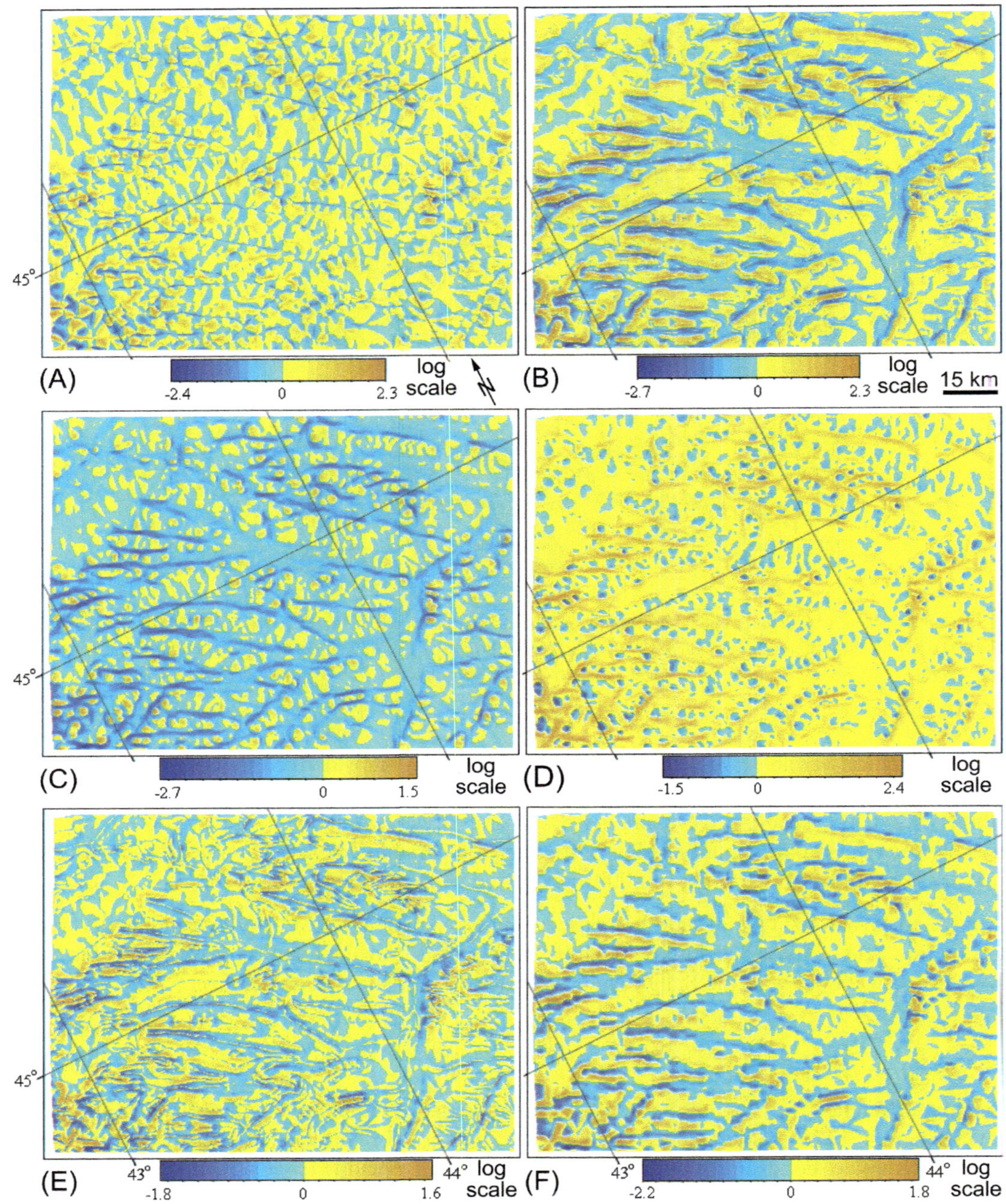

FIG. 4.5 The Stavropol Upland: (A) Horizontal curvature. (B) Vertical curvature. (C) Minimal curvature. (D) Maximal curvature. (E) Difference curvature. (F) Mean curvature.

(Continued)

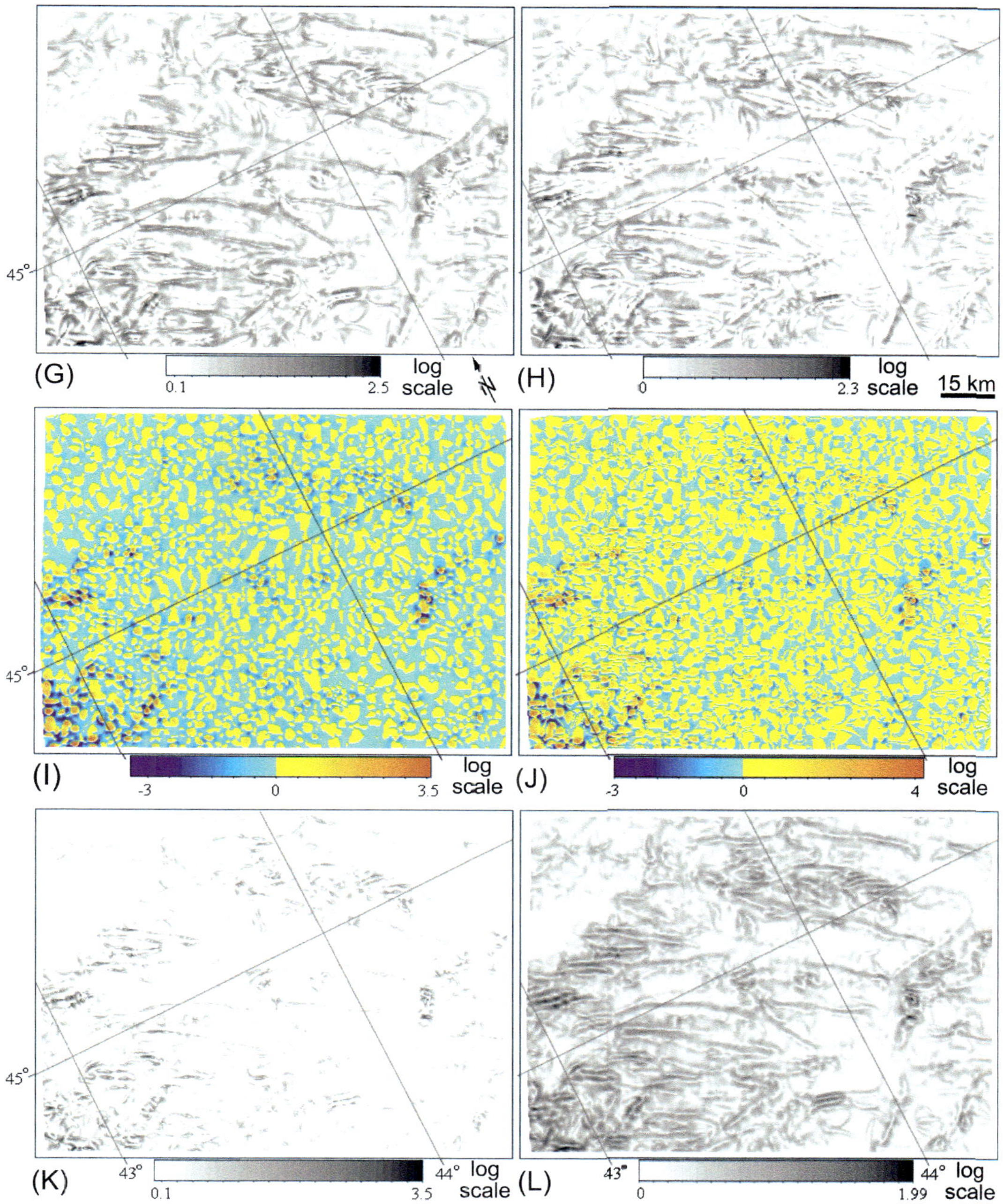

FIG. 4.5, CONT'D (G) Horizontal excess curvature. (H) Vertical excess curvature. (I) Gaussian curvature. (J) Accumulation curvature. (K) Ring curvature. (L) Unsphericity curvature. Calculations were conducted on the square grid with the 5 × 5 moving window (see details in the text). For the elevation map, see Fig. 4.3. *(Modified from Florinsky, I.V., 2010. Theory and Applications of Mathematical Cartographic Modeling of Topography (D.Sc. thesis). Institute of Mathematical Problems of Biology, Russian Academy of Sciences, Pushchino, Russia (in Russian), Fig. 1.5.)*

Indeed, comparing k_h maps derived by two methods (Fig. 4.8), one can see that local changes in map patterns enclose from one to about eight pixels. This is a level of denoising rather than generalization. One can find the same level of changes in the k_h dynamic range: it varied from −2.55 to 2.44 for the Evans-Young method (Fig. 4.8A) and from −2.49 to 2.31 for the author's method (Fig. 4.8B). Thus, the change in the k_h dynamic range was only about 4%. This is an insignificant alteration.

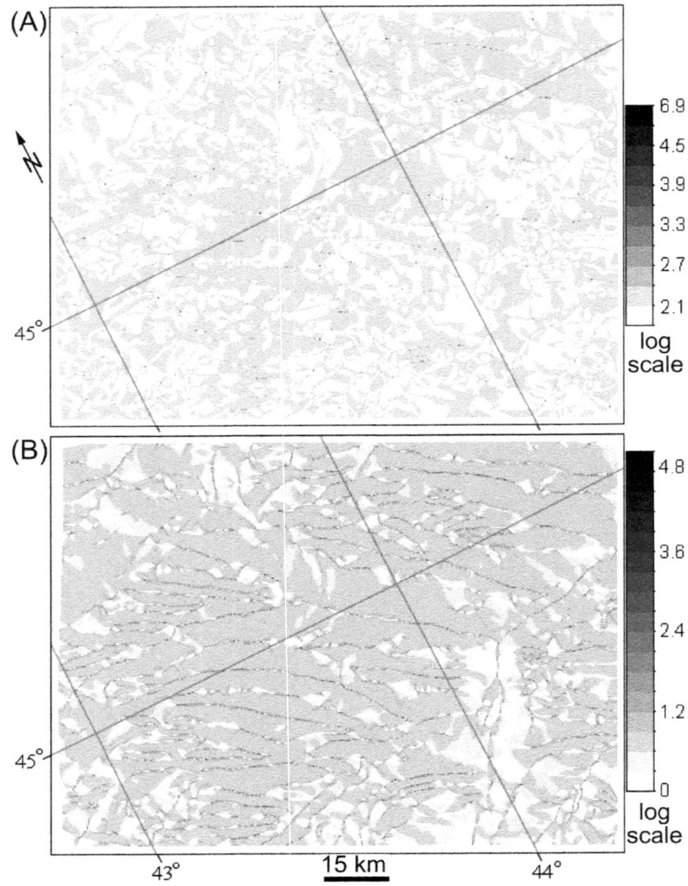

FIG. 4.6 The Stavropol Upland: RMSE maps of k_h calculated by the Evans-Young method (A) and the author's method (B). For the elevation map, see Fig. 4.3. *(Modified from Florinsky, I.V., 2009a. Accurate method for derivation of local topographic variables. Geodesy Cartogr. (4), 19–23 (in Russian), Fig. 3A.)*

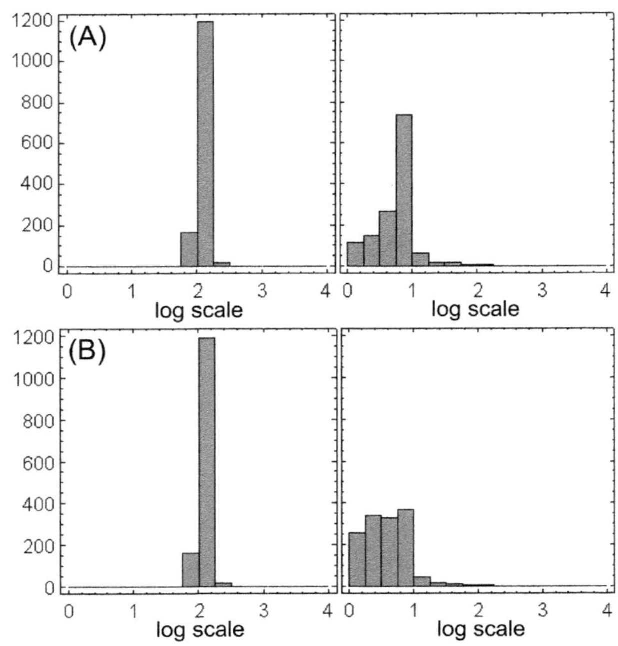

FIG. 4.7 Statistical distribution of logarithmically transformed RMSE values of curvatures calculated by the Evans-Young method (*left*) and the author's method (*right*): (A) RMSE of horizontal curvature. (B) RMSE of vertical curvature. *(Modified from Florinsky, I.V., 2009a. Accurate method for derivation of local topographic variables. Geodesy Cartogr. (4), 19–23 (in Russian), Fig. 3B and C.)*

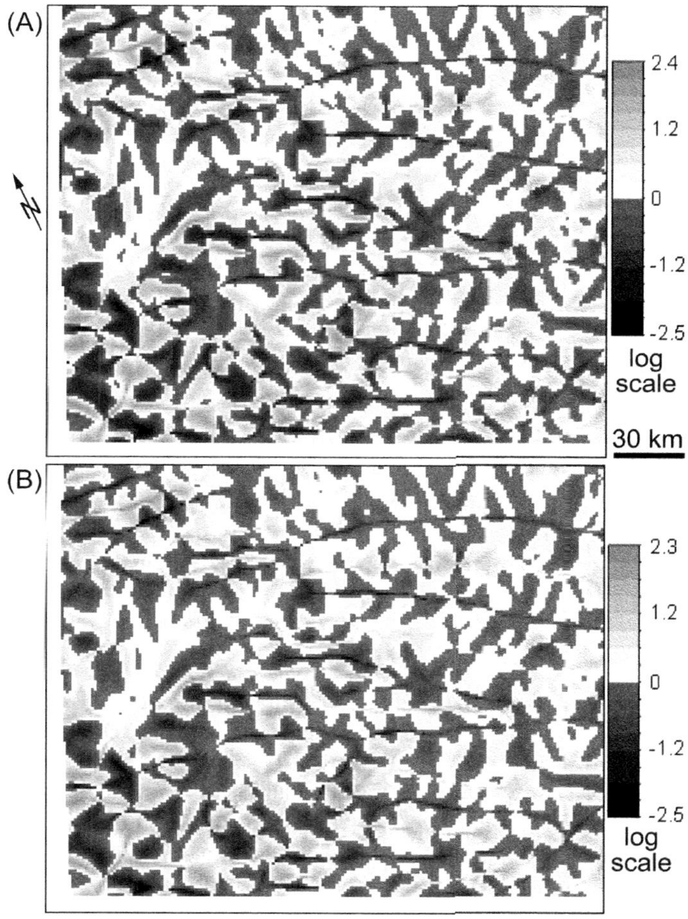

FIG. 4.8 The Stavropol Upland: enlarged lower left portions of the horizontal curvature maps derived by the Evans-Young method (A) and the author's method (B). For the entire map, see Fig. 4.5A. *(Modified from Florinsky, I.V., 2009a. Accurate method for derivation of local topographic variables. Geodezy Cartogr. (4), 19–23 (in Russian), Fig. 3D.)*

Histograms of the two k_h samples are very similar (Fig. 4.9A). Most changes between k_h models derived by the Evans-Young and author's methods were in the range of −0.1 to 0.1 (Fig. 4.9B). Again, this is 4% of the dynamic range of k_h derived by the Evans-Young method. Results of the Kolmogorov–Smirnov test for paired samples derived by different methods demonstrated that there is no statistically significant difference between the two distributions at the 95% confidence level. For k_h, for example, the estimated overall statistic DN = 0.025, the two-sided large sample K-S statistic is 0.67, and the *P*-value = 0.77. Therefore, the author's method provides additional denoising of a DEM without undue generalization of the surface.

A stronger suppression of high-frequency noise in DEMs by the author's method can be associated with two factors. First, we use the Taylor polynomial of a higher order than the Evans-Young method employs. Second, we use the 5×5 moving window, while the Evans-Young method utilizes the 3×3 one. Indeed, the increase of the moving window size may decrease the influence of high-frequency noise and interpolation errors on the computation of topographic variables (Albani et al., 2004).

Thus, compared to the Evans-Young method, the described method has essentially higher accuracy of calculation and stronger suppression of high-frequency noise. Therefore, it is advisable to use this method as a standard tool for data processing in digital terrain modeling.

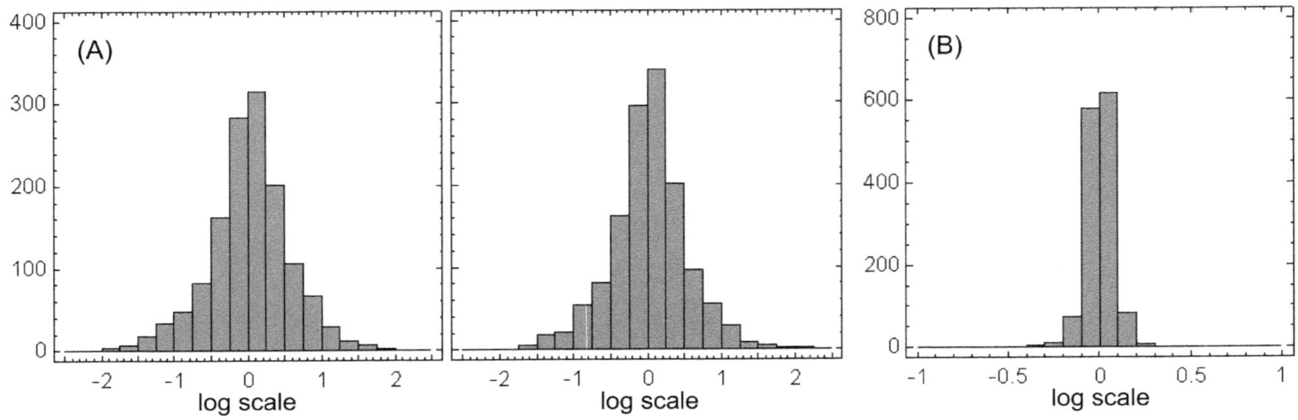

FIG. 4.9 Statistical distribution of logarithmically transformed values of horizontal curvature: (A) Horizontal curvature derived by the Evans-Young method (*left*) and the author's method (*right*). (B) Difference between horizontal curvature values derived by the Evans-Young and the author's methods. (*Modified from Florinsky, I.V., 2009a. Accurate method for derivation of local topographic variables. Geodesy Cartogr. (4), 19–23 (in Russian), Fig. 3E and F.*)

4.3 Calculation of local morphometric variables on a spheroidal equal angular grid on a sphere and an ellipsoid of revolution

4.3.1 Motivation

Many DEMs are based on spheroidal equal angular grids for an ellipsoid of revolution or a sphere (Section 3.2.2). Among these are global and quasi-global DEMs of the Earth (Table 3.1) as well as global DEMs of other planetary bodies, such as Mars (Smith et al., 2003), Venus (Ford et al., 1992; Sjogren et al., 1997), and the Moon (Zuber et al., 1996; Neumann 2008). It is obvious that processing and analysis of spheroidal equal angular DEMs should be done considering the grid geometry.

As discussed earlier (Sections 4.1 and 4.2), digital models of local morphometric variables are usually derived by methods based on the approximation of the partial derivatives of elevation by finite differences. Such methods have been developed and intended to calculate the partial derivatives from DEMs based on plane square grids. A plane square grid and a spheroidal equal angular grid have principally different geometry (Section 3.2.2). In spheroidal equal angular grids for an ellipsoid of revolution and a sphere, a grid/cell size, measured in linear units, depends on the latitude. For such grids, a grid spacing, approximately equal in linear units along both meridians and parallels, exists only at and near the equator. For example, the $3'' \times 3''$ cell of the SRTM3 DEM (Rabus et al., 2003; Farr et al., 2007) has the linear sizes of about $52\,\text{m} \times 93\,\text{m}$ at the latitude of 55°N.

Therefore, it is obvious that methods developed for plane square grids cannot be applied to spheroidal equal angular grids. This obstacle can be gotten around by reprojection using interpolation of elevation values from a spheroidal equal angular grid to a plane square grid (e.g., USGS, 2000; Ehsani and Quiel, 2008). However, this is an additional data-processing step, and any interpolation introduces new errors into a new DEM (Willmott et al., 1985). A detailed discussion of computational errors caused by the application of plane square-grid algorithms to spheroidal equal angular DEMs can be found elsewhere (Florinsky, 2017b).

Next, we describe the method for calculating local morphometric variables on spheroidal equal angular grids for an ellipsoid of revolution and a sphere (Florinsky, 1998c, 2017b).

4.3.2 Formulas

Let elevation be given as $z=f(x,y)$, where x and y are the orthogonal spheroidal coordinates. Let there be a 3×3 moving window formed by four adjacent spheroidal equal angular trapezoids (Fig. 4.1C). a, b, and c are linear lengths of three angularly equal arcs of parallels, while d and e are linear lengths of two angularly equal arcs of meridians. The orthogonal spheroidal coordinates and elevations are known for the window points $(-c, e, z_1)$, $(0, e, z_2)$, (c, e, z_3), $(-b, 0, z_4)$, $(0, 0, z_5)$, $(b, 0, z_6)$, $(-a, -d, z_7)$, $(0, -d, z_8)$, and $(a, -d, z_9)$. Let us assume that the curvature of the planet can be ignored within the moving window; that is, the length of the window diagonal is less than at most 0.1 of the average radius of the planet.

To estimate the first- and second-order partial derivatives r, t, s, p, and q (Eq. 2.2) at the central point $(0, 0, z_5)$, let us fit the polynomial (Eq. 4.1) to the nine points of the 3×3 spheroidal equal angular window by the least-squares method

(Bjerhammar, 1973). By writing the polynomial (Eq. 4.1) for all points of the window, we obtain a system of nine conditional linear equations. Let us write it in the form of Eq. (4.8). In this case, $\boldsymbol{\alpha}$ is a 9×1 matrix of the nine measured values of z:

$$\boldsymbol{\alpha} = \begin{pmatrix} z_1 \\ \vdots \\ z_9 \end{pmatrix}, \tag{4.23}$$

$\boldsymbol{\beta}$ is a 6×1 matrix of the six unknown coefficients of the polynomial (Eq. 4.1):

$$\boldsymbol{\beta} = \begin{pmatrix} r \\ t \\ s \\ p \\ q \\ u \end{pmatrix}, \tag{4.24}$$

and $\mathbf{F}$ is a 9×6 matrix of the known coefficients of the equation system (Eq. 4.8):

$$\mathbf{F} = \begin{pmatrix} \frac{c^2}{2} & \frac{e^2}{2} & -ce & -c & e & 1 \\ 0 & \frac{e^2}{2} & 0 & 0 & e & 1 \\ \frac{c^2}{2} & \frac{e^2}{2} & ce & c & e & 1 \\ \frac{b^2}{2} & 0 & 0 & -b & 0 & 1 \\ 0 & 0 & 0 & 0 & 0 & 1 \\ \frac{b^2}{2} & 0 & 0 & b & 0 & 1 \\ \frac{a^2}{2} & \frac{d^2}{2} & ad & -a & -d & 1 \\ 0 & \frac{d^2}{2} & 0 & 0 & -d & 1 \\ \frac{a^2}{2} & \frac{d^2}{2} & -ad & a & -d & 1 \end{pmatrix}. \tag{4.25}$$

To determine the unknown coefficients of the polynomial (Eq. 4.1), it is necessary to solve Eq. (4.12). Using matrix operations, we find a 6×9 matrix $(\mathbf{F}^{\mathrm{T}}\mathbf{F})^{-1}\mathbf{F}^{\mathrm{T}}$:

$$(\mathbf{F}^{\mathrm{T}}\mathbf{F})^{-1}\mathbf{F}^{\mathrm{T}} = \left\| \mathbf{J}_1 \ \mathbf{J}_2 \ \mathbf{J}_3 \ \mathbf{J}_4 \ \mathbf{J}_5 \ \mathbf{J}_6 \ \mathbf{J}_7 \ \mathbf{J}_8 \ \mathbf{J}_9 \right\|, \tag{4.26}$$

where

$$\mathbf{J}_1 = \left\| \begin{matrix} \dfrac{c^2}{a^4 + b^4 + c^4} \\ \dfrac{2\left[d\left(a^4 + b^4 + b^2c^2\right) - c^2e\left(a^2 - b^2\right)\right]}{3de(d+e)\left(a^4 + b^4 + c^4\right)} \\ -\dfrac{c\left[a^2(d+e) + b^2e\right]}{2\left[a^2c^2(d+e)^2 + b^2\left(a^2d^2 + c^2e^2\right)\right]} \\ -\dfrac{a^2cd(d+e)}{2\left[a^2c^2(d+e)^2 + b^2\left(a^2d^2 + c^2e^2\right)\right]} \\ \dfrac{d^2\left(a^4 + b^4 + b^2c^2\right) + c^2e^2\left(a^2 - b^2\right)}{3de(d+e)\left(a^4 + b^4 + c^4\right)} \\ -\dfrac{b^2c^2}{3\left(a^4 + b^4 + c^4\right)} \end{matrix} \right\|, \tag{4.26a}$$

$$\mathbf{J}_2 = \left\| \begin{matrix} -\dfrac{2c^2}{a^4+b^4+c^4} \\ \dfrac{2\left[d\left(a^4+b^4+3c^4-2b^2c^2\right)+2c^2e\left(a^2-b^2\right)\right]}{3de(d+e)\left(a^4+b^4+c^4\right)} \\ 0 \\ 0 \\ \dfrac{d^2\left(a^4+b^4+3c^4-2b^2c^2\right)-2c^2e^2\left(a^2-b^2\right)}{3de(d+e)\left(a^4+b^4+c^4\right)} \\ \dfrac{2b^2c^2}{3\left(a^4+b^4+c^4\right)} \end{matrix} \right\|, \tag{4.26b}$$

$$\mathbf{J}_3 = \left\| \begin{matrix} \dfrac{c^2}{a^4+b^4+c^4} \\ \dfrac{2\left[d\left(a^4+b^4+b^2c^2\right)-c^2e\left(a^2-b^2\right)\right]}{3de(d+e)\left(a^4+b^4+c^4\right)} \\ \dfrac{c\left[a^2(d+e)+b^2e\right]}{2\left[a^2c^2(d+e)^2+b^2\left(a^2d^2+c^2e^2\right)\right]} \\ \dfrac{a^2cd(d+e)}{2\left[a^2c^2(d+e)^2+b^2\left(a^2d^2+c^2e^2\right)\right]} \\ \dfrac{d^2\left(a^4+b^4+b^2c^2\right)+c^2e^2\left(a^2-b^2\right)}{3de(d+e)\left(a^4+b^4+c^4\right)} \\ -\dfrac{b^2c^2}{3\left(a^4+b^4+c^4\right)} \end{matrix} \right\|, \tag{4.26c}$$

$$\mathbf{J}_4 = \left\| \begin{matrix} \dfrac{b^2}{a^4+b^4+c^4} \\ -\dfrac{2\left[d\left(a^4+c^4+b^2c^2\right)+e\left(a^4+c^4+a^2b^2\right)\right]}{3de(d+e)\left(a^4+b^4+c^4\right)} \\ -\dfrac{b\left(a^2d-c^2e\right)}{2\left[a^2c^2(d+e)^2+b^2\left(a^2d^2+c^2e^2\right)\right]} \\ -\dfrac{b\left(a^2d^2+c^2e^2\right)}{2\left[a^2c^2(d+e)^2+b^2\left(a^2d^2+c^2e^2\right)\right]} \\ -\dfrac{d^2\left(a^4+c^4+b^2c^2\right)-e^2\left(a^4+c^4+a^2b^2\right)}{3de(d+e)\left(a^4+b^4+c^4\right)} \\ \dfrac{a^4+c^4}{3\left(a^4+b^4+c^4\right)} \end{matrix} \right\|, \tag{4.26d}$$

$$\mathbf{J}_5 = \left\| \begin{array}{c} -\dfrac{2b^2}{a^4 + b^4 + c^4} \\ -\dfrac{2\left[\left(a^4 + 3b^4 + c^4\right)(d + e) - 2b^2\left(a^2 e + c^2 d\right)\right]}{3de(d + e)\left(a^4 + b^4 + c^4\right)} \\ 0 \\ 0 \\ -\dfrac{\left(a^4 + 3b^4 + c^4\right)\left(d^2 - e^2\right) + 2b^2\left(a^2 e^2 - c^2 d^2\right)}{3de(d + e)\left(a^4 + b^4 + c^4\right)} \\ \dfrac{a^4 + 3b^4 + c^4}{3\left(a^4 + b^4 + c^4\right)} \end{array} \right\|, \tag{4.26e}$$

$$\mathbf{J}_6 = \left\| \begin{array}{c} \dfrac{b^2}{a^4 + b^4 + c^4} \\ -\dfrac{2\left[d\left(a^4 + c^4 + b^2 c^2\right) + e\left(a^4 + c^4 + a^2 b^2\right)\right]}{3de(d + e)\left(a^4 + b^4 + c^4\right)} \\ \dfrac{b\left(a^2 d - c^2 e\right)}{2\left[a^2 c^2 (d + e)^2 + b^2\left(a^2 d^2 + c^2 e^2\right)\right]} \\ \dfrac{b\left(a^2 d^2 + c^2 e^2\right)}{2\left[a^2 c^2 (d + e)^2 + b^2\left(a^2 d^2 + c^2 e^2\right)\right]} \\ -\dfrac{d^2\left(a^4 + c^4 + b^2 c^2\right) - e^2\left(a^4 + c^4 + a^2 b^2\right)}{3de(d + e)\left(a^4 + b^4 + c^4\right)} \\ \dfrac{a^4 + c^4}{3\left(a^4 + b^4 + c^4\right)} \end{array} \right\|, \tag{4.26f}$$

$$\mathbf{J}_7 = \left\| \begin{array}{c} \dfrac{a^2}{a^4 + b^4 + c^4} \\ \dfrac{2\left[e\left(b^4 + c^4 + a^2 b^2\right) + a^2 d\left(b^2 - c^2\right)\right]}{3de(d + e)\left(a^4 + b^4 + c^4\right)} \\ \dfrac{a\left[c^2 (d + e) + b^2 d\right]}{2\left[a^2 c^2 (d + e)^2 + b^2\left(a^2 d^2 + c^2 e^2\right)\right]} \\ -\dfrac{a c^2 e (d + e)}{2\left[a^2 c^2 (d + e)^2 + b^2\left(a^2 d^2 + c^2 e^2\right)\right]} \\ -\dfrac{e^2\left(b^4 + c^4 + a^2 b^2\right) - a^2 d^2\left(b^2 - c^2\right)}{3de(d + e)\left(a^4 + b^4 + c^4\right)} \\ -\dfrac{a^2 b^2}{3\left(a^4 + b^4 + c^4\right)} \end{array} \right\|, \tag{4.26g}$$

$$\mathbf{J}_8 = \left\| \begin{matrix} -\dfrac{2a^2}{a^4+b^4+c^4} \\ \dfrac{2\left[e\left(3a^4+b^4+c^4-2a^2b^2\right)-2a^2d\left(b^2-c^2\right)\right]}{3de(d+e)\left(a^4+b^4+c^4\right)} \\ 0 \\ 0 \\ -\dfrac{e^2\left(3a^4+b^4+c^4-2a^2b^2\right)+2a^2d^2\left(b^2-c^2\right)}{3de(d+e)\left(a^4+b^4+c^4\right)} \\ \dfrac{2a^2b^2}{3\left(a^4+b^4+c^4\right)} \end{matrix} \right\|, \tag{4.26h}$$

$$\mathbf{J}_9 = \left\| \begin{matrix} \dfrac{a^2}{a^4+b^4+c^4} \\ \dfrac{2\left[e\left(b^4+c^4+a^2b^2\right)+a^2d\left(b^2-c^2\right)\right]}{3de(d+e)\left(a^4+b^4+c^4\right)} \\ -\dfrac{a\left[c^2(d+e)+b^2d\right]}{2\left[a^2c^2(d+e)^2+b^2\left(a^2d^2+c^2e^2\right)\right]} \\ \dfrac{ac^2e(d+e)}{2\left[a^2c^2(d+e)^2+b^2\left(a^2d^2+c^2e^2\right)\right]} \\ -\dfrac{e^2\left(b^4+c^4+a^2b^2\right)-a^2d^2\left(b^2-c^2\right)}{3de(d+e)\left(a^4+b^4+c^4\right)} \\ -\dfrac{a^2b^2}{3\left(a^4+b^4+c^4\right)} \end{matrix} \right\|, \tag{4.26i}$$

Solving Eq. (4.12), we obtain the formulas for r, t, s, p, and q:

$$r = \frac{c^2(z_1+z_3-2z_2)+b^2(z_4+z_6-2z_5)+a^2(z_7+z_9-2z_8)}{a^4+b^4+c^4}, \tag{4.27}$$

$$\begin{aligned} t = {} & \frac{2}{3de(d+e)(a^4+b^4+c^4)} \times \left\{\left[d\left(a^4+b^4+b^2c^2\right)-c^2e\left(a^2-b^2\right)\right](z_1+z_3)\right. \\ & -\left[d(a^4+c^4+b^2c^2)+e(a^4+c^4+a^2b^2)\right](z_4+z_6) \\ & +\left[e(b^4+c^4+a^2b^2)+a^2d(b^2-c^2)\right](z_7+z_9) \\ & +d\left[b^4(z_2-3z_5)+c^4(3z_2-z_5)+(a^4-2b^2c^2)(z_2-z_5)\right] \\ & +e\left[a^4(3z_8-z_5)+b^4(z_8-3z_5)+(c^4-2a^2b^2)(z_8-z_5)\right] \\ & \left.-2\left[a^2d(b^2-c^2)z_8-c^2e(a^2-b^2)z_2\right]\right\}, \end{aligned} \tag{4.28}$$

$$s = \frac{c\left[a^2(d+e)+b^2e\right](z_3-z_1)-b(a^2d-c^2e)(z_4-z_6)+a\left[c^2(d+e)+b^2d\right](z_7-z_9)}{2\left[a^2c^2(d+e)^2+b^2\left(a^2d^2+c^2e^2\right)\right]}, \tag{4.29}$$

$$p = \frac{a^2cd(d+e)(z_3-z_1)+b\left(a^2d^2+c^2e^2\right)(z_6-z_4)+ac^2e(d+e)(z_9-z_7)}{2\left[a^2c^2(d+e)^2+b^2\left(a^2d^2+c^2e^2\right)\right]}, \tag{4.30}$$

$$
\begin{aligned}
q=&\frac{1}{3de(d+e)(a^4+b^4+c^4)}\times\{[d^2(a^4+b^4+b^2c^2)+c^2e^2(a^2-b^2)](z_1+z_3)\\
&-[d^2(a^4+c^4+b^2c^2)-e^2(a^4+c^4+a^2b^2)](z_4+z_6)\\
&-[e^2(b^4+c^4+a^2b^2)-a^2d^2(b^2-c^2)](z_7+z_9)\\
&+d^2[b^4(z_2-3z_5)+c^4(3z_2-z_5)+(a^4-2b^2c^2)(z_2-z_5)]\\
&+e^2[a^4(z_5-3z_8)+b^4(3z_5-z_8)+(c^4-2a^2b^2)(z_5-z_8)]\\
&-2[a^2d^2(b^2-c^2)z_8+c^2e^2(a^2-b^2)z_2]\}.
\end{aligned}
\tag{4.31}
$$

Moving the 3×3 window along a DEM, one can compute values of r, t, s, p, and q (and so values of local morphometric variables) for all points of the spheroidal equal angular DEM, except for boundary rows and columns. If one processes a virtually closed, global spheroidal equal angular DEM, it is possible to estimate values of topographic variables at each point of such a DEM.

We do not present a formula for the residual term u of the polynomial (Eq. 4.1) because it is not used in calculations. Eqs. (4.27)–(4.31) were developed with the Maple software (© Waterloo Maple Inc., 1981–1997).

4.3.3 Linear sizes of a spheroidal equal angular window

Values of a, b, c, d, and e vary depending on the latitude. Since geographic coordinates are known for every point of a spheroidal equal angular grid, so a, b, c, d, and e can be estimated using mathematical expressions for a solution to the inverse geodetic problem. For a sphere and an ellipsoid of revolution, there are well-developed theories and computational algorithms of the inverse geodetic problem (Helmert, 1880, chap. 5; Bagratuni, 1967, chaps. V and VI; Vincenty, 1975; Morozov, 1979, chap. IV; Bespalov, 1980, chaps. VI and VII; Sjöberg, 2006b; Karney, 2013; Panou, 2013). Next, we use formulas for the solution of the inverse geodetic problem at short distances (Morozov, 1979, pp. 178–179).

On an ellipsoid of revolution or a sphere, the distance L (measured in meters) between two points (φ_1, λ_1) and (φ_2, λ_2), where φ is latitude and λ is longitude, can be found by the following equation:

$$
L=\sqrt{Q^2+P^2},
\tag{4.32}
$$

where

$$
Q=\Theta \mathrm{M}_m\left[1-\left(E'^2-2\eta_m^2\right)\frac{\Theta^2}{8}-\left(1+\eta_m^2\right)\frac{(\Lambda\cos\varphi_m)^2}{12}-\frac{(\Lambda\sin\varphi_m)^2}{8}\right],
\tag{4.33}
$$

$$
P=\Lambda\cos\varphi_m \mathrm{N}_m\left[1+\left(1-9E'^2+8\eta_m^2\right)\frac{\Theta^2}{24}-\frac{(\Lambda\sin\varphi_m)^2}{24}\right],
\tag{4.34}
$$

where

$$
\Theta=\varphi_2-\varphi_1 \text{ (measured in radians)},
\tag{4.35}
$$

$$
\Lambda=\lambda_2-\lambda_1 \text{ (measured in radians)},
\tag{4.36}
$$

$$
\varphi_m=\frac{1}{2}(\varphi_1+\varphi_2),
\tag{4.37}
$$

$$
E'=\frac{\sqrt{A^2-B^2}}{B},
\tag{4.38}
$$

where E' is the second eccentricity, and A and B are semimajor and semiminor axes of the ellipsoid of revolution, respectively;

$$
\eta_m^2=E'^2\cos^2\varphi_m,
\tag{4.39}
$$

$$
\mathrm{N}_m=\frac{C}{\sqrt{1+\eta_m^2}},
\tag{4.40}
$$

$$C = \frac{A^2}{B}, \tag{4.41}$$

$$\mathrm{M}_m = \frac{\mathrm{N}_m}{1 + \eta_m^2}. \tag{4.42}$$

In a similar manner, one can determine sizes of the moving window and weights of its points (Eq. 6.9) for DEM smoothing. Alternative formulas to solve the inverse geodetic problem can be found elsewhere (Sodano, 1965; Vincenty, 1975; Bowring, 1996).

4.3.4 Discussion

In the described method, as in the Evans-Young method (Section 4.1), the polynomial (Eq. 4.1) is approximated to elevation values of the 3 × 3 window rather than passing exactly through them. This leads to a local suppression of high-frequency noise and may enhance the calculation of the partial derivatives sensitive to a high-frequency component of a signal (Florinsky, 2002) (Section 5.4.1). For the method's accuracy in terms of RMSE of a function of measured variables, see Section 5.3.3.2.

Differentiation operations are commonly applied to analyze digital models of various geophysical characteristics based on spheroidal equal angular grids. For example, Ekman (1988) calculated the Gaussian curvature of the "surface" of the isostatic uplift while exploring the origin of caves in Fennoscandia. Grachev et al. (2001) studied the gradients and curvatures of a "surface" for velocity of vertical neotectonic movements in Northern Eurasia. The described method offers possibilities for the direct use of such operations on spheroidal equal angular grids, without data interpolation to plane square grids.

Fig. 4.10 illustrates derivation of local topographic variables by the described method. The global SRTM30_PLUS DEM (Sandwell et al., 2008) was used to extract a spheroidal equal angular DEM of the Sinai Peninsula and adjacent territories (Fig. 4.10A). The area measuring 4° × 4° is located between 27.5°N and 31.5°N, and 32°E and 36°E. The DEM includes 230,880 points (the elevation matrix 480 × 481); the grid spacing is 30″. To suppress the high-frequency noise in bathymetric portions of the DEM, it was smoothed one time with the 3 × 3 moving window. Some local topographic variables were finally calculated by the described method (Fig. 4.10B–F). To deal with the large dynamic range of k_h, k_v, k_{min}, and k_{max}, we logarithmically transformed their values by Eq. (8.1) with $n = 6$. Data were processed by the LandLord software (Florinsky, 2016).

4.4 Calculation of nonlocal morphometric variables

Unlike local topographic variables calculated with the finite-difference approximation of elevation partial derivatives, logical procedures called "flow routing algorithms" are commonly applied to estimate key nonlocal morphometric variables, such as catchment and dispersive areas. These algorithms determine a route along which a flow is distributed from a given point of the topographic surface to downslope points.

4.4.1 Plane square grid

There are several flow routing algorithms to derive *CA* and *DA* from DEMs based on plane square grids. These algorithms can be grouped into two main types:

1. Eight-node single-flow direction (D8) algorithms
2. Multiple-flow direction (MFD) algorithms

D8 algorithms use only one of the eight possible directions separated by 45° to model a flow from a given point (O'Callaghan and Mark, 1984; Jenson and Domingue, 1988; Martz and de Jong, 1988). The flow direction is determined by estimating a value of *G* to each neighbor of the point; it corresponds to the direction for which *G* is the greatest (Fig. 4.11A). *CA* at a downslope point is determined as the number of upslope points, passed by flows to reach this point, multiplied by the cell unit area (Fig. 4.11B). The advantages of these algorithms are the simplicity and speed. Their disadvantage is oversimplification of the flow route modeling due to the limited number of possible flow directions.

MFD algorithms use flow partitioning (Freeman, 1991; Quinn et al., 1991). In this case, a flow is directed to every adjacent downslope point of a 3 × 3 moving window using slope weights (Fig. 4.11C). Thus, *CA* at a downslope point is composed of partial contributions from various upslope points. The advantage of these algorithms is their more natural

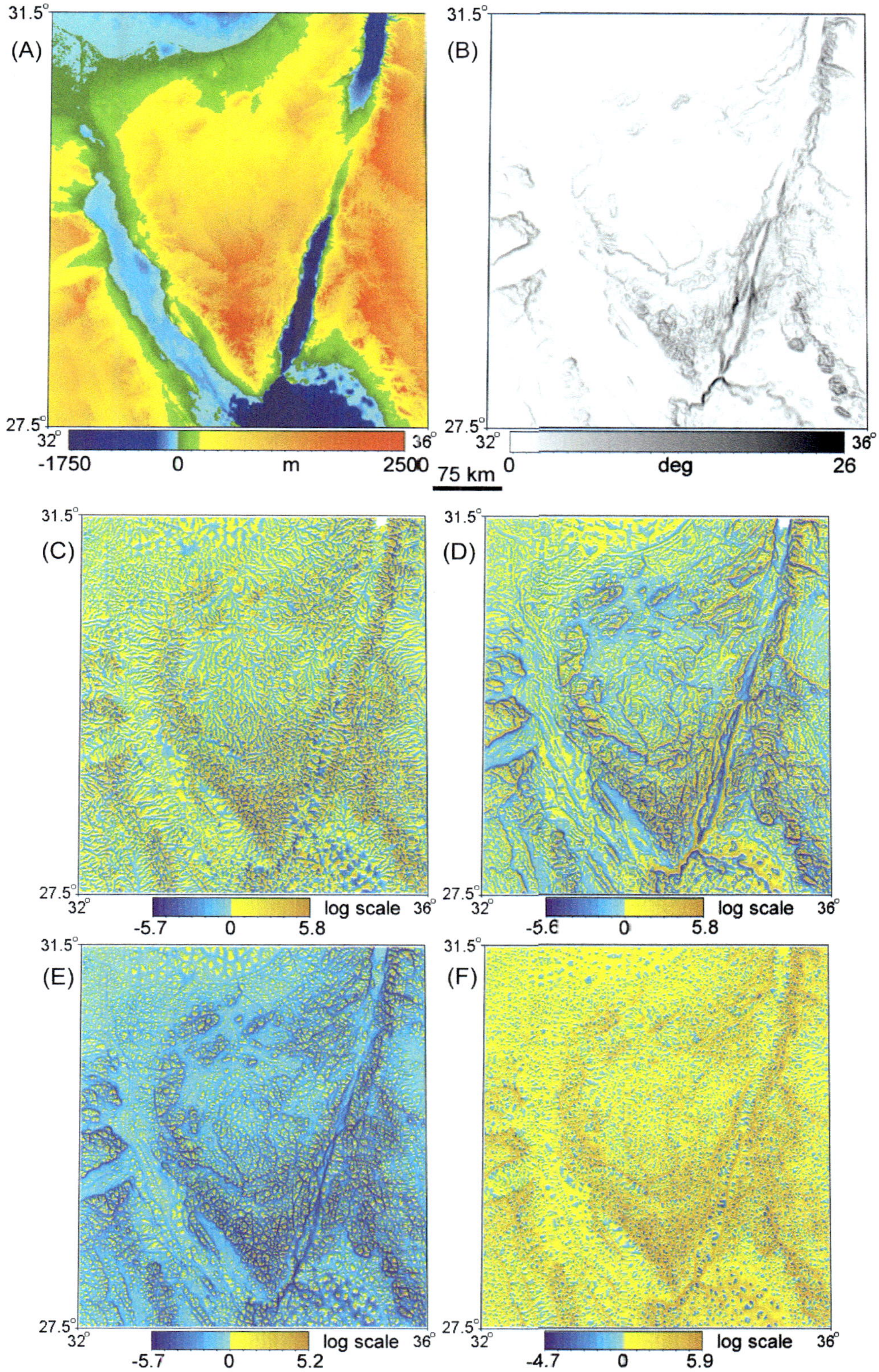

FIG. 4.10 Sinai Peninsula: (A) Elevation. (B) Slope gradient. (C) Horizontal curvature. (D) Vertical curvature. (E) Minimal curvature. (F) Maximal curvature. Calculations were conducted on the spheroidal equal angular grid (see details in the text).

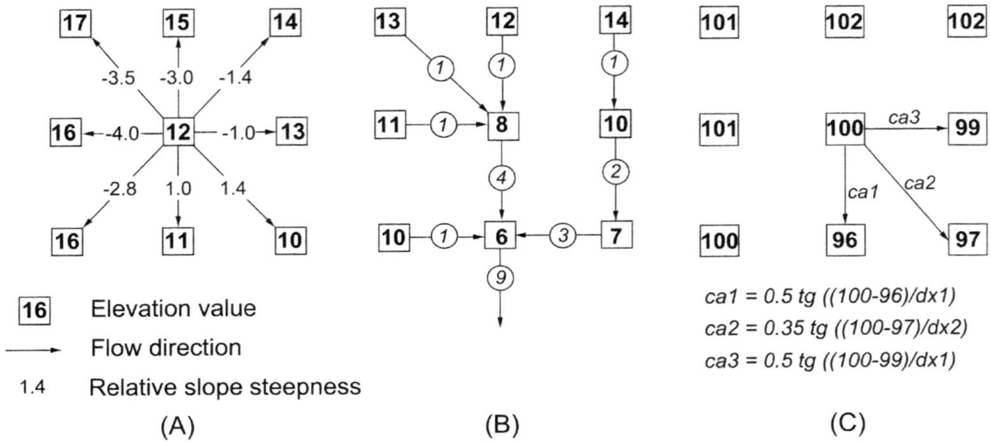

FIG. 4.11 Principles of calculation of nonlocal morphometric variables: (A) Gradient estimation at a given point of the DEM by the Martz-de Jong method: a flow line is going from the central point to the right bottom point. (B) Estimation of the catchment area by the Martz-de Jong method; *italic*, number of grid points contributing to the flow. (C) Flow partitioning: *ca*1, *ca*2, *ca*3, weights for the flow partitioning; *dx*1, *dx*2, flow line lengths. *((A) and (B) From Martz, L.W., de Jong E., 1988. CATCH: a Fortran program for measuring catchment area from digital elevation models. Comput. Geosci. 14 (5), 627–640, Figs. 1 and 4, reproduced with permission from Elsevier; (C) from Quinn, P., Beven, K., Lamb, R., 1995. The ln(a/tanβ) index: how to calculate it and how to use it within the TOPMODEL framework. Hydrol. Process. 9 (2), 161–182, Fig. 3, reproduced with permission from John Willey & Sons.)*

description of flow routing. Their disadvantage is a strong dependence of the flow dispersion on the DEM grid spacing value (Orlandini et al., 2012).

There are also some methods that combine the advantages of D8 and MFD algorithms and minimize their drawbacks (Tarboton, 1997; Seibert and McGlynn, 2007; Orlandini and Moretti, 2009).

DA can be determined in a similar manner, but a DEM should be inverted before the processing (e.g., by multiplying the DEM by −1). As a result, ridges "become" thalwegs, and thalwegs "become" ridges.

Before *CA* or *DA* calculation, it is reasonable to fill closed pits or depressions (Martz and de Jong, 1988; Lindsay and Creed, 2005; Arnold, 2010), which can be both errors of DEM generation and actual elements of the topographic surface. In this connection, Martz and de Jong (1988) distinguished two sorts of *CA* and *DA*: minimal and maximal one. Minimal *CA* is computed without pit filling. This terminates all simulated flows in closed depressions. In this case, *CA* values at points located downslope of a close pit do not consider *CA* values at points located upslope of the pit. To calculate maximal *CA*, all closed pits should be filled. In this case, *CA* values at each downslope point consider *CA* values at all connected upslope points.

A detailed description and comparison of flow routing algorithms can be found elsewhere (Band, 1993; Desmet and Govers, 1996; Rieger, 1998; Wilson et al., 2008; Orlandini et al., 2012; Qin et al., 2013; Huang and Lee, 2016).

Fig. 4.12 illustrates the derivation of *CA* and *DA* by the Martz-de Jong method. We used the square-gridded DEM of the Stavropol Upland (Fig. 4.3) described in Section 4.2.3.1. To deal with the large dynamic range of nonlocal topographic variables, we logarithmically transformed their values by Eq. (8.1). Data were processed by the LandLord software (Florinsky, 2016).

4.4.2 Spheroidal equal angular grid on a sphere and an ellipsoid of revolution

All methods mentioned previously are intended for the calculation of nonlocal topographic variables from a DEM based on a plane square grid, which has the same cell unit area (Π) at every point of the DEM (Fig. 4.13A). In the case of spheroidal equal angular grids on a sphere or an ellipsoid of revolution, Π depends on the latitude (Fig. 4.13B). For a sphere and an ellipsoid of revolution, there are well-developed theories and computational algorithms for estimating areas of spheroidal trapeziums (Helmert 1880, pp. 61–63; Bagratuni 1967, § 10; Morozov 1979, pp. 33–34; Sjöberg 2006a). Here we use the following expression (Morozov, 1979, p. 34):

$$\Pi = B^2(\lambda_2 - \lambda_1)\left| \sin\varphi + \frac{2}{3}E^2\sin^3\varphi + \frac{3}{5}E^4\sin^5\varphi + \frac{4}{7}E^6\sin^7\varphi + \cdots \right|_{\varphi_1}^{\varphi_2}, \tag{4.43}$$

where *E* is the first eccentricity of an ellipsoid of revolution:

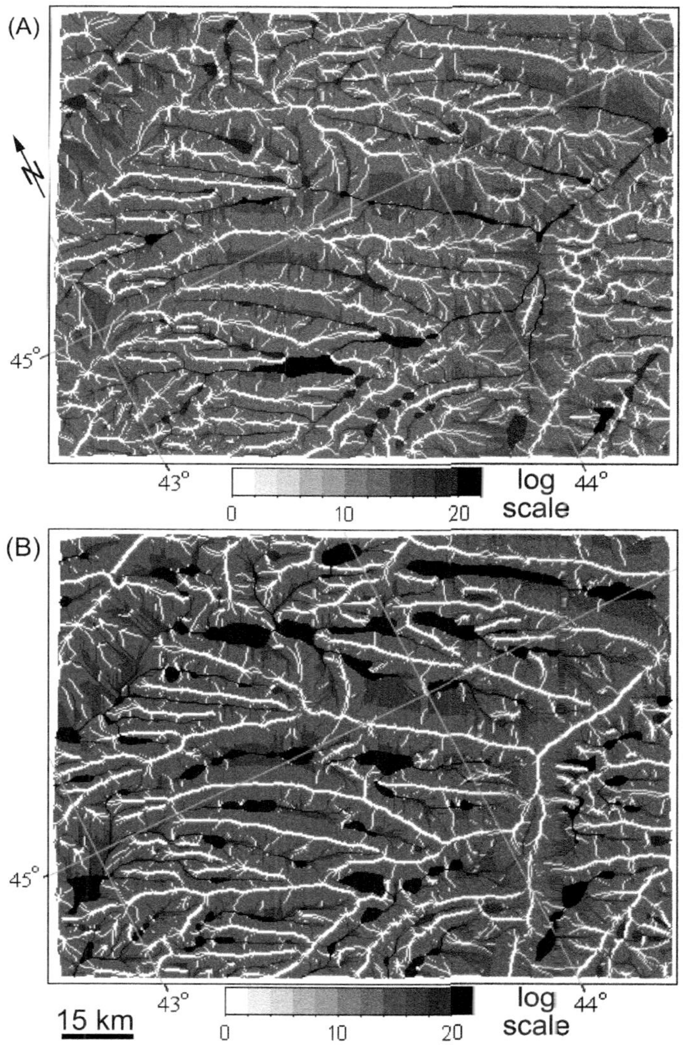

FIG. 4.12 The Stavropol Upland, nonlocal morphometric variables: (A) Catchment area. (B) Dispersive area. Calculations were conducted on the square grid (see details in the text). For the elevation map, see Fig. 4.3. *(Modified from Florinsky, I.V., 2010. Theory and Applications of Mathematical Cartographic Modeling of Topography (D.Sc. thesis). Institute of Mathematical Problems of Biology, Russian Academy of Sciences, Pushchino, Russia (in Russian), Fig. 1.7.)*

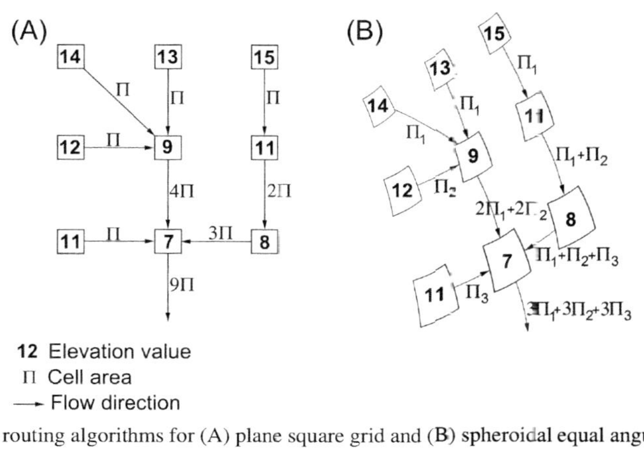

FIG. 4.13 Differences in the D8 flow routing algorithms for (A) plane square grid and (B) spheroidal equal angular grid on a sphere and an ellipsoid of revolution. Π_1, Π_2, Π_3, cell unit areas at different latitudes. *(From Florinsky, I.V., 2017. Spheroidal equal angular DEMs: the specificity of morphometric treatment. Trans. GIS. 21 (6), 1115–1129, Fig. 3, reproduced with permission from John Willey & Sons.)*

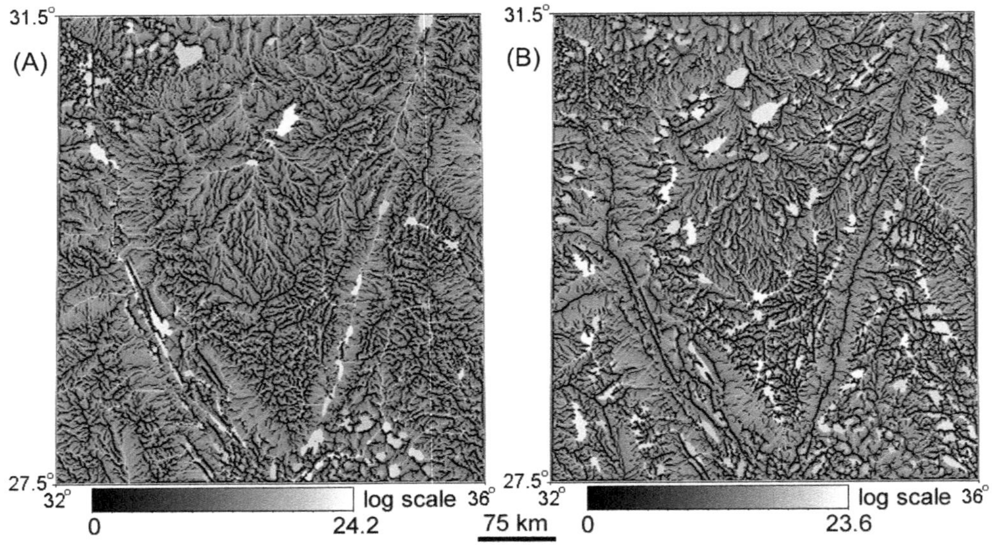

FIG. 4.14 Sinai Peninsula, nonlocal morphometric variables: (A) Catchment area. (B) Dispersive area. Calculations were conducted on the spheroidal equal angular grid (see details in the text). For the elevation map, see Fig. 4.10A.

$$E = \frac{\sqrt{A^2 - B^2}}{A}, \tag{4.44}$$

where A and B are semimajor and semiminor axes of the ellipsoid of revolution, respectively.

For a spheroidal equal angular grid, determination of the flow direction by estimating G value is easily performed by the method described in Section 4.3. If necessary, one can use formulas of the inverse geodetic problem at short distances (Section 4.3.3) to determine direction weights and distances between DEM cells. Eq. (4.43) and this principle can be used in any D8 or MFD algorithm.

The described approach offers possibilities for the direct calculation of nonlocal topographic variables on spheroidal equal angular grids on a sphere or an ellipsoid of revolution, without additional data interpolation to plane square grids (or data reprojection into an equal area projection).

Fig. 4.14 illustrates the derivation of CA and DA for the Sinai Peninsula by the Martz-de Jong method adapted to the spheroidal equal angular grid on a sphere or an ellipsoid of revolution. We used the DEM (Fig. 4.10A) described in Section 4.3.4. To deal with the large dynamic range of these nonlocal topographic variables, we logarithmically transformed their values by Eq. (8.1). Data were processed by the LandLord software (Florinsky, 2016).

4.5 Calculation of structural morphometric variables

Digital terrain modeling is closely allied to image processing. Thus, delineation of ridges and thalwegs is one of the topical tasks of both digital terrain modeling and image processing (Koenderink and van Doorn, 1993, 1994). In digital terrain modeling, researchers focus attention on the detection of thalwegs. The delineation of ridges receives much consideration in image processing. There is a principal difference between the topographic surface, whose properties are in many respects controlled by gravity, and the image intensity function. However, from the technical point of view, the algorithms of digital terrain modeling may be applied in image processing, and vice versa, algorithms of image processing are used in digital terrain analysis.

4.5.1 Conventional algorithms

The twofold nature of structural morphometric variables (Section 2.4) is responsible for the existence of two fundamentally different groups of techniques to delineate ridges and thalwegs.

The methods of the first group are based on principles of differential geometry. They are predominantly used in image processing (Haralick, 1983; Eberly et al., 1994; López and Serrat, 1996; Rieger, 1997; López et al., 1998). In these approaches, differential geometric criteria are applied to detect loci of extreme values of mean, principal, and plan curvatures (Section 4.5.2).

The methods of the second group are based on logical processing of data, such as cell-to-cell flow routing and *CA* calculation (Section 4.4), with subsequent thresholding of its values. These approaches are commonly used in digital terrain modeling (Mark, 1984; O'Callaghan and Mark, 1984; Band, 1986; Douglas, 1986; Skidmore, 1990; Tribe, 1992; Jones, 2002).

In scale terms, the methods of the first group are local because they concern the vicinity of a point within a moving window. The methods of the second group are nonlocal because determining the flow routing requires an analysis of rather large portions of the topographic surface.

The third group of algorithms combines local differential geometric and nonlocal logical approaches. In this case, differential geometric criteria are used to detect saddle points or points of extreme contour curvature, and then logical procedures are applied to link the points for visualization of ridges and thalwegs (Gauch and Pizer, 1993; Kweon and Kanade, 1994; Steger, 1999).

Detailed comparisons of the methods can be found elsewhere (Skidmore, 1990; Tribe, 1992; Gauch and Pizer, 1993; López et al., 1999).

Fig. 4.15 illustrates the revealing of crests and thalwegs using a method belonging to the second group. We used the Stavropol Upland DEM (Fig. 4.3) described in Section 4.2.3.1. The map was obtained with the CatchmentSIM software (Ryan and Boyd, 2003). Flats and pits were removed from the DEM by the breaching algorithm (Jones, 2002). In flow routing, a downslope flow angle was determined by a modified multiple-direction algorithm (Lea, 1992). Thalwegs were defined as pixels with *CA* values greater than a threshold (Jones, 2002); we used the threshold of 25 pixels. When treating an inverted DEM, the ridge network was delineated in the same way.

4.5.2 Horizontal deflection

One method of the first group is based on the derivation of the horizontal deflection (Shary and Stepanov, 1991). It is known that k_h is the measure of convergence and divergence of slope lines (Section 2.2.4.2). Negative values of k_h indicate convergence areas of slope lines, while positive values of k_h correspond to their divergence areas. At the same time, k_h is the product of the curvature of contour line and slope factor (Section 2.2.4.2) (Shary, 1991; Shary et al., 2002b). This means that k_h contains information on the behavior of both contours and slope lines. Thus, to find loci of extreme curvature of both families of spatial curves, one needs to find loci of extreme values of k_h. They correspond to zero values of the k_h derivative, that is, horizontal deflection D_{kh} (Section 2.2.5.1). Theoretically, ridges and convex break lines relate to the locus of $D_{kh}=0$ within divergence areas, while thalwegs and concave break lines correspond to the locus of $D_{kh}=0$ within convergence areas (Shary and Stepanov, 1991).

Fig. 4.16 illustrates the revealing of crest and thalweg lines for calculating horizontal deflection. We used the Stavropol Upland DEM (Fig. 4.3) described in Section 4.2.3.1. To estimate the first-, second-, and third-order partial derivatives of elevation, we applied the author's method described in Section 4.2. To deal with the large dynamic range of D_{kh}, we logarithmically transformed its values by Eq. (8.1) with $n=5$. Data were processed by the LandLord software (Florinsky, 2016).

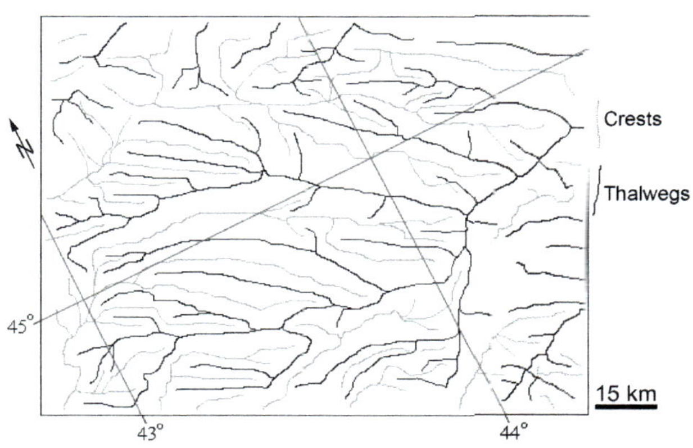

FIG. 4.15 The Stavropol Upland: ridge and thalweg lines. Calculations were conducted on the square grid (see details in the text). For the elevation map, see Fig. 4.3. *(Modified from Florinsky, I.V., 2010. Theory and Applications of Mathematical Cartographic Modeling of Topography (D.Sc. thesis). Institute of Mathematical Problems of Biology, Russian Academy of Sciences, Pushchino, Russia (in Russian), Fig. 1.9A.)*

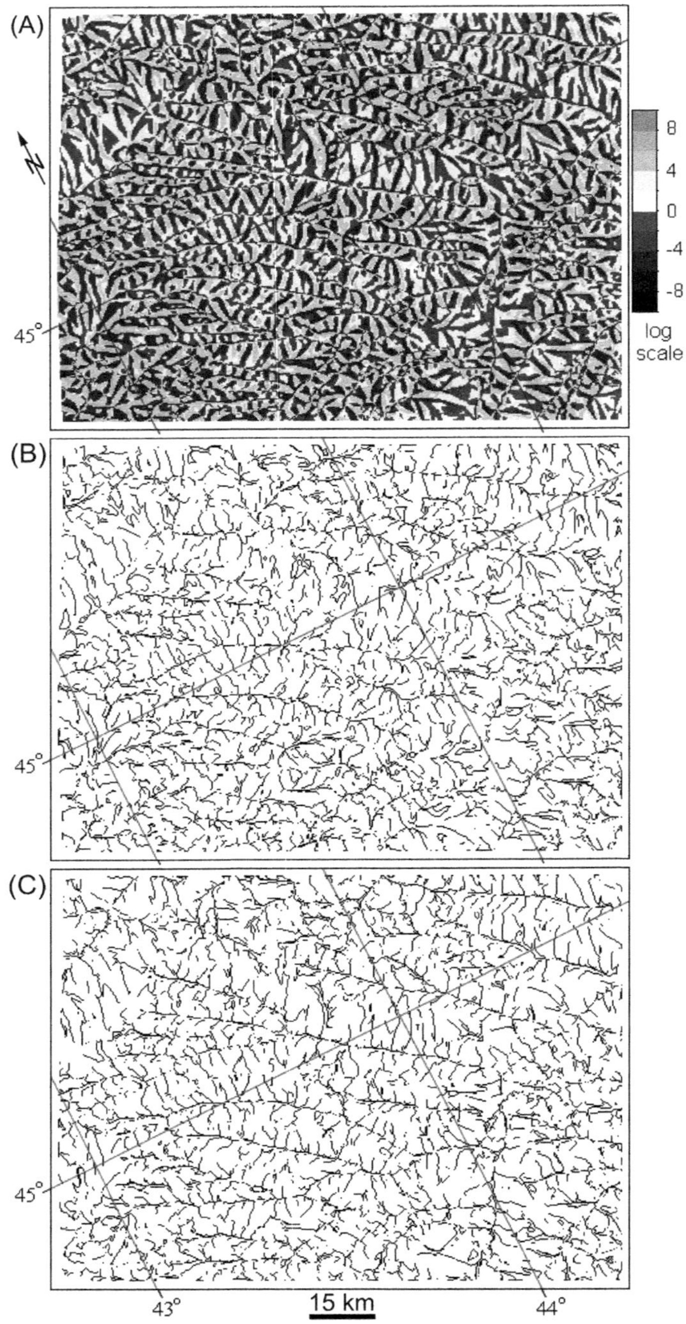

FIG. 4.16 The Stavropol Upland, horizontal deflection: (A) Spatial distribution of D_{kh} values. (B) Zero values of D_{kh} within divergence areas, the locus of the positive extreme curvature of the topographic surface. (C) Zero values of D_{kh} within convergence areas, the locus of the negative extreme curvature of the topographic surface. Calculations were conducted on the square grid with the 5×5 moving window (see details in the text). For the elevation map, see Fig. 4.3. *(Modified from Florinsky, I.V., 2018a. Geomorphometry on the surface of a triaxial ellipsoid: towards the solution of the problem. Int. J. Geogr. Inf. Sci. 32 (8), 1558–1571, Fig. 4.)*

The map of D_{kh} (Fig. 4.16A) includes all loci of the extreme curvature of the topographic surface. They appear as borders between areas of positive and negative values of D_{kh} (light and dark patterns, respectively). The map of zero values of horizontal deflection, located within flow divergence areas ($D_{kh}=0$ with $k_h>0$), displays the locus of the positive extreme curvature (Fig. 4.16B). The map of zero values of horizontal deflection, located within flow convergence areas ($D_{kh}=0$ with $k_h<0$), displays the locus of the negative extreme curvature (Fig. 4.16C).

Theoretically, the positive extreme curvature map (Fig. 4.16B) should display ridges and convex break lines, while the negative extreme curvature map (Fig. 4.16C) should represent thalwegs and concave break lines (Shary and Stepanov, 1991). However, even a cursory examination of the maps showed that they failed to detect these structural lines correctly. This inference is supported by comparison with the maps of ridge/thalweg networks delineated using the nonlocal logical technique (cf. Figs. 4.15 and 4.16). The main problem is that the loci of the extreme curvature are interrupted in many places (Fig. 4.16B and C). This can be caused by interpolation and smoothing errors, the remains of the high-frequency noise of the DEM, and loci of special points, namely local maxima and minima, saddles, and minor flat horizontal areas.

Therefore, in terms of ridge/thalweg detection, D_{kh} mapping has some disadvantages: First, loci of the extreme curvature may only partially define thalweg/ridge lines. Second, the calculation of the partial derivatives is sensitive to errors and noise existing copiously in DEMs (Section 5.4.1) (Florinsky, 2002). Third, D_{kh} mapping cannot reveal thalweg/ridge networks as hierarchical structures of tributary orders.

4.6 Calculation of two-field-specific morphometric variables

4.6.1 Calculation of solar morphometric variables

Reflectance (Section 2.5.1.1) and insolation (Section 2.5.1.2) are functions of the first-order partial derivatives of elevation and two solar angles (Table 2.2). Therefore, there are no special methods to calculate these variables. Depending on the DEM grid type, one can choose a particular calculation method for local morphometric variables (Sections 4.1–4.3).

Fig. 4.17 illustrates the derivation of insolation by the author's method (Section 4.2) for the Stavropol Upland. Its square-gridded DEM (Fig. 4.3) is described in Section 4.2.3.1. Data were processed by the LandLord software (Florinsky, 2016).

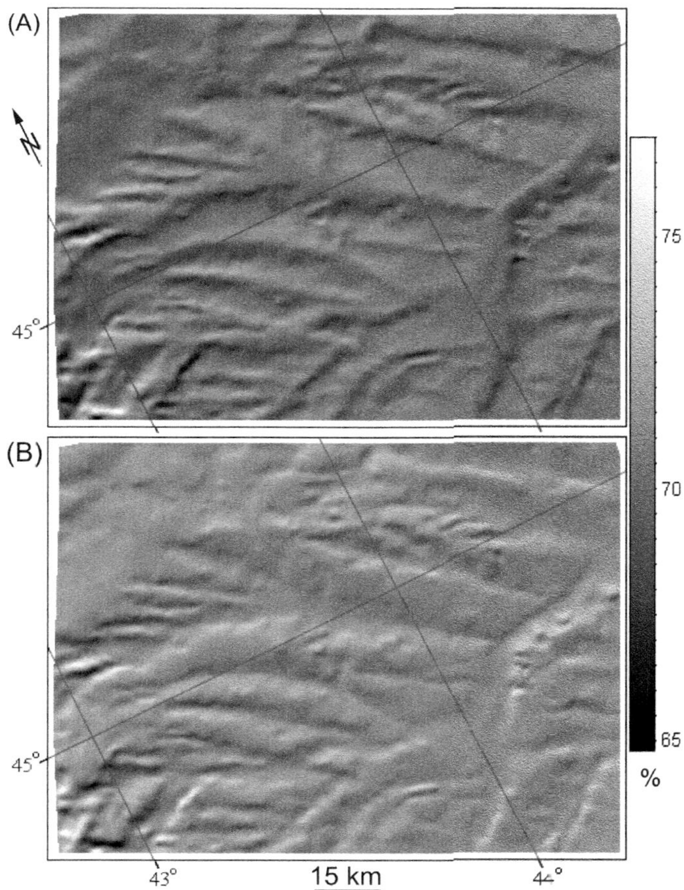

FIG. 4.17 The Stavropol Upland: insolation for various values of elevation and azimuth solar angles: (A) 35° and 180° (south), (B) 45° and 0° (north). Calculations were conducted on the square grid with the 5 × 5 moving window (see details in the text). For the elevation map, see Fig. 4.3.

4.6.2 Calculation of wind morphometric variables

Computation of the wind effect index *WEf* and wind exposition index *WEx* (Section 2.5.2) can be done as follows (Böhner, 2004). For a particular wind direction sector with two user-defined parameters—a central angle (w_α) and a search distance (L) (Fig. 4.18A)—mean weighted horizon angles for windward and leeward directions (σ_W and σ_L, respectively) can be estimated by the next expressions:

$$\sigma_W = \frac{\sum_{i=1}^{n}\left[\frac{1}{d_{Wi}}\arctan\left(\frac{\Delta z_{Wi}}{\sqrt{d_{Wi}}}\right)\right]}{\sum_{i=1}^{n}\frac{1}{d_{Wi}}}, \quad (4.45)$$

$$\sigma_L = \frac{\sum_{i=1}^{n}\left[\frac{1}{\ln(1+d_{Li})}\arctan\left(\frac{\Delta z_{Li}}{\sqrt{d_{Li}}}\right)\right]}{\sum_{i=1}^{n}\frac{1}{\ln(1+d_{Li})}}, \quad (4.46)$$

where d_{Wi} and d_{Li} are horizontal distances from a given point P to a point P_i located within a wind direction sector in windward and leeward directions, respectively; Δz_{Wi} and Δz_{Li} are elevation differences between the given point P and the point P_i located within the wind direction sector in windward and leeward directions, respectively; $i=1, \ldots n$; n is the number of points within the wind direction sector (Fig. 4.18).

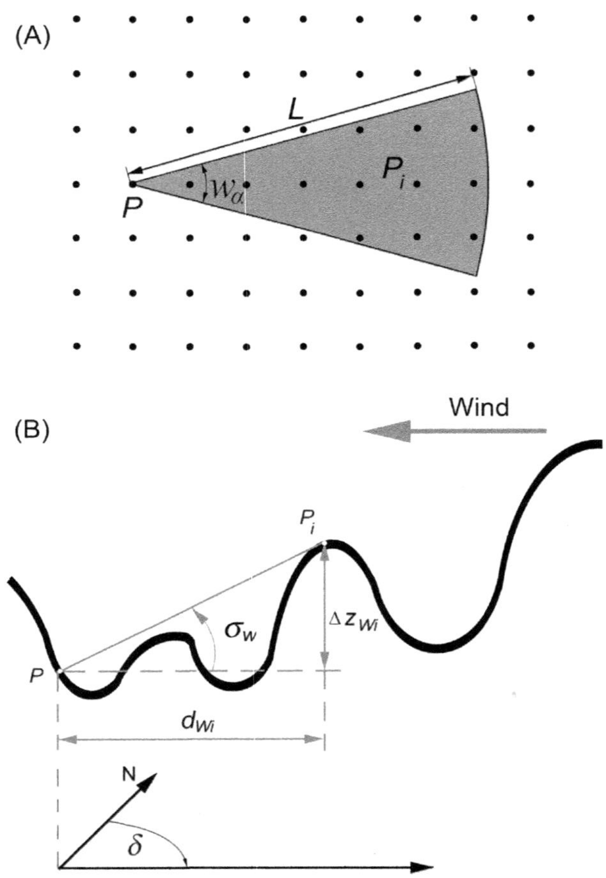

FIG. 4.18 Principles of calculation of wind morphometric variables: (A) Wind direction sector at a given point P against a background of a square-gridded DEM; L, search distance; P_i, points located within a wind direction sector; w_α, central angle. (B) Key values against a terrain profile (*bold curve*): d_{Wi}, horizontal distance from a given point P of the topographic surface to a point P_i located in a wind direction sector in windward direction; N, north direction; Δz_{Wi}, elevation difference between the given point P and the point P_i located in a wind direction sector in windward direction; δ, wind direction; σ_w, horizon angle in the wind direction for the pair of points P and P_i.

Then Eqs. (4.45) and (4.46) are converted into dimensionless functions:

$$\begin{cases} \sigma'_W = 1 + \ln(1 + \sigma_W) & \text{for} \quad \sigma_W > 0 \\ \sigma'_W = \dfrac{1}{1 + \ln(1 - \sigma_W)} & \text{for} \quad \sigma_W < 0, \end{cases} \tag{4.47}$$

$$\begin{cases} \sigma'_L = \sqrt{1 + \ln(1 + \sigma_L)} & \text{for} \quad \sigma_L > 0 \\ \sigma'_L = \dfrac{1}{\sqrt{1 + \ln(1 - \sigma_L)}} & \text{for} \quad \sigma_L < 0. \end{cases} \tag{4.48}$$

Finally, for the given point P of the topographic surface, the wind effect index is estimated by the following expression:

$$WEf = \sqrt[4]{\sigma'_W \cdot \sigma'_L}. \tag{4.49}$$

The wind exposition index is an average value of WEf estimated for all wind direction sectors of the full unit circle:

$$WEx = \frac{1}{m}\sum_{j=1}^{m}\left(\sqrt[4]{\sigma'_W \cdot \sigma'_L}\right)_j, \tag{4.50}$$

where j is the number of a wind direction sector; m is the number of wind direction sectors representing the full unit circle, $j=1, \ldots m$; w_α is an angular step, that is, $w_\alpha = 360°/m$.

Fig. 4.19 illustrates the derivation of WEf and WEx for the Stavropol Upland. Its square-gridded DEM (Fig. 4.3) is described in Section 4.2.3.1. Data were processed by the SAGA software (Conrad et al., 2015).

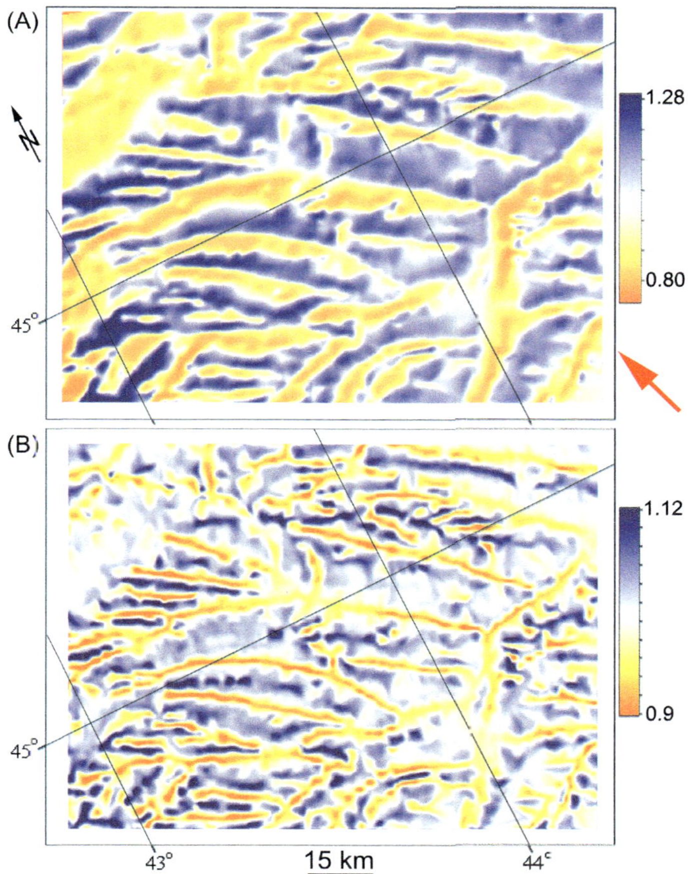

FIG. 4.19 The Stavropol Upland, wind morphometric variables: (A) Wind effect index. Search distance is 3 km; *red arrow*, wind direction, 198°. (B) Wind exposition index. Search distance is 3 km, $w_\alpha = 15°$. Calculations were conducted on the square grid (see details in the text). Boundary effects are removed. For the elevation map, see Fig. 4.3.

Notice that *WEf* was originally developed by Böhner (2004) as a proxy of wind speed distribution to model orographic precipitation, which results from the uplift of moisture-bearing air masses at windward slopes of mountain ranges and the leeward rain shadow. In that case, Eq. (4.45) included a second summand to account for the leeward impact of previously traversed mountain chains (Böhner, 2004, 2005; Gerlitz et al., 2015).

4.6.3 Calculation of view morphometric variables

Calculation of terrain visibility (singular viewshed) and total viewshed (Section 2.5.3) is based on line-of-sight (LoS) algorithms (De Floriani et al., 1994; Nagy, 1994; Fisher, 1995; Sanchez-Fernandez et al., 2021). In this group of algorithms, rays are usually projected from the view point toward the DEM boundaries to determine which DEM points are visible or invisible. A typical LoS algorithm includes the following main steps (Sanchez-Fernandez et al., 2021):

1. Performing an azimuthal partition of the DEM into n azimuthal equiangular sectors with central angles α starting at a view point P_v (Fig. 4.20); $n = 360°/\alpha$.
2. Searching visible points within an nth azimuthal sector. A given point is visible from the view point P_v if its angular elevation is higher than all the previous ones along the axis of the azimuthal sector (Fig. 4.20). Points satisfying this condition form a set of visible points of the nth azimuthal sector. This step is repeated until all points of the nth azimuthal sector are checked.
3. Projecting all visible segments within the nth azimuthal sector onto a horizontal plane produces visible sections of the nth azimuthal sector (Fig. 4.20).
4. Repeating steps 1–3 for all n azimuth sectors of the view point P_v.

The result of steps 1–4 is a binary viewshed map representing visible/invisible areas for the view point P_v.

Fig. 4.21A illustrates the derivation of terrain visibility for the Stavropol Upland. Its square-gridded DEM (Fig. 4.3) is described in Section 4.2.3.1. Data were processed by the SAGA software (Conrad et al., 2015).

If there is a need to estimate the size of the area visible from the view point P_v, three additional steps should be performed:

5. Measuring distances r_i and R_i from the view point P_v to the starting (P_{si}) and ending (P_{ei}) points, respectively, of all visible sections (Fig. 4.20).
6. Computing areas of all visible sections A_{VSi} of the view point P_v:

$$A_{VSi} = \frac{\alpha\pi}{360}\left(R_i^2 - r_i^2\right). \tag{4.51}$$

7. Summing the areas of all visible sections A_{VSi} of the view point P_v.

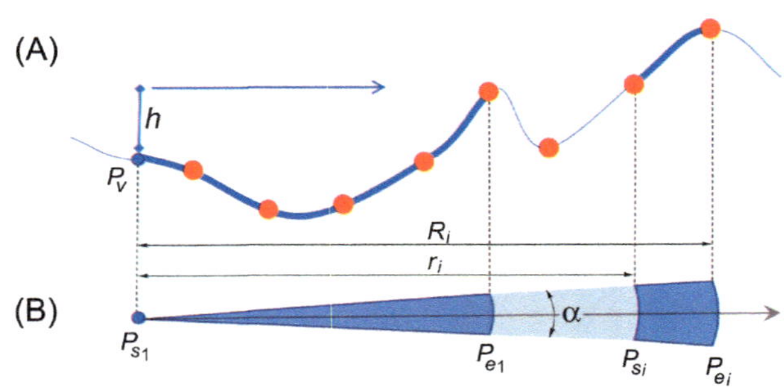

FIG. 4.20 A scheme of a line-of-sight algorithm for estimating terrain visibility (singular viewshed). (A) Side projection: *blue dot*, a view point P_v with a relative height h; *red dots*, target points; *dark blue thick curves*, visible segments. (B) Zenith projection: nth azimuthal sector; α, central angle of the nth azimuthal sector; *gray line with arrow*, axis of the nth azimuthal sector; *dark and light blue areas*, visible and invisible sections of the nth azimuthal sector, respectively; P_{si} and P_{ei}, starting and ending points of visible sections, respectively. *(Modified from Sanchez-Fernandez, A.J., Romero, L.F., Bandera, G., Tabik S., 2021. A data relocation approach for terrain surface analysis on multi-GPU systems: a case study on the total viewshed problem. Int. J. Geogr. Inf. Sci. 35 (8), 1500–1520, Fig. 2, reproduced with permission from Taylor & Francis Ltd.)*

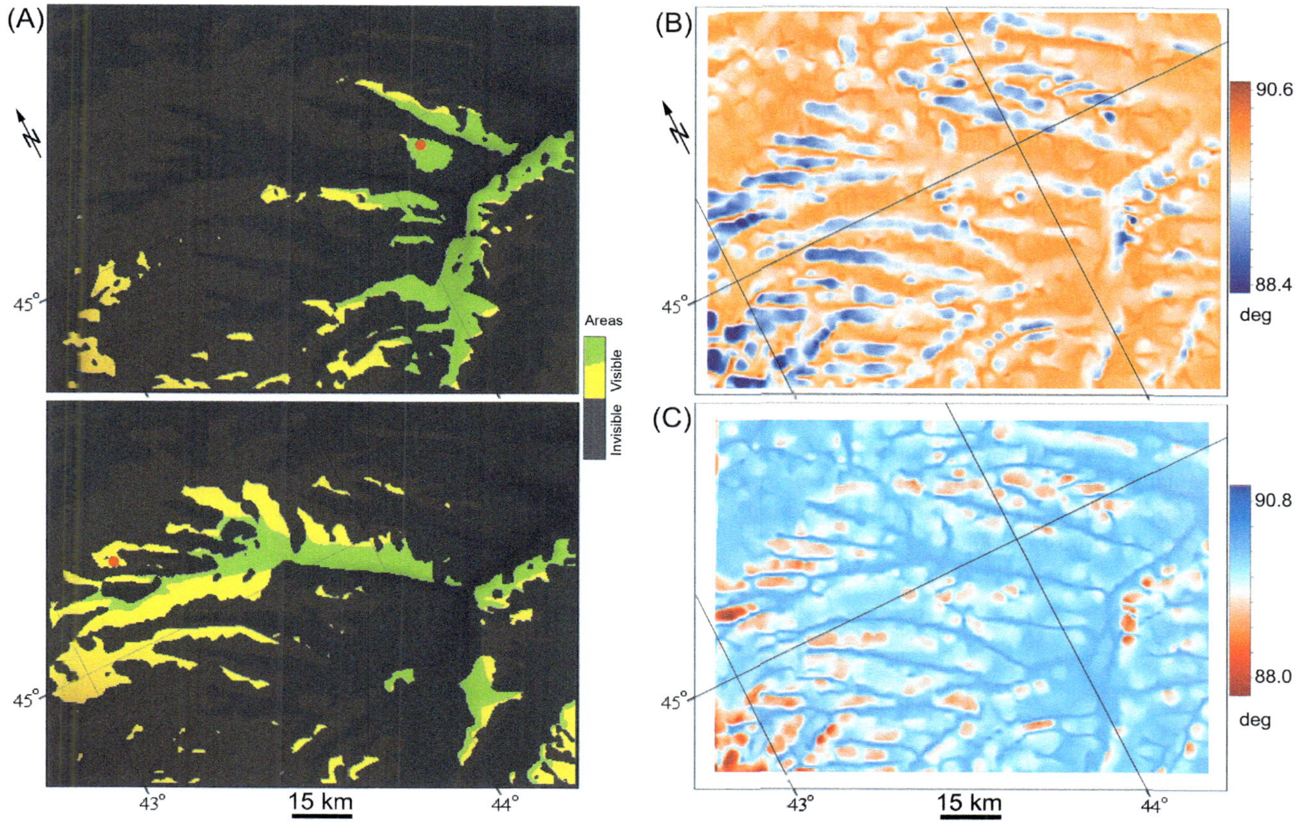

FIG. 4.21 The Stavropol Upland, view morphometric variables: (A) Terrain visibility (singular viewshed) for two different view points with relative heights of 10m (*red dots*). The binary viewshed map is draped over the elevation map (Fig. 4.3). Invisible areas are darkened. (B) Positive openness and (C) negative openness calculated with an 8-direction line tracing and a radial distance of 3 km. Calculations were conducted on the square grid (see details in the text). Boundary effects are removed. For the elevation map, see Fig. 4.3.

Estimation of the total viewshed V_Σ (Section 2.5.3) is carried out by repeating the singular viewshed estimation (steps 1–7, as discussed earlier) for each point of a DEM (Tabik et al., 2013; Sanchez-Fernandez et al., 2021).

Topographic openness *TO* (Section 2.5.3) is usually computed by line tracing algorithms based on determination of maximal slope gradients G_z and G_n (and so zenith and nadir angles ϕ and ψ) between a central point of a moving window $n \times n$ (n are odd numbers) and window's points located along all directions available for a given window size n within a radial distance $L = (n - 1)/2$. For a $n \times n$ window, there are $4(n - 1)$ possible directions (Yokoyama et al., 2002; Alonso-Sarría et al., 2018).

Fig. 4.21B and C illustrates the derivation of positive and negative *TO*, respectively, for the Stavropol Upland. Its square-gridded DEM (Fig. 4.3) is described in Section 4.2.3.1. Data were processed by the SAGA software (Conrad et al., 2015).

4.7 Calculation of combined morphometric variables

First, let us discuss calculation of combined topographic variables, which are functions of nonlocal and local morphometric attributes. Such combined variables include *TWI* and *SPI*, which are functions of *CA* and *G* (Section 2.6.1, Table 2.2). Thus, *TWI* and *SPI* derivation consist of (1) calculations of *CA* and *G* by related methods described previously in this chapter and (2) subsequent arithmetic combinations of two DTMs obtained.

It is obvious that selection of both methods depends strongly on the DEM grid type. It would be incorrect to mix methods intended for different grid types, for instance, to calculate *CA* by a method for square grids while computing *G* by a method for spheroidal equal angular grids (or vice versa). A detailed discussion of computational errors caused by application of plane square-grid algorithms to spheroidal equal angular DEMs can be found elsewhere (Florinsky, 2017b).

Fig. 4.22 illustrates the derivation of *TWI* and *SPI* for the Stavropol Upland. Its square-gridded DEM (Fig. 4.3) is described in Section 4.2.3.1. The *CA* model (Fig. 4.12A) was computed by the Martz-de Jong method (Section 4.4.1), while the *G* model (Fig. 4.4A) was calculated by the author's method (Section 4.2).

Fig. 4.23 illustrates the derivation of *TWI* and *SPI* for the Sinai Peninsula and adjacent territories. The spheroidal equal angular DEM (Fig. 4.10A) is described in Section 4.3.4. The *CA* model (Fig. 4.14A) was computed by the Martz–de Jong method adapted to spheroidal equal angular grids on a sphere or an ellipsoid of revolution (Section 4.4.2), while the *G* model (Fig. 4.10B) was calculated by the author's method (Section 4.3).

In both cases, to deal with the large dynamic range of *TWI* and *SPI*, we logarithmically transformed their values by Eq. (8.1). Data were processed by the LandLord software (Florinsky, 2016).

Thus, *TWI* and *SPI* can be directly calculated on spheroidal equal angular grids, without additional procedures of data interpolation to plane square grids (or data reprojection into an equal area projection). Such reprojections have been performed to provide correct computation of *G*, *CA*, *TWI*, and some other attributes for HYDRO1k, the quasi-global hydrographic database (Verdin and Greenlee, 1998; USGS, 2000) derived from GTOPO30 (USGS, 1996). However, Marthews et al. (2015) reported the preparation of a quasi-global model of *TWI* using HydroSHEDS, the SRTM3-enhanced and derived hydrographic database (Lehner et al., 2008). The computation of *TWI* was performed directly on the spheroidal equal angular grid, but the dependence of cell sizes on the latitude was ignored. Such incorrectness could be avoided by applying the approach described in this section, especially since it was used earlier to calculate the global model of *TWI* for Mars (Florinsky, 2008b).

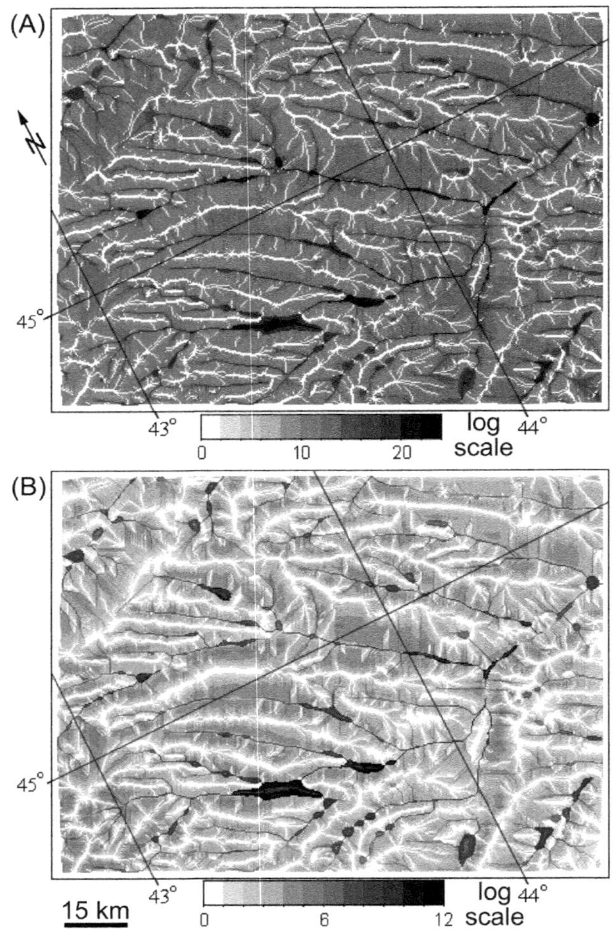

FIG. 4.22 The Stavropol Upland, combined morphometric variables: (A) Topographic wetness index. (B) Stream power index. Calculations were conducted on the square grid (see details in the text). For the elevation map, see Fig. 4.3. *(Modified from Florinsky, I.V., 2010. Theory and Applications of Mathematical Cartographic Modeling of Topography (D.Sc. thesis). Institute of Mathematical Problems of Biology, Russian Academy of Sciences, Pushchino, Russia (in Russian), Fig. 1.10.)*

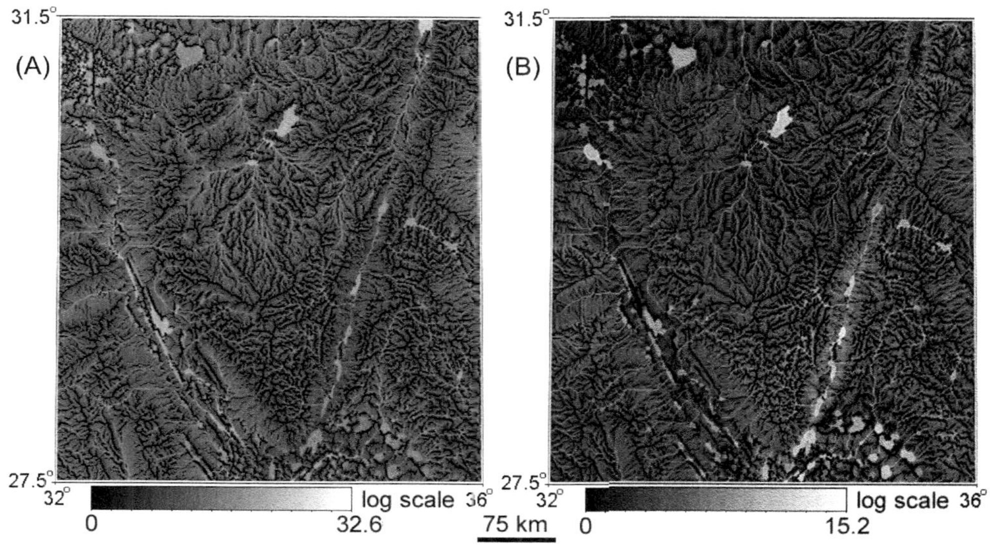

FIG. 4.23 Sinai Peninsula, combined morphometric variables: (A) Topographic wetness index. (B) Stream power index. Calculations were conducted on the spheroidal equal angular grid (see details in the text). For the elevation map, see Fig. 4.10A.

Finally, let us consider combined topographic variables, which are functions of several local morphometric attributes. Such combined variables include, for example, *IS* and *C*, which are functions of local morphometric attributes, namely, *H* and *K* as well as k_{max} and k_{min}, respectively, which in turn are functions of the first- and second-order partial derivatives of elevation (Section 2.6.2, Table 2.2). Therefore, there are no special methods to calculate these variables. Depending on the DEM grid type, one can select a calculation method for local morphometric variables (Sections 4.1–4.3).

4.8 Calculation of landform classifications

Most landform classifications are based on signs or value intervals of local morphometric variables as classification criteria (Section 2.7). Therefore, at the first stage, these morphometric variables are calculated using the methods described in Sections 4.1–4.3, depending on the grid geometry of the initial DEM. At the second stage, landform classification itself occurs, that is, quantitative characteristics of the topographic surface turn into qualitative categorical concepts of geomorphology (Section 2.7). Standard logical procedures are used for this purpose.

An algorithm for the geomorphon classification (Section 2.7.4) includes the coding of elevation values within a nine-point moving window (Fig. 2.24B). At a selected search distance from the central point of the window, the relative elevation of eight other points is determined along eight main compass directions. Relative elevation can take three values: higher, lower or same. The order of these codes determines the position of the central point and allows it to be assigned to 1 of the 10 most common landform classes (Fig. 2.24B). Calculations can be performed at different distances from the central point of the window to classify landforms of different scales (Jasiewicz and Stepinski, 2013).

For examples of calculated landform classification maps, see Figs. 2.23 and 2.25.

4.9 Calculations on a triaxial ellipsoid

4.9.1 Motivation

In the previous sections of this chapter, we demonstrated that existing methods and algorithms of geomorphometry can be applied to DEMs based on plane square grids or spheroidal equal angular grids on a sphere or an ellipsoid of revolution. However, it is advisable to use a triaxial ellipsoid for describing forms of small moons and asteroids (Duxbury, 1974; Soter and Harris, 1977; Thomas, 1989; Stooke, 1998; Drummond and Christou, 2008).

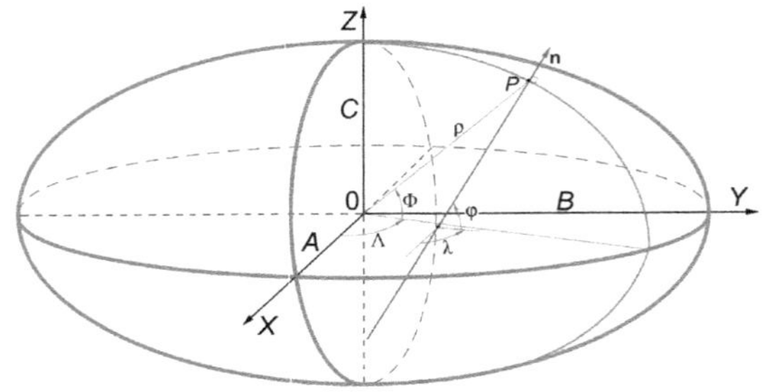

FIG. 4.24 A triaxial ellipsoid. A, B, C, semiaxes; **n**, normal to the ellipsoid surface in the given point P; X, Y, Z, the Cartesian axes; φ and λ, geodetic latitude and longitude, respectively; Φ and Λ, planetocentric latitude and longitude, respectively; ρ, radius vector from the ellipsoid center to the given point P. *(From Florinsky I.V., 2018a. Geomorphometry on the surface of a triaxial ellipsoid: towards the solution of the problem. Int. J. Geogr. Inf. Sci. 32 (8), 1558–1571, Fig. 1, reproduced with permission from Taylor & Francis Ltd.)*

A triaxial ellipsoid is a closed central surface of the second order (Fig. 4.24). In the Cartesian coordinates X, Y, and Z, its canonical implicit equation has the form:

$$\frac{X^2}{A^2} + \frac{Y^2}{B^2} + \frac{Z^2}{C^2} = 1, \tag{4.52}$$

where A and B are the large and small equatorial semiaxes, and C is the polar semiaxis; $A > B > C$.

In geodesy and differential geometry, various sections of the theory of a triaxial ellipsoid have been studied with varying depth and success. Coordinate systems of a triaxial ellipsoid and relationships between them have been repeatedly discussed (Shebuev, 1896; Kagan, 1947, § 11; Bespalov, 1980, chaps. I and III; Bugaevsky, 1999, § 1.1.1; Feltens, 2009; Ligas, 2012; Bektaş, 2015; Panou et al., 2016). Progress has been made in the development of cartographic projections for a triaxial ellipsoid (Snyder, 1985; Stooke and Keller, 1990; Bugaevsky, 1999; Nyrtsov et al., 2015). There is an analytical basis for describing geodesics, solving the direct and inverse geodetic problems, determining the length of meridian and parallel arcs, as well as the area of an ellipsoidal trapezium on a triaxial ellipsoid (Jacobi, 1839; Shebuev, 1896, 1898; Krasovsky, 1902, 1936; Bespalov, 1980, §§ 12 and 47; Karney, 2023; Panou, 2013). However, related computational issues are still insufficiently developed (Baillard, 2013).

It is clear that the geometry of a triaxial ellipsoid differs significantly from the geometry of an ellipsoid of revolution and especially a sphere (Fig. 4.24). For a triaxial ellipsoid, linear sizes of equal angular spheroidal trapezoids depend not only on the latitude (as for a sphere and an ellipsoid of revolution—Fig. 4.1C), but also on the longitude (Fig. 4.25A). Therefore,

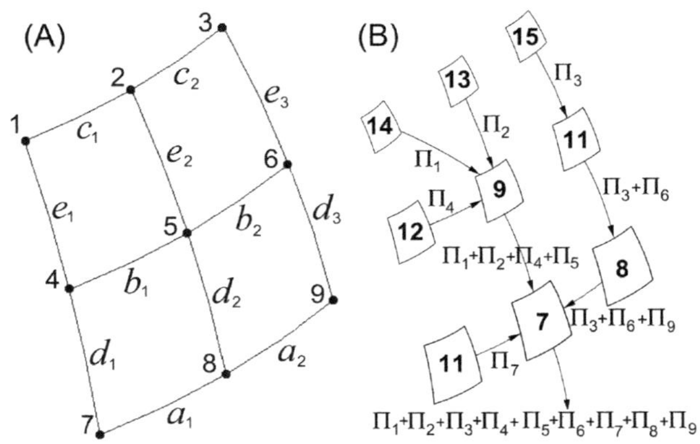

FIG. 4.25 Schemes for calculation of morphometric variables on the triaxial ellipsoid surface. (A) A spheroidal trapezoidal moving window is used to calculate local morphometric variables; a_i, b_i, c_i, d_i, e_i, linear sizes of window elements; 1, …, 9, node numbers. For a triaxial ellipsoid, $a_1 \neq a_2$, $b_1 \neq b_2$, $c_1 \neq c_2$, $d_1 \neq d_2 \neq d_3$, and $e_1 \neq e_2 \neq e_3$, while for an ellipsoid of revolution and a sphere, $a_1 = a_2 = a$, $b_1 = b_2 = b$, $c_1 = c_2 = c$, $d_1 = d_2 = d_3 = d$, and $e_1 = e_2 = e_3 = e$ (cf. Fig. 4.1C). (B) A scheme for a flow routing algorithm; *arrows*, flow directions; *bold numbers*, elevation values; Π_i, cell areas ($i = 1$, …, 9) at grid nodes located on different latitudes and longitudes. For a triaxial ellipsoid, $\Pi_1 \neq \Pi_2 \neq \Pi_3$, $\Pi_4 \neq \Pi_5 \neq \Pi_6$, and $\Pi_7 \neq \Pi_8 \neq \Pi_9$, while for an ellipsoid of revolution and a sphere, $\Pi_1 = \Pi_2 = \Pi_3$, $\Pi_4 = \Pi_5 = \Pi_6$, and $\Pi_7 = \Pi_8 = \Pi_9$ (cf. Fig. 4.13B). *(From Florinsky I.V., 2018a. Geomorphometry on the surface of a triaxial ellipsoid: towards the solution of the problem. Int. J. Geogr. Inf. Sci. 32 (8), 1558–1571, Fig. 3, reproduced with permission from Taylor & Francis Ltd.)*

methods and algorithms designed for an ellipsoid of revolution or a sphere (Sections 4.3 and 4.4.2) cannot be used for a triaxial ellipsoid. All these particular issues complicate the correct application of geomorphometric approaches in studies of small moons and asteroids.

Next, we formulate the problem of geomorphometric modeling on a triaxial ellipsoid as well as present its solution in analytical and computational forms (Florinsky, 2018a).

4.9.2 Coordinate systems

Global and quasi-global DEMs of planets and large moons, based on spheroidal equal angular grids, are commonly presented using the geodetic coordinate system. At the same time, available DEMs of small celestial bodies can be described with the planetocentric coordinate system. However, some calculations on the surface of a triaxial ellipsoid may be easier to perform using the ellipsoidal coordinate system. In this regard, let us recall definitions and formulas for coordinate systems of a triaxial ellipsoid and their transformation.

In the geodetic coordinate system (Krasovsky, 1902; Feltens, 2009), the coordinate lines are $\varphi=\text{const}$ and $\lambda=\text{const}$, where φ is the geodetic latitude and λ is the geodetic longitude. φ is an angle between the equatorial plane XY and a normal **n** to the ellipsoid surface at the given point P. λ is an angle measured in the XY plane between a line parallel to the X axis and the projection of the normal **n** to the ellipsoid surface on the plane XY (Fig. 4.24).

The Cartesian and geodetic coordinates are interrelated as follows (Bespalov, 1980, p. 55; Feltens, 2009):

$$\begin{aligned} X &= v\cos\varphi\cos\lambda \\ Y &= v\left(1-e_e^2\right)\cos\varphi\sin\lambda \\ Z &= v\left(1-e^2\right)\sin\varphi, \end{aligned} \tag{4.53}$$

where v is the length of the radius of the prime vertical:

$$v = \frac{A}{\sqrt{1-e^2\sin^2\varphi - e_e^2\cos^2\varphi\sin^2\lambda}}, \tag{4.54}$$

e and e_e are the first polar and equatorial eccentricities:

$$e = \frac{\sqrt{A^2-C^2}}{A}, \quad e_e = \frac{\sqrt{A^2-B^2}}{A}. \tag{4.55}$$

The inverse relation between the geodetic and Cartesian coordinates has the form (Feltens, 2009):

$$\varphi = \arctan\left(\frac{\left(1-e_e^2\right)}{\left(1-e^2\right)}\frac{Z}{\sqrt{\left(\left(1-e_e^2\right)^2X^2+Y^2\right)}}\right), \tag{4.56}$$

$$\lambda = \arctan\left(\frac{1}{\left(1-e_e^2\right)}\frac{Y}{X}\right). \tag{4.57}$$

In the planetocentric coordinate system (Bugaevsky, 1999, p. 13), the planetocentric latitude Φ is an angle between the equatorial plane XY and the radius vector ρ from the ellipsoid center to the given point P on the ellipsoid surface. The planetocentric longitude Λ is an angle between the plane XZ and a plane passing through the ellipsoid axis Z and the given point P on the ellipsoid surface (Fig. 4.24). The Cartesian and planetocentric coordinates are interrelated as follows (Bespalov, 1980, pp. 30–31):

$$\begin{aligned} X &= \rho\cos\Phi\cos\Lambda \\ Y &= \rho\cos\Phi\sin\Lambda \\ Z &= \rho\sin\Phi, \end{aligned} \tag{4.58}$$

$$\rho = \left[\left(\frac{\cos\Phi\cos\Lambda}{A}\right)^2 + \left(\frac{\cos\Phi\sin\Lambda}{B}\right)^2 + \left(\frac{\sin\Phi}{C}\right)^2\right]^{-1/2}. \tag{4.59}$$

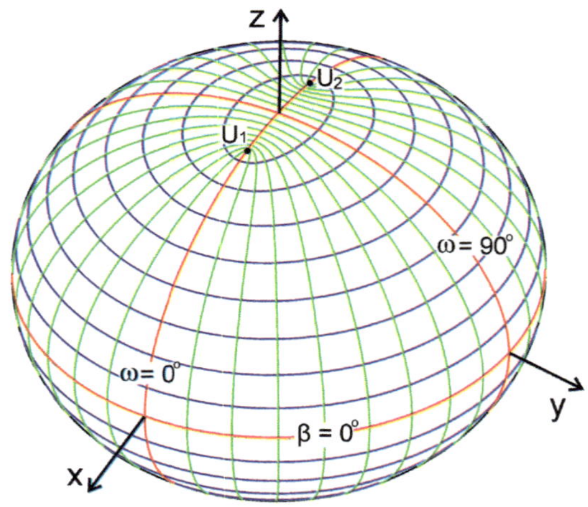

FIG. 4.26 A triaxial ellipsoid. *Blue and green lines*, ellipsoidal coordinates β and ω, respectively; *red lines*, principal sections; U_1, U_2, umbilical points. *(From Panou, G., 2013. The geodesic boundary value problem and its solution on a triaxial ellipsoid. J. Geodet. Sci. 3 (3), 240–249, Fig. 1, © Versita, courtesy of De Gruyter Open Ltd.)*

In the ellipsoidal coordinate system (Kagan, 1947, § 11), the coordinate lines are $\beta=$const and $\omega=$const, where β is the ellipsoidal latitude and ω is the ellipsoidal longitude (Fig. 4.26). Closed curves β are intersection loci of a triaxial ellipsoid and a confocal system of two-sheeted hyperboloids. Closed curves ω are intersection loci of a triaxial ellipsoid and a confocal system of one-sheeted hyperboloids.

The following formulas describe the relationship between the Cartesian and ellipsoidal coordinates (Jacobi, 1839; Bespalov, 1980, pp. 53–55; Karney, 2023):

$$\begin{aligned} X &= A\cos\omega\sqrt{\frac{A^2 - B^2\sin^2\beta - C^2\cos^2\beta}{A^2 - C^2}} \\ Y &= B\cos\beta\sin\omega \\ Z &= C\sin\beta\sqrt{\frac{A^2\sin^2\omega + B^2\cos^2\omega - C^2}{A^2 - C^2}}, \end{aligned} \tag{4.60}$$

To calculate the ellipsoidal coordinates from the known Cartesian and geodetic ones, it is necessary to apply the following iterative formulas (Bektaş, 2015):

$$\begin{aligned} \omega_{i+1} &= \arctan\left(\frac{AY\sqrt{(m^2-1)\sin^2\beta_i + 1}}{BX\cos\beta_i}\right) \\ \beta_{i+1} &= \arctan\left(\frac{BZ\sin\omega_i}{CY\sqrt{1 - m^2\cos^2\omega_i}}\right), \end{aligned} \tag{4.61}$$

where $i=0, 1, 2, \ldots$; $\beta_0=\varphi$, $\omega_0=\lambda$;

$$m = \sqrt{\frac{A^2 - B^2}{A^2 - C^2}}. \tag{4.62}$$

4.9.3 Solution for the inverse geodetic problem

Recall that to derive digital models of local morphometric variables (Table 2.2) from a DEM, one should estimate values of the partial derivatives of elevation r, t, s, p, and q (Eq. 2.2). For an ellipsoid of revolution or a sphere, the partial derivatives of elevation are functions of a, b, c, d, and e (Fig. 4.1C, Eqs. 4.27–4.31). For a triaxial ellipsoid, the partial derivatives of elevation are analogically functions of a_i, b_i, c_i, d_i, and e_i, which are linear distances between nodes of a 3 × 3 spheroidal equal angular moving window (Fig. 4.25A).

A distance S between two points $P_1(\beta_1, \omega_1)$ and $P_2(\beta_2, \omega_2)$ on a triaxial ellipsoid surface can be determined by the solution of Jacobi (1839) in the ellipsoidal coordinate system. The following equation can be applied (Karney, 2023):

$$dS = \frac{\sqrt{B^2 \sin^2\beta + C^2 \cos^2\beta}\sqrt{(B^2 - C^2)\cos^2\beta - \gamma}}{\sqrt{A^2 - B^2 \sin^2\beta - C^2 \cos^2\beta}} d\beta + \frac{\sqrt{A^2 \sin^2\omega + B^2 \cos^2\omega}\sqrt{(A^2 - B^2)\sin^2\omega + \gamma}}{\sqrt{A^2 \sin^2\omega + B^2 \cos^2\omega - C^2}} d\omega, \tag{4.63}$$

where

$$\gamma = (B^2 - C^2)\cos^2\beta \sin^2\alpha - (A^2 - B^2)\sin^2\omega \cos^2\alpha, \tag{4.64}$$

where α is an angle between a geodesic from $P_1(\beta_1, \omega_1)$ to $P_2(\beta_2, \omega_2)$ and a line $\omega = \text{const}$.

4.9.4 Length of meridian and parallel arcs

a_i, b_i, c_i, d_i, and e_i values, necessary for computing the partial derivatives of elevation, are arcs of parallels and meridians (Fig. 4.25A). Thus, to derive local morphometric variables, one can alternatively determine lengths of parallel and meridian arcs.

Bespalov (1980, chap. III) developed analytical formulas for the lengths of parallel and meridian arcs on a triaxial ellipsoid as well as published several particular expressions for numerical solutions. Later, we give general computational formulas.

To estimate the length S_p of a parallel arc located between two points $P_1(\varphi, \lambda_1)$ and $P_2(\varphi, \lambda_2)$ on a latitude φ between longitudes λ_1 and λ_2, the following expression can be used:

$$S_p = \frac{A \cos\varphi(1 - e^2 \sin^2\varphi - e_e^2 \cos^2\varphi)}{\sqrt{(1 - e^2 \sin^2\varphi - e_e^2 \cos^2\varphi \sin^2\lambda_m)^3}}(\lambda_2 - \lambda_1), \tag{4.65}$$

where $\lambda_m = (\lambda_1 + \lambda_2)/2$; $(\lambda_2 - \lambda_1)$ is measured in radians.

To determine the length S_m of a meridian arc located between two points $P_1(\varphi_1, \lambda)$ and $P_2(\varphi_2, \lambda)$ on a longitude λ between latitudes φ_1 and φ_2, one can apply the following formula:

$$S_m = A(1 - e^2)\sqrt{\frac{1 - 2e_e^2 \cos^2\lambda - e_e^4 \sin^2\lambda(1 - \cos^2\varphi_m \cos^2\lambda)}{(1 - e^2 \sin^2\varphi_m - e_e^2 \cos^2\varphi_m \sin^2\lambda)^3}}(\varphi_2 - \varphi_1), \tag{4.66}$$

where $\varphi_m = (\varphi_1 + \varphi_2)/2$; $(\varphi_2 - \varphi_1)$ is measured in radians.

One can see that the computations of the length of meridian/parallel arcs are straightforward and are performed in the commonly used geodetic coordinate system. In the geomorphometric context, these computations are easier, considerably shorter, and so preferred over the solution of the inverse geodetic problem (Section 4.9.3) because the latter requires the transformation from geodetic coordinates to ellipsoidal ones (Eqs. 4.61 and 4.62) as well as the determination of α for Eq. (4.64).

Table 4.1 represents the lengths of 1-arc-minute parallel and meridian arcs computed by Eqs. (4.65) and (4.66), respectively, for some latitudes and longitudes of an arbitrary triaxial ellipsoid. The following semiaxes values were used: $A = 6{,}378{,}245$ m, $B = 6{,}368{,}245$ m, and $C = 6{,}356{,}863$ m. A and C values are semiaxes of the Krasovsky ellipsoid of revolution, and B value was chosen to accentuate the triaxial ellipsoidal effect on lengths of parallel/meridian arcs and areas of ellipsoidal trapeziums (Section 4.9.5). One can clearly see, in particular, that linear lengths of parallel/meridian arcs depend not only on the latitude (as for an ellipsoid of revolution and a sphere), but also on the longitude.

TABLE 4.1 Lengths (m) of 1′ parallel and meridian arcs as well as areas (m²) of 1″ × 1″ ellipsoidal trapeziums for selected latitudes (φ) and longitudes (λ) of the triaxial ellipsoid (for the semiaxes values, see Section 4.9.4).[a]

φ	λ							
	0	45	90	135	180	225	270	315
65	785.83	786.16	786.49	786.16	785.83	786.16	786.49	786.16
	1852.41	1856.12	1859.82	1856.12	1852.41	1856.12	1859.82	1856.12
	404.58	404.81	405.04	404.81	404.58	404.81	405.04	404.81
35	1518.29	1520.69	1523.10	1520.69	1518.29	1520.69	1523.10	1520.69
	1843.24	1849.07	1854.91	1849.07	1843.24	1849.07	1854.91	1849.07
	779.07	780.72	782.37	780.72	779.07	780.72	782.37	780.72
15	1787.29	1791.22	1795.16	1791.22	1787.29	1791.22	1795.16	1791.22
	1838.39	1845.34	1852.31	1845.34	1838.39	1845.34	1852.31	1845.34
	915.42	918.10	920.80	918.10	915.42	918.10	920.80	918.10
−5	1842.60	1846.90	1851.22	1846.90	1842.60	1846.90	1851.22	1846.90
	1837.29	1844.49	1851.71	1844.49	1837.29	1844.49	1851.71	1844.49
	943.33	946.27	949.22	946.27	943.33	946.27	949.22	946.27

[a]*Value triplets:* upper, *the length of parallel arc;* center, *the length of meridian arc;* lower, *the area of ellipsoidal trapezium.*

4.9.5 Cell area

To derive digital models of nonlocal and combined morphometric variables from a DEM (Fig. 4.25B), one should estimate areas of ellipsoidal trapezoidal cells (Section 4.4.2) (Florinsky, 2017b). Bespalov (1980, chap. III) developed an analytical formula for the area of a triaxial ellipsoidal trapezium as well as published a particular expression for the numerical solution. A general computational formula is presented later.

To estimate the area Π of a triaxial ellipsoidal trapezium, located between two parallels with latitudes φ_1 and φ_2 as well as two meridians with longitudes λ_1 and λ_2, one can apply the following expression:

$$\Pi = \frac{A^2(1-e^2)\left(1-e_e^2\right)(\lambda_2-\lambda_1)(\sin\varphi_2 - \sin\varphi_1)}{\left(1-e^2\sin^2\varphi_m - e_e^2\cos^2\varphi_m\sin^2\lambda_m\right)^2}. \tag{4.67}$$

This computation is conducted in the geodetic coordinate system.

Table 4.1 represents the areas of 1″ × 1″ ellipsoidal trapeziums computed by Eq. (4.67) for some latitudes and longitudes of a triaxial ellipsoid with the parameters specified in Section 4.9.5. One can clearly see that areas of ellipsoidal trapeziums depend not only on the latitude (as for an ellipsoid of revolution and a sphere), but also on the longitude.

4.9.6 General algorithms for geomorphometric calculations

A general algorithm for the derivation of local morphometric variables on a triaxial ellipsoid can be set up as follows:

1. Take a DEM.
2. If a DEM is given in the planetocentric coordinate system, transform it into the geodetic coordinate system using Eqs. (4.56)–(4.58).
3. Read a set of the geodetic coordinates and elevation values (φ_i, λ_i, z_i), $i = 1, \ldots, 9$, for the nine nodes of a spheroidal trapezoidal moving window (Fig. 4.25A).
4. Estimate linear sizes of the window elements a_i, b_i, c_i, d_i, and e_i (Fig. 4.25A) using Eqs. (4.65) and (4.66) and the known values of φ_i and λ_i.
5. Compute values of the first- and second-order partial derivatives r, t, s, p, and q (Eq. 2.2) for the central node of the moving window using the known values of z_i as well as the estimated values of a_i, b_i, c_i, d_i, and e_i. Considering that within a moving window, $a_1 = a_2 = a$, $b_1 = b_2 = b$, $c_1 = c_2 = c$, $d_1 = d_2 = d_3 = d$, and $e_1 = e_2 = e_3 = e$ with a millimeter-range accuracy (Florinsky, 2018a), Eqs. (4.27)–(4.31) can be applied to calculate r, t, s, p, and q.

6. Compute the value of a local morphometric variable for the central node of the moving window using the estimated values of r, t, s, p, and q.
7. Move the window to the next position and repeat ## 3–6.

A general algorithm for derivation of nonlocal morphometric variables on a triaxial ellipsoid can be set up as follows:

1. Take a DEM.
2. If a DEM is given in the planetocentric coordinate system, transform it into the geodetic coordinate system using Eqs. (4.56)–(4.58).
3. Read the geodetic coordinates (φ_i, λ_i) for all DEM cells.
4. Calculate areas of all DEM cells Π_i by Eq. (4.67).
5. Apply one of the flow routing methods (Section 4.4.1), which should be adapted to the geometry of spheroidal equal angular grids (Section 4.4.2) (Florinsky, 2017b). Specifically, at a downslope cell, *CA* is determined as the sum of areas Π_i of upslope cells, which are passed by flows to reach the downslope cell (Fig. 4.25B).

Computation of combined variables, which are functions of nonlocal and local morphometric attributes (e.g., *TWI*), is performed by the consecutive application of two algorithms for nonlocal and local variables (as discussed earlier) and then by the arithmetic combination of two derived models.

Chapter 5

Errors and accuracy

5.1 Sources of DEM errors

DEM accuracy and errors depend on the type of topography, method for DEM generation, type of DEM grid, and DEM resolution or DEM grid spacing (Isaacson and Ripple, 1990; Li, 1994; Hunter and Goodchild, 1995; Aguilar et al., 2005; Reuter et al., 2009). Rieger (1996) proposed the following classification of DEM errors:

- Low-frequency systematic errors resulting, for example, from procedures of the photogrammetric orientation. These errors are critical for combined processing and analysis of DTMs and remotely sensed data.
- Medium-frequency systematic and random errors caused, for instance, by errors in photogrammetric measurements and interpolation procedures. These errors are critical for DTM-based soil, geological, and other applied studies.
- High-frequency random errors (noise), which may be, for example, connected with image defects and sensor instability.

5.1.1 DEMs from conventional topographic surveys

The accuracy of DEMs, produced by conventional topographic surveys using total station instruments, depends essentially on systematic instrumental errors and random operator errors (Ghilani, 2022, § 8.20). Instrumental errors are mainly caused by maladjustment of reference axes of a total station instrument.

5.1.2 DEMs from kinematic GNSS topographic surveys

The accuracy of DEMs created using kinematic GNSS survey is, in particular, influenced by ionospheric and tropospheric refraction, satellite ephemeris errors, satellite and receiver clock biases, limited access to satellites in forested areas, and so on (Ghilani, 2022, §§ 13.6, 14.8, and 15.11). The accuracy of a kinematic GNSS survey based on the combined use of several global navigation satellite systems, for example, GPS and GLONASS, can be higher than that based on the GPS or GLONASS satellites only (Pirti et al., 2013; Kaartinen et al., 2015).

5.1.3 DEMs from stereophotogrammetry of optical remote sensing images

The accuracy of DEMs generated by photogrammetric techniques is impacted by the following factors (Lobanov, 1984; Makarovic, 1984b; Kraus, 2007):

- systematic and random operator errors (Fig. 5.1)
- systematic and random instrumental errors
- resolution of aerial or satellite images depending mainly on the camera or sensor characteristics
- vertical resolution of images dictated essentially by the base:height ratio
- snow cover, vegetation coverage, and clouds (Fig. 5.2)
- type of topography
- image accuracy depending on the curvature of the planet, atmospheric refraction, air- or spacecraft stability, camera distortion or sensor stability, image processing, errors of image transformation for removing geometric distortion, and the like

In digital photogrammetry, errors in elevation measurements can account for 5%–20% of the total measurements, depending on the image quality (Akovetsky, 1994). Aguilar et al. (2005) demonstrated that in digital photogrammetry, DEM accuracy depends essentially on three factors: type of topography, DEM grid spacing, and interpolation technique.

Digital Terrain Analysis. https://doi.org/10.1016/B978-0-443-24798-9.00025-X

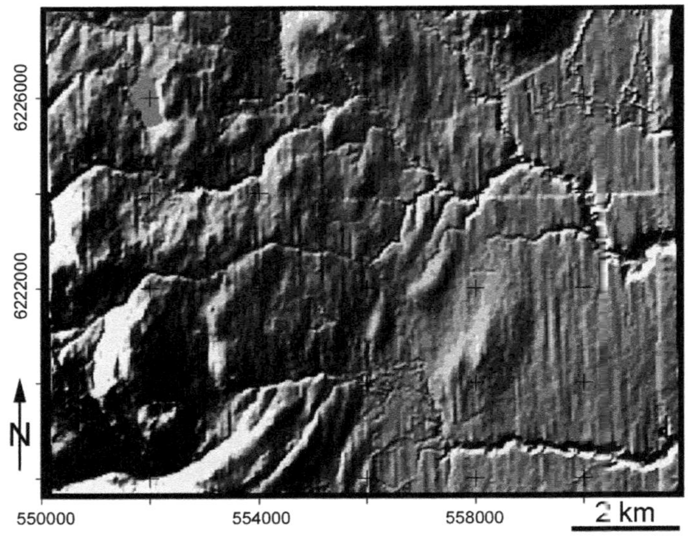

FIG. 5.1 North-south-oriented striping artifacts on the hill-shading map, the result of the DEM generation using a semiautomatic profiling of aerial stereo images. Terrain Resource Information Management DEM for the Fort St. John Forest District, British Columbia, Canada; DEM grid spacing is 30m. *(From Albani, M., Klinkenberg, B., 2003. A spatial filter for the removal of striping artifacts in digital elevation models. Photogrammet. Eng. Remote Sens. 69 (7), 755–765, Fig. 1, reproduced with permission from American Society for Photogrammetry and Remote Sensing.)*

FIG. 5.2 Samples of cloud patterns in the ASTER GDEM. Tasmania, DEM grid spacing is 1″. *(From Hirt, C., Filmer, M.S., Featherstone, W. E., 2010. Comparison and validation of the recent freely available ASTER-GDEM ver1, SRTM ver4.1 and GEODATA DEM-9S ver3 digital elevation models over Australia. Aust. J. Earth Sci. 57 (3), 337–347, Fig. 2b, © Geological Society of Australia, reproduced with permission from Taylor & Francis Ltd.)*

Fully automated digital photogrammetric processing of stereo images (without further editing) produces, as a rule, less accurate DEMs than analog photogrammetric techniques (Gong et al., 2000).

5.1.4 UAV- and SfM-based DEMs

UAV- and SfM-based DEMs can be characterized by serious systematic broad-scale errors such as vertical "doming" of the surface. This can be caused by the use of relatively simple, consumer-grade cameras with high distortion and inappropriate calibration procedures (James and Robson, 2014; James et al., 2017b, 2020).

Automatic ground filtering of the dense point cloud (Section 3.1.4) can produce artifacts in DEMs. Qin et al. (2023) identified the following sources of such artifacts:

- outliers under/above the ground
- objects with large coverage (e.g., large buildings)
- objects close to the ground (e.g., bridges connected to the ground, low vegetation cover, fences beside the road)
- areas with dense vegetation cover and sharply changing topography (e.g., cliffs and shores)
- areas characterized by low and high slopes (e.g., slope areas with scatted buildings)
- disconnected terrain (e.g., jagged hilly edges)

Application of different methods for ground filtering leads to different resulted DEMs and different artifacts (Tan et al., 2018; Yilmaz et al., 2018). Moreover, some of such artifacts can be seen only after derivating models of local morphometric variables (that is, after calculating partial derivatives of elevation) due to the sensitivity of partial derivatives to sharp, step-like changes in elevation function values (Section 5.4.1).

As an example, consider the ground filtering of UAV- and SfM-based DEMs of the Zaoksky testing ground (Florinsky et al., 2018b). It was impossible to remove a low grass cover with an average height of about 9 cm from the dense point cloud. Although these artifacts are not visible on the elevation map (Fig. 5.3A), there are barely noticeable traces of several trails on the horizontal curvature map (Fig. 5.3C). In this case, step-like changes in elevation values are located along the edges of trails and the low grass cover. Also, there is a pronounced polygonal artifact located in the southwestern corner of the horizontal curvature map (Fig. 5.3D) related to a forested zone. It was impossible to filter the dense point cloud because true values of ground elevation within the forested zone cannot be derived from aerial images. Such artifacts are unrecoverable without ancillary elevation data obtained, for instance, from aerial lidar or ground-based GNSS surveys. On the left side of the horizontal curvature map, one can also see several pronounced linear artifacts (Fig. 5.3D). These are traces of the roads occurring for the same reason as the trail traces (as discussed earlier). There are trees and shrubs along the edges of the roads; that is, there are sharp, step-like changes in elevation values along the road edges. Notice that the discussed polygonal and linear artifacts are not visible on the elevation map (Fig. 5.3B).

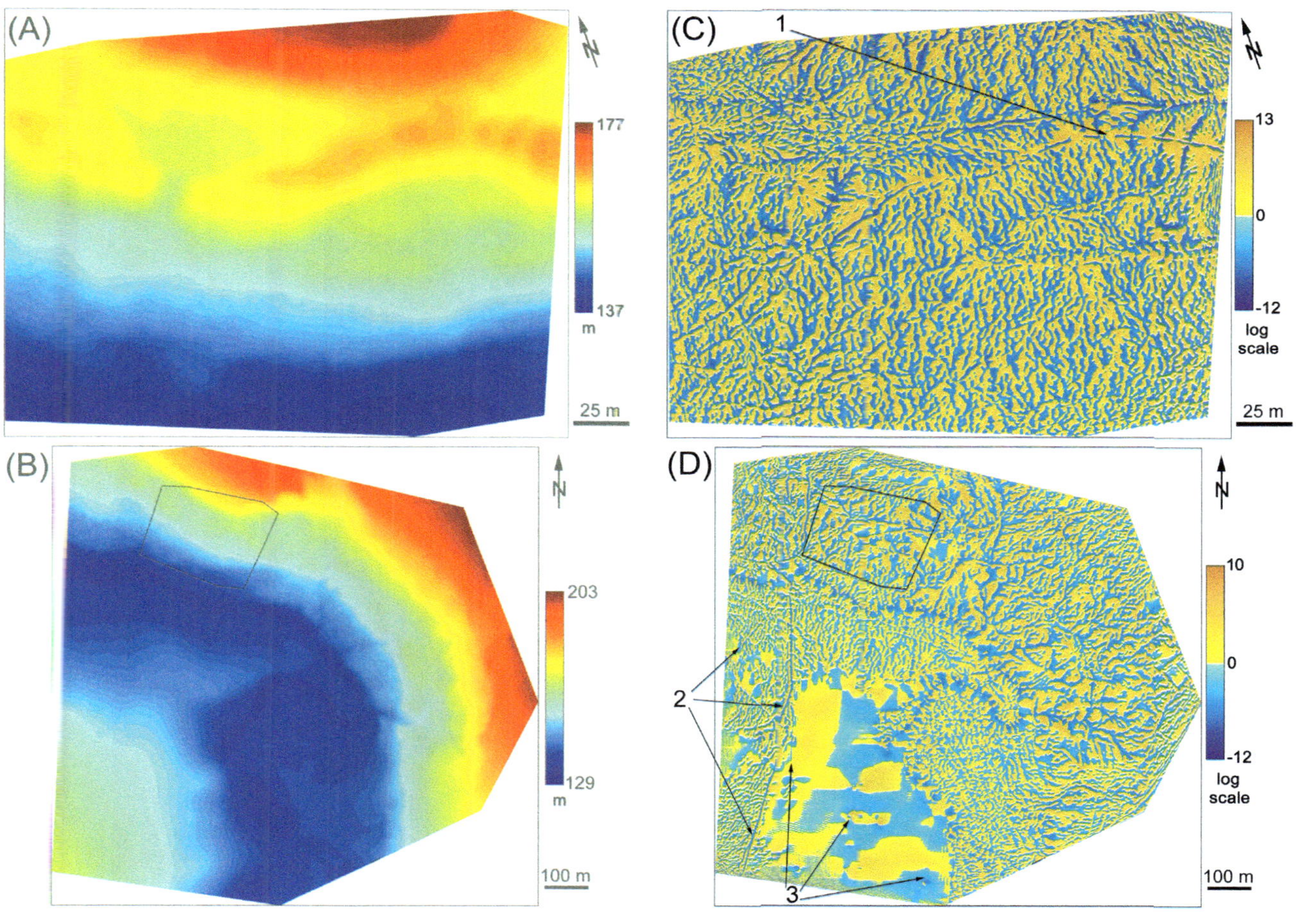

FIG. 5.3 Artifacts in UAV- and SfM-based DEMs caused by incomplete ground filtering. The Zaoksky testing ground, Tula Region, Russia. Elevation: (A) 20-cm grid spacing. (B) 1-m grid spacing. Horizontal curvature: (C) 20-cm grid spacing. (D) 1-m grid spacing. *Black frames in maps (B) and (D)* indicate the location of maps (A) and (C), respectively. *Arrows*, artifacts: 1, trail; 2, roads; 3, forest. *(From Florinsky, I.V., Kurkov, V.M., Bliakharskii, D.P., 2018b. Geomorphometry from unmanned aerial surveys. Trans. GIS 22 (1), 58–81, Figs. 7 and 9, reproduced with permission from John Wiley & Sons.)*

5.1.5 Lidar DEMs

The accuracy of lidar-derived DEMs depends on sensor characteristics, vertical and horizontal error in the positioning of the laser platform, laser scan angle, flight speed and altitude, as well as physical properties of the surface, such as surface reflectivity and slope gradient (Huising and Gomes-Pereira, 1998; Wehr and Lohr, 1999; Brenner et al., 2007) (Fig. 5.4). In lidar survey, systematic errors can range from 5 cm for flat, paved areas to 200 cm for areas with grass and shrub cover, while random errors can range from 10 cm for flat areas to 200 cm for hilly areas (Huising and Gomes-Pereira, 1998). Errors also occur during the processing of the raw data, including data filtering, gridding, segmentation, and object reconstruction (Aguilar and Mills, 2008).

5.1.6 Interferometric DEMs

The accuracy of InSAR-derived DEMs depends on radar characteristics, the type of topography, and the physical properties of the surface (Jarvis et al., 2004; Falorni et al., 2005; Brenner et al., 2007). For example, more than 15% of SRTM DEM tiles have voids for 1% of the covered area, and more than 5% of those have voids for 5% of the covered area.

Large voids are typical for mountainous terrains; geometric artifacts are here caused by layover, foreshortening, and radar shadow (Fig. 5.5). Large voids are also observed in sand deserts; a low dielectric constant of the surface is responsible for a low level of the signal reflectance (Elsner and Bonnici, 2007). In forested areas, the SRTM elevation is on average located ~40% of the distance from the top of the vegetation canopy to the ground (Carabajal and Harding, 2006).

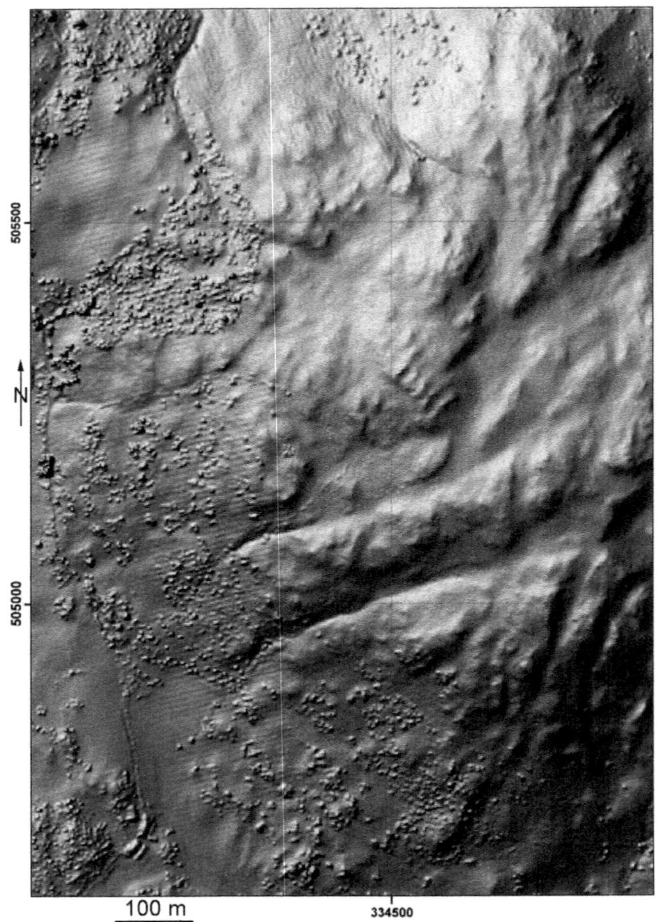

FIG. 5.4 Hill-shaded map of the lidar DEM for Loughrigg Fell, Lake District, United Kingdom. The artificial diagonal banding in two dominant directions (~75° and 143°) is associated with aircraft pitch motion. Illumination from the northwest; DEM grid spacing is 2 m. *(Lidar data © Environment Agency Science Enterprise Centre. From Arrell, K., Wise, S., Wood, J., Donoghue, D., 2008. Spectral filtering as a method of visualising and removing striped artefacts in digital elevation data. Earth Surf. Process. Landf. 33 (6), 943–961, Fig. 4, reproduced with permission from John Wiley & Sons.)*

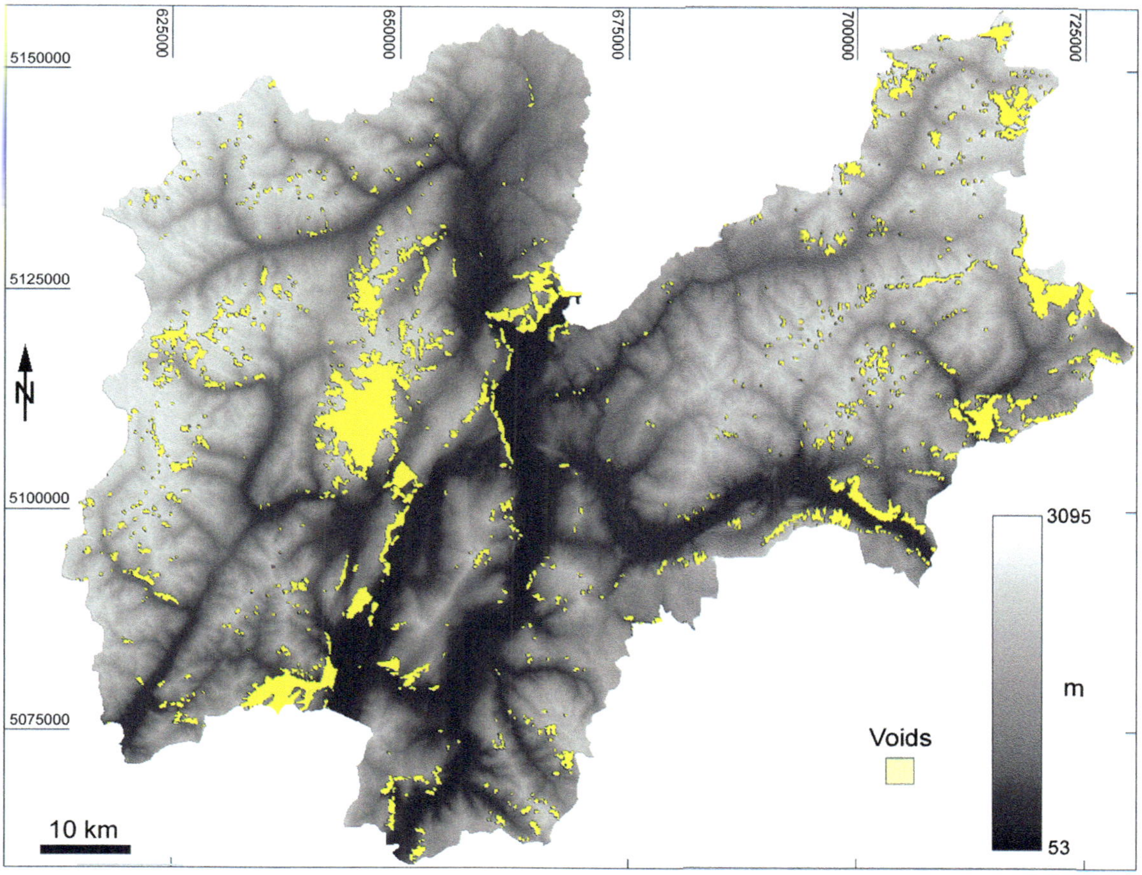

FIG. 5.5 Voids (yellow areas) in the SRTM1 DEM for Trentino, Italian Alps. DEM grid spacing is 25 m.

SRTM DEMs also suffer from striping artifacts: "the stripes have a typical wavelength of about 800 m with an amplitude of about 0.2–4 m and are aligned diagonally in a pattern that suggests a close relationship to the orbital paths … Striping at varying levels occurred everywhere except where it was obscured by high topographic relief" (Gallant and Read, 2009, pp. 149 and 151).

It is necessary to stress that the use of fully automated approaches of digital photogrammetry and SfM as well as lidar and InSAR data processing leads, in fact, to "blind sampling" of elevations (Lemmens, 1999). Such DEMs need editing and filtering (Jarvis et al., 2004; Sithole and Vosselman, 2004; Falorni et al., 2005; Evans and Hudak, 2007; Karkee et al., 2008; Gallant and Read, 2009) because they describe not the topographic surface but a surface consisting of portions of the land surface, buildings, vegetation cover, and the like (Fig. 5.6), that is, digital surface models (DSMs) (Egels and Kasser, 2002, chap. 3). This kind of data per se can be useful for the recognition of man-made objects and urban mapping (e.g., Gruen et al., 1997).

5.1.7 Bathymetric DEMs

Errors in bathymetric DEMs produced by echo sounding may be caused by instrumental errors of an echo sounder and navigation errors. The errors are common for flat seafloor areas and steep deep-water slopes (Smith, 1993; Bourillet et al., 1996; Abramova, 2020a,b, 2021a,b).

Errors in bathymetric DEMs created with optical sensing depend on instrumental errors in sensors, the correctness of regression models for relationships between image brightness and water depth, water turbidity, and bottom reflectance relating to bottom characteristics and water depth (Gao, 2009). The accuracy of lidar-based bathymetric DEMs is influenced by water clarity and bottom features (Pe'eri et al., 2011).

The accuracy of bathymetric DEMs produced with satellite radar altimetry data depends on the platform and radar characteristics. Also, error sources include water surface roughness (waves), tide model errors, ocean currents, and signal delay

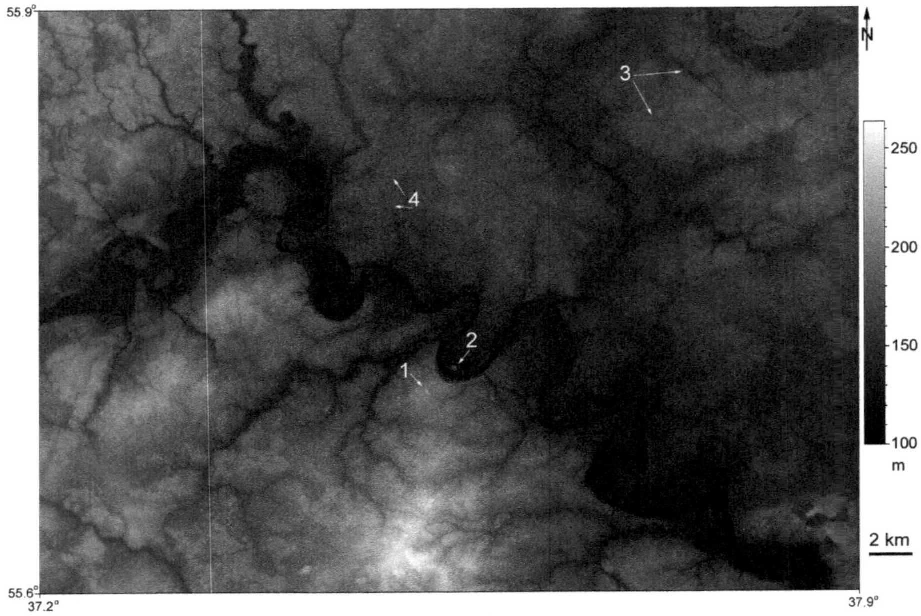

FIG. 5.6 Examples of human-made objects contained in the SRTM1 DEM for the city of Moscow: 1, the main building of the Lomonosov Moscow State University; 2, Luzhniki Stadium; 3, portion of the Moscow Perimeter Highway and other roads crossing the forested area of the Losiny Ostrov National Park; 4, some wide streets. The DEM includes 1,363,141 points (the elevation matrix 1261 × 1081); the DEM grid spacing is 2″ × 1″.

in the ionosphere and troposphere (Sandwell and Smith, 2001). However, most errors are associated with limitations of physical mathematical models linking gravitational anomalies with seafloor topography, which are applied to convert altimetric data into seafloor DEMs. One may obtain adequate results for wavelengths of 20–200 km within areas with a sedimentary cover up to 200 m thick.

5.1.8 Subsurface DEMs

The accuracy of auger hole or borehole data-derived DEMs of soil horizon surfaces or stratigraphic surfaces is mostly affected by the absence of information on the shape of subsurface horizons between auger holes or boreholes (Jones and Johnson, 1983; Groshong Jr., 2006, chap. 10). There are no reasons to assume that interpolation of subsurface DEMs may provide an adequate picture of stratigraphic surfaces.

The accuracy of DEMs of stratigraphic surfaces produced with 3D seismic survey is, in particular, influenced by limitations in the survey design, coherent noise, and systematic errors in the processing (Brown, 2011; Liner, 2016). Groshong Jr. (2006, pp. 30–31) made the following observations:

> *"The geometry of a structure that is even moderately complex displayed in travel time is likely to be significantly different from the true geometry of the reflecting boundaries because of the distortions introduced by steep dips and laterally and vertically varying velocities. Reflections from steeply dipping units may return to the surface beyond the outer limit of the recording array and so are not represented on the seismic profile. The structural interpretation of seismic reflection data requires the conversion of the travel times to depth. This requires an accurate model for the velocity distribution, something not necessarily well-known for a complex structure. The most accurate depth conversion is controlled by velocities measured in nearby wells."*

5.1.9 Contour-based DEMs

The accuracy of contour-based DEMs depends on the following factors (Carter, 1988; Kumler, 1994):

- accuracy of topographic maps
- random operator errors (up to 90% of the total number of DEM errors)
- systematic instrumental errors

If a DEM has been created using any interpolation technique, then the DEM obviously includes systematic and random errors of the interpolator (Carrara et al., 1997; Bater and Coops, 2009).

5.1.10 Fused DEMs

Two problems are typical for a DEM compiled from a set of other DEMs:

(1) Different portions of the DEM may have different accuracy.
(2) Boundaries between the portions may clearly be seen on all morphometric maps.

These problems are familiar to users of existing global and quasi-global DEMs (Table 3.1). An example of such errors is discussed in Section 6.3.3.

5.2 Estimation of DEM accuracy

DEM accuracy is commonly estimated by the criterion of RMSE of elevation computed by comparing DEM points and reference (ground-truth) points (Ivanov and Kruzhkov, 1992; Bolstad and Stowe, 1994). Elevation RMSE is usually presented as the integral metric for the accuracy of the entire DEM. However, different portions of a DEM may have distinct accuracy (Hunter and Goodchild, 1995; Carlisle, 2005; Fisher and Tate, 2006), in particular, depending on the terrain type and shape.

In this regard, the accuracy of global and nearly global DEMs is evaluated in terms of common error statistics (e.g., mean error, mean absolute error, RMSE) for particular regions or continents (Rodríguez et al., 2006; Farr et al., 2007; Sharma et al., 2010; Slater et al., 2011; Bulatović et al., 2012). Ground-truth points are commonly used to estimate these statistics. Also, data from more accurate DEMs are utilized as reference points; for example, Carabajal and Harding (2006) estimated the accuracy of SRTM DEMs by comparing them with ICESat laser altimetry elevations. In some cases, a comparative analysis of several DEMs for a particular region can be useful to understand the accuracy and applicability of these DEMs (Hirt et al., 2010; Grohmann, 2014).

Rieger (1996) proposed comparing a target DEM with a reference DEM. Recently, for accuracy assessment and control of the spatial distribution of errors in a target DEM, a matrix of residuals is often used. This matrix represents the difference between the target DEM and the reference DEM (Fig. 5.7). Lidar DEMs are usually used as reference ones (Florinsky et al., 2019; Guth and Geoffroy, 2021; Trevisani et al., 2023a). Sometimes a matrix of residuals is called a DEM of difference (DoD).

Various approaches can be used to analyze the spatial distribution of DEM errors. In particular, Makarovič (1972) proposed estimating DEM accuracy by calculating the mean deviation of sinusoidal amplitudes obtained after interpolation from those existing before interpolation. Tempfli (1980) evaluated DEM accuracy by spectral analysis. Frederiksen (1981) developed a method to predict DEM accuracy based on summing the Fourier spectrum for high-frequency components of topographic profiles.

Polidori et al. (1991) attempted to estimate DEM interpolation errors by calculating the DEM fractal dimension at different scales and in different directions. Ackermann (1978) and Li (1994) proposed formulas for elevation RMSE as a function of slope gradient. However, the accuracy of the gradient calculation also depends on elevation RMSE (Felicísimo, 1995; Florinsky, 1998a) (Section 5.3). Hannah (1981) developed a technique to detect DEM random errors. This method was based on a point comparison with adjacent points, using slope gradient and change-gradient values with specified thresholds.

Felicísimo (1994b) proposed an elegant method utilizing differences between two elevation values for each DEM point: the first is recorded in the DEM, and the second is interpolated from its four neighboring points. A mean, a standard deviation, and the Student's t-criterion are then calculated. Relatively high t-values indicate possible errors.

Kraus (1994) and Hunter and Goodchild (1995) described two approaches for mapping the horizontal accuracy of elevation data. In the first approach, "the epsilon band" is drawn around the position of a contour line defining the elevation z_i. The borders of the epsilon band are two contours defining elevations $(z_i - \Delta z/2)$ and $(z_i + \Delta z/2)$, where Δz is a contour interval. The epsilon band approach means that the true position of the z_i contour is somewhere between the band borders. In the second approach, the researcher should calculate and map the likelihood of the true value of elevation exceeding a threshold.

López (1997) described a method that was based on the decomposition of a regular DEM into strips, considering it as a multivariate table. Principal component analysis was applied to select unlikely elevations.

Detailed reviews of methods to analyze and model spatial distribution of DEM errors can be found elsewhere (Carlisle, 2005; Mesa-Mingorance and Ariza-López, 2020; Polidori and El Hage, 2020).

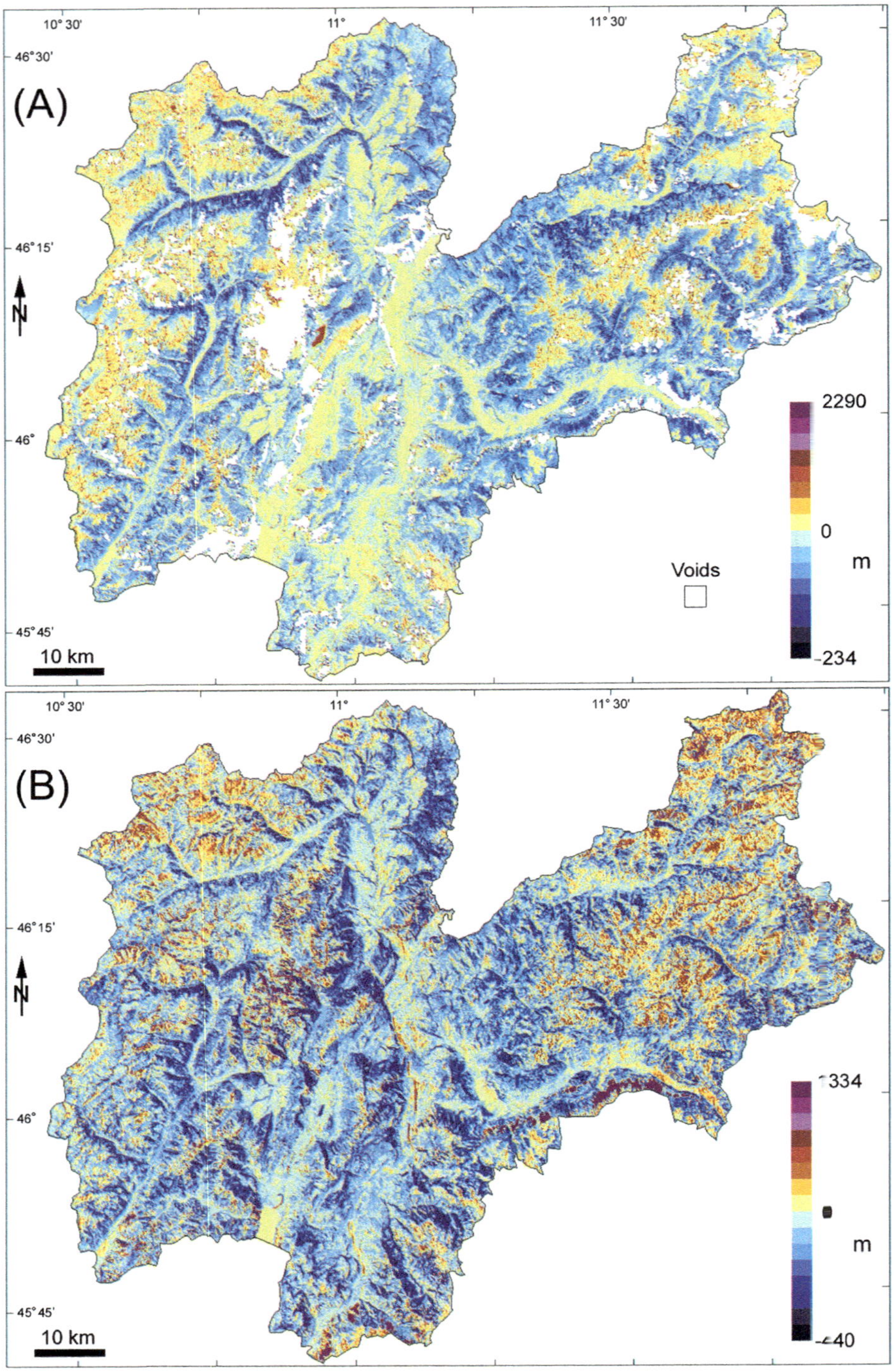

FIG. 5.7 Residuals (DoD) between two target DEMs and a lidar-based reference DEM for Trentino, Italian Alps: (A) SRTM1 DEM and (B) ASTER GDEM. DEM grid spacing is 25 m. *(From Florinsky, I.V., Skrypitsyna, T.N., Trevisani, S., Romaikin, S.V., 2019. Statistical and visual quality assessment of nearly-global and continental digital elevation models of Trentino, Italy. Remote Sens. Lett. 10 (8), 726–735, Fig. 3A and B, reproduced with permission from Taylor & Francis Ltd.)*

5.3 Calculation accuracy of local morphometric variables

5.3.1 Motivation

Since the inception of digital terrain modeling, considerable attention has been given to various issues of DTM accuracy (Fisher and Tate, 2006). This is connected with the fact that errors in digital models of morphometric attributes influence the accuracy and objectivity of DTM-based studies and the modeling of natural processes and phenomena (Desmet, 1997;

Holmes et al., 2000; Vörösmarty et al., 2000; Thompson et al., 2001; Van Niel et al., 2004; Bishop et al., 2006; Callow et al., 2007; Erskine et al., 2007; Wise, 2007).

In the early years of digital geomorphometry, calculation accuracy of local topographic variables was estimated by comparing their computed and "reference" values. For reference data, researchers used hand measurements of G and A from topographic maps (Evans, 1980; Skidmore, 1989), field measurements of G, A, and k_v (Bolstad and Stowe, 1994; Giles and Franklin, 1996), as well as values of morphometric variables derived from reference DEMs of actual terrains (Chang and Tsai, 1991; Sasowsky et al., 1992) and simulated surfaces (Carter, 1992; Felicísimo, 1995; Hodgson, 1995; Zhou and Liu, 2004).

However, the accuracy of digital models of local topographic attributes cannot be determined by the comparison of computed and "reference" values due to the accuracy paradox of geomorphometry. Indeed, it is well known that measurement accuracy is the difference between a measured value and an actual value of a variable. However, the actual land surface is not mathematically smooth. So, it cannot have derivatives and, hence, local morphometric variables, which are functions of the partial derivatives of elevation (Section 2.2). For the land surface, notions of derivatives arise only during its measurements (Shary, 1991, pp. 57–58). Therefore, the calculation accuracy of local topographic variables cannot be determined by a comparison of computed and "reference" values because there are no actual values.

Moreover, such a strategy can lead to artifacts as well as subjective and conflicting conclusions. For instance, it was reported that errors of A calculation are typical for flat areas (Chang and Tsai, 1991; Carter, 1992), while errors of G calculation are predominantly positioned on steep slopes (Chang and Tsai, 1991; Sasowsky et al., 1992; Bolstad and Stowe, 1994). However, Carter (1992) found that both G and A errors grow to high values within flat areas. At the same time, Davis and Dozier (1990) reported that G and A errors concentrate within zones of rapid change in G and A values (e.g., nearby crests or thalwegs).

Also, it is well known that the map accuracy of G, A, k_h, k_v, and other curvatures depends on DEM grid spacing (Evans, 1979, 1980; Stewart and Podolski, 1998; Shary et al., 2002b; Bergbauer and Pollard, 2003; Zhu et al., 2008). For instance, the increase of DEM grid spacing can transform small, steep zones into broad areas marked by medium G values. It was reported that G and A calculation errors increase as grid spacing increases (Chang and Tsai, 1991). However, Carter (1992) found that G and A computed values more closely correspond to their "reference" values as grid spacing increases.

Therefore, the calculation accuracy of local morphometric variables cannot be adequately studied by a comparison of calculated and "reference" values. At the same time, it is obvious that it principally depends on the accuracy of initial data, that is, DEMs, and the accuracy of a calculation algorithm. So, attention should be focused on these two factors. Thus, Felicísimo (1995) found that G errors increase with increasing elevation RMSE (m_z). Brown and Bara (1994) and Giles and Franklin (1996) intimated that the number of errors in calculations of the partial derivatives of elevation depends on the DEM noise level. Skidmore (1989) and Hodgson (1995) compared the accuracy of several methods for G and A derivation, that is, the calculation of p and q (Section 2.2). Skidmore (1989) found that p- and q-computation algorithms based on six points of a 3×3 moving window are more accurate than the four-point algorithms. At the same time, Hodgson (1995) argued that the four-point algorithms are more accurate than the six-point ones. This contradiction may be explained by the fact that those two studies were also carried out with a comparison of calculated and "reference" values of topographic variables.

It is clear that local morphometric variables are functions of measured variables $F = \varphi(x, y, \ldots, u)$, where $x, y, \ldots, u$ are measured variables. To estimate the accuracy of calculation of local morphometric attributes, Kuryakova (1996) proposed using the criterion of RMSE of a function of measured variables m_F (Merriman, 1899, p. 32; Bolshakov and Gaidaev, 1977, p. 117):

$$m_F = \sqrt{\left(\frac{\partial F}{\partial x}\right)^2 m_x^2 + \left(\frac{\partial F}{\partial y}\right)^2 m_y^2 + \cdots + \left(\frac{\partial F}{\partial u}\right)^2 m_u^2} \tag{5.1}$$

where $m_x, m_y, \ldots, m_u$ are RMSE of $x, y, \ldots, u$. Next, we describe the method for estimating the accuracy of calculation of local morphometric variables by this criterion (Florinsky, 1998a,c, 2008e, 2009b).

5.3.2 RMSE formulas for local morphometric variables

Using Eq. (5.1), let us develop formulas for RMSE of calculation of local morphometric attributes. Functions of measured variables are $G, k_h, k_v, K, k_{min}, k_{max}, H, E, K_a, M, K_r, k_{ve}$, and k_{he} (Table 2.2); measured variables are p, q, r, s, and t (Eq. 2.2). After differentiation and simple algebraic operations, we obtain the following expressions (Florinsky, 1998a, 2008e):

$$m_G = \sqrt{\left(\frac{\partial G}{\partial p}\right)_0^2 m_p^2 + \left(\frac{\partial G}{\partial q}\right)_0^2 m_q^2} = \dots = \frac{1}{(1+p^2+q^2)}\sqrt{\frac{p^2 m_p^2 + q^2 m_q^2}{p^2+q^2}}, \tag{5.2}$$

$$\begin{aligned} m_{k_h} &= \sqrt{\left(\frac{\partial k_h}{\partial r}\right)_0^2 m_r^2 + \left(\frac{\partial k_h}{\partial t}\right)_0^2 m_t^2 + \left(\frac{\partial k_h}{\partial s}\right)_0^2 m_s^2 + \left(\frac{\partial k_h}{\partial p}\right)_0^2 m_p^2 + \left(\frac{\partial k_h}{\partial q}\right)_0^2 m_q^2} \\ &= \dots = \frac{1}{p^2+q^2}\left(\frac{1}{1+p^2+q^2}\left\{m_p^2\left[p\left(q^2 r - 2pqs + p^2 t\right)\left(\frac{2}{p^2+q^2} + \frac{1}{1+p^2+q^2}\right) + 2(qs - pt)\right]^2\right.\right. \\ &\quad + m_q^2\left[q\left(q^2 r - 2pqs + p^2 t\right)\left(\frac{2}{p^2+q^2} + \frac{1}{1+p^2+q^2}\right) + 2(ps - qr)\right]^2 \\ &\quad \left.\left. + m_r^2 q^4 + 4m_s^2 p^2 q^2 + m_t^2 p^4\right\}\right)^{\frac{1}{2}}, \end{aligned} \tag{5.3}$$

$$\begin{aligned} m_{k_v} &= \sqrt{\left(\frac{\partial k_v}{\partial r}\right)_0^2 m_r^2 + \left(\frac{\partial k_v}{\partial t}\right)_0^2 m_t^2 + \left(\frac{\partial k_v}{\partial s}\right)_0^2 m_s^2 + \left(\frac{\partial k_v}{\partial p}\right)_0^2 m_p^2 + \left(\frac{\partial k_v}{\partial q}\right)_0^2 m_q^2} \\ &= \dots = \frac{1}{(p^2+q^2)(1+p^2+q^2)} \\ &\quad \times\left(\frac{1}{1+p^2+q^2}\left\{m_p^2\left[p\left(p^2 r + 2pqs + q^2 t\right)\left(\frac{2}{p^2+q^2} + \frac{3}{1+p^2+q^2}\right) - 2(pr + qs)\right]^2\right.\right. \\ &\quad + m_q^2\left[q\left(p^2 r + 2pqs + q^2 t\right)\left(\frac{2}{p^2+q^2} + \frac{3}{1+p^2+q^2}\right) - 2(ps + qt)\right]^2 \\ &\quad \left.\left. + m_r^2 p^4 + 4m_s^2 p^2 q^2 + m_t^2 q^4\right\}\right)^{\frac{1}{2}}, \end{aligned} \tag{5.4}$$

where m_p, m_q, m_r, m_s, and m_t are RMSE of calculation of p, q, r, s, and t, respectively; m_G, m_{k_h}, and m_{k_v} are RMSE of calculation of G, k_h, and k_v, respectively. Similarly, we obtain equations for other local topographic attributes:

$$m_K = \frac{\sqrt{16\left(p^2 m_p^2 + q^2 m_q^2\right)\left(rt - s^2\right)^2 + \left(r^2 m_t^2 + 4s^2 m_s^2 + t^2 m_r^2\right)\left(1+p^2+q^2\right)^2}}{\left(1+p^2+q^2\right)^3}, \tag{5.5}$$

$$m_{k_{max}} = m_{k_{min}} = \sqrt{m_H^2 + m_M^2}, \tag{5.6}$$

$$m_H = m_E = \frac{1}{2}\sqrt{m_{k_h}^2 + m_{k_v}^2}, \tag{5.7}$$

$$m_{K_a} = \sqrt{k_v^2 m_{k_h}^2 + k_h^2 m_{k_v}^2}, \tag{5.8}$$

$$m_M = \frac{1}{2}\sqrt{\frac{4H^2 m_H^2 + m_K^2}{H^2 - K}}, \tag{5.9}$$

$$m_{K_r} = \sqrt{k_{ve}^2 m_{k_{he}}^2 + k_{he}^2 m_{k_{ve}}^2}, \tag{5.10}$$

$$m_{k_{he}} = \sqrt{m_{k_h}^2 + m_{k_{min}}^2}, \tag{5.11}$$

$$m_{k_{ve}} = \sqrt{m_{k_v}^2 + m_{k_{min}}^2}, \tag{5.12}$$

where m_K, $m_{k_{min}}$, $m_{k_{max}}$, m_H, m_E, m_{K_a}, m_M, m_{K_r}, $m_{k_{he}}$, and $m_{k_{ve}}$ are RMSE of calculation of K, k_{min}, k_{max}, H, E, K_a, M, K_r, k_{he}, and k_{ve}, respectively.

5.3.3 RMSE formulas for the partial derivatives

According to Eqs. (5.2)–(5.12), one should calculate p, q, r, s, and t as well as m_p, m_q, m_r, m_s, and m_t to derive a digital model of m_F. Functions of measured variables are p, q, r, s, and t; measured variables are elevation values at nodes of a moving window z_i, where $i = 1, 2, \ldots, 9$ for the Evans-Young method (Section 4.1) and the author's method for the spheroidal equal angular grid (Section 4.3); $i = 1, 2, \ldots, 25$ for the author's method for the plane square grid (Section 4.2). So, formulas for m_p, m_q, m_r, m_s, and m_t can also be determined by the general expression for RMSE of a function of measured variables (Eq. 5.1).

5.3.3.1 *Calculation on the plane square grid*

Using Eq. (5.1), let us develop formulas for RMSE of calculation of p, q, r, s, and t for the Evans-Young method (Eqs. 4.2–4.6), and g, h, k, m, r, t, s, p, and q for the author's method (Eqs. 4.14–4.22).

For the Evans-Young method, we obtain, in particular:

$$m_r = \sqrt{\left(\frac{\partial r}{\partial z_1}\right)_0^2 m_{z_1}^2 + \left(\frac{\partial r}{\partial z_2}\right)_0^2 m_{z_2}^2 + \cdots + \left(\frac{\partial r}{\partial z_9}\right)_0^2 m_{z_9}^2}, \tag{5.13}$$

where m_{z_1}, m_{z_2}, …, m_{z_9} are RMSE of z_1, z_2, …, z_9, respectively. In the strict sense, $m_{z_i} = \psi(x, y)$ depends on the type of topography as well as methods of DEM generation and interpolation (Hunter and Goodchild, 1995). There are expressions to estimate m_z as a function of G (Li, 1994). However, this approach is incorrect because the accuracy of G calculation also depends on m_z (Eq. 5.2) (Florinsky, 1998a). At the same time, Li (1994) demonstrated that for contour-based DEMs, one can assume:

$$m_z = const = B\Delta z, \tag{5.14}$$

where Δz is the contour interval, $B = (0.16–0.33)$ depending on the type of topography. So, let us consider $m_{z_1} = m_{z_2} = \cdots = m_{z_9} = m_z$ and assume that this simplification is applicable to any type of DEM. Substituting m_z into Eq. (5.13), we obtain:

$$m_r = m_z \sqrt{\left(\frac{\partial r}{\partial z_1}\right)_0^2 + \left(\frac{\partial r}{\partial z_2}\right)_0^2 + \left(\frac{\partial r}{\partial z_3}\right)_0^2 + \cdots + \left(\frac{\partial r}{\partial z_9}\right)_0^2}. \tag{5.15}$$

After differentiation and simple algebraic operations, we obtain expressions of m_r, m_t, m_s, m_p, and m_q for the Evans-Young method (Table 5.1). Similarly, for the author's method, we obtain formulas of m_r, m_t, m_s, m_p, and m_q as well as m_g, m_h, m_k, and m_m (viz., RMSE of calculation of g, h, k, and m, respectively) (Table 5.1).

The formulas (Table 5.1) show that RMSEs of the partial derivatives of elevation are in direct proportion to m_z and in inverse proportions to w (for m_p and m_q), w^2 (for m_r, m_t, and m_s), and w^3 (for m_g, m_h, m_k, and m_m). Thus, the third-order partial derivatives are the most sensitive to the grid spacing value, while the first-order ones are the least sensitive to that.

The Evans-Young method has the highest accuracy among its analogues based on the approximation of the partial derivatives by finite differences on the 3 × 3 moving window (Florinsky, 1998a). Let us compare the formulas of m_r, m_t, m_s, m_p, and m_q related to the Evans-Young and the author's methods.

The formulas of m_r, m_t, m_s, m_p, and m_q (Table 5.1) show that under the same m_z and w values, application of the author's method results in significantly lower values of m_r, m_t, and m_s, than those related to the Evans-Young method. Indeed, m_r and m_t for the author's method are almost six times less than those for the Evans-Young method. m_s for the author's method is five times less than that for the Evans-Young method. m_p and m_q for the author's method are a mere 10% higher than those

TABLE 5.1 RMSE of partial derivatives calculated by the Evans-Young method (Section 4.1) and the author's method (Section 4.2).

RMSE	m_p and m_q	m_r and m_t	m_s	m_g and m_h	m_k and m_m
The Evans-Young method	$\frac{m_z}{\sqrt{6}w}$	$\frac{\sqrt{2}m_z}{w^2}$	$\frac{m_z}{2w^2}$	–	–
The author's method	$\sqrt{\frac{527}{70}}\frac{m_z}{6w}$	$\sqrt{\frac{2}{35}}\frac{m_z}{w^2}$	$\frac{m_z}{10w^2}$	$\frac{m_z}{\sqrt{2}w^3}$	$\frac{m_z}{\sqrt{35}w^3}$

for the Evans-Young method. This means that the author's method can provide a higher level of accuracy in the derivation of topographic variables than the Evans-Young method.

However, G is calculated using two first-order partial derivatives, p and q. Table 5.1 shows that m_p and m_q for Eqs. (4.21) and (4.22) are 10% higher than those for Eqs. (4.5) and (4.6). Therefore, the Evans-Young method is more accurate than the author's method for G calculation.

5.3.3.2 Calculation on the spheroidal equal angular grid

Similarly, we obtain formulas of m_p, m_q, m_r, m_s, and m_t for calculation of the partial derivatives of elevation on the spheroidal equal angular grid (Eqs. 4.27–4.31) (Florinsky, 1998c, 2008e):

$$m_p = m_z\sqrt{\frac{a^2d^2 + c^2e^2}{2\left[a^2c^2(d+e)^2 + b^2\left(a^2d^2 + c^2e^2\right)\right]}}, \tag{5.16}$$

$$\begin{aligned} m_q = {} & \frac{m_z}{3de(d+e)\left(a^4+b^4+c^4\right)} \\ & \times\Big(2\Big\{\left[d^2\left(a^4+b^4+b^2c^2\right)+c^2e^2\left(a^2-b^2\right)\right]^2+\left[e^2\left(a^4+c^4+a^2b^2\right)-d^2\left(a^4+c^4+b^2c^2\right)\right]^2 \\ & +\left[e^2\left(b^4+c^4+a^2b^2\right)+a^2d^2\left(c^2-b^2\right)\right]^2\Big\}+\left[d^2\left(a^4+b^4+3c^4-2b^2c^2\right)+2c^2e^2\left(b^2-a^2\right)\right]^2 \\ & +\left[e^2\left(a^4+3b^4+c^4-2a^2b^2\right)-d^2\left(a^4+3b^4+c^4-2b^2c^2\right)\right]^2 \\ & +\left[2a^2d^2\left(c^2-b^2\right)-e^2\left(3a^4+b^4+c^4-2a^2b^2\right)\right]^2\Big)^{\frac{1}{2}}, \end{aligned} \tag{5.17}$$

$$m_r = m_z\sqrt{\frac{6}{a^4+b^4+c^4}}, \tag{5.18}$$

$$m_s = m_z\frac{\sqrt{c^2\left[a^2(d+e)+b^2e\right]^2+b^2\left(a^2d-c^2e\right)^2+a^2\left[c^2(d+e)+b^2d\right]^2}}{\sqrt{2}\left[a^2c^2(d+e)^2+b^2\left(a^2d^2+c^2e^2\right)\right]}, \tag{5.19}$$

$$\begin{aligned} m_t = {} & \frac{2m_z}{3de(d+e)\left(a^4+b^4+c^4\right)} \\ & \times\Big(2\Big\{\left[d\left(a^4+b^4+b^2c^2\right)+c^2e\left(b^2-a^2\right)\right]^2+\left[d\left(a^4+c^4+b^2c^2\right)+e\left(a^4+c^4+a^2b^2\right)\right]^2 \\ & +\left[e\left(b^4+c^4+a^2b^2\right)+a^2d\left(b^2-c^2\right)\right]^2\Big\}+\left[d\left(a^4+b^4+3c^4-2b^2c^2\right)+2c^2e\left(a^2-b^2\right)\right]^2 \\ & +\left[d\left(a^4+3b^4+c^4-2b^2c^2\right)+e\left(a^4+3b^4+c^4-2a^2b^2\right)\right]^2 \\ & +\left[e\left(3a^4+b^4+c^4-2a^2b^2\right)+2a^2d\left(c^2-b^2\right)\right]^2\Big)^{\frac{1}{2}}, \end{aligned} \tag{5.20}$$

where a, b, c, d, and e are linear lengths of elements of the 3×3 spheroidal equal angular moving window (Fig. 4.1C).

The development of formulas (Eqs. 5.2–5.5, 5.16–5.20, Table 5.1) was carried out with the Maple software (© Waterloo Maple Inc., 1981–1997). A generalization of the described method was proposed by Zhou and Liu (2004).

Notice that the RMSE formula (Eq. 5.1) is not complete. In this particular form, it is usually applied to functions of independent measured variables. For a function of correlated or dependent variables, the RMSE formula includes residual terms of the Taylor series (Bolshakov and Gaidaev, 1977, p. 116):

$$\begin{aligned} m_F = {} & \left[\left(\frac{\partial F}{\partial x}\right)^2 m_x^2 + \left(\frac{\partial F}{\partial y}\right)^2 m_y^2 + \cdots + \left(\frac{\partial F}{\partial u}\right)^2 m_u^2 + 2\left(\frac{\partial F}{\partial x}\right)\left(\frac{\partial F}{\partial y}\right)m_x m_y \right. \\ & \left. + 2\left(\frac{\partial F}{\partial x}\right)\left(\frac{\partial F}{\partial u}\right)m_x m_u + \cdots\right]^{\frac{1}{2}}. \end{aligned} \tag{5.21}$$

It is obvious that residual terms can influence RMSE values. In some cases, elevation values at the nodes of a moving window may be correlated. Thus, it is reasonable to use Eq. (5.21) to estimate RMSE of calculation of the partial derivatives and morphometric variables. An interested reader may wish to develop related RMSE formulas as well as to estimate the influence of residual terms.

5.3.4 RMSE mapping

Eqs. (5.2)–(5.12) can be easily applied to produce digital models of m_F. Indeed, values of r, t, s, p, q, m_r, m_t, m_s, m_p, and m_q (and, hence, m_F) are estimated for the central point of the 3×3 (or 5×5) moving window. Moving the window along a DEM, it is possible to calculate m_F values for each point of the DEM except for boundary rows and columns.

Mapping is a convenient and pictorial strategy for visualizing the propagation of errors in spatial modeling (Heuvelink et al., 1989; Kraus, 1994; Hunter and Goodchild, 1995). Let us illustrate m_F mapping with two examples:

1. RMSE estimation on the plane square grid is exemplified by a portion of the Kursk Region, Russia.
2. RMSE estimation on the spheroidal equal angular grid is exemplified by the Moscow Region, Russia.

A DEM of the Kursk Region (Fig. 5.8A) is described in Section 14.4.2.2. A DEM of the Moscow Region (Fig. 5.9A) was extracted from the GLOBE DEM (GLOBE Task Team and Others, 1999). This DEM includes 59,400 points (the matrix 330×180); the grid spacing is $1'$.

To suppress high-frequency noise, both DEMs were smoothed three times with the 3×3 moving window. Using the author's method (Section 4.2), we calculated some morphometric attributes including H (Fig. 5.8B) for the Kursk Region; the grid spacing of 150 m was utilized. Using the other author's method (Section 4.3), we calculated k_h (Fig. 5.9B) and some other topographic variables for the Moscow Region; the grid spacing of $1'$ was applied. Using the formulas (Eqs. 5.3, 5.8, 5.16–5.20, and Table 5.1) with $m_z = 5$ m, we derived the m_H digital model for the Kursk Region (Fig. 5.8C) and the m_{k_h} digital model for the Moscow Region (Fig. 5.9C). To deal with the large dynamic range of H, k_h, m_H, and m_{k_h}, we logarithmically transformed their digital models by Eq. (8.1) with $n = 5$ for H and m_H, and $n = 7$ for k_h and m_{k_h}. Data were processed by the LandLord software (Florinsky, 2016).

Analysis of m_F maps allowed us to determine some regularity for the spatial distribution of m_F values. In a general case, the flatter the topography, the higher the m_F value (Florinsky, 1998a,c). m_F values can be in excess of the maximum absolute values of F within flat areas. However, this does not mean that there are errors in DTMs within such areas: m_F is a statistical property of a function F; that is, m_F values represent a possibility for errors.

To improve the impartiality of DTM-based studies, researchers have to treat data on local topographic variables with criticism as well as consider the mentioned effects in data interpretation. Models and maps of m_F can be used, for instance, as follows:

- To account for a spatial distribution of m_F in the analysis and interpretation of F map.
- To refine a DEM within areas having high m_F values, and then to recalculate F within these areas.

5.4 Ignoring the sampling theorem

5.4.1 Motivation

Let us recall the third consequence of the sampling theorem (Section 3.3): "landforms" with typical planar sizes less than $\lambda_{x,y}$, occurring in a DEM after its interpolation, should be considered interpolator-induced high-frequency noise. Every so often such a noise arises in a DEM with "enhanced" resolution, or overresolution, that is, a DEM produced by interpolation using a grid spacing essentially less than (1) an average density of points in an initial irregular DEM or (2) a grid spacing in an original regular DEM.

As this takes place, high-frequency noise may not be clearly seen in elevation maps derived from DEMs with "enhanced" resolution. This is because vertical amplitude of noise may be too low compared with a contour interval. However, these minor false "landforms" may dramatically be magnified in the subsequent derivation of topographic variables using differentiation procedures.

It is common knowledge of signal and image processing that the differentiation of a signal increases noise manifestation in a derivative (Baker, 1982; Rosenfeld and Kak, 1982, p. 238). In other words, differentiation can impair the signal-to-noise ratio. In a general case, the noise can have a higher derivative than the signal because the noise is less "smooth" than the signal and can fluctuate more randomly than the signal.

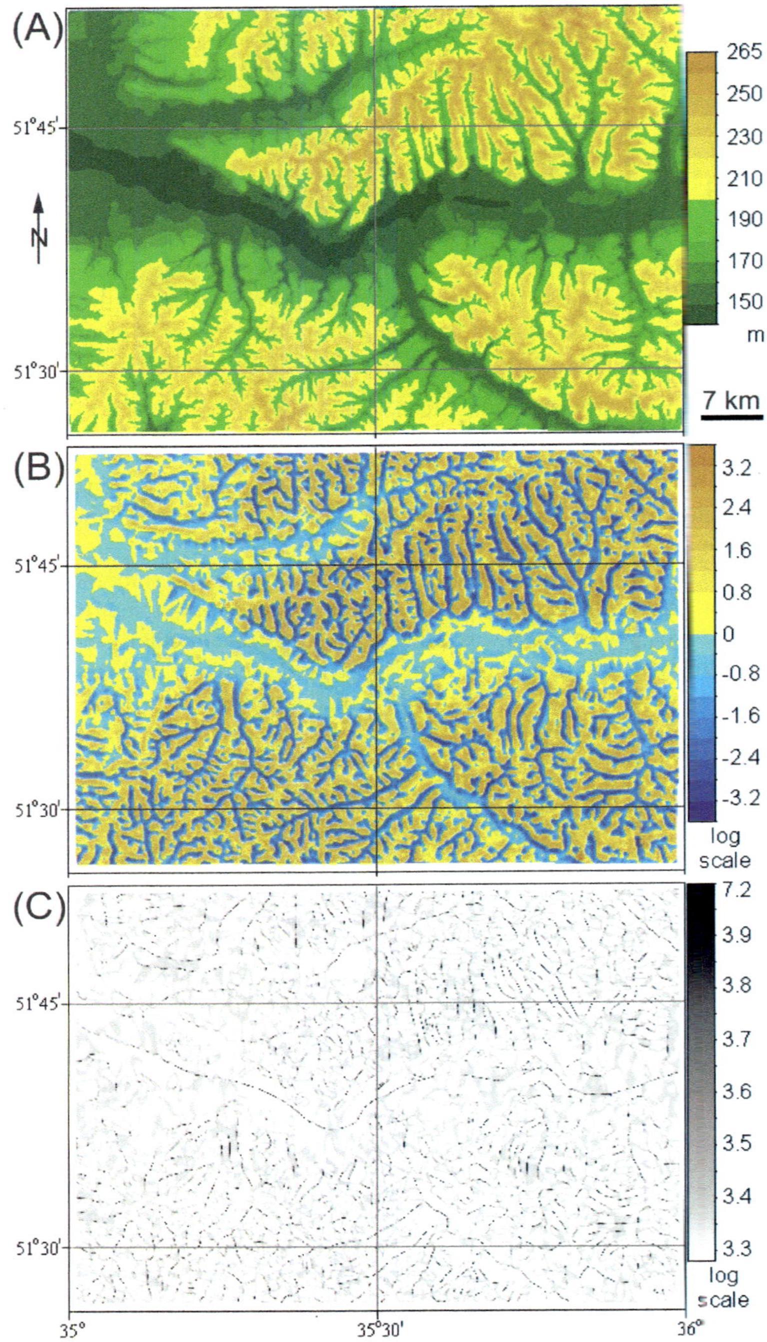

FIG. 5.8 The Kursk Region: (A) Elevation. (B) Mean curvature. (C) RMSE of the mean curvature calculation. *(Modified from Florinsky, I.V., 2010. Theory and Applications of Mathematical Cartographic Modeling of Topography (D. Sc. thesis). Institute of Mathematical Problems of Biology, Russian Academy of Sciences, Pushchino, Russia (in Russian), Fig. 2.6.)*

The higher the order of the derivative, the higher the noise manifestation. The following example can clarify this point for the one-dimensional case (Fig. 5.10). For an edge of the signal (Fig. 5.10A), its first-order derivative (Fig. 5.10B) has one extremum (a peak or a pit), while the second-order derivative (Fig. 5.10C) has two extrema (a peak and a pit). Also, for a peak or a pit of the signal, the first-order derivative has two extrema (a peak and a pit), while the second-order derivative has three extrema (a peak surrounded by two pits, or a pit surrounded by two peaks).

Derivation of local topographic variables from DEMs is carried out using the first- and second-order partial derivatives of the elevation function (Section 2.2). As any DEM includes noise (i.e., random and systematic errors), its propagation

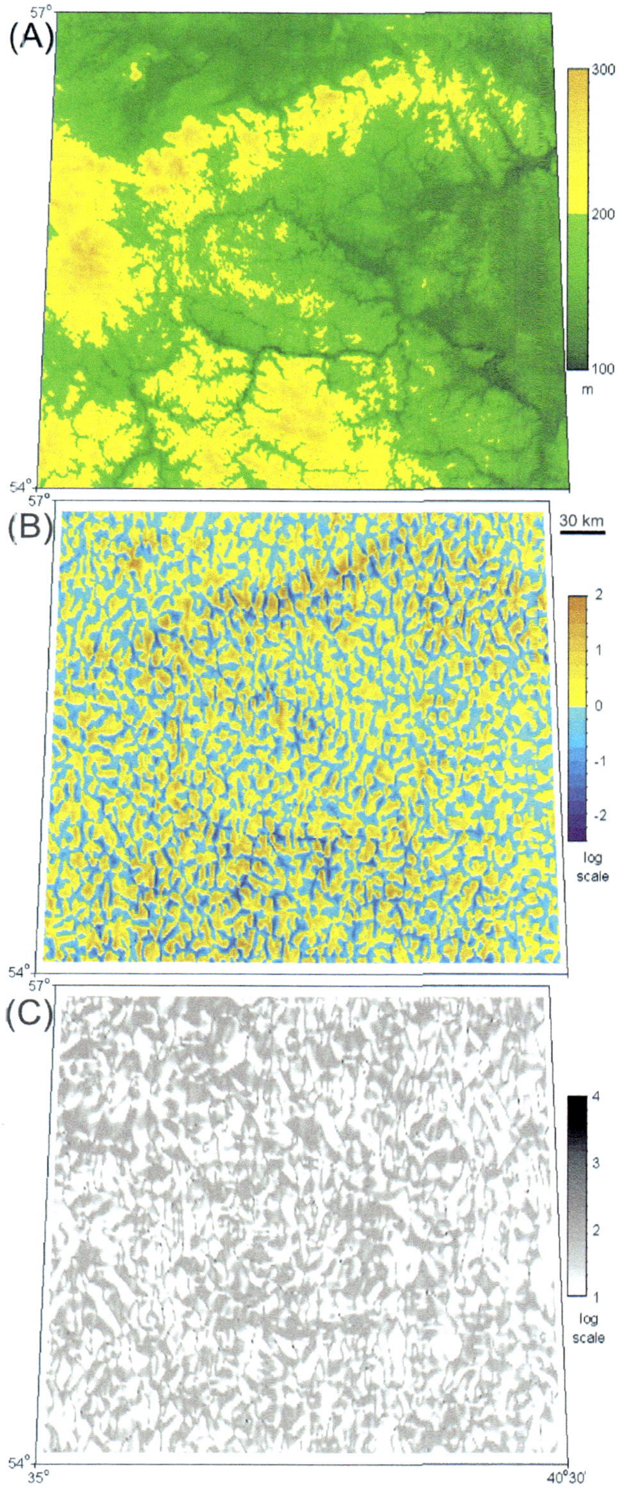

FIG. 5.9 The Moscow Region and adjacent territories: (A) Elevation. (B) Horizontal curvature. (C) RMSE of the horizontal curvature calculation. *(From Florinsky, I.V., 2008e. On calculation accuracy in digital terrain modeling. Geodesy Cartogr. (6), 28–32 (in Russian), Figs. 1, 2A and C.)*

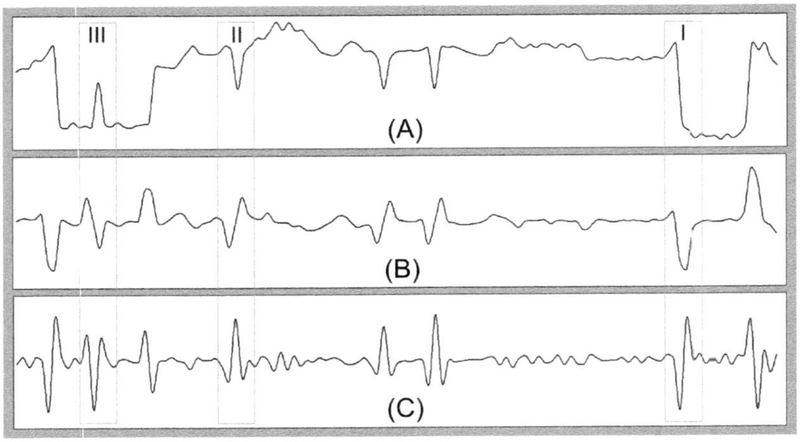

FIG. 5.10 Differentiation of a one-dimensional signal: (A) Signal. (B) First-order derivative. (C) Second-order derivative. *I*, *II*, *III*, edge, pit, and peak of the signal, respectively.

with a differentiation-induced magnification is typical for digital terrain modeling (Brown and Bara, 1994; Giles and Franklin, 1996; Desmet, 1997; Florinsky, 2002; Oksanen and Sarjakoski, 2005) (see, for example, illustrations in Sections 6.2.5 and 6.3.3).

In particular, scholars noticed errors typical of DEMs with "enhanced" resolution, such as false terraces, "traces" of contours, and triangular patterns (see later) (Batson et al., 1975; Wood and Fisher, 1993; Eklundh and Mårtensson, 1995; Desmet, 1997). However, these artifacts were explained by the lack of interpolation methods and were not associated with the overresolution of DEMs. In fact, such errors are the result of ignoring the sampling theorem. In the following section, we discuss this type of DTM error.

5.4.2 Materials and data processing

To illustrate errors caused by overresolution, we used DEMs of two areas: (1) the south coast of Crimea and adjacent sea bottom and (2) the Severny Gully in the city of Pushchino. The generation of the DEMs is described in Sections 14.4.1.2 and 10.3.2, respectively.

Using the inverse distance weighting interpolation (Watson, 1992, pp. 114–115) of the irregular DEM of Crimea, we produced two regular DEMs with $w=500$ m (Fig. 5.11A) and $w=3000$ m. Application of $w=500$ m gives an "enhanced" resolution since such w value is less than average distances between points within all parts of the irregular DEM (Fig. 5.11A). The grid spacing of 3000 m corresponds approximately to average distances between points describing the seafloors of the Black Sea. The Crimean Mountains can be treated with a smaller grid spacing, such as 1000 m, but we applied the single grid spacing value fitting all parts of the DEM. Digital models of G (Fig. 5.11B and C) and k_v (Fig. 5.11D and E) were derived from the two regular DEMs of Crimea by the Evans-Young method (Section 4.1).

Applying the Delaunay triangulation and a piecewise quadric polynomial interpolation with matching derivatives along triangle edges (Agishtein and Migdal, 1991) to the irregular DEM of the Severny Gully, we produced two regular DEMs with $w=0.25$ m (Fig. 5.12A) and $w=3$ m. Application of $w=0.25$ m certainly gives "enhanced" resolutions since this w value is well less than average distances between points of the irregular DEM. The grid spacing of 3 m approximately relates to average distances between points of the irregular DEM (Fig. 5.12A). Digital models of G (Fig. 5.12B and C) and k_h (Fig. 5.12D and E) were derived from the two regular DEMs of the Severny Gully by the Evans-Young method (Section 4.1).

To highlight artifacts on the k_v and k_h maps, we subdivided k_v and k_h values into two intervals relative to the zero value (Figs. 5.11D and E, 5.12D and E). Data were processed by the LandLord software (Florinsky, 2016).

5.4.3 Results and discussion

On the G and k_v maps of the Crimea's south coast derived from the DEM with $w=500$ m, one can see "traces" of contours (viz., curvilinear artifacts in form and location related to the contour lines) within the seafloor (Fig. 5.11B and C). These maps are obviously unfit for use for any purpose since they display a system of false curvilinear "landforms." However, G and k_v maps derived from the DEM with $w=3000$ m (Fig. 5.11C and E) can be useful in regional geomorphic and

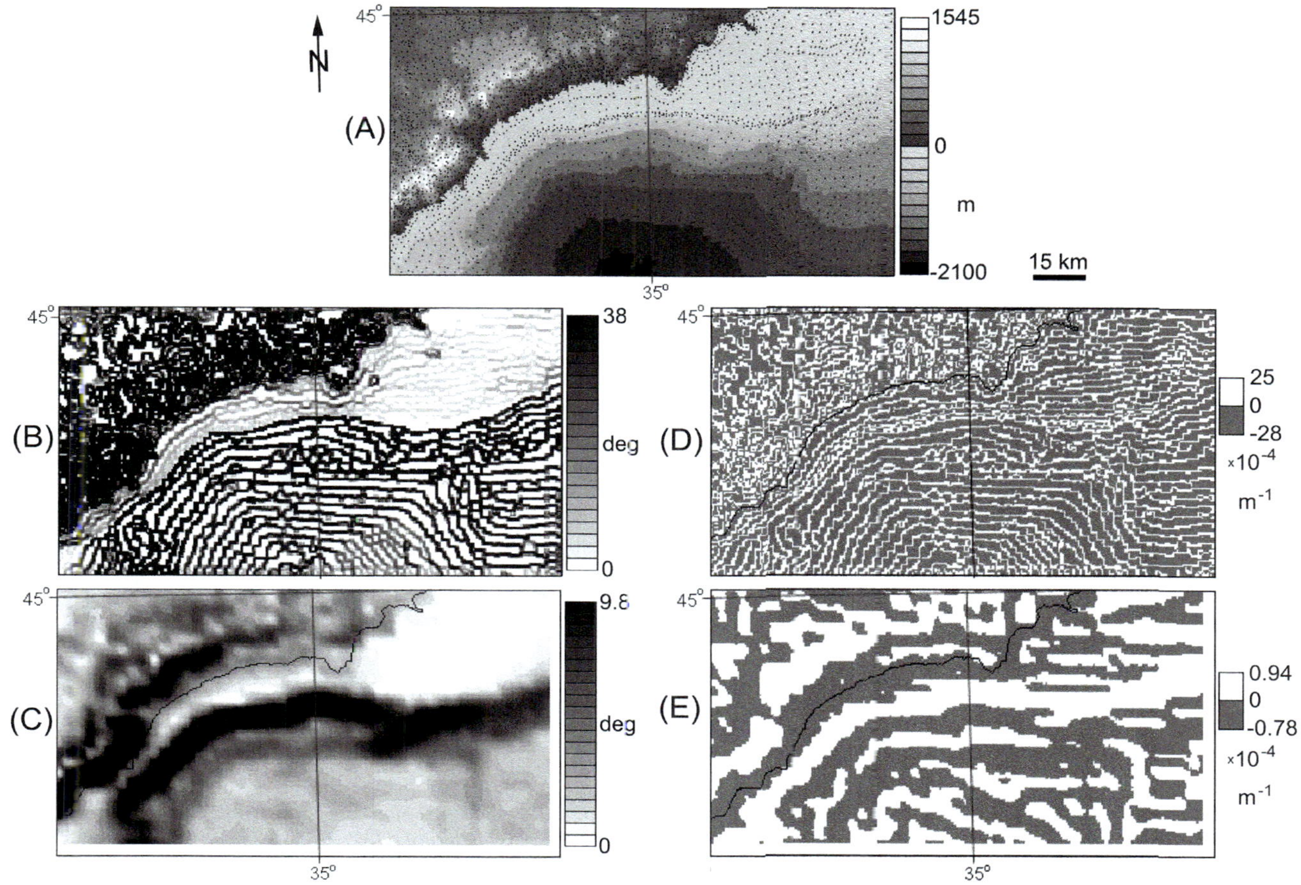

FIG. 5.11 South coast of Crimea and the adjacent sea bottom: (A) Elevation ($w = 500$ m) and distribution of the DEM points. (B) Slope gradient, $w = 500$ m. (C) Slope gradient, $w = 3000$ m. (D) Vertical curvature, $w = 500$ m. (E) Vertical curvature, $w = 3000$ m. *(Modified from Florinsky, I.V., 2010. Theory and Applications of Mathematical Cartographic Modeling of Topography (D.Sc. thesis). Institute of Mathematical Problems of Biology, Russian Academy of Sciences, Pushchino, Russia (in Russian), Fig. 1.14.)*

geological studies. For example, the k_v map (Fig. 5.11E) was successfully employed to recognize and classify lineaments (Florinsky, 1996) (Section 14.4.1).

On the G and k_h maps of the Severny Gully derived from the DEM with $w = 0.25$ m, one can see "marks" of triangulation, such as triangles and other relatively regular features (Fig. 5.12B and D). These maps are also unfit for use in any DTM-based application, as they reveal a set of false triangular "landforms." At the same time, G and k_h data derived from the DEM with $w = 3$ m (Fig. 5.12C and E) were successfully applied to study the dependence of soil moisture on topography (Florinsky and Kuryakova, 2000) (Section 10.3).

Traces of contours (Fig. 5.11B and D) are portraits of slender upright artificial "escarps" of broad flat false "terraces." "Escarps" may arise along contour lines, while "terraces" may occur between contours in DEM interpolation by the inverse distance weighting method. This method is sensitive to clustering of points: the search for nearest neighbors finds a lot of points along contours but none between ones. As a result, an elevation of a contour is assigned to regular DEM points along the contour, while circa average elevation of two neighboring contours is assigned to regular DEM points between these contours. It has been suggested that the method is not appropriate to interpolate contour-based DEMs (Wood and Fisher, 1993; Eklundh and Mårtensson, 1995). Triangular artifacts (Fig. 5.12B and D) are results of some inaccuracy of the matching derivatives along the edges of triangles in the piecewise interpolation (Agishtein and Migdal, 1991). However, these imperfections of the interpolation algorithms make themselves evident in the case of overdetailed interpolation only. Therefore, the actual root of the artifacts is a lack of information about elevation between contours and triangulation nodes.

Such inaccuracies lead to relatively minor errors in the regular DEMs with "enhanced" resolution. At least, it is impossible to see them on the elevation maps (Figs. 5.11A and 5.12A). Therefore, one can ignore these artifacts in some cases; for example, an elevation map derived from a DEM with "enhanced" resolution can be used for illustration goals. However,

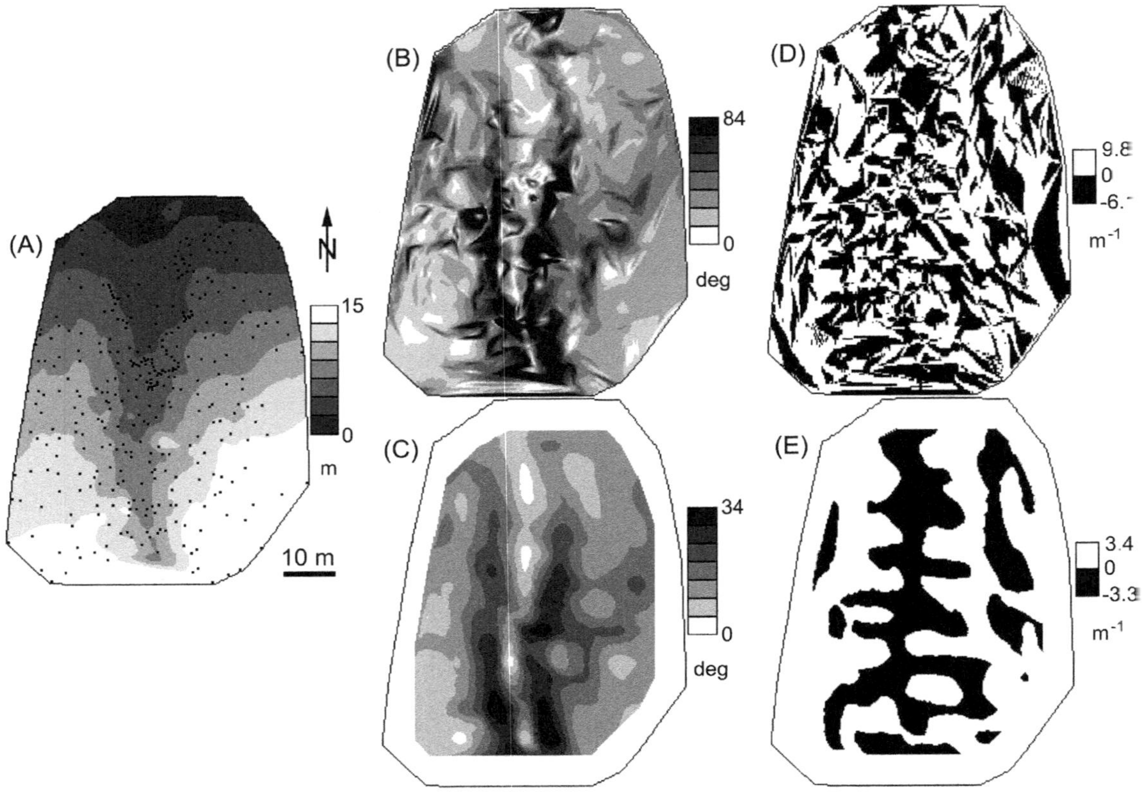

FIG. 5.12 The Severny Gully: (A) Elevation ($w = 0.25$ m) and distribution of the DEM points. (B) Slope gradient, $w = 0.25$ m. (C) Slope gradient, $w = 3$ m. (D) Horizontal curvature, $w = 0.25$ m. (E) Horizontal curvature, $w = 3$ m. *(Modified from Florinsky, I.V., 2010. Theory and Applications of Mathematical Cartographic Modeling of Topography (D.Sc. thesis). Institute of Mathematical Problems of Biology, Russian Academy of Sciences, Pushchino, Russia (in Russian), Fig. 1.15.)*

these errors are dramatically magnified after derivation of local morphometric variables, as they are calculated by differentiation (Section 2.2).

There are several ways to avoid these artifacts. First, it has been suggested that more sophisticated and smoother interpolators should be used to prevent the formation of "terraces" in DEMs (Eklundh and Mårtensson, 1995), contour "traces" on G, A, k_v, and hill-shaded maps (Batson et al., 1975; Wood and Fisher, 1993; Desmet, 1997), and triangle patterns on G and A maps (Desmet, 1997).

Second, Eklundh and Mårtensson (1995) proposed generalizing contours in order to prevent terraces in DEMs. Third, high-frequency filtering and smoothing are applied before differentiation to reduce the noise level in signal and image processing (Baker, 1982; Rosenfeld and Kak, 1982, § 6.4). Similar filtering procedures are used in digital terrain modeling (Chapter 6) to suppress high-frequency noise in DTMs after their derivation from DEMs (Horn, 1981) or in DEMs before DTM derivation (Brown and Bara, 1994; Giles and Franklin, 1996; Desmet, 1997; Wise, 2000).

All these approaches may improve the results of DEM interpolation and hide the structure of an original irregular DEM grid. However, this cannot clearly increase the actual resolution of a DEM (conversely, the second option can decrease it). It is incorrect to use DEMs with "enhanced" resolution for DTM derivation because an overresolution of a DTM does not relate to information on actual landforms kept in a DEM before interpolation. Treating DTMs beyond a resolution limit can lead just to some abstract investigation of the geometry of a matrix of interpolated values $z = f(x, y)$ rather than the geometry of the topographic surface.

Common sense leads us to assume that there is only one appropriate solution to the problem discussed. Let the spatial resolution of an irregular DEM correspond to $\tilde{\lambda}_{x,y}$ (Section 3.3). In this case, one should use w value relating, at least, to $\tilde{\lambda}_{x,y}$ in DEM interpolation, if an interpolated DEM will be then utilized to derive other DTMs. Practically, this w value should relate to the average distance between points in the original irregular DEM. Other w values chosen from considerations of the study tasks (Chapter 10) (Florinsky and Kuryakova, 2000) should be higher than this minimal w value.

5.5 The Gibbs phenomenon

5.5.1 Motivation

The Gibbs phenomenon is a specific behavior of some functions manifested as over- and undershoots around a jump discontinuity (Nikolsky, 1977b, § 15.9; Hewitt and Hewitt, 1980; Jerri, 1998). The Gibbs phenomenon is typical for the Fourier series, orthogonal polynomials, splines, wavelets, and some other approximation functions. It appears in many scientific problems and applications involving signal and image processing (Rosenfeld and Kak, 1982, p. 158).

For the one-dimensional case, the simplest mathematical illustration of the Gibbs phenomenon is an approximation of a square wave function (Fig. 5.13A)

$$F(x) = \begin{cases} 1, & 0 < x < \pi \\ 0, & x = 0, \pm\pi \\ -1, & -\pi < x < 0 \end{cases} \tag{5.22}$$

by a trigonometric polynomial of the form

$$f_{2k-1}(x) = \frac{4}{\pi}\sum_{k=1}^{\infty}\frac{\sin(2k-1)x}{2k-1} = \frac{4}{\pi}\left(\sin x + \frac{\sin 3x}{3} + \cdots + \frac{\sin(2k-1)x}{2k-1}\right). \tag{5.23}$$

The polynomial (Eq. 5.23) converges uniformly to the square wave function (Eq. 5.22), except for points of discontinuity of $F(x)$, $x=0, \pm\pi$. This means that as k increases, the graphs of the partial sums $f_{2k-1}(x)$ approach arbitrarily closely to the graphs of $F(x)=\pm1$, except for the vicinities of the points $x=0, \pm\pi$ (Fig. 5.13B–E). This convergence defect manifests itself as over- and undershoots around the points $x=0, \pm\pi$. It is important that as k increases, the vertical size of the over- and undershoots does not change (Fig. 5.13B–E). It is equal to 17.9% of half the jump size, or 8.95% of the jump size. With $k \to \infty$ the ultimate graph of $f_{2k-1}(x)$ is a jagged line including extended vertical legs as big as 8.95% of the jump size

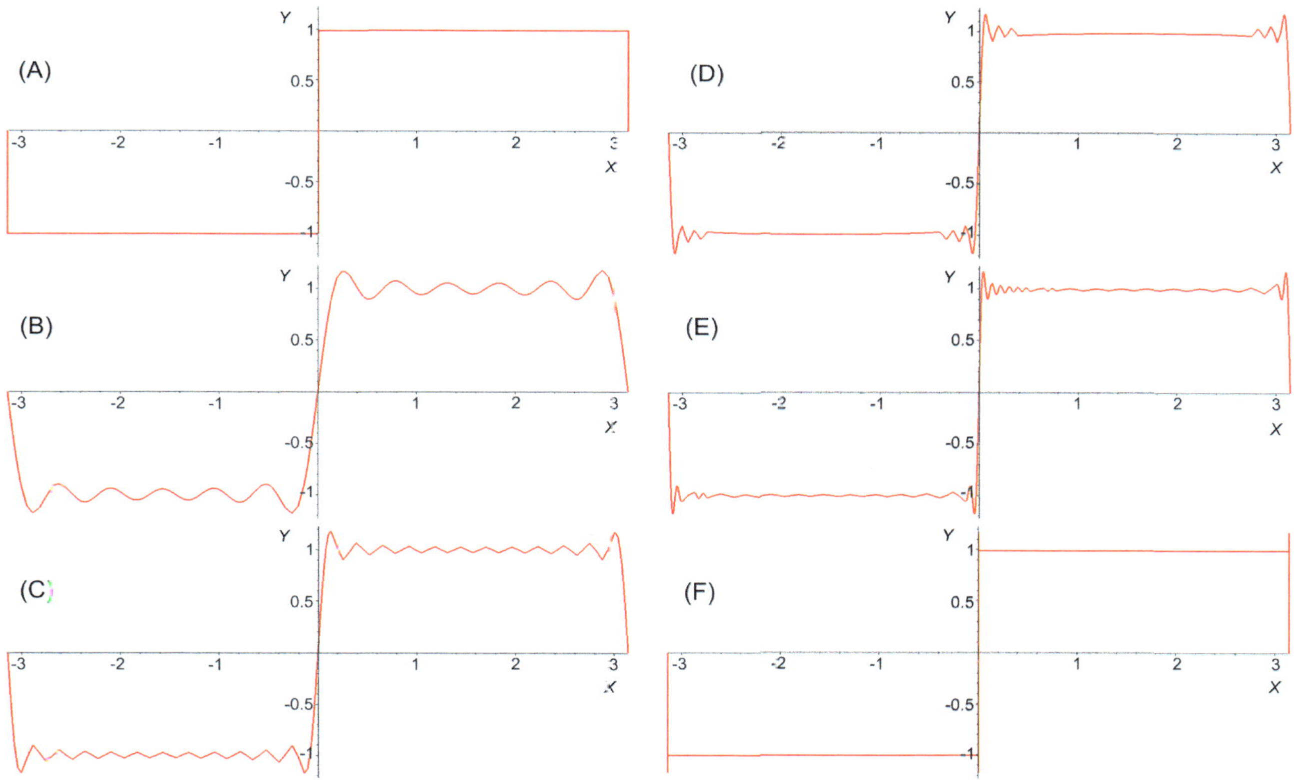

FIG. 5.13 Approximation of a square wave function by a trigonometric polynomial: (A) Square wave function. Partial sums: (B) $k=6$, (C) $k=12$, (D) $k=24$, (E) $k=36$, and (F) $k \to \infty$. (Approximation was done by the Maple software (© Waterloo Maple Inc., 1981–1997.)

(Fig. 5.13F). By increasing k, it is possible to reduce the horizontal extension of the Gibbs phenomenon only (Fig. 5.13) (Fichtenholz, 1966b, pp. 490–497).

In the Gibbs phenomenon, an over- and undershoot size depends on the size of a jump discontinuity and a sort of function. In the classical case of the Fourier series, the over- and undershoot size is equal to 8.95% of the jump size (Jerri, 1998). Fichtenholz (1966b) established that the Gibbs phenomenon takes place for any piecewise smooth function (e.g., splines). The over- and undershoot size is 8.95% of the jump size for a spline of the order $k \to \infty$. Richards (1991) demonstrated that the over- and undershoot size increases if a spline order decreases. For example, it is 9.49% and 13.39% of the jump size for $k=8$ and $k=2$, respectively. The higher the jump, the higher the over- and undershoots.

Since splines and other approximating functions are widely used for DEM interpolation (Watson, 1992), understanding of the Gibbs phenomenon becomes substantial to provide correct digital terrain modeling, for instance, to choose an appropriate interpolator for reducing or eliminating over- and undershoots. Practically, any DEM includes many jump discontinuities in the elevation function wherein the Gibbs phenomenon can arise after interpolation. These are areas with sharp changes in elevation (or steep gradient), such as terraces, escarpments, abrupt slopes, peaks, and pits. Jump discontinuities can also occur near pronounced systematic and random DEM errors, for example, near an abrupt linear step in elevations resulting from processing images as separate patches and subsequently joining them to assemble a DEM (Hunter and Goodchild, 1995), or near a point with false elevation of 100 m within an area with an average altitude of 10 m.

For instance, the following two artificial landforms may appear near a steep escarpment with a relative altitude of 50 m: a "bank" (an overshoot) along the escarpment edge and a "ditch" (an undershoot) along the escarpment foot. In this case, the relative height of the "bank" as well as the relative depth of the "ditch" can be about 4.5 m (8.95% of 50 m) if the interpolation is carried out with high-order splines. Artificially closed "banks" (overshoots) and "ditches" (undershoots) may arise around isolated pits and peaks, respectively.

DEM interpolation has been commonly used in the past five decades, so researchers have noted over- and undershoots near steep gradients in interpolated DEMs and have also tried to solve this problem (Akima, 1974; McCullagh, 1981; Nielson and Franke, 1984; Mitášová and Mitáš, 1993); see detailed discussion in Section 5.5.3. However, these over- and undershoots have not been recognized as manifestations of the Gibbs phenomenon. In the following sections, we discuss the properties of the Gibbs phenomenon as applied to digital terrain modeling.

5.5.2 Materials and data processing

First, we produced a DEM of a simulated surface. The surface measured 7 × 7 m. The square-gridded DEM consisted of 64 points. The surface included two jump discontinuities: (1) a scarp with an elevation difference of about 26 m and a horizontal distance between edge and foot of about 1 m and (2) a false elevation of −100 m within an area with an average elevation of −4 m.

Second, we derived two DEMs with a grid spacing of 0.03 m (Fig. 5.14A) from the initial DEM by two interpolation methods: (1) the Delaunay triangulation and a linear interpolation and (2) the Delaunay triangulation and a piecewise quadric polynomial interpolation with matching derivatives along triangle edges (Agishtein and Migdal, 1991). The linear interpolation was performed by the Surfer software (© 1993–1996, Golden Software Inc.). The smooth interpolation and data visualization were carried out by the LandLord software (Florinsky, 2016).

Finally, cross sections A–A' and B–B' were constructed through jump discontinuities (Fig. 5.14B and C).

5.5.3 Results and discussion

On the elevation map produced with the smooth interpolation, one can see three marks ("knolls") of the Gibbs phenomenon around the false elevation (Fig. 5.14A). With the exception of a small "pit," the elevation map does not demonstrate the Gibbs phenomenon marks along the scarp due to too-high contour interval value. At the same time, there are no the Gibbs phenomenon marks on the elevation map produced with the linear interpolation (Fig. 5.14A).

Fig. 5.14B presents the cross sections A–A' with and without the Gibbs phenomenon marks. The cross section A–A' without marks of the Gibbs phenomenon relates to the DEM produced with the linear interpolation. At the same time, there are the Gibbs phenomenon marks on the cross section A–A' corresponding to the DEM produced with the smooth interpolation. These are over- and undershoots to the left and to the right of the jump discontinuity. So, two "landforms" arose due to the Gibbs phenomenon after the smooth interpolation: the "bank" is along the edge, and the "ditch" is along the foot.

The cross sections B–B' with and without the Gibbs phenomenon marks are presented in Fig. 5.14C. The cross section B–B' without marks of the Gibbs phenomenon relates to the DEM produced with the linear interpolation. At the same time, there are two marks of the Gibbs phenomenon on the cross section B–B' corresponding to the DEM produced with the

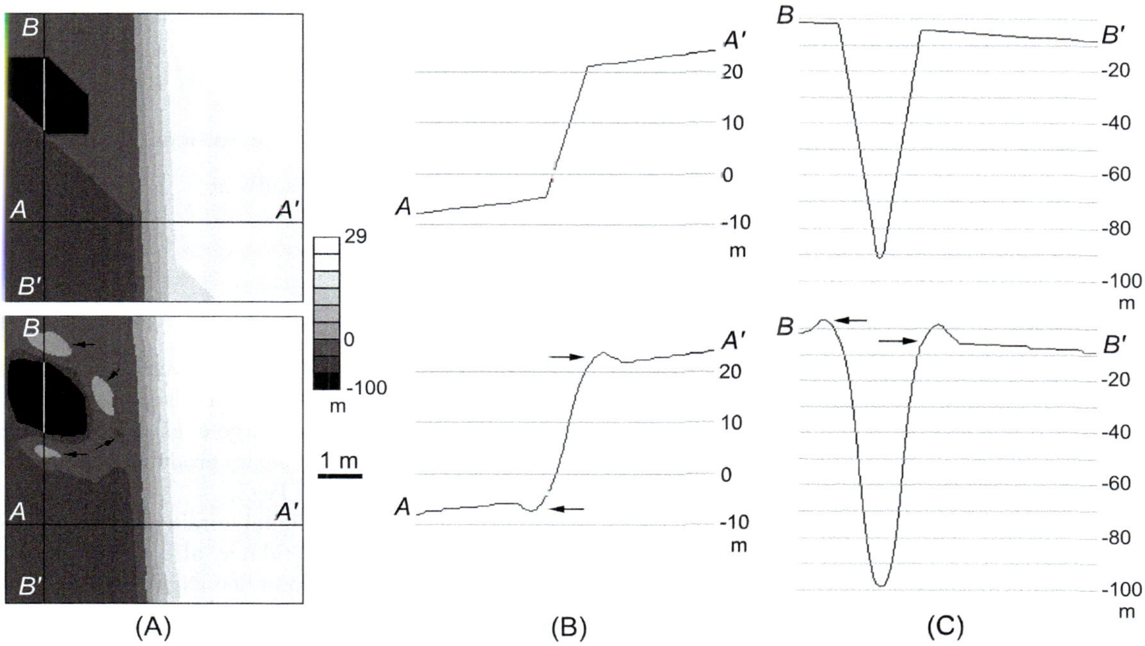

FIG. 5.14 The Gibbs phenomenon in DEM interpolation: (A) Elevation; *A–A′* and *B–B′*, cross sections through areas with jump discontinuities. (B) Cross sections *A–A′*. (C) Cross sections *B–B′*. *Upper*, linear interpolation; *lower*, smooth interpolation; *arrows*, manifestations of the Gibbs phenomenon. *(Modified from Florinsky, I.V., 2010. Theory and Applications of Mathematical Cartographic Modeling of Topography (D.Sc. thesis). Institute of Mathematical Problems of Biology, Russian Academy of Sciences, Pushchino, Russia (in Russian), Fig. 1.17.)*

smooth interpolation. These overshoots are part of the "bank" around the "hole" associated with the false elevation of −100 m. The results demonstrate that the Gibbs phenomenon can arise after a smooth interpolation rather than a linear one.

It is obvious that DEM errors caused by the Gibbs phenomenon can propagate through the processing and can produce new errors in secondary DTMs derived from a DEM. Indeed, differentiation increases the manifestation of noise in derivation of local morphometric variables (Section 5.4.1). For catchment area and combined local-nonlocal topographic attributes (Sections 2.3.1 and 2.6.1), errors can arise at each point located downslope from the Gibbs phenomenon rather than near jump discontinuities only. As vertical errors adversely affect flow-path determination (Veregin, 1997), the Gibbs phenomenon can disturb the design of crest and thalweg network maps (Section 2.4).

There are four key ways to prevent or reduce DEM errors caused by the Gibbs phenomenon:

1. *Decreasing the jump discontinuity before DEM interpolation*. This can be carried out by DEM refining, that is, inserting additional points into an irregular DEM within the area with a jump discontinuity. This should be done before DEM interpolation. For example, the irregular DEM includes an escarp described by two sets of points along its edge and foot. To reduce the jump discontinuity, it would be rational to insert at least one additional set of points along the midslope of the escarp. This sort of DEM refining can decrease over- and undershoots or may hold an interpolation function in position without over- and undershoots because the function passes through (or is approximated to) additional points. Since the refining is carried out before interpolation, it neither violates a regular grid nor leads to a refining of the entire regular DEM. A "drawback" of this way is the necessity of careful control of the generation of the irregular DEM, including a choice of a proper distribution of irregular grid nodes. However, this is a common problem of DEM generation.
2. *Omitting over- and undershoots after DEM interpolation*. This option can be carried out by interactive editing of DEMs (Weibel and Brändli, 1995). Its drawback is the necessity of very careful control of the regular DEM produced by interpolation. An operator should search over- and undershoots near steep terraces, escarpments, abrupt slopes, peaks, pits, and so on. To do this correctly, the operator should be well aware of the actual topography of a modeled area. Notice that the Gibbs phenomenon can be useful for DEM editing because, as a rule, it introduces additional "banks" and "ditches" around pronounced random and systematic errors in DEM production. These over- and undershoots may assist the operator in finding and eliminating such errors.

3. *Using the Gibbs phenomenon-free interpolators*. This option may be performed using algorithms of smooth interpolation based on splines with tension (Nielson and Franke, 1984; Franke, 1985; Mitášová and Mitáš, 1993) (Section 3.4). They were specially developed to protect interpolation of bivariate functions from over- and undershoots near jump discontinuities.
4. *Applying global polynomial approximation with the Fejér summation*. For details, see Sections 7.2 and 7.5.

The choice of a particular method depends on the user's qualifications and available software.

5.6 Grid displacement

5.6.1 Motivation

It is obvious that somewhat different point sets can be sampled using different positions of a discretization grid about the original signal or image. Subsequently, one can produce slightly different reconstructions of the signal or image for different positions of the discretization grid. As a rule, these minor discretization errors can be considered high-frequency noise, and one may, in most cases, ignore them. However, they may be increased in secondary products derived from these slightly different signals or images using differentiation (Section 5.4.1) (Zlatopolsky, 1992).

Discretization of the two-dimensional function of elevation is a key procedure of digital terrain modeling (Makarovič, 1973, 1977; Stefanovic et al., 1977) (Sections 3.2 and 3.3). The geometry of a DEM grid may affect patterns on maps of topographic variables (Mark, 1975a; Carter, 1988; Wilson et al., 1998). It is also apparent that slightly different DEMs of the same area can be produced by displacement (in a general case, by rotation) of a grid, wherein elevation values are measured or interpolated. We mean that the geometry of an irregular grid, or the spacing of a regular grid, remains constant; it is only the grid position and/or orientation that change. Such DEMs can have minor differences, but all of them may be used to describe the area. Moreover, grid displacement may sometimes improve the DEM accuracy: Endreny et al. (2000) found that rotation of a SPOT-derived DEM about the axis may improve the DEM RMSE.

The following question arises: what differences could be found in digital models of a topographic attribute derived from a DEM after displacement of a DEM grid? In other words, how does the displacement of a DEM grid influence the derivation of topographic variables? Of particular interest is the effect of different positions or orientations of a DEM grid on the revealing of topographically expressed geological features such as lineaments (Chapter 14). In the following sections, we discuss the influence of DEM grid displacement on map patterns of two topographic variables used in geological studies.

5.6.2 Materials and data processing

The study area—a portion of the Stavropol Upland, Russia—measures about 172 × 143 km (Fig. 5.15). An irregular DEM including 4459 points was produced by digitizing a topographic map (Central Board of Geodesy and Cartography, 1968).

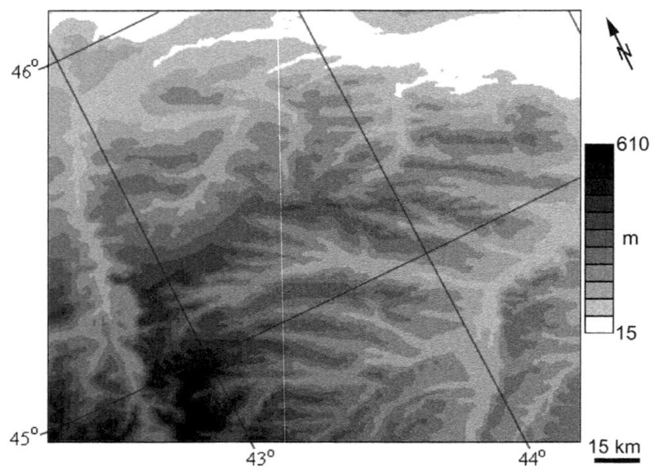

FIG. 5.15 The Stavropol Upland: elevation. *(Modified from Florinsky, I.V., 2010. Theory and Applications of Mathematical Cartographic Modeling of Topography (D.Sc. thesis). Institute of Mathematical Problems of Biology, Russian Academy of Sciences, Pushchino, Russia (in Russian), Fig. 1.18.)*

Using the Delaunay triangulation and a piecewise quadric polynomial interpolation (Agishtein and Migdal, 1991), we produced three regular DEMs with $w = 2500$ m. Regular grids of the DEMs had distinct orientations relative to the irregular DEM: rotation angles of the reference axes of the regular grid are 0°, 25°, and 90° in the counterclockwise direction in relation to the reference axes of the irregular grid (Fig. 5.16A).

Digital models and maps of k_h and k_v (Fig. 5.16B and C) were derived from the three regular DEMs by the Evans-Young method (Section 4.1). We subdivided k_h and k_v values into two intervals relative to the zero value because this type of presentation of curvature data is used to reveal lineaments (Florinsky, 1992, 1996) (Chapter 14). Data were processed by the LandLord software (Florinsky, 2016).

Using samples extracted from k_h and k_v models, we conducted a pairwise comparison of the DTMs obtained. The sample size was 3600 points. Statistical analysis was carried out by the Statgraphics software (© Statistical Graphics Corp., 1994–1997).

5.6.3 Results and discussion

DEM rotation leads to the revealing of slightly different topographic patterns. Although k_h and k_v maps derived from the variously oriented DEMs have many similarities, they also include a number of distinctions (Fig. 5.16B and C). In particular, some patterns break or merge, and other patterns change their width and length. Most of the patterns remain in

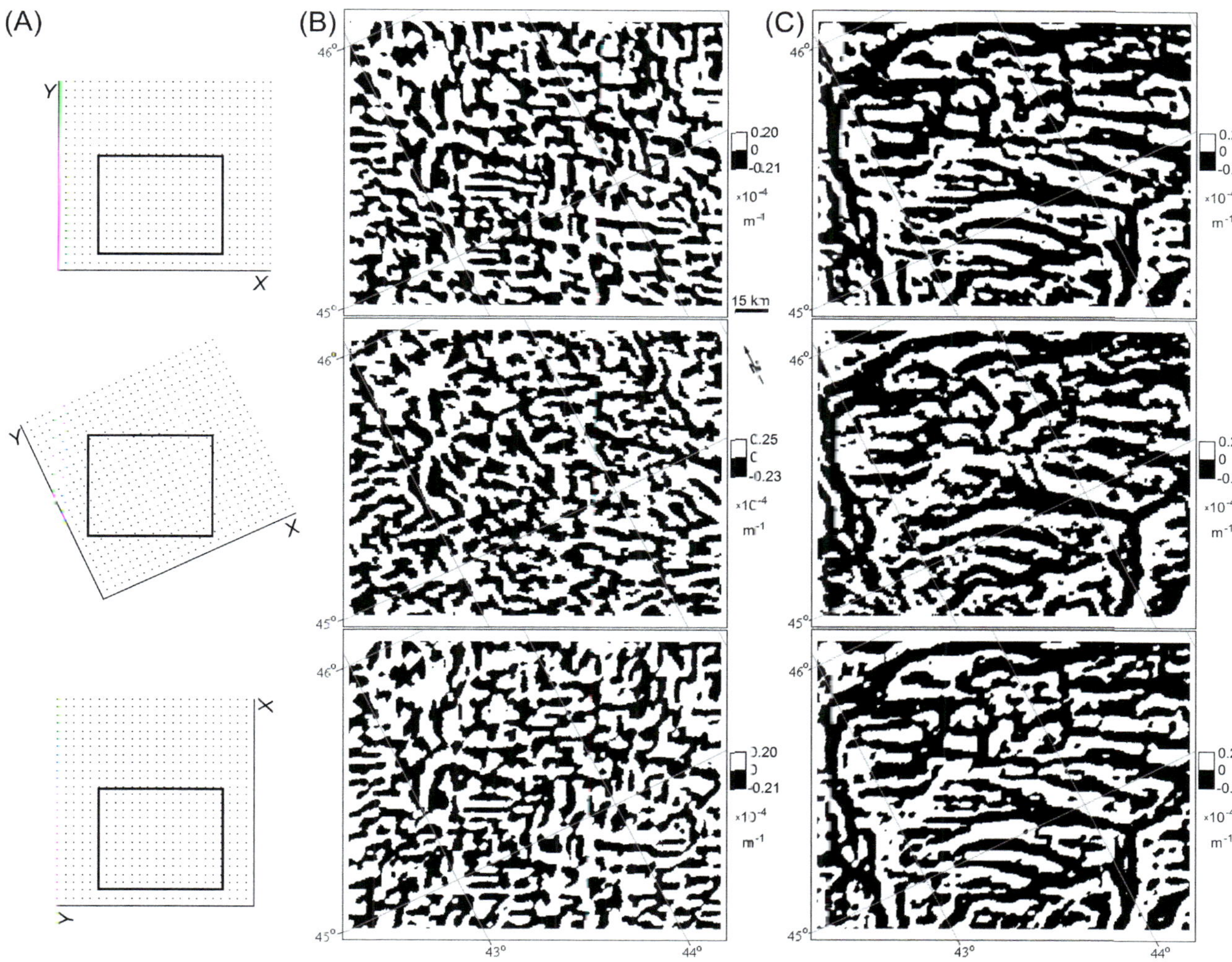

FIG. 5.16 Effect of the rotation of the discretization grid: (A) Scheme of the rotation; *black frame*, the irregular DEM. (B) Horizontal curvature derived from the rotated DEMs. (C) Vertical curvature derived from the rotated DEMs. *Upper*, 0°-; *center*, 25°-; *lower*, 90°-rotation of the discretization grid. For the initial DEM, see Fig. 5.15. *(Modified from Florinsky, I.V., 2010. Theory and Applications of Mathematical Cartographic Modeling of Topography (D. Sc. thesis). Institute of Mathematical Problems of Biology, Russian Academy of Sciences, Pushchino, Russia (in Russian), Figs. 1.19 and 1.20.)*

their position. However, small dots, narrow lines, and small particles of big patterns appear and disappear on maps. Notice that the typical sizes of these patterns are less than the grid spacing. Therefore, they fit into a high-frequency noise. These artifacts are, within certain limits, manifestations of DEM discretization errors increased by differentiation (Section 5.4.1).

Similar effects arise in lineament revealed by digital processing of satellite images: some linear structures can appear and disappear owing to the image rotation in relation to the discretization grid (Zlatopolsky, 1992). There is no way to prevent such effects completely when dealing with discrete functions, which describe the surface continuum.

Results of the DTM pairwise comparison are presented in Fig. 5.17 and Table 5.2. Histograms of both k_h and k_v samples are very similar (Fig. 5.17). The Kolmogorov-Smirnov test applied to k_h and k_v pair samples demonstrated that no

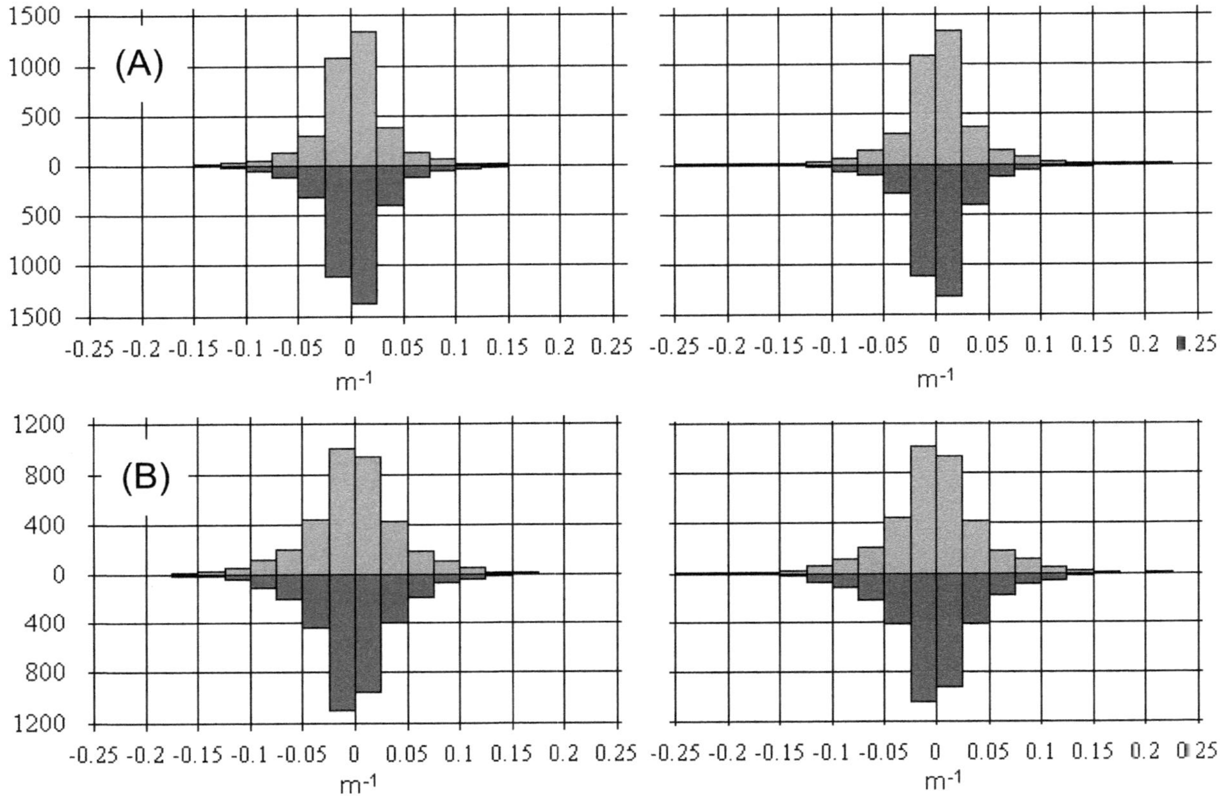

FIG. 5.17 Statistical distribution of (A) horizontal curvature and (B) vertical curvature values derived from the regular DEMs interpolated with various rotation angles of the discretization grid: all upper histograms are for 0°-rotation, and lower histograms are for 25°- (*left*) and 90°-rotation (*right*). *(From Florinsky, I.V., 2010. Theory and Applications of Mathematical Cartographic Modeling of Topography (D.Sc. thesis). Institute of Mathematical Problems of Biology, Russian Academy of Sciences, Pushchino, Russia (in Russian), Fig. 1.21.)*

TABLE 5.2 Pairwise Kolmogorov-Smirnov test of the k_h and k_v samples derived from the regular DEMs interpolated with various rotation angles of the discretization grid (0°-, 25°-, and 90°-rotation).

Sample pairs		Statistics	
		K-S	P
k_h	0° and 25°	0.71	0.70
	0° and 90°	0.52	0.95
k_v	0° and 25°	1.20	0.11
	0° and 90°	0.60	0.86

(From Florinsky, I.V., 2010. Theory and Applications of Mathematical Cartographic Modeling of Topography (D.Sc. thesis). Institute of Mathematical Problems of Biology, Russian Academy of Sciences, Pushchino, Russia (in Russian), Table 1.3.)

statistically significant difference exists between the distributions of k_h (or k_v) calculated with different positions of the discretization grid at the 95% confidence level (Table 5.2). In other words, visually observable discretization errors are statistically insignificant.

5.7 Linear artifacts

5.7.1 Motivation

In DTM-based revealing of geological lineaments, indicators of these structures are linear patterns on maps of local morphometric variables (Chapter 14). Approximately north-, west-, northeast-, and northwest-striking linear patterns are commonly observed on such maps. These orthogonal and diagonal lineaments can reflect topographically expressed geological structures (Chebanenko, 1963; Moody, 1966; Shults, 1971; Katterfeld and Charushin, 1973; Trifonov et al., 1983; Bryukhanov and Mezhelovsky, 1987; Anokhin and Odesskii, 2001; Maslov and Anokhin, 2006) (Fig. 5.18). However, they may also be artifacts. There are several possible causes for orthogonal and diagonal lineaments: (1) DTM grid geometry, (2) DEM production errors, and (3) DEM interpolation errors. Let us briefly discuss these sources of artifacts.

1. It is obvious that DTM mapping can, more or less, visualize the DTM grid geometry, that is, orthogonal and diagonal (cardinal) directions. However, it is possible to ignore this effect using DTMs with a sufficiently high resolution, when the human visual system does not perceive the grid geometry. It is also obvious that the grid geometry may affect calculations (e.g., interpolation, see later).
2. There are, at least, five sources for linear artifacts in DEM generation:
   (a) Photogrammetric generation of DEMs can lead to linear artifacts if one uses a semiautomatic profiling of stereo images (Brown and Bara, 1994; Albani and Klinkenberg, 2003) (Fig. 5.1).
   (b) There are abrupt linear altitude steps resulting from the processing of images as separate patches and joining them to assemble a DEM (Hunter and Goodchild, 1995).
   (c) Artificial orthogonal linear scarps can be found in small-scale and global DEMs, the result of assembling several DEMs varying in origin, scale, and accuracy (see illustrations in Section 6.3).
   (d) Artificial lineaments may occur in detailed DEMs produced by laser and radar surveys. In particular, such artifacts relate to human-made linear elevation differences along wide and straight streets in cities or straight highways crossing forested territories (Fig. 5.6).
   (e) Artificial lineaments may relate to striping artifacts of lidar and InSAR DEMs associated with air- or spacecraft motion, for example, aircraft pitch motion or orbital paths (Fig. 5.4 and Section 5.1) (Arrell et al., 2008; Gallant and Read, 2009). Linear artifacts (roll errors) can be found in deep-water areas in bathymetric DEMs produced by echo sounding (Bourillet et al., 1996).

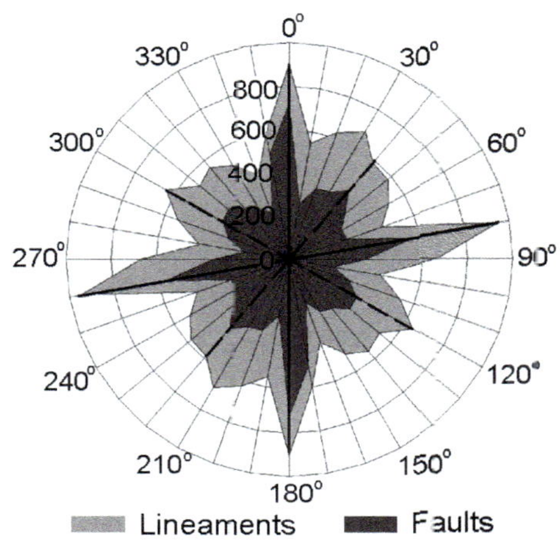

FIG. 5.18 Rose diagram for direction azimuths of the global distribution of lineaments and faults: *radial coordinate*, number of lineaments and faults; *solid lines*, orthogonal system of lineaments and faults; *dashed lines*, the diagonal system of lineaments and faults. *(From Maslov, L.A., Anokhin, V.M., 2006. The Earth's decelerated rotation and regularities in orientation of its surface lineaments and faults. Planet. Space Sci. 54 (22), 216–218, Fig. 1, reproduced with permission from Elsevier.)*

3. DEM interpolation along a series of profiles of the cardinal directions, orthogonal and diagonal, can lead to linear artifacts striking in these directions (Wood and Fisher, 1993). For sparsely sampled areas, interpolation based on polynomials, splines, and triangulation with linear interpolation may also produce artificial linear patterns of the cardinal directions (Declercq, 1996).

It is common knowledge in the theory of image processing that a derivative operator for transformation of two-dimensional signals may be isotropic or anisotropic (Rosenfeld and Kak, 1982, § 6.3.1). By isotropic operators are meant rotation invariants: rotating a function $z=f(x,y)$ by an angle φ about the z-axis and then applying an operator Ω to $z=F(x' y')$ give the same result as applying Ω to $z=f(x,y)$ and then rotating a result by the angle φ about the z-axis; that is, $F(x',y')=f(x,y)$, where x, y and x', y' are the unrotated and rotated Cartesian coordinates, respectively. An operator is anisotropic if this condition is not met. An example of isotropic operators is the Laplacian, whereas anisotropic operators are exemplified by compass operators measuring gradients in several directions.

Derivation of a digital model of a local topographic variable Θ from a DEM by related formulas (Table 2.2) can be considered as applying an operator Ω of the variable Θ to the elevation function z. The operator Ω transforms the elevation function $z=f(x,y)$ into the function of the local topographic variable $\Theta=\psi(x,y)$, or, what is the same, Ω transforms a DEM into the digital model of Θ.

Local morphometric variables are functions of the partial derivatives of elevation (Section 2.2). It would therefore be reasonable to suggest that operators of these variables may also be isotropic or anisotropic. We should stress that from mathematical definitions of local morphometric attributes (Table 2.1), it is clear that all of them, except for A, are independent of the orientation of the x- and y-axes. These variables are associated with directions related to intrinsic properties of the surface (Section 2.2.1) rather than directions of the coordinate axes. However, some researchers prefer to ascribe an occurrence of linear patterns of cardinal directions in maps of topographic variables to probable anisotropic properties of their operators, ignoring mathematical definitions of these variables. This suggests that it is reasonable to present the proof of the isotropy for operators of local topographic variables (Florinsky, 2005).

5.7.2 Isotropy of local morphometric variables

To prove the isotropy of operators of local topographic variables (Table 2.2), we apply a principle for testing derivative operators in image processing (Rosenfeld and Kak, 1982, pp. 238–239). First, let us write the first- and second-order partial derivatives of $z=F(x',y')$, that is, p', q', r', s', and t', in terms of the first- and second-order partial derivatives of $z=f(x,y)$, that is, p, q, r, s, and t (Eq. 2.2). On differentiation, simple algebraic operations, and substitutions of the well-known expressions of coordinate rotation

$$x = x'\cos\varphi - y'\sin\varphi, \quad y = x'\sin\varphi + y'\cos\varphi, \tag{5.24}$$

we obtained the following formulas (Florinsky, 2005):

$$p' = \frac{\partial z}{\partial x'} = \frac{\partial z}{\partial x}\frac{\partial x}{\partial x'} + \frac{\partial z}{\partial y}\frac{\partial y}{\partial x'} = p\cos\varphi + q\sin\varphi, \tag{5.25}$$

$$q' = \frac{\partial z}{\partial y'} = \frac{\partial z}{\partial x}\frac{\partial x}{\partial y'} + \frac{\partial z}{\partial y}\frac{\partial y}{\partial y'} = q\cos\varphi - p\sin\varphi, \tag{5.26}$$

$$r' = \frac{\partial^2 z}{\partial x'^2} = r\cos^2\varphi + 2s\cos\varphi\sin\varphi + t\sin^2\varphi, \tag{5.27}$$

$$s' = \frac{\partial^2 z}{\partial x'\partial y'} = (t-r)\cos\varphi\sin\varphi + s\left(\cos^2\varphi - \sin^2\varphi\right), \tag{5.28}$$

$$t' = \frac{\partial^2 z}{\partial y'^2} = r\sin^2\varphi - 2s\cos\varphi\sin\varphi + t\cos^2\varphi. \tag{5.29}$$

As $p\neq p'$, $q\neq q'$, $r\neq r'$, $s\neq s'$, and $t\neq t'$, these partial differential operators are not rotation invariants.

Second, let us write a k_h formula (Eq. 2.55) for the rotated Cartesian coordinates, that is, in terms of p', q', r', s', and t' (Eqs. 5.25–5.29) (Florinsky, 2005):

$$
\begin{aligned}
k_h' &= -\frac{q'^2 r' - 2p'q's' + p'^2 t'}{(p'^2 + q'^2)\sqrt{1 + p'^2 + q'^2}} \\
&= \Big\{-(q\cos\varphi - p\sin\varphi)^2\left(r\cos^2\varphi + 2s\cos\varphi\sin\varphi + t\sin^2\varphi\right) \\
&\quad -2(p\cos\varphi + q\sin\varphi)(q\cos\varphi - p\sin\varphi)\left\{(t-r)\cos\varphi\sin\varphi + s\left(\cos^2\varphi - \sin^2\varphi\right)\right\} \\
&\quad + (p\cos\varphi + q\sin\varphi)^2\left(r\sin^2\varphi - 2s\cos\varphi\sin\varphi + t\cos^2\varphi\right)\Big\}/ \\
&\qquad \Big\{\left[(p\cos\varphi + q\sin\varphi)^2 + (q\cos\varphi - p\sin\varphi)^2\right] \\
&\quad \times\sqrt{1 + (p\cos\varphi + q\sin\varphi)^2 + (q\cos\varphi - p\sin\varphi)^2}\Big\} \\
&= -\frac{q^2 r - 2pqs + p^2 t}{(p^2 + q^2)\sqrt{1 + p^2 + q^2}}
\end{aligned}
\tag{5.30}
$$

Comparing the k_h formulas for the rotated and unrotated coordinates (cf. Eqs. 5.30 and 2.55), it is possible to see that $k_h' = k_h$. These mean that the k_h operator is isotropic. Similarly, it was proved that operators of G, k_v, H, K, K_a, K_r, M, E, k_{min}, k_{max}, k_{he}, and k_{ve} (Table 2.2) are isotropic, while the operator of A is anisotropic (Florinsky, 2005). This means that:

- Rotating an elevation function about the z-axis and then applying operators of G, k_h, k_v, H, K, K_a, K_r, M, E, k_{min}, k_{max}, k_{he}, and k_{ve} as well as any of their linear transformations, cannot lead to variations in both values of the variables and patterns on their maps, compared with results of applying these operators to an unrotated elevation function.
- Application of these operators to DEMs cannot be responsible for the occurrence of artificial lineaments of cardinal directions in morphometric maps.

Chapter 6

Filtering

6.1 Tasks of DTM filtering

A surface can be viewed as a sum of surfaces. This triviality forms the basis for solving a variety of problems in digital terrain analysis and DTM-based geoscientific studies. The most popular tasks are:

- Separation of low- and high-frequency components of the topographic surface to study regularities in its structure and its elements of different scales.
- DTM denoising.
- DTM generalization, that is, removal of nonnoise high-frequency components of the topographic surface from DTMs.

6.1.1 Decomposition of the topographic surface

It is well-known that topography is a result of the interaction of both endogenous and exogenous processes of different scales. One of the problems involved in geomorphology is reconstructing the history and characteristics of these processes from the current properties of topography. Another important problem of geomorphology is determining the hierarchical structure of topography (Phillips, 1988; De Boer, 1992), which can be used as a basis for describing the hierarchical structure of ecosystems (Puzachenko et al., 2002).

Researchers usually try to solve these two problems by decomposing the topographic surface into its components of different scales. Mathematically, such a task is not complex and can be performed by various methods. In particular, these are trend-surface analysis (Section 6.2.1), the Filosofov method (Section 6.2.2), two-dimensional (2D) discrete Fourier transform (Section 6.2.3), 2D discrete wavelet transform (Section 6.2.4), smoothing (Section 6.2.5), data elimination from DTMs (Section 6.2.6), 2D singular spectrum analysis (Section 6.3), and so on.

It is obvious, however, that some low-frequency harmonics of the elevation function do not necessarily relate, for instance, to some tectonic structure, while some high-frequency harmonics do not relate to some exogenic (e.g., eolian) landforms. The decomposition of the topographic surface may nevertheless be useful for the initial estimation of a hierarchical structure of the terrain and the origin of its component.

6.1.2 Denoising

Any DEM contains many random errors and artifacts (Bourillet et al., 1996; Hastings and Dunbar, 1998; Arabelos, 2000; Holmes et al., 2000; Bergbauer et al., 2003; Smith and Sandwell, 2003; Rodríguez et al., 2006; Reuter et al., 2009). The occurrence of high-frequency noise in a DEM leads to the production of much more noisy models of morphometric attributes derived from the DEM (Brown and Bara, 1994; Giles and Franklin, 1996; Steen et al., 1998; Stewart and Podolski, 1998; Roberts, 2001; Florinsky, 2002; Bergbauer and Pollard, 2003; Bergbauer et al., 2003; Oksanen and Sarjakoski, 2005). For example, curvature maps derived from nondenoised DEMs are almost unreadable and cannot be used in research. This is associated with the increase of high-frequency noise by the computation of partial derivatives of elevation (Section 5.4.1).

Suppression of high-frequency noise in DEMs can be performed by low-frequency spatial filtering based on the 2D discrete Fourier transform (Section 6.2.3), 2D discrete wavelet transform (Section 6.2.4), smoothing (Section 6.2.5), cutting method (Section 6.2.7), 2D singular spectrum analysis (Section 6.3.1), depression filling (Section 4.4), and some other techniques (Milledge et al., 2009).

Digital Terrain Analysis. https://doi.org/10.1016/B978-0-443-24798-9.00002-9

6.1.3 Generalization

Any research usually includes special stages of data processing dedicated to reducing excess information and revealing the most important features of objects, phenomena, and processes under study. In order to do this, each science has qualitative and quantitative approaches for analyzing and transforming initial data. In sciences dealing with spatially distributed data as well as using a cartographic representation of data and a cartographic method of investigation (Salichtchev, 1955; Dury, 1962; Berlyant, 1986), a set of such approaches is known as generalization (Eckert, 1908; Filippov, 1955; Steward, 1974; Brassel and Weibel, 1988; Florinsky, 1991; McMaster and Shea, 1992; Weibel, 1997; Mackaness et al., 2007). Because a map is one of the key outputs of digital terrain modeling, next we briefly discuss the main notions of cartographic generalization.

6.1.3.1 Generalization in cartography

The term *cartographic generalization* usually refers to the generalization of map content according to the map's purpose and scale. There are two types of generalization: the scale-oriented generalization and the aim-oriented one (Shiryaev, 1977).

The scale-oriented generalization is a necessary condition for producing any map. This is the cartographic manifestation of reducing excess information. "Generalization is immanent for a map; in fact, a cartographer performs generalization during an initial survey and production of the very first map (which can be as detailed as one likes) without thinking of any generalization" (Baransky, 1946, p. 188). There are two reasons for this: (1) the complexity of the form, configuration, or structure of mapped features does not allow for generating their perfect images and (2) any graphical tool for representing information should have clear and laconic language (Bowman, 1968).

The scale-oriented generalization deals with the following three tasks:

1. Simplification of the map structure, preserving a general similarity to the initial map.
2. Preservation of maximum precision for the location of mapped features.
3. A compromise between the maximum informativeness of a map and its readability.

The aim-oriented generalization is a necessary condition for revealing the most important or particular features of objects, phenomena, and processes concerned. Perception and understanding of such information using a nongeneralized map may be difficult due to the integrative nature of data representation.

The aim-oriented generalization deals with the following two tasks:

1. Revealing and representation of particular attributes, properties, and relations of objects, phenomena, and processes under study.
2. Transition from individual characteristics to group ones.

Resolution of these tasks can improve the results of the cartographic analysis (Lawrence, 1971) as well as allow one to obtain qualitatively new information (Salichtchev, 1955; Berlyant, 1986) that can be of fundamental and practical importance. A prominent example of the role of cartographic generalization in fundamental research is the production of a small-scale hypsometric map of the East European Plain; its analysis allowed Tillo (1890) to reveal regularities in the topographic structure at the subcontinental scale. The generalized nature of small-scale maps and globes made it possible to develop geological theories of continental drift (Wegener, 1915) and the expanding Earth (Carey, 1988).

To carry out the aim-oriented generalization, one uses concepts and algorithms of the scale-oriented generalization.

6.1.3.2 DTM generalization

The topographic surface is an extremely complicated natural object. Thus, its investigation must incorporate data generalization. Otherwise, a researcher risks obtaining informationally overloaded, poorly readable maps. Recasting Baransky (1946) (Section 6.1.3.1), generalization is immanent for a DTM. Indeed, discretization of the continuous function of elevation is data generalization. Models of any topographic variables derived from a DEM consider the generalized nature of any DEM.

Weibel (1992) recognized three classes of DTM generalization techniques: global filtering, selective filtering, and heuristic generalization:

1. *Global filtering*. This class includes trend-surface analysis (Section 6.2.1), a low-frequency spatial filtering based on the 2D discrete Fourier transform (Section 6.2.3) and wavelet transform (Section 6.2.4), mean, median, and Gaussian smoothing (Sections 6.2.5.1 and 6.2.5.2), row and column elimination from DTMs (Section 6.2.6), cutting method

(Section 6.2.7), 2D singular spectrum analysis (Section 6.3.1), fractal interpolation schemes (Bindlish and Barros, 1995), and so on. From the standpoint of traditional cartography, these approaches are too formal because they do not consider important features of the topography (e.g., structural lines) in data processing.

2. *Selective filtering*. Methods of this class remove insignificant elevation points from a DEM or modify such elevation values using some thresholds. Among these techniques are feature-preserving smoothing approaches (Section 6.2.5.3) as well as adaptive triangular mesh filtering (Heller, 1990), which allows preserving structural lines if their points are assigned higher weights than other points (Weibel, 1997).
3. *Heuristic generalization*. Techniques of this class are based on an emulation of principles used in manual cartography. In particular, they consider important topographic features (e.g., structural lines, peaks, and pits), which are generalized and preserved in output DEMs (Weibel, 1997). Among these are the methods proposed by Leonowicz et al. (2009), Ai and Li (2010), Samsonov (2010, 2011), Zhou and Chen (2011), Duan et al. (2017), Behrens et al. (2018a), and Wu et al. (2019).

For example, the method of Leonowicz et al. (2009) uses upper and lower quartile filters that tend to preserve the original elevations of crests and thalwegs (Fig. 6.1). First, an initial DEM is filtered with an upper-quartile filter to produce a model appropriate for ridge areas. Second, the initial DEM is filtered with a lower quartile filter to produce a model appropriate for

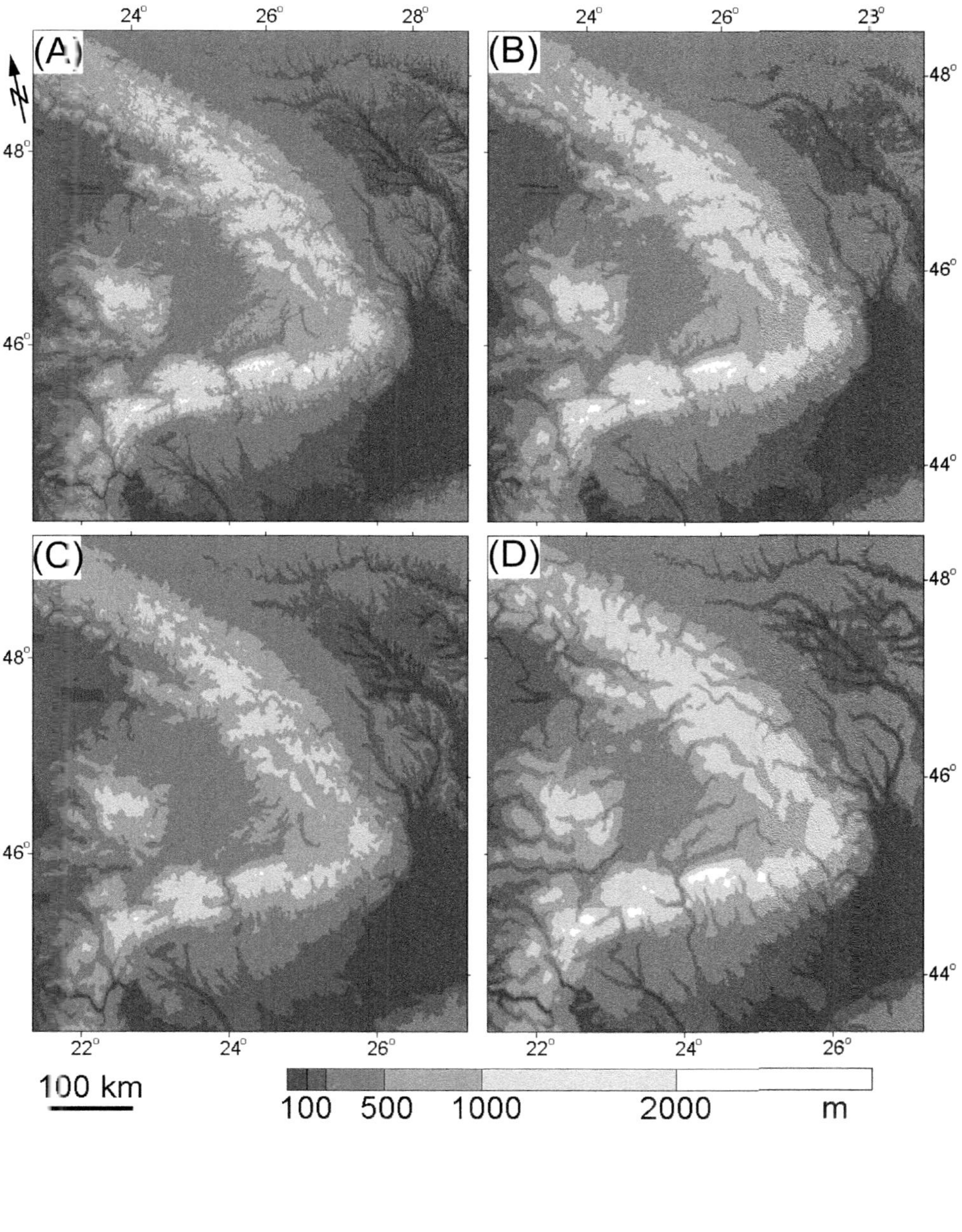

FIG. 6.1 DEM generalization by the method of Leonowicz et al. (2009). Elevation maps of the Carpathian Mountains: (A) Without generalization. (B) Application of an upper-quartile filter. (C) Application of a lower quartile filter. (D) Combination of filtered maps with simplified drainage network and removing pits. A DEM was extracted from the GTOPO30 DEM (USGS, 1996). The grid spacing is 30″. *(From Leonowicz, A.M., Jenny, B., Hurni, L., 2009. Automatic generation of hypsometric layers for small-scale maps. Comput. Geosci. 35 (10), 2074–2083, Figs. 4, 6, and 12, reproduced with permission from Elsevier.)*

valley bottoms. Third, the two filtered DEMs are combined. Finally, a generalization level is adjusted by simplifying the drainage network (the shortest streams are removed).

There are several peculiarities of the scale-oriented generalization in the context of digital terrain analysis, notably:

- The grid density of an irregular DEM or the grid spacing of a regular DTM formally determines the generalization level or resolution of a DTM. If different portions of an irregular DEM have different grid densities, such distinctions are transmitted to all DTMs derived from the DEM; their generalization level will also be a function of coordinates.
- Interpolation of an irregular DEM may play a role of generalization if the grid spacing of a regular grid is much larger than the density of the irregular grid (Rhind, 1971).
- There are minor errors (high-frequency noise) in any DEM. This dictates a need for low-frequency spatial filtering or smoothing of a DEM (Sections 6.2.3–6.2.5) that can be considered a form of generalization (Florinsky, 1991). In the derivation of local morphometric variables, some denoising takes place if one uses the Evans-Young or the author's finite-difference methods (Sections 4.1–4.3), since these approximate polynomials (Eqs. 4.1 and 4.7) to elevation values of the moving windows.
- Numerical results of morphometric calculations depend on the DEM grid spacing, that is, on the DEM generalization level (Evans, 1972, 1979, 1980; Zhang and Montgomery, 1994; Stewart and Podolski, 1998; Shary et al., 2002b; Bergbauer and Pollard, 2003; Wynn and Stewart, 2003; Hancock, 2005; Smith et al., 2006; Sørensen and Seibert, 2007; Zhu et al., 2008). This is because the DEM grid spacing is included in the expressions defining partial derivatives of elevation (Chapter 4). The accuracy of such calculations also depends on the DEM grid spacing (Section 5.3).
- To generalize maps of morphometric variables, one can aggregate (decrease the number of) quantitative classes of mapped surfaces (Jenks, 1963). For example, by classifying values of horizontal curvature into two levels—more or less than zero—it is possible to clearly represent flow divergence and convergence areas (cf. Figs. 10.5B and 5.9F).
- The generalization of topographic segmentation maps (Section 2.7) is carried out by the aggregation of landform classes. For example, mapping of accumulation, transit, and dissipation zones is the generalization in reference to segmentation using the Efremov-Krcho classification. Topographic segmentation by the Gaussian classification can be considered a generalization with respect to segmentation by the Shary classification. Similar approaches can be used for other segmentation models (MacMillan and Pettapiece, 2000; MacMillan et al., 2000).
- DTMs are commonly used in statistical analyses of relations between topographic and soil/plant characteristics and DTM-based prediction of soil or vegetation properties (Chapters 10–12). It is known that data generalization may increase the absolute values of correlation coefficients between related processes or phenomena owing to the removal of minor variations (Berlyant, 1986). For example, the correlation coefficient describing the association between soil moisture content and mean curvature changed from −0.58 to −0.88 after the DEM smoothing (Kuryakova et al., 1992). Thus, it is natural that DTM smoothing may improve the prediction of soil taxonomic classes in medium-scale soil mapping (Moran and Bui, 2002; Grinand et al., 2008; Behrens et al., 2010). The same can be true for DTM-based, medium-scale plant or ecological mapping.
- Although one of the key goals of generalization is to avoid the creation of informationally overloaded, poorly readable maps, an equally important goal is to avoid the creation of overgeneralized, primitive maps. In the context of digital terrain modeling, the question is: When should surface simplification be stopped? Samsonov (2022) demonstrated that the typical landform size should be 5–6 mm at the mapping scale to achieve the clarity and readability in topography representation usual for manually drawn maps.

6.2 Methods of DTM filtering

6.2.1 Trend-surface analysis

Trend-surface analysis, a form of regression analysis, is in regular use in the geosciences (Krumbein, 1959; Chorley and Haggett, 1965; Tobler, 1969; Aronov, 1977, chap. 2; Davis, 2002, chap. 5; Agterberg, 2014, chap. 7). Using the least-squares approach, a digital model of some property is approximated by a global function of the form

$$z = f_1(x,y) + f_2(x,y) + \cdots + f_k(x,y) + R(x,y), \tag{6.1}$$

where $f_k(x,y)$ are trend components and $R(x, y)$ is a residual. Trend-surface analysis is traditionally based on low-order polynomials.[a]

a. The use of high-order polynomials in digital terrain analysis is discussed in Chapter 7.

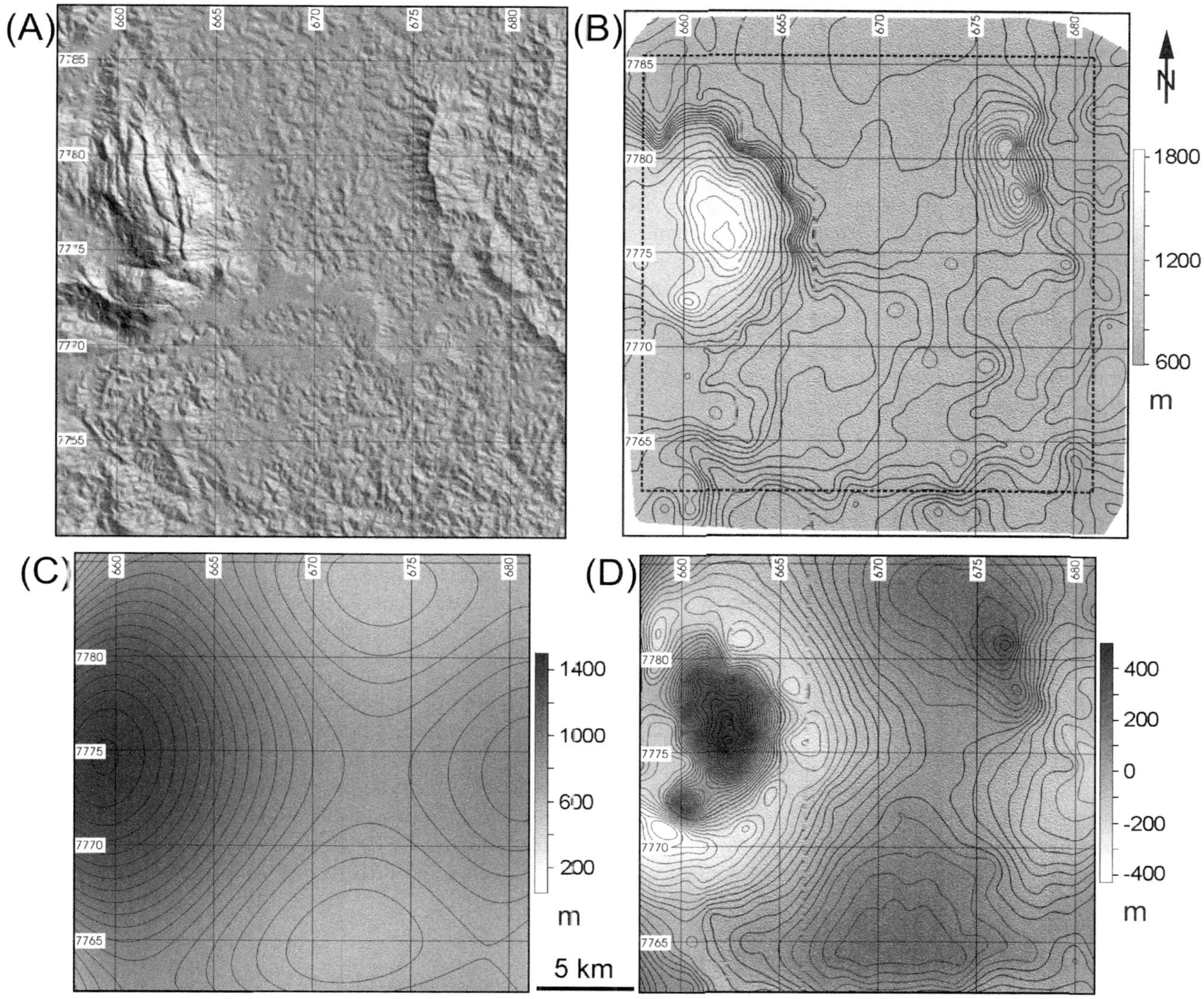

FIG. 6.2 Use of the Filosofov method and trend-surface analysis in geological studies of the Quadrilátero Ferrífero region, the São Francisco Craton, Brazil: (A) Hill shading ($\Theta = 45°$, $\psi = 30°$). A DEM was generated by contour interpolation from a topographic map, scale 1:50,000. The DEM includes 292,357 points (the elevation matrix 559×523); the grid spacing is 50 m. (B) Original isobase surface; dotted line indicates the area used for the trend-surface analysis. (C) The sixth-order trend surface. (D) The residual of the sixth-order trend surface. *((A) From Grohmann, C.H., 2004. Técnicas de Geoprocessamento Aplicadas a Análise Morfométrica (M.Sc. thesis). Institute of Geosciences, University of São Paulo, São Paulo, Brazil, Fig. 2, reproduced with permission from the author; (B)–(D) from Grohmann, C.H., 2005. Trend-surface analysis of morphometric parameters: a case study in southeastern Brazil. Comput. Geosci. 31 (8), 1007–1014, Fig. 3A, C, and D, reproduced with permission from Elsevier.)*

Trend components are mapped separately and are then visually analyzed and interpreted (Fig. 6.2). Trend-surface analysis of the land surface DEMs was applied to study geological ring structures (Gosteva et al., 1983; Zverev and Strykov, 1985) and lineaments (Fedorov, 1991).

6.2.2 The Filosofov method

The method of Filosofov (1960) was initially designed for manual analysis of topographic maps. The method allows the revealing of some tectonic structures as well as modeling the interaction of erosion and vertical tectonic movements. Over many years, structural and petroleum geologists have used this manual method on a wide range of scales, from 1:10,000 to 1:2,500,000 (Filosofov, 1960, 1975; Gvin, 1963; Golbreich et al., 1968; Zuchiewicz, 1989; Golts and Rosenthal, 1993;

Verkhovtsev, 2008). It makes sense to provide a reader with a brief description of the original technique because its digital versions were also developed (Grohmann, 2004a, b; Jordan, 2007; Grohmann et al., 2007, 2011a) (Fig. 6.2).

The Filosofov method uses the hypothesis that valleys of different orders were formed at different times and are associated with tectonic motions and structures of different spatial scales. The method is based on the graphical decomposition of the topographic surface via manual transformations of the topographic contours. It may be considered a sort of manual trend-surface analysis. The method includes the following basic steps (Filosofov, 1960):

1. Preparation of a map of thalweg orders using the Strahler numbering system (Strahler, 1952, 1957): the first-order thalweg is a thalweg without tributaries, the second-order thalweg is a thalweg formed by the joining of two first-order thalwegs, and so on. In a similar manner, a map of crest orders can be produced.
2. Generation of a map of a base-level surface of some order by interpolating elevation values related to thalwegs of this and higher orders. For example, to produce a map of the third-order base-level surface, one should interpolate elevations associated with the third- and higher-order thalwegs. The higher the order of the base-level surface, the higher the generalization level of topography. It is recommended to utilize the second-order base-level maps for revealing local anticlinal structures, the third- and fourth-order base-level maps for recognizing platform structures (e.g., flexures and depressions), and the fifth- or sixth-order base-level maps for analyzing anteclises and syneclises.
3. Preparation of a map of a residual topography of some order by subtracting the base-level surface of this order from the topographic surface. These maps can be used to reveal local tectonic structures.
4. Production of a map of a summit-level surface of some order by interpolating elevation values related to crests of this and higher orders. Summit-level maps are generated similarly to base-level maps (as discussed earlier). They can also be used to detect areas of neotectonic uplift and/or subsidence of different scales.
5. Generation of a map of a difference between two base-level surfaces of different orders. Such maps may characterize a change of elevations in a period between the formation of valleys of different orders. The greater the difference in elevations between neighboring base-level surfaces, the more intensive the vertical tectonic movements.

Simplified computerized approaches are available to calculate digital models of the base- and summit-level surfaces as well as relative relief. In this case, the relative relief is the difference between summit level (the highest altitude for a given area) and base level (the lowest altitude for a given area) (Dury, 1962, p. 174). Calculations are performed with moving windows. Relative relief may be used as an index of the relative velocity of vertical tectonic movements. Digital models of such attributes have been applied to reveal active tectonic structures (Ioffe and Kozhurin, 1997; Szynkaruk et al., 2004), to recognize paleosurfaces (Johansson, 1999), to estimate seismic activity (Zamani and Hashemi, 2000), and to study the interaction between endogenous and exogenous processes of orogenesis (Kühni and Pfiffner, 2001a, b; Zhang et al., 2006).

6.2.3 Two-dimensional discrete Fourier transform

The 2D discrete Fourier transform (Sundararajan, 2001) is widely used in the geosciences. Within the framework of the 2D Fourier transform, the topographic surface is represented as a linear combination of sinusoidal surfaces of various frequencies, phase shifts, and amplitudes. The problem of the 2D Fourier transform is determining the coefficients of these sinusoidal surfaces. The problem of the 2D inverse Fourier transform is reconstructing the topographic surface using the coefficients of sinusoidal surfaces.

Let a 2D discrete function of elevation $z=f(x,y)$, that is, a DEM is a matrix $N \times M$ of $z(k,l)$; $k=0, 1, 2, \ldots, N-1$; $l=0, 1, 2, \ldots, M-1$. For this function, the 2D discrete Fourier transform is

$$Z(n,m) = \sum_{k=0}^{N-1} \sum_{l=0}^{M-1} z(k,l) e^{-i2\pi(nk/N+ml/M)}, \tag{6.2}$$

where $Z(n, m)$ are the coefficients of sinusoidal surfaces, i is the imaginary unit, and n and m are frequencies: $n=0, 1, 2, \ldots, N-1$; $m=0, 1, 2, \ldots, M-1$.

The $Z(n, m)$ coefficients represent the frequency domain of the terrain. A matrix $N \times M$ of the modules of the $Z(n, m)$ coefficients is the power spectrum of the terrain.

The 2D inverse discrete Fourier transform is

$$z(k,l) = \frac{1}{N} \frac{1}{M} \sum_{n=0}^{N-1} \sum_{m=0}^{M-1} Z(n,m) e^{i2\pi(nk/N+ml/M)}. \tag{6.3}$$

Wavelengths of nth and mth harmonics ω_n and ω_m are

$$\omega_n = \frac{Nw}{n}, \quad 0 < n \le N/2; \quad \omega_n = \frac{Nw}{N-n}, \quad N/2 \le n < N; \tag{6.4}$$

$$\omega_m = \frac{Mw}{m}, \quad 0 < m \le M/2; \quad \omega_m = \frac{Mw}{M-m}, \quad M/2 \le m < M, \tag{6.5}$$

where w is a DEM grid spacing (Papo and Gelbman, 1984).

To filter a DEM, $Z(n, m)$ coefficients of the harmonics concerned should be considered in Eq. (6.3) during the reconstruction of the topographic surface. For low-pass filtering, for example, $Z(n, m)$ coefficients corresponding to wavelengths shorter than some threshold should be set to zero. Calculations are usually performed by the well-known algorithm of the fast Fourier transform (Cooley and Tukey, 1965).

The 2D discrete Fourier transform has been periodically used to generalize DTMs (Papo and Gelbman, 1984; Rudy, 1999), to describe the hierarchical structure of the topography (Harrison and Lo, 1996; Gallant and Hutchinson, 1997; Puzachenko et al., 2002), and to denoise DEMs (Arrell et al., 2008; Gallant and Read, 2009). In structural geological analysis, the 2D discrete Fourier transform of DEMs has been applied to both stratigraphic surfaces (Bergbauer et al., 2003) and the land surface (Chigirev, 1976; Dhont and Chorowicz, 2006).

For instance, the 2D discrete Fourier transform was utilized to reconstruct the topography of the ancient Ethiopian dome comprising the Ethiopian and Yemeni Plateaus (Fig. 6.3) (Collet et al., 2000a, b). The DEM of the current topographic surface of the region (Fig. 6.3A) was used as the initial data. First, those authors "closed" three rifts (the Red Sea, Gulf of Aden, and Ethiopian rifts), which had been initiated by mantle-plume activity during the Oligocene times. To do this, portions of the initial DEM, corresponding to the Arabian and Somalian plates, were shifted relative to the stationary Nubian plate (Fig. 6.3B). Second, elevation values were masked and recalculated for the areas of the recent volcanic edifices and gaps between the rift shoulders remaining after the first step. Then the topography was smoothed using the fast Fourier transform. As a result, a model of the prerift topographic surface of the Ethiopian dome was obtained (Fig. 6.3C).

6.2.4 Two-dimensional discrete wavelet transform

The wavelet transform is a signal decomposition using a system of wavelets, that is, functions, each of which is a shifted and scaled copy of some function called the mother wavelet. The term *wavelet transform* defines a class of decomposition methods. Existing types of the wavelet transform differ widely in definitions and properties (Chui, 1992; Daubechies, 1992; Pereberin, 2001).

The wavelet transform of a one-dimensional discrete signal $\mathbf{s} = \{s_j\}, j \in Z$ is carried out by the following recursive formulas (Pereberin, 2001):

$$\mathbf{v}_i = \downarrow_2 [\mathbf{v}_{i+1} * \tilde{\mathbf{h}}], \quad \mathbf{w}_i = \downarrow_2 [\mathbf{v}_{i+1} * \tilde{\mathbf{g}}], \quad i = i_1 - 1, i_1 - 2, \ldots, i_0, \tag{6.6}$$

where $\mathbf{v}_i$ is the approximation of the signal with a lower resolution (a low-frequency component); $\mathbf{w}_i$ is the detailing information (a high-frequency component), $\tilde{\mathbf{h}}$ and $\tilde{\mathbf{g}}$ are low- and high-frequency filters, respectively; $\downarrow_2[\mathbf{x}]$ is the operator removing every second element from the signal $\mathbf{x}$; * denotes convolution; i is the scale.

Some scale value i_1 is related to the initial signal $\mathbf{s}: \mathbf{v}_{i_1} = \mathbf{s}$. On each step of the wavelet transform, the signal is separated into two components, $\mathbf{v}_i$ and $\mathbf{w}_i$. Elements of high-frequency components are wavelet coefficients. The lowest scale i_0 corresponds to the case when a low-frequency component consists of the only element.

The inverse wavelet transform or reconstruction of the signal $\mathbf{s}$ is carried out by the following formula (Pereberin, 2001):

$$\mathbf{v}_{i+1} = \uparrow_2 [\mathbf{v}_i] * \mathbf{h} + \uparrow_2 [\mathbf{w}_i] * \mathbf{g}, \quad i = i_0, i_0 + 1, \ldots, i_1 - 1, \tag{6.7}$$

where $\mathbf{h}$ and $\mathbf{g}$ are low- and high-frequency synthesis filters, respectively; $\uparrow_2[\mathbf{x}]$ is the operator inserting the zero element between elements of the signal $\mathbf{x}$.

Eqs. (6.6) and (6.7) are general expressions for the direct and inverse wavelet transforms of discrete signals. A particular type of the transform is defined by the four filters: $\tilde{\mathbf{h}}$, $\tilde{\mathbf{g}}$, $\mathbf{h}$, and $\mathbf{g}$. Mathematical descriptions of particular filters (e.g., Haar, D_4, biorthogonal, B-spline) can be found elsewhere (Chui, 1992; Daubechies, 1992; Pereberin, 2001).

The 2D wavelet transform can be performed by combining one-dimensional (1D) wavelet transforms. Let there be a finite-size matrix. Let us apply one step of the 1D wavelet transform to each row of the matrix. We obtain two matrices, whose rows include low- and high-frequency components of rows of the initial matrix. Also, let us apply one step of the 1D

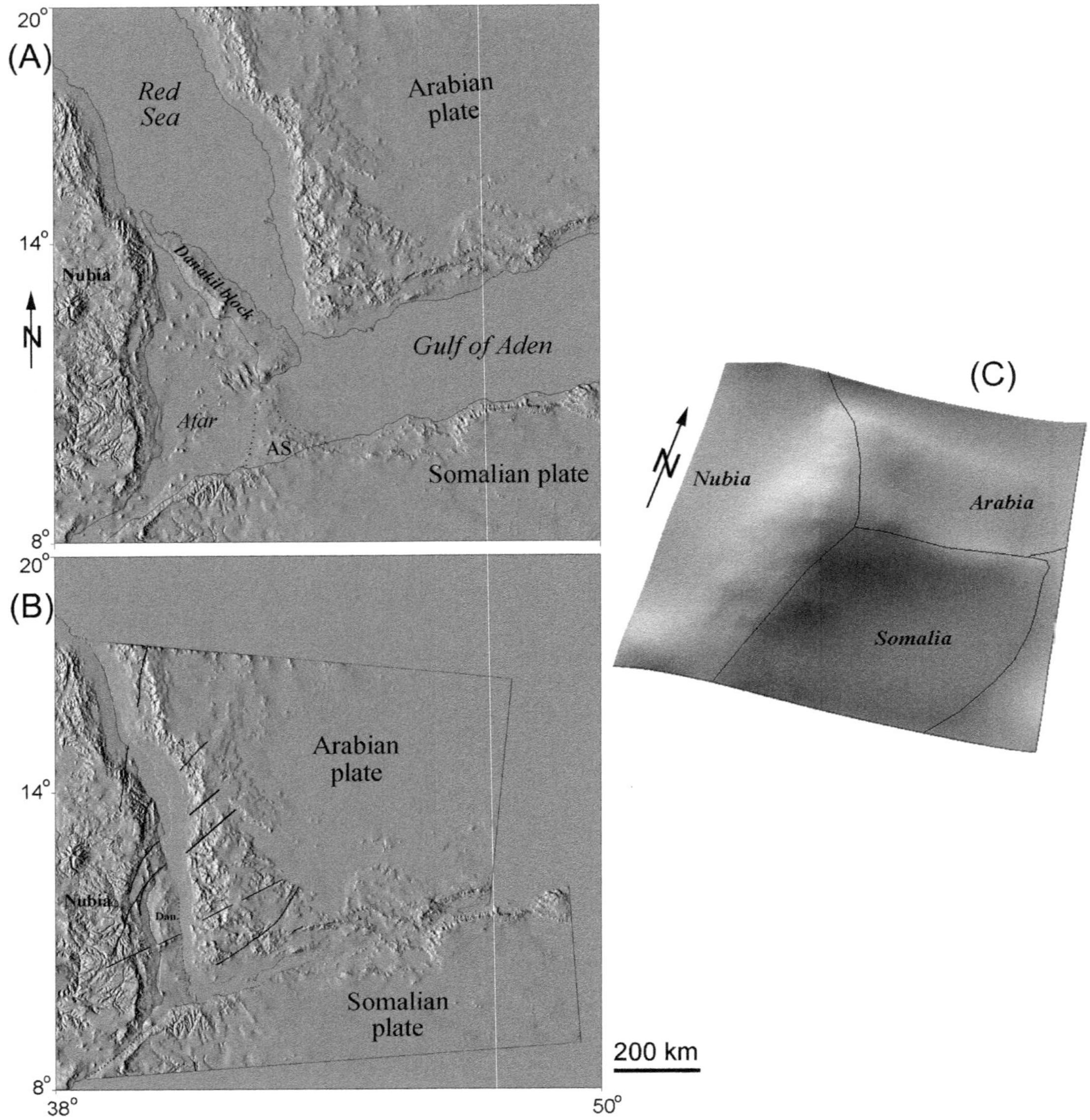

FIG. 6.3 Application of the 2D discrete Fourier transform to model the ancient Ethiopian dome: (A) Hill shading of the current location of tectonic blocks: *black lines*, block limits, *dotted lines*, and supposed limit of the Ali-Sabieh (*AS*) block. A DEM was derived from the Digital Chart of the World (US DMA, 1992) by contour interpolation. The grid spacing is 2000 m. (B) Reconstitution of the territory for the Oligocene time (prerifting period): *dotted lines*, the Ethiopian rift axis and the Arabia/Danakil limit; *solid lines*, lineaments; *Dan.*, Danakil block. (C) Three-dimensional model of the prerifting topography of the Ethiopian dome after applying the fast Fourier transform to the DEM. *((A) and (B) From Collet, B., Taud, H., Parrot, J.F., Bonavia, F., Chorowicz, J., 2000b. A new kinematic approach for the Danakil block using a digital elevation model representation. Tectonophysics 316 (3–4), 343–357, Figs. 1 and 3, reproduced with permission from Elsevier; (C) from Collet, B., Parrot, J.F., Taud, H., 2000a. Orientation of absolute African plate motion revealed by tomomorphometric analysis of the Ethiopian dome. Geology 28 (12), 1147–1149, Fig. 2, reproduced with permission from Geological Society of America.)*

wavelet transform to each column of both matrices. As a result, we obtain four matrices. The first of them is the low-frequency component of the initial 2D signal, while the other three include detailed information.

In geomorphometry, the 2D wavelet transform has been used for the denoising of DEMs (Fig. 6.4) (Bergbauer et al., 2003; Falorni et al., 2005; Lark, 2007), decomposition of the topographic surface into components of different scales and description of the hierarchical terrain structure (Gallant and Hutchinson, 1996, 1997; McArthur et al., 2000; Sulebak and Hjelle, 2003), and DTM generalization (Bjørke and Nilsen, 2003; Wu, 2003). Audet (2011) applied spherical wavelets to global topographic and gravitational data to analyze the lithospheric isotropy for the Earth, Venus, Mars, and the Moon.

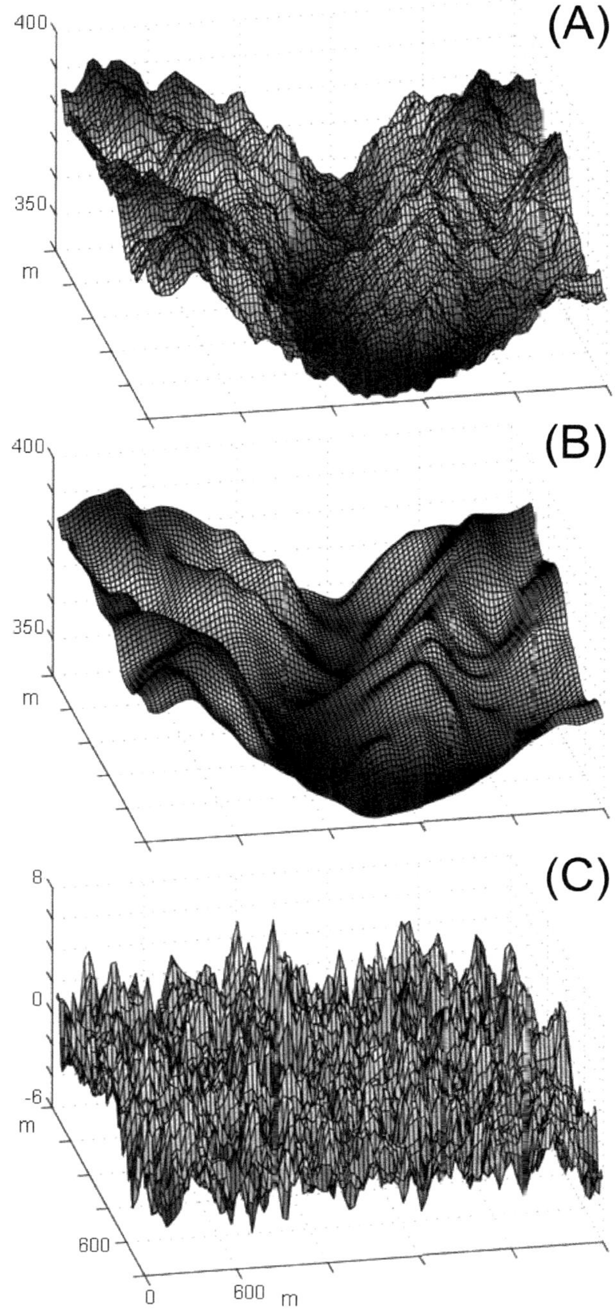

FIG. 6.4 Application of a wavelet filter to denoise an SRTM DEM of a portion of the Little Washita Basin, Oklahoma, USA: (A) Raw SRTM DEM. (B) Denoised DEM. (C) Noise removed from the SRTM DEM. A biorthogonal wavelet with five levels of decomposition was used. *(From Falorni, G., Teles, V., Vivoni, E.R., Bras, R.L., Amaratunga, K.S., 2005. Analysis and characterization of the vertical accuracy of digital elevation models from the Shuttle Radar Topography Mission. J. Geophys. Res. Earth 110 (F2), F02005, Fig. 11A, C and D, © American Geophysical Union, reproduced with permission from John Wiley & Sons.)*

6.2.5 Smoothing

DEM smoothing with moving windows is one of the most popular approaches for removal of artifacts from DTMs, DEM denoising, and DEM generalization (Tobler, 1966; Evans, 1972; Florinsky, 1991; Weibel, 1992; Schmid-McGibbon, 1995; MacMillan and Pettapiece, 2000; Albani and Klinkenberg, 2003; Leonowicz et al., 2009; Reuter et al., 2009).

DEM smoothing is widely used in DTM-based geological (Stewart and Podolski, 1998; Korsakova, 2002; Bergbauer and Pollard, 2003; Florinsky, 2007a) and soil (Moran and Bui, 2002; Grinand et al., 2008; Behrens et al., 2010) studies to enhance results of geological or soil modeling.

There are various types of smoothing filters, such as mean, median, and Gaussian ones. These well-known techniques came to geomorphometry from the field of image processing (Pratt, 2001).

6.2.5.1 Mean smoothing

The general formula for the mean, or weighted moving average smoothing, is as follows:

$$z'_{(n+1)/2} = \frac{\sum_{i=1}^{n} W_i z_i}{\sum_{i=1}^{n} W_i}, \tag{6.8}$$

where $z'_{(n+1)/2}$ is the smoothed elevation value in the central point of the moving window, $n = 9, 25, 49, \ldots$ for $3 \times 3, 5 \times 5, 7 \times 7, \ldots$ windows, respectively; z_i is the elevation value at the ith point of the window, $i = 1, \ldots n$; W_i is the weight of the ith point (Wood, 1996, § 4.4.5):

$$W_i = \frac{1}{(1 + d_i)^m}, \tag{6.9}$$

where d_i is the distance from the ith point to the central point, $m = 0, 1, 2$. The size of a moving window and the number of smoothing iterations are selected empirically.

Fig. 6.5 illustrates the mean smoothing. A DEM of Moscow city and its suburbs was extracted from the SRTM3 DEM (Rabus et al., 2003; Farr et al., 2007). The DEM includes 284,544 points (the elevation matrix 624×456); the DEM grid spacing is 3″. Horizontal and vertical curvatures were derived from the DEM by the author's method (Section 4.3). To deal with the large dynamic range of k_h and k_v, we logarithmically transformed their digital models by Eq. (8.1) with $n = 4$.

One can see that maps of horizontal and vertical curvatures derived from the nonsmoothed initial DEM (Fig. 6.5A) are almost unreadable. It is unlikely that they may be used in DTM-based research. This is because the DEM includes high-frequency noise increased by the calculation of derivatives (Section 5.4.1). Note, however, that the elevation map derived from the initial DEM is quite readable and can be used as an illustration (Fig. 6.5A). Maps of horizontal and vertical curvatures became readable after several iterations of the DEM smoothing. The greater the number of smoothing iterations, the more generalized maps can be obtained (Fig. 6.5B and C).

Mean smoothing has to be used with caution as it may produce high-frequency artifacts such as phase inversion (Chigirev, 1976). For instance, a pit may replace a hill, or a blockage may arise in a valley (Armstrong and Martz, 2003).

6.2.5.2 Median and Gaussian smoothing

Median smoothing uses an $n \times n$ window, where n is an odd number. For the central point of the moving window, a smoothed elevation value is the median value of the elevations within the window (Pratt, 2001, § 10.3.2; Zhu and Huang, 2012).

A Gaussian filter modifies the input DEM by its convolution with a Gaussian function (Lin et al., 1996; Pratt, 2001, § 9.4.3)

$$G(x, y) = \frac{1}{2\pi\sigma^2} e^{-\frac{x^2 + y^2}{2\sigma^2}}, \tag{6.10}$$

where x and y are the Cartesian coordinates of a point; σ is the standard deviation of the Gaussian distribution. The latest is the spread parameter determining the width of the Gaussian function (Eq. 6.10), that is, the window size. As σ decreases, the smoothing decreases, and vice versa, as σ increases, the smoothing increases. For the central point of a moving window, a smoothed elevation value is the weighted average of elevation values of points within the moving window. The central point receives the largest weight, which is the highest Gaussian function value, and neighboring points receive smaller weights (lower values of the Gaussian function) depending on their distances from the central point.

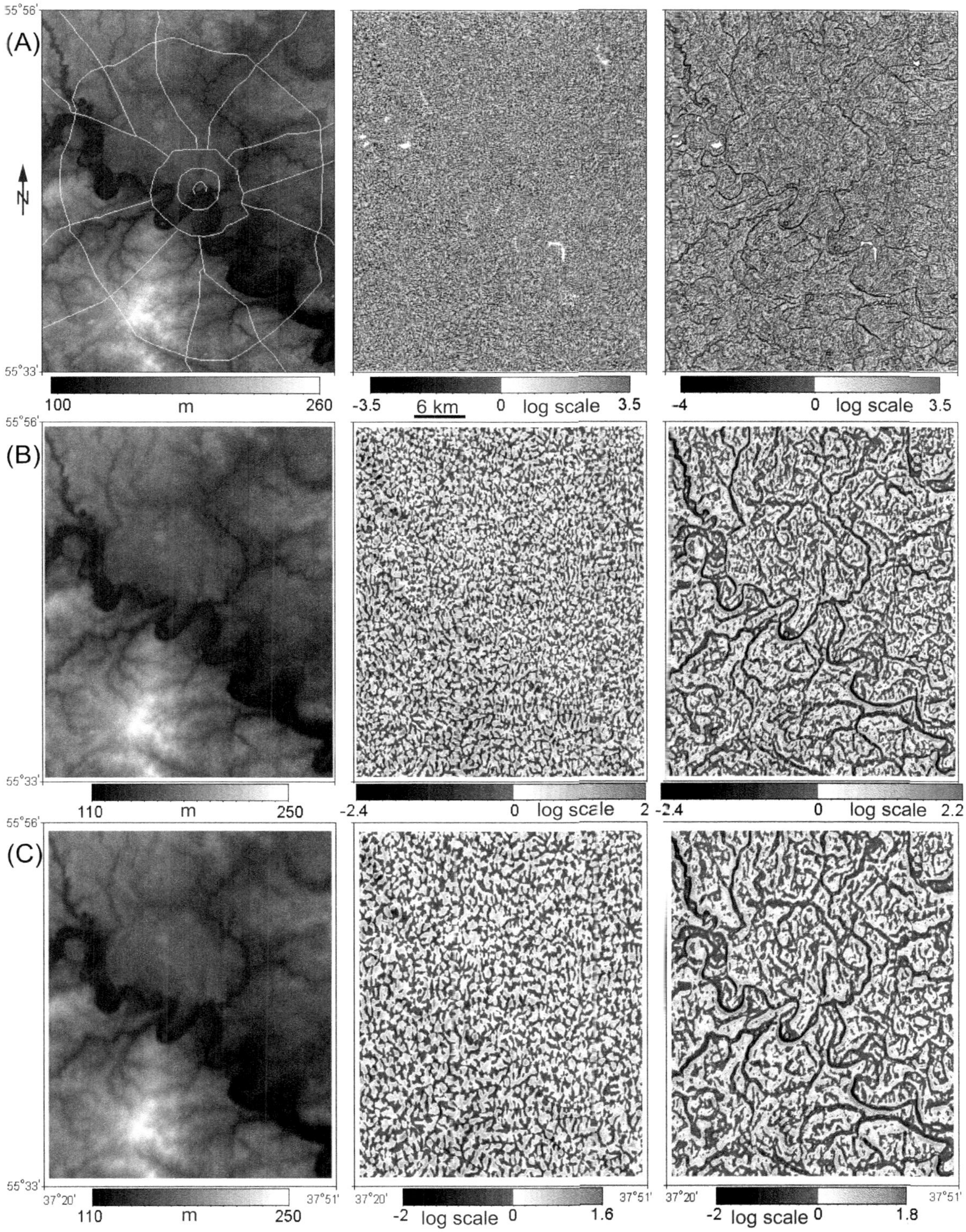

FIG. 6.5 Application of a mean filter for smoothing a DEM of Moscow city and adjacent areas, Russia: (A) Nonsmoothed DEM. (B) 5-times smoothed DEM. (C) 10-times smoothed DEM. *Left*, elevation (*white lines*, main highways); *center*, horizontal curvature; *right*, vertical curvature. A 3 × 3 moving window was used. *(From Florinsky, I.V., 2010. Theory and Applications of Mathematical Cartographic Modeling of Topography (D.Sc. thesis). Institute of Mathematical Problems of Biology, Russian Academy of Sciences, Pushchino, Russia (in Russian), Fig. 1.22.)*

6.2.5.3 Feature-preserving smoothing

Mean, median, and Gaussian filters (Sections 6.2.5.1 and 6.2.5.2) suppress both noise and signal. To circumvent the problem of signal corruption, several filtering techniques preserving topographic features were developed and applied to denoise SRTM DEMs (Stevenson et al., 2010; Gallant, 2011).

Lindsay et al. (2019) presented a feature-preserving DEM smoothing (FPDEMS) method based on the modification of surface normals at each point of the topographic surface described by a square-gridded DEM. A prototype of the FPDEMS method was an approach for denoising three-dimensional (3D) meshes in computer graphics (Sun et al., 2007). The FPDEMS method includes two main stages.

First, the normal vector field is modified using an $n \times n$ weighted moving window. The weights are larger for the window's points with normals aligned with the normal of the central point of the window. The weights decrease as the angle between the central and neighboring normals increases. A user specifies a window size: the larger the window, the stronger the smoothing.

The smoothed normal $\mathbf{n}'_i$ at the central point of the window

$$\mathbf{n}'_i = \sum_j^N w_j \mathbf{n}_j, \tag{6.11}$$

where N is the number of points within the window; $\mathbf{n}_j$ is the normal of the jth neighboring point; w_j is the weight of the jth neighboring point:

$$w_j = \begin{cases} \dfrac{\left(\cos\left(\mathbf{n}_i, \mathbf{n}_j\right) - \cos\Theta_t\right)^2}{\sum_j^N \left(\cos\left(\mathbf{n}_i, \mathbf{n}_j\right) - \cos\Theta_t\right)^2}, & \left(\mathbf{n}_i, \mathbf{n}_j\right) < \Theta_t \\ 0, & \left(\mathbf{n}_i, \mathbf{n}_j\right) \geq \Theta_t \end{cases}, \tag{6.12}$$

where $(\mathbf{n}_i, \mathbf{n}_j)$ is an angle between the central and neighboring normals; Θ_t is a user-defined threshold angle.

Since w_j are zero for neighboring points with $(\mathbf{n}_i, \mathbf{n}_j)$ larger than Θ_t (Eq. 6.12), such normals $\mathbf{n}_j$ are excluded from $\mathbf{n}_i'$ calculation by Eq. (6.11). In other words, neighboring but slope-break-separated points are not involved in the local smoothing.

Second, after smoothing the normal vector field, elevation values are iteratively modified (the number of iterations is determined empirically). For the central point of the window, a modified (smoothed) elevation value is calculated from the smoothed tangent planes of eight neighboring points using their smoothed normals (Lindsay et al., 2019).

Comparing with mean, median, and Gaussian filters, the FPDEMS method can provide better suppression of noise, preserve sharp slope breaks, and retain the topographic complexity. The method may be useful for denoising ultra-high-resolution UAS- and lidar-based DEMs (Fig. 6.6).

6.2.6 Data elimination

Point, row, and column elimination from a DEM is one of the most popular approaches to DTM generalization (Rhind, 1971; Florinsky, 1991; Stewart and Podolski, 1998; Bergbauer and Pollard, 2003). The simplest technique of the scale-oriented DEM generalization is the generation of a new DEM from every nth row and every nth column of an original regular DEM; n is an integer (Rhind, 1971). This approach can be utilized as an aim-oriented generalization because the choice of DTM grid spacing depends on a particular problem in geoscientific research (Chapter 10).

Although elevation values in the DEM grid nodes are unchanged by row and column elimination, such an approach generates artifacts in a generalized DEM; for example, neighboring valleys belonging to different basins may merge into one valley (Armstrong and Martz, 2003).

Obviously, row and column elimination is the equivalent of changing the grid spacing value. This approach has been used to decompose the topographic surface for the description of the hierarchical structure of the terrain (Schmidt and Andrew, 2005) as well as to determine the adequate resolution for DTMs in soil studies (Florinsky and Kuryakova, 2000; Zhu et al., 2008; Roecker and Thompson, 2010); see details in Chapter 10. The change in the grid spacing value was applied to suppress high-frequency noise and reveal geological structures of different scales using digital models of principal curvatures of stratigraphic surfaces (Stewart and Podolski, 1998; Bergbauer and Pollard, 2003; Wynn and Stewart, 2003). Notice that the grid spacing value influences the calculation accuracy of morphometric variables (Section 5.3).

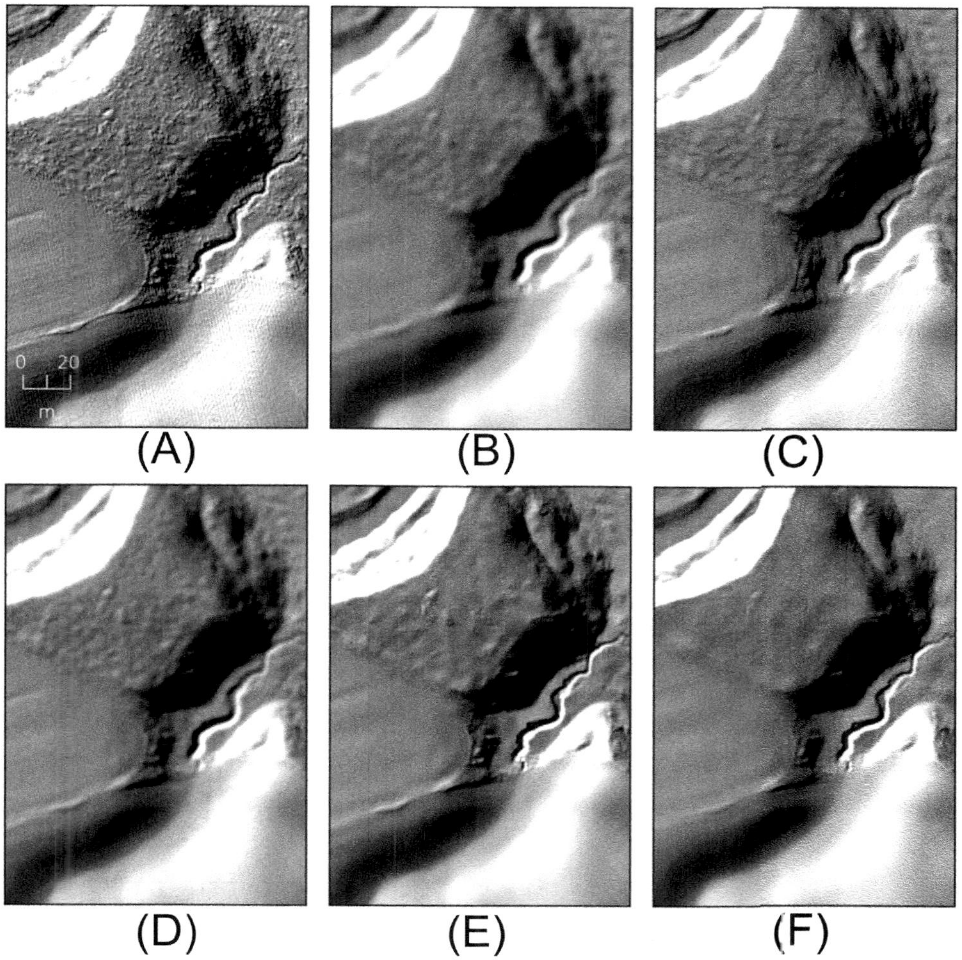

FIG. 6.6 A comparison of mean, median, and Gaussian filters and the FPDEMS method exemplified by a portion of 0.5-m resolution lidar DEM for an agricultural area in southwestern Ontario, Canada: (A) Initial DEM. (B) DEM smoothed by a mean filter with a 7×7 moving window. (C) DEM smoothed by a median filter with a 7×7 moving window. (D) DEM smoothed by a Gaussian filter with $\sigma = 1.7$ cells. (E) DEM smoothed by the FPDEMS method with an 11×11 moving window, $\Theta_t = 15°$, and 10 smoothing iteration. (F) DEM smoothed by the FPDEMS method with a 17×17 moving window, $\Theta_t = 20°$, and 15 smoothing iteration. Hill-shaded maps are presented. *(From Lindsay, J.B., Francioni, A., Cockburn, J.M. H., 2019. LiDAR DEM smoothing and the preservation of drainage features. Remote Sens. (Basel) 11 (16), 1926, Fig. 9, © Authors, distributed under the terms and conditions of the Creative Commons Attribution (CC BY) license.)*

Multiresolution DEMs have the built-in possibility of generalizing a model by removing particular hierarchical layers. Among such models are DEMs based on hierarchical TINs (De Floriani et al., 1993, 1997, 2000; Abdelguerfi et al., 1998; Pajarola and Gobbetti, 2007; Duan et al., 2017; Feciskanin and Minár, 2021) and spheroidal DEMs based on the hierarchical tessellation of a sphere by spherical polygons (Dutton, 1999; Sahr et al., 2003; Mahdavi-Amiri et al., 2015).

DTM generalization by row and column elimination is illustrated in Fig. 6.7. Two DEMs of the East European Plane and adjacent territories (Fig. 6.7A) with different resolutions (4′ and 10′) were compiled using several sources. Land elevations were extracted from the GLOBE DEM (GLOBE Task Team and Others, 1999), the bathymetry of all oceans and most seas was taken from the ETOPO2 DEM (NOAA, 2001), and the bathymetry of the Caspian Sea and some large lakes was digitized from topographic maps (Florinsky, 2007a).

Models of vertical curvature (Fig. 6.7B) were derived from three times smoothed DEMs by the author's method (Section 4.3). To deal with the large dynamic range of k_v, we logarithmically transformed its digital models by Eq. (8.1) with $n = 7$ and 8 for the DEM grid spacing of 4′ and 10′, respectively. The generalization effect is especially seen on maps of vertical curvature.

6.2.7 Cutting method

Generalization of curvature maps and visualization of landforms or geological structures of a higher hierarchical level can be achieved using the curvature threshold (k_t). It specifies an absolute value of a curvature below which the curvature is set to zero.

Koshkarev (1982) introduced a method of cutting or lopping off negligible or minor curvature values. Mynatt et al. (2007a) applied this approach to classify forms of a geological structure (Fig. 6.8) by signs of the Gaussian and mean

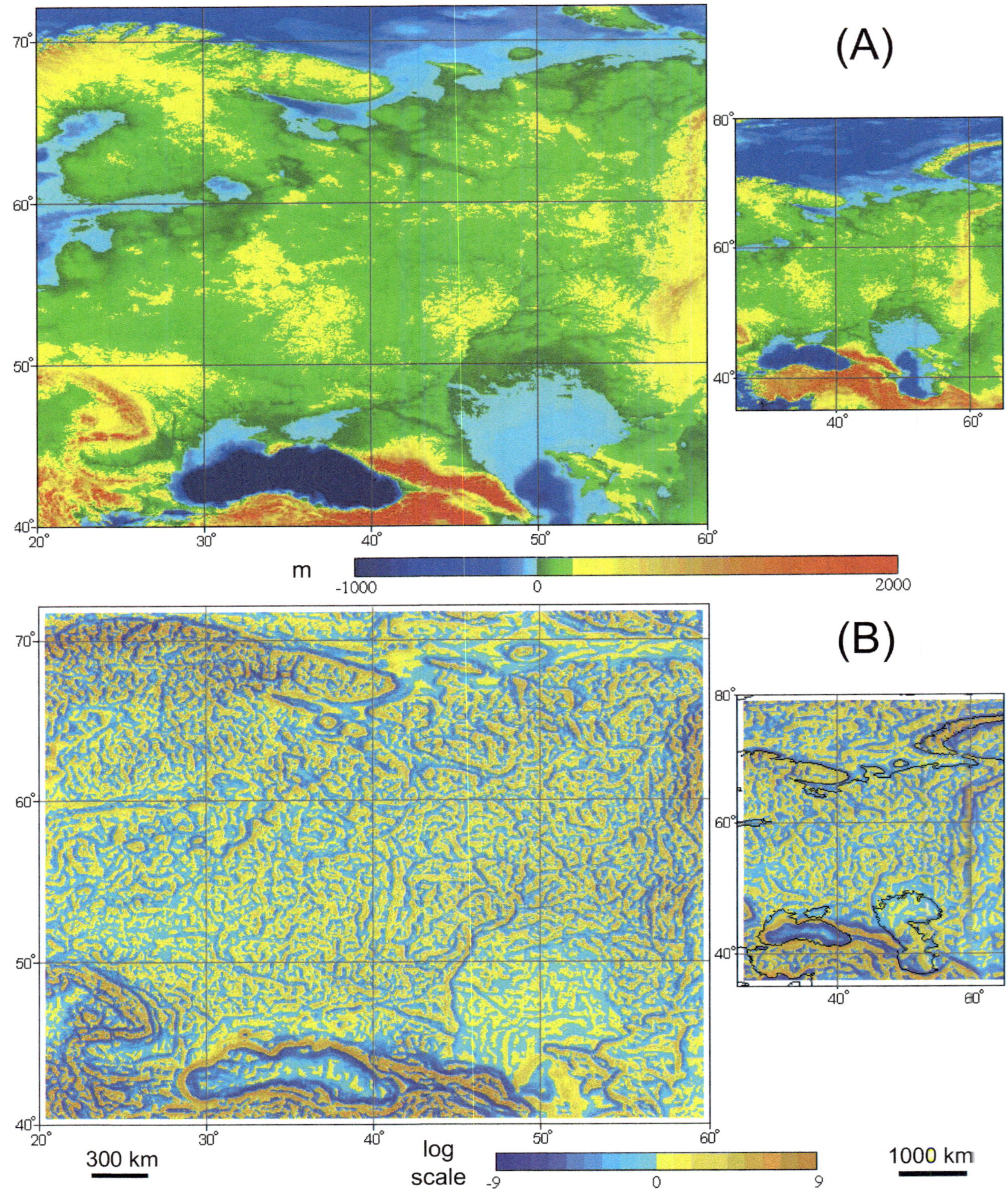

FIG. 6.7 The East European Plane and adjacent territories: (A) Elevation. (B) Vertical curvature. Grid spacing is 4′ (*left*) and 10′ (*right*). DEMs include 289,081 points (the elevation matrix 601 × 481) and 64,530 points (the elevation matrix 239 × 270), respectively. *(From Florinsky, I.V., 2010. Theory and Applications of Mathematical Cartographic Modeling of Topography (D.Sc. thesis). Institute of Mathematical Problems of Biology, Russian Academy of Sciences, Pushchino, Russia (in Russian), Fig. 1.23.)*

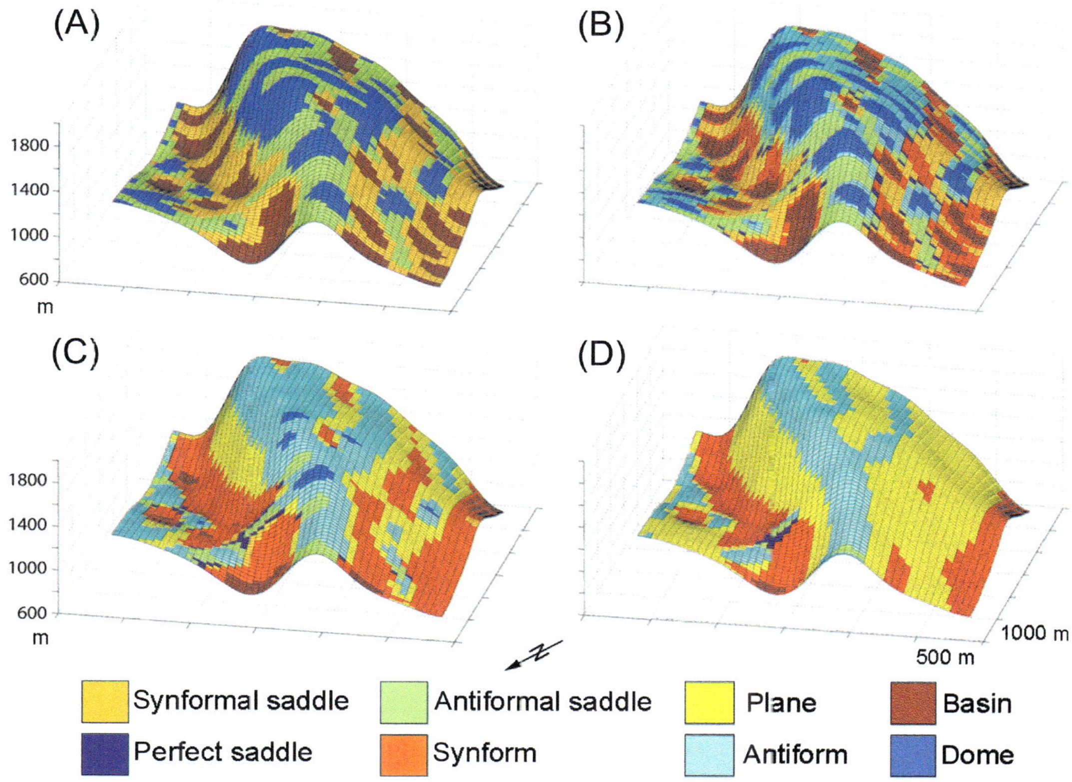

FIG. 6.8 DTM generalization by the cutting method. Classification of the Sheep Mountain Anticline, Wyoming, United States, by signs of the Gaussian and mean curvatures with varying values of the curvature threshold: (A) $k_t = 0\,\text{m}^{-1}$. (B) $k_t = 10^{-4}\,\text{m}^{-1}$. (C) $k_t = 5 \times 10^{-4}\,\text{m}^{-1}$. (D) $k_t = 10^{-3}\,\text{m}^{-1}$. A DEM was created by digitizing a structure contour map. The regular DEM includes 4641 points; the grid spacing is 50m. *(From Mynatt, I., Bergbauer, S., Pollard, D.D., 2007. Using differential geometry to describe 3-D folds. J. Struct. Geol. 29 (7), 1256–1266, Fig. 5, reproduced with permission from Elsevier.)*

curvatures (Section 2.7.1). The cutting was applied to the principal curvatures, which were then used to estimate K and H (Section 2.2.4.5). Those authors noted a generalization effect, which should be considered in interpretation of such maps:

For $k_t = 0\,\text{m}^{-1}$ (k_{min} and k_{max} were not set to zero), forms of the geological surface were classified as domes, basins, and antiformal and synformal saddles (Fig. 6.8A). There were no cylindrical or planar forms.[b] "The anticlinal hinge of the fold and the subparallel undulations on the backlimb are identified as convex up domes and antiformal saddles, while the synclinal hinge is composed of basins and synformal saddles" (Mynatt et al., 2007a, p. 1261). For $k_t = 10^{-4}\,\text{m}^{-1}$, minor values of k_{min} and k_{max} were set to zero, and small cylindrical areas and local perfect saddles appeared on the map (Fig. 6.8B). For $k_t = 5 \times 10^{-4}\,\text{m}^{-1}$ and $1 \times 10^{-3}\,\text{m}^{-1}$, larger cylindrical and planar areas appeared (Fig. 6.8C and D). "The anticlinal hinge and adjacent backlimb undulations are composed almost entirely of antiformal points and the synclinal hinge is composed primarily of synclinal points. Both limbs are approximated as composed predominately of planar points" (Mynatt et al., 2007a, p. 1262).

6.3 Two-dimensional singular spectrum analysis

Singular spectrum analysis (SSA) originated as a model-free technique to analyze one-dimensional time series (Elsner and Tsonis, 1996; Danilov and Zhigljavsky, 1997; Golyandina et al., 2001; Golyandina and Zhigljavsky, 2013). The SSA can be used to decompose a time series into a sum of a trend, oscillations, and noise, to detect periodicities, to smooth and denoise signals, to forecast time series, to impute missing data, and so on.

---

b. This is the expected result: cylindrical and planar landforms are "rare forms"; they do not practically occur in nature (Shary, 1995). For details, see Section 2.7.3.

There are several multidimensional extensions of the SSA. For example, a multichannel SSA is intended to simultaneously analyze a set of time series with common features (Elsner and Tsonis, 1996; Danilov and Zhigljavsky, 1997; Golyandina and Stepanov, 2005). The multichannel SSA can formally be applied to 2D scalar fields if one dimension is considered to be time. The two-dimensional singular spectrum analysis (2D-SSA) was specially designed to process 2D scalar fields (Danilov and Zhigljavsky, 1997; Golyandina and Usevich, 2010). The 2D-SSA is invariant regarding field rotation.

In the next section, the 2D-SSA algorithm developed by N.E. Golyandina is described according to our joint papers (Golyandina et al., 2007, 2008).

6.3.1 Formulas

Let us consider a 2D discrete field $f: \{1,\ldots,N_r\} \times \{1,\ldots,N_c\} \mapsto \mathbf{R}$ given by a matrix

$$\mathrm{F} = \begin{pmatrix} f(1,1) & f(1,2) & \ldots & f(1,N_c) \\ f(2,1) & f(2,2) & \ldots & f(2,N_c) \\ \vdots & \vdots & \ddots & \vdots \\ f(N_r,1) & f(N_r,2) & \ldots & f(N_r,N_c) \end{pmatrix}. \tag{6.13}$$

Algorithm parameters are window sizes (L_r, L_c), where $1 \le L_r \le N_r$, $1 \le L_c \le N_c$, $1 < L_r L_c < N_r N_c$. Set $K_r = N_r - L_r + 1$ and $K_c = N_c - L_c + 1$. The algorithm includes two stages—decomposition and reconstruction—each of which consists of two steps. Details on matrix calculus can be found elsewhere (Magnus and Neudecker, 2007, chap. 1).

6.3.1.1 *Decomposition*

The first step—embedding—consists of constructing the trajectory matrix of the field F by moving $L_r \times L_c$ windows. In the 1D case, one transforms a 1D object into a 2D matrix (Elsner and Tsonis, 1996; Golyandina et al., 2001; Golyandina and Zhigljavsky, 2013). Here, a 2D object is embedded into a four-dimensional (4D) space. To flatten the 4D object, we transform moving windows

$$\mathrm{F}_{i,j} = \begin{pmatrix} f(i,j) & f(i,j+1) & \ldots & f(i,j+L_c-1) \\ f(i+1,j) & f(i+1,j+1) & \ldots & f(i+1,j+L_c-1) \\ \vdots & \vdots & \ddots & \vdots \\ f(i+L_r-1,j) & f(i+L_r-1,j+1) & \ldots & f(i+L_r-1,j+L_c-1) \end{pmatrix}, \tag{6.14}$$

where $1 \le i \le K_r$, $1 \le j \le K_c$, to columns of the flattened trajectory matrix $\mathbf{W}$ ($\mathrm{F}_{i,j}$ transfers to the $(i+(j-1)K_r)$th column). For example, if $L_r = L_c = 2$, then the window

$$\begin{pmatrix} f(1,1) & f(1,2) \\ f(2,1) & f(2,2) \end{pmatrix} \tag{6.15}$$

is transformed into the first column $(f(1,1), f(2,1), f(1,2), f(2,2))^{\mathrm{T}}$; ^{T} denotes the matrix transpose.

It is appropriate to use vectorizing and matricizing operations: for an $M \times N$ matrix $\mathbf{B}$, $\mathrm{vec}\mathbf{B} \in \mathbf{R}^{MN}$ is the vector constructed from stacked columns of $\mathbf{B}$. If we fix matrix sizes M and N, then (M, N)-matricizing will be opposite to vectorizing: $\mathrm{matr}\ \mathrm{vec}\mathbf{B} = \mathbf{B}$. Thus, the trajectory matrix $\mathbf{W}$ of the field F consists of $K_r K_c$ columns $\mathrm{vec}\mathrm{F}_{i,j}$, where $1 \le i \le K_r$, $1 \le j \le K_c$. Furthermore, the matrix $\mathbf{W}$ of the size $L_r L_c \times K_r K_c$ can be presented in a more structured form:

$$\mathbf{W} = \begin{pmatrix} \mathbf{H}_1 & \mathbf{H}_2 & \ldots & \mathbf{H}_{K_c} \\ \mathbf{H}_2 & \mathbf{H}_3 & \ldots & \mathbf{H}_{K_{c+1}} \\ \vdots & \vdots & \ddots & \vdots \\ \mathbf{H}_{L_c} & \mathbf{H}_{L_{c+1}} & \ldots & \mathbf{H}_{N_c} \end{pmatrix}, \tag{6.16}$$

where

$$\mathbf{H}_i = \begin{pmatrix} f(1,i) & f(2,i) & \dots & f(K_r,i) \\ f(2,i) & f(3,i) & \dots & f(K_r+1,i) \\ \vdots & \vdots & \ddots & \vdots \\ f(L_r,i) & f(L_r+1,i) & \dots & f(N_r,i) \end{pmatrix}, \tag{6.17}$$

The matrix $\mathbf{W}$ has the block-Hankel structure with the same blocks along secondary diagonals. Each block $\mathbf{H}_i$ is Hankel itself; it is the trajectory matrix of the 1D series $f(, i)$ (the ith column of the initial field F). The matrix $\mathbf{W}$ will be further called the block-Hankel trajectory matrix of the field F. Note that there is a one-to-one correspondence between $N_r \times N_c$ fields and block-Hankel matrices with $L_r \times K_r$ Hankel blocks.

The next step is a singular value decomposition of the block-Hankel trajectory matrix:

$$\mathbf{W} = \sum_{i=1}^{d} \mathbf{W}_i = \sum_{i=1}^{d} \sqrt{\lambda_i} U_i V_i^{\mathrm{T}}, \tag{6.18}$$

where $\lambda_1, \ldots, \lambda_d$ are nonzero eigenvalues of the matrix $\mathbf{W}\mathbf{W}^{\mathrm{T}}$ arranged in decreasing order of magnitudes ($\lambda_1 \geq \lambda_2 \geq \cdots \geq \lambda_d > 0$); $\{U_1, \ldots, U_d\}$; $U_i \in \mathbf{R}^{L_r L_c}$ is the corresponding orthonormal system of the eigenvectors; and $\{V_1, \ldots, V_d\}$, $V_i \in \mathbf{R}^{K_r K_c}$ is the orthonormal system of the corresponding factor vectors

$$V_i = \mathbf{W}^{\mathrm{T}} U_i / \sqrt{\lambda_i}. \tag{6.19}$$

By analogy with principal component analysis, the vectors $\sqrt{\lambda_i} V_i$ are called principal component vectors. They are conveniently considered as matrices: the (L_r, L_c)-matricizing of an eigenvector is called an eigenfield; the (K_r, K_c)-matricizing of a factor vector is called a factor field; and the (K_r, K_c)-matricizing of a vector of principal components is called a principal component field. A set of square roots of ith eigenvalue, ith eigenfield, and ith factor field is called ith eigentriple (ET).

6.3.1.2 Reconstruction

The next step consists of the grouping of addends in the decomposition (Eq. 6.18), that is, the corresponding eigentriples. Let us divide the set $\{1, \ldots, d\}$ into m disjoint subsets $\mathrm{I}_1, \ldots, \mathrm{I}_m$. Summing $\mathbf{W}_i$, $i \in \mathrm{I}_k$, we come to the expansion

$$\mathbf{W} = \sum_{k=1}^{m} \mathbf{W}_{\mathrm{I}_k}. \tag{6.20}$$

Grouped matrices $\mathbf{W}_{\mathrm{I}_k}$ do not necessarily have the block-Hankel form. Therefore, one needs an additional step to transfer the decomposition (Eq. 6.20) of the block-Hankel trajectory matrix $\mathbf{W}$ into a decomposition of the initial field F. This can be done by the orthogonal projection (in the Frobenious norm) of the matrices $\mathbf{W}_{\mathrm{I}_k}$ on the set of the block-Hankel matrix with Hankel blocks, like Eq. (6.16). After projection, we obtain

$$\mathbf{W} = \sum_{k=1}^{m} \tilde{\mathbf{W}}_k, \tag{6.21}$$

where $\tilde{\mathbf{W}}_k$, $k = 1, \ldots, m$, have a form of Eq. (6.16). Using the one-to-one correspondence between block-Hankel trajectory matrices and 2D fields, we come to the final decomposition of the initial field:

$$\mathrm{F} = \sum_{k=1}^{m} \tilde{\mathrm{F}}_k. \tag{6.22}$$

The algorithm decomposes the initial field into a sum of components. It is expected that if the field F is a sum of a smooth surface, oscillations, and noise, there exists such a grouping that the resultant decomposition (Eq. 6.22) is close to the initial field decomposition. This gives an opportunity for smoothing, denoising, removing periodic noise, and the like.

Rules to select 2D-SSA parameters are mostly similar to those for 1D-SSA (Golyandina et al., 2001; Golyandina and Zhigljavsky, 2013). In particular, small window sizes produce adaptive smoothing. The most detailed decomposition and the better separation of the field components can be obtained if window sizes are close to ($N_r/2$, $N_c/2$). However, the use of large window sizes can lead to a mixing problem caused by too-detailed decomposition of components.

6.3.2 Materials and data processing

To demonstrate the abilities of the 2D-SSA as a tool of DTM filtering, we selected a portion of the Northern Andes measuring 4° × 4°, located between 2°S and 2°N, and 78°30′W and 74°30′W (Fig. 6.9A). The area covers some regions of Ecuador, Colombia, and Peru, including parts of the Coastal Plain, the Andean Range, and the Upper Amazon Basin.

A DEM of the study area was extracted from the GTOPO30 DEM (USGS, 1996). The DEM includes 230,880 points (the elevation matrix 480 × 481); the grid spacing is 30″ (Fig. 6.9B).

We selected this area and GTOPO30 on two counts. First, it is well known that this DEM incorporates high-frequency noise caused by interpolation errors and inaccurate merging of topographic charts having different accuracy. The spatial distribution of the noise in GTOPO30 is uneven and depends on the accuracy of cartographic sources. In particular, the potent noise is typical for forested regions of Africa and South America because reasonably detailed and accurate topographic data were unavailable for such areas. Thus, interpolation of sparse contours has been used to compile these portions of GTOPO30. Although DEM noise is no obstacle to producing realistic maps of elevation, it leads to the derivation of noisy and unreadable maps of local morphometric variables (computation of the partial derivatives of elevation dramatically

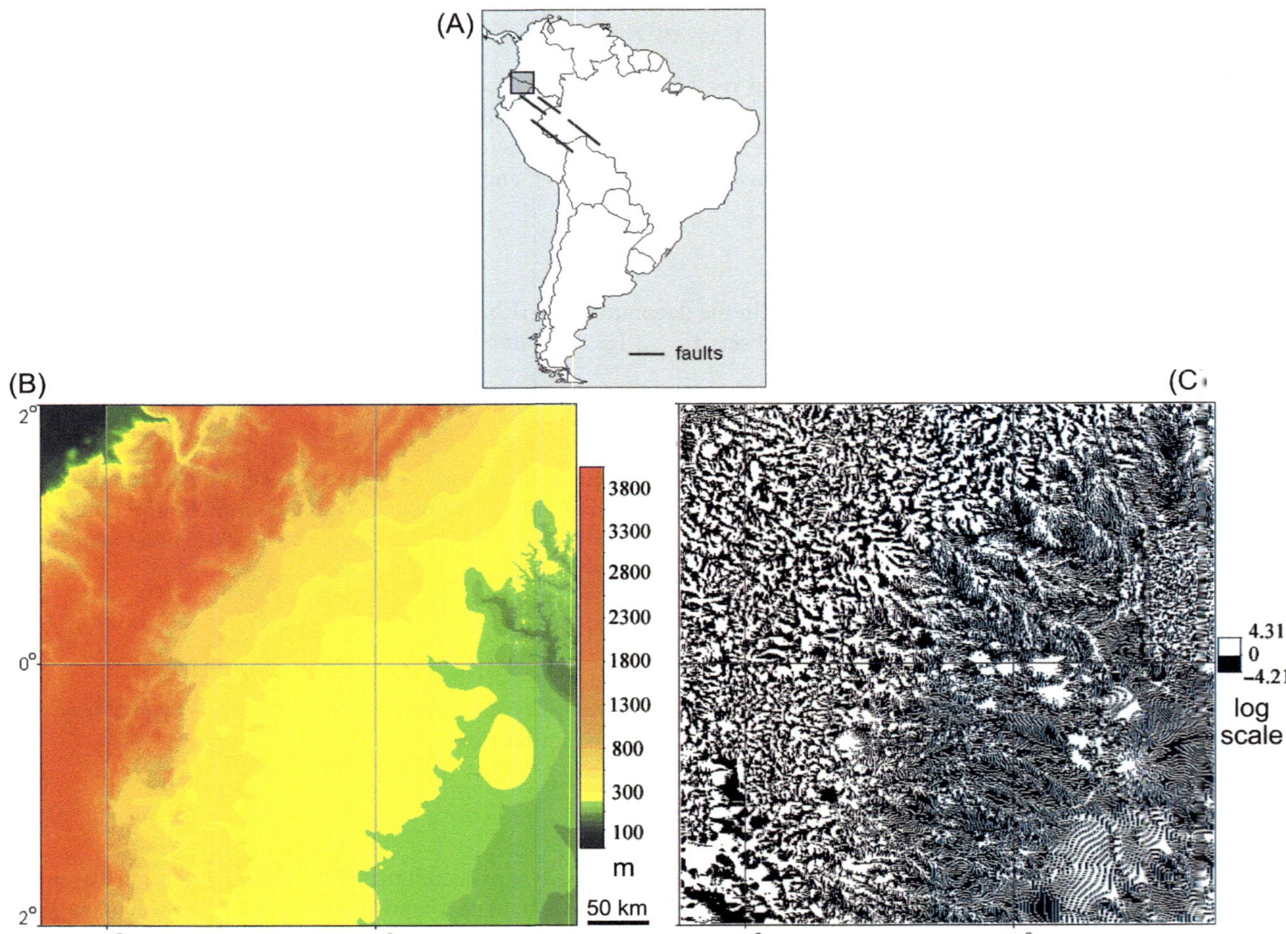

FIG. 6.9 The Northern Andes: (A) Geographical location of the study area. Some deep-seated faults are shown after the global fault scheme (Chebanenko, 1963, Fig. 28). (B) Elevation. (C) Horizontal curvature derived from the initial DEM. *Publisher's note: Map lines delineate the study area and do not necessarily depict accepted national boundaries. ((A) From Florinsky, I.V., 2010. Theory and Applications of Mathematical Cartographic Modeling of Topography (D.Sc. thesis). Institute of Mathematical Problems of Biology, Russian Academy of Sciences, Pushchino, Russia (in Russian), Fig. 1.24; (B) and (C) from Golyandina, N.E., Usevich, K.D., Florinsky, I.V., 2008. Singular spectrum analysis for filtering of digital terrain models. Geodesy Cartogr. (5), 21–28 (in Russian), Fig. 2A and D.)*

increases the noise, see Section 5.4.1). The study area, consisting of two main zones—high mountains and forested foothills—which have different signal-to-noise ratios, is ideally suited for validating the 2D-SSA as a tool to denoise DEMs. Second, the use of this DEM allowed us to demonstrate the ability of the 2D-SSA to decompose the topographic surface into components of different scales under complicated geomorphic conditions.

To reduce the huge range of elevations (about 6080 m, see Fig. 6.9B), the initial DEM was logarithmically transformed. Logarithmic DEMs were used in the further processing and mapping.

Using the window size of 30×30, the initial DEM was decomposed into 900 eigentriples (Fig. 6.10). We evaluated various combinations of eigentriples to reconstruct DEMs. Finally, we selected ET combinations producing interpretable reconstructed DEMs. To denoise the initial DEM (Fig. 6.9B), a DEM was reconstructed from the ET 1–100 (Fig. 6.11A). To exemplify DTM generalization, a DEM was reconstructed from the ET 1–50 (Fig. 6.11B) and ET 1–25. To separate continental, regional, and local components of topography, six DEMs were reconstructed from the ET 1 (Fig. 6.12A), ET 2, ET 3, ET 2–3 (Fig. 6.12C), ET 4–25 (Fig. 6.12D), and ET 51–100 (Fig. 6.13A). To visualize a noise component, a DEM was reconstructed from the ET 101–900 (Fig. 6.13B). The DEM was processed by the 2D-SSA software (© K. Usevich and N. Golyandina, 2005–2007).

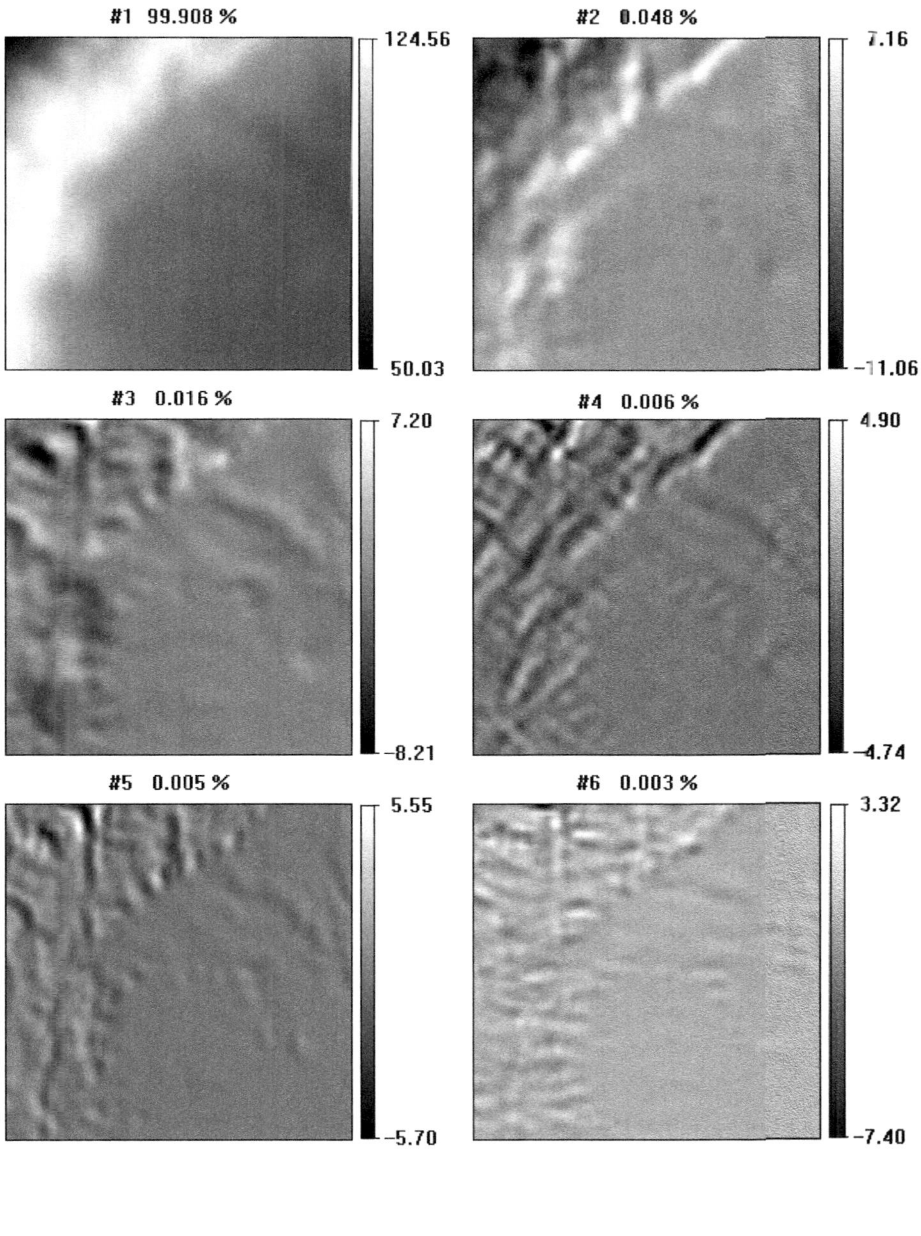

FIG. 6.10 The Northern Andes, principal component fields 1–6 of the DEM decomposition. The percentage reflects shares of the corresponding eigentriples in the singular value decomposition. Logarithmic scale is used. *(From Golyandina, N.E., Usevich, K.D., Florinsky, I.V., 2008. Singular spectrum analysis for filtering of digital terrain models. Geodesy Cartogr. (5), 21–28 (in Russian), Fig. 1.)*

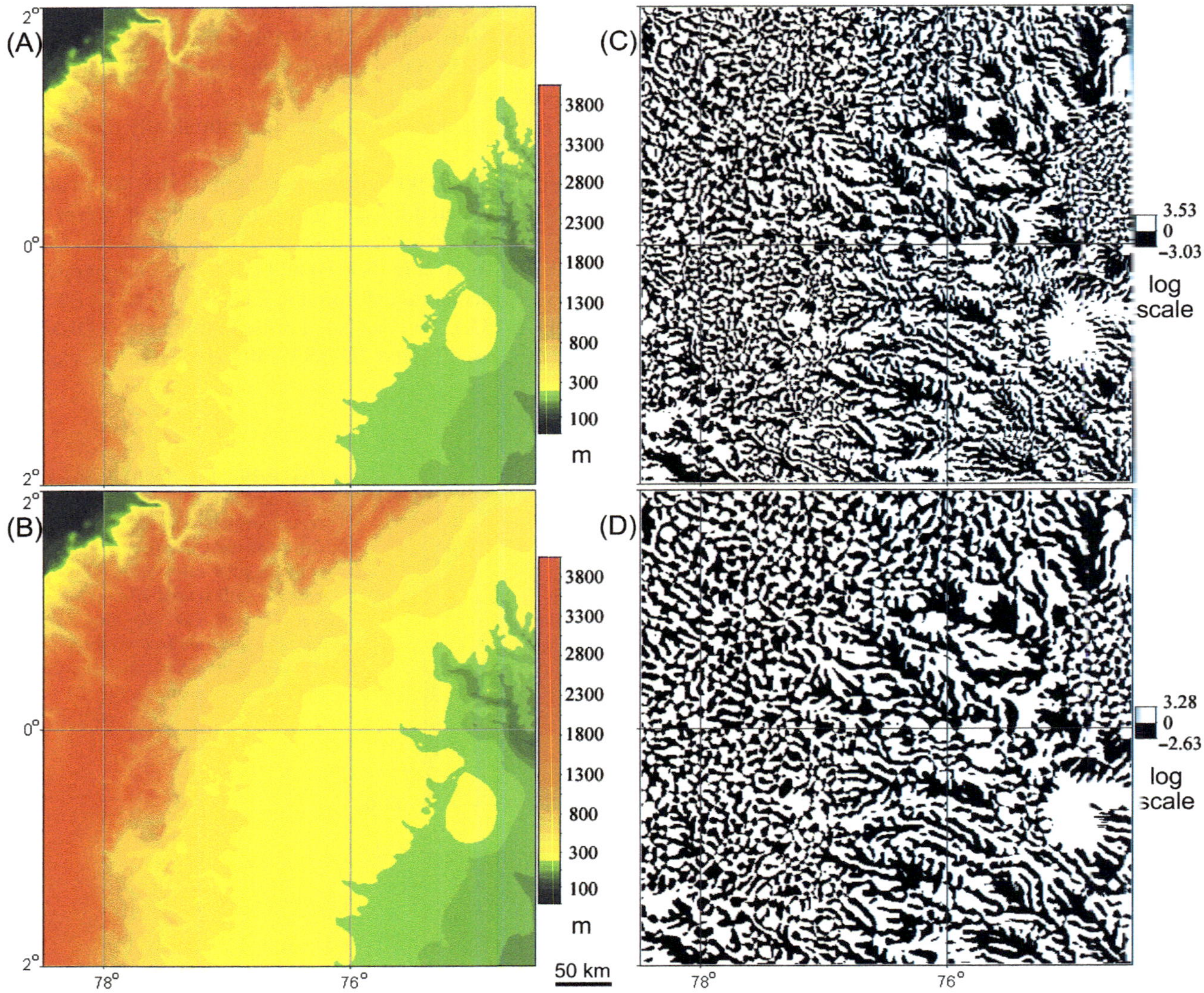

FIG. 6.11 The Northern Andes, DTM denoising. Elevation reconstructed from: (A) ET 1–100 and (B) ET 1–50. Horizontal curvature derived from: (C) ET 1–100 DEM and (D) ET 1–50 DEM. *(From Golyandina, N.E., Usevich, K.D., Florinsky, I.V., 2008. Singular spectrum analysis for filtering of digital terrain models. Geodesy Cartogr. (5), 21–28 (in Russian), Fig. 2B, C, E, and F.)*

To return to elevation values, reconstructed logarithmic DEMs were exponentiated. Using the author's method (Section 4.3), we derived k_h models from the initial DEM and some reconstructed DEMs. To deal with the large dynamic range of k_h, we logarithmically transformed its digital models by Eq. (8.1) with $n=8$. To clarify the effects of denoising and generalization, we mapped k_h subdividing its values into two levels: $k_h>0$ and $k_h<0$ (flow divergence and convergence areas, respectively).

The DTMs produced had a grid spacing of 30″. Exponentiation, k_h calculation, and mapping were done by the LandLord software (Florinsky, 2016).

Using samples extracted from each DEM and k_h model, we performed a pairwise comparison of the DTMs obtained. The sample size was 2209 points (the elevation matrix 47 × 47); the grid spacing was 5′. Statistical analysis was carried out with the Statgraphics software (© Statistical Graphics Corp., 1994–1997).

6.3.3 Results and discussion

A visual comparison of elevation maps derived from the initial DEM (Fig. 6.9B) and two DEMs reconstructed from the ET 1–100 and ET 1–50 (Fig. 6.11A and B) permits one to see nothing but minor changes in image patterns. A cursory examination may lead to underestimation of the DEM denoising and generalization results.

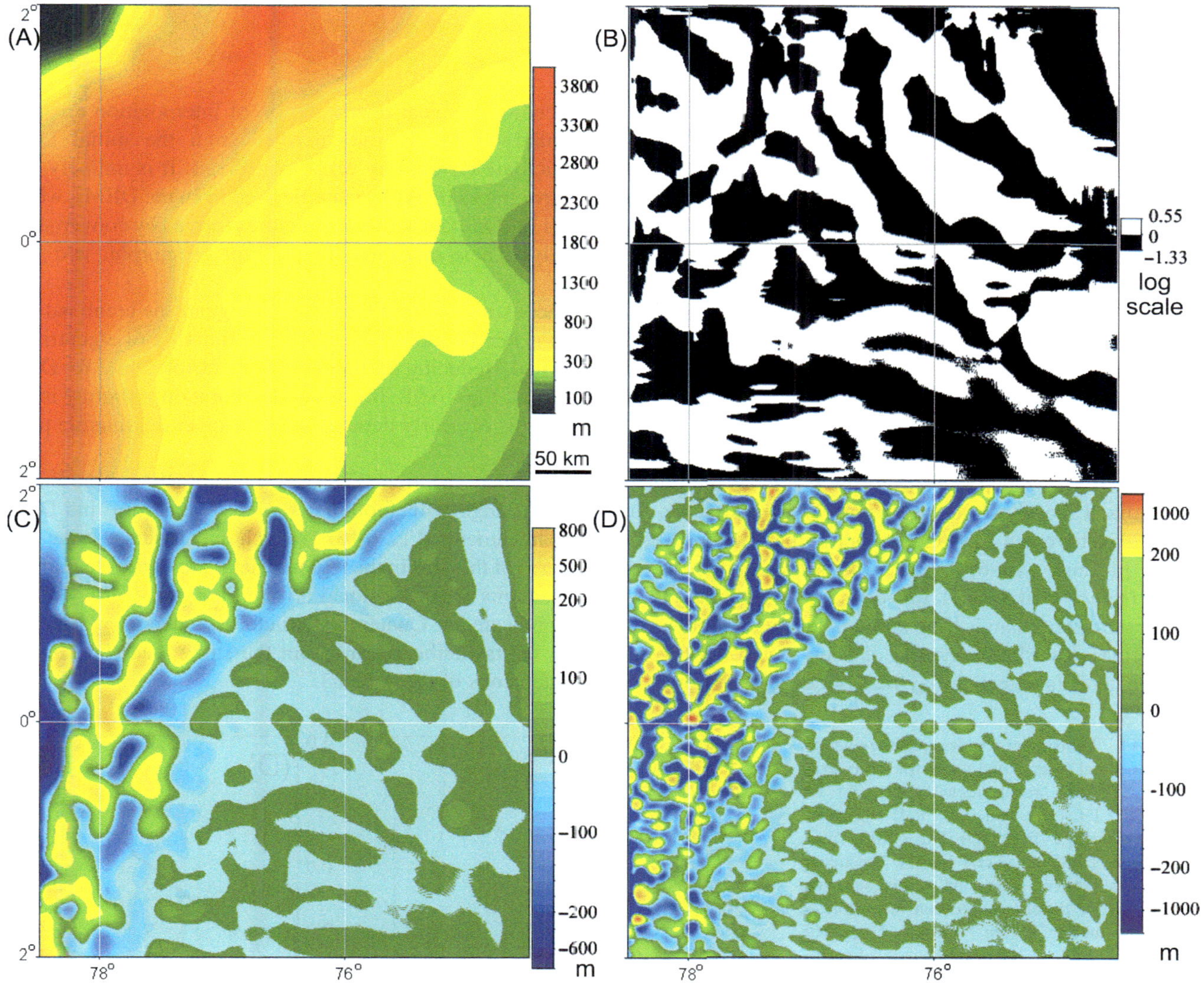

FIG. 6.12 The Northern Andes, low-frequency components. (A) Elevation reconstructed from the ET 1. (B) Horizontal curvature is derived from ET 1 DEM. Elevation reconstructed from: (C) ET 2–3 and (D) ET 4–25. *(From Golyandina, N.E., Usevich, K.D., Florinsky, I.V., 2008. Singular spectrum analysis for filtering of digital terrain models. Geodesy Cartogr. (5), 21–28 (in Russian), Fig. 3A–D.)*

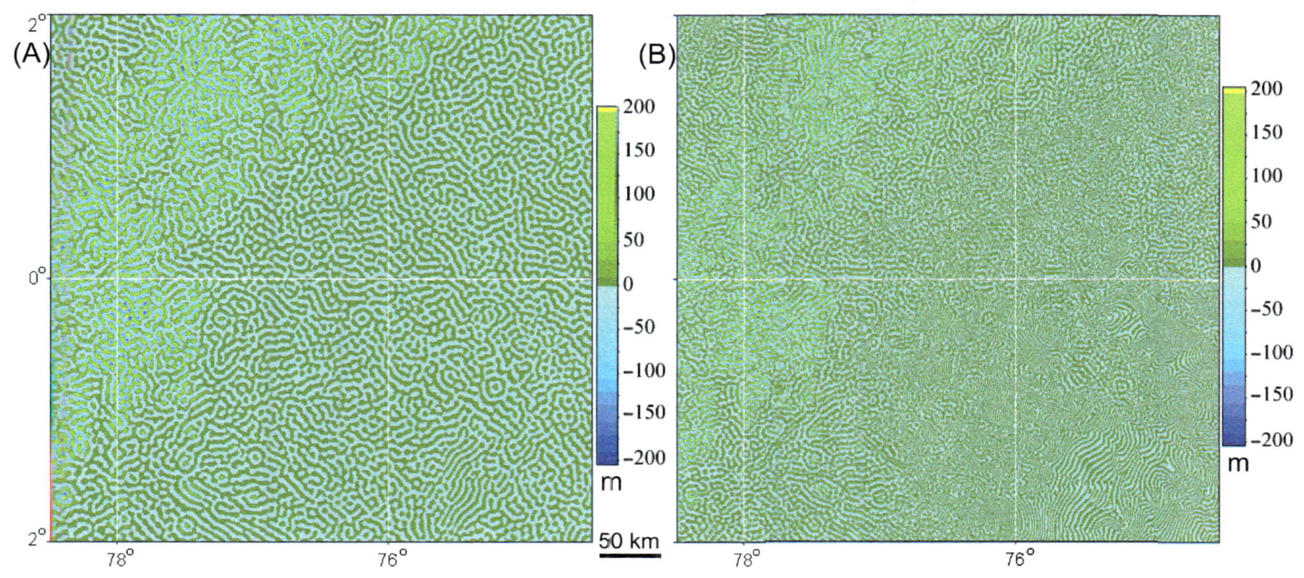

FIG. 6.13 The Northern Andes, high-frequency components. Elevation reconstructed from (A) ET 51–100 and (B) ET 101–900. *(From Golyandina, N.E., Usevich, K.D., Florinsky, I.V., 2008. Singular spectrum analysis for filtering of digital terrain models. Geodesy Cartogr. (5), 21–28 (in Russian), Fig. 3E and F.)*

The k_h maps provide better insight into these outcomes. Indeed, the typical manifestation of interpolation errors—contour "tracks" (Florinsky, 2002) (Section 5.4)—can be found on the k_h map derived from the initial DEM (Fig. 6.9C). These tracks are typical of the Andean foothills, which are covered by dense rainforests. It is unlikely that this map may be used for any application. However, there are no error tracks on the k_h maps derived from the DEMs, which were reconstructed from the ET 1–100 and ET 1–50 (Fig. 6.11C and D). One can see flow structures formed by convergence and divergence areas (black and white image patterns, respectively). These maps may be used for geomorphic and geological interpretation.

A comparison between the k_h maps, derived from the DEMs reconstructed from the different eigentriple combinations (Figs. 6.9B, 6.11C and D), shows a pronounced effect of the map generalization. The smaller the number of eigentriples used to reconstruct a DEM, the smoother and simpler k_h maps obtained. Reducing the number of eigentriples used for DEM reconstruction leads to a marked reduction in the range of k_h values (cf. Figs. 6.9B, 6.11C and D), but it only slightly influences the range of elevation values (cf. Figs. 6.9A, 6.11A and B). This is clearly demonstrated by quantile-quantile plots (Fig. 6.14).

The DEM reconstructed from the ET 1 has the highest level of generalization (Fig. 6.12A). This DEM represents a generalized morphostructure of the continental scale, the Andean Range with foothills. k_h derivation from this DEM allowed us to reveal a system of approximately northwest-striking lineaments (Fig. 6.12B), which may indicate strike-slip faults (Florinsky, 1996) (Chapter 14). Although there are no structures of this sort in the regional database of the Quaternary faults (Eguez et al., 2003), this does not demonstrate that the lineaments are of erosional origin. First, the geology of the Upper Amazon Basin is little known. Second, the lineaments may indicate pre-Quaternary structures. Indeed, Chebanenko (1963) has described a system of deep-seated northwest-striking faults southeastward from the study area (Fig 6.9A). The lineaments detected may be associated with northwestern extensions of these faults.

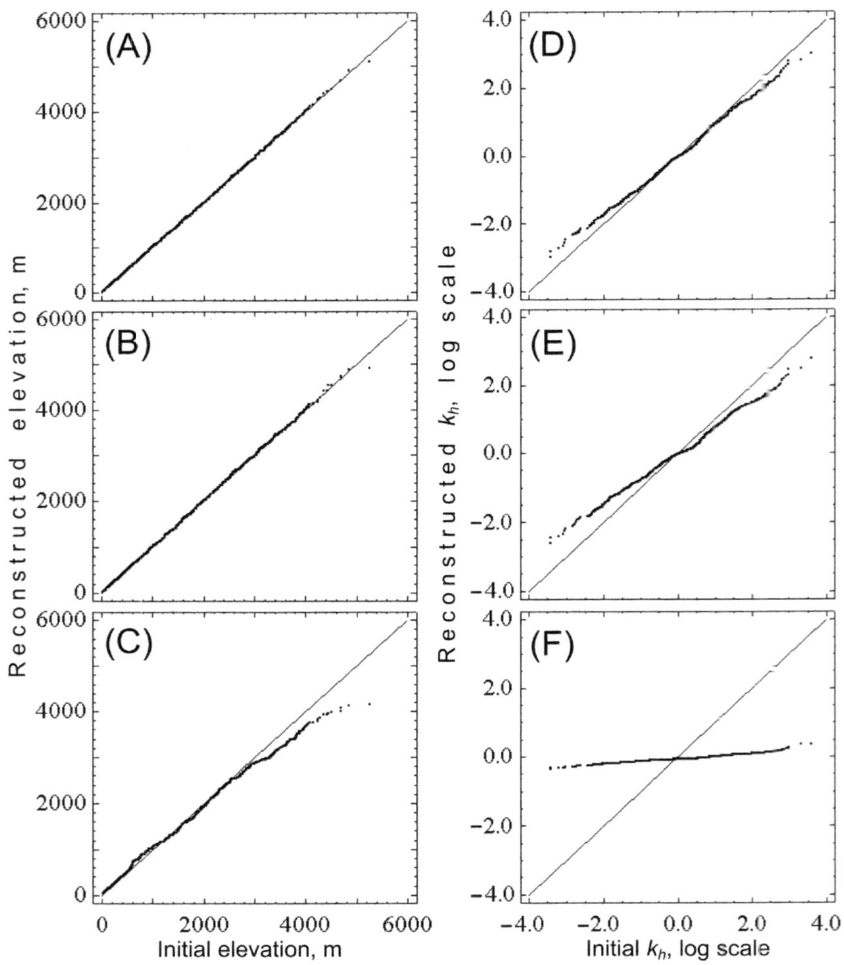

FIG. 6.14 The Northern Andes, quantile-quantile plots for initial (x-axis) and reconstructed (y-axis) values. Elevation reconstructed from (A) ET 1–100, (B) ET 1–50, and (C) ET 1. Horizontal curvature derived from (D) ET 1–100 DEM, (E) ET 1–50 DEM, and (F) ET 1 DEM. The greater the distance between points and the diagonal line, the greater the difference between reconstructed and initial models. *(From Florinsky, I.V., 2010. Theory and Applications of Mathematical Cartographic Modeling of Topography (D.Sc. thesis). Institute of Mathematical Problems of Biology, Russian Academy of Sciences, Pushchino, Russia (in Russian), Fig. 1.29.)*

The DEM reconstructed from the ET 2–3 (Fig. 6.12C) represents landforms that are probably connected with regional tectonic structures. For the DEM reconstructed from the ET 2 only, the elevation map shows approximately northeast-striking patterns, which agree closely with the direction of the Andean Range within the study area. For the DEM reconstructed from the ET 3 only, the elevation map represents approximately northwest-striking structures probably related to the k_h lineaments (Fig. 6.12B). It is conceivable that these structures control the spatial distribution of river valleys in the Upper Amazon Basin. The DEM reconstructed from the ET 4–25 represents landforms that are probably associated with geomorphic processes of a regional scale. For example, one can see generalized side valleys of the Amazon River (Fig. 6.12D).

The DEM reconstructed from the ET 51–100 (Fig. 6.13A) represents high-frequency components, which cannot be considered as noise. However, it is impossible to find familiar patterns in the drainage network on this map. These components might represent topographic manifestation of local (exogenous) geomorphic processes. This DEM is the difference between two DEMs reconstructed from the ET 1–100 and ET 1–50 (Fig. 6.11A and B).

The DEM reconstructed from the ET 101–900 (Fig. 6.13B) is a residual of the initial DEM (Fig. 6.9B) after its reconstruction from the ET 1–100 (Fig. 6.11A). Among other high-frequency components, this model includes the noise inherent in the initial DEM. This is precisely the noise that rendered the k_h map derived from the initial DEM unsuitable for applications. Among contour "tracks" one can see a rectangular feature measured $1° \times 30'$, which is located eastward from 75° 30′W and northward from the equator (Fig. 6.13B). The "rectangle" can also be seen on three maps of k_h (Figs. 6.9B, 6.11C and D). This is a trace of the inaccurate merging of adjacent topographic charts with different accuracies during the GTOPO30 compilation.

The DEM reconstructions from the ET 1, ET 2, ET 3, and ET 2–3 may be considered an application of low-pass filters to the initial DEM, while the DEM reconstructions from the ET 101–900 and ET 51–100 can be regarded as an application of high-pass filters.

The results demonstrate that the 2D-SSA is a powerful method to denoise DTMs. Indeed, 2D-SSA-based denoising leads to very fine changes in elevations. These changes cannot be captured except through the computation of partial derivatives. This suggests that the 2D-SSA is an important method for the preliminary treatment of noisy DEMs. This opens the way to utilizing them for the derivation of local and nonlocal morphometric variables.

The 2D-SSA can be used to decompose the topographic surface into components of continental, regional, and local scales. At the same time, eigentriple selection to reconstruct topographic components of different scales is nonunique and ambiguous. Notice that a similar problem arises if trend-surface analysis or the 2D Fourier transform is used to extract regional or local topographic components. However, this problem is largely associated with the qualitative nature of the geomorphological concepts—scale and hierarchical level—rather than with mathematical features of the methods.

Chapter 7

Universal spectral analytical modeling

7.1 Motivation

As the previous chapters show, mathematical issues of digital terrain analysis can be summarized into three key problems:

1. Interpolation of irregularly or regularly spaced DEMs. This task is commonly carried out by various local interpolation methods (Section 3.4).
2. Filtering is used to denoise, generalize, smooth, and decompose DEMs into components of different spatial scales. These tasks can be attacked by 2D discrete Fourier transform, 2D discrete wavelet transform, and 2D singular spectrum analysis (Chapter 6).
3. Derivation of local morphometric variables (Chapter 2.2) from regularly spaced DEMs using local approximations of the partial derivatives of elevation by finite differences (Sections 4.1–4.3).

Traditionally, these three problems of geomorphometry have been solved separately by means of procedures implemented in different methods and algorithms. Exceptions were very rare. For example, Mitášová and Mitáš (1993) developed a method for local interpolation of DEMs using regularized splines with tension, which allowed simultaneous estimation of the partial derivatives of elevation (and so, morphometric variables).

However, all three problems may be resolved within a single, end-to-end framework of DTM treatment based on the global approximation[a] of a bivariate function by high-order orthogonal polynomials. In the 1960s–80s, there were attempts to apply high-order orthogonal polynomials for DEM global approximation (Heifetz, 1964; Van Rossel, 1972; Yagodina, 1972; Segu, 1984). These attempts faced challenges (McCullagh, 1988, p. 753). First, global polynomial approaches require considerable computer resources. Second, practical tasks demanded working with increasingly large DEMs containing tens of thousands to a few million points. Methods and computers that existed at that time could not handle such data amounts. As a result, an opinion was formed that the topographic surface is too complicated for the application of global polynomial approximations. The prevailing view was that "high quality global interpolation methods … are computationally impractical" for digital terrain modeling (Hutchinson, 2008, p. 150). Only low-order orthogonal polynomials were commonly utilized in trend-surface analysis, a technique for decomposing the topographic surface (Section 6.2.1).

Progress in the theory and computational practice of polynomial approximation (Press et al., 1992, chaps. 3–5; Dedus A.F. et al., 1995; Dedus F.F. et al., 1995, 1999, 2002, 2004; Gautschi, 2004; Pankratov, 2004a, b; Pankratov and Britenkov, 2004; Tetuev and Dedus, 2007; Pankratov et al., 2011, 2015), as well as advances in computer technology, allowed us to develop a universal spectral analytical method based on high-order orthogonal expansions using the Chebyshev polynomials (Florinsky and Pankratov, 2015, 2016). Next, we describe the method, which is intended for the processing of regularly spaced DEMs within a single, end-to-end framework, including DEM global approximation, denoising, smoothing, generalization, calculation of the partial derivatives of elevation, and derivation of local morphometric variables.

7.2 Method

Consider the function $z=f(x,y)$ (Eq. 2.1) defined in a rectangular domain. This function can be approximated using the bivariate Chebyshev series expansion (Clenshaw and Hayes, 1965; Fox and Parker, 1968, § 6.22):

---

a. Two problems are considered in the approximation theory (Davis, 1975; Phillips, 2003): interpolation and approximation. A polynomial $Q_{mn}(x, y)$ is the approximating function of the actual function $f(x, y)$ if $Q_{mn}(x_i, y_j) \approx f(x_i, y_j)$ at the grid nodes (x_i, y_j), where $i=0, 1, \dots, m$; $j=0, 1, \dots, n$. The polynomial $Q_{mn}(x, y)$ is the interpolating function of $f(x, y)$ if $Q_{mn}(x_i, y_j)=f(x_i, y_j)$ at the grid nodes. Thus, the first problem is a particular case of the second one.

Digital Terrain Analysis. https://doi.org/10.1016/B978-0-443-24798-9.00022-4

$$z(x,y) = \sum_{i=0}^{l-1}\sum_{j=0}^{l-1} c_{ij}T_i(x)T_j(y), \tag{7.1}$$

where c_{ij} are expansion coefficients; $T_i(x)$ and $T_j(y)$ are the Chebyshev polynomials of the first kind in the variables x and y of degrees i and j, respectively; l is the maximum degree of expansion. It is assumed that the domain of the original function is translated into the domain of the orthogonal polynomials by a linear transformation.

The method consists of four main stages: (1) calculation of expansion coefficients; (2) the Fejér summation; (3) reconstruction of the approximated function; and (4) calculation of derivatives. The 2D approximation of the original function (Eq. 2.1) can be computationally performed as a superposition of two one-dimensional approximations by the variables x and y. Later, we describe the stages for the variable x only; for the variable y, the notations are mathematically identical.

7.2.1 Calculation of expansion coefficients

Let us apply numerical techniques for the approximation of a univariate function $u(x)$ using the Chebyshev series expansion (Press et al., 1992, § 5.8; Pankratov, 2004a, b):

$$u(x) = \sum_{i=0}^{l-1} c_i T_i(x), \tag{7.2}$$

where c_i are expansion coefficients; $T_i(x)$ are the Chebyshev polynomials of the first kind in the variable x of degree i; l is the maximum degree of expansion. The Chebyshev polynomials are defined as follows (Rivlin, 1974, § 1.1):

$$T_i(x) = \cos(i \arccos x), \tag{7.3}$$

where $i=0, \ldots, \infty$. We normalize the first polynomial of the function (Eq. 7.3) so that $T_0(x) = \frac{1}{\sqrt{2}}$. This allows making the norm of each polynomial the same and equal to $\sqrt{\pi/2}$.

In the continuous form, the function $T_i(x)$ (Eq. 7.3) satisfies the orthogonality condition under the scalar product:

$$\left(T_i, T_j\right) = \int_{-1}^{1} \frac{T_i(x)T_j(x)}{\sqrt{1-x^2}}dx = \begin{cases} 0 \text{ if } i \neq j \\ \pi/2 \text{ if } i = j \end{cases}, \tag{7.4}$$

where $i, j=0, \ldots, \infty$.

In the discrete form, the orthogonality condition is also satisfied under the scalar product on a Gaussian quadrature grid of k nodes

$$(T_n, T_m) = \frac{2}{k}\sum_{i=1}^{k} T_n(\xi_i)T_m(\xi_i) = \begin{cases} 0 \text{ if } n \neq m \\ 1 \text{ if } n = m \end{cases}, \tag{7.5}$$

where $n, m=0, \ldots, k-1$; ξ_i are nodes of the Gaussian quadrature grid:

$$\xi_i = \cos\left(\frac{\pi(i-1/2)}{k}\right), \quad i = 1, 2, \ldots, k, \tag{7.6}$$

with k zeros of the orthogonal polynomial $T_k(x)$.

The expansion coefficients c_j are calculated by the expression:

$$c_j = \left(u, T_j\right) = \frac{2}{k}\sum_{i=1}^{k} u(\xi_i)T_j(\xi_i), \tag{7.7}$$

where u is the original univariate function on the Gaussian quadrature grid; $j=0, \ldots, l-1$; $l \leq k$; the round brackets in the central part of the expression denote the scalar product.

7.2.2 The Fejér summation

Approximations based on orthogonal polynomials always lead to oscillatory artifacts due to the Gibbs phenomenon (Jerri, 1998, chap. 3; Florinsky, 2002, § 2) (Section 5.5). In our case, this means that the Chebyshev series expansion (Eq. 7.2) does not converge uniformly to the original function $u(x)$ at all points. According to the Fejér theorem, the arithmetic mean of the partial sums of the Fourier series of a continuous function converges uniformly to the function (Courant and Hilbert, 1989, pp. 102–103; Jerri, 1998, § 1.4). The Fejér theorem is also true for other orthogonal series (Jerri, 1998, § 3.5). In other words, a uniform convergence of the approximation can be achieved by replacing the initial expansion of $u(x)$ (Eq. 7.2) with the arithmetic mean of the partial sums of the Chebyshev series:

$$\tilde{u}_n(x) = \frac{1}{n}\sum_{l=1}^{n} u_l(x), \tag{7.8}$$

where $\tilde{u}_n(x)$ is the Fejér sum. The Fejér summation is a powerful method to suppress oscillatory artifacts in polynomial approximation (Jerri, 1998; Pankratov and Kulikova, 2006).

In our case, the transformation of the expansion coefficients c_j, corresponding to the Fejér summation, has the form:

$$\tilde{c}_j = \frac{l-j}{l}c_j, \tag{7.9}$$

where $\tilde{c}_j$ are coefficients of an orthogonal series $\tilde{u}_n(x)$; $j=0, \ldots, l-1$.

7.2.3 Reconstruction of the approximated function

Reconstruction of the original function $u(x)$ is conducted by a simple summation of the orthogonal series. Generalization, smoothing, and/or denoising of the function are performed, decreasing the number of expansion coefficients $\tilde{c}_j$ used for the reconstruction.

7.2.4 Calculation of derivatives

Let us calculate the first and second derivatives of the reconstructed function. For a Chebyshev-approximated univariate function, Press et al. (1992, § 5.9) described the calculation of the expansion coefficients of a derivative from the expansion coefficients of the original function. For an arbitrary basis, Tetuev and Dedus (2007, pp. 32–34) published a generalized conversion scheme for the expansion coefficients.

For the present case, the recurrence relations for the expansion coefficients c'_j of the first derivative of an orthogonal series are as follows:

$$\begin{cases} c'_{l-1} = 0 \\ c'_{l-2} = 2(l-2)\tilde{c}_{l-1} \\ \ldots \\ c'_j = c'_{j+2} + 2j\tilde{c}_{j+1}, \quad j = l-3, \ldots, 0 \\ \ldots \\ c'_0 = \dfrac{c'_0}{\sqrt{2}}. \end{cases} \tag{7.10}$$

In the case of a linear transformation of function domain, expansion coefficients of the derivative should be scaled:

$$c'_j = \frac{2}{W}c'_j, \tag{7.11}$$

where W is the length of a function interval, $j=0, \ldots, l-1$.

The expansion coefficients c''_j of the second derivative are calculated in the same manner as c'_j (Eq. 7.10):

$$\begin{cases} c''_{l-1} = 0 \\ c''_{l-2} = 2(l-2)c'_{l-1} \\ \ldots \\ c''_j = c''_{j+2} + 2jc'_{j+1}, \quad j = l-3, \ldots, 0 \\ \ldots \\ c''_0 = \dfrac{c''_0}{\sqrt{2}}. \end{cases} \tag{7.12}$$

After repeating stages 1–4 for the variable y, we can obtain expressions for the partial derivatives of elevation p, q, r, s, and t (Eq. 2.2) in the form of a bivariate orthogonal series similar to Eq. (7.1):

$$p(x,y) = \sum_{i=0}^{l-1}\sum_{j=0}^{l-1} c_{ij}^{p} T_i(x) T_j(y), \tag{7.13}$$

$$q(x,y) = \sum_{i=0}^{l-1}\sum_{j=0}^{l-1} c_{ij}^{q} T_i(x) T_j(y), \tag{7.14}$$

$$r(x,y) = \sum_{i=0}^{l-1}\sum_{j=0}^{l-1} c_{ij}^{r} T_i(x) T_j(y), \tag{7.15}$$

$$s(x,y) = \sum_{i=0}^{l-1}\sum_{j=0}^{l-1} c_{ij}^{s} T_i(x) T_j(y), \tag{7.16}$$

$$t(x,y) = \sum_{i=0}^{l-1}\sum_{j=0}^{l-1} c_{ij}^{t} T_i(x) T_j(y), \tag{7.17}$$

where c_{ij}^{p}, c_{ij}^{q}, c_{ij}^{r}, c_{ij}^{s}, and c_{ij}^{t} are expansion coefficients for p, q, r, s, and t (Eq. 2.2), respectively.

Now we can calculate values of the first- and second-order partial derivatives of elevation for all points of the reconstructed DEM by a simple summation of the related orthogonal series (Eqs. 7.13–7.17). Finally, to derive local morphometric variables, all the calculated values of the derivatives are substituted into the related expressions (Table 2.2).

One can similarly calculate the third-order partial derivatives of elevations g, h, k, and m (Eq. 2.2), which are necessary for derivation of third-order morphometric variables (Section 2.2.1).

The described method can be used if one can ignore the planetary curvature within a DEM area. In other words, the diagonal length of a DEM rectangle should be less than at most 0.1 of the average radius of the planet.

The described method is analytical because (1) a single analytical function (Eq. 7.1) defines elevations within the entire study area and (2) each calculated partial derivative of elevation is defined by a single analytical function (Eqs. 7.13–7.17) for the entire study area. This is the fundamental difference between the described method and the traditional techniques for DEM processing (Sections 4.1–4.3) in which one uses a DEM—a discrete function of elevation—to calculate discrete functions of the partial derivatives of elevation. The method is universal because it combines all three key mathematical procedures of geomorphometry.

7.3 Algorithm

To describe an algorithm implementing the universal spectral analytical method, we use a polynomial-oriented matrix approach (Pankratov, 2004b).

Let the original function of elevation z (Eq. 2.1) be specified as an $m \times n$ matrix $\mathbf{A}$ with the elevation values at the nodes of a square grid; $x_1, \ldots, x_m$ and $y_1, \ldots, y_n$ are grid nodes in the interval $[-1, 1]$ along the x- and y-axes, respectively; $\xi_1, \ldots, \xi_k$ are nodes of the Gaussian quadrature grid.

The k values determine the maximum degree of expansion l in Eq. (7.1), that is, $k \geq l$. Because $m \approx n$, the Gaussian quadrature grid is identical for abscissas and ordinates. For a more accurate calculation of the expansion coefficients, it is recommended to choose the k value much greater than m and n: $k = N \bullet \max(m, n)$, $N \geq 8$.

Let us introduce the following notations: $\mathbf{L}_{\mathbf{x}\xi}$ and $\mathbf{L}_{\mathbf{y}\xi}$ are $k \times m$ and $k \times n$ matrices of linear interpolation from the grids x and y, respectively, to the grid ξ; $\mathbf{T}_{\xi}$ is an $l \times k$ matrix of the Chebyshev polynomial values $T_i(\xi_j)$ in the grid ξ, where $i = 0, \ldots, l-1$, and $j = 1, \ldots, k$; $\mathbf{F}$ is an $l \times l$ diagonal matrix with diagonal elements $(l - i)/l$ for the Fejér summation.

The calculation of the expansion coefficients is carried out in two steps. First, all columns of the matrix $\mathbf{A}$ are transformed into an $l \times n$ intermediate matrix of the expansion coefficients $\mathbf{C}_{\mathbf{y}}$ that corresponds to the expansion in the variable y:

$$\mathbf{C}_{\mathbf{y}} = \frac{2}{k}\mathbf{F}\mathbf{T}_{\xi}\mathbf{L}_{\mathbf{y}\xi}\mathbf{A}. \tag{7.18}$$

Second, the matrix $\mathbf{C}_{\mathbf{y}}$ is transposed, and the transformation is repeated, which corresponds to the expansion in the variable x:

$$\mathbf{C} = \frac{2}{k}\mathbf{F}\mathbf{T}_{\xi}\mathbf{L}_{\mathbf{x}\xi}\mathbf{C}_{\mathbf{y}}^{\mathrm{T}}, \tag{7.19}$$

where $\mathbf{C}$ is an $l \times l$ resultant square matrix of the expansion coefficients; ^{T} denotes the matrix transpose.

Reconstruction of the approximated function is carried out by a summation of the orthogonal series:

$$\mathbf{Z} = \mathbf{T}_{\mathbf{x}}^{\mathrm{T}}\mathbf{C}^{\mathrm{T}}\mathbf{T}_{\mathbf{y}}, \tag{7.20}$$

where $\mathbf{Z}$ is an $m \times n$ matrix of the approximated values of the original function z (Eq. 2.1) at the nodes of the square grid; $\mathbf{T}_{\mathbf{x}}$ and $\mathbf{T}_{\mathbf{y}}$ are $l \times m$ and $l \times n$ matrices of the Chebyshev polynomial values $T_i(x_j)$ and $T_i(y_j)$ in the grids x and y, respectively.

Let $\mathbf{D}$ be an $l \times l$ differentiation operator in the space of expansion coefficients. In this case, $\mathbf{C_p}$, $\mathbf{C_q}$, $\mathbf{C_r}$, $\mathbf{C_s}$, and $\mathbf{C_t}$—$l \times l$ square matrices of expansion coefficients of the first- and second-order partial derivatives p, q, r, s, and t (Eq. 2.2)—are as follows:

$$\begin{aligned}
\mathbf{C_p} &= \mathbf{DC}, \\
\mathbf{C_q} &= \left(\mathbf{DC}^{\mathrm{T}}\right)^{\mathrm{T}} = \mathbf{CD}^{\mathrm{T}}, \\
\mathbf{C_r} &= \mathbf{DC_p}, \\
\mathbf{C_s} &= \left(\mathbf{DC_p^{\mathrm{T}}}\right)^{\mathrm{T}} = \mathbf{C_pD}^{\mathrm{T}}, \\
\mathbf{C_t} &= \left(\mathbf{DC_q^{\mathrm{T}}}\right)^{\mathrm{T}} = \mathbf{C_qD}^{\mathrm{T}}.
\end{aligned} \tag{7.21}$$

Reconstruction of the partial derivatives is carried out by a summation of the orthogonal series:

$$\begin{aligned}
\mathbf{P} &= \mathbf{T_x^{\mathrm{T}}C_p^{\mathrm{T}}T_y}, \\
\mathbf{Q} &= \mathbf{T_x^{\mathrm{T}}C_q^{\mathrm{T}}T_y}, \\
\mathbf{R} &= \mathbf{T_x^{\mathrm{T}}C_r^{\mathrm{T}}T_y}, \\
\mathbf{S} &= \mathbf{T_x^{\mathrm{T}}C_s^{\mathrm{T}}T_y}, \\
\mathbf{T} &= \mathbf{T_x^{\mathrm{T}}C_t^{\mathrm{T}}T_y},
\end{aligned} \tag{7.22}$$

where $\mathbf{P}$, $\mathbf{Q}$, $\mathbf{R}$, $\mathbf{S}$, and $\mathbf{T}$ are $m \times n$ matrices of the reconstructed values of the first- and second-order partial derivatives of elevation p, q, r, s, and t (Eq. 2.2), respectively.

Matrices of the reconstructed values of the third-order partial derivatives of elevation g, h, k, and m (Eq. 2.2) can similarly be obtained.

7.4 Materials and data processing

To show the possibilities of the universal spectral analytical method, we selected a portion of the East African rift system measuring 4° × 4°, located between 2°S and 2°N, and 35°E and 39°E (Fig. 7.1A). The area covers central and western regions of Kenya including a portion of the Great Rift Valley and central Kenyan highlands. A DEM of this territory was extracted from the global SRTM30_PLUS DEM (Sandwell et al., 2008). The DEM includes 230,880 points (the elevation matrix 480 × 481); the grid spacing is 30″ (Fig. 7.1B).

First, using the author's finite-difference method (Section 4.3), we derived k_h and k_v models from the initial DEM (Fig. 7.1C and D).

Second, we evaluated various sets of expansion coefficients to approximate and reconstruct DEMs. In particular, DEMs were reconstructed using 7000, 6000, 5000, 4000, 3400, 3000, 2880, 2400, 1920, 1200, 960, 480, 360, 240, 120, 60, and 30 expansion coefficients. Some of the sets of expansion coefficients were selected to illustrate the capabilities of the method (Fig. 7.2). The first- and second-order partial derivatives of elevation were calculated from each reconstructed DEM. Digital models of k_h and k_v were then computed using the derivatives (Figs. 7.3 and 7.4). The DTMs produced had a grid spacing of 30″.

Third, to evaluate the approximation accuracy, digital models of residuals—that is, differences between the reconstructed DEMs and the initial ones—were calculated (Fig. 7.5). Fourth, we performed a pairwise comparison of the reconstructed DTMs with the initial ones using samples extracted from each DTM. The sample size was 2209 points (the matrices 47 × 47); the grid spacing was 5′ (Fig. 7.6).

Fifth, to evaluate the role of the Fejér summation in the DTM treatment, we repeated DEM approximation as well as k_h and k_v derivations with the same sets of expansion coefficients but without the Fejér summation (Figs. 7.7 and 7.8). Digital models of residuals for the Fejér-free reconstructed DEMs were also calculated (Fig. 7.9).

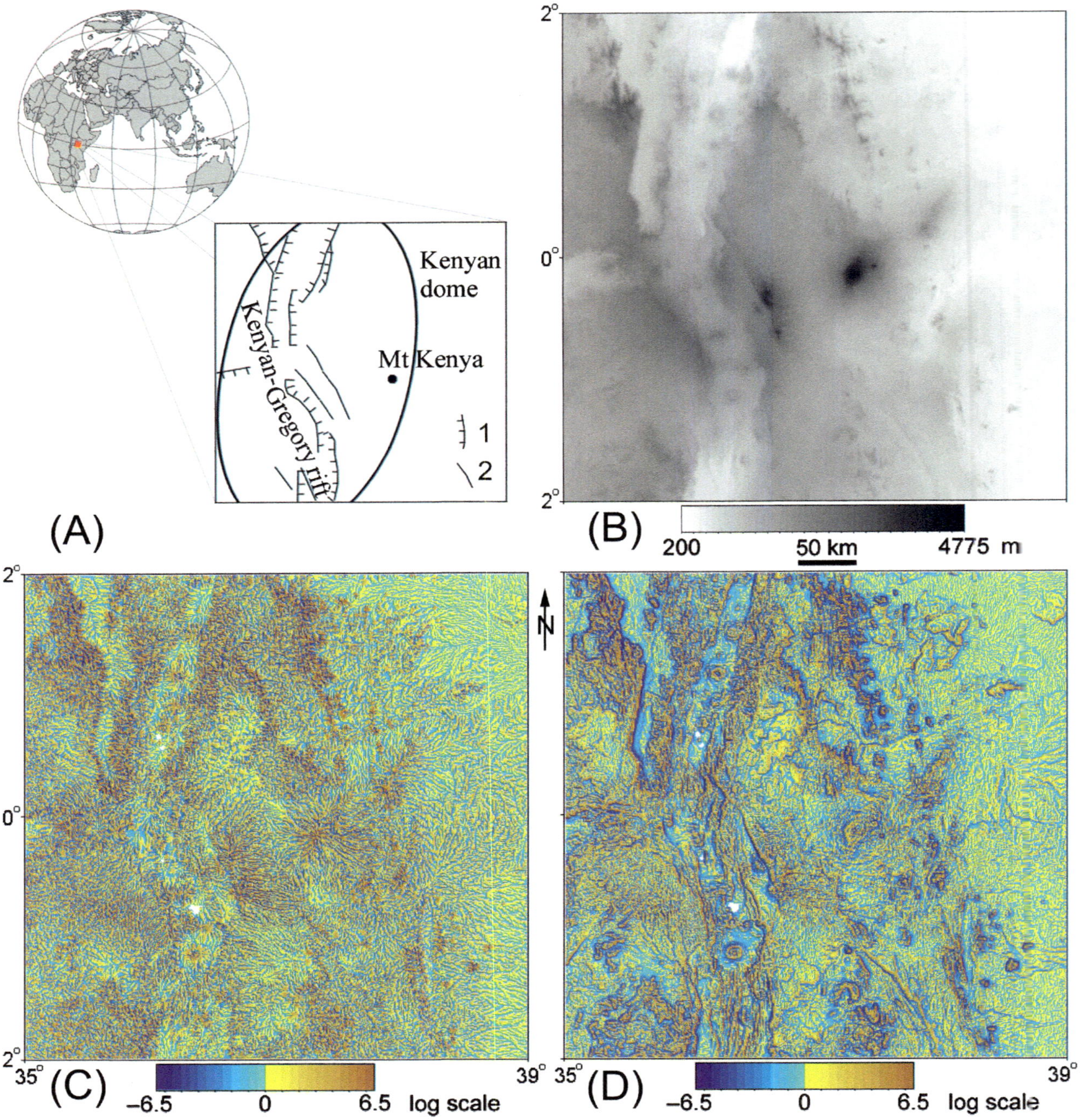

FIG. 7.1 Central and western Kenya: (A) Geographical location of the study area and key tectonic features: *1*, normal faults; *2*, strike-slip faults. (B) Elevations from the initial DEM. (C) Horizontal curvature derived from the initial DEM. (D) Vertical curvature derived from the initial DEM. *Publisher's note: Map lines delineate the study area and do not necessarily depict accepted national boundaries. ((A) Modified from Chorowicz, J., 2005. The East African rift system. J. Afr. Earth Sci. 43 (1–3), 379–410, Fig. 7, reproduced with permission from Elsevier.)*

Finally, statistical characteristics of the residuals for the DEMs, reconstructed with and without Fejér summation, were obtained using samples from the digital models of residuals (Table 7.1). The sample size was 2209 points (the matrices 47×47); the grid spacing was 5′.

To deal with the large dynamic range of k_h and k_v, we logarithmically transformed their digital models by Eq. (8.1) with $n=6$. Data were processed by the MATLAB software (© MathWorks Inc., 1984–2008). Statistical analysis was carried out with the Statgraphics software (© Statistical Graphics Corp., 1994–1997). Mapping was conducted with the LandLord software (Florinsky, 2016).

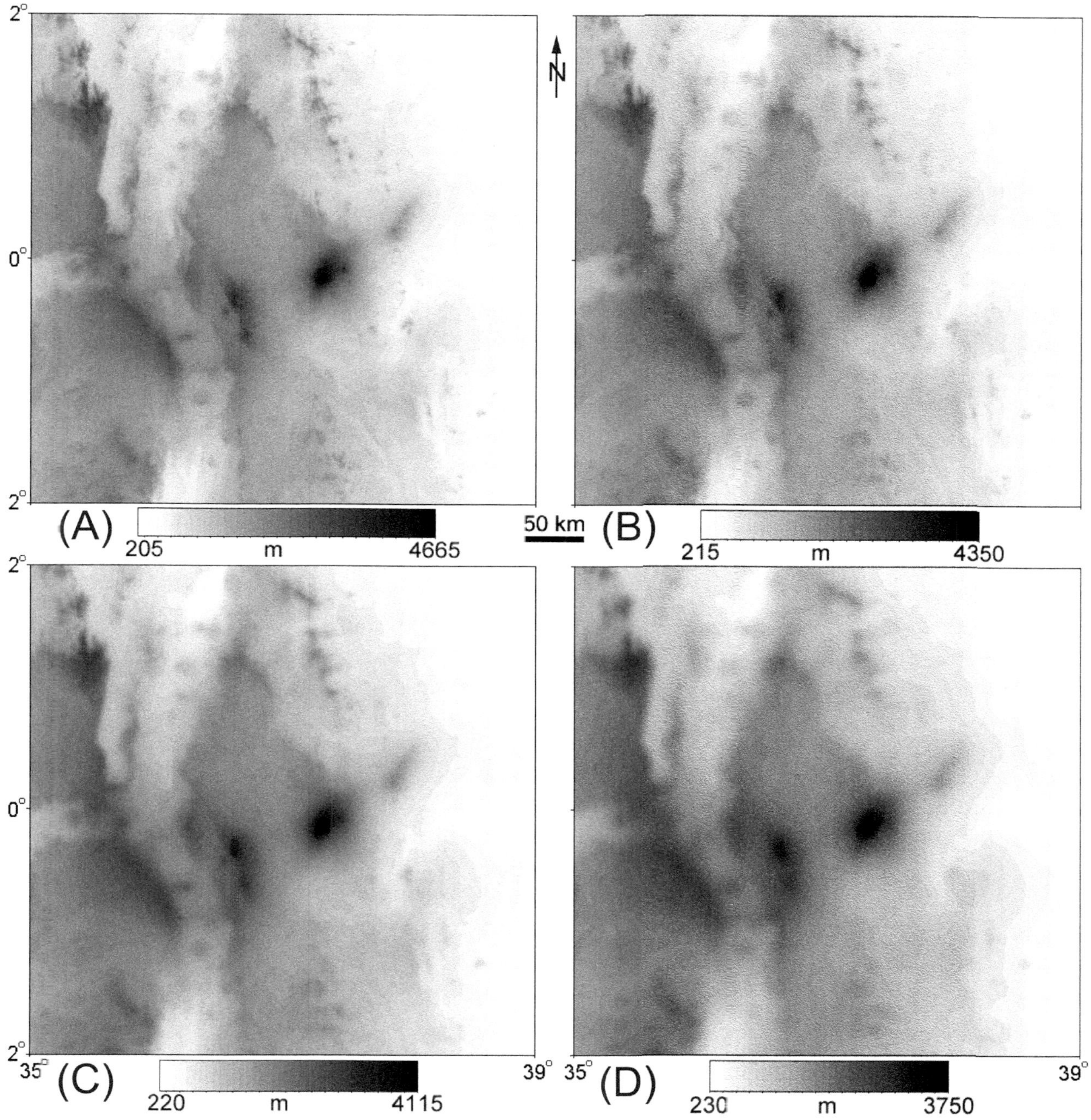

FIG. 7.2 Central and western Kenya: elevations reconstructed with (A) 2880, (B) 480, (C) 240, and (D) 120 expansion coefficients.

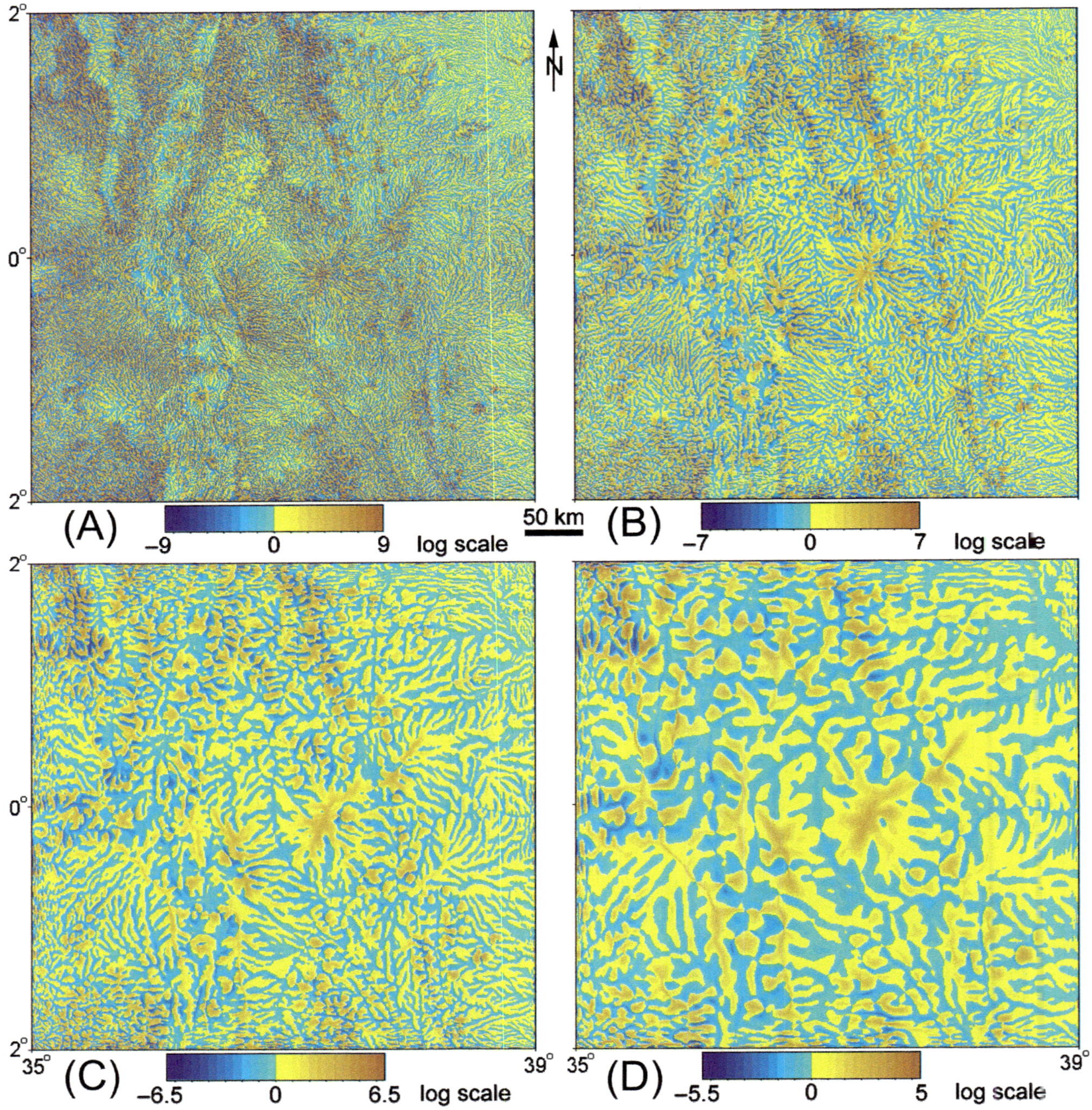

FIG. 7.3 Central and western Kenya: horizontal curvature derived from the DEMs reconstructed with (A) 2880, (B) 480, (C) 240, and (D) 120 expansion coefficients.

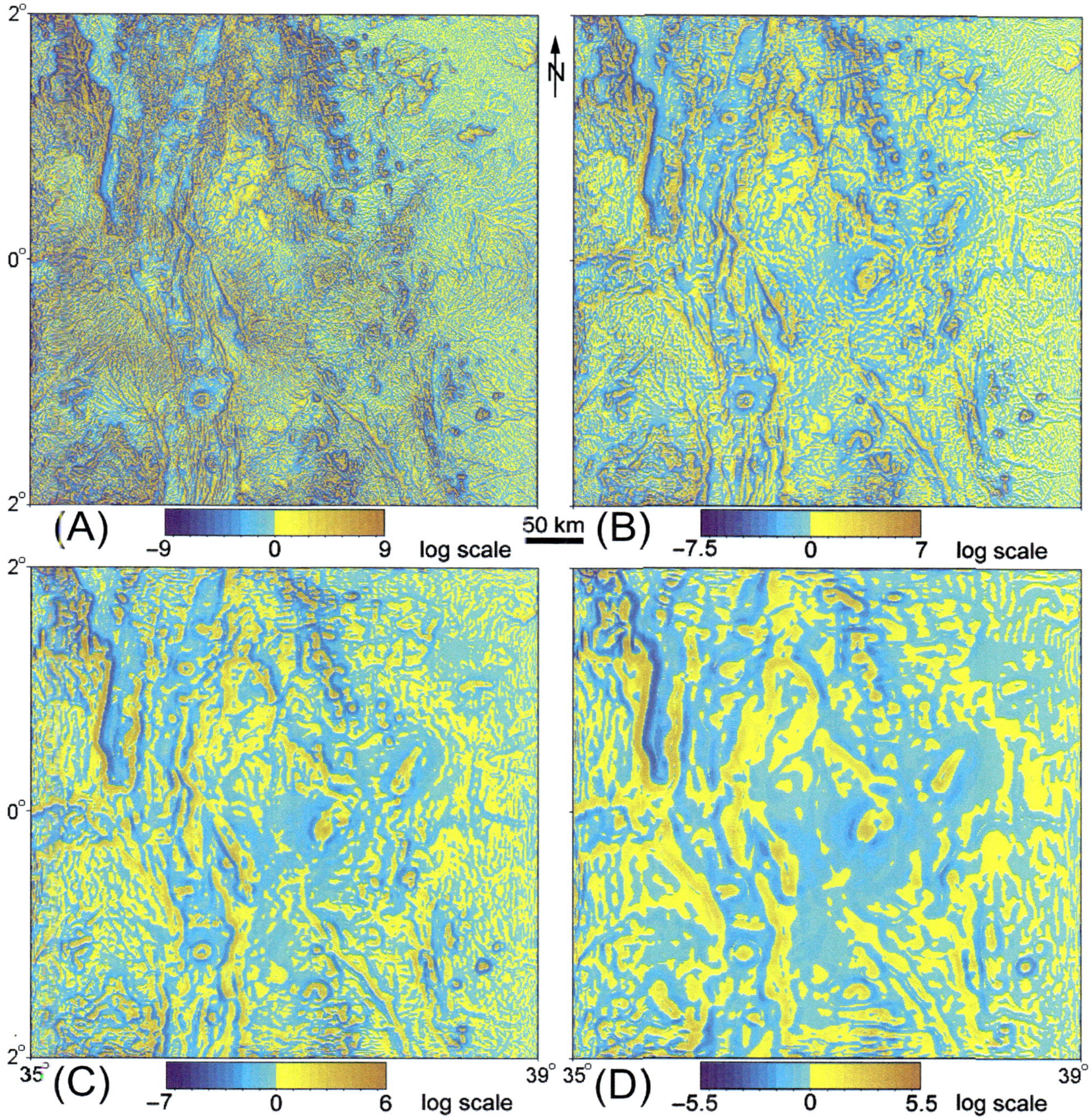

FIG. 7.4 Central and western Kenya: vertical curvature derived from the DEMs reconstructed with (A) 2880, (B) 480, (C) 240, and (D) 120 expansion coefficients.

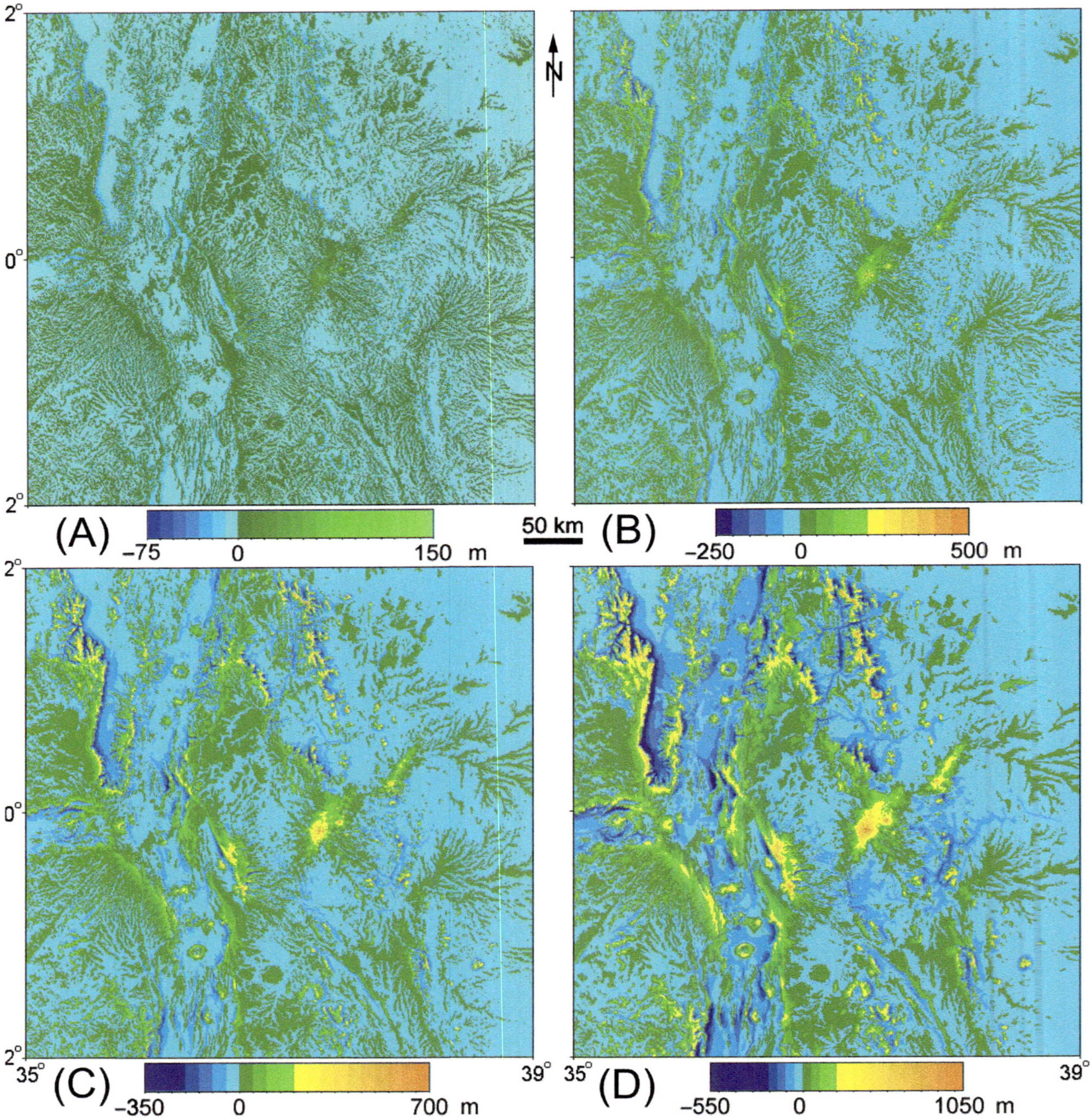

FIG. 7.5 Central and western Kenya: residuals between the initial DEM and DEMs reconstructed with (A) 2880, (B) 480, (C) 240, and (D) 120 expansion coefficients.

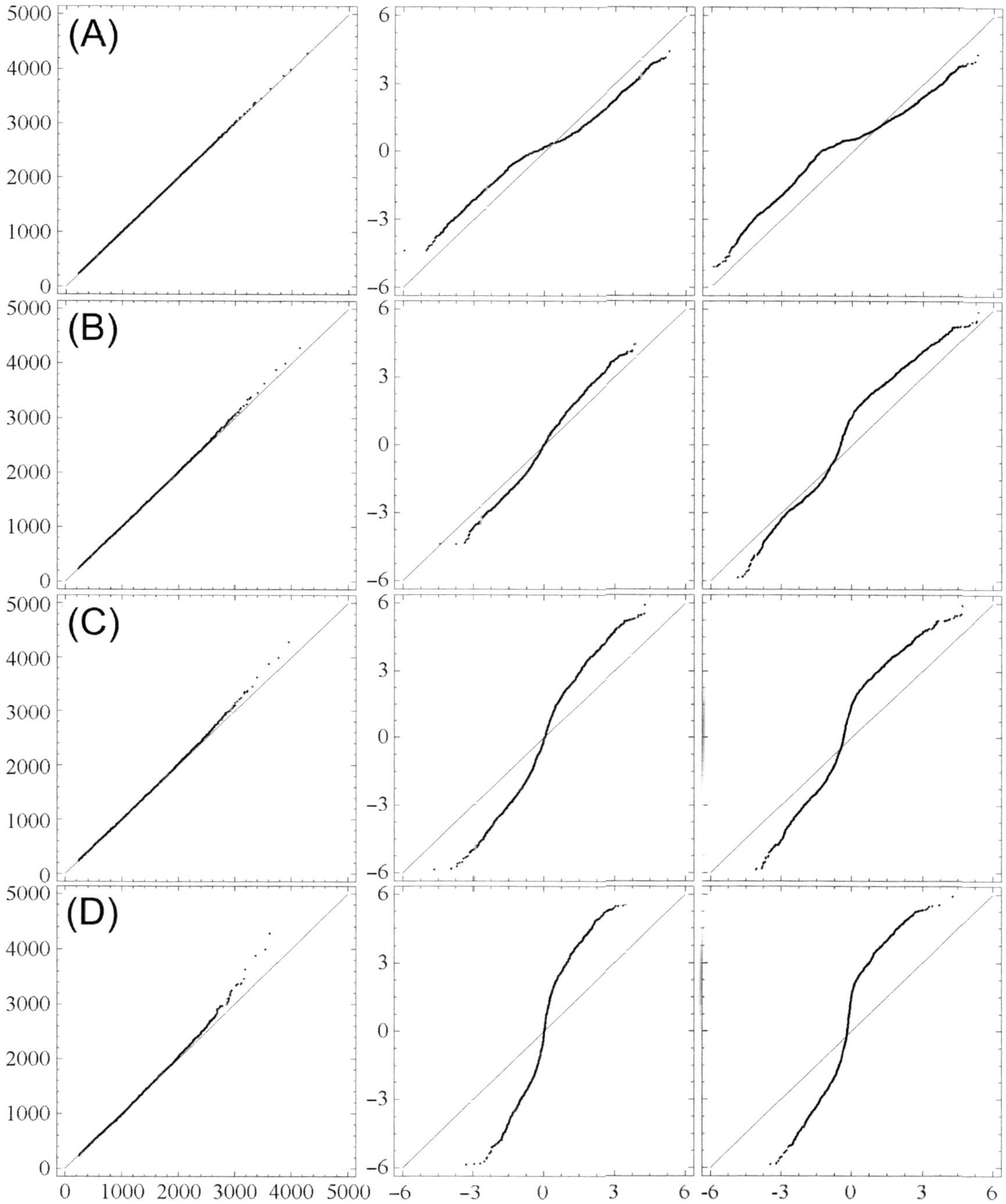

FIG. 7.6 Quantile-quantile plots for elevation (*left*), horizontal curvature (*center*), and vertical curvature (*right*), initial (*Y*-axis) and reconstructed (*X*-axis) with (A) 2880, (B) 480, (C) 240, and (D) 120 expansion coefficients. The greater the distance between points and the diagonal line, the greater the difference between reconstructed and initial models. *(From Florinsky, I.V., Pankratov A.N., 2016. A universal spectral analytical method for digital terrain modeling. Int. J. Geogr. Inf. Sci. 30 (12), 2506–2528, Fig. 6, reproduced with permission from Taylor & Francis Ltd.)*

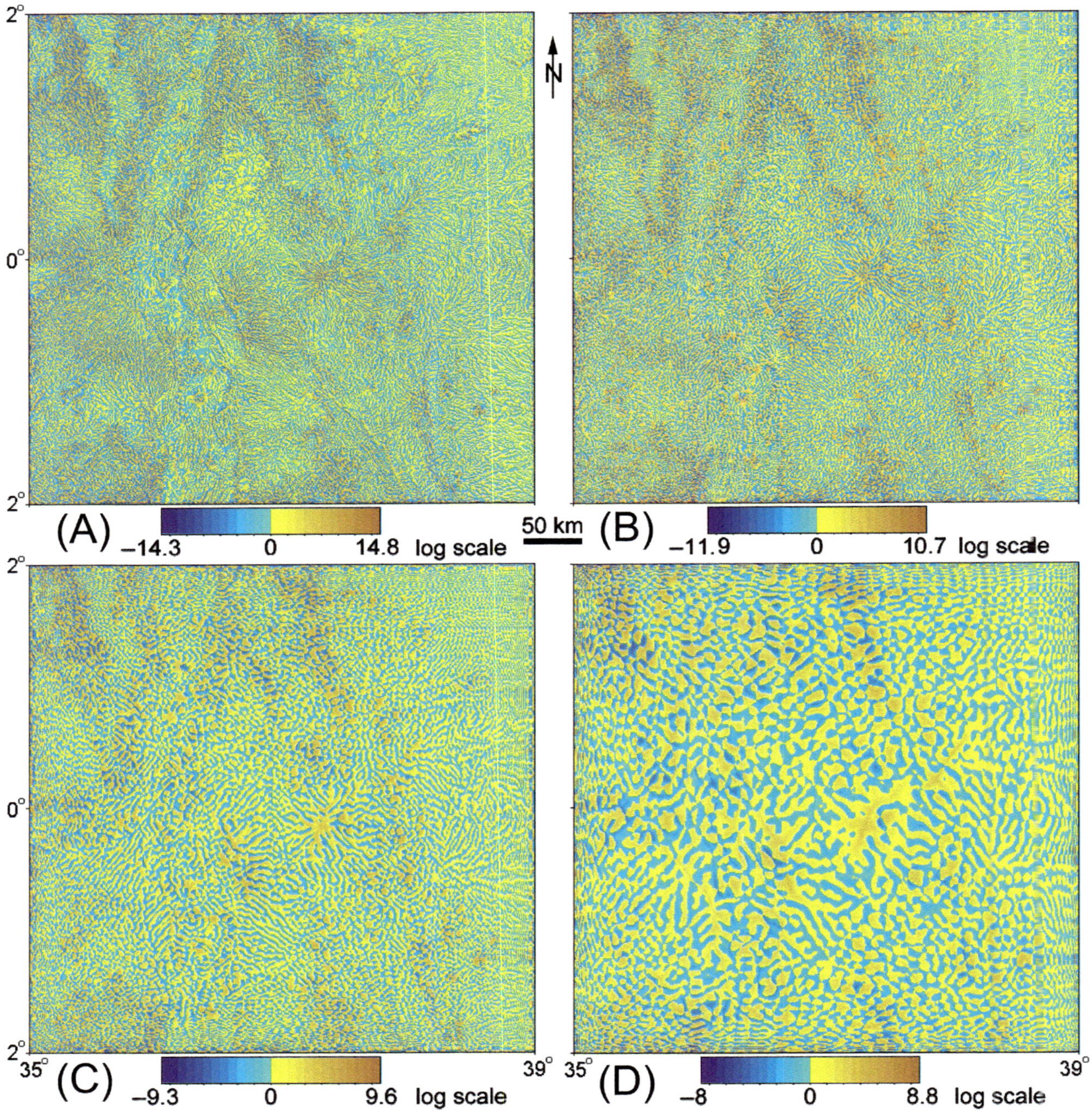

FIG. 7.7 Central and western Kenya: horizontal curvature derived from the DEMs reconstructed without the Fejér summation using (A) 2880, (B) 480, (C) 240, and (D) 120 expansion coefficients.

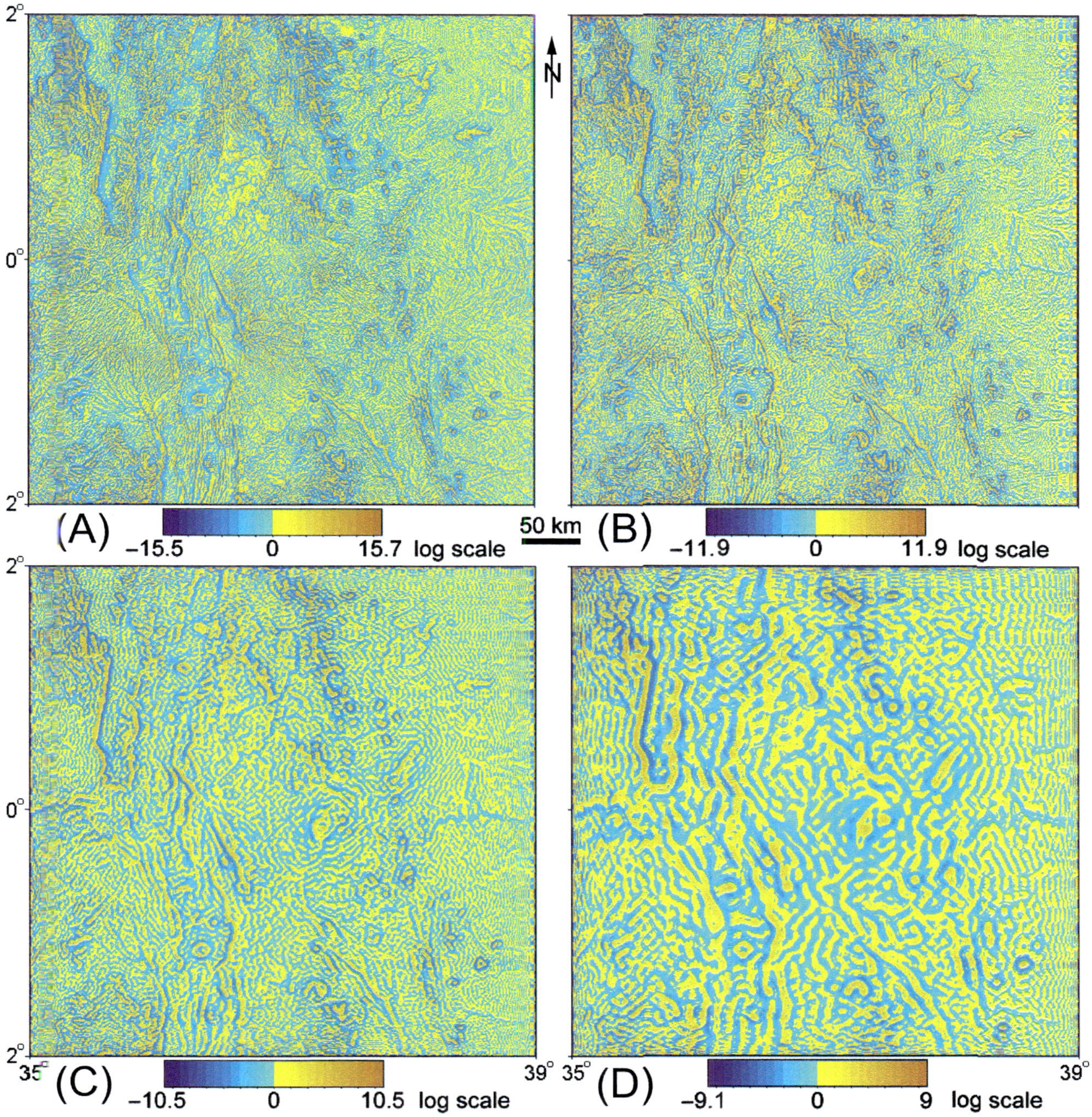

FIG. 7 8 Central and western Kenya: vertical curvature derived from the DEMs reconstructed without the Fejér summation using (A) 2880, (B) 480, (C) 240, and (D) 120 expansion coefficients.

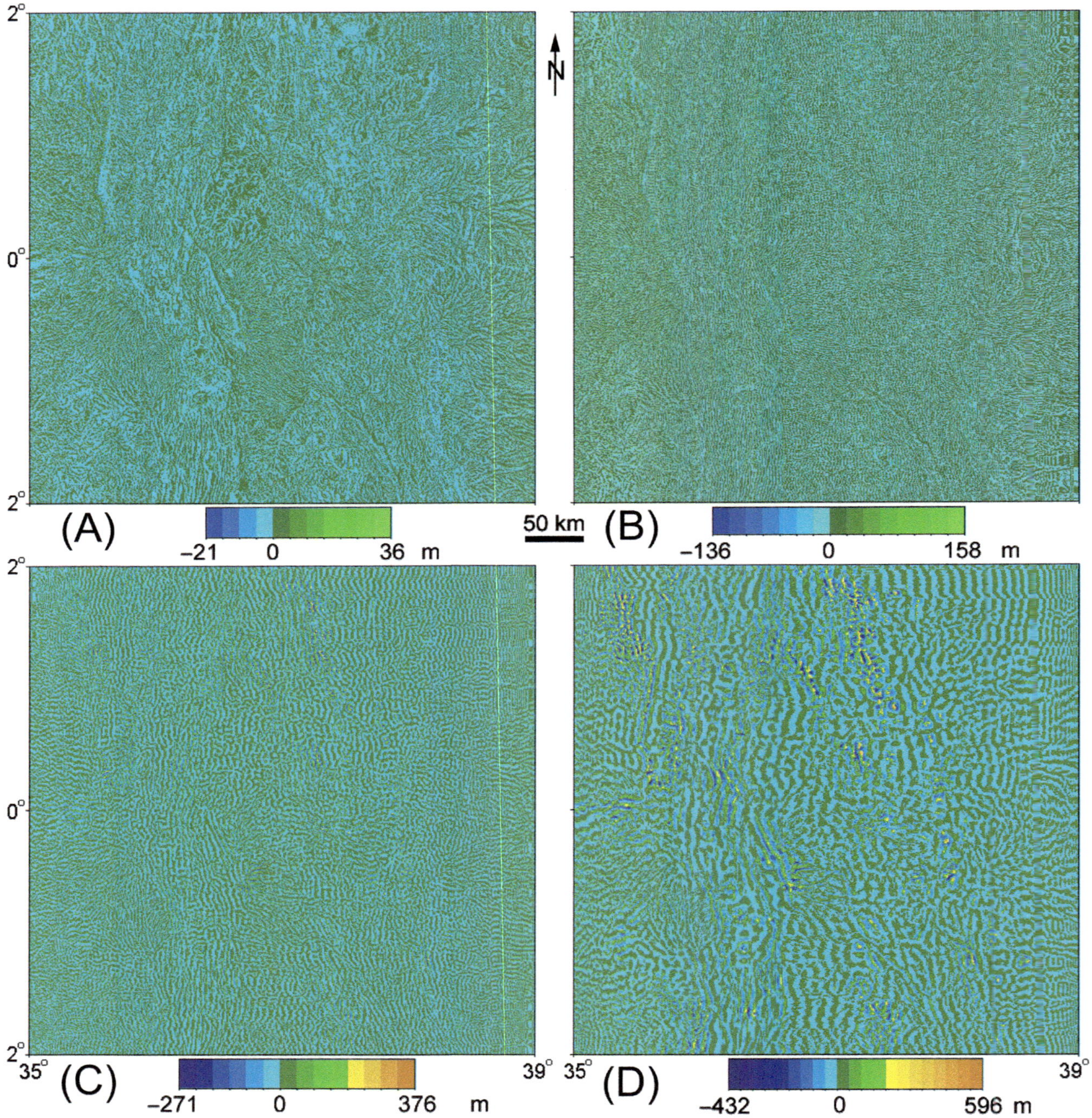

FIG. 7.9 Central and western Kenya: residuals between the initial DEM and DEMs reconstructed without the Fejér summation using (A) 2880, (B) 480, (C) 240, and (D) 120 expansion coefficients.

TABLE 7.1 Statistical characteristics of the residuals between the DEMs reconstructed with/without Fejér summation and the initial DEM.

Statistics	Number of expansion coefficients											
	60	120	240	480	960	1920	2880	3400	4000	5000	6000	7000
	With the Fejér summation											
Minimum	−418.27	−287.46	−198.30	−146.08	−94.71	−57.51	−44.35	−39.97	−36.06	−31.54	−28.84	−28.20
Maximum	1163.49	669.29	427.11	265.23	158.20	90.72	64.26	55.43	48.49	41.61	37.50	36.09
Mean	6.57	3.40	1.41	0.66	0.33	0.17	0.12	0.10	0.09	0.08	0.07	0.08
Standard deviation	131.69	84.94	53.50	32.94	20.19	12.07	8.83	7.75	6.82	5.72	4.97	4.47
	Without the Fejér summation											
Minimum	−369.48	−251.56	−178.15	−80.92	−37.86	−20.73	−13.06	−13.41	−7.32	−8.47	−7.96	−5.38
Maximum	679.34	322.94	219.00	70.32	40.10	18.14	13.65	12.98	11.26	8.48	8.15	7.12
Mean	0.45	−0.66	−0.37	0.03	−0.01	0.00	0.01	0.02	0.01	−0.02	−0.01	0.01
Standard deviation	69.58	42.46	24.12	11.40	5.29	2.71	1.83	1.63	1.31	1.05	0.89	0.80

The sample size is 2209.

(From Florinsky, I.V., Pankratov A.N., 2016. A universal spectral analytical method for digital terrain modeling. Int. J. Geogr. Inf. Sci. 30 (12), 2506–2528, Table 1, reproduced with permission from Taylor & Francis Ltd.)

7.5 Results and discussion

A set of elevation maps reconstructed with different numbers of expansion coefficients (Fig. 7.2) demonstrates a process of DEM generalization, from its zero level (Fig. 7.1B) to the highest one (Fig. 7.2D). Similarly, a set of k_h and k_v maps calculated from reconstructed DEMs (Figs. 7.3 and 7.4) shows k_h and k_v generalization, from the zero levels (Fig. 7.1C and D) to the highest ones (Figs. 7.3D and 7.4D). Generally, the lower the number of expansion coefficients used to reconstruct a DEM, the more smooth and simplified maps are obtained.

On the initial k_h and k_v map, one can find numerous flow structures descending from the Kenyan highlands to the coast of the Indian Ocean (Fig. 7.1C), as well as scarps and terraces of the mountainous topography (Fig. 7.1D). However, these nongeneralized maps are overloaded with information. It is hard to see any generalization effect on maps calculated with 2880 expansion coefficients (cf. Figs. 7.1B and 7.2A; 7.1C and 7.3A; 7.1D and 7.4A). Readable and interpretable k_h and k_v maps were derived from the DEMs reconstructed with 480 and 240 expansion coefficients; one can clearly see flow structures (Fig. 7.3B and C) as well as scarps and terraces (Fig. 7.4B and C). On k_h and k_v maps derived from the DEM reconstructed with 120 expansion coefficients, one can recognize patterns probably connected with the evolution of the Kenya Dome (cf. Figs. 7.3D and 7.1A) as well as approximately north-striking large structures associated with elements of the Kenyan and Gregory rifts (cf. Figs. 7.4D and 7.1A).

The decrease in the number of expansion coefficients in DEM reconstruction can act as high-frequency filtering. However, the visual analysis of the initial models of k_h and k_v (Figs. 7.3A and 7.4A) shows that the initial DEM does not include a pronounced high-frequency noise, which would inevitably be increased after calculating derivatives and curvatures (Section 5.4.1). An example of the high-frequency filtering of a noisy DEM using the method can be found elsewhere (Florinsky and Pankratov, 2015).

It is well known that orthogonal polynomial approximation produces boundary effects that cannot be completely eliminated. In our case, these effects appear as narrow linear artifacts at the boundaries of k_h and k_v maps (Figs. 7.3D and 7.4D). A problem with boundaries of curvature maps also arises using the finite-difference algorithms: it is impossible to estimate the partial derivatives for boundary columns and rows of a DEM (Fig. 7.1C and D) because they are estimated for the center point of the moving window (Sections 4.1–4.3).

The analysis of Fig. 7.5 shows that maximum residuals of the approximation predictably occur in areas of the contrast mountainous topography with dramatic elevation changes. However, an estimation of the approximation accuracy can be relevant only for DEMs reconstructed with a high number of expansion coefficients, for instance, with the set of 2880 coefficients (Fig. 7.2A). In this particular case, the residuals range from −73 to 146 m (Fig. 7.5A), while the original elevation ranges from 206 to 4774 m (Fig. 7.1B), that is, the range of calculation errors (219 m) is 4.8% of the elevation range in the initial DEM (4568 m). At the same time, the high residual values are typical for relatively small areas adjacent to the scarps of the Great Rift Valley, Mount Kenya, and the like. For most of the territory, the residuals are 0.1 ± 8.8 m (Fig. 7.5A and Table 7.1). This value can be considered a permissible computational error comparable with the vertical error of the initial DEM. Indeed, the land portion of the SRTM30_PLUS is based on the SRTM30 DEM (Becker et al., 2009), characterized by an absolute vertical error of 5–9 m for this part of Africa (Farr et al., 2007).

It is clear that the smaller the number of expansion coefficients used to reconstruct a DTM, the greater the residual between the reconstructed DTM and the initial one (Figs. 7.5 and 7.6). However, we purposely generalize the topographic surface reducing the number of expansion coefficients, so the approximation accuracy problem becomes irrelevant for generalized DEMs. In this case, the key criterion is the sufficient plausibility of reconstructed maps.

Figs. 7.7–7.10 and Table 7.1 demonstrate the role of the Fejér summation in the universal spectral analytical method. On the one hand, although both approximation versions—with and without the Fejér summation—are characterized by a distinct monotonic convergence, the convergence rate of the Fejér-free approximation is markedly increased (Fig. 7.10 and Table 7.1). For example, the residuals for the DEM reconstructed with the set of 2880 expansion coefficients without the Fejér summation range from −21 to 36 m (Fig. 7.9A); that is, the range of calculation errors (57 m) does not exceed 1.3% of the elevation range in the initial DEM (4568 m). Moreover, the residuals are ±1.8 m for most of the territory (Fig. 7.9A and Table 7.1), which is less than the absolute vertical error of the SRTM DEM, 5–9 m, for this part of Africa (Farr et al., 2007). For the set of 7000 expansion coefficients, the residuals for the DEM reconstructed without the Fejér summation range from −11 to 15 m; that is, the range of calculation errors does not exceed 0.6% of the elevation range in the initial DEM. In this case, the residuals are ±0.8 m for most of the territory (Table 7.1). Thus, the Fejér-free approximation can lead to much better results in terms of elevation errors compared with the approximation with the Fejér summation.

Notice that zero residuals are impossible to achieve because the polynomial series can theoretically completely converge to the original function on the Gaussian quadrature grid, which does not coincide with the DEM grid (these are errors of interpolation from the Gaussian quadrature grid to the DEM grid). Furthermore, there are unavoidable rounding errors in the computations.

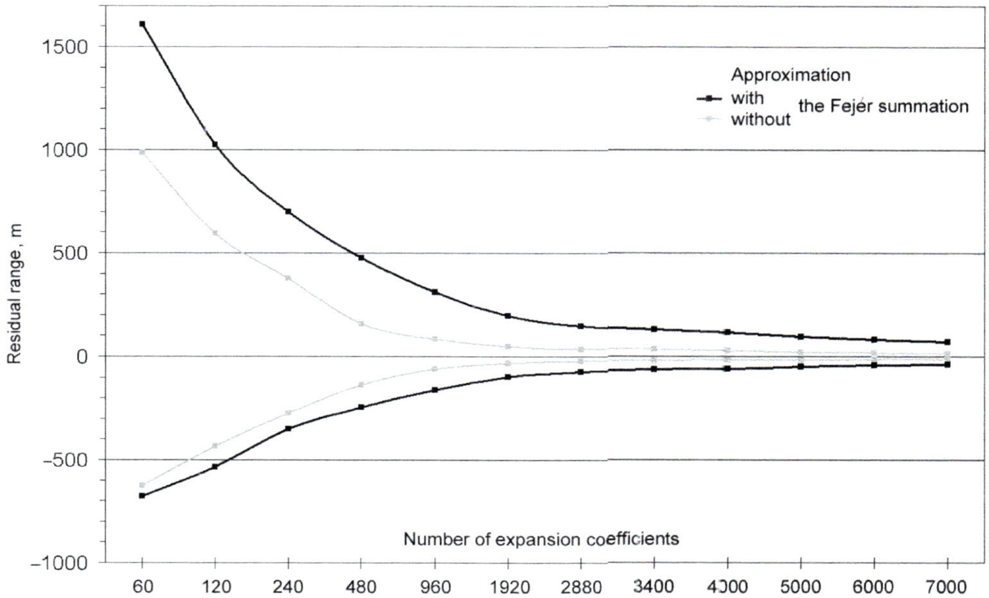

FIG. 7.10 Convergence rate for the DEM approximation with and without the Fejér summation in terms of the residual range. *(From Florinsky, I.V., Pankratov A.N., 2016. A universal spectral analytical method for digital terrain modeling. Int. J. Geogr. Inf. Sci. 30 (12), 2506–2528, Fig. 9, reproduced with permission from Taylor & Francis Ltd.)*

On the other hand, pronounced ripple-like artifacts emerged on the maps of horizontal and vertical curvatures derived from the Fejér-free approximated DEMs (i.e., in the eastern, relatively flat part of the study area, see Figs. 7.7B–D and 7.8B–D). Even for the set of 2880 expansion coefficients, one can see minor ripples in the northeastern corner, the flattest portion of the study area (Figs. 7.7A and 7.8A). These are oscillatory artifacts typical for approximations based on orthogonal polynomials. Classic boundary effects are also clearly visible on these maps. It is obvious that the reconstructed DEMs also include these oscillations, but they are weakly expressed and therefore cannot be seen on the elevation maps (but can be found on the residual maps, see Fig. 7.9). Differentiation amplifies their manifestation on the curvature maps (Section 5.4.1). At the same time, these artifacts do not appear on the curvature maps if the DEM approximation includes the Fejér summation (cf. Figs. 7.3 and 7.4 with Figs. 7.7 and 7.8, respectively). Thus, the Fejér summation, suppressing this computational noise, is the necessary stage of the described method.

The results demonstrated a good performance of the presented method. It can be utilized as a universal tool for analytical treatment of DEMs, including DEM global approximation, denoising, generalization, smoothing, calculation of the partial derivatives of elevation, and derivation of local morphometric variables. The method is an effective alternative to common techniques of DTM processing.

A further development of the method may include (1) its generalization to other orthogonal polynomials, such as the Chebyshev of the second kind, Legendre, Fourier, Jacobi, and Hermite ones and (2) the implementation of a fast recurrence algorithm (Pankratov et al., 2011) for calculating the expansion coefficients.

Chapter 8

Mapping and visualization

Relief representation was and remains one of the central problems of cartography (Du Carla, 1782; Lehmann, 1799; Peucker, 1898; Skvortsov, 1950, 1976; Koldaev, 1956; Imhof, 1965, 1982; Hurni, 2010; Kovaleva, 2012). On the one hand, mapping of topography should be mathematically rigorous, but on the other, cartographic tools should facilitate the adequate, quick, and easy perception of topography by the human eye. In this chapter, relief representation is briefly considered in the context of digital terrain modeling.

8.1 Peculiarities of morphometric mapping

One should consider several general rules to produce readable and interpretable morphometric maps:

1. As early as in the late 19th century, Tillo (1890, p. 10) noted that "contours by themselves do not form the integral or even partial picture of topography," although they describe mathematically spatial distribution of elevations.[a] It is also undesirable to use isoline (or isopleth) maps to represent other morphometric attributes; see, for example, poorly readable morphometric maps made in the early days of computer graphics by Krcho (1973). This is connected with very complicated shapes of morphometric surfaces. As a result, human eyes perceive isopleth-based morphometric maps with difficulty. For the correct perception of topographic peculiarities by the human visual system, there is a need to use layer or hypsometric tinting. In this well-known cartographic method (Peucker, 1898; Imhof, 1965, 1982; Skvortsov, 1976; Patterson and Jenny, 2011; Huffman and Patterson, 2013; Kovaleva, 2014), a surface $z = f(x, y)$ is subdivided into a set of z-value intervals. Then each interval is assigned a color, shade of color, or shade of gray from a color scale, monochromic scale, or gray scale, respectively (Fig. 8.1). This enables surface peculiarities to be clearly displayed. In this book, all maps of morphometric attributes are produced with color or grayscale tinting.
2. For all morphometric variables whose values can be both positive and negative, it is necessary to subdivide their values relative to the zero value. This is in part connected with the fact that values of different signs can measure different directions of a physical process associated with a particular variable (Table 2.1). For example, positive values of k_h describe flow divergence, while negative ones define flow convergence (Section 2.2.4.2).
3. Correct selection of a tint scale is essential for morphometric mapping (Roberts, 2001). Our experience shows that there is a need to use contrast color schemes for all morphometric variables, whose values can be both positive and negative. This offers a means to distinguish between areas of positive and negative values (e.g., areas of flow divergence and flow convergence, respectively). For gray scales, it is reasonable to apply darker halftones to negative values, while lighter ones to positive values. In this book, we systematically use such color or gray scales.
4. Wide dynamic ranges usually characterize topographic variables. To avoid loss of information on the spatial distribution of values of morphometric attributes in mapping, it makes sense to apply a logarithmic transform using the following expression (Shary et al., 2002b):

$$\Theta' = \operatorname{sign}(\Theta) \ln(1 + 10^{nm}|\Theta|), \tag{8.1}$$

where Θ is a value of a morphometric variable; $n = 0$ for nonlocal variables and $n = 2, \ldots, 9$ for local variables; $m = 2$ for K, K_a, and K_r, $m = 1$ for other variables. Selection of the n value depends on the DTM grid spacing; recommended values are presented in Table 8.1. Such a form of logarithmic transformation considers that dynamic ranges of many topographic attributes include both positive and negative values.

a. The relief representation using contour lines was theoretically described by Du Carla (1782). Lacroix (1811) established the mathematical basis for this technique (see historical details elsewhere—Konvitz, 1987, pp. 77–81, Lamandé, 2009).

Digital Terrain Analysis. https://doi.org/10.1016/B978-0-443-24798-9.00009-1

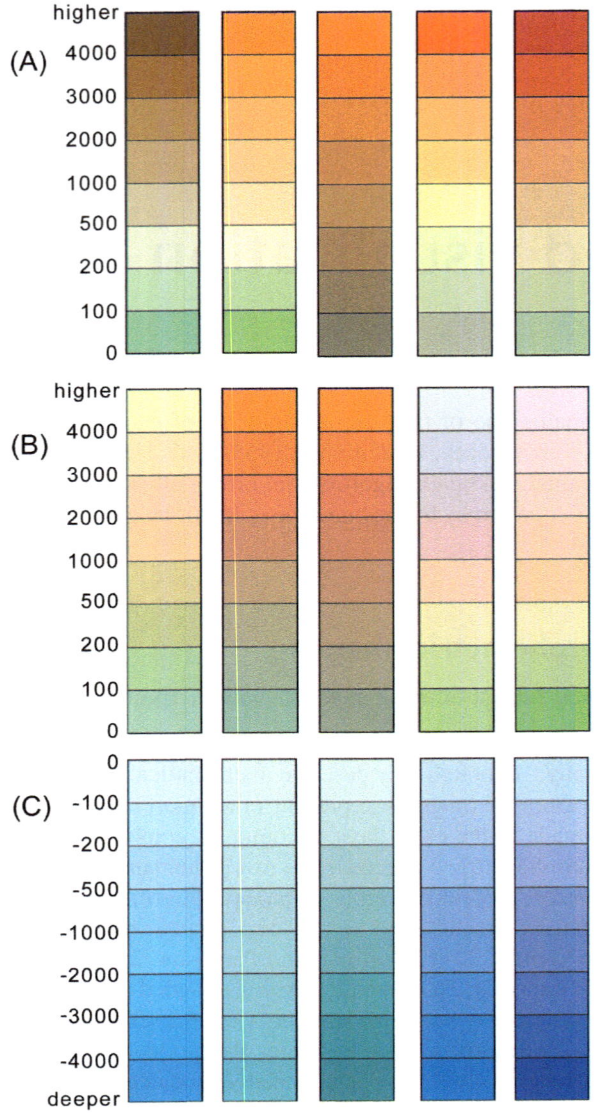

FIG. 8.1 Examples of color hypsometric tint scales: (A) Scales for allocating major elevation zones (*two left*) and scales of color plasticity (*three right*). (B) Scales of aerial perspective. (C) Scales for bathymetry. *(From Kovaleva, O.V., 2012. Perfecting the Representation of Topography on Small-Scale Maps (Ph.D. thesis). Moscow State University of Geodesy and Cartography, Moscow, Russia (in Russian), Figs. 3.37–3.39 and 3.44, reproduced with permission from the author.)*

TABLE 8.1 Values of n for Eq. (8.1) depending on the DEM grid spacing.

DEM grid spacing, m	n
< 1	2
1–10	3
10–100	4
100–1000	5
1000–5000	6
5000–10,000	7
10,000–75,000	8
$> 75,000$	9

8.2 Combined visualization of morphometric variables

Sometimes researchers try to represent several morphometric variables on a single map (Chu et al., 2019). This is a difficult task both from a technical point of view (to avoid the overlap of layers) and in the context of the human visual system's perception of multilayer graphical information.

In the precomputer era, Kvietkauskas (1963–1964) proposed a method to produce four-color morphometric maps with the following color scheme: a yellow scale was used for elevation values, a gray scale for slope gradient, a red scale for positive values of vertical curvature, and a blue scale for its negative values. Elevation, gradient, and vertical curvature were mapped together using four nonoverlapping color raster grids.

Currently, two morphometric layers—elevation and hill shading—are most commonly combined for color relief shading (Edwards and Davis, 1994; Sterner, 1995; Jenny and Hurni, 2006). Input elevation maps are displayed by color hypsometric tinting, while input hill-shaded data are presented as monochrome gray-scale maps (Fig. 8.2). Input elevation can also be described using grayscale hypsometric tinting (Huffman and Patterson, 2013). In this case, the outputs are also grayscale maps.

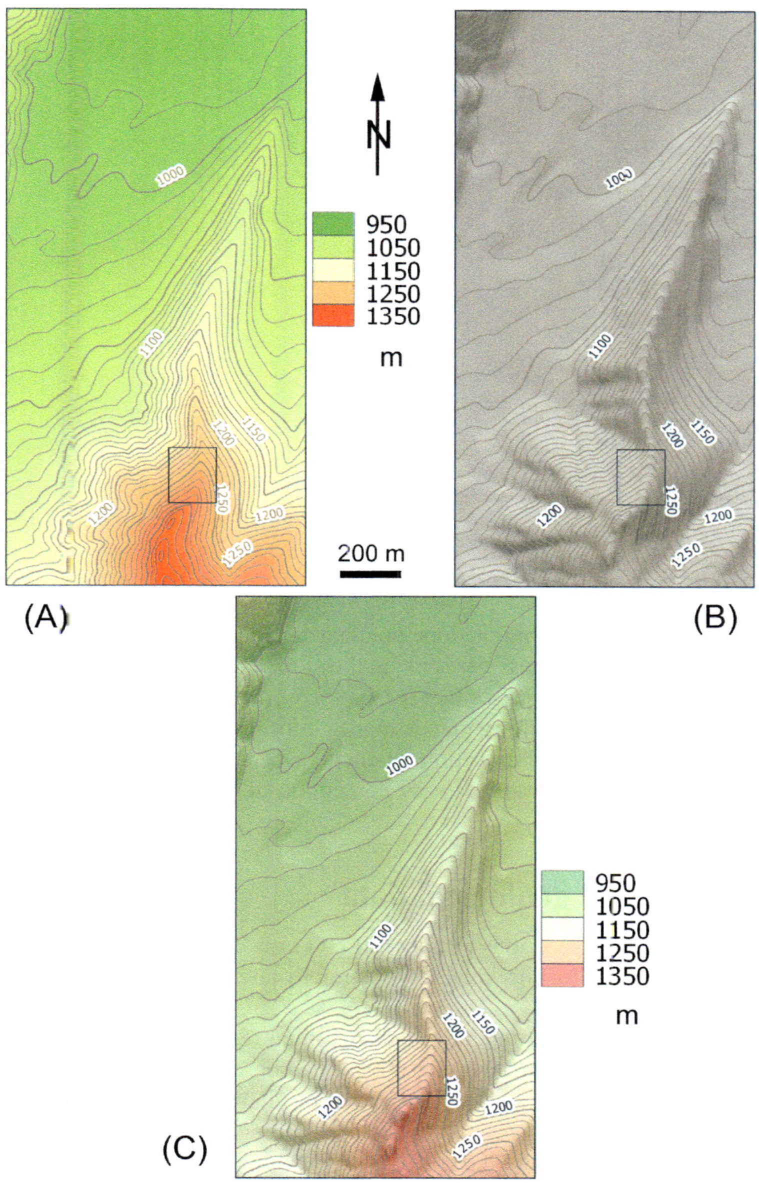

FIG. 8.2 Color relief shading. The Molnieboi Spur, Altai Mountains, Russia (50°10′30″N, 85°42′30″E): (A) Elevation. (B) Hill-shaded map. (C) Color-shaded map produced by combination of the elevation and hill-shaded maps. A DEM with a grid spacing of 2 m was created by digitizing a detailed topographic map. Maps were produced by the QGIS software (QGIS, 2009–2023). *(Modified from Florinsky, I.V., Selezneva, E.V., Kulikova, A.I., 2015. GIS-based support for the complex botanical studies at the Molnieboi Spur, Altai. arXiv Preprint 1512.05143, Fig. 4.)*

FIG. 8.3 Combined visualization of slope gradient and aspect by colored hachures. *(From Samsonov, T., 2014. Morphometric mapping of topography by flowline hachures. Cartogr. J. 51 (1), 63–74, Fig. 15, © The British Cartographic Society, reproduced with permission from Taylor & Francis Ltd.)*

Samsonov (2010, 2014) developed an approach for combined mapping of slope gradient and aspect. According to this method, both G and A are represented by hachures. The hachure thickness depends on the slope gradient value, as in the classical hachure technique (Lehmann, 1799), creating a hill-shading effect. Unlike the monochromatic classical technique, the hachure color depends on the slope aspect value (Fig. 8.3). The resulting map is highly informative. Quantitatively, it integrates the properties of several morphometric maps, such as maps of elevation, gradient, aspect, and flow lines. Qualitatively, such a topographic image has a high degree of plasticity that allows easy perception of the shape of the topographic surface.

8.3 Combining hill-shaded maps with soil and geological data

Technically, combining thematic and morphometric layers on a single map is a rather difficult task. The readability of such a map governs the effectiveness of its analysis in studying interrelationships between topography and other components of a geosystem.

One of the popular approaches to geoscientific visualization is to drape a hill-shaded map (Section 2.5.1.1) with a thematic—geological, soil, plant, etc.—map (Vigil et al., 2000a, b; Barton et al., 2003; Giasson et al., 2008; Marchetti et al., 2010; Masoud and Koike, 2011b). This technique can enhance the visual perception of relationships between geological/soil/plant and geomorphic features. Fig. 8.4 illustrates this technique: the small-scale geological map of the United States (King et al., 1974) is integrated with a hill-shaded map (Thelin and Pike, 1991a, b). The draping of soil data on hill-shaded maps is shown in Figs. 11.3 and 11.5.

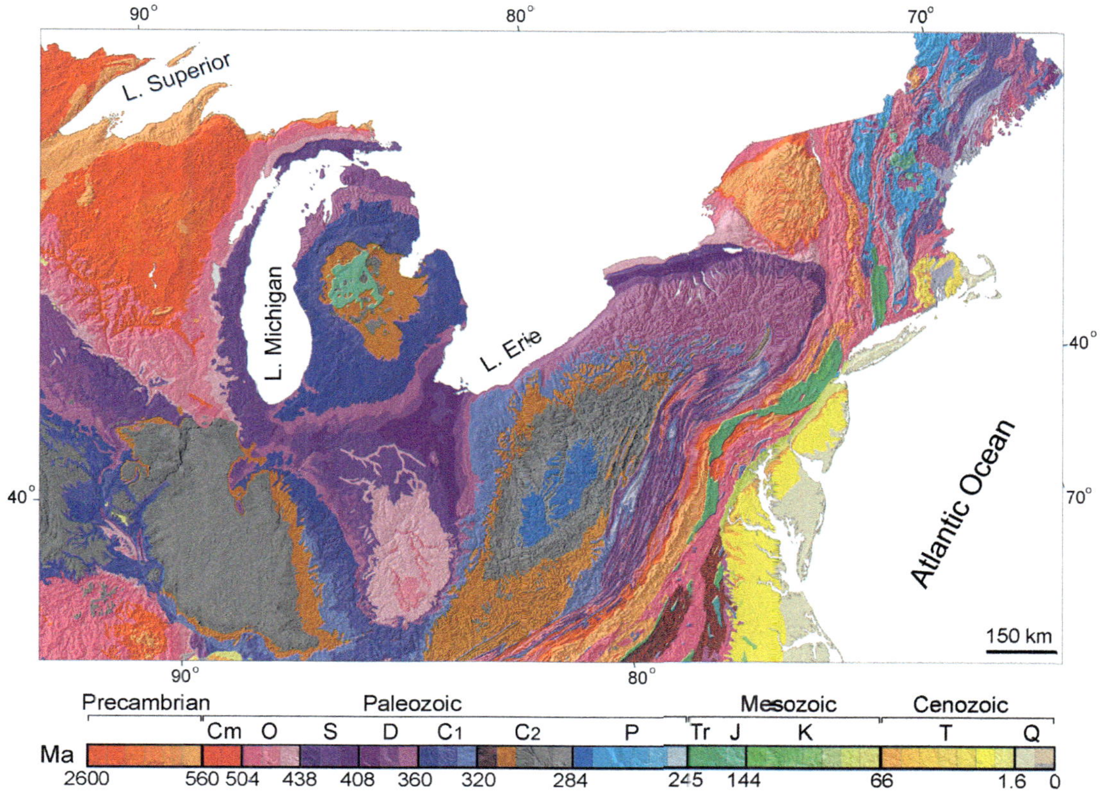

FIG. 8.4 Combining geological information and hill-shaded maps: a northeastern portion of the USA geological map. *(From Vigil, J.F., Pike, R.J., Howell, D.G., 2000. A Tapestry of Time and Terrain. Geologic Investigations Series Map I-2720, scale 1 : 3,500,000. US Geological Survey, Denver, CO, courtesy of the U.S. Geological Survey.)*

8.4 Cross sections

Cross-section construction using DEMs of both the land and stratigraphic surfaces is a standard procedure (Ichoku et al., 1994; Groshong Jr., 2006, chap. 6). Cross sections allow one to display the peculiarities of relations between the land surface and the surfaces of soil or stratigraphic horizons. Groshong Jr. (2006, p. 133) distinguishes two types of cross sections:

> *"[They are] illustrative or predictive. The purpose of an illustrative cross section is to illustrate the cross section view of an already-completed map or 3D interpretation. A slice through a 3D interpretation is a perfect example. The purpose of a predictive cross section is to assemble scattered information and, utilizing appropriate rules, predict the geometry between control points. A predictive cross section can be used to predict the geometry of a horizon for which little or no information is available … Cross sections constructed for the purpose of structural interpretation are usually oriented perpendicular to the fold axis, perpendicular to a major fault, or parallel to these trends."*

To construct interpretable cross sections, one should use an exaggeration of the vertical scale depending on the type of topography and the size of the territory concerned. Fig. 8.5 illustrates the use of DEM-derived cross sections in geological studies.

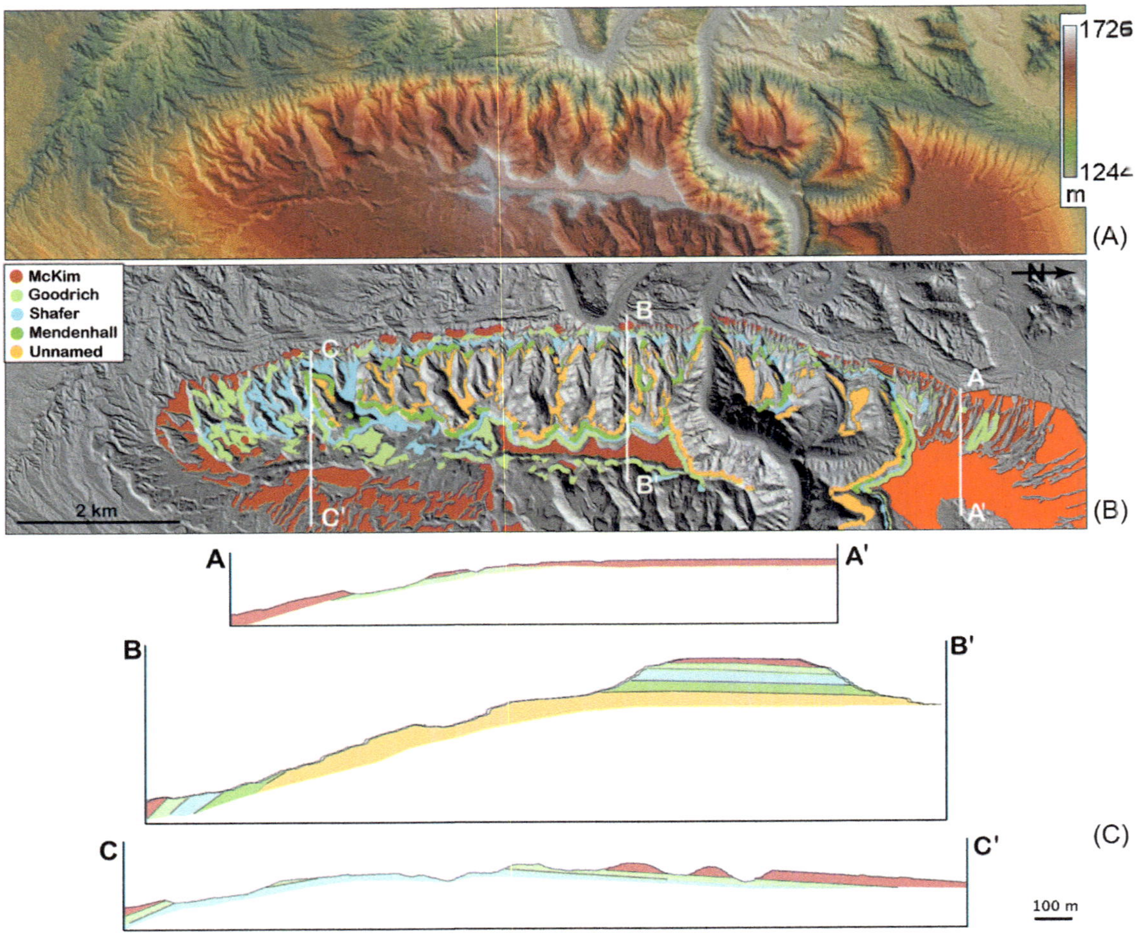

FIG. 8.5 DEM-derived cross sections in geological studies. The Raplee Ridge monocline, Utah, USA: (A) Elevation. (B) Hill shading with the spatial distribution of the five bedding-plane surfaces exhumed within the fold. (C) Cross sections (no vertical exaggeration). The DEM was derived from the airborne laser swath mapping data. The grid spacing is 1 m. *(From Hilley, G.E., Mynatt, I., Pollard, D.D., 2010. Structural geometry of Raplee Ridge monocline and thrust fault imaged using inverse Boundary Element Modeling and ALSM data. J. Struct. Geol. 32 (1), 45–58, Figs. 2a and 3, reproduced with permission from Elsevier.)*

8.5 Three-dimensional modeling

Three-dimensional (3D) modeling is a useful data-processing step concerning the human perception of spatially distributed phenomena and information. Viewing 3D models from different points can enhance the visual interpretation of remotely sensed images and maps as well as understanding of the relationships between elements of a geosystem.

8.5.1 Three-dimensional terrain modeling

There are two key types of DEM-based 3D terrain modeling:

1. 3D modeling of the land surface (Evans, 1972; Peucker, 1980b; Eyton, 1986; Raper, 1989; Patterson, 2001; Mach and Petschek, 2007). Currently, 3D terrain modeling with single-source illumination and simple orthographic or perspective projections is a standard procedure (Fig. 8.6A). Also, there are more complex techniques, such as perspective panoramic 3D terrain modeling (Patterson, 2001; Jenny et al., 2010), plan oblique relief mapping (Jenny and Patterson, 2007), hierarchical 3D terrain modeling (Chen et al., 2018), as well as modeling with multiple-source illumination (Section 8.5.1.1). As for cross sections, to construct interpretable 3D models, one should as a rule use an exaggeration of the vertical scale depending on the type of topography and the size of territory to be modeled.

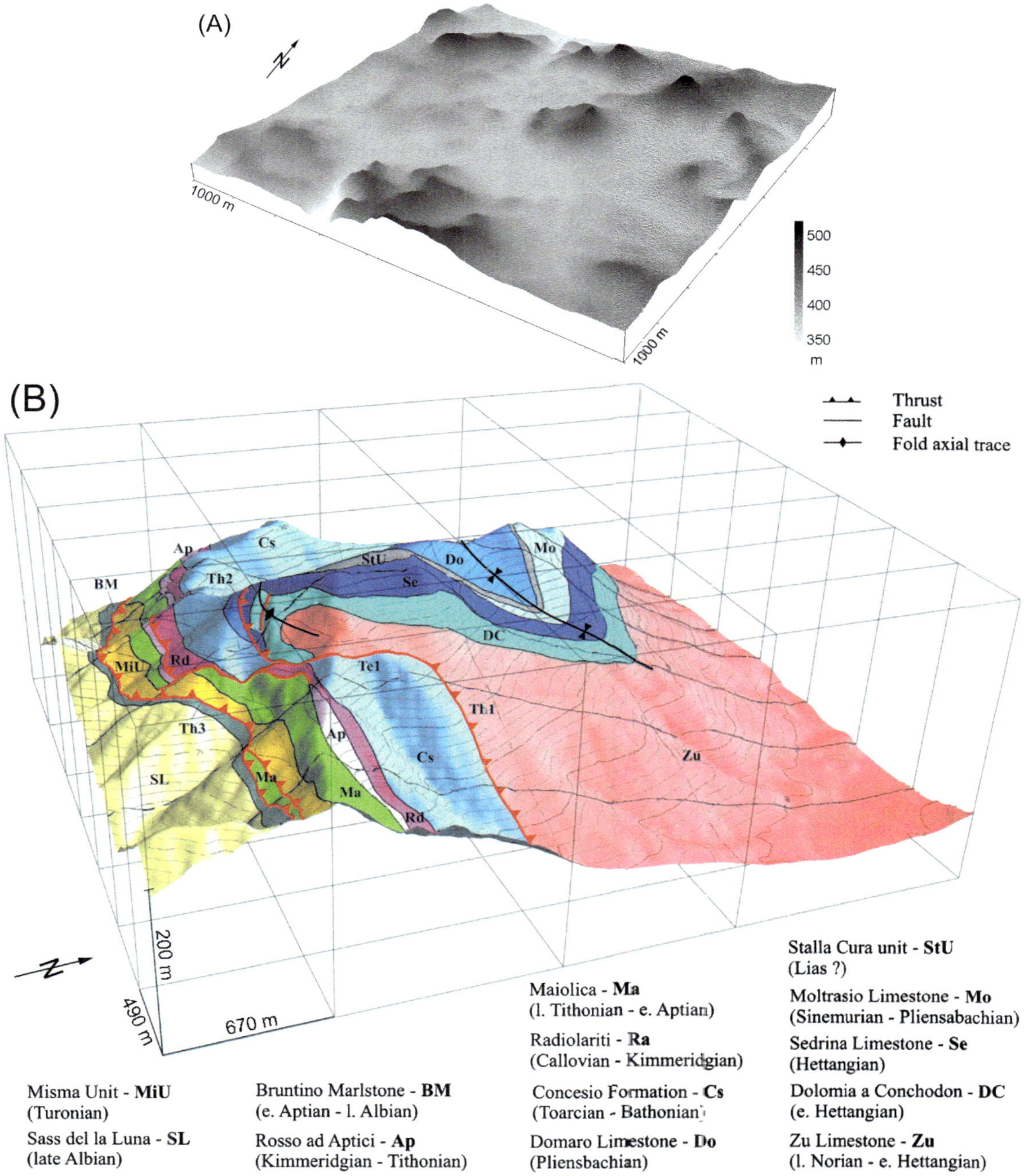

FIG. 8.6 3D terrain models: (A) 3D model of the Tunguska Event area, Krasnoyarsk Region, Russia (60°53′09″N, 101°53′40″E). A DEM was derived from a topographic map, scale 1:100,000. The DEM includes 182,213 points (the elevation matrix 431 × 423); the grid spacing is 20 m. The lower Triassic Kulikovsky paleovolcanic complex is visible due to the $5^{\times}$ vertical exaggeration. (B) 3D geological map of Mt. Misma, Italian Alps. A DEM was derived from [illegible]0-m contour interval topographic maps. The grid spacing is 2.5 m. Geological information was derived from a 1:5000 scale geological map. *((A) From Florinsky, I.V., 2008. 100 years anniversary of the Tunguska Event: digital terrain modeling of the epicenter area. Geodesy. Cartogr. (8), 20–22 (in Russian), Fig. 1. (B). From Zanchi, A., Francesca, S., Stefano, Z., Simone, S., Graziano, G., 2009. 3D reconstruction of complex geological bodies: examples from the Alps. Comput. Geosci. 35 (1), 49–69, Fig. 10, reproduced with permission from Elsevier.)*

2. Draping thematic data (orthorectified remotely sensed scenes, soil and geological maps, etc.) over 3D terrain models (Eyton, 1986; Ziadat et al., 2003; Palyvos et al., 2006; Qi and Zhu, 2006; Mach and Petschek, 2007; Kukowski et al., 2008) (Figs. 8.6B) as well as draping morphometric layers over 3D terrain models (e.g., Figs. 2.8, 2.9, 2.11, 2.15). This approach is also a standard procedure.

Sequential spatiotemporal series of undraped or draped 3D terrain models can be used to create 3D terrain animation (Mitasova et al., 2012). In the first case, these can be multitemporal 3D models of rapidly changing topography, for example, loose sand dunes. In the second case, one can, for instance, illustrate the dynamics of grass cover depending on the season using multitemporal remote images superimposed on the same 3D terrain model.

The vast majority of DEM-based 3D terrain models are used to present and analyze them on a computer screen. However, 3D digital terrain models are input data for the production of 3D physical terrain models using 3D printing technology (Hasiuk et al., 2019; Solla et al., 2020; Harding et al., 2021).

Note that in this book we do not consider methods for creating 3D models of synthetic terrains, landforms, and landscapes used in computer graphics, games, etc. (Galin et al., 2019; Lochner et al., 2023).

8.5.1.1 Modeling with multiple-source illumination

A usual, single-source illumination of 3D models may be deficient for topographically complex terrains. Here we describe an approach for 3D terrain modeling with multiple-source illumination using the Blender package, free and open-source software for 3D modeling, visualization, and animation (Blender Foundation, 2003–2023; Kent, 2015; Filippov, 2018).

The approach includes the following stages (Florinsky and Filippov, 2019c): (1) creating a polygonal object; (2) selecting an algorithm to model the 3D geometry; (3) selecting a vertical exaggeration scale; (4) selecting types, parameters, a number, and positions of light sources; (5) selecting methods for generating shadows; (6) selecting a shading method for the 3D model surface; (7) selecting a material for the 3D model surface; (8) overlaying a texture on the 3D model; (9) setting a virtual camera; and (10) rendering the 3D model.

Stage 1. Creating a polygonal object

In Blender, a 3D terrain model is a polygonal mesh object. To create it, one should prepare lists of vertices, edges, and faces of the object. A vertex is a three-dimensional vector, that is, a list of three numbers corresponding to the coordinates (x, y, z). The vertex array is a list of vectors. One can similarly prepare lists for indices of vertices forming edges and faces of a polygonal lattice. A list defining an edge contains two vertex indices. A list defining a face contains three or more vertex indices.

Stage 2. Selecting an algorithm to model the 3D geometry

The aim of this stage is to produce a high-resolution virtual mesh without modifying the original one. This is performed by recursive subdivision of the mesh's faces into smaller faces. It leads to a smooth appearance of the model (Blender Foundation, 2018). One can use two subdivision algorithms (Fig. 8.7):

1. A simple subdivision. It maintains the original shape by subdividing the faces and edges without smoothing. It can be utilized to increase the original mesh resolution.
2. The Catmull–Clark subdivision (Catmull and Clark, 1978). It subdivides and smoothes the faces and edges producing a smooth surface.

Stage 3. Selecting a vertical exaggeration scale

To produce readable and interpretable 3D terrain models, exaggeration of the vertical scale should as a rule be applied depending on the type of topography, area size, and elevation difference within the area to be modeled. Although there are no direct mathematical relationships, one should be guided by the following principle: the larger the area and the smaller the elevation difference, the greater the vertical scale exaggeration. The choice of a specific value is made empirically and interactively, considering the correct perception and appearance of the 3D terrain model.

Stage 4. Selecting types, parameters, a number, and position of light sources

To illuminate a 3D terrain model, three types of Blender light sources can be applied: the Sun, Spot, and Point (Fig. 8.8). The Sun emits light in a given direction and is located infinitely far away. The Spot emits light in a particular direction within a cone. The Point emits light equally in all directions (Blender Foundation, 2018).

There are two key parameters of a light source: (1) energy, the illumination intensity; and (2) falloff, the decrease in illumination intensity depending on the distance from a light source. For the Spot light source, one can use additional parameters: light cone angle and light attenuation zone at the cone periphery.

The number and position of light sources depend on the type and complexity of the topography to be modeled. It usually makes sense to apply one or two main light sources illuminating the entire area as well as an unlimited number of additional light sources highlighting shaded landforms (e.g., deep depressions and basins) if they are important for perception and interpretation of the 3D model.

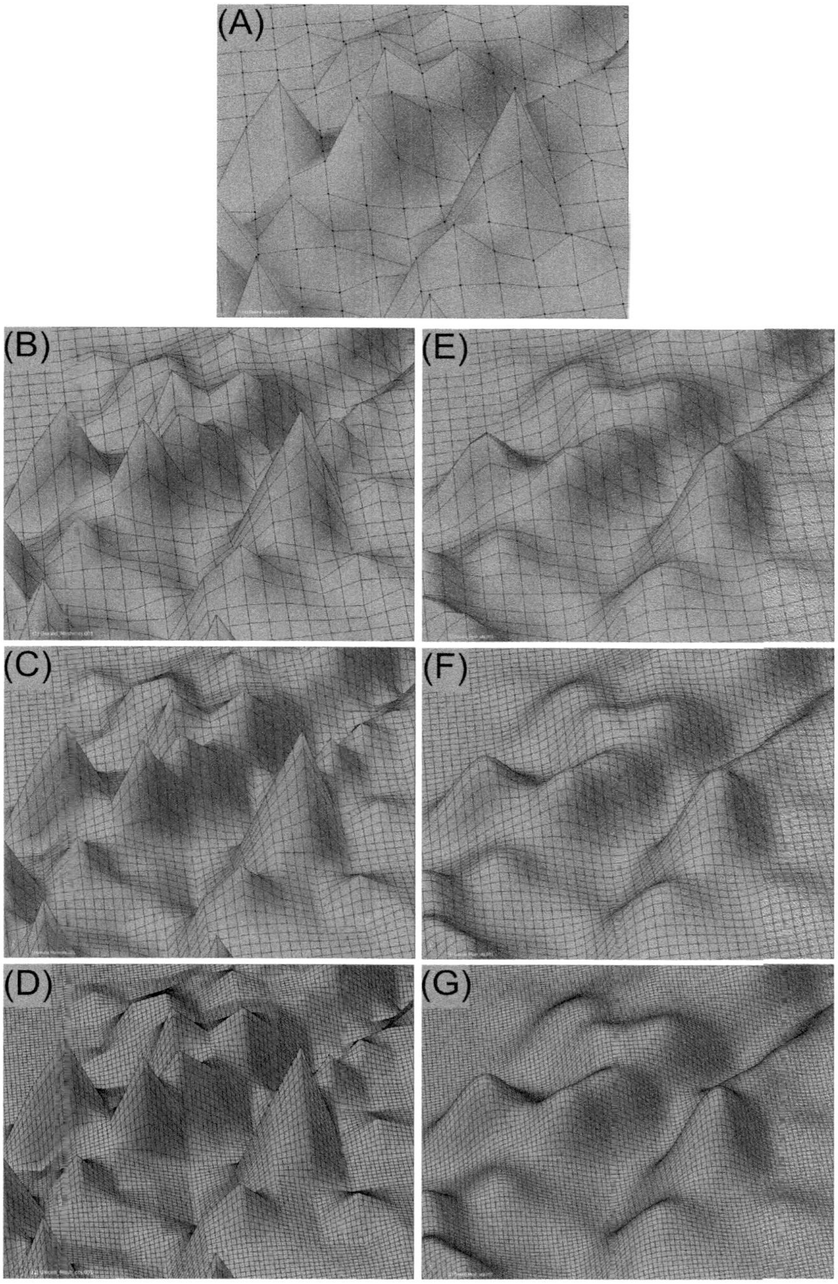

FIG. 8.7 Modeling the 3D geometry by different algorithms of mesh subdivision: (A) Original mesh. Simple subdivisions without smoothing: (B) single, (C) double, and (D) triple subdivisions of faces. The Catmull–Clark subdivisions with smoothing: (E) single, (F) double, and (G) triple subdivision of faces. The models were produced in the Blender viewport (real-time, OpenGL visualization). *(From Florinsky, I.V., Filippov, S.V., 2019. Three-dimensional terrain modeling with multiple-source illumination. Trans. GIS 23 (5), 937–959, Fig. 1, reproduced with permission from John Wiley & Sons.)*

Stage 5. Selecting methods for generating shadows

Shadows are inseparably linked with light sources. To generate shadows, one can choose two major approaches: ray tracing and buffering (Fig. 8.9). The Sun and Point light sources are capable of casting ray-traced shadows; the Spot light source is able to cast both ray-traced and buffered shadows (Blender Foundation, 2018).

The ray-tracing approach is used to determine shaded areas of a mesh. During the final rendering at Stage 10, the tracer draws rays from a focal point of the virtual camera (see Stage 9) through each pixel of its image plane to a 3D surface point corresponding to the pixel. Then reflected rays are drawn to light sources. The point is in shadow if a ray cannot be traced to a light source without intersection with some part of the model. In the buffering approach, areas that cannot be seen from a light source are considered in shadow (Simonds, 2013).

The ray-tracing approach is more accurate and slower than the buffering one. It can generate both crisp and soft shadows by the constant and adaptive Quasi-Monte Carlo (QMC) methods (Caflisch, 1998). The constant QMC is slower, but it

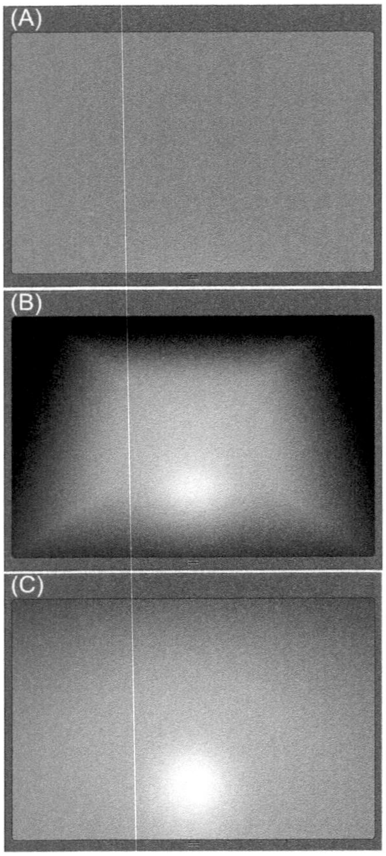

FIG. 8.8 Brightness distribution for different light sources: (A) Sun. (B) Spot. (C) Point. *(From Florinsky, I.V., Filippov, S.V., 2019. Three-dimensional terrain modeling with multiple-source illumination. Trans. GIS 23 (5), 937–959, Fig. 2, reproduced with permission from John Wiley & Sons.)*

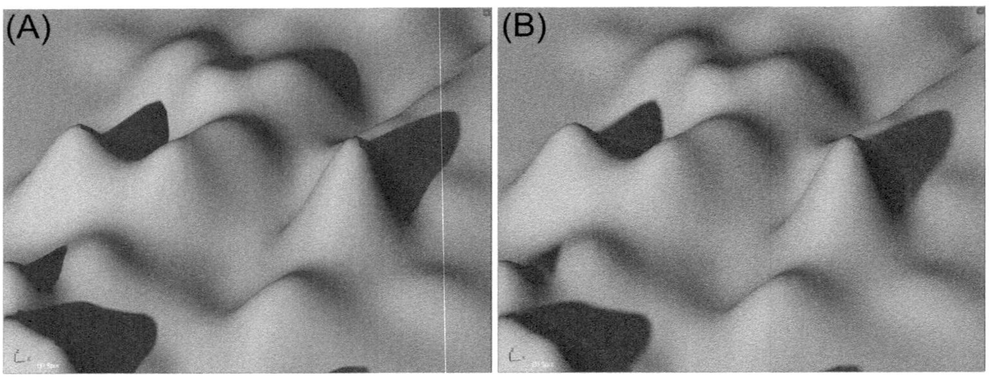

FIG. 8.9 Shadows generated by different methods: (A) Ray-traced shadows generated by the adaptive QMC method. (B) Buffered shadows are generated by the classic-halfway buffering method with the box anti-aliased filtering. The models were produced in the Blender viewport (real-time, OpenGL visualization) for a 3D model smoothed by the Catmull–Clark double subdivision. *(From Florinsky, I.V., Filippov, S.V., 2019. Three-dimensional terrain modeling with multiple-source illumination. Trans. GIS 23 (5), 937–959, Fig. 3, reproduced with permission from John Wiley & Sons.)*

produces less noisy shadows than the adaptive QMC. The buffering approach is faster but less accurate than the ray-tracing one. It makes sense to choose between the classic-halfway and irregular methods for generating buffered shadows (Simonds, 2013).

Two filters can be applied to anti-alias buffered shadows: the box and Gauss filters. The box filter creates sharp anti-aliasing and is usually applied to low-resolution models. The Gauss filter creates soft anti-aliasing and is utilized to handle high-resolution models.

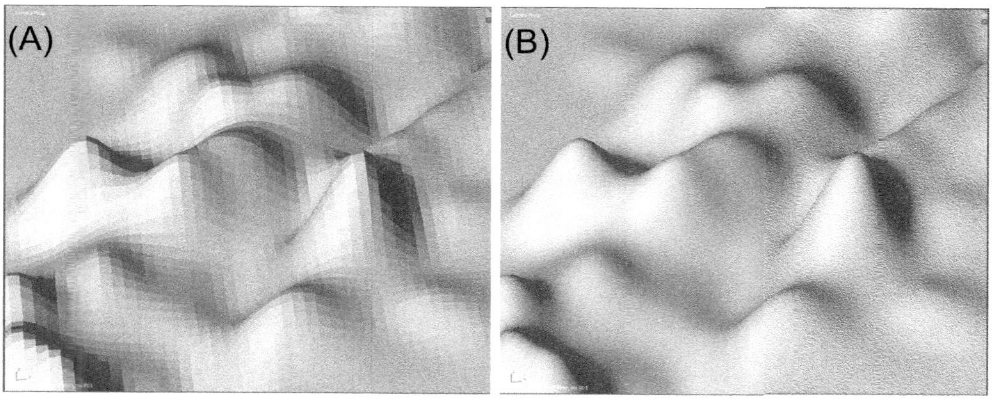

FIG. 8.10 Shading performed by different methods: (A) Flat. (B) Smooth. The models were produced in the Blender viewport (real-time, OpenGL visualization). *(From Florinsky, I.V., Filippov, S.V., 2019. Three-dimensional terrain modeling with multiple-source illumination. Trans. GIS 23 (5), 937–959, Fig. 4, reproduced with permission from John Wiley & Sons.)*

Stage 6. Selecting a shading method for the 3D model surface

A 3D model is approximated by a polygon mesh (Fig. 8.7A). When rendered, the 3D model appears as a rasterized surface consisting of a plurality of small flat faces. To smooth such an appearance, shading should be applied (Blender Foundation, 2018). Two shading methods are used: the flat and smooth ones (Fig. 8.10).

In flat shading, a color or grayscale brightness value is determined for one point within each polygon using a normal to the flat polygon's face. Then this color/brightness value is attributed to the entire face of the polygon.

In the smooth shading, color/brightness values are determined at the polygon's vertices. For pixels between the vertices, color/brightness values are bilinearly interpolated. This procedure creates the illusion of a smooth surface because the smooth shading does not modify the 3D model geometry (i.e., the mesh). The shading algorithm of Phong (1975) is used for smooth shading.

Stage 7. Selecting a material for the 3D model surface

Selecting a material means selecting reflection properties for the 3D surface to be further rendered (Blender Foundation, 2018). One can apply two diffuse reflectance models: the Lambertian model (Horn, 1981) (Section 2.5.1.1) (Fig. 8.11A–G) and the Oren–Nayar (1994) model (Fig. 8.11E–H). These models can be used with or without the application of additional specular reflection models by Phong (1975) (Fig. 8.11B and F), Blinn (1977) (Fig. 8.11C and G), Cook and Torrance (1982), as well as Ward and Simmons (1999) (Fig. 8.11D and H).

The choice of a specific reflection model is generally quite subjective and depends on the aesthetic taste of a user and goals of a particular 3D model.

Stage 8. Overlaying a texture on the 3D model

This stage may be skipped if a user constructs a grayscale, hill-shaded 3D terrain model. Otherwise, a color thematic (hypsometric, bathymetric, morphometric, etc.) texture should be produced as an image file. Then the image file is projected to the mesh. To preserve the consistency of a texture with the 3D terrain model, it is necessary to use the flat projection type in UV coordinates (U and V are axes of a two-dimensional (2D) texture of a 3D object), zero offset, and 1:1 scaling factor. One should also set values for the color and emit channels. These determine a balance between the texture color intensity and the surface luminosity.

Stage 9. Setting a virtual camera

A virtual camera is needed to observe a 3D model from a particular point. A user should specify the coordinates of the camera position, its rotation angles, focal length, and resolution of the image to be rendered. For 3D terrain modeling, a perspective projection is used. To visualize a 3D model from various angles, one can either move the same camera to desired positions or set several cameras.

Stage 10. Rendering the 3D model

Renderers are programs that produce images from meshes, materials, and light sources. Blender includes real-time and offline renderers (Blender Foundation, 2018).

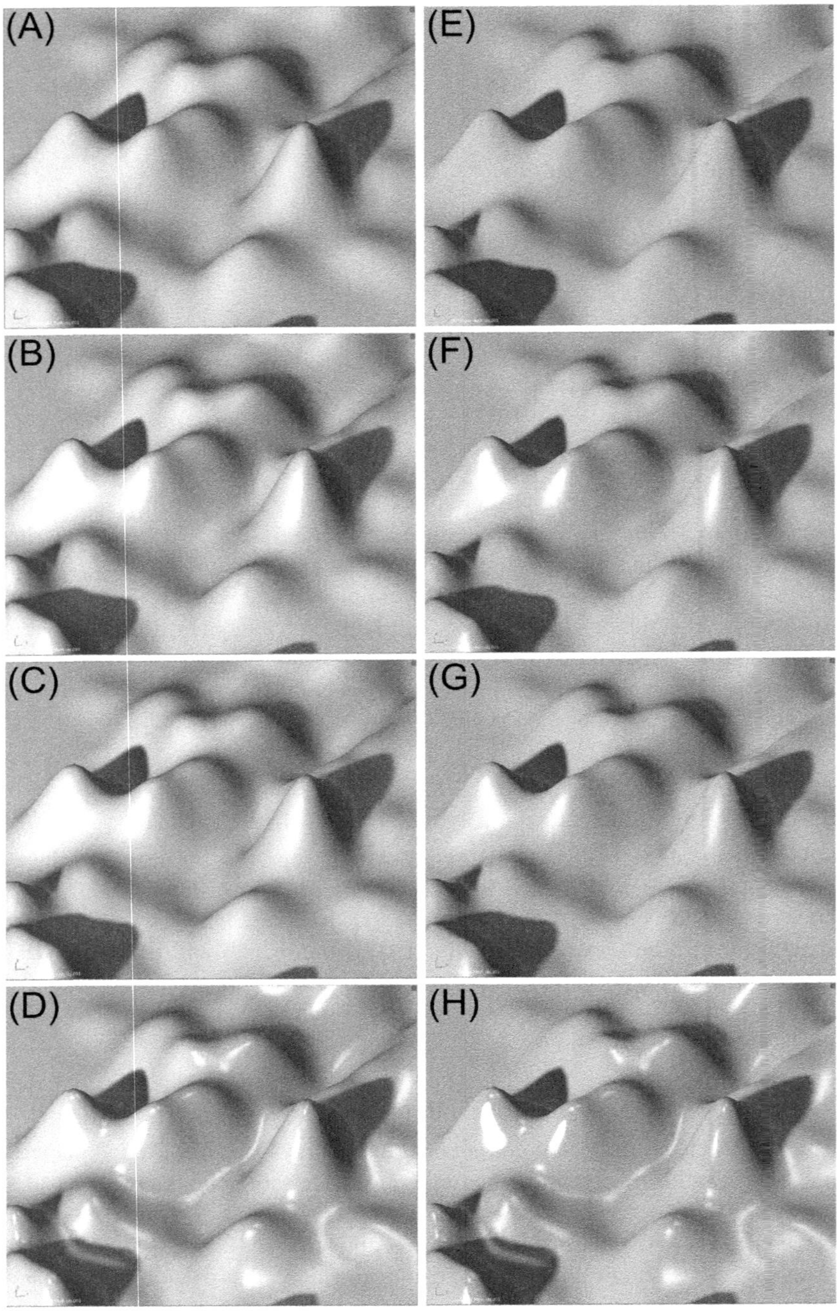

FIG. 8.11 Applying materials with different reflection properties for a 3D model surface. The Lambertian diffuse reflectance model: (A) No specular reflection. (B) Phong specular reflection. (C) Blinn specular reflection. (D) Ward specular reflection. The Oren–Nayar diffuse reflectance model: (E) No specular reflection. (F) Phong specular reflection. (G) Blinn specular reflection. (H) Ward–Simmons specular reflection. The models were produced in the Blender viewport (real-time, OpenGL visualization). *(From Florinsky, I.V., Filippov, S.V., 2019. Three-dimensional terrain modeling with multiple-source illumination. Trans. GIS 23 (5), 937–959, Fig. 5, reproduced with permission from John Wiley & Sons.)*

Engines of the first type render 3D scenes in real time allowing a user to interact with graphical objects during the rendering. Real-time rendering is carried out with the Blender Game Engine (BGE). It is marked by limited visualization capabilities, hardly supporting reflections, transparency, shadows, etc.

Engines of the second type render 3D scenes offline. It is impossible to modify them after the rendering. There are two offline renderers:

1. *Blender Render*. This is a fast, non-photorealistic, legacy-rendering engine. It does not use graphics processing unit (GPU), but it well supports multicore processors, that is, central processing unit (CPU).
2. *Cycles*. This is a slow, physically based engine. It is perfectly accelerated on GPU. Its main advantage is photorealism, which is not always necessary for 3D terrain modeling.

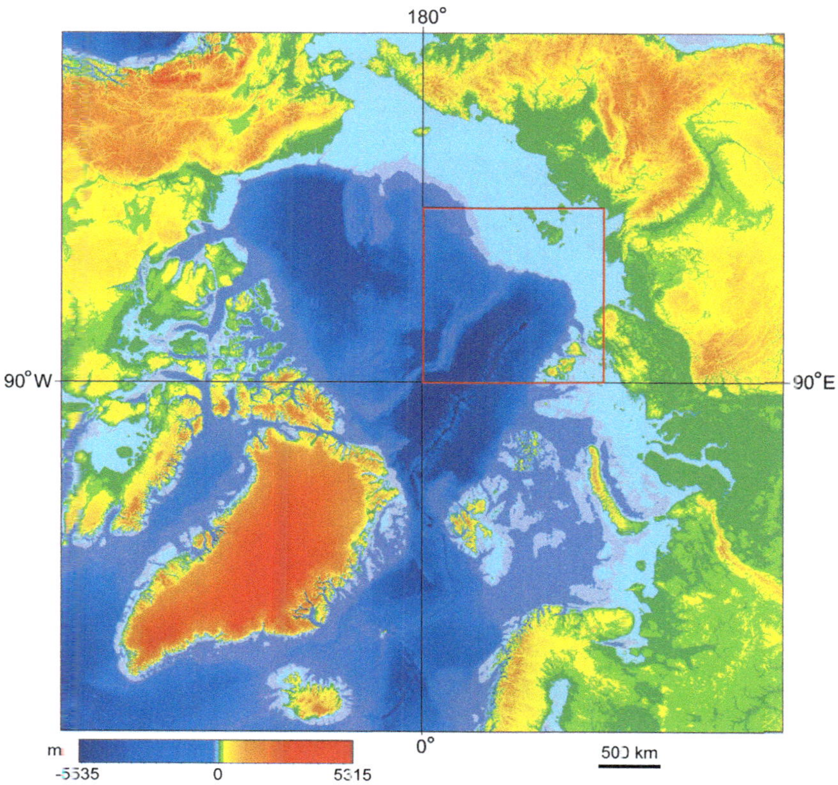

FIG. 8.12 IBCAO 3.0 depth/elevation map with a 5-km grid spacing. *Red frame*, the DEM location. *(From Florinsky, I.V., Filippov, S.V., 2019. Three-dimensional terrain modeling with multiple-source illumination. Trans. GIS 23 (5), 937–959, Fig. 6, reproduced with permission from John Wiley & Sons.)*

To exemplify the described approach, we use a small, low-resolution DEM for a portion of the Arctic Ocean floor (Fig. 8.12). It was extracted from the 500-m gridded DEM belonging to the data set of the International Bathymetric Chart of the Arctic Ocean version 3.0 (IBCAO 3.0) (IBCAO Compilation Group, 2012; Jakobsson et al., 2012). The DEM (Fig. 8.12) describes a territory measuring about 1440 km × 1440 km. It is based on a 15-km square grid (a 97 × 97 elevation matrix; 9409 points). Elevation/depth values range from 860 m to −5179 m. The DEM is presented in the polar stereographic projection. Its left lower corner coincides with the North Pole, while its bottommost row and leftmost column coincide with the meridians of 90°E and 180°E, respectively.

Fig. 8.13 illustrates a progressive smoothing of the 3D model by combination of the mesh subdivision and shading. Fig. 8.13A represents a created unsmoothed polygonal object consisting of 9409 vertices and 9216 faces. The 3D model with the unsmoothed mesh and flat shading can be seen in Fig. 8.13B. Next, the unsmoothed mesh of the 3D model with the smooth shading is shown in Fig. 8.13C. Finally, the geometry of the 3D model was smoothed by the Catmull–Clark subdivision algorithm with the double subdivision (Fig. 8.13D). This resulted in a smoothed polygonal object with 147,456 faces. Here we applied the Lambertian diffuse reflection model with the Phong specular reflection.

We selected the $40^{\times}$ vertical scale exaggeration to enhance the visual perception of topographic features. Such a large exaggeration factor was set because the elevation difference is only around 6 km within the study area, measuring about 1440 km by 1440 km.

Three light sources were used to illuminate the 3D model (Fig. 8.14A):

1. The Sun light source was located at the zenith above the study area center (Fig. 8.14B). It was intended for the uniform illumination of the entire area. It also highlights deep shadows from other light sources. The ray-tracing approach was applied to calculate shadows.
2. The Point light source was located above a deep depression and was intended for its illumination (Fig. 8.14B). Shadows were calculated by the ray-tracing approach.
3. The Spot light source was located outside the study area and was directed at a slight angle to the horizontal plane (Fig. 8.14C). It was intended for the visual enhancement of relief plasticity. Shadows were calculated by the classic-halfway buffering method and smoothed using the box filter.

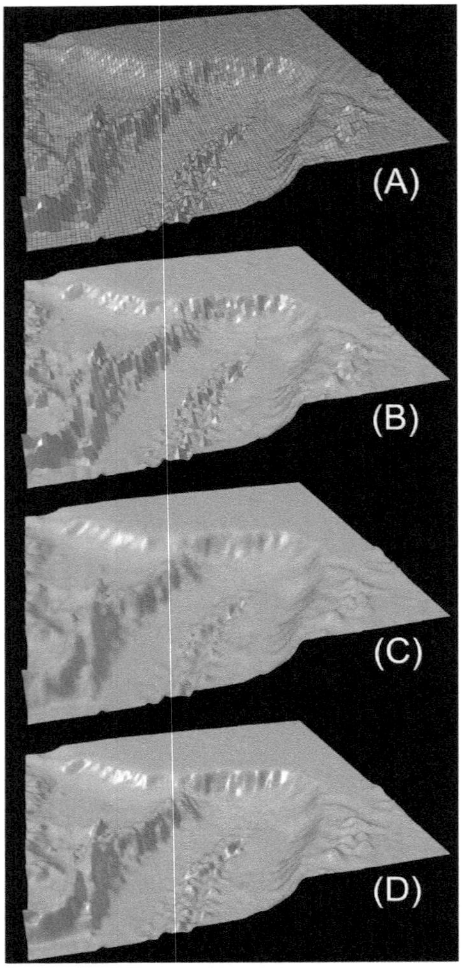

FIG. 8.13 Smoothing the geometry and shading the 3D model: (A) The original polygon mesh. (B) No subdivision and flat shading. (C) No subdivision and smooth shading. (D) The Catmull–Clark double subdivision and the smooth shading. The models were produced in the Blender viewport (real-time, OpenGL visualization). *(From Florinsky, I.V., Filippov, S.V., 2019. Three-dimensional terrain modeling with multiple-source illumination. Trans. GIS 23 (5), 937–959, Fig. 7, reproduced with permission from John Wiley & Sons.)*

Fig. 8.15 shows the results of applying three individual light sources and their combinations to illuminate the 3D model. Here, we utilized the Lambertian diffuse reflection model without specular reflection. One can notice that the application of individual light sources or their pairs is not enough for adequate 3D representation of even such a small and simple model.

To exemplify an overlay of a texture on the 3D surface (Fig. 8.16B), we used a bathymetric map (Fig. 8.16A) derived from the initial DEM by the LandLord software (Florinsky, 2016).

Fig. 8.17 shows one of the camera positions used for further rendering. In the final, offline visualization with the Blender Render engine (Fig. 8.18), we applied the Lambertian diffuse reflection model without specular reflection. Although the rendered 3D terrain models (Fig. 8.18) are quite generalized, they are well perceived and provide a general, small-scale visual description of submarine topography within this area of the Arctic Ocean.

Fig. 8.19 demonstrates examples of 3D modeling with larger DEMs using the described approach (Florinsky and Filippov, 2019a, b, 2021).

8.5.2 Three-dimensional subsurface modeling

In 3D subsurface modeling, researchers use various combinations of DEM-derived 3D models of the land surface, geological surfaces, or surfaces of soil horizons, as well as other data on spatially distributed geological and soil features. Caumon et al. (2009, p. 928) explained the importance of 3D geological modeling as follows:

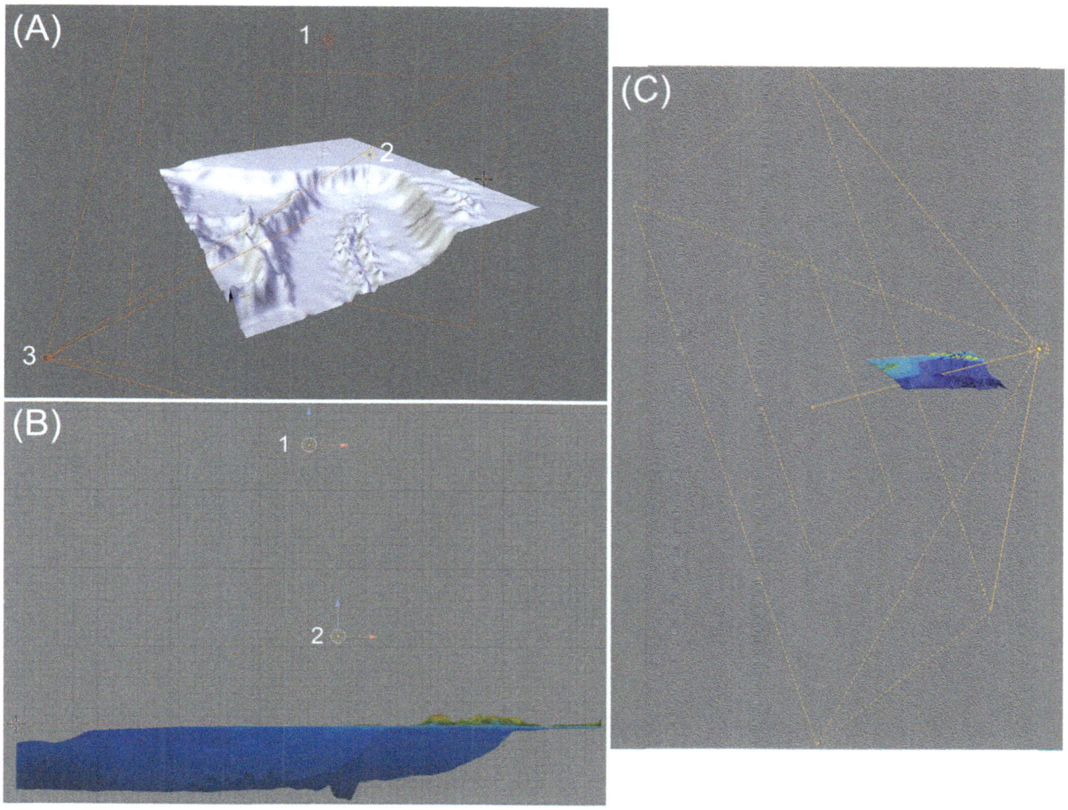

FIG. 8.14 Position of light sources relative to the 3D model: (A) General view: *1*, Sun; *2*, Point; *3*, Spot. (B) Side view: *1*, Sun; *2*, Point. (C) Side view: Spot. The general and side views are presented using the model without and with an overlaid bathymetric texture, respectively. *(From Florinsky, I.V., Filippov, S.V., 2019. Three-dimensional terrain modeling with multiple-source illumination. Trans. GIS 23 (5), 937–959, Fig. 8, reproduced with permission from John Wiley & Sons.)*

> *"Understanding the spatial organization of subsurface structures is essential for quantitative modeling of geological processes. It is also vital to a wide spectrum of human activities, ranging from hydrocarbon exploration and production to environmental engineering. Because it is not possible to directly access the subsurface except through digging holes and tunnels, most of this understanding has to come from various indirect acquisition processes. 3D subsurface modeling is generally not an end, but a means of improving data interpretation through visualization and confrontation of data with each other and with the model being created as well as a way to generate support for numerical simulations of complex phenomena (i.e., earthquakes, fluid transport) in which structures play an important role. As the interpretation goes, the 3D framework forces us to make interpretive decisions that would be left on the side in map or cross section interpretations. Skilled geologists know how to translate 3D into 2D and vice versa, but, no matter how experienced one can be, this mental translation is bound to be qualitative, hence inaccurate and sometimes incorrect. 3D model building calls for a complex feedback between the interpretation of the data and the model."*

A general scheme for construction of 3D geological models, or block diagrams consists of the following steps (Van Driel, 1989; Pflug and Harbaugh, 1992; Breunig, 1996; Lemon and Jones, 2003; Masumoto et al., 2004; Dhont et al., 2005; Kaufmann and Martin, 2008; Fernandez et al., 2009; Mateo Lázaro et al., 2014; Thornton et al., 2018):

1. Geo-referencing and digitizing data in 3D space:
   (a) Deriving a DEM of the land surface (Section 3.1).
   (b) Digitizing geological borders from geological maps and field data.
   (c) Digitizing faults and other structural data from geological maps and field data.
   (d) Deriving DEMs of geological (stratigraphic) surfaces from geological maps, cross sections, borehole data (Section 3.1.10), and 3D seismic survey data (Section 3.1.11).
2. For each geological surface and fault, estimate local strikes and dips by geometrical analysis of areal extents of geological contacts (Section 14.3.3).

FIG. 8.15 Illumination of the 3D model by individual light sources and their combinations: (A) Sun. (B) Spot. (C) Point. (D) Spot and Sun. (E) Sun and Point. (F) Spot and Point. (G) All three light sources. The models were produced by the offline scene visualization with the Blender Render engine. *(From Florinsky, I.V., Filippov, S.V., 2019. Three-dimensional terrain modeling with multiple-source illumination. Trans. GIS 23 (5), 937–959, Fig. 9, reproduced with permission from John Wiley & Sons.)*

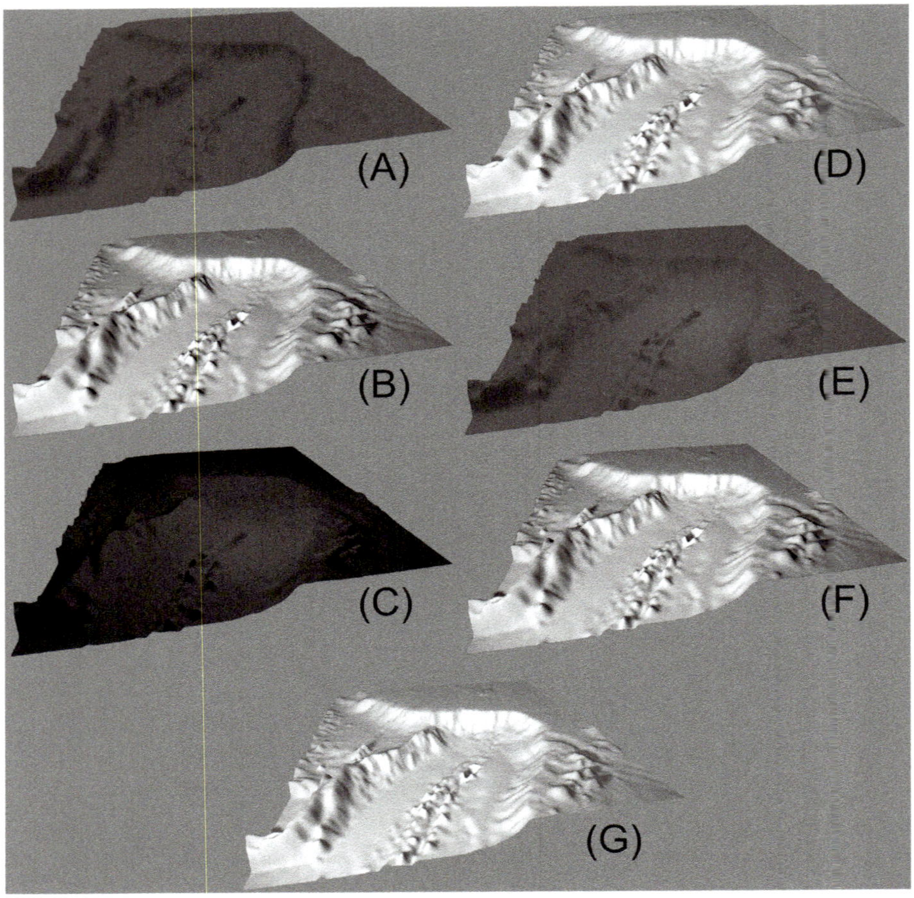

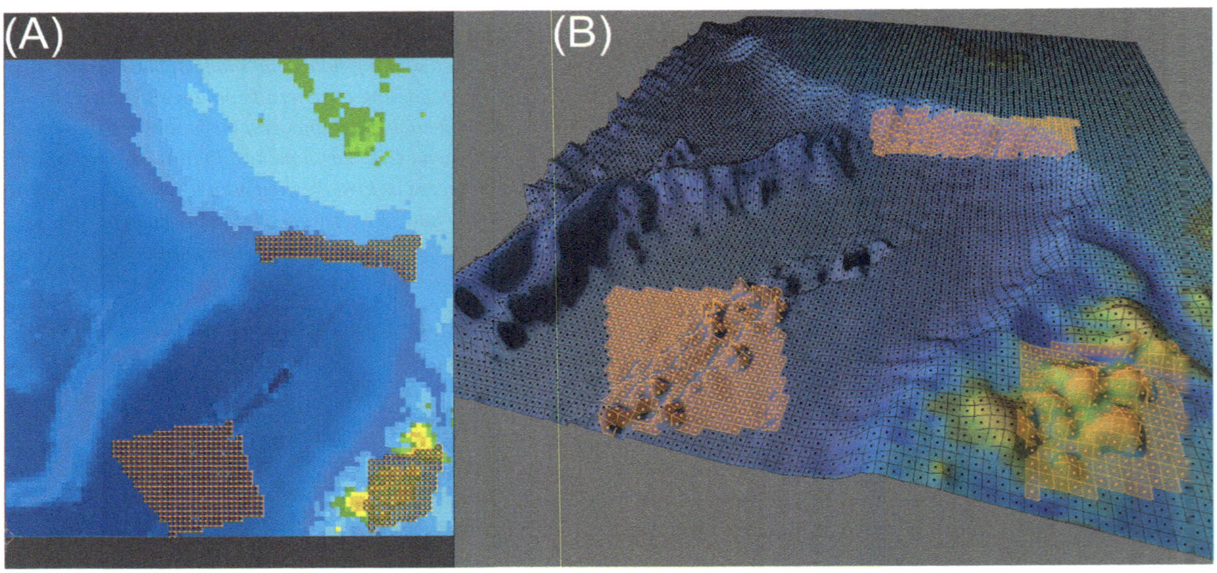

FIG. 8.16 Overlaying the bathymetric texture (A) on the 3D surface (B). One can see the underlying, smoothed polygonal grid in three locations. *(From Florinsky, I.V., Filippov, S.V., 2019. Three-dimensional terrain modeling with multiple-source illumination. Trans. GIS 23 (5), 937–959, Fig. 10, reproduced with permission from John Wiley & Sons.)*

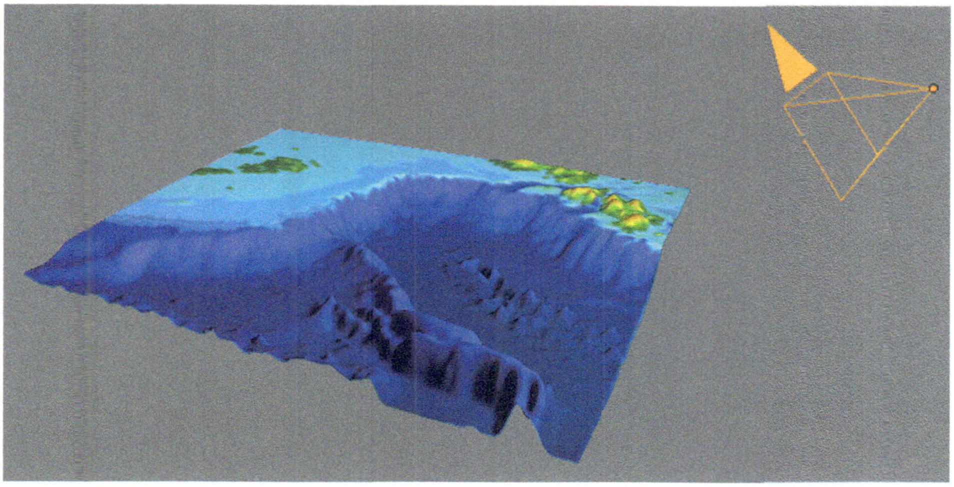

FIG. 8.17 One of the virtual camera positions with respect to the 3D model. *(From Florinsky, I.V., Filippov, S.V., 2019. Three-dimensional terrain modeling with multiple-source illumination. Trans. GIS 23 (5), 937–959, Fig. 11, reproduced with permission from John Wiley & Sons.)*

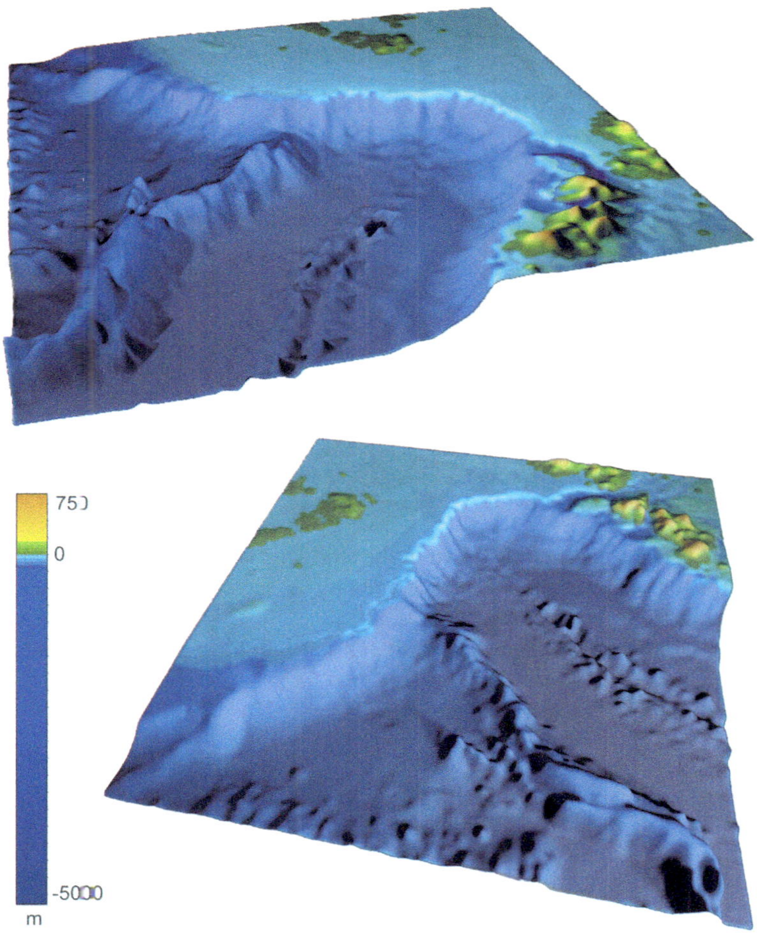

FIG. 8.18 Examples of the final 3D model produced by the offline scene visualization with the Blender Render engine. *(From Florinsky, I.V., Filippov, S.V., 2019. Three-dimensional terrain modeling with multiple-source illumination. Trans. GIS 23 (5), 937–959, Fig. 12, reproduced with permission from John Wiley & Sons.)*

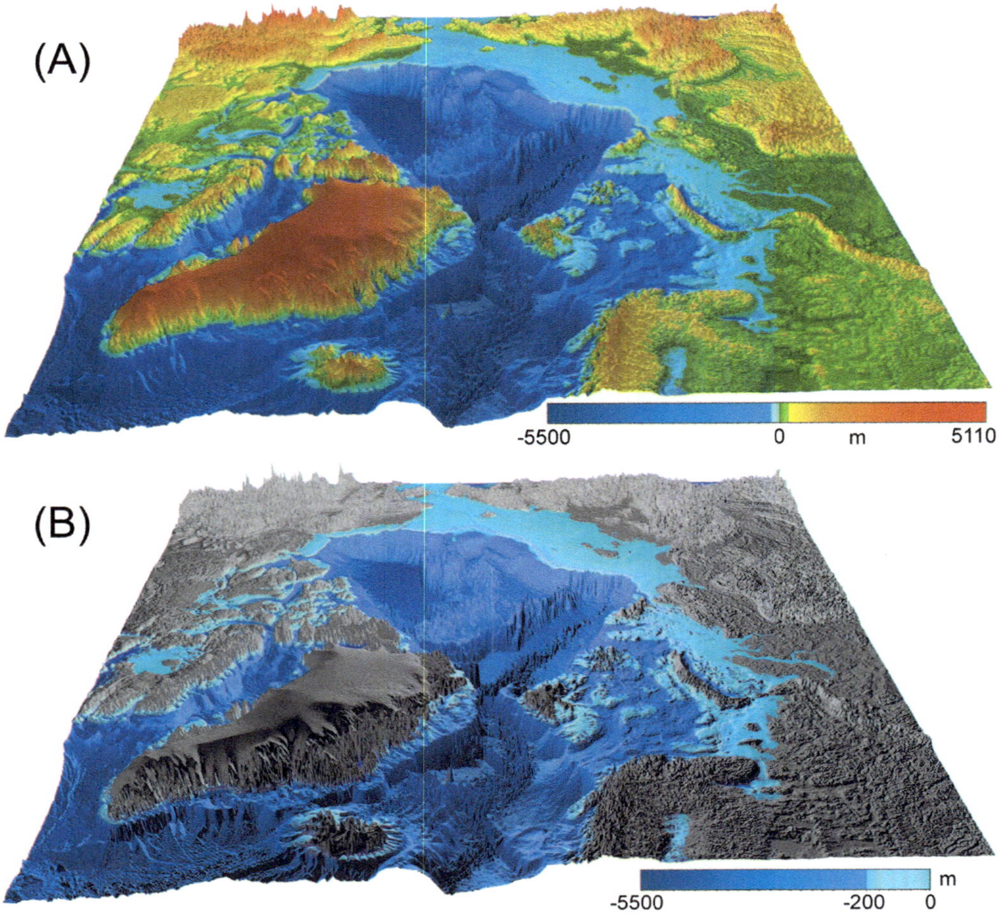

FIG. 8.19 Examples of 3D models of the Arctic Ocean floor and adjacent territories: (A) 10-km gridded DEM (581 × 581 elevation matrix; 337,561 points). (B) 5-km gridded DEM (1161 × 1161 elevation matrix; 1,347,911 points). Perspective views from the Atlantic. The DEMs were extracted from IBCAO 3.0 (IBCAO Compilation Group, 2012). *((A) From Florinsky, I.V., Filippov, S.V., 2019. Three-dimensional geomorphometric modelling of the Arctic Ocean submarine topography (a testing, low-resolution desktop system). InterCarto InterGIS 25 (1), 275–289 (in Russian), Fig. 6A. (B) From Florinsky, I.V., Filippov, S.V., 2019. Three-dimensional desktop morphometric models for the Arctic Ocean floor. Proc. Int. Cartogr. Assoc. 2, 32, Fig. 1.)*

3. Generating a set of digital models for all geological surfaces and faults known in the study area. To do this, one should create partial surfaces honoring local strikes and dips for each geological surface and fault. Assembling partial surfaces associated with each geological surface or fault should also be done.
4. Analyzing relative positions of obtained models of geological surfaces and faults, in particular, to detect their intersections. Correcting positional errors.
5. Assembling a complete 3D model using constructed geological surfaces and faults (Fig. 8.20).

3D models are widely used in geological studies of various kinds and scales (Morris, 1991; Maerten et al., 2001; Guillaume et al., 2008; De Donatis et al., 2009; Jones et al., 2009; Zanchi et al., 2009; Quiroga et al., 2022; García-Gil et al., 2023), in particular, in oil and gas exploration (McMahon and North, 1993; Durham, 1999; Clarke and Phillips, 2003; Liu et al., 2022).

It is well-known that soil is a 3D body, and the thicknesses of soil horizons are spatially variable (Hartemink et al., 2020). Smith et al. (2008, p. 190) noted that "spatial soil horizon modeling can use methods and software developed originally for shallow geological modelling, when soil horizons follow a unique super positional order. The modelling of horizons that do not have one single super positional order or are even overturned and convoluted e.g., in periglacially disturbed soils, can be solved using techniques developed for modelling faulted and overturned geology." However, compared with geology, 3D soil modeling has not yet found wide acceptance in soil science (Pereira and FitzPatrick, 1998; Grunwald et al., 2000, 2007; Mendonça Santos et al., 2000; Chaplot and Walter, 2003; Ramasundaram et al., 2005; Smith et al., 2008; Delarue et al., 2009; Gerke et al., 2022).

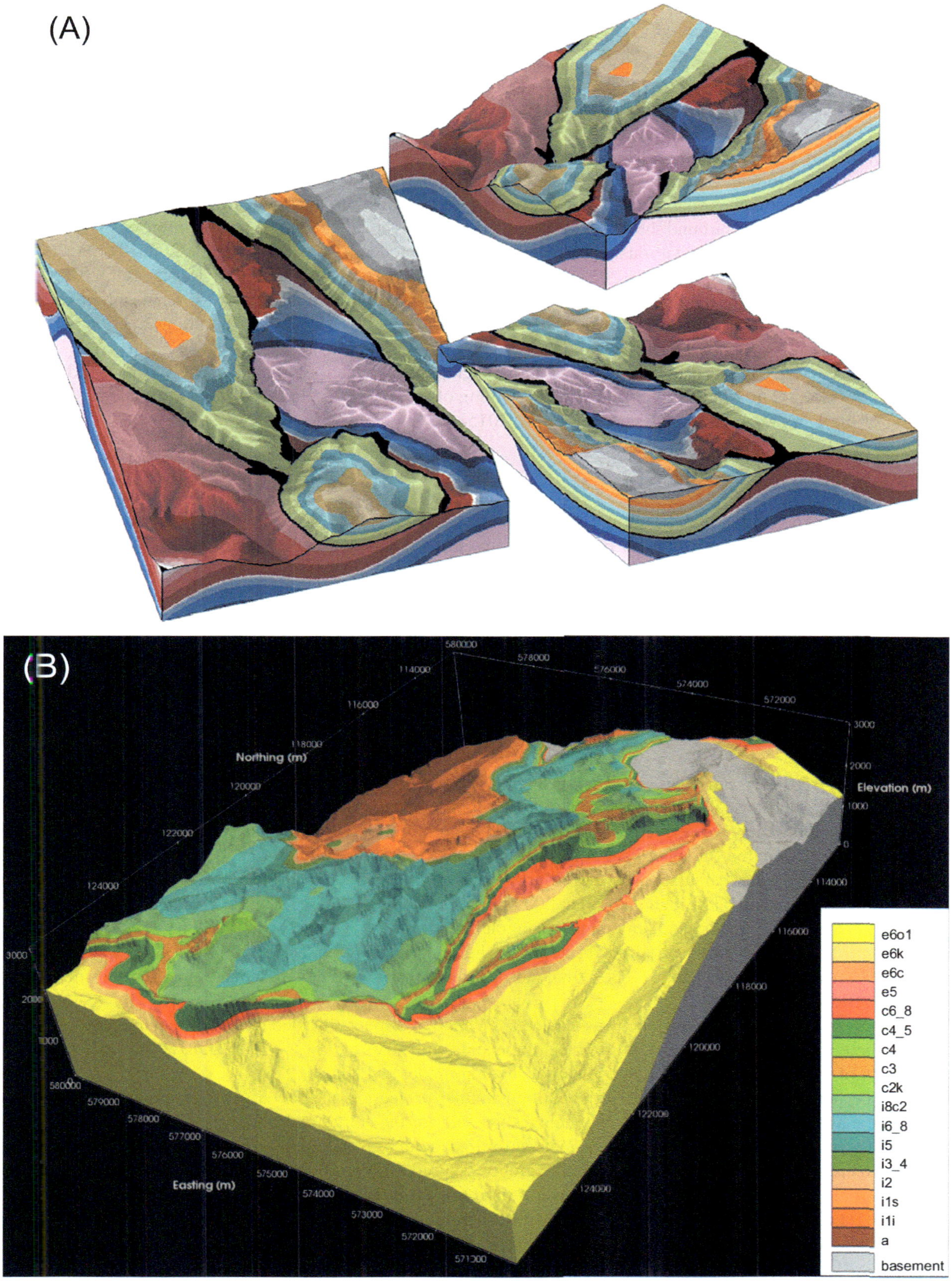

FIG. 8.20 3D geological models: (A) 3D model of simulated geological strata. (B) 3D model of the bedrock geology of the Nappe de Morcles, western Swiss Alps. The legend refers to the coding of the formations of the stratigraphic pile. *((A) From: Mateo Lázaro, J., Sánchez Navarro, J.Á., García Gil, A., Edo Romero, V., 2014. 3D-geological structures with digital elevation models using GPU programming. Comput. Geosci. 70, 138–146, Fig. 13, reproduced with permission from Elsevier. (B) From Thornton, J.M., Mariethoz, G., Brunner, P., 2018. A 3D geological model of a structurally complex Alpine region as a basis for interdisciplinary research. Sci. Data 5, 180238, Fig. 8, © Authors, distributed under the terms and conditions of the Creative Commons Attribution (CC BY) license.)*

8.6 Virtual globes

In the last twenty years, significant progress has been made in the development and application of desktop and web-based virtual globes (Tuttle et al., 2008; Kennedy, 2009; Blaschke et al., 2012). Virtual globes are programs implementing interactive 3D models of planetary bodies. The list of virtual globes includes World Wind (NASA, 2003–2023), Earth3D (Gunia, 2004–2018), Google Earth (Google, 2005–2023), Marble (KDE, 2007–2023), Cesium (Cesium GS, 2012–2023), and others. Virtual globes enable to perform 3D multiscale visualization of complex, spatially distributed multilayer data with capabilities to move around the globe and to change the user's viewing angle and position relative to the globe.

Virtual globes are used to solve various multiscale tasks of geosciences (Ryakhovsky et al., 2003; Tooth, 2006; Chen and Bailey, 2011; Paraskevas, 2011; Guth, 2012; Whitmeyer et al., 2012; Yu and Gong, 2012; Zhu et al., 2014; Scheffers et al., 2015; Müller et al., 2016; Bladin et al., 2018). In virtual globes, it is a common practice to visualize topography using three well-known cartographic approaches (Cozzi and Ring, 2011, Pt. 4): (1) hypsometric tinting (Section 8.1), (2) analytical hill shading (Section 2.5.1.1), and (3) color relief shading (Section 8.2) (Fig. 8.21).

Virtual globes utilize engines usually developed with 3D graphics application programming interfaces (APIs), such as OpenGL and WebGL (Cozzi and Ring, 2011). Rapid zooming of massive datasets is commonly provided by hierarchical tessellation of the globe surface (Mahdavi-Amiri et al., 2015).

8.6.1 Desktop globes

Here we describe a desktop system of virtual morphometric globes for the Earth, Mars, and the Moon (Florinsky and Filippov, 2015, 2017) developed with the Blender, the free and open-source software for 3D modeling, visualization, and animation (Blender Foundation, 2003–2023; Kent, 2015; Filippov, 2018). The system development consisted of two key steps: (1) derivation of a set of global low-resolution morphometric maps of the Earth, Mars, and the Moon, and (2) 3D modeling of morphometric globes.

As the initial data, we used the following three 15′-gridded global DEMs:

- A DEM of the Earth extracted from the SRTM30_PLUS DEM (Sandwell et al., 2008).
- A DEM of Mars extracted from the Mars Orbiter Laser Altimeter (MOLA) data archive (Smith et al., 1999, 2003).
- A DEM of the Moon extracted from the Lunar Orbiter Laser Altimeter (LOLA) data archive (Neumann, 2008; Smith D.E. et al., 2010).

To suppress high-frequency noise, the DEMs were smoothed using a 3×3 moving window. The DEM of Mars was smoothed two times, and the DEMs of the Earth and the Moon were smoothed three times.

For these celestial bodies, we derived digital models of local morphometric attributes (Table 2.1) from the smoothed DEMs by the author's method for spheroidal equal angular grids (Section 4.3). Digital models of nonlocal morphometric variables (Table 2.1) were calculated by the Martz–de Jong method adapted to spheroidal equal angular grids (Section 4.4.2). Each global DTM included 1,036,800 points (the matrix 1440×720); the grid spacing was 15′.

Then we derived global maps of all calculated topographic attributes.[b] To deal with the large dynamic range of morphometric variables, we logarithmically transformed their digital models by Eq. (8.1) with $n = 8$. Data were processed by the LandLord software (Florinsky, 2016). The global morphometric maps were saved as image files for their subsequent use as textures in Blender.

To construct the 3D model of the globe, we selected a UV map of a sphere divided into 1152 rectangular polygons; that is, the 3D sphere is tessellated into 48×24 spherical trapezoids with sizes of $7.5° \times 7.5°$ (Fig. 8.22). A reasonably smooth representation of such a sphere was achieved using a shading model of Phong (1975).

Global morphometric maps, presented in the equirectangular projection and saved as raster image files, were draped over the sphere as textures influencing two channels responsible for the color and luminosity of an object. It was done to avoid (1) glares on the surface of the globe and (2) need for light sources traditionally complicating real-time rendering of the scenes. To create a latitude/longitude grid on the globe, we used a UV map of such a grid with a grid spacing of 7.5°. This UV map was draped on the sphere surface over a morphometric texture as the second semi-transparent texture.

The globe rotation was performed using a mouse actuator embedded in BGE. The globe rotation was also implemented using cursor movement keys of the keyboard. The scaling (zooming) of the globe was performed using both mouse and keyboard. The BGE camera was connected with two logical chains including two mouse sensors as well as two key sensors. They were linked with motion actuators, providing the BGE camera motion along the Y-axis (Fig. 8.23).

b. Examples of similar morphometric maps derived from global DEMs with a resolution of 30′ can be found in Figs. 16.2–16.9.

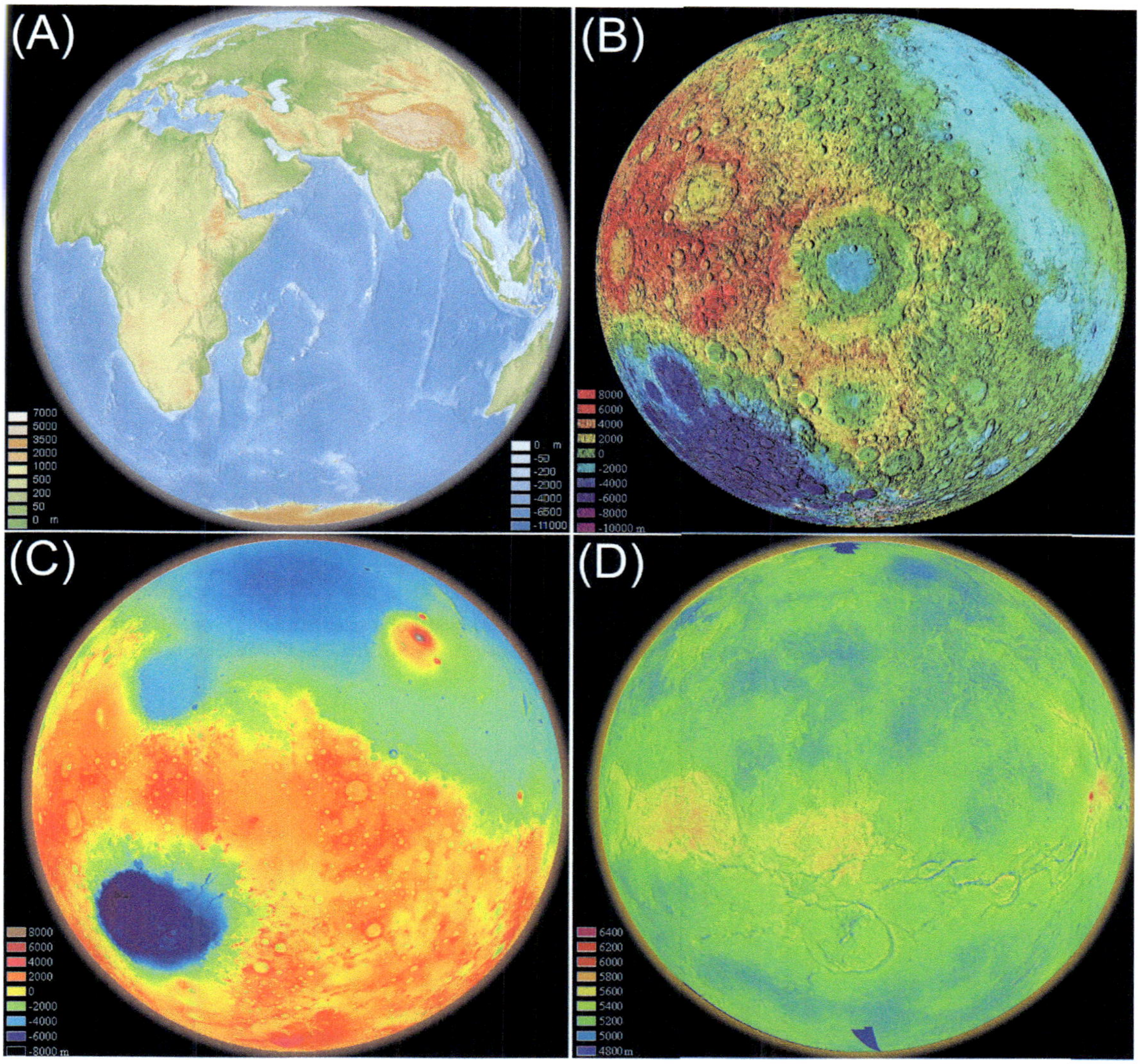

FIG. 8.21 Visualization of topography using hypsometric tinting and hill shading in the Marble Virtual Globe (KDE, 2007–2014): (A) The Earth. (B) The Moon; elevation data are from the Clementine data archive (Zuber et al., 1996) draped on the USGS hill-shaded map. (C) Mars; elevation data are from the MOLA data archive (Smith et al., 2003). (D) Venus; elevation data are from the Magellan data archive (Ford et al., 1992). *(Elevation maps are courtesy of the U.S. Geological Survey Astrogeology Science Center. From Florinsky, I.V., Filippov, S.V., 2015. Development of virtual morphometric globes using Blender. arXiv Preprint 1512.08511, Fig. 1.)*

All morphometric globes were generated as individual scenes within a project of the Blender Editor. To switch between scenes, a corresponding logic was created. Finally, the Blender project was assembled into a standalone desktop program using the Blender tools.

Figs. 8.24–8.26 show screenshotted examples of the developed morphometric globes for the Earth, Mars, and the Moon, respectively. For each celestial body, the screenshots were made for two same positions of the six globes relative to the observer.

8.6.2 Web globes

Here we describe a web system of virtual morphometric globes for Mars and the Moon (Florinsky et al., 2018a). The web system was designed as a separate unit of a 3D web geoportal for storage and access to planetary data (Karachevtseva et al., 2019). To implement the web system of morphometric globes, a distributed application was developed. The application

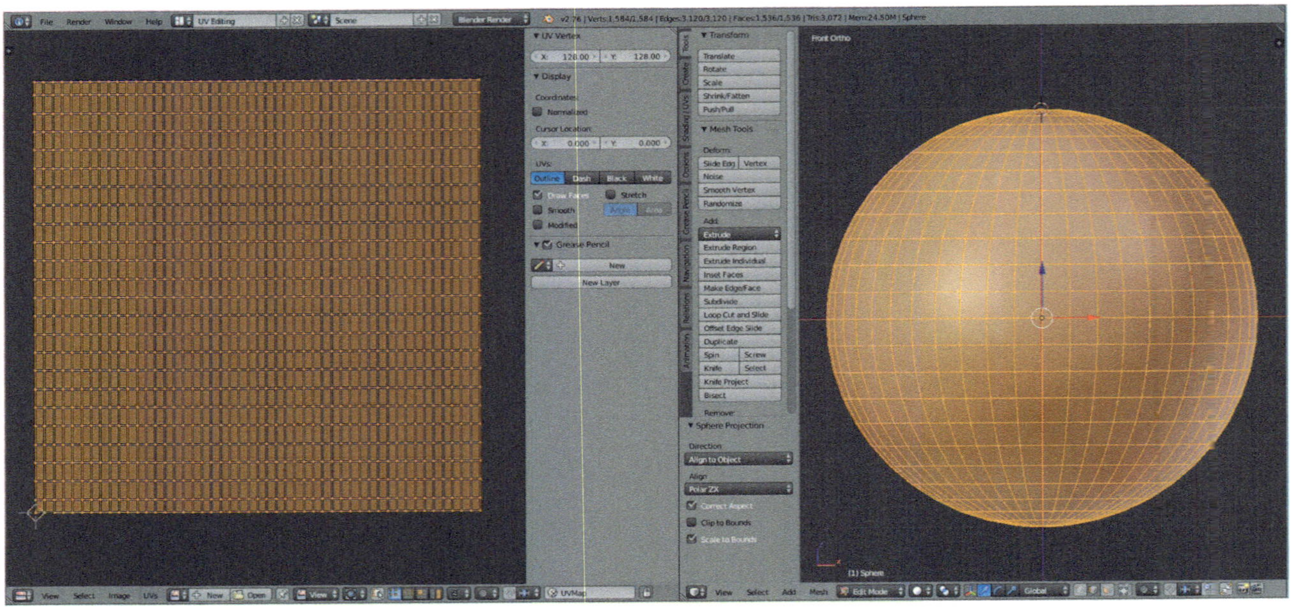

FIG. 8.22 A geometric model for a virtual globe: UV map of a sphere (*left*) and 3D model of a sphere (*right*). *(From Florinsky, I.V., Filippov, S.V., 2017. A desktop system of virtual morphometric globes for Mars and the Moon. Planet. Space Sci. 137, 32–39, Fig. 2, reproduced with permission from Elsevier.)*

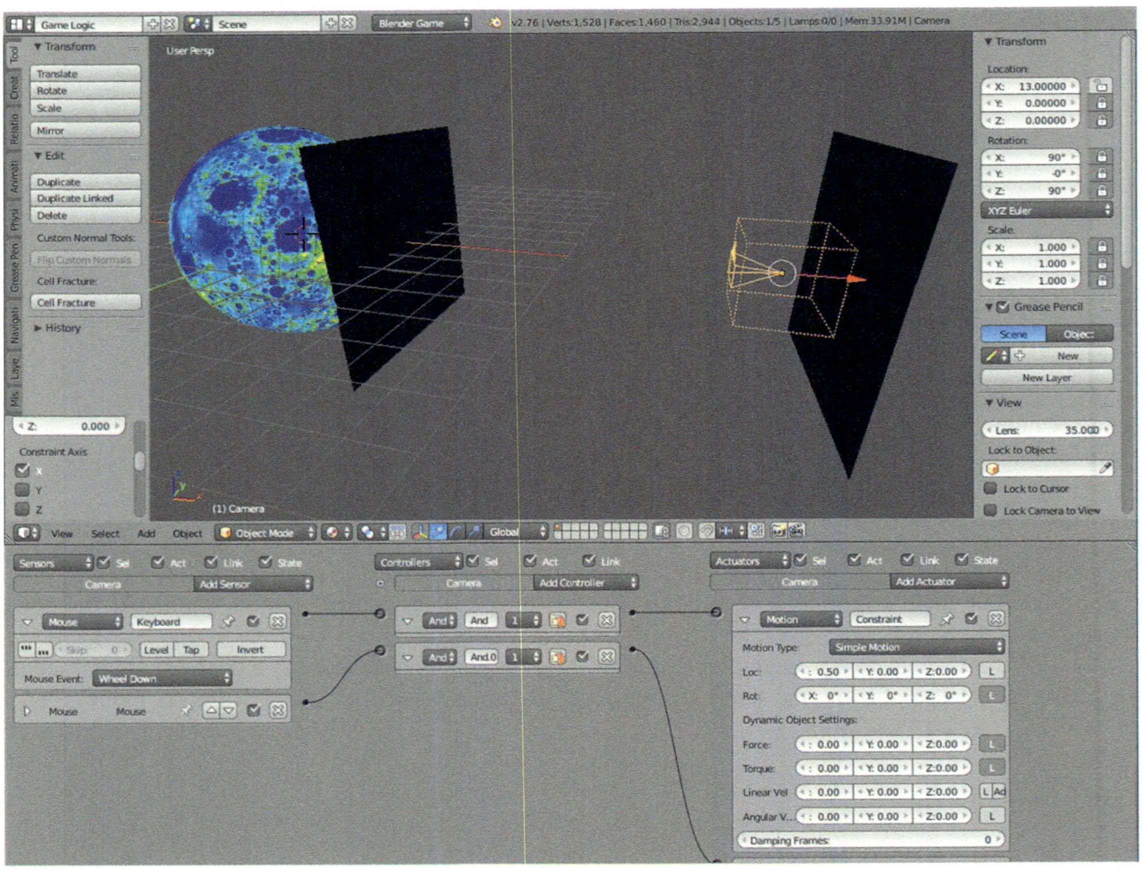

FIG. 8.23 The general view of a scene (*upper*) and the logic of zooming (*lower*). *(From Florinsky, I.V., Filippov, S.V., 2017. A desktop system of virtual morphometric globes for Mars and the Moon. Planet. Space Sci. 137, 32–39, Fig. 3, reproduced with permission from Elsevier.)*

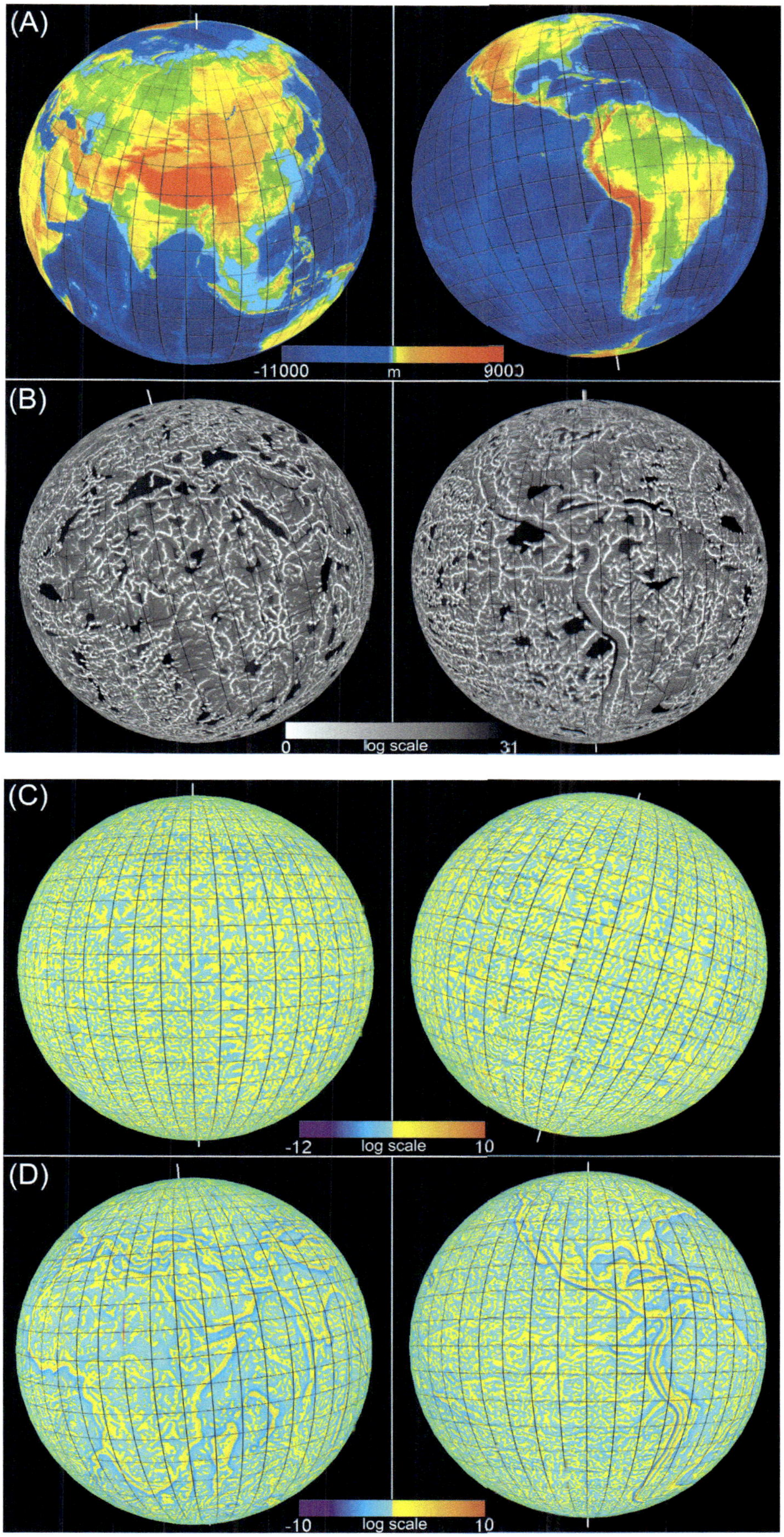

FIG. 8.24 The Earth, desktop virtual morphometric globes: (A) Elevation. (B) Catchment area. (C) Horizontal curvature. (D) Vertical curvature.

(Continued)

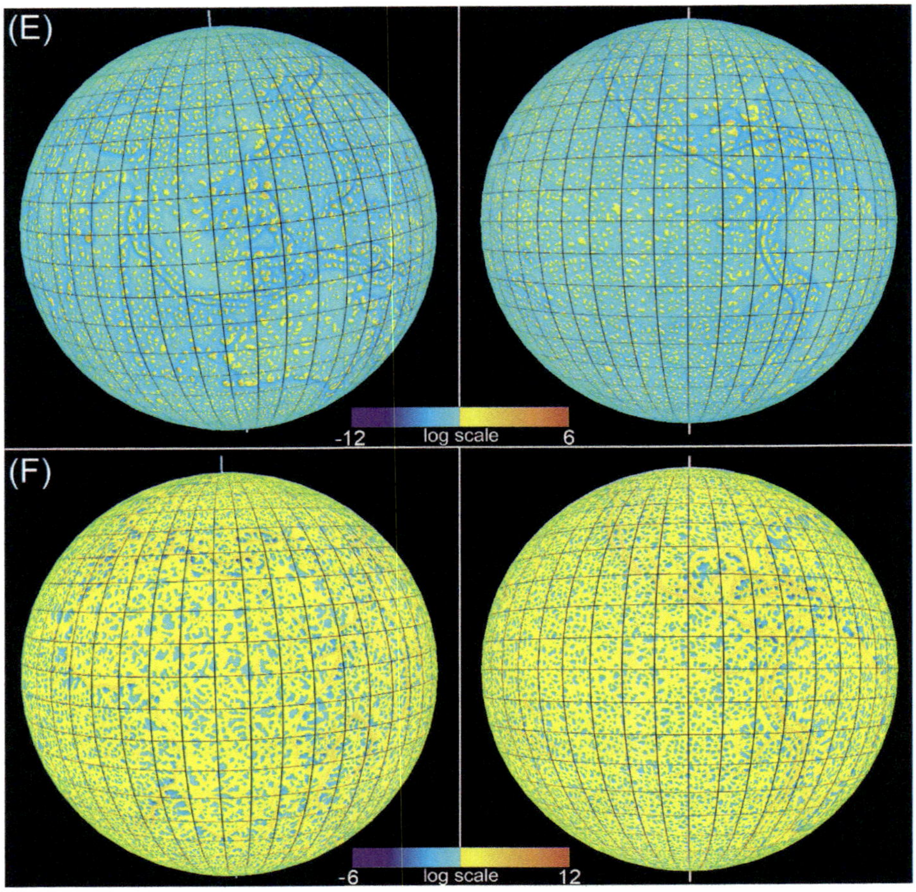

FIG. 8.24, CONT'D (E) Minimal curvature. (F) Maximal curvature. The globe screenshots (*left* and *right*) are made for two same positions of the six globes relative to the observer. *(From Florinsky, I.V., Filippov, S.V., 2015. Development of virtual morphometric globes using Blender. arXiv Preprint 1512.08511, Fig. 5.)*

consisted of (1) a client front-end running in a browser and providing a user with an interface and visualization of a globe and morphometric parameters; and (2) a server back-end supporting a client and feeding it with all required data

The application back-end represented a layered structure consisting of data storage, web services, and hypertext transfer protocol (HTTP) server layers.

The data storage layer was implemented as a structured query language (SQL) database. The type of data, split into many small-keyed tiles, makes the use of data tables a viable and fit solution. We chose a relational database management system to ensure fast-key search and effective thread utilization and reuse.

For the web service layer, we used a previously developed application server (Garov et al., 2016), providing an API similar to the OpenGIS Web Map Tile Service standard (OGC, 2010). The application server extracts data tiles from the database and sends them to a client as responses to the client's HTTP *post* or *get* requests.

The HTTP server was used for the static site hosting. No site templating or dynamic server-side site rendering was used. Session state was not used at the server; we opted for a fully stateless approach.

For the front-end implementation, we used an approach utilizing a graphic API in which GPU helps to build 3D scenes (i.e., a globe surface for visualization) considering an observer position, dynamically transforming via OpenGL Shading Language (GLSL) (Khronos Group, 2008–2021) shaders, and mutually adjusting a set of standard quad geometries into sphere segments. This set is re-used, for example, when changing the observer position.

Models of morphometric variables were presented in the form of data layers with the possibility of separate as well as joint visualization. When displayed, the layer data were aligned with a correspondent sphere segment and loaded into the GPU memory. We implemented scaling with automatic adjustment of level of detail (LOD).

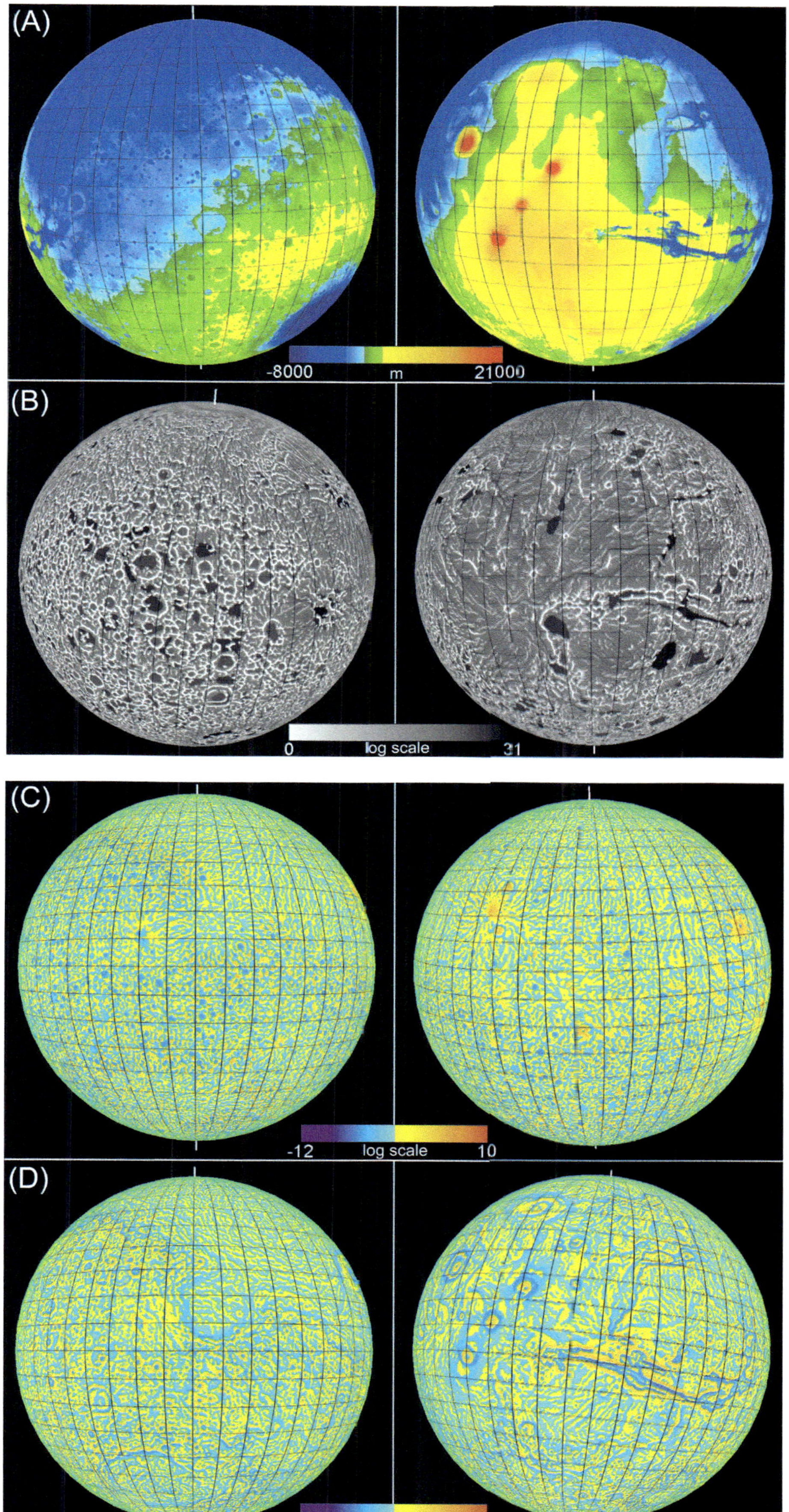

FIG. 8.25 Mars, desktop virtual morphometric globes: (A) Elevation. (B) Catchment area. (C) Horizontal curvature. (D) Vertical curvature.

(Continued)

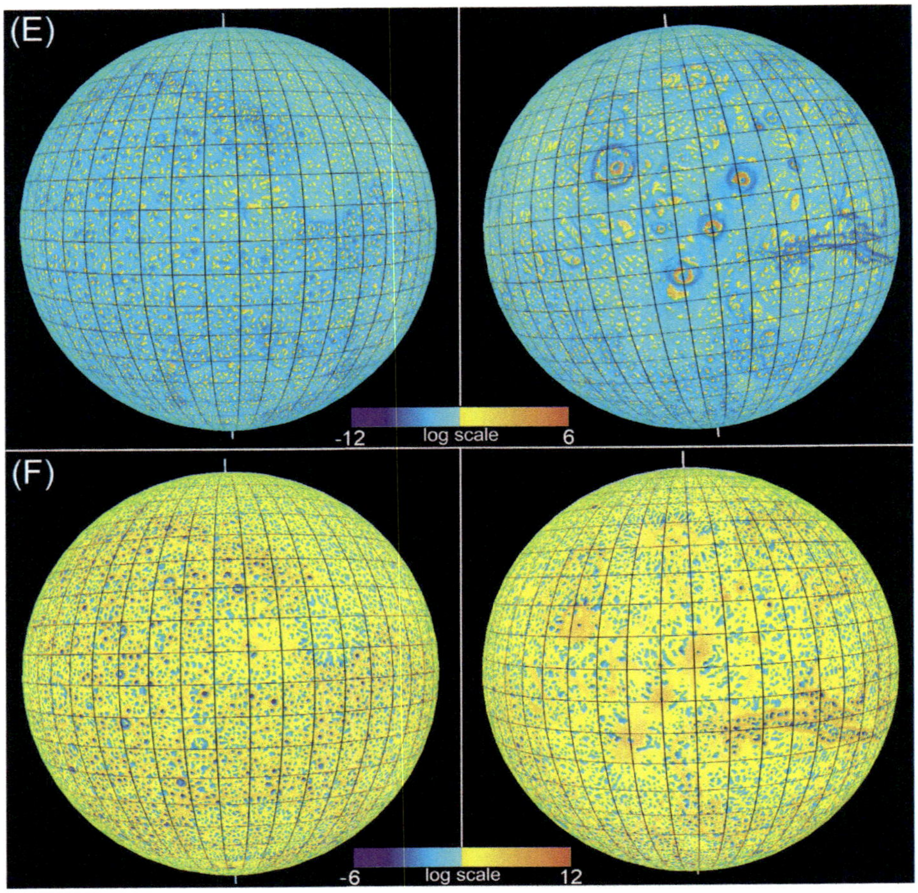

FIG. 8.25, CONT'D (E) Minimal curvature. (F) Maximal curvature. The globe screenshots (*left* and *right*) are made for two same positions of the six globes relative to the observer. *(From Florinsky, I.V., Filippov, S.V., 2015. Development of virtual morphometric globes using Blender. arXiv Preprint 1512.08511, Fig. 6.)*

To provide performance of 3D visualization, we selected a solution with cross-compilation. The code responsible for 3D rendering—fast extensible graphics "engine"—was written in C++ language and then cross-compiled to a compact static code form, which can be run by a browser.

The client gets morphometric data from the server in the form of tiles. Rendering of a morphometric model is performed directly in a browser using one of the color-changeable legends.

Figs. 8.27 and 8.28 show screenshotted examples of virtual morphometric globes for Mars and the Moon, respectively. In both cases, the screenshots were made for the same position of the six globes relative to the observer. One can compare these globes with their desktop analogs (Section 8.6.1).

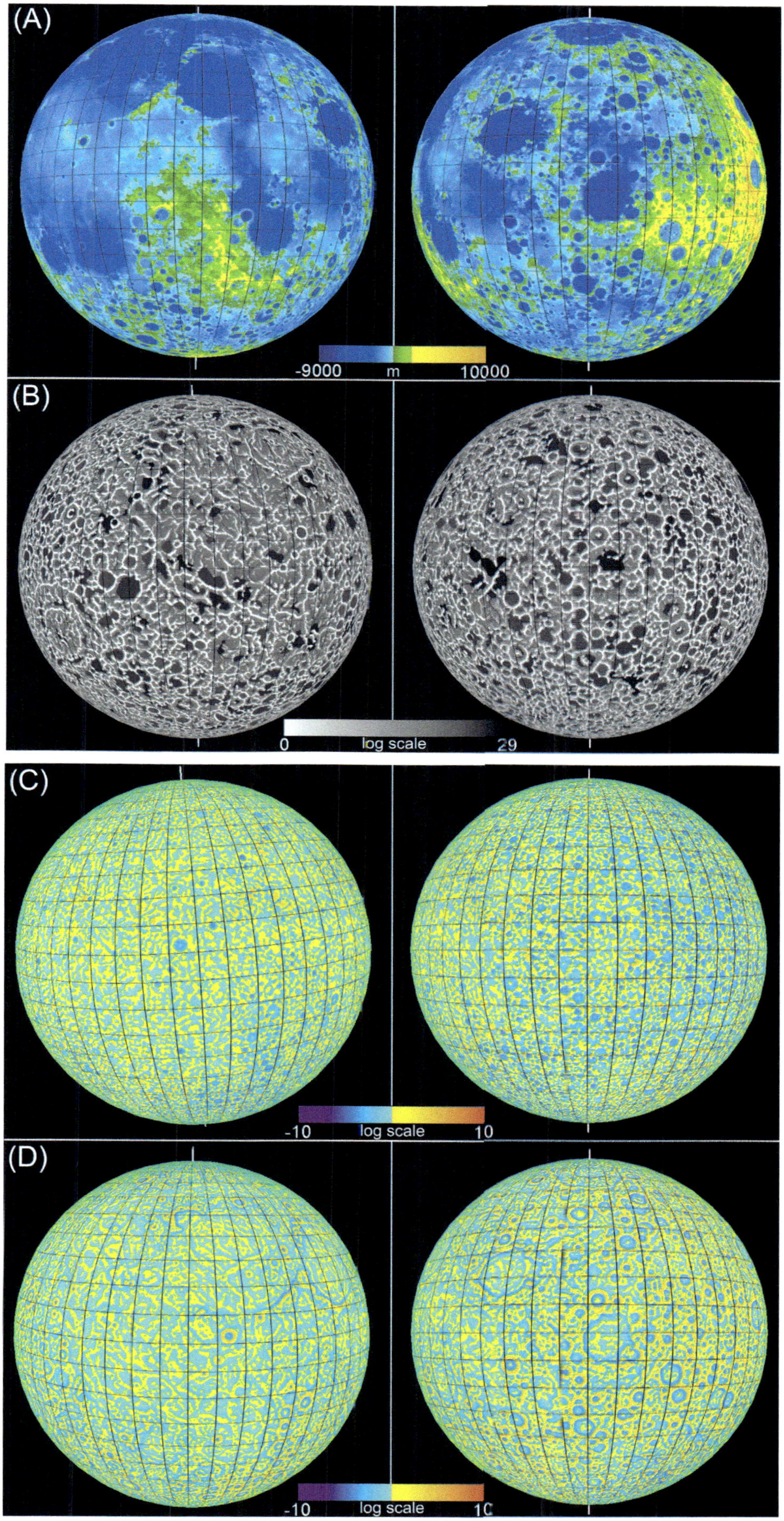

FIG. 8.26 The Moon, desktop virtual morphometric globes: (A) Elevation. (B) Catchment area. (C) Horizontal curvature. (D) Vertical curvature. (E) Minimal curvature.

(Continued)

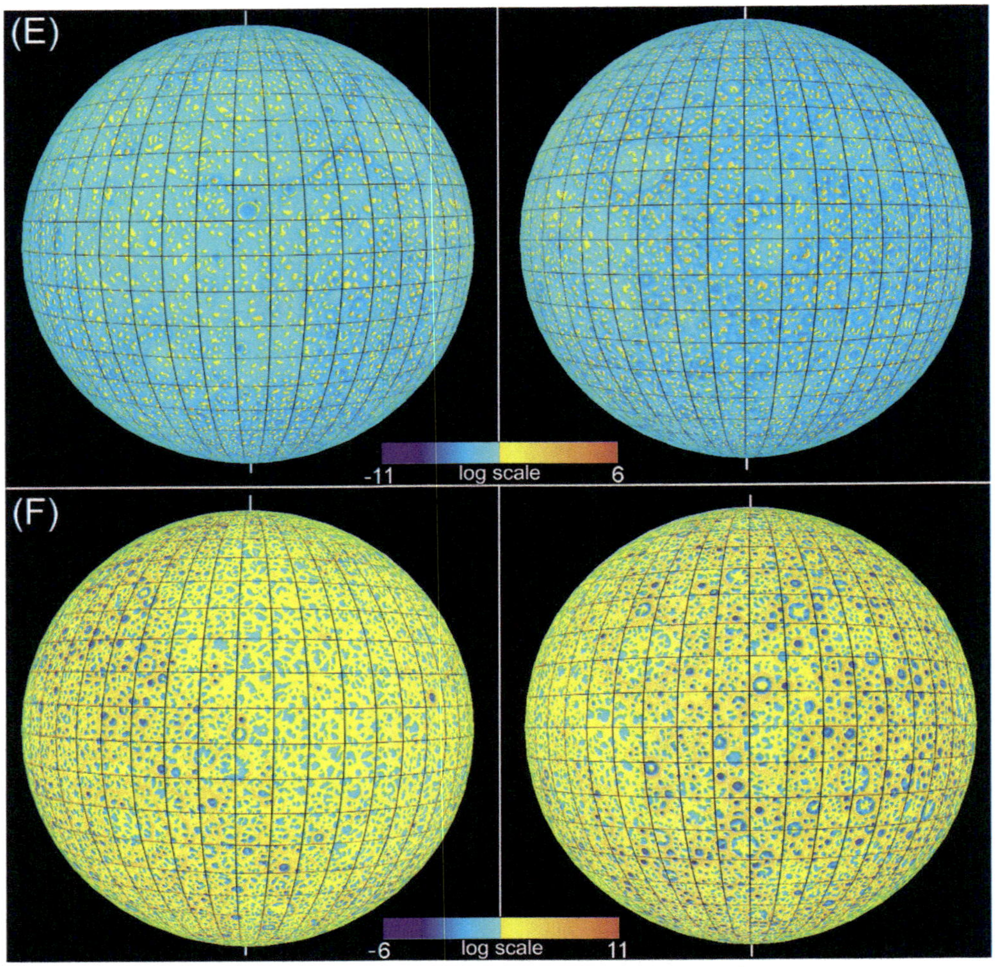

FIG. 8.26, CONT'D (F) Maximal curvature. The globe screenshots (*left* and *right*) are made for two same positions of the six globes relative to the observer. *(From Florinsky, I.V., Filippov, S.V., 2015. Development of virtual morphometric globes using Blender. arXiv Preprint 1512.08511. Fig. 7.)*

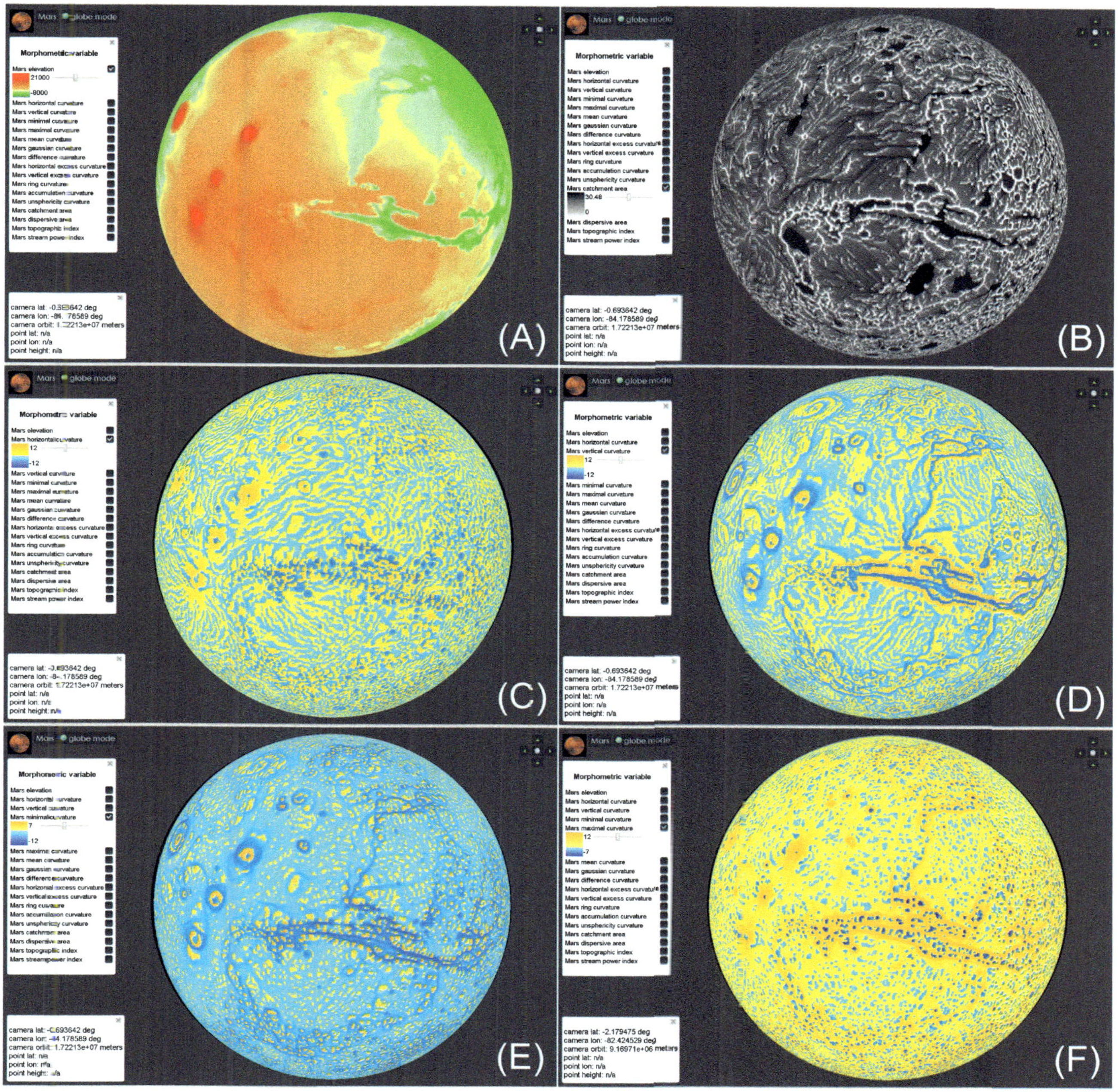

FIG. 8.27 Mars, web virtual morphometric globes: (A) Elevation. (B) Catchment area. (C) Horizontal curvature. (D) Vertical curvature. (E) Minimal curvature. (F) Maximal curvature. *(From Florinsky, I.V., Garov, A.S., Karachevtseva, I.P., 2018. A web system of virtual morphometric globes for Mars and the Moon. Planet. Space Sci. 159, 105–109, Fig. 1, reproduced with permission from Elsevier.)*

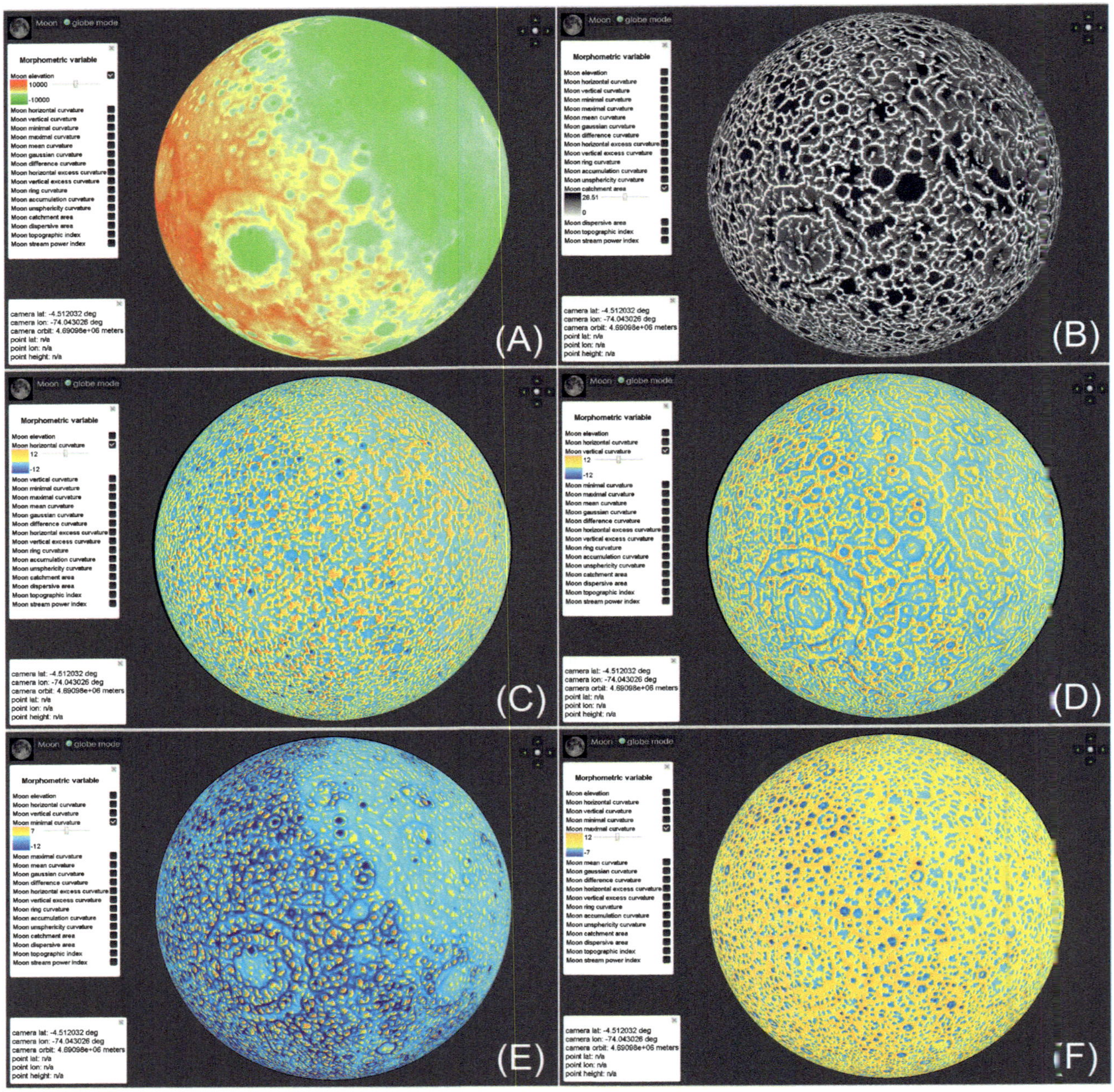

FIG. 8.28 The Moon, web virtual morphometric globes: (A) Elevation. (B) Catchment area. (C) Horizontal curvature. (D) Vertical curvature. (E) Minimal curvature. (F) Maximal curvature. *(From Florinsky, I.V., Garov, A.S., Karachevtseva, I.P., 2018. A web system of virtual morphometric globes for Mars and the Moon. Planet. Space Sci. 159, 105–109, Fig. 2, reproduced with permission from Elsevier.)*

Part II

Digital terrain modeling in soil science

Chapter 9

Influence of topography on soil properties

9.1 Introduction

It is well known that topography is one of the soil-forming factors (Dokuchaev, 1883, 1891; Sibirtsev, 1899; Vysotsky, 1906; Zakharov, 1911, 1913; Neustruev, 1915, 1927, 1930; Jenny, 1941; Huggett, 1975; Fridland, 1976; Gerrard, 1981; Schaetzl and Anderson, 2005). Topography influences (micro)climatic and meteorological characteristics, which affect hydrological and temperature regimes of soils (Neustruev, 1927, 1930; Geiger, 1927; Romanova, 1977; Kondratyev et al., 1978; Raupach and Finnigan, 1997; Böhner and Antonić, 2009; Emeis and Knoche, 2009), the prerequisites of the gravity-driven lateral overland and intrasoil transport of water and other substances (Kirkby and Chorley, 1967; Young, 1972; Speight, 1980), as well as the spatial distribution of the vegetation cover (Yaroshenko, 1961; Franklin, 1995; Florinsky and Kuryakova, 1996). Thus, it is natural that topography directly or indirectly controls the spatial distribution of physical, chemical, and biological soil properties (Moore et al., 1991; Florinsky et al., 2002; Shary et al., 2002b; Schaetzl and Anderson, 2005, chap. 13). An in-depth understanding of this control is required for further modeling and mapping of soil properties based on topographic data (Chapter 11).

Topography influences soil properties through two key "tools":

1. The gravity-driven lateral migration and accumulation of water.
2. Spatial differentiation of the temperature regime of slopes.

Vysotsky (1906, pp. 3–4) noted:

> *"Water for soil is like blood for a living organism. It dissolves and transports various more or less soluble and organic substances, which are created by the weathering of parent rocks or decomposition of organic residuals, or moved through the atmosphere as dust and precipitation. But apart from such an internal action, water is an important external actor conducting (in known cases, together with wind action) a lot of work, such as erosion, denudation, and accumulation as well as separation of soil matter .... The direction of such actions depends, first and foremost, on topography ... because its forms influence not only overland water circulation but also intrasoil one."*

Legates et al. (2011, p. 65) stressed the importance of soil moisture:

> *"Soil moisture is a critical component of the earth system and plays an integrative role among the various subfields of physical geography ... Soil moisture affects atmospheric, geomorphic, hydrologic, and biologic processes ... it lies at the intersection of these areas of scientific inquiry. Soil moisture impacts earth surface processes in such a way that it creates an obvious synergistic relationship among the various subfields of physical geography. The dispersive and cohesive properties of soil moisture also make it an important variable in regional and microclimatic analyses, landscape denudation and change through weathering, runoff generation and partitioning, mass wasting, and sediment transport."*

Thus, it is reasonable to use the example of soil moisture to discuss the principal aspects of the topographic influence on soil properties.

The following quantitative topographic characteristics are responsible for the spatial distribution and redistribution of water in the landscape: slope gradient, slope aspect, horizontal, vertical, and mean curvatures (local morphometric variables), as well as catchment area (the nonlocal topographic attribute) (Table 2.1). We will discuss their role in detail.

9.2 Local morphometric variables and soil

Slope gradient (Section 2.2.2) controls soil moisture content as follows: as G increases, the slope area and velocity of water flow increase, so the precipitation received per unit area and its infiltration decrease, the runoff and evaporation area increase, and hence soil moisture content decreases (Zakharov, 1913, 1940). This leads to the usual negative correlations between the soil moisture content and G.

Digital Terrain Analysis. https://doi.org/10.1016/B978-0-443-24798-9.00021-2

Slope aspect (Section 2.2.3) influences the soil-water balance since A, in association with G, affects insolation (Kondratyev et al., 1978) and evapotranspiration (Romanova, 1977). It is well known that in the northern hemisphere, soil moisture content tends to be the highest on north slopes, intermediate on west and east slopes, and least on south slopes (Sibirtsev, 1899; Zakharov, 1913; Ponagaibo, 1915; Neustruev, 1927, 1930). Also, A affects soil moisture content by controlling the impact of neighboring geographical objects (i.e., mountains, seas, lakes, and deserts), which determine the character and direction of atmospheric flows (Neustruev, 1915, 1930; Zakharov, 1940).

Slope gradient and aspect affect the redistribution of snow over the land surface. Thus, these morphometric variables influence the spatial differentiation and dynamics of freezing and melting of soils and, in turn, the spatial differentiation of soil-water storage (Taychinov and Fayzullin, 1958; Kuz'min, 1963; Lapen and Martz, 1996; DeWalle and Rango, 2008).

Horizontal and vertical curvatures (Sections 2.2.4.2 and 2.2.4.3) are the key topographic factors determining overland and intrasoil water dynamics (Kirkby and Chorley, 1967), being the measures of flow convergence/divergence and flow deceleration/acceleration, respectively. Lateral intrasoil flow of the saturation zone and soil moisture content increase when flows converge (k_h takes negative values) and decrease when flows diverge (k_h takes positive values) (Kirkby and Chorley, 1967; Carson and Kirkby, 1972). Moreover, horizontal curvature influences hydrological processes in unsaturated soils: infiltration flux diverges when $k_h > 0$ and converges when $k_h < 0$ (Zaslavsky and Rogowski, 1969). It was found experimentally that the dynamics of lateral flows in the saturation zone and soil moisture content depend essentially on horizontal curvature (Anderson and Burt, 1978). This topographic attribute can play a key role in the formation of saturation zones: they are the most stable in areas of flow convergence (O'Loughlin, 1981). Soil moisture content also increases when flows decelerate (k_v takes negative values) and decreases when flows accelerate (k_v takes positive values) (Kirkby and Chorley, 1967; Kuryakova et al., 1992). These facts explain the usual negative correlations of the soil moisture content with k_h and k_v.

For the arid climatic conditions and relatively flat topography of a southern desert in Israel, Sinai et al. (1981) observed a high correlation (−0.9) between soil moisture content of the root zone with mean curvature approximated by the Laplacian (Section 2.2.4.13). Such a relation results from the microtopographic control of the lateral intrasoil water transport rather than from the redistribution of overland water flows (they are not typical for that landscape). A strong dependence of soil moisture on mean curvature was also found in Russia's southern Moscow Region, which has a contrasting topography and is located in a moderate continental climate zone (Kuryakova et al., 1992) (Section 10.3).

Saturation zones are often correlated with landforms for which both horizontal and vertical curvatures are negative (Feranec et al., 1991). These are relative accumulation zones, where both flow convergence and relative deceleration of flows act together (for details, see Section 2.7.2). There is landsliding (Lanyon and Hall, 1983), soil gleying, maximum thickness of the A horizon, and maximum depth of calcium carbonate (Pennock et al., 1987) in these zones due to increased water content in soils and grounds.

9.3 Nonlocal morphometric variables and soil

Relationships between soil moisture content and a relative slope position (i.e., upslope, midslope, and downslope) were qualitatively understandable in the early 20th century (Zakharov, 1913). Quantitatively, the dependence of soil moisture content on catchment area (which in fact describes the relative position of a point on the topographic surface—Section 2.3) was probably first described by Zakharov (1940, p. 384) as follows: "water amount per unit area increases from upslope to downslope due to additional water supply." Thus, as CA increases, soil moisture content also increases. This explains the usual positive correlations between soil moisture content and CA. Speight (1980) argued that catchment area rather than horizontal curvature is of first importance for the soil moisture dynamics because catchment area considers the relative position of a point in the landscape.

The topographic wetness index (Section 2.6), combining the metrics of slope gradient with catchment area, can further improve the description of morphometric prerequisites for the spatial distribution of soil moisture (Moore et al., 1986). This is because TWI takes into account both the local geometry of a slope and the relative location of the point in the landscape. As CA increases and G decreases, TWI and soil moisture content increase. This resulted in higher absolute values of correlation coefficients of soil moisture content with TWI than with CA or G (Thompson and Moore, 1996).

However, TWI and horizontal curvature cannot individually offer the prospect of predicting soil moisture dynamics. Saturation zone depth may have higher correlations with some other empirically determined variables, such as a product of horizontal curvature and catchment area (Burt and Butcher, 1985). This generates a need for the use of a representative set of morphometric attributes in soil studies (Section 11.4).

9.4 Discussion

It is obvious that soil moisture content depends not only on topography but also on some physical and hydraulic characteristics of soils, such as soil texture and soil-water retention (Lin, 2012). However, the spatial distribution of these parameters also depends on morphometric variables (Moore et al., 1993; Pachepsky et al., 2001) because they are, in one way or another, controlled by the intensity and direction of gravity-driven overland and intrasoil transport of substances.

Soils and sediments on dense clay or rock formations are affected not only by surface flows of water and dissolved or suspended substances but also by intrasoil flows along the bedrock surface. To study and model the properties of such soils, it is necessary to analyze the microtopography of both the land surface and the top surface of an underlying clay or rock horizon (Florinsky and Arlashina, 1998; Freer et al., 2002; Chaplot and Walter, 2003; Chaplot et al., 2004; Kim, 2009; Carless et al., 2021) (Fig. 9.1).

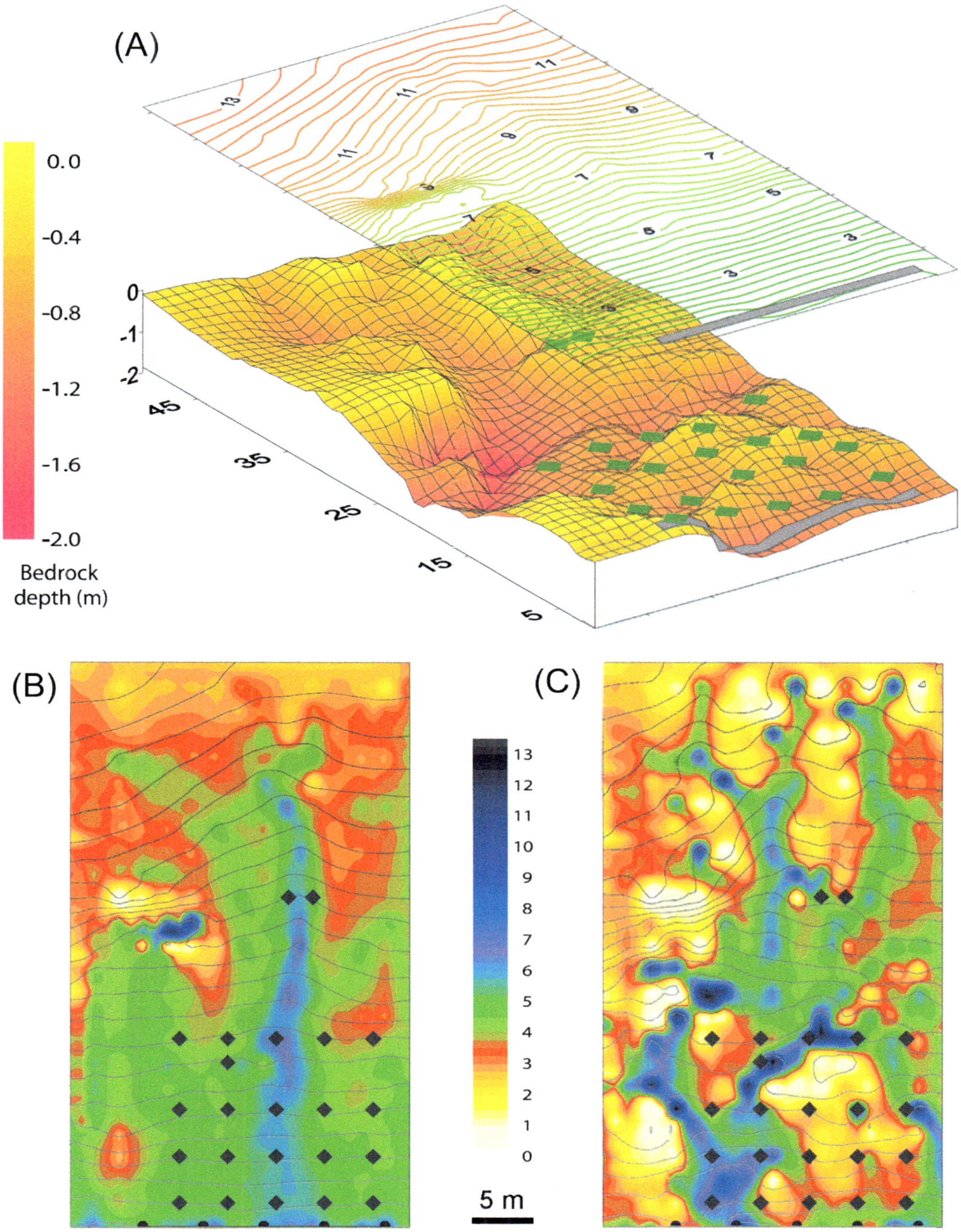

FIG. 9.1 Topographic indices derived for the ground and bedrock surfaces in soil-hydrological modeling: (A) Elevation of the ground surface (contours) and depth to the bedrock surface (3D model). (B) Topographic wetness index of the ground surface. (C) Topographic wetness index of the bedrock surface. *Diamonds*, tensiometer positions. *(From Freer, J., McDonnell, J.J., Beven, K.J., Peters, N.E., Burns, D.A., Hooper, R.P. et al., 2002. The role of bedrock topography on subsurface storm flow. Water Resour. Res. 38 (12), 1269, Fig. 2, © American Geophysical Union, reproduced with permission from John Wiley & Sons.)*

Finally, we should note that the influence of topography on soil properties depends on the management or tillage practice; for instance, zero tillage versus conventional tillage (Farenhorst et al., 2003; Senthilkumar et al., 2009). Most of the works dealing with relationships between topography and soil properties in agricultural landscapes have been conducted in Canada, the United States, and Australia on fields tilled over a 50–150-year period without dramatic modifications of the land surface and soil cover. This may be one of the reasons why high correlations have been systematically observed for the topography-soil system in agrolandscapes. A strong, long-term (several hundred years) agricultural load can seriously reduce the topographic control of soil properties (Venteris et al., 2004; Samsonova et al., 2007).

Chapter 10

Adequate resolution of models

10.1 Motivation

In landscape studies, the following question usually arises: What density of sampling points should be used to adequately depict the spatial distribution of a landscape property concerned for a given scale, measurement accuracy, and minimum number of samples? (Lidov, 1949; Mueller-Dombois and Ellenberg, 1974; Campbell, 1979; Kershaw and Looney, 1985; Burrough, 1993). If one carries out a study with a regular grid of sampling points, the problem reduces to the determination of grid spacing. The success of a study depends on the correct solution to this problem.

At first sight, the problem would seem to be a technical task. However, this is the fundamental problem because it is connected with determining the spatial scale of an object, phenomenon, or process under study. This is a critical question because different physical laws and landscape processes dominate at different spatial scales. Extension of one or other concept or model to all scales can result in an invalid description of actual relationships (Haggett et al., 1965; Schumm and Lichty, 1965; Klemeš, 1983; Phillips, 1988; De Boer, 1992; Band and Moore, 1995; Bierkens et al., 2000; Pachepsky and Hill, 2017).

An adequate description of a landscape property with a minimum number of samples implies that a grid spacing corresponds to an area wherein property values vary smoothly or are assumed to have a constant value for a given scale and measurement accuracy. Different sciences use different terms for such an area. In this book, an area and a grid spacing are denoted as *an adequate area* of a landscape property and *an adequate grid spacing*, respectively (see details in Section 10.2).

In DTM-based geomorphic and geological studies, one can determine an adequate grid spacing (w) of a DTM from a typical size S of topographic features, landforms, or geological structures concerned. According to the sampling theorem (Section 3.3), the adequate w cannot be more than $S/2$. In studies of other landscape components (e.g., soils and plants) as well as landscape processes (e.g., lateral intrasoil transport of substances), determination of the adequate w is less trivial. This is due to the high spatial variability of landscape properties (Campbell, 1979; Kershaw and Looney, 1985; Oliver and Webster, 1986; Trangmar et al., 1987). In addition, a landscape property can have several adequate areas associated with different natural processes (Sitnikov, 1978, 1980; Kershaw and Looney, 1985) (Section 10.2). There are some closely related methods for determining adequate areas of a landscape property. In these methods, an indicator of an adequate area is a smooth portion of a graph describing the dependence of a property or its statistical parameter on the area or grid spacing used to measure the property (Mead, 1974; Mueller-Dombois and Ellenberg, 1974; Kershaw and Looney, 1985; Oliver and Webster, 1986; Trangmar et al., 1987; Wood et al., 1988; Famiglietti and Wood, 1995).

The problem can be further complicated if one analyzes data on two or more landscape properties in combination or predicts an attribute through an analysis of other attributes. This is because different landscape properties may have different adequate areas (Trangmar et al., 1987; Phillips, 1988, 1995). Moreover, existing a priori relationships between two landscape properties (justified theoretically or observed at other places) can be manifested at only certain adequate areas of these properties (Phillips, 1988). For instance, it is well known that a water table may look like a generalized land surface. Such regularity may be observed by an analysis of water table depths and a DEM with w omitting minor landforms. However, one can find a low correlation of the water table with the topography using an excessively detailed or overgeneralized DEM (Thompson and Moore, 1996).

Thus, one may establish invalid statistical regularities (e.g., low correlation coefficients) as well as incorrect predictions and conclusions for a priori interrelated landscape properties using high-quality data. This problem is typical in the analysis of data sets that have different resolutions and incongruous grids (Lidov, 1949; Band and Moore, 1995). However, the problem cannot be solved merely by using data sets with the same resolutions and grids. The situation is complicated by the fact that adequate areas may differ from "basic" landscape units, such as stow and facies (Phillips, 1988).

Digital Terrain Analysis. https://doi.org/10.1016/B978-0-443-24798-9.00019-4

Determining grid spacing is always required in combined analyses of DTMs and data on other landscape components, as well as in DTM-based modeling of landscape properties. The grid spacing was and is commonly estimated by expert opinion (Anderson and Burt, 1980; Burt and Butcher, 1985; Moore et al., 1993; Quinn et al., 1995; Florinsky and Kuryakova, 1996), hence such a solution can be subjective. Arbitrary choice of w can lead to incorrect results and artifacts. For example, Speight (1980) did not find a relationship between soil moisture content and k_h because w was too small (Anderson and Burt, 1980). Sinai et al. (1981) did not observe a correlation between soil salinization and H because w was too large. In *TWI*-based hydrological modeling, the average depth of the water table decreases, and the overland runoff increases when w increases (Wolock and Price, 1994). For forest ecosystems of the British Columbia, the use of *TWI* calculated with $w = 4$ m resulted in the correct prediction of the water table depth, while the invalid prediction was obtained using $w = 16$ m (Thompson and Moore, 1996). Slopes of DTM-based hydrographs essentially depend on w (Zhang and Montgomery, 1994; Da Ros and Borga, 1997). DTMs with $w < 30$ m can provide the correct prediction of soil hydromorphy, while a coarser DEM resolution deteriorates the prediction (Chaplot et al., 2000). Other examples can be found elsewhere (Penížek et al., 2016; Mondal et al., 2017; Goyal et al., 2018; Jacobs et al., 2018; Chang et al., 2019; Guo et al., 2019; Maxwell and Warner, 2019; Riihimäki et al., 2021; Piedallu et al., 2022).

Therefore, a correct choice of w is one of the key problems of DTM-based soil studies. This issue has become even more relevant due to the widespread use of high-resolution, UAS- and lidar-based DEMs (Sections 3.1.4 and 3.1.5). Many researchers use more and more detailed DTMs, ignoring the fact that a high resolution does not necessarily correspond to an adequate grid spacing, and it certainly does not determine the accuracy of DTM-based soil prediction (Cavazzi et al., 2013; Pradervand et al., 2014; Samuel-Rosa et al., 2015; Han et al., 2018; Larson et al., 2022).

In this chapter, we describe the method for determining the adequate grid spacing of DTMs used to analyze and model soil properties (Florinsky and Kuryakova, 2000). The method is exemplified by studying relationships between topography and soil moisture.

10.2 Theory

The method is largely based on the concept of a representative elementary volume used in mass transfer modeling, in particular, in hydrodynamics (Bear, 1972, chap. 1; Bear and Braester, 1972; Sitnikov, 1978, chap. 2, 1980). Some principles of this concept can be applied to solve a wide range of two-dimensional problems in geosciences if one switches from a representative elementary volume to a representative elementary area (Wood et al., 1988; Famiglietti and Wood, 1995).

Assume that a response presented in Fig. 10.1A describes the variation in the value of some landscape property (e.g., a rock property) with the volume or area for which the property is observed (Sitnikov, 1978, § 2.1, 1980). A pronounced spatial heterogeneity (e.g., pore effects) can explain most of the variability of property values within the interval I. Property values vary smoothly within the interval II. Then another distinct spatial heterogeneity (e.g., macrocracks) leads to abrupt variations of property values within the interval III. Again, property values vary smoothly within the interval IV.

By elementary volume or area, we mean a minimum volume or area in which a property is independent of heterogeneities; that is, the property values vary smoothly or have low variability (Sitnikov, 1978, § 2.1, 1980). For instance, V_1 and V_3 are elementary volumes or areas for the given property (Fig. 10.1A). For a landscape property, one can therefore distinguish a set of elementary volumes or areas associated with several intervals of smooth relationships between property values and volume or area. Intervals of smooth variation of a landscape property can sometimes not be found due to peculiarities of natural processes or measurement errors.

An elementary volume or area can also depend on time due to temporal variability of landscape properties (Fig. 10.1B) (Sitnikov, 1978, § 2.1, 1980). In addition, the elementary volume or area of a property can depend on specific natural conditions. If a terrain includes heterogeneous patches, a landscape property can have different elementary volumes or areas at adjacent patches. For example, assume that a terrain consists of three heterogeneous patches A, B, and C. Assume also that the dependencies of property values on volume or area are smooth within intervals I, II, and III in patches A, B, and C, respectively (Fig. 10.1C). Therefore, a property has elementary volumes or areas V_2, V_1, and V_3 in patches A, B, and C, respectively. If one can find borders between the patches, it is desirable to study this property separately at each patch. However, one can observe interval IV, wherein property values have low variability at all three patches (Fig. 10.1C). The elementary volume or area of this interval is V_3. It can be used as a common elementary volume or area for the entire terrain (Sitnikov, 1978, § 2.1, 1980; Kershaw and Looney, 1985).

By adequate interval of volumes or areas, we mean an interval of volumes or areas wherein property values vary smoothly or are constant. A variability of values may be ignored if it is within the accuracy of measurement (Sitnikov, 1980). For example, intervals II and IV are adequate intervals of volumes or areas for the property discussed (Fig. 10.1A). It is mandatory to use adequate intervals of volumes or areas. Otherwise, one can obtain nonreproducible

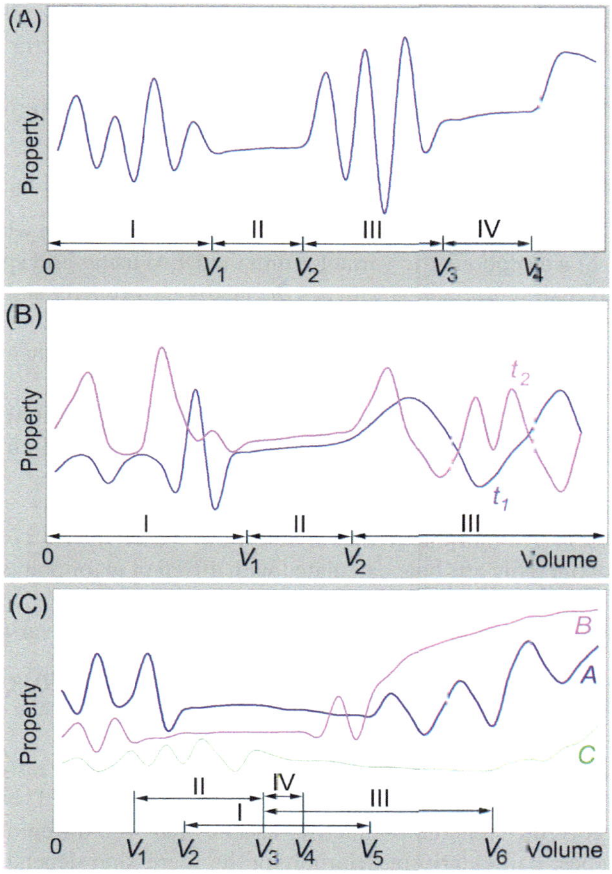

FIG. 10.1 Possible dependence of a landscape property on volume or area of observation (Sitnikov 1978, 1980): (A) Dependence on volume or area. (B) Dependence on volume or area at points in time t_1 and t_2. (C) Dependence on volume or area for heterogeneous patches *A*, *B*, and *C*.

and uninterpretable results due to the high and unpredictable variability of property values within inadequate intervals of volumes or areas.

By adequate volume or area, we mean a volume or area belonging to an adequate interval of volumes or areas (Sitnikov, 1978, § 2.1, 1980). By adequate grid spacing, we mean a grid spacing corresponding to an adequate area. According to the sampling theorem (Section 3.3), if s^2 is the adequate area of a landscape property, then $s/2$ is the adequate grid spacing, which relates to s^2 and provides an adequate description of the property.

An adequate grid spacing can be determined by (1) measuring a landscape property at grid nodes using different values of w and (2) plotting values measured against w (similarly to Fig. 10.1A). Smooth portions of this graph will indicate intervals of adequate w (Sitnikov, 1980).

In a combined analysis of two landscape properties and prediction of one property through analysis of another property, one should work with an adequate interval of areas common to two properties. Values of both properties are constant or vary smoothly within this interval, by definition of the adequate interval (see above). Therefore, correlation coefficients between values of two properties can also vary smoothly within the common adequate interval. At the same time, one can observe a high variability of correlation between two properties in adjacent inadequate intervals marked by a high variability of property values. So to determine a common adequate interval of areas and grid spacings providing the correct combined study of two properties, one should (1) analyze the correlation between these properties observed with different grid spacings and (2) plot correlation coefficients versus w (similarly to Fig. 10.1A). A smooth portion of this graph can indicate an interval of adequate grid spacings and, hence, an adequate interval of areas common to both properties.

By adequate w, we mean adequate grid spacing if a morphometric variable (described as a DTM) is one of two landscape properties under study. The adequate area corresponding to the adequate w determines the typical planar size of landforms, providing topographic control of the landscape property concerned. To define adequate w, one should carry out the following procedures:

- Derive a set of DTMs using a series of w.
- Perform a correlation analysis of data on a landscape property and a morphometric variable estimated with various values of w.
- Plot correlation coefficients between the landscape property and the morphometric variable versus w.
- Determine smooth portions of the graph indicating intervals of adequate w.

There are three main ways to implement the method described, depending on initial data types:

1. A DEM and data on a landscape property are obtained using the same square grids with a grid spacing w. One can then derive a set of digital models of a morphometric variable from the DEM using grid spacings w, $2w$, $3w$, …, nw; n is an integer. Then one should analyze the correlations between the landscape property and the morphometric variable using samples with these grid spacings.
2. A DEM and data on a landscape property are obtained using distinct grids and are then interpolated to a common square grid with a grid spacing w. Further steps are as in the first case.
3. A DEM and data on a landscape property are obtained using distinct grids, and spatial interpolation of data on the landscape property is undesirable or impossible (e.g., landscape data are collected along a transect or a contour, see Section 10.3.2). In this case, one should:
   a. Produce several DEMs by interpolation with different w.
   b. Derive a set of digital models of a morphometric variable from these DEMs.
   c. Interpolate values of the topographic attribute calculated with different w for points in which the landscape property was observed.
   d. Analyze correlations between the landscape property and the morphometric variable computed with different w.

To minimize interpolation errors, the first way is the best because an interpolation is not used. Next is the third way in which one should interpolate only DTMs (Section 10.3.2). The second way may result in the greatest number of interpolation errors.

Different morphometric variables can be connected with landscape processes of different scales (Anderson and Burt, 1980; Florinsky and Kuryakova, 1996). Thus, a landscape property can be "controlled" by different topographic attributes at different adequate areas. Therefore, different morphometric variables can offer different adequate w suited for each landscape property. In landscape studies, particularly in determining the regression dependence of a landscape property on morphometric variables, it is desirable to find the "main" interval of an adequate w common to all morphometric variables under consideration. It makes sense to perform a regression analysis only with an adequate w of this interval.

It is also obvious that one should only consider statistically significant correlations. Therefore, it is advisable to use a sample size of about 40 as well as to ignore smooth portions of correlation coefficient graphs if these portions include statistically insignificant coefficients.[a]

10.3 Field study

10.3.1 Study site

The study site is located in the center of the East European Plain, south of the Moscow Region, Russia, near the city of Pushchino (Fig. 10.2A). It is characterized by a moderate continental climate with warm summer and prolonged cold winter. Average temperatures in January and July are −10 and 18.6 °C, respectively. Precipitation is about 640 mm per year, 350–450 mm of which are rainfall.

The site is situated on the soddy, partially forested landslide slope of the Oka River valley. Elevations are about 130 m above sea level. The slope has a gradient of about 10° and a northerly aspect. Quaternary loams cover fractured and karstified Middle Carboniferous limestones, which lie at a depth of about 6 m and comprise the unconfined aquifer (Lyubashin and Lisitsin, 1981).

The study site includes a portion of the north-striking Severny Gully and adjacent crests. It measures about 58 × 77 m; the elevation difference is about 15 m (Fig. 10.2B). The soil complex includes gray forest soils at crests and slopes and meadow hydromorphic soils at the gully bottom. Vegetation cover consists of birches and herbs on crests, hazels and common horsetails on slopes, and nettles in the gully bottom.

a. It is commonly recommended to use a minimum sample size of 50–60. The sample size of 40 is chosen as a compromise: a collection of 50–60 soil samples may be difficult due to laboriousness and the cost of field and laboratory work. At the same time, correlation coefficients of 0.3 and higher are statistically significant for the sample size of 40. Notice that in geostatistical studies, a sample size should be >100 (Webster and Oliver, 1992).

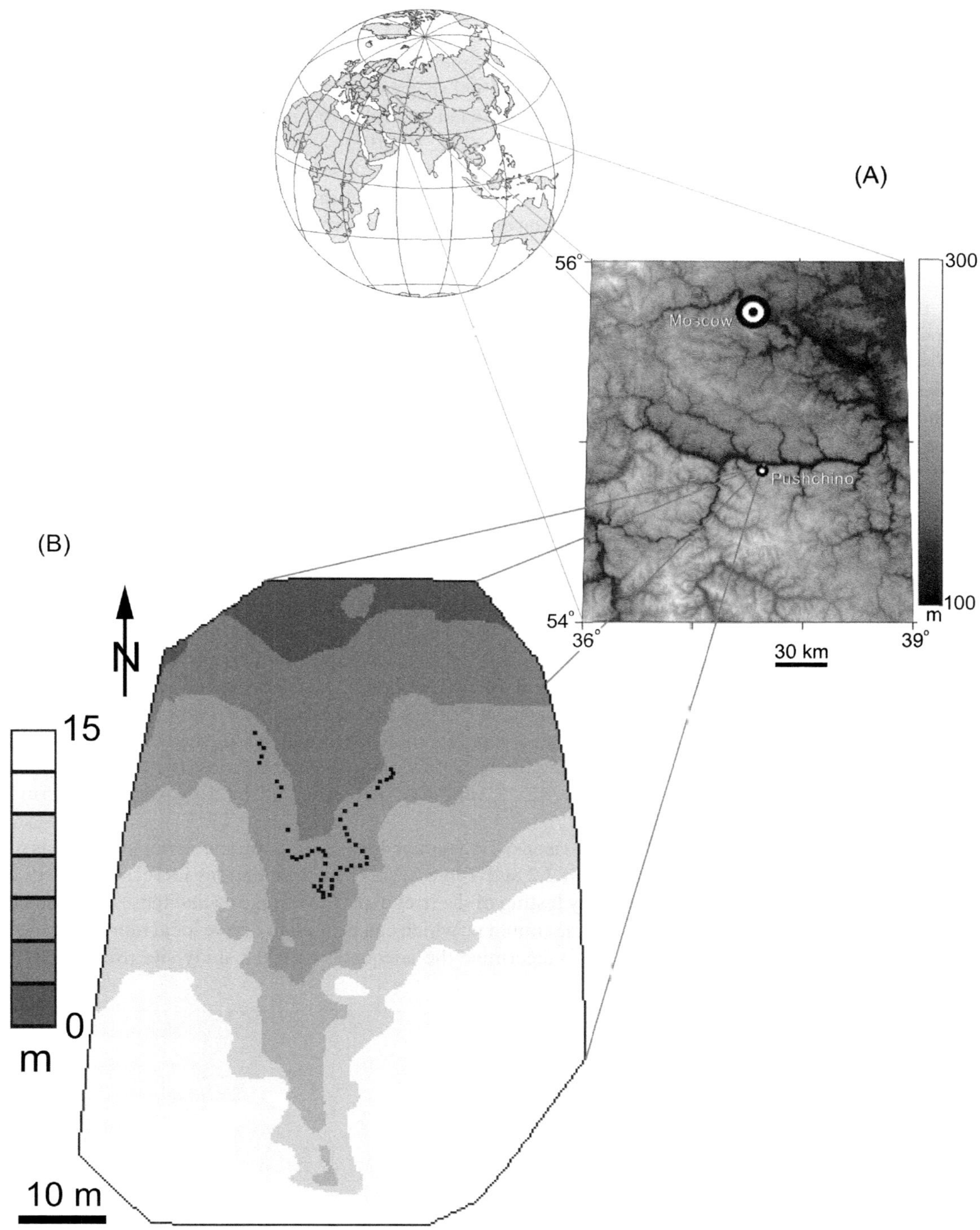

FIG. 10.2 The Severny Gully: (A) Geographical location (37°34′30″ N, 54°41′10″ E). on the background of the SRTM30_PLUS elevation map. (B) Elevations. *Dots*, locations of soil sampling sites along the 4.25-m contour. *(Modified from Florinsky I.V., 2010. Theory and Applications of Mathematical Cartographic Modeling of Topography (D.Sc. thesis). Institute of Mathematical Problems of Biology, Russian Academy of Sciences, Pushchino, Russia (in Russian), Fig. 3.3.) Publisher's note: Map lines delineate the study area and do not necessarily depict accepted national boundaries.*

10.3.2 Materials and methods

We obtained an irregular DEM of the study site through a tacheometric survey with an optical tacheometer TaN. The survey was performed by G.A. Kuryakova and the author in June 1990 (Kuryakova et al., 1992).

The irregular DEM consisted of 374 points. It was constructed in a relative Cartesian coordinate system. As the local datum, we used the minimum elevation value within the site (Fig. 10.2B).

To estimate the surficial soil moisture content (*Moist*), we carried out soil sampling on June 20, 1990. Precipitation was about 60 mm from 1 to 20 June 1990 in Pushchino and the suburb. There was drizzling rain (about 2 mm) on the eve of the soil sampling. To prevent a strong influence of evaporation on *Moist*, we performed the soil sampling in the morning, within 1 h, at an air temperature of 20–22 °C.

The soil sampling was performed at 62 points (included in the irregular DEM), which were located along the 4.25-m contour (Fig. 10.2B). These points were rather evenly distributed along the contour, with a distance ranging from 0.5 to 1.5 m. On the west side of the gully, three larger distances corresponded to two tree falls and a track. We chose the 4.25-m contour because (1) it passed over the main landforms within the site: two crests, slopes, and the gully bottom (Fig. 10.2B) and (2) it was suitable for the soil sampling because higher elevation slopes were too steep.

We took three soil samples at each of the 62 points at a depth of about 0.1 m. All samples had similar soil texture (loam). The soil sampling was conducted by G.A. Kuryakova, P.A. Shary, and the author (Kuryakova et al., 1992).

We evaluated *Moist* for each of these samples by weighing pre- and postdrying samples on an analytical damper balance ADV-200. The drying of samples was carried out by a drying box 2V-151 for 24 h at 105 °C (Carter, 1993). To reduce the influence of random deviations in *Moist* values, we used the average of *Moist* for the three samples collected at each of the 62 points as net values of *Moist* (Fig. 10.3). Laboratory analysis was carried out by the author.

We performed the soil sampling along one contour in order to eliminate from consideration the apparent influence of elevation on soil properties. Obviously, the actual influence of elevation on soil moisture observed in mountainous regions (due to altitudinal zonation) cannot occur within the study site. G, A, k_h, k_v, H, and CA are responsible for the physical mechanisms of distribution and redistribution of moisture in landscapes (Chapter 9). Elevation is not responsible by itself for the physical mechanisms of gravity-driven transport of soil moisture. Dependence of soil moisture on elevation observed in plain landscapes is in fact a result of the influence of CA on the spatial distribution of water in soils.

Using the Delaunay triangulation and a piecewise quadric polynomial interpolation (Agishtein and Migdal, 1991) of the irregular DEM, we produced 13 square-gridded DEMs with the following w (the numbers in parentheses are point numbers in the regular DEMs): 1 m (3312), 1.5 m (1421), 2 m (742), 2.5 m (461), 3 m (301), 3.5 m (202), 4 m (152), 4.5 m (110), 5 m (85), 5.5 m (63), 6 m (48), 6.5 m (35), and 7 m (31).

We decided on this range of w for the following reasons: To estimate local morphometric variables, one has to calculate the first and second derivatives of elevation (Section 2.2 and Table 2.2). These calculations increase errors in DEM generation and interpolation (Section 5.4.1). Preliminary testing of the irregular DEM suggests that these errors can be ignored for $w \geq 1$ m. Considering the site size, $w = 7$ m is the maximum w, which can be used to derive local morphometric variables. $\Delta w = 0.5$ m was of interest, and its use allowed us to determine the adequate w for the study site (Section 10.3.3).

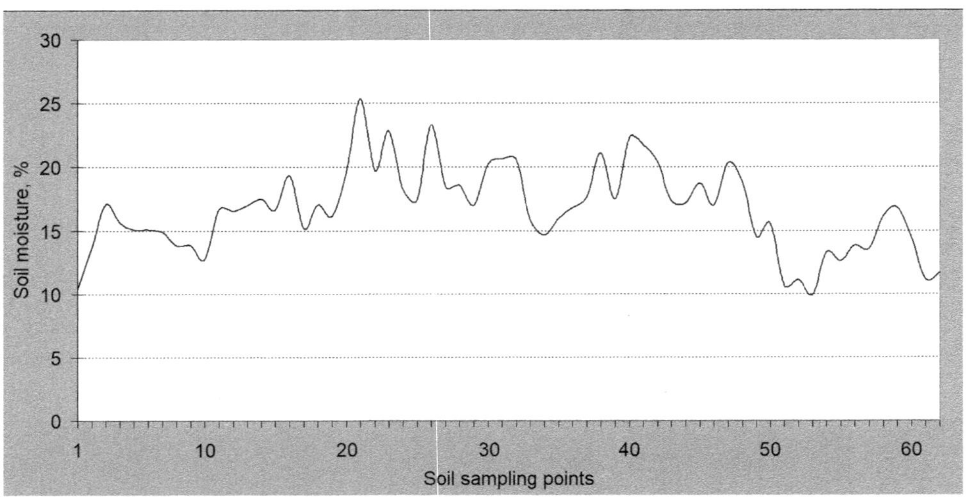

FIG. 10.3 Distribution of the soil moisture content along the 4.25-m contour.

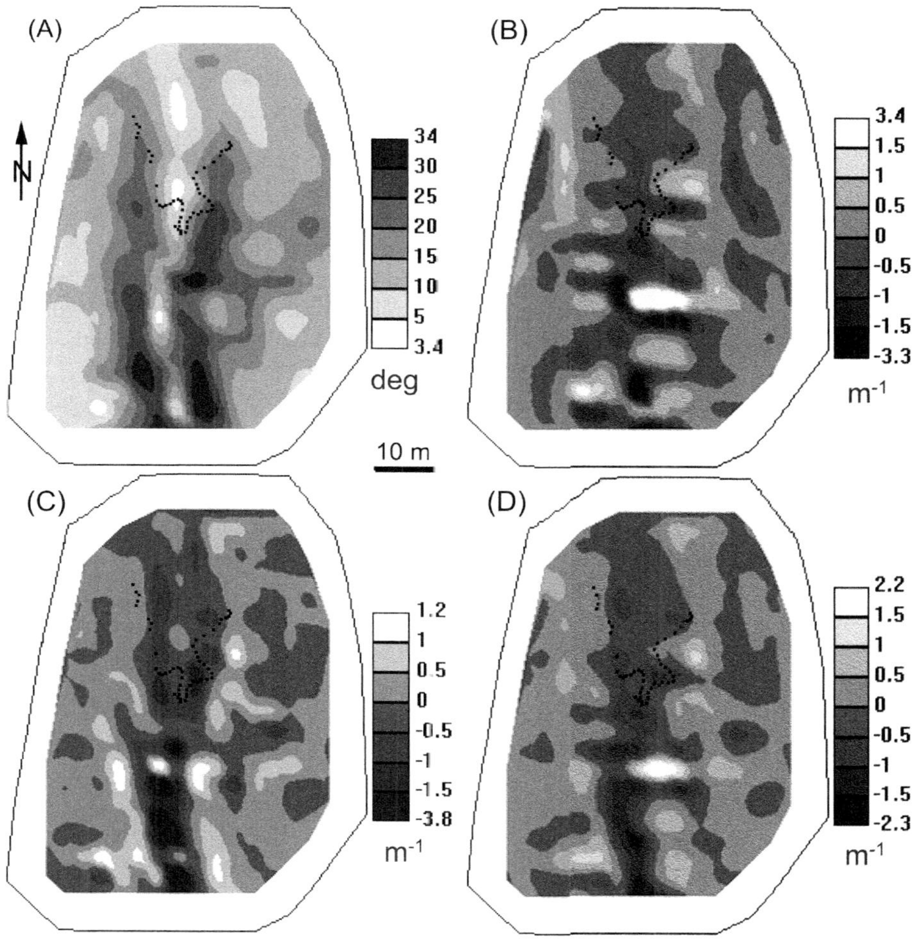

FIG. 10.4 The Severny Gully, morphometric variables derived with $w = 3$ m: (A) Slope gradient. (B) Horizontal curvature. (C) Vertical curvature. (D) Mean curvature. *Dots*, locations of soil sampling sites along the 4.25-m contour. *(From Florinsky, I.V., Kuryakova, G.A., 2000. Determination of grid size for digital terrain modelling in landscape investigations—exemplified by soil moisture distribution at a micro-scale. Int. J. Geogr. Inf. Sci. 14 (8) 815–832, Fig. 5, reproduced with permission from Taylor & Francis Ltd.)*

Digital models of G, k_h, k_v, and H (Fig. 10.4) were derived from all regular DEMs by the Evans-Young method (Section 4.1). Altogether, we produced 52 regular DTMs. Then we used the Delaunay triangulation and a piecewise quadric polynomial interpolation (Agishtein and Migdal, 1991) of these regular DTMs to determine the values of G, k_h, k_v, and H at the 62 sampling points.

To determine an adequate w and to estimate the dependence of *Moist* on morphometric variables, we carried out a linear multiple correlation analysis of *Moist* with G, k_h, k_v, and H calculated for the 13 values of w. Sixty-two-point samples were used for DTMs with $w = 1, \ldots, 5$ m; 59-point samples for DTMs with $w = 5.5$ and 6.5 m; 56-point samples for DTMs with $w = 6$ m; and 53-point samples for DTMs with $w = 7$ m.[b] Then we presented the dependence of correlation coefficients on w in the form of graphs for G, k_h, k_v, and H. To describe the effect of topography on the distribution of *Moist*, the "best" combination of topographic variables was chosen by the stepwise linear regression (Kleinbaum et al., 2014). We used the 62-point samples corresponding to two adequate w determined for regression analysis.

We did not study the influence of A on *Moist* because the study site is located (1) on the slope with a uniform northerly aspect and (2) in the forest where sunlight is diffused by trees. Therefore, we can consider A as a background at the site, although the dependence of soil moisture on A is well known (Section 9.2). We did not study the influence of CA and TWI on

b. The gradual increase of w led to the gradual increase of the width of the DEM boundary rows and columns. Local morphometric variables cannot be calculated for points in these rows and columns (Sections 4.1 and 4.2). Thus, some sampling points close to the DEM edges were lost. As a result, the sample size gradually decreased.

Moist because we did not perform the tacheometric survey for the gully head; derivation of *CA* and *TWI* from DEMs without data on catchment heads can lead to invalid results.

Statistical analysis was carried out by the Statgraphics software (© Statistical Graphics Corp., 1994–1997). Digital terrain modeling was performed by the LandLord software (Florinsky, 2016).

10.3.3 Results and discussion

Table 10.1 presents the results of the correlation analysis, and Fig. 10.5 provides the graphs of correlation coefficients between *Moist* and G, k_h, k_v, and H versus w. As expected, these graphs include some portions of smooth variation of correlation coefficients with w as well as some portions of high variability for these relationships (Fig. 10.5). All four graphs include two main portions: the left portion (between $w = 1$ m and $w \cong 4$ m) with relatively smooth variations, and the right portion (between $w \cong 4$ m and $w = 7$ m) with pronounced fluctuations.

Using smooth portions of the graphs as indicators of adequate intervals of w, we can distinguish:

- Adequate interval W_1 for H, k_v, and k_h ranging between $w \cong 2.25$ and $w \cong 4$ m.
- Adequate interval W_2 for G, H, k_v, and k_h ranging between $w \cong 2.5$ and $w \cong 3$ m.

W_2 is the main adequate interval of w because all correlations peak there except spurious correlations of *Moist* with G for $w = 4.5$ and 6 m (see explanation later). For example, for $w = 3$ m correlation coefficients of *Moist* with G, k_h, k_v, and H are

TABLE 10.1 The Severny Gully: correlation coefficients between the soil moisture content and some morphometric variables depending on the DEM grid spacing.

Morphometric variable	DEM grid spacing (m)												
	1	1.5	2	2.5	3	3.5	4	4.5	5	5.5	6	6.5	7
G	−0.29	−0.24	–	−0.32	−0.28	–	–	0.34	0.24	–	0.58	–	–
k_h	–	–	−0.34	−0.38	−0.52	−0.36	−0.36	–	−0.35	–	−0.29	–	−0.42
k_v	−0.32	−0.40	−0.37	−0.53	−0.50	−0.44	−0.47	–	−0.52	–	−0.53	–	−0.33
H	−0.27	−0.29	−0.46	−0.58	−0.60	−0.53	−0.49	–	−0.46	–	−0.41	–	−0.39

$P \leq 0.05$ for statistically significant correlations; dashes are statistically insignificant correlations.
(From Florinsky, I.V., Kuryakova, G.A., 2000. Determination of grid size for digital terrain modelling in landscape investigations—exemplified by soil moisture distribution at a micro-scale. Int. J. Geogr. Inf. Sci. 14 (8) 815–832, Table 2, reproduced with permission from Taylor & Francis Ltd.)

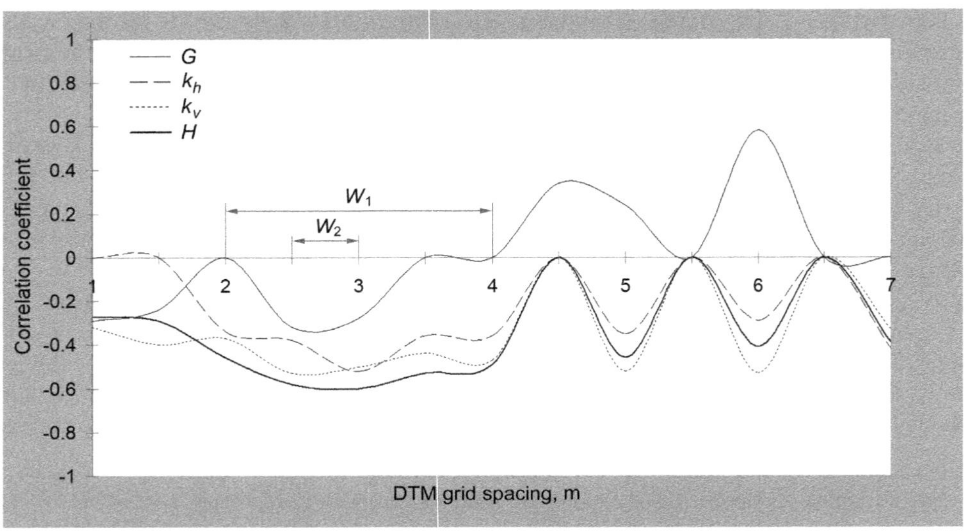

FIG. 10.5 Correlation between the soil moisture content and morphometric variables versus the DTM grid spacing.

TABLE 10.2 The Severny Gully: parameters of the regression equations for the dependence of the soil moisture content on morphometric variables under adequate *w*.

w (m)	Independent variables	Coefficient	*P*	R^2
2.5	*G*	−0.20	0.00	0.45
	H	−4.77	0.00	
	Constant	18.05	0.00	
3.0	*G*	−0.14	0.02	0.39
	H	−5.16	0.00	
	Constant	17.13	0.00	

(From Florinsky, I.V., Kuryakova, G.A., 2000. Determination of grid size for digital terrain modelling in landscape investigations—exemplified by soil moisture distribution at a micro-scale. Int. J. Geogr. Inf. Sci. 14 (8) 815–832, Table 3, reproduced with permission from Taylor & Francis Ltd.)

−0.28, −0.52, −0.50, and −0.60, respectively (Table 10.1). These results generally conform to previous findings on the topographic control of soil moisture (see details in Chapter 9).

Intervals of inadequate *w* differ greatly in appearance from adequate intervals of *w*: there are drastic variations in the dependence of correlation coefficients on *w* (Fig. 10.5). Correlations of *Moist* with k_h and *H* are much lower within the inadequate interval of *w* (between $w \cong 4$ m and $w = 7$ m) than within the main adequate interval (Table 10.1 and Fig. 10.5). Moreover, we observed positive correlations between *Moist* and *G* within this inadequate interval (Table 10.1 and Fig. 10.5). This is the evident artifact caused by the use of the inadequate *w*. Indeed, the soil moisture content decreases when *G* increases (Section 9.2). So if there is a dependence of *Moist* on *G*, correlation coefficients between *Moist* and *G* can take negative values, as we found for $w = 1$–3.5 m (Table 10.1 and Fig. 10.5). This demonstrates the importance of the correct determination of *w*. This is also a pictorial example of how one may obtain invalid conclusions using statistics alone, without understanding the physical senses of morphometric variables and the relationships between topography and landscape processes.

The results of the regression analysis are presented in Table 10.2. We obtained regression equations for the two adequate *w*, 2.5 and 3 m. In both cases, *G* and *H* entered the equations as independent variables. The regression equations describe 45% and 39% of *Moist* variability for $w = 2.5$ m and $w = 3$ m, respectively. These R^2 values are not high. Possibly, this is because we did not analyze the dependence of *Moist* on *CA*, essentially affecting the spatial distribution of soil moisture (Zakharov, 1940; Speight, 1980) (Section 9.3).

The method allows one to estimate not only the adequate *w* but also an adequate area of landforms supporting a topographic control of landscape properties. From the relationship between an adequate area and an adequate grid spacing (Section 10.2), the adequate area of landforms controlling *Moist* lies in the range of 20–40 m^2 within the study site. Notice that the adequate $w \cong 3$ m is close to $w = 4$ m, providing the *TWI*-based prediction of the water table depth in forests of the British Columbia (Thompson and Moore, 1996). This may indicate that topographic control of some soil hydrological processes occurs on landforms with an adequate area ranging between 36 and 64 m^2 in subboreal forested regions. Therefore, 3–4 m are suitable values of the adequate *w* for DTM-based soil and hydrological studies in these climatic conditions.

Using the method described, some difficulties can arise in choosing the minimum *w* and Δw. It is easy to see (Fig. 10.5) that for $\Delta w = 1$ m, it would be impossible to find fluctuations in the dependence of correlation coefficients on *w* within the right inadequate interval (between $w \cong 4$ m and $w = 7$ m). Using $\Delta w = 1$ m, these fluctuations cannot be found with the minimum *w* of 1 m or 1.5 m. We suppose that as a minimum *w* one should use such a *w* value that DTM errors caused by calculation of derivatives can be ignored (Section 5.4). Δw should be less than half of the minimum *w*.

Chapter 11

Predictive soil mapping

11.1 The Dokuchaev hypothesis as a central idea of soil predictions

In the mid-1880s, Vasily Dokuchaev (1886, pp. 352–353) formulated the following hypothesis[a]:

> "Any … soil is always and everywhere a mere *function*[b] of the following factors of soil formation[c]: 1) the nature (content and structure) of the parent rock, 2) the climate of the given terrain, 3) the mass and character of vegetation, 4) the age of the terrain, and finally, 5) the terrain topography. It immediately follows that (a) if the mentioned *factors* are the same in two different localities (however far apart they might be), the soils in the two localities should also be similar, and vice versa; consequently, (b) if we have thoroughly studied these factors, *we may predict in advance* what the soil itself should be like. Next, (c) it is well known that *momentum* should not change if one force component increases or decreases by some value, while another force component changes by the opposite value; thus there should be a similar, *to some extent*, relationship between *character of soil* and *character* of its forming factors. Hence, it is clear that it is theoretically possible to state and solve, for example, the following problem: How would a given soil change if there is an increase of the terrain temperature by say 1–2°, with a synchronous increase of meteoric water by 1–2 inches? How would soil change if there is an annual increase of the vegetative mass by 20 poods per desyatina[d] with a decrease of the temperature by 1–2°?
>
> Again, all these are so clear, so logically legitimate and necessarily, that *nobody* will probably argue with that.
>
> *For now* unfortunately, it is difficult to prove all these propositions *using facts* and *with desirable completeness*, particularly to answer the last question (c) *in details*. The reasons are quite obvious. Firstly, there is a great complexity of conditions affecting soil; secondly, these conditions have no absolute values, and, therefore, it is difficult to express them numerically; finally, we possess very few data for some factors and none whatever for others. Nevertheless, we may hope that all these difficulties will be overcome with time, and then soil science will truly become an exact science."

Later, Dokuchaev (1899, p. 3) carried out the first step towards the formalization of the problem. He proposed the first soil formation equation:

$$\Pi = f(\text{К}, \text{О}, \Gamma)\text{В}, \tag{11.1}$$

where П is soil, К is climate, О is organisms, Г is parent material, and В is the age of the soil (topography was not included in the expression due to a stenographer's mistake). Zakharov (1927, p. 8) presented a general soil formation equation, ideally describing the first law of soil science:

$$\pi = f(\text{М.Г.П.}, \text{Р.Ж.Орг.}, \text{Кл.}, \text{Возр.стр.}, \text{Р}-\text{ф}), \tag{11.2}$$

where π is soil, М.Г.П. is parent rock material, Р.Ж.Орг. is plant and animal organisms, Кл. is climate, Возр.стр. is the age of the terrain, and Р-ф is the relief.

In 1927, the Dokuchaev hypothesis and Eqs. (11.1) and (11.2) became known in the West due to two circumstances. First, Afanasiev (1927, p. 10) presented an English translation of the hypothesis at the First International Congress of Soil Science held in Washington, DC, in June 1927. Second, Sergey Zakharov took part in the Congress. Western soil scientists,

---

a. See details elsewhere (Florinsky, 2012).

b. Italics are by V. Dokuchaev.

c. The fundamental idea of soil science that soil is a result of the interaction of the five principal forming factors was first postulated by Dokuchaev (1883, p. III): "Soils being a result of a very complicated interaction between local climate, plant and animal organisms, content and structure of parent rocks, topography, and, finally, age of the terrain." Zakharov (1927, p. 8) called this idea "the first basic law of soil science."

d. Pood is the unit of weight of the Russian Imperial measurement system, 1 pood = 16.38 kg; desyatina is the unit of area of the Russian Imperial measurement system, 1 desyatina = 1.0925 ha.

Digital Terrain Analysis. https://doi.org/10.1016/B978-0-443-24798-9.00004-2

including Hans Jenny (Waksman and Deemer, 1928), had the opportunity to learn about the hypothesis and the equations during the Congress and the transcontinental soil excursion. Jenny (1941, p. 16) adopted the Zakharov Eq. (11.2) replacing Russian abbreviations with English ones:

$$S = f(cl, o, r, p, t, \ldots), \tag{11.3}$$

where S is soil; cl, o, r, p, and t are soil-forming factors: climate, organisms, relief, parent material, and time, respectively. The ellipsis indicates that additional soil formers may be included in Eq. (11.3).

McBratney et al. (2003) proposed a Zakharov-like formulation for empirical quantitative descriptions of relationships between soil and other spatially distributed factors using them as soil predictors:

$$S_c, S_a = f(s, c, o, r, p, a, n), \tag{11.4}$$

where S_c is soil classes, S_a is soil attributes, s is soil, other properties of the soil at a point; c is climate, local climatic properties; o is organisms, vegetation, fauna, and human activity; r is relief, morphometric variables; p is parent material, lithology; a is age, time; and n is space, spatial position.

One of the main obstacles to implementing such models is the complexity involved in producing reasonably accurate, detailed quantitative descriptions of most soil-forming factors (i.e., climate, parent material, and biota) and properties of the soil. However, a detailed and relatively accurate quantitative description of topography—the most temporally stable component of geosystems—can be quickly and easily performed. Indeed, there are quasi-global DEMs with a grid spacing of 1″ (~30 m) and smaller (Table 3.1). Moreover, advances in unmanned aerial surveys led to the situation in which DEMs with a meter- and decimeter-range resolution can be relatively easily and inexpensively created for virtually any terrain (Section 3.1.4). However, it is unlikely that similar (in terms of resolution and coverage) quantitative descriptions of soil properties will be available even in the long term. This is associated with spatial and temporal variability of soil properties (Beckett and Webster, 1971; Campbell, 1979; Burrough, 1993; Heuvelink and Webster, 2001), as well as labor-intensive and high-cost soil surveys or samplings and soil laboratory analyses.

Thus, in techniques of predictive digital soil mapping (DSM)[e] (McBratney et al., 2000, 2003; Scull et al., 2003; Lagacherie and McBratney, 2007; Walter et al., 2007; Lagacherie, 2008; MacMillan, 2008; Grunwald, 2009; Heung et al., 2016; Minasny and McBratney, 2016; Malone et al., 2018; Ma Y. et al., 2019; Shary, 2023), morphometric variables are commonly used as key predictors of soil properties[f] (for details, see Section 11.3). Such techniques are based on the following general assumption: Let us assume that (1) dependence of a soil property on morphometric variables is found using a relatively small sample set of soil and topographic variables and (2) statistical relationships between two subsets of variables—a soil property and morphometric factors prerequisites—are reasonably strong. In this case, it is possible to predict the spatial distribution of a soil property within a field, landscape, and, sometimes, a region using quantitative topographic data.

The objects of predictive soil mapping are morphological, physical, chemical, and biological quantitative properties of soil as well as soil taxonomic units. As predictors, researchers use different sets of morphometric variables, from the only elevation (e.g., Leenaers et al., 1990; Baxter and Oliver, 2005) up to 69 (!) morphometric indices[g] (Behrens et al., 2005). The most popular set includes elevation, G, k_h, k_v, and *TWI*.

As predictors, one may use morphometric variables of not only the land surface but also the top surface of parent material. This can be useful if rock properties (e.g., density) may influence the intrasoil lateral migration and accumulation of water and, hence, the spatial distribution of soil properties (Section 9.4).

It is obvious that digital soil mapping should be considered in the broader context of modeling and mathematization of soil science (Bockheim and Gennadiyev, 2010; Stockmann et al., 2011; McBratney et al., 2018).

11.2 Early models

In early predictive soil models, topographic maps were used as a source of input topographic data.

Fedoseev (1959) used (1) a coefficient describing the water storage in the root zone for different landscape positions relative to a reference hillslope and (2) data on the seasonal dynamics of soil moisture content depending on G, A, and the

e. Do not confuse this abbreviation with DSM as digital surface model (Chapter 3).

f. Besides DTMs the list of soil predictors includes remotely sensed data (Mulder et al., 2011), data on soil electrical conductivity, and other thematic layers (e.g., bedrock lithology, climate, and vegetation).

g. For independent predictors, Behrens et al. (2005) used k_h, k_v, k_{max}, and k_{min}; each of them was estimated by several calculation methods. It is unlikely that such an approach is reasonable. Selection of (topographic) predictors was, for example, discussed by Lark et al. (2007).

slope shape (convex, concave, and flat) to predict the spatial distribution of soil moisture content. Kirkby and Chorley (1967) proposed semi-empirical mathematical models of intrasoil lateral migration of water using data on slope gradient and horizontal curvature. Romanova (1970, 1971) developed a method for predictive mapping of seasonal moisture distribution using maps of G, A, and slope shape as well as derived empirical graphs describing the dependence of soil moisture content on these attributes. However, these approaches took into account only local peculiarities of topography. They did not consider the contribution of upslope areas to the formation and dynamics of soil moisture at downslope areas (i.e., the role of catchment area).

Romanova (1963) also developed a method to predict and map the redistribution of meteoric waters, considering the catchment area and slope gradient. On a topographic map, a slope was separated into sections having similar soil properties and G values. Water input to a slope section was the sum of water precipitated to this section and runoff from upslope sections. The runoff was calculated as a difference between the water precipitated there and the water infiltrated into the soil, depending on the upslope area. Each slope section was assigned a value of the empirical runoff coefficient, which considered the dependence of runoff and meteoric water infiltration on G, initial soil moisture, and vertical transmissivity of the soil. The main obstacle to this method was its empirical basis, which hindered its formalization and further development. Eventually, Svetlitchnyi et al. (2003) computerized the methods of Romanova (1963, 1970, 1971).

The method of Romanova (1963) is close to TOPMODEL, a concept of distributed hydrological modeling at a scale of small watersheds. The first version of TOPMODEL was based on the manual analysis of topographic maps (Beven and Kirkby, 1979). Later, several computerized versions were developed (Quinn and Beven, 1993; Kirkby, 1997; Beven and Freer, 2001). TOPMODEL used three main approximations:

1. Saturation zone dynamics can be approximated by a sequence of stable states.
2. At the given point on the land surface, the hydraulic gradient of the saturation zone can be approximated by the local slope gradient.
3. The dependence of lateral soil transmissivity on the depth can be described by an exponent function of water table depth (or soil storage deficit).

The first two approximations allowed the use of simple relations between the moisture capacity (or soil storage deficit) of a watershed and local water table depths (or local soil storage deficits). A key variable in these relations was the topographic wetness index (Section 2.6.1). Various versions of TOPMODEL have been used in digital modeling of overland flow dynamics, evapotranspiration, soil moisture, and water table depth in subboreal zones (Beven and Kirkby, 1979; Beven et al., 1984; Quinn and Beven, 1993; Thompson and Moore, 1996; Blazkova et al., 2002), the Mediterranean (Durand et al., 1992; Piñol et al., 1997), the humid tropics (Moličová et al., 1997), and so on.

Like the Romanova method, the basic TOPMODEL did not consider the spatial variability of soil moisture caused by the influence of the slope aspect. To solve this problem, (1) catchment slopes were classified into several groups depending on A, (2) TOPMODEL was calibrated for each slope group, and (3) the soil hydrological characteristics of each slope group were separately modeled (Band et al., 1993). In addition, TOPMODEL was a pretty simple model developed for the natural conditions particular to central England and Wales. Therefore, utilization of TOPMODEL in other soil and hydrological conditions may produce incorrect results (Quinn et al., 1995; Beven, 1997).

Information on plan curvature (Section 2.2.4.1) has been widely used to produce soil maps by the Relief Plasticity method, a manual geomorphometric technique (Anisimov et al., 1977; Stepanov et al., 1984, 1987; Stepanov and Loshakova, 1998). According to this method, boundaries of soil polygons were determined by a morphoisograph[h]—the isoline of $k_p = 0$—separating divergence zones of overland and intrasoil flows ($k_p > 0$) from their convergence zones ($k_p < 0$). Then soils of crests and upslopes were attributed to divergence zones; soils of valleys, downslopes, and depressions were assigned to convergence zones; midslope soils were set to intermediate positions. Soil data was usually derived from existing soil maps. The Relief Plasticity method has been applied to produce a lot of medium- and small-scale soil maps for vast territories of Central Asia, the East European Plain, and West Siberia[i] (Stepanov, 1984, 1989; Satalkin, 1988).

h. A morphoisograph is the approximation for the isoline of $k_h = 0$ (Section 2.2.4.2). A manual graphical technique to derive the isoline of $k_p = 0$ from topographic maps was first described by Sobolevsky (1932).

i. Stepanov et al. (1998, p. 30) presented a list of 17 published and 13 unpublished soil maps of Russian regions (mostly, at 1: 200,000 and 1: 300,000 scales), which were produced by the Relief Plasticity method.

11.3 Current predictive methods

MacMillan (2008, p. 114) made the following observations about digital soil mapping methods:

> *"Almost all efforts to develop and apply DSM techniques can be seen to follow approximately the same basic steps .... Differences in protocols arise from differences in the kinds of outputs that are to be predicted, the kinds of input data layers selected to support predictions and the kinds of equations or rules developed to make predictions. These steps are inter-related such that decisions on what to predict (individual soil properties or soil classes) influence both the selection of input variables and the development of predictive equations and vice versa."*

The six basic steps of digital soil mapping are as follows (MacMillan, 2008):

1. Conceptualization of (a) a discrete (categorical or quantitative) area-class entity of a soil map or (b) a quantitative continuous individual soil property to be predicted.
2. Identification and obtaining or derivation of relevant input data sets including DTMs, remotely sensed data, thematic maps, and data from field surveys.
3. Development of classification rules or predictive equations using related mathematical approaches (Section 11.3.2).
4. Application of classification rules or predictive equations to input data.
5. Evaluation of accuracy or efficiency of prediction in terms of root mean square error (RMSE), positional error, and so on.
6. Production of final maps.

11.3.1 Classification of methods

DTM-based methods of predictive digital soil mapping are naturally classified into two main groups:

1. Methods applied to predict quantitative soil properties. There are two subgroups (Florinsky et al., 1999, 2002; Pachepsky et al., 2001):
   - **(a)** Methods for prediction of a quantitative soil property at each point of the land surface. They are used to produce quantitative *continuous* maps of quantitative continuous soil properties. Such methods are usually based on multiple regression analysis, hybrid geostatistical techniques, regression kriging, etc. (for details, see Section 11.3.2).
   - **(b)** Methods for prediction of a quantitative soil property for distinct landforms of the land surface (e.g., upper and lower slopes, shoulders, footslopes). Such methods commonly use landform classification approaches (Section 2.7). In modeling, a transition occurs from a quantitative continuous scale of a soil property to a quantitative discrete (categorical) one. Such methods are utilized to derive quantitative *discrete unit* maps of quantitative continuous soil properties, such as soil horizon thickness, depth to calcium carbonate, soil organic carbon content, etc. (e.g., Pennock et al., 1987; Florinsky et al., 1999, 2002; Park et al., 2001; Pennock and Corre, 2001; Florinsky and Eilers, 2002; Park and van de Giesen, 2004).
2. Methods applied to predict categorical soil variables (e.g., soil classes or units). There are three subgroups (McBratney et al., 2003; MacMillan, 2008):
   - **(a)** Unsupervised classification approaches. These approaches are used to derive new soil classes from the observed soil properties. The type and nature of predicted classes are greatly dependent on the type of predictors and sample locations rather than on local expert knowledge. Such methods can, for example, be based on decision tree models (Section 11.3.2).
   - **(b)** Supervised classification approaches. These approaches are used to develop classification rules for recognizing soil classes. The rules describe relationships between each output class and a set of environmental predictors from representative sites (so-called training data). Such approaches can, for instance, be based on decision tree models, fuzzy logic, discriminant analysis, and Bayesian analysis (Section 11.3.2). Also, supervised classification approaches include deterministic methods (Walter et al., 2007). They are used if a simplified physical law may explain the spatial distribution of a soil property. For example, the genesis and spatial distribution of hydric-like soils are associated with periodically waterlogged areas, which may be derived from DTMs. Thus, digital models of *TWI* and *DA* (Sections 2.6.1 and 2.3.1) have been applied to predict the spatial distribution of such soils (e.g., Merot et al., 1995; Bedard-Haughn and Pennock, 2002; Mourier et al., 2008; Penizek and Boruvka, 2008).
   - **(c)** Heuristic classification approaches. These approaches can be used if one is able to identify (i) all soil classes to be predicted and (ii) conditions under which all classes may occur. Such expert knowledge may arise from (i) local field experience or (ii) theoretical considerations about relationships between morphometric variables (or quantitatively segmented landforms—Section 2.7) and soil units (e.g., MacMillan et al., 2000; Zhu et al., 2001; Buivydaite and Mozgeris, 2007; Barringer et al., 2008; Smith S. et al., 2010).

11.3.2 Mathematical approaches

DTM-based methods of predictive digital soil mapping can use a variety of mathematical approaches. Below we list the most common of them:

1. Multiple regression analysis, a classical least-squares regression technique (Kleinbaum et al., 2014). A dependent variable is a soil property, and independent predictors are morphometric variables.[j] Hybrid regression approaches are sometimes applied, such as combination of stepwise linear regression, regression trees, and robust linear regression (Thompson and Kolka, 2005) as well as combination of stepwise linear regression and principal component regression (Huang et al., 2007).
2. Multinomial logistic regression, which is used to predict a categorically distributed predictand from a set of real-, binary-, or categorical-valued predictors (Kleinbaum et al., 2014).
3. Geographically weighted regression, which is used to examine spatial heterogeneity or nonstationarity providing spatially varying coefficient estimates by calibrating a location-specific weighted regression model (Brunsdon et al., 1998).
4. Generalized linear models, an extension of linear regression allowing models to fit data that follow a nonnormal probability distribution (McCullagh and Nelder, 1999).
5. Hybrid geostatistical methods (cokriging, kriging with external drift, etc.), which use morphometric attributes as ancillary variables in kriging of a predictand (Goovaerts, 1997; Webster and Oliver, 2007).
6. Regression kriging, the combination of a multiple linear regression model with kriging of the regression residuals (Odeh et al., 1995).
7. Multiple-point geostatistics, an alternative approach to two-point geostatistic techniques utilizing a conceptual training image of the expected spatial structure of the predicted variable. The idea behind training images is that there may be another site—the training image—from which the necessary spatial information can be obtained (Guardiano and Srivastava, 1993; Meerschman et al., 2013).
8. Fuzzy logic (Zadeh, 1965; McBratney and Odeh, 1997). The central idea of fuzzy logic is that a function, defining if an element belongs to a set, can take any value between 0 and 1, not only 0 or 1. As soil landscapes are characterized by continuous nature, fuzzy logic is useful in predictive soil mapping. Sometimes the fuzzy logic approach is combined with regression kriging (Carré and Girard, 2002; Carré and McBratney, 2005) and discriminant analysis (Sorokina and Kozlov, 2009).
9. Generalized additive models, a regression approach in which nonlinearity can be introduced by replacing linear combinations of the predictors with combinations of nonparametric smoothing functions (Hastie and Tibshirani, 1990).
10. Artificial neural networks (Graupe, 2007), a mathematical model consisting of an interconnected "neurons." Such models can change their structures using information flowing through the network in training.
11. Support vector machine, a supervised machine learning technique providing a non-linear solution for classification and regression tasks (Vapnik, 2000). The key idea of the technique is to transfer original vectors into a higher-dimensional space and to search for a separating hyperplane with the largest gap in this space. Two parallel hyperplanes are constructed on both sides of the hyperplane separating the classes. The technique is based on the assumption that the greater the distance between parallel hyperplanes, the smaller the average error of the classifier.
12. Decision trees (Clark and Pregibon, 1997). These methods work by splitting data into homogeneous subsets. Two main types of decision tree analysis are used in digital soil mapping: classification tree analysis (a predictand is a categorical variable) and regression tree analysis (a predictand is a numeric variable).
13. Random forest regression, a machine-learning classification and regression method combining several randomized decision trees and aggregating their predictions by averaging. It shows the best performance in settings where the number of observations is much smaller than the number of variables (Biau and Scornet, 2016; Hengl et al., 2018).
14. Gradient boosting, a method in which classification or regression trees are sequentially constructed from the residuals of the preceding trees. At each stage, boosting occurs by selecting a random sample of data that leads to a gradual improvement of the model (Friedman, 2002).
15. Structural equation modeling, a multivariate statistical analysis technique used to analyze relations between measured and latent variables (Maruyama, 1997).
16. Discriminant analysis, a classical statistical technique to predict group membership using knowledge of interval variables (Webster and Burrough, 1974).

---

j. Section 11.4 includes a detailed description of an algorithm to predict soil properties using multiple regression.

17. Bayesian analysis of evidence (Ghosh et al., 2006), an approach quantifying a predictive strength of an input subset of data relative to all other subsets.
18. Ensemble modeling, approaches creating a single predictive model of a variable from its several primary predictive models produced by using different modeling algorithms (Heuvelink and Bierkens, 1992; Malone et al., 2014; Román Dobarco et al., 2017).

Detailed, soil-mapping oriented description of these approaches can be found elsewhere (McBratney et al., 2000, 2003; Bishop and McBratney, 2001; Scull et al., 2003; Walter et al., 2007; MacMillan, 2008; Heung et al., 2016; Hengl et al., 2018; Padarian et al., 2019; Wadoux et al., 2020). Tables 11.1 and 11.2 summarize the long-term experience of applying the listed techniques for spatial prediction, modeling, and mapping of quantitative soil properties and categorical soil variables, respectively. Figs. 11.1–11.5 represent examples of digital soil predictive maps derived by some of the listed techniques.

TABLE 11.1 Studies focused on prediction and modeling of quantitative soil properties by DTM-based techniques of pedometrics and digital soil mapping.

Quantitative soil property	References
Total thickness and thickness/depth of soil horizons	Kachanoski et al. (1985a,b), Pennock et al. (1987), Moore et al. (1993), Bell et al. (1994b), Odeh et al. (1994, 1995), Gessler et al. (1995, 2000), Florinsky et al. (1999, 2002), McKenzie and Ryan (1999), Bourennane et al. (2000), Park et al. (2001), Thompson et al. (2001), Hengl et al. (2004, 2017b), Shi et al. (2004), Herbst et al. (2006), Puzachenko et al. (2006), Sumfleth and Duttmann (2008), Ballabio (2009), Kuriakose et al. (2009), Tesfa et al. (2010), Zhu and Lin (2010), Angelini et al. (2016), Drăguţ and Dornik (2016), Libohova et al. (2016), Mulder et al. (2016a), Li A. et al. (2017), Poggio and Gimona (2017), Han et al. (2018), Nussbaum et al. (2018), Patton et al. (2018), Piccini et al. (2019), Aitkenhead and Coull (2020), Wang et al. (2021), Zhang et al. (2021), and Khanifar and Khademalrasoul (2022)
Organic matter content	Moore et al. (1993), Cook et al. (1996), Hengl et al. (2004, 2007a), Dobos et al. (2007), Chai et al. (2008), Marchetti et al. (2010), Pei et al. (2010), and Zhu and Lin (2010)
pH	Moore et al. (1993), Hengl et al. (2004, 2017b, 2021), Zhu and Lin (2010), Mulder et al. (2016a), Dharumarajan et al. (2017), Li X. et al. (2017), Poggio and Gimona (2017), Sun et al. (2017), Marques et al. (2018), Nussbaum et al. (2018), Ramcharan et al. (2018), Shepard et al. (2018), Poppiel et al. (2019, 2021), Wadoux (2019), Baltensweiler et al. (2020), and Piedallu et al. (2022)
Bulk density	Kachanoski et al. (1985a,b), Ballabio et al. (2016), Hengl et al. (2017b, 2021), Poggio and Gimona (2017), Ramcharan et al. (2018), and Aitkenhead and Coull (2020)
Texture (sand, silt, and clay content)	Moore et al. (1993), Odeh et al. (1994, 1995, 2007), De Bruin and Stein (1998), Lark (1999), McBratney et al. (2000), Shi et al. (2004), Bishop et al. (2006), Qi and Zhu (2006), Lark et al. (2007), Minasny and McBratney (2007), Mora-Vallejo et al. (2008), Sumfleth and Duttmann (2008), Zhu and Lin (2010), Angelini et al. (2016, 2017), Ballabio et al. (2016), Drăguţ and Dornik (2016), Mulder et al. (2016a), Forkuor et al. (2017), Hengl et al. (2017b, 2021), Poggio and Gimona (2017), Stumpf et al. (2017), Sun et al. (2017), Behrens et al. (2018a,b), Marques et al. (2018), Nussbaum et al. (2018), Ramcharan et al. (2018), Shepard et al. (2018), Caubet et al. (2019), Piccini et al. (2019), Poppiel et al. (2019, 2021), Wadoux (2019), Riza et al. (2021), Dornik et al. (2022), Mello et al. (2022), and Siqueira et al. (2023)
Moisture content	Sinai et al. (1981), Burt and Butcher (1985), Kuryakova et al. (1992), Florinsky et al. (1999, 2002), Lark (1999), Sulebak et al. (2000), Chaplot and Walter (2003), Park and van de Giesen (2004), Lark et al. (2007), Lei et al. (2016), Baldwin et al. (2017), Li X. et al. (2017), Martínez-Murillo et al. (2017), Guevara and Vargas (2019), Luo W. et al. (2019), Fan et al. (2020), Mallet et al. (2020), and Larson et al. (2022)
Soil temperature	Fan et al. (2020)
Available water capacity	Malone et al. (2009) and Ballabio et al. (2016)
Soil hydraulic properties	Herbst et al. (2006) and Dornik et al. (2022)
Total carbon content	McKenzie and Ryan (1999), Huang et al. (2007), Sumfleth and Duttmann (2008), Malone et al. (2009), Kaleita et al. (2017), and Hengl et al. (2021)
Organic carbon content	Florinsky et al. (1999, 2002), Gessler et al. (2000), Chaplot et al. (2001), Mitusov (2001), Pennock and Corre (2001), Florinsky and Eilers (2002), Terra et al. (2004), Thompson and Kolka (2005), Lark et al. (2007), Mora-Vallejo et al. (2008), Ballabio (2009), Mendonça-Santos et al. (2010), Shary and

TABLE 11.1 Studies focused on prediction and modeling of quantitative soil properties by DTM-based techniques of pedometrics and digital soil mapping—cont'd

Quantitative soil property	References
	Pinskii (2013), Akpa et al. (2016), Angelini et al. (2016, 2017), Lin et al. (2016), Mulder et al. (2016a,b), Song et al. (2016, 2017), Dharumarajan et al. (2017), Dokuchaev (2017), Fissore et al. (2017), Forkuor et al. (2017), Hengl et al. (2017a,b, 2021), Kaleita et al. (2017), Li X. et al. (2017, 2018a,b), Malone et al. (2017), Poggio and Gimona (2017), Roudier et al. (2017), Sun et al. (2017), Bogunovic et al. (2018), Hounkpatin et al. (2018), Nussbaum et al. (2018), Ramcharan et al. (2018), Shepard et al. (2018), Zhu et al. (2018), Guo et al. (2019), Padarian et al. (2019), Piccini et al. (2019), Poppiel et al. (2019, 2021), Wadoux (2019), Wadoux et al. (2019), Adhikari et al. (2020), Aitkenhead and Coull (2020), Guevara et al. (2020), Mallick et al. (2022), and Mello et al. (2022)
Total nitrogen content	Florinsky et al. (1999, 2002), Sumfleth and Duttmann (2008), Lin et al. (2016), Forkuor et al. (2017), Hengl et al. (2017a, 2021), Kaleita et al. (2017), Li X. et al. (2017), Ramcharan et al. (2018), and Wadoux (2019)
Mineral nitrogen content	Lark (1999) and Baxter and Oliver (2005)
Phosphorus content	Moore et al. (1993), McKenzie and Ryan (1999), Florinsky et al. (1999, 2002), Mitusov (2001), Hengl et al. (2017a, 2021), Li X. et al. (2017), and Shen et al. (2019)
Macro- and trace-element content	Leenaers et al. (1990), Lark et al. (2007), Ballabio (2009), Ballabio and Comolli (2010), Zhu and Lin (2010), Hengl et al. (2017a, 2021), Behrens et al. (2018b), Marques et al. (2018), Shepard et al. (2018), and Dvornikov et al. (2022)
Microbial biomass carbon and nitrogen content	Florinsky et al. (2004) and Kaleita et al. (2017)
Microbial enzyme activity	Florinsky et al. (2004) and Fairbanks et al. (2020)
Denitrification rate	Pennock et al. (1992), Florinsky et al. (2004), and Li et al. (2018c)
Methane flux	Kaiser et al. (2018)
Nitrous oxide flux	Corre et al. (1996) and Pennock and Corre (2001)
Cation exchange capacity	McBratney et al. (2000), Bishop and McBratney (2001), Lark et al. (2007), Mulder et al. (2016a), Angelini et al. (2017), Forkuor et al. (2017), Hengl et al. (2017b, 2021), Marques et al. (2018), Nussbaum et al. (2018), Piccini et al. (2019), and Poppiel et al. (2019)
Electrical conductivity	Odeh et al. (2007), Dharumarajan et al. (2017), and Taghizadeh-Mehrjardi et al. (2021)
Gamma radiation	Malone et al. (2016) and Wadoux et al. (2018)

TABLE 11.2 Studies focused on prediction and modeling of categorical soil variables by DTM-based techniques of digital soil mapping.

Categorical soil variable	References
Soil taxonomic classes	Skidmore et al. (1991), Odeh et al. (1992a,b), Zhu et al. (1996, 2001), Lagacherie and Holmes (1997), Thomas et al. (1999), Dobos et al. (2000), McBratney et al. (2000), Zhu (2000), Bui and Moran (2001, 2003), Bedard-Haughn and Pennock (2002), Carré and Girard (2002), Moran and Bui (2002), Zhou et al. (2004), Behrens et al. (2005, 2010), Carré and McBratney (2005), MacMillan et al. (2005), Scull et al. (2005), Behrens and Scholten (2007), Buivydaite and Mozgeris (2007), Cole and Boettinger (2007), Hengl et al. (2007b, 2017b), Hollingsworth et al. (2007), Saunders and Boettinger (2007), Barringer et al. (2008), Giasson et al. (2008), Grinand et al. (2008), Mendonça-Santos et al. (2008), Mourier et al. (2008), Debella-Gilo and Etzelmüller (2009), Kozlov (2009), Sorokina and Kozlov (2009), Moonjun et al. (2010), Smith S. et al. (2010), Chaney et al. (2016), Drăguţ and Dornik (2016), Penížek et al. (2016), Silva et al. (2016), Dokuchaev (2017), Sarmento et al. (2017), Zeraatpisheh et al. (2017), Dornik et al. (2018), Horáček et al. (2018), Ramcharan et al. (2018), Vincent et al. (2018), Atkinson et al. (2020), and Mello et al. (2021)
Soil drainage classes	Bell et al. (1992, 1994a), Merot et al. (1995), Cialella et al. (1997), and Kravchenko et al. (2002)
Soil gleying degrees	Sorokina and Kozlov (2009)
Soil salinity risk	Florinsky et al. (2000), MacMillan and Marciak (2001), Taylor and Odeh (2007), and Vermeulen and van Niekerk (2017)
Soil texture classes	Hengl et al. (2007b), Ballabio et al. (2016), Wu et al. (2018), and Atkinson et al. (2020)

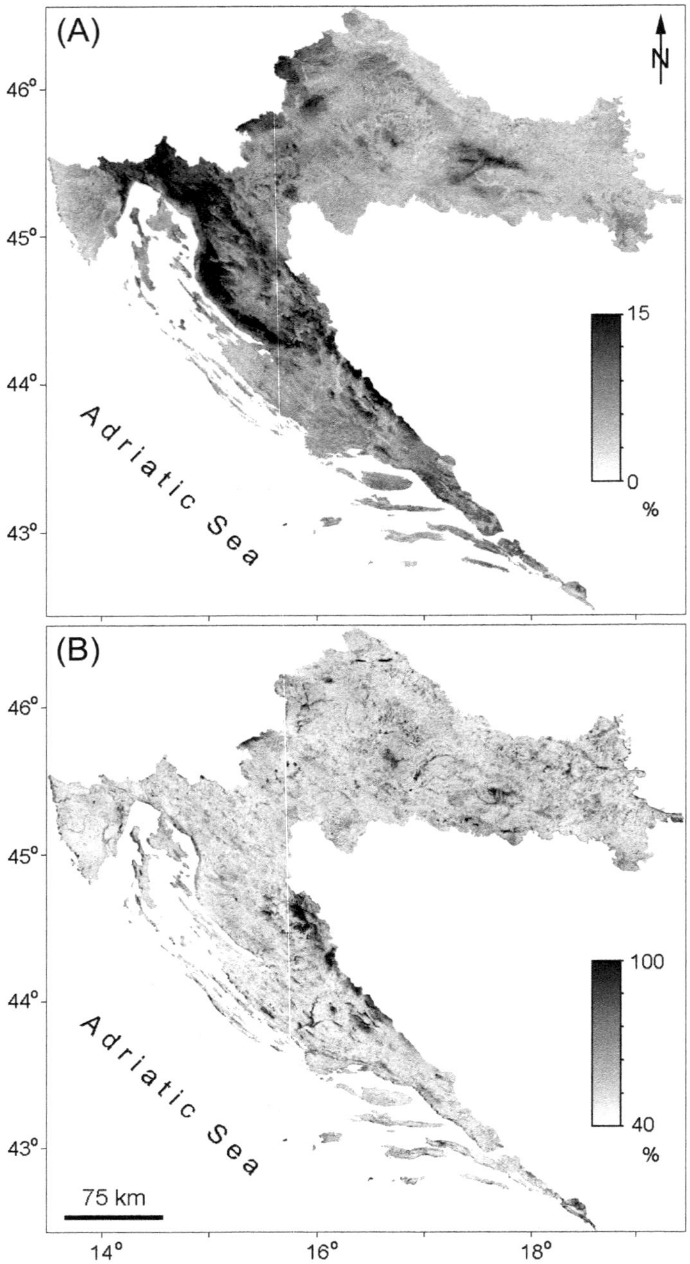

FIG. 11.1 Small-scale predictive mapping of topsoil organic matter content using regression kriging for Croatia: (A) Predicted values. (B) Prediction error in terms of regression kriging variance. Original sampled values (2087 measurements) were obtained from the Croatian Soil Database. Sixteen environmental predictors were derived from a DEM, generated using the Shuttle Radar Topography Mission (SRTM) data (Rabus et al., 2003; Farr et al., 2007) and 1:25,000 topographic maps, and multitemporal satellite radiometric images of the Moderate Resolution Imaging Spectroradiometer (MODIS). Among six predictors selected by the stepwise regression, the best topographic predictors were *TWI* and elevation. Predictors explained 40% of the variation in soil organic matter. *(From Hengl, T., Heuvelink, G.B.M., Rossiter, D.G., 2007a. About regression kriging: from equations to case studies. Comput. Geosci. 33 (10), 1301–1315, Fig. 3C and D, reproduced with permission from Elsevier.)*

11.3.3 Small-scale predictive models and upscaling

DTM-based predictive mapping of quantitative soil properties (Table 11.1) is typical for field and small watershed scales. DTM-based predictive mapping of categorical soil entities (Table 11.2) is common for the production of soil maps at field, watershed, and regional scales. However, DTM-based predictive mapping is sometimes used at large regional/national (Florinsky et al., 2000; Dobos et al., 2007; Hengl et al., 2007a; Mulder et al., 2016a,b), continental (Hengl et al., 2015,

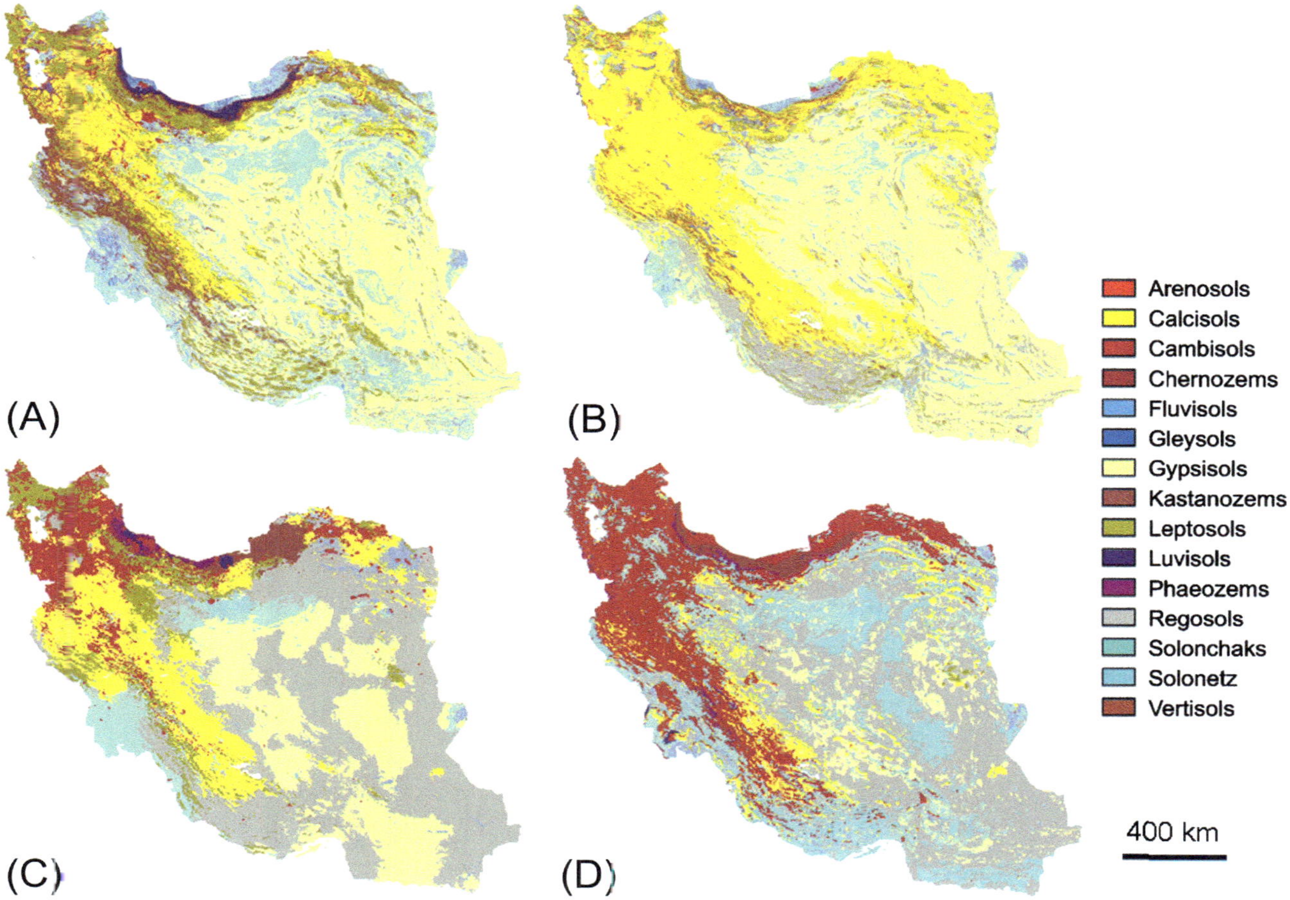

FIG. 11.2 Small-scale predictive mapping of World Reference Base soil groups for the territory of Iran. Soil groups were interpolated by (A) supervised classification using maximum likelihood, (B) multinomial logistic regression, (C) regression kriging on memberships, and (D) per-pixel classification of interpolated taxonomic distances. A DEM was produced by combining SRTM data (Rabus et al., 2003; Farr et al., 2007) and GTOPO30 (USGS, 1996). The DEM grid spacing was 1 km. *G*, *H*, *TWI*, and some other topographic attributes were used as predictors. *(From Hengl, T., Toomanian, N., Reuter, H.I., Malakouti, M.J., 2007b. Methods to interpolate soil categorical variables from profile observations: lessons from Iran. Geoderma 140 (4), 417–427, Fig. 5, reproduced with permission from Elsevier.)*

2017a, 2021; Ballabio et al., 2016; Chaney et al., 2016; Ramcharan et al., 2018; Guevara and Vargas, 2019; Guevara et al., 2020) and even global (Hengl et al., 2017b) scales.

Mallavan et al. (2010) proposed the Homosoil method, a technique for deriving predictive soil maps of vast territories whose soils are poorly known. The Homosoil was intended for small-scale extrapolation of soil taxonomic classes across the globe. Using available data on parent materials, topography (elevation, slope gradient, and topographic wetness index), and climate (solar radiation, rainfall, temperature, and evapotranspiration), the Homosoil identifies a so-called homosoil area with similar soil-forming factors for an area with no soil data. If soils of the homosoil area are known, models describing relationships between soils and selected lithologic, topographic, and climatic predictors can be applied to the area with no soil data. In fact, the Homosoil directly implemented the idea of Dokuchaev (1886) on predictive soil mapping (Section 11.1).

Upscaling procedures (Bierkens et al., 2000) may be applied to produce predictive soil maps for vast territories. However, it is obvious that such procedures are complex and ambiguous, at least in the context of DTM-based soil prediction (Thomas et al., 1999; Lagacherie et al., 2001). For instance, Florinsky and Eilers (2002) used a concept of accumulation, transit, and dissipation zones (Section 2.7.2) to predict the spatial distribution of the topsoil organic carbon content at a field scale and to upscale the prediction to a regional and subcontinental scales within the Black Soils zone of the Canadian Prairies. First, relationships between the organic carbon content of the A horizon and accumulation, transit, and dissipation zones of microtopography were found at the field scale. Second, the field-scale relations were applied to

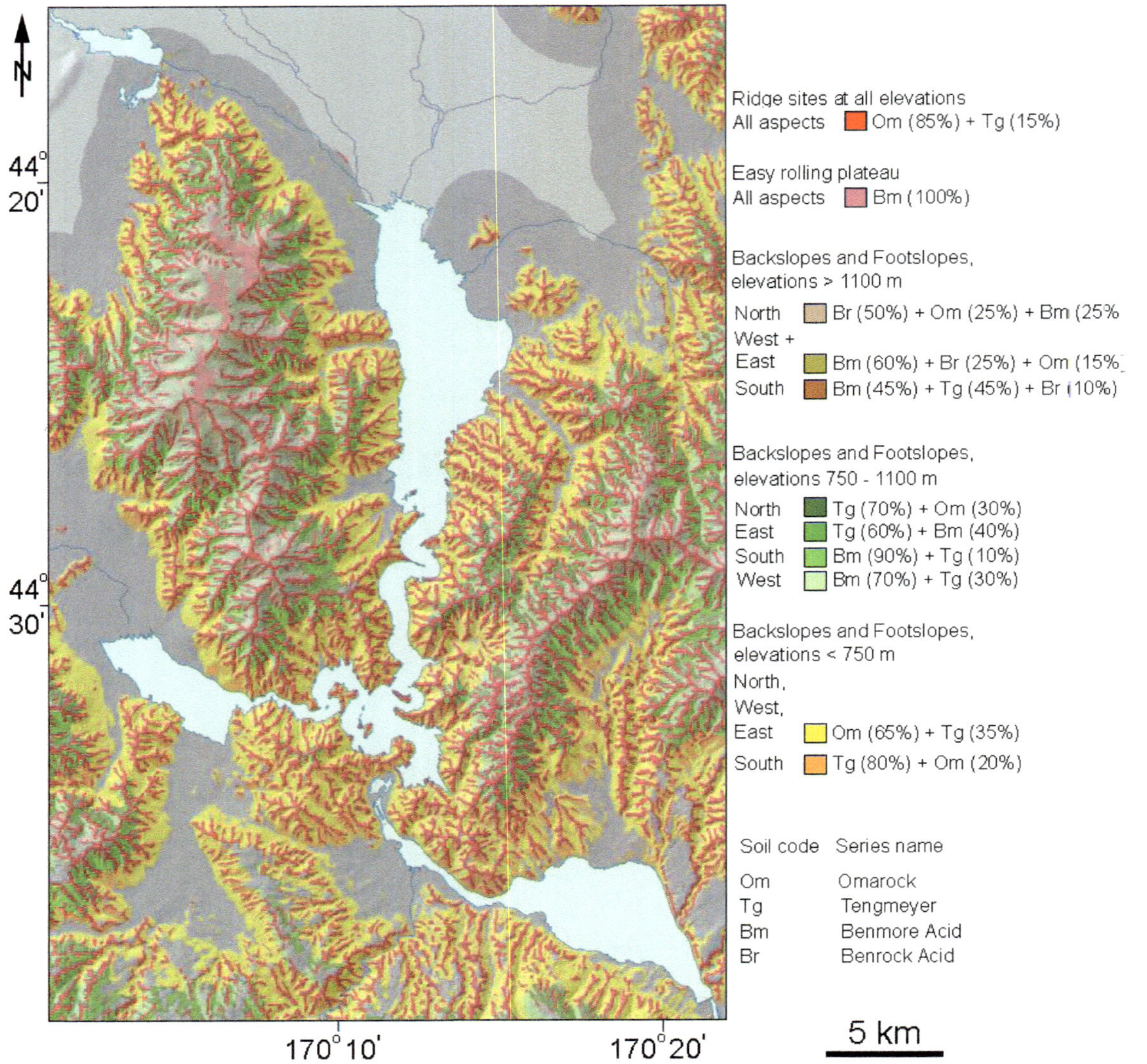

FIG. 11.3 Large-scale predictive soil mapping by means of landscape segmentation and fuzzy logic (exemplified by the Benmore Range, South Island, New Zealand). The national DEM of New Zealand was used; the DEM grid spacing was 25 m. *(From Barringer, J.R.F., Hewitt, A.E., Lynn, I.H., Schmidt, J., 2008. National mapping of landform elements in support of S-Map, a New Zealand soils database. In: Zhou, Q., Lees, B., Tang, G.-A. (Eds.), Advances in Digital Terrain Analysis. Lecture Notes in Geoinformation and Cartography Book Series. Springer, Berlin, Germany, pp. 443–458, Fig. 3, reproduced with permission from Springer Nature.)*

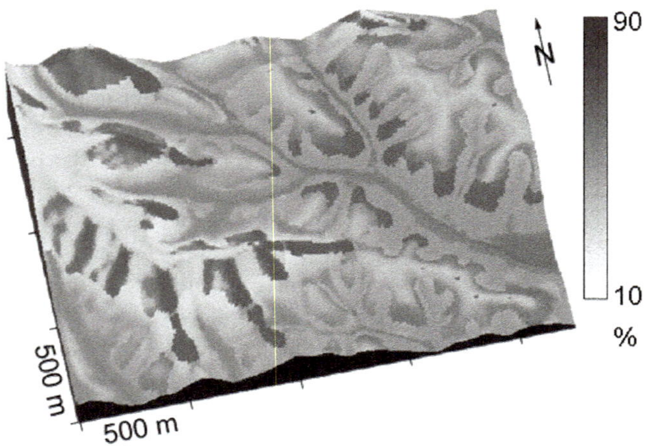

FIG. 11.4 Large-scale predictive mapping of the distribution of the A horizon silt percentage using fuzzy logic (exemplified by the Raffelson Watershed, La Crosse, Wisconsin, USA). A DEM was produced by digitizing a 1:24,000 topographic map; the grid spacing was 30 m. Elevation, G, k_a, k_v, and TWI were used for prediction. *(From Qi, F., Zhu, A., 2006. Fuzzy soil mapping based on prototype category theory. Geoderma 136 (3–4), 774–787, Fig. 5-right, reproduced with permission from Elsevier.)*

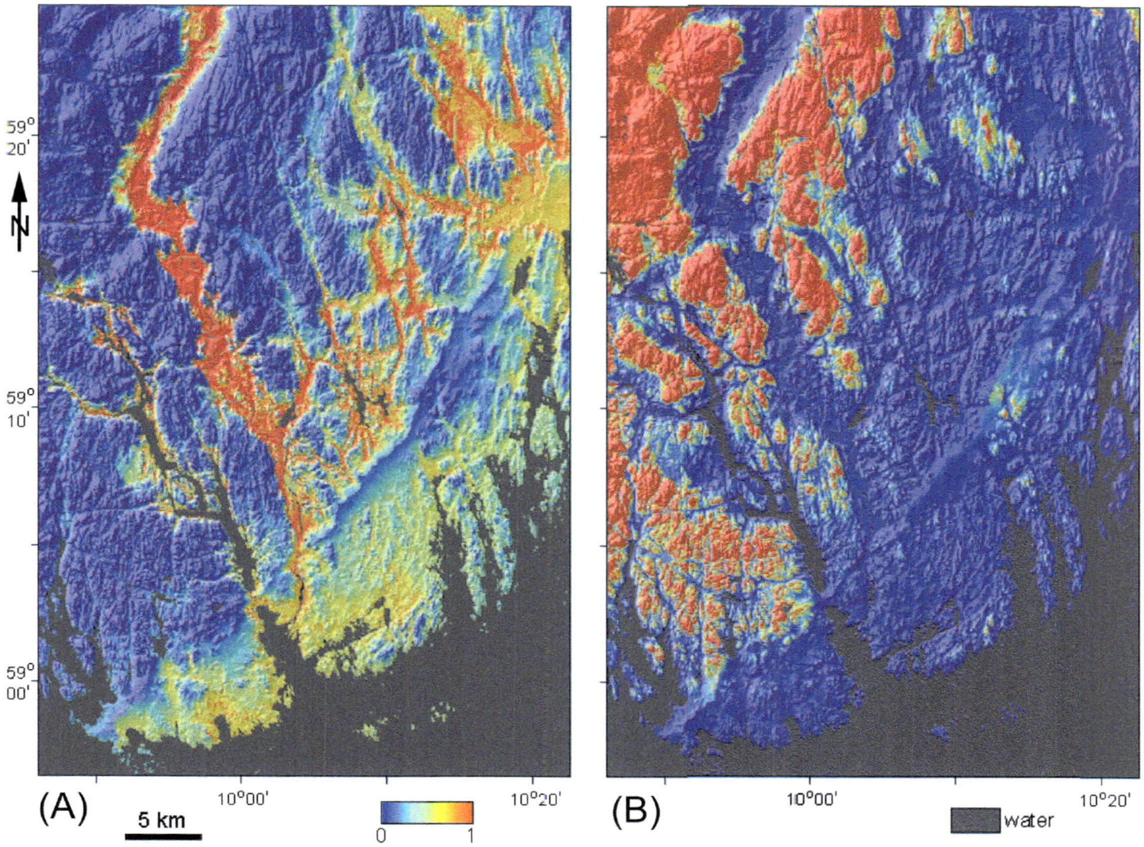

FIG. 11.5 Large-scale mapping of the probability distribution, *p*-value, of (A) Cambisols and (B) Leptosols in the Vestfold County, southeastern Norway. Prediction was carried out by multinomial logistic regression. Leptosols occupy hills and rocky areas, and Cambisols occupy valleys. A DEM was created using 20-m contour topographic maps; the DEM grid spacing was 25 m. *(From Debella-Gilo, M., Etzelmüller, B., 2009. Spatial prediction of soil classes using digital terrain analysis and multinomial logistic regression modeling integrated in GIS: examples from Vestfold County, Norway. Catena 77 (1), 8–18, Fig. 5, reproduced with permission from Elsevier.)*

meso- and macrotopography of ecological districts with similar soils, parent materials, and landforms. Finally, DEM-derived maps of accumulation, transit, and dissipation zones were used to produce predictive maps at three different scales (Fig. 11.6). Generalization of organic carbon content polygons came directly from generalization of accumulation, transit, and dissipation zones due to the use of DEMs of various resolution and DEM smoothing.

Carré et al. (2008) compared two principally different approaches to generalize soil polygons on DTM-based predictive soil maps in transition from 1:25,000 scale to 1:250,000 scale. The first, bottom-up approach was based on (1) a taxonomic aggregation of soil properties (carbonate rate, hydromorphic rate, texture, parent material) and topographic attributes (elevation and *TWI*) to define taxonomic units and (2) generalization of their polygons to produce pedolandscape mapping units. The second, top-down approach included (1) a taxonomic aggregation of the topographic attributes to delineate taxonomic units and (2) generalization of their polygons to produce pedolandscape mapping units. These approaches produced principally distinct compositions of soil polygons. The first approach produced patchwork-like soil polygons (Fig. 11.7A), while soil polygons constructed by the second approach reflected the main peculiarities of the regional topography and landscape structure (Fig. 11.7B). Although there were no objective criteria to choose a particular approach to generalize soil polygons, the second approach was intuitively preferred. Carré et al. (2008, p. 210) also accepted that the second "approach can be recommended for digital soil mapping with limited data since" topography "is the most relevant factor of the pedogenesis."

This is not a new problem for soil mapping. There was a discussion on the manual delineation of soil polygons using the Relief Plasticity method[k] versus conventional soil mapping approaches (Stepanov et al., 1984, 1987; Dmitriev, 1998;

k. For a description of the Relief Plasticity method, see Section 11.2.

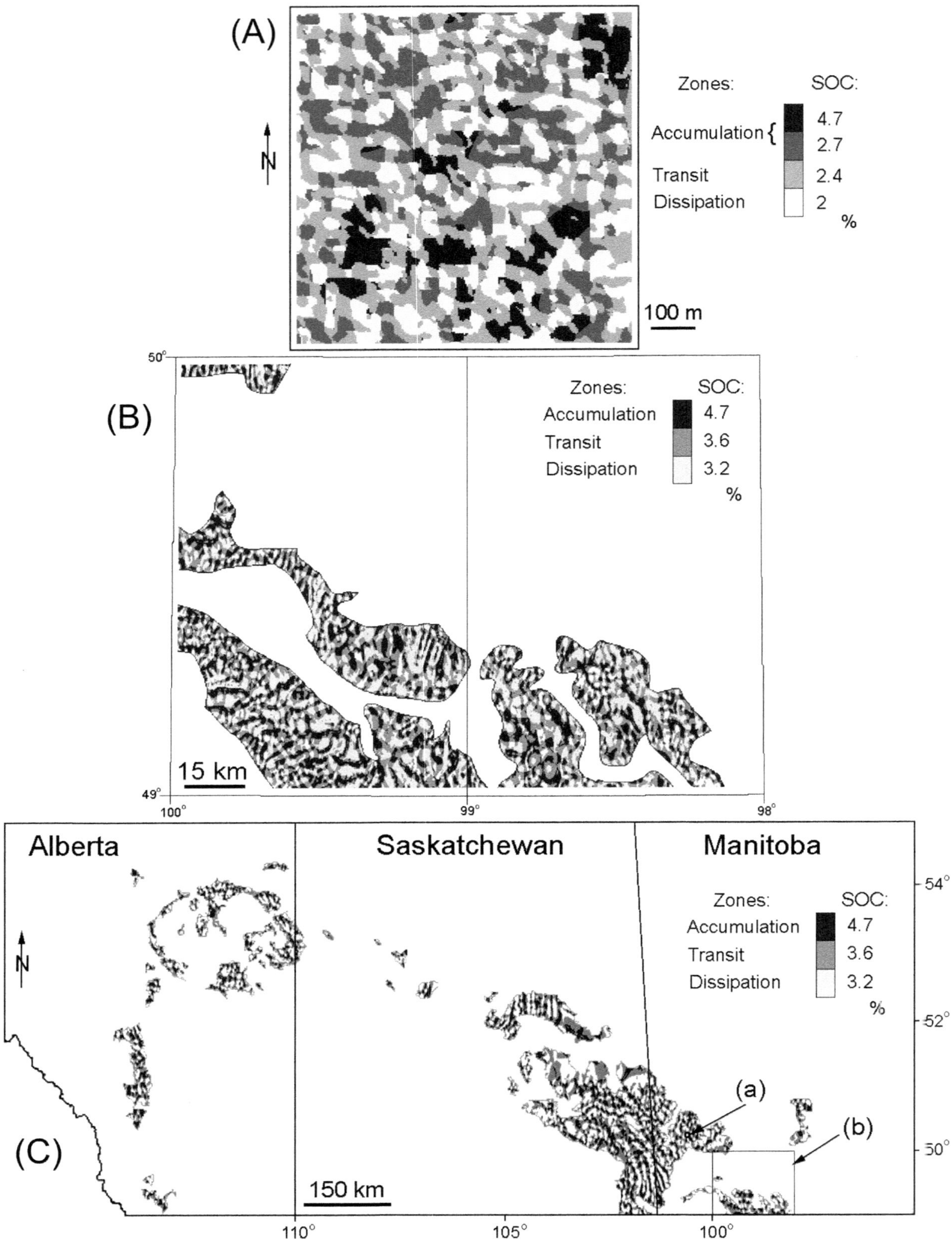

FIG. 11.6 Upscaling prediction of the topsoil organic carbon content within the Black Soils Zone of the Canadian prairies: (A) Field scale, the Miniota Precision Agriculture Research Site. (B) Regional scale, a part of southern Manitoba. (C) Subcontinental scale, southern portions of Alberta, Saskatchewan, and Manitoba provinces, Canada. *Circle (a)*, location of the field-scale study site; *frame (b)*, location of the regional-scale study area. For the field scale, a DEM was produced by the kinematic GNSS survey (see details in Section 11.3.1); the DEM grid spacing was 15 m. For the regional scale, we used a DEM derived from the Canadian Digital Elevation Data files based on the 62 G topographic chart, 1:250,000 scale (Natural Resources Canada, 1997). The DEM included 321,201 points (the elevation matrix 801 × 401); the grid spacing was 9″. For the subcontinental scale, we used a DEM derived from the GLOBO DEM (GLOBE Task Team and Others, 1999). The DEM included 1,080,000 points (the elevation matrix 1800 × 600); the grid spacing was 1′. To suppress high-frequency noise, the regional and subcontinental DEMs were smoothed. *(Modified from Florinsky, I. V., Eilers, R.G., 2002. Prediction of the soil organic carbon content at micro-, meso- and macroscales by digital terrain modelling. Transactions of the 17th World Congress of Soil Science, Bangkok, Thailand, 14–21 Aug. 2002, Symposium 52. International Society of Soil Science, Bangkok, Thailand, # 24, Figs. 2C, 3C, and 4C.)*

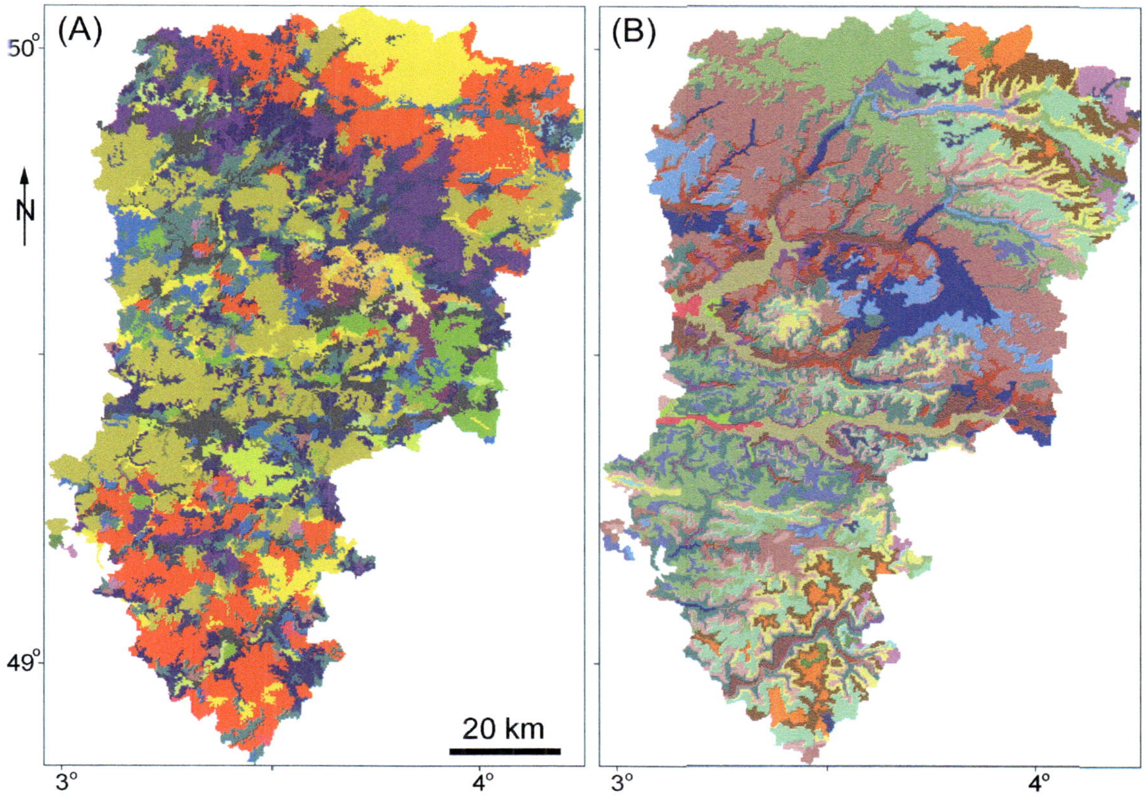

FIG. 11.7 Upscaling soil polygons for medium-scale soil mapping (exemplified by the Aisne Department, France): pedolandscape units resulting from the (A) bottom–up and (B) top–down approaches. The DEM grid spacing was 50 m. Map legends are not displayed as we discuss the geometry of soil polygons only. *(From Carré, F., Reuter, H.I., Daroussin, J., Scheurer, O., 2008. From a large to a small scale soil map: top-down against bottom-up approaches. Application to the Aisne soil map (France). In: Hartemink, A.E., McBratney, A., Mendonça-Santos, M.L. (Eds.), Digital Soil Mapping with Limited Data. Springer, Dordrecht, the Netherlands, pp. 203–212, Fig. 17.2, reproduced with permission from Springer Nature.)*

Stepanov and Loshakova, 1998; Volkova and Zhuchkova, 2000) (Fig. 11.8). Soil maps compiled by this method were positively different from conventional patchwork-like soil maps because of the clear representation of soil-topography relationships (cf. Fig. 11.8A and B).

11.3.4 Prediction accuracy

The problem of accuracy or effectiveness of predictive soil mapping is rather ambiguous. On the one hand, all mathematical approaches discussed in Section 11.3.2 have some criteria of prediction quality. Various validation indices can be used to measure prediction accuracy (e.g., RMSE and mean error) (Fig. 11.1B). On the other hand, the accuracy of quantitative methods in soil survey (pedometric techniques) can be described as follows (McBratney et al., 2000):

> *"Application of each of the pedometric techniques depends on the purpose, resolution and setup of the survey as the ultimate use of soil survey information determines the accuracy required. Different techniques produce different error of interpolation … In all the cases, since the purposes are different, the risk of taken wrong decisions due to survey error is also different. Therefore, the pedometric techniques … cannot just be applied to any situation without consideration of the specific needs and appropriateness of the inherent assumptions of the techniques."*

Particular attention should be given to pronounced differences in predictive soil patterns produced by different digital soil-mapping techniques (e.g., Fig. 11.2). However, when different soil scientists use conventional soil-mapping techniques, they also usually produce significantly different maps in terms of soil polygons (patterns) and soil taxonomic classes (Fig. 11.8). This effect is associated with (1) a high spatial variability of the soil cover, (2) spatiotemporal variability of the most soil properties, and (3) application of different concepts of soil formation and classification.

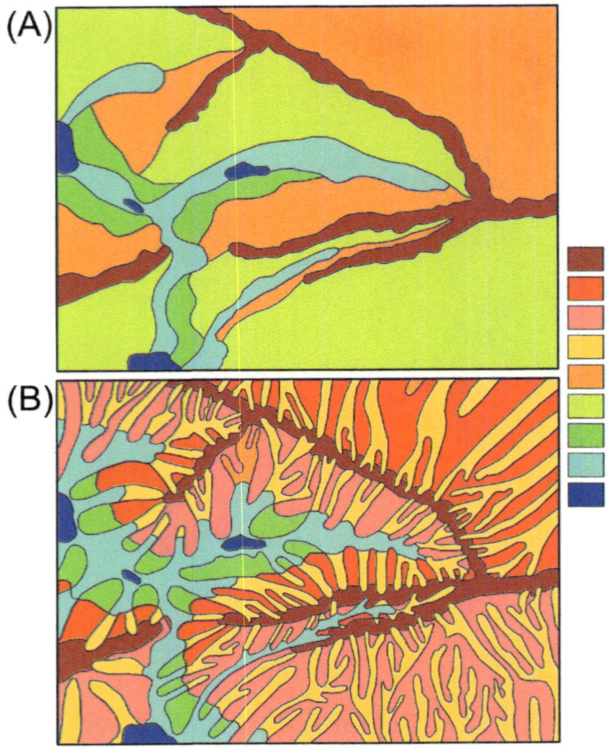

FIG. 11.8 Different geometry of soil polygons on two small-scale soil maps for a semi-desert region of the Central Asia (Stepanov et al., 1984) produced by two methods: (A) A conventional manual technique when a soil scientist implicitly considers topographic peculiarities. (B) The Relief Plasticity method when a soil scientist considers topographic peculiarities using a map of zero isolines of plan curvature (Stepanov et al., 1984, 1987). Map scale, coordinates, and legends are not displayed as we discuss the geometry of soil polygons only.

Comparative analyses of prediction accuracy or effectiveness of various methods of digital predictive soil mapping can be found elsewhere (Odeh et al., 1994, 2007; McBratney et al., 2000; Bishop and McBratney, 2001; Terra et al., 2004; Behrens and Scholten, 2007; Hengl et al., 2007b; Minasny and McBratney, 2007; Taylor and Odeh, 2007).

In closing this section, we should stress that although "a range of data analysis methods can be applied to develop models for spatial prediction using environmental correlation, … the success of environmental correlation depends on the strength of relationships between soil and environmental variables" (McKenzie and Ryan, 1999, pp. 81 and 92).

11.4 Topographic multivariable approach

The author's approach (Florinsky, 2009c) described in this section relates to predictive soil-mapping methods based on multiple regression analysis (Section 11.3.2). The approach includes the following steps:

1. Generation of a DEM of a site (field). One can use any appropriate technique to derive a DEM, such as unmanned aerial survey, kinematic global navigation satellite system (GNSS) survey, lidar aerial survey, or conventional topographic survey (Section 3.1). It is obvious that a particular type of survey affects the accuracy and resolution of DTMs and soil maps to be derived, but it does not influence the sequence of further data processing.
2. Derivation of a representative set of digital models of morphometric variables from the DEM. The recommended set includes elevation, 14 local attributes (k_h, k_v, H, K, k_{min}, k_{max}, K_a, E, K_r, k_{he}, k_{ve}, M, G, A), two nonlocal attributes (CA and DA, or SCA and SDA) as well as two combined variables (TWI and SPI). Definitions, interpretations, and formulas of these variables can be found in Chapter 2. It is a priori impossible to know which particular morphometric variables control a soil property under given natural conditions. Thus it is reasonable to utilize this representative set of attributes.
3. Selection of a plot within the site, which is representative relative to the entire site in terms of the spatial distribution of morphometric variables. The location and size of a representative plot are defined by expert opinion. The author's experience shows that the plot area should be no more than 10% of the entire site area. Validation of the morphometric representativeness of the plot (that is, estimation if there is a significant difference between two distributions of

topographic variables, within the plot and the entire site) can, for example, be performed by the Kolmogorov-Smirnov two-sample test (Daniel, 2000, chap. 8).

4. Soil sampling within the plot and soil laboratory analyses.
5. Correlation and multiple regression analyses of soil properties versus topographic variables for samples collected at the plot (the sample size should be no less than 40; see Footnote a in Chapter 10). Regression equations for soil properties are derived using morphometric variables as independent predictors. Notice that one may want to normalize topographic attributes in the 0–1 range in correlation and regression analyses. This allows estimation of the relative contribution of topographic variables directly from values of regression coefficients (Tomer and Anderson, 1995).
6. Derivation of predictive maps of soil properties for the entire site using regression equations obtained for the plot and DTMs of the entire site.

Notice that validation of the prediction using an independent data set is important, but it may be problematic because of labor and cost involved in field and laboratory works. Validation can be done through the following steps:

- Collect soil test samples at several points of the site located outside the plot.
- Measure soil properties under study for the test samples.
- Predict values of soil properties for the test points by DTMs and regression equations obtained previously.
- Perform statistical analysis to compare measured and predicted values of soil properties.

Jackknifing (Efron, 1982; Bishop and McBratney, 2001) can alternatively be applied to validate regression models. To provide an independent assessment of the prediction accuracy, the data set is randomly split into prediction and validation subsets. The first subset is used to create a model, which is validated with the second subset.

Several peculiarities are associated with the use of morphometric variables in predictive soil modeling:

- Slope aspect is a circular variable: its values range from 0° to 360°, and both of these values correspond to the north direction. So A cannot be used in linear statistical analysis. To estimate statistical relations between A and other variables, one can use either the approach of circular statistics (Mardia, 1972; Batschelet, 1981) or A_E and A_N instead A (King et al., 1999) (Section 2.2.3.1).
- Many soil and morphometric characteristics have nonnormal distributions. Therefore, it is correct to use the Spearman rank correlation coefficients (Daniel, 2000, chap. 9) to estimate statistical relationships between topography and soil properties.
- Some topographic attributes are linear combinations of others. For example, mean curvature is a combination of horizontal and vertical curvatures (Section 2.2.4.4), while topographic wetness and stream power indices are combinations of catchment area and slope gradient (Section 2.6.1). In this connection, it is incorrect to use together horizontal, vertical, and mean curvatures (or catchment area, slope gradient, and topographic wetness index) in multiple linear regression analysis.
- Every so often, a dynamic range of a topographic variable for a plot can be less than a dynamic range of the variable for the entire site. This is typical for catchment and dispersive areas. However, it is technically difficult to select a plot satisfying dynamic ranges of all morphometric attributes within the site. In the case of an imperfect representativeness of the plot, regression equations cannot be applied to zones, where values of a topographic predictor differ significantly from its dynamic ranges within the plot.

Examples of applying the described method are presented in Chapter 12.

Chapter 12

Analyzing relationships in the topography-soil system

12.1 Motivation

The correctness of DTM-based modeling and predictive mapping of soil properties depends not only on the spatial and temporal variability of soil properties by themselves (Campbell, 1979; Burrough, 1993) but also on two related factors.

First, there exists temporal variability in soil-topography relationships. For example, Burt and Butcher (1985) described temporal variability in the dependence of saturation depth and slope discharge on k_h and *TWI*, but there were no explanations for this phenomenon. Heddadj and Gascuel-Odoux (1999) found seasonal variations in the dependence of unsaturated hydraulic conductivity on slope position.

Second, there exist variations in the topographic control of soil properties depending on the depth of the soil layer. It is essential to recognize an effective soil layer wherein relationships between soil and topography are observable and statistically significant. For example, TOPMODEL (Section 11.2) uses the assumption that soil moisture content decreases with the soil depth due to a decline in hydraulic conductivity (Beven and Kirkby, 1979).

Study of denitrification—a process of biological conversion of nitrate (NO_3^-) into nitrous oxide (N_2O) and nitrogen (N_2) gases (Payne, 1981)—is important for understanding two fundamental interdisciplinary problems: (1) the nitrogen dynamics at regional and global scales (Mishustin and Shilnikova, 1971; Khalil and Rasmussen, 1992) and (2) the contribution of N_2O emission to global warming, stratospheric ozone depletion, and photochemical air pollution (Conrad, 1996; Meixner and Eugster, 1999). Denitrification is also of concern in agricultural practice because it is responsible for most of the nitrogen loss from fall-applied fertilizers: about 30% of nitrogen fertilizers applied to agricultural soils are lost to the atmosphere as a result of the activity of denitrifying bacteria (Murray et al., 1989). It is well known that denitrification is influenced by soil water content; the hydrological differences in the landscape control patterns of the spatial distribution of denitrifying activity (Groffman and Hanson, 1997). Thus, it is not surprising that denitrification processes depend on topography (Pennock et al., 1992; Van Kessel et al., 1993; Corre et al., 1996; Whelan and Gandolfi, 2002; Florinsky et al., 2004).

In this chapter, we demonstrate the possibilities of DTM-based analysis, modeling, and predictive mapping of the spatial distribution of soil properties using the approach described in Section 11.4. We studied:

- Temporal variability in the influence of topography on dynamic properties of soil (Florinsky et al., 1999, 2002).
- Changes in the influence of topography on dynamic properties of soil depending on the soil layer depth (Florinsky et al., 1999, 2002).
- The effect of topography on the activity of denitrifiers under different soil moisture conditions (Florinsky et al., 2004, 2009c).

12.2 Study sites

The studies were carried out at two sites, Miniota and Minnedosa, situated in southern Manitoba, Canada (Fig. 12.1). The sites are located about 260–280 km west of the city of Winnipeg. The Miniota site measures 809 × 820 m; the maximum elevation difference is about 6 m (Fig. 12.2A). The Minnedosa site measures 1680 × 820 m; the maximum elevation difference is about 13 m (Fig. 12.3A).

The study sites are located in the continental climate zone with warm summers and prolonged, cold winters. The mean annual temperature is 2.5°C, the mean summer temperature is 16°C, the mean winter temperature is −11°C. The mean annual precipitation is 460 mm including 310 mm of rainfall and 150 mm of snowfall (Fitzmaurice et al., 1999).

Digital Terrain Analysis. https://doi.org/10.1016/B978-0-443-24798-9.00017-0

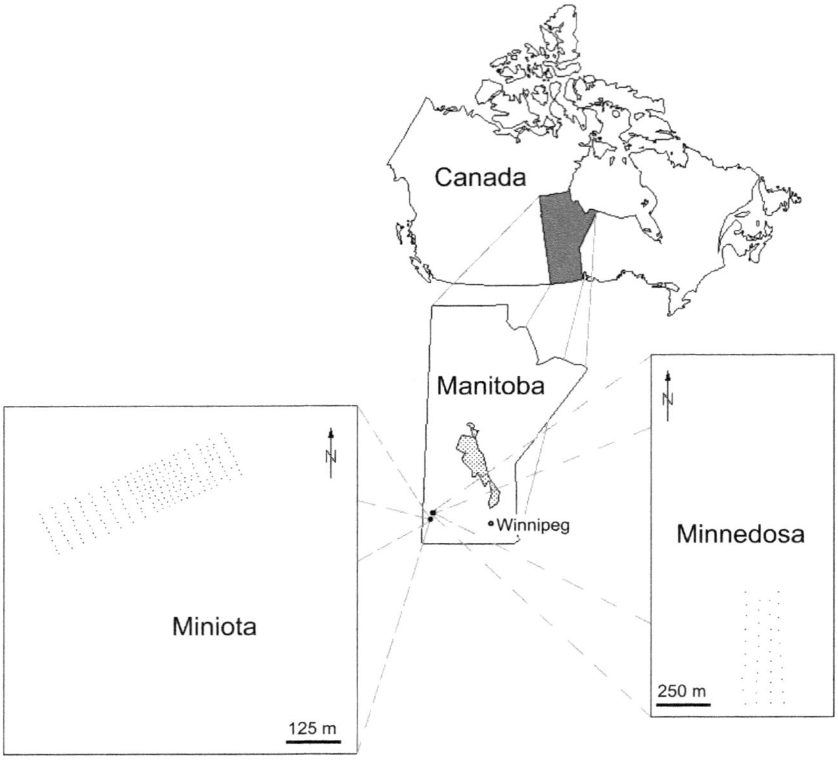

FIG. 12.1 Geographical location of the study sites: the Miniota site (50°13′40″ N 100°51′20″ W) and the Minnedosa site (50°14′43″ N 99°5[illegible]′34″ W). *Dots*, soil-sampling sites. *Publisher's note: Map lines delineate study areas and do not necessarily depict accepted national boundaries. (From Florinsky, I.V., 2010. Theory and Applications of Mathematical Cartographic Modeling of Topography (D.Sc. thesis). Institute of Mathematical Problems of Biology, Russian Academy of Sciences, Pushchino, Russia (in Russian), Fig. 3.7.)*

The study sites are representative of a broad region of undulating-to-hummocky glacial till landscapes in Western Canada. They are situated on the Newdale Plain, at an elevation of 500–580 m above sea level. The parent material consists of loamy-textured glacial till deposits (Clayton et al., 1977). For most part of the sites, soils are Black Chernozems and Gleysols (Soil Classification Working Group, 1998). The Newdale Orthic Black Chernozems series is predominant on well-drained crests and midslopes. The Varcoe Gleyed Carbonated Rego series is indicative of imperfectly drained downslopes, often in association with the Angusville Gleyed Eluviated series. The Penrith, Hamiota, and Drokan Gleysols series predominate in poorly drained depressions (Fitzmaurice et al., 1999; Bergstrom et al., 2001b).

The sites are located in the aspen parkland of the Canadian prairies, the northern extension of open grasslands in the Great Plains of North America. Native vegetation of willows (*Salix* sp.), aspen (*Populus tremuloides*), and sedges (*Carex* sp.) surrounds water-saturated depressions. Most of the sites have been cropped for more than 50 years. Before 1976, the Miniota site was farmed in a wheat-fallow rotation. In 1976, continuous cropping was initiated, with a cereal-broadleaf rotation. Since 1988, a zero-tillage management system has been employed. The Minnedosa site is a conventional tilled field cultivated by a deep-tiller, with one pass in autumn and one or two passes in spring. Crops have historically included wheat, barley, oats, rape, and flax (Bergstrom et al., 2001a). The site was cropped to rape in 2000 and wheat in 2001.

12.3 Materials and methods

Field works and laboratory analyses were performed by employees and students of the Manitoba Land Resource Unit, Agriculture and Agri-Food Canada, as well as the Department of Soil Science, University of Manitoba, Winnipeg, Canada (Manning, 1999; Bergstrom et al., 2001b; McMahon, 2001). Standard field and laboratory techniques were utilized (see later).

12.3.1 Field work

Irregular DEMs of the study sites were constructed using kinematic global navigation satellite system (GNSS) surveys (Section 3.1.2) with single-frequency Trimble 4600LS Surveyors mounted on all-terrain vehicles. The irregular DEMs of the Miniota and Minnedosa sites include 4211 and 7193 points, respectively.

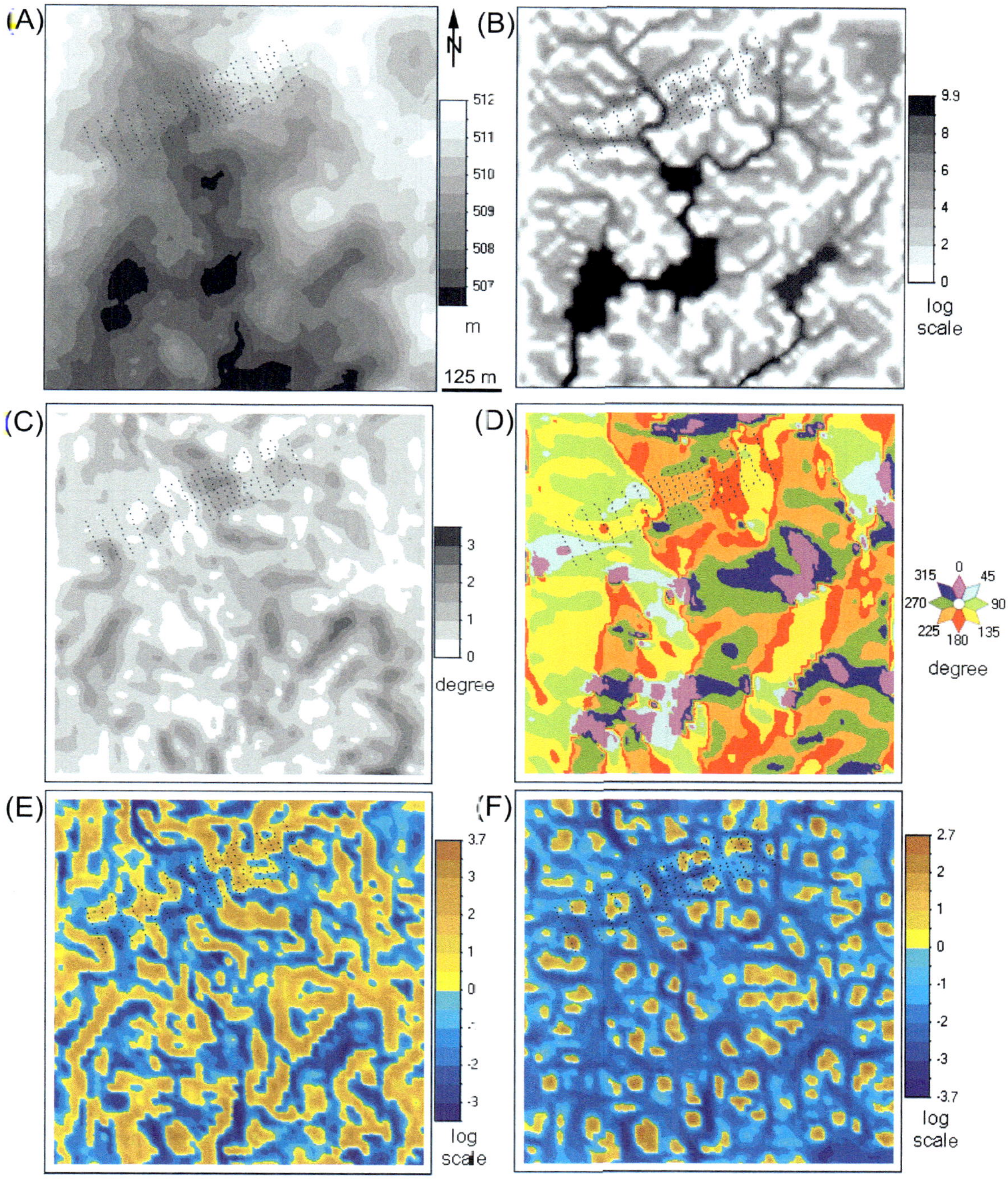

FIG. 12.2 The Miniota site, some morphometric variables: (A) Elevation. (B) Specific catchment area. (C) Slope gradient. (D) Slope aspect. (E) Vertical curvature. (F) Minimal curvature. *Dots*, soil-sampling sites. *(From Florinsky, I.V., Eilers, R.G., Burton, D.L., McMahon, S.K., Monreal, C.M., Farenhorst, A., 2009a. Predictive soil mapping based on digital terrain modeling. Geoinformatika (1), 22–32 (in Russian), Fig. 1.)*

A plot was selected within each study site to include a typical soil catena (Fig. 12.1).

On the Miniota site, the plot measured about 450 × 150 m with a difference in elevation of about 4.2 m (Fig. 12.2A). Soil samples were collected along 10 parallel transects about 450 m long, each about 11 m apart. There were 21 sampling points along each transect: points spaced at intervals of approximately 30 m on the crest and northeastern slope and at intervals of approximately 15 m on the southwestern steeper slope. In total, there were 210 sampling points in the plot.

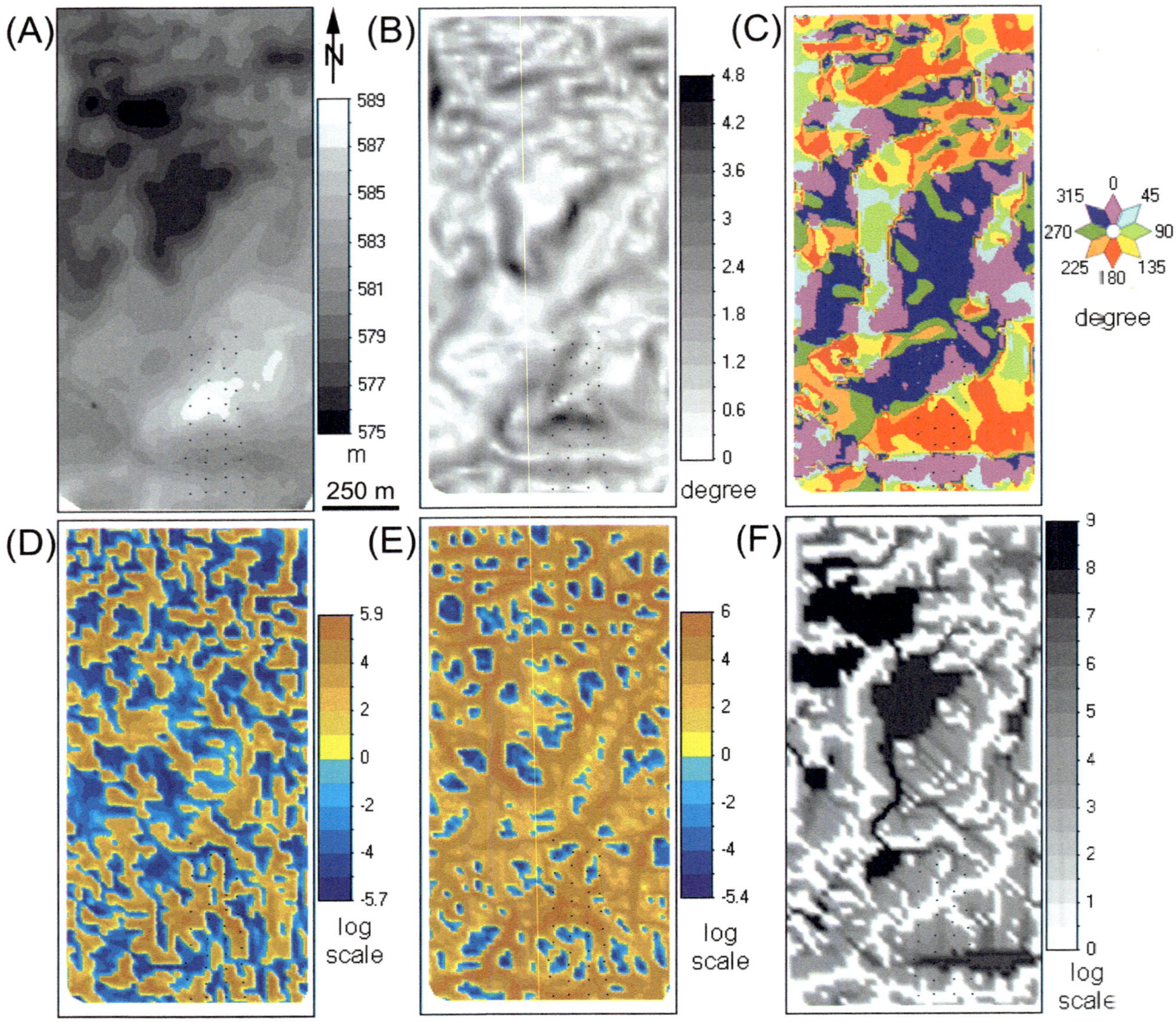

FIG. 12.3 The Minnedosa site, some morphometric variables: (A) Elevation. (B) Slope gradient. (C) Slope aspect. (D) Horizontal curvature. (E) Maximal curvature. (F) Specific catchment area. *Dots*, soil-sampling sites. *(From Florinsky, I.V., Eilers, R.G., Burton, D.L., McMahon, S.K., Monreal, C.M., Farenhorst, A., 2009a. Predictive soil mapping based on digital terrain modeling. Geoinformatika (1), 22–32 (in Russian), Fig. 2.)*

On the Minnedosa site, the plot measured about 500 × 200 m with a difference in elevation of about 8 m (Fig. 12.3A). Soil samples were collected along four parallel transects about 500 m long, each about 50 m apart. There were 10 sampling points along transects spaced at intervals of approximately 50 m, for a total of 40 sampling points in the plot.

All sampling points were georeferenced by GNSS and flagged.

On the Miniota site, soil samples were collected at 210 points from four depth increments—0–0.3, 0.3–0.6, 0.6–0.9, and 0.9–1.2 m—using a soil auger (Carter, 1993). Among other soil properties, these samples were used to determine gravimetric soil moisture content (%). For each depth increment, the soil moisture content was determined six times: in early May, early July, and late August in both 1997 and 1998 (in August 1997, for depth increments of 0–0.3 and 0.3–0.6 m).

On the Minnedosa site, two sets of soil attributes were sampled:

1. Two environmental properties influencing soil microbial activity: gravimetric soil moisture content (%) and soil bulk density ($g\,cm^{-3}$).
2. Six indices of soil microbial activity (Table 12.1): most probable number of denitrifiers, microbial biomass carbon content, denitrifier enzyme activity, denitrification rate, microbial respiration rate, and N_2O flux.

TABLE 12.1 Interpretations of soil microbial variables.

Variable, unit	Interpretation
Most probable number of denitrifiers, $\#_{organisms}g_{soil}^{-1}$	A measure of the number of denitrifiers in the soil
Microbial biomass carbon, $\mu g_C g_{soil}^{-1}$	A measure of the microbial biomass expressed as carbon
Denitrifier enzyme activity, $\mu g_N g_{soil}^{-1} h^{-1}$	A measure of the amount of denitrifying enzymes in the soil
Denitrification rate, $\mu g_N g_{soil}^{-1} h^{-1}$	A measure of the total gas N production from the soil
Microbial respiration rate, $\mu g_{CO_2} g_{soil}^{-1} h^{-1}$	A measure of the rate of the total microbial respiration in the soil
N_2O flux, $ng_{N_2O} m^{-2} s^{-1}$	A measure of the rate of N_2O emission from the soil

(From Florinsky, I.V., McMahon, S., Burton, D.L., 2004. Topographic control of soil microbial activity: a case study of denitrifiers. Geoderma 119 (1–2), 33–53, Table 2, reproduced with permission from Elsevier.)

Soil samples were collected at 40 points using aluminum soil cores 5 cm in diameter and 5 cm in height. The sampling depth was approximately 10 cm because each core pressed into the ground passed through surface litter and discontinuities (about 2 cm).

To minimize the impact of temporal variability in denitrification and storage of samples, the collection of soil cores and N_2O flux measurements occurred simultaneously. N_2O flux was estimated using vented static chambers (Hutchinson and Mosier, 1981). Chambers were inserted within 1 m of each sample point. After 1 h of accumulation, a 15-mL gas sample was taken from each chamber by a syringe, injected into 10-mL vacutainers, and returned to the laboratory for analysis (Burton et al., 2000).

Sampling of all soil attributes took place at two times, July 2000 and July 2001, to assay the effect of topography on the activity of denitrifiers at different humidity levels. The 2000 sampling occurred during a period of increased rainfall, while the 2001 sampling occurred following a period of decreased rainfall. Monthly precipitation of 133 and 26 mm was observed in July 2000 and July 2001, respectively, at the nearest weather station in the city of Brandon, located 40 km southward from the site.

12.3.2 Laboratory analyses

The soil moisture content was determined by drying 10-g soil subsamples at 105°C for 24 h. The bulk density was calculated from the moist weight of soil, water content, and core volume (Carter, 1993).

To allow direct comparison of microbial activity indices, analyses were conducted on a single subsample taken from each soil core within 1 h of collection.

The microbial biomass carbon content was determined using a fumigation-direct extraction technique (Carter, 1993). Two 15-g samples were weighed into square French bottles. One sample was extracted immediately using 30 mL of 0.5 M K_2SO_4. The second sample was fumigated for 24 h under a chloroform atmosphere and then extracted. Filtrate was analyzed for carbon using a Technicon Autoanalyzer.

The most probable number of denitrifiers was determined with the modified method of Tiedje (1994). Aliquots (0.5 mL) of serial dilutions from 10^{-3} to 10^{-6} were added to 4.5 mL of sterile nutrient broth in 10-mL Vacutainers. They were incubated at 25°C for approximately 7 days. Denitrifier presence was determined by measuring N_2O accumulation in the headspace using a Varian 3800 gas chromatograph.

Measurements were made sequentially (microbial respiration rate, denitrification rate, and denitrifier enzyme activity) within 48 h of sampling.

To measure the microbial respiration rate, the modified method of Zibilske (1994) was used. Soil (5 g) was incubated in 20-mL headspace vials at ambient temperature for 2 h. Then the headspace was sampled, and CO_2 concentration was determined by a Varian 3800 gas chromatograph.

To measure the denitrification rate, the modified method of Beauchamp and Bergstrom (1993) was used. Soil (5 g) was incubated at ambient temperature in 20-mL headspace vials containing atmospheric air and 10% acetylene. After 24 h, the headspace was analyzed for N_2O concentration using a Varian 3800 gas chromatograph. The acetylene blocks the conversion of N_2O to N_2, so the N_2O accumulation rate can be used as a proxy for total denitrification (N_2O+N_2).

To measure the denitrifying enzyme activity, the modified method of Beauchamp and Bergstrom (1993) was applied. Soil (5 g) was incubated at ambient temperature in 20-mL headspace vials with 4 mL of a buffer solution under a helium/acetylene atmosphere. The buffer solution consisted of 10 mM glucose, 10 mM KNO_3, and 50 mM K_2HPO_4. N_2O in the headspace was measured at 30-min intervals using a Varian 3800 gas chromatograph.

The N_2O flux samples collected in situ were analyzed using a Varian 3800 gas chromatograph.

12.3.3 Data processing

12.3.3.1 *Topographic modeling*

The irregular DEMs (Section 12.3.1) were converted into regular ones (Figs. 12.2A and 12.3A) by the Delaunay triangulation and a piecewise quadric polynomial interpolation (Agishtein and Migdal, 1991). The DEM grid spacings were 15 and 20 m for the Miniota and Minnedosa sites, respectively.

All recommended morphometric variables (Section 11.4) were derived from the regular DEMs (Figs. 12.2B–F and 12.3B–F). Local topographic attributes were calculated by the author's method (Section 4.2); nonlocal ones were estimated by the Martz-de Jong method (Section 4.4.1). Each DTM obtained included 2743 and 3193 points for the Miniota and Minnedosa sites, respectively. To deal with the large dynamic range of morphometric variables, we logarithmically transformed DTMs by Eq. (8.1) with $n=4$.

Then we used the Delaunay triangulation and a piecewise quadric polynomial interpolation (Agishtein and Migdal, 1991) of these DTMs to determine the values of all derived morphometric attributes at the sampling points.

Digital terrain analysis was performed by the LandLord software (Florinsky, 2016).

12.3.3.2 *Statistical analysis*

Tables 12.2 and 12.3 represent the statistical characteristics of morphometric variables for the plots and the entire sites. Tables 12.4 and 12.5 represent statistical characteristics of soil properties for the plots. On the Miniota site, the sample sizes were 210 and 2743 for the plot and the entire site, respectively. On the Minnedosa site, the sample sizes were 40 and 3193 for the plot and the entire site, respectively.

To estimate the topographic representativeness of the plots relative to the entire sites (Tables 12.2 and 12.3), we applied the Kolmogorov-Smirnov two-sample test (Daniel, 2000, chap. 8). Based on the test results, there are no statistical differences between curvature distributions within the plots and the entire sites. However, distributions of elevation, slope gradient, specific catchment area, as well as topographic and stream power indices within the plots are statistically different from those within the entire sites.

We also applied the Kolmogorov-Smirnov two-sample test to estimate statistical differences between two distributions of each soil property measured in different years (1997 and 1998 for the Miniota site; 2000 and 2001 for the Minnedosa site). Results are presented in Tables 12.4 and 12.5.

To evaluate the statistical relationships between soil properties and topographic attributes on the plots, we carried out multiple rank correlation analysis between all measured soil properties on the one hand and all morphometric variables on the other. The sample sizes were 210 and 40 for the Miniota and Minnedosa sites, respectively. The results of the correlation analysis are presented in Tables 12.6 and 12.7.

To describe the relationships between measured soil properties and topographic variables on the plots, the best combinations of these variables were chosen by stepwise linear regression (Kleinbaum et al., 2014). The sample sizes were 210 and 40 for the Miniota and Minnedosa sites, respectively. Results of regression analysis are presented in Tables 12.8–12.10 (we show regression equations with $R^2 \geq 0.40$).

Using the regression equations (Tables 12.8–12.10) and DTMs (Figs. 12.2 and 12.3) inserted into the equations as predictors, we derived predictive digital models and maps of some soil properties (Fig. 12.4). These models include 2756 and 3193 points for the Miniota and Minnedosa sites, respectively.

Notice that we did not derive predictive maps for areas of the sites wherein the values of morphometric predictors differ significantly from the dynamic ranges of the predictors within the plots (see Kolmogorov-Smirnov statistics in Tables 12.2 and 12.3). For the Miniota site, we did not perform predictive mapping for areas with values of $SCA > 6028\,\text{m}^2/\text{m}$, $z < 507.7\,\text{m}$, and $G > 2.5°$. For the Minnedosa site, we did not map areas with values of $SCA > 1754\,\text{m}^2/\text{m}$, $z < 580.8\,\text{m}$, and $G > 3.7°$. Hatching indicates areas omitted from the prediction (Fig. 12.4). Predictive patterns of soil properties (Fig. 12.4) generally resemble the structure of some morphometric maps (Figs. 12.2 and 12.3). Standard errors of the estimate may be used to assess roughly the prediction accuracy outside the plots (Tables 12.8–12.10).

Statistical analysis was carried out by Statgraphics software (© Statistical Graphics Corp., 1994–1997). Predictive soil maps were derived using the LandLord software (Florinsky, 2016).

TABLE 12.2 The Miniota Site: density traces and statistics for morphometric variables at the plot and the entire site.[a]

Density trace		Plot	Site	Density trace		Plot	Site
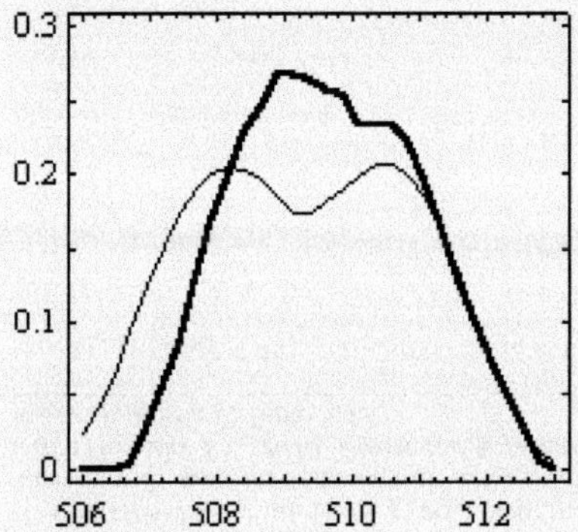	z, m			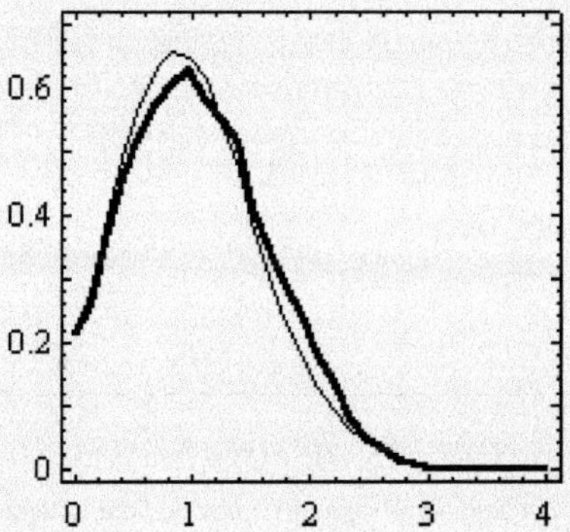	G, °		
	min	507.73	506.31		min	0.07	0.03
	max	511.82	512.13		max	2.34	3.43
	$\bar{x}$	509.66	509.31		$\bar{x}$	1.02	0.96
	s	1.13	1.47		s	0.51	0.50
	Dn (P)	0.211 (0.00)			Dn (P)	0.082 (0.14)	
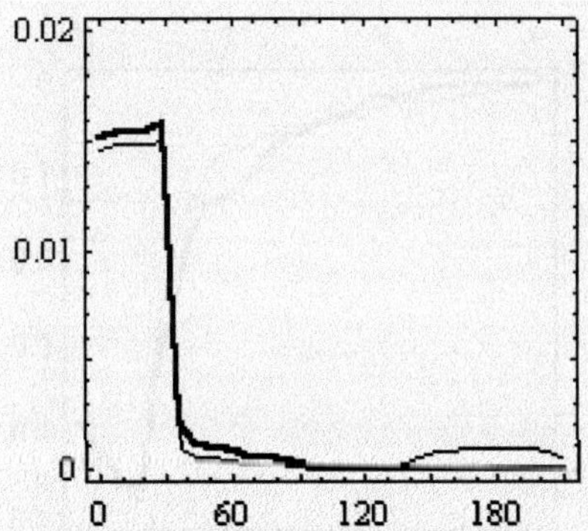	SCA, $\times 10^2$, m			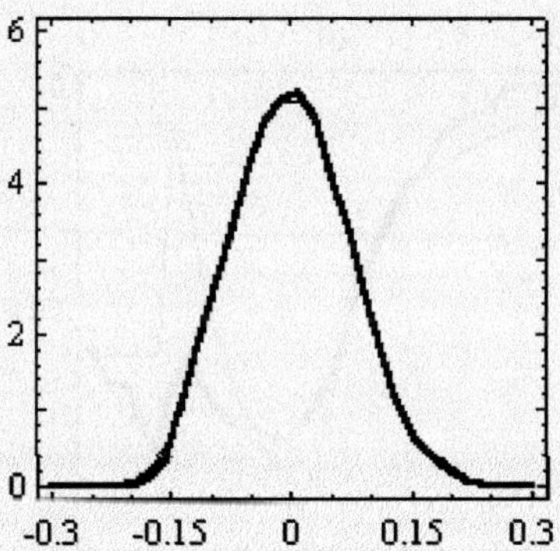	H, $\times 10^{-2}$, m^{-1}		
	min	0.00	0.00		min	−0.097	−0.250
	max	60.28	201.15		max	0.160	0.291
	$\bar{x}$	3.84	11.66		$\bar{x}$	0.003	0.000
	s	10.74	41.17		s	0.051	0.053
	Dn (P)	0.386 (0.00)			Dn (P)	0.053 (0.64)	
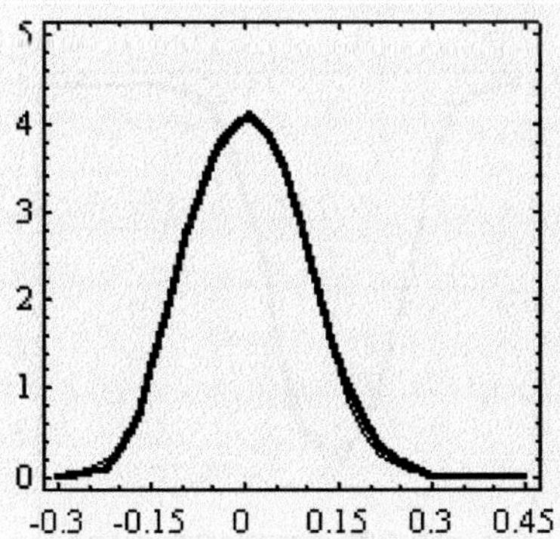	k_h, $\times 10^{-2}$, m^{-1}			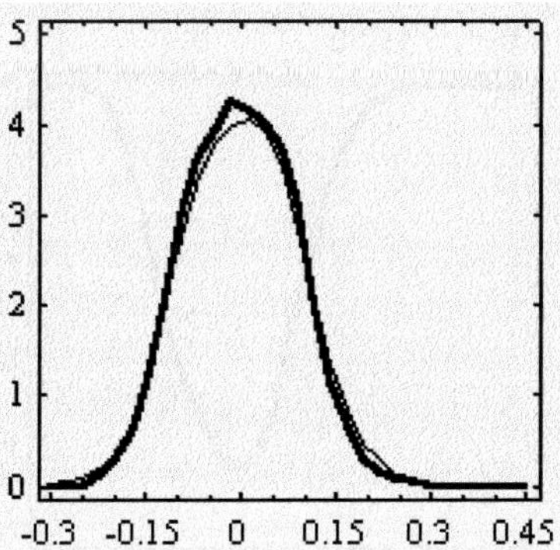	k_v, $\times 10^{-2}$, m^{-1}		
	min	−0.173	−0.296		min	−0.130	−0.288
	max	0.184	0.446		max	0.182	0.403
	$\bar{x}$	0.006	0.001		$\bar{x}$	0.000	−0.001
	s	0.066	0.066		s	0.055	0.066
	Dn (P)	0.056 (0.59)			Dn (P)	0.076 (0.21)	

Continued

TABLE 12.2 The Miniota Site: density traces and statistics for morphometric variables at the plot and the entire site—cont'd

Density trace		Plot	Site	Density trace		Plot	Site
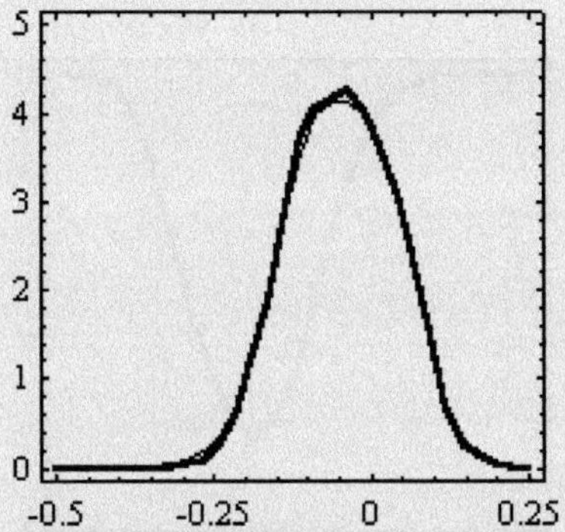	k_{min}, $\times 10^{-2}$, m^{-1}			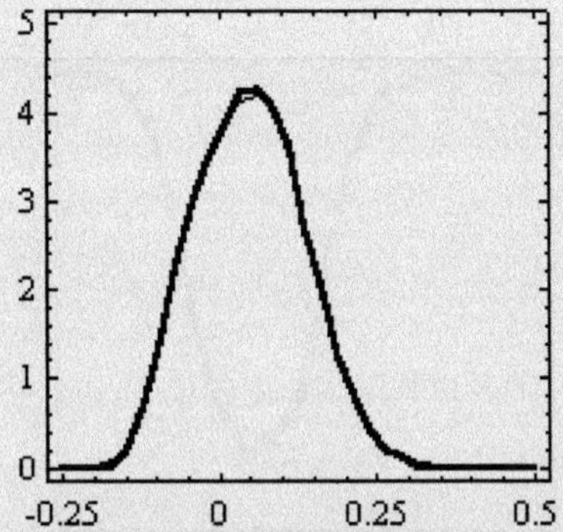	k_{max}, $\times 10^{-2}$, m^{-1}		
	min	−0.203	−0.431		min	−0.064	−0.185
	max	0.127	0.142		max	0.193	0.454
	$\overline{x}$	−0.044	−0.047		$\overline{x}$	0.050	0.047
	s	0.055	0.060		*s*	0.056	0.062
	Dn (*P*)	0.060 (0.49)			Dn (*P*)	0.051 (0.70)	
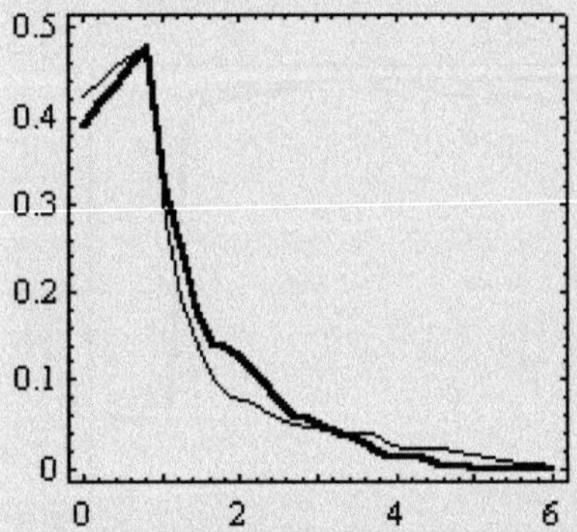	*SPI*			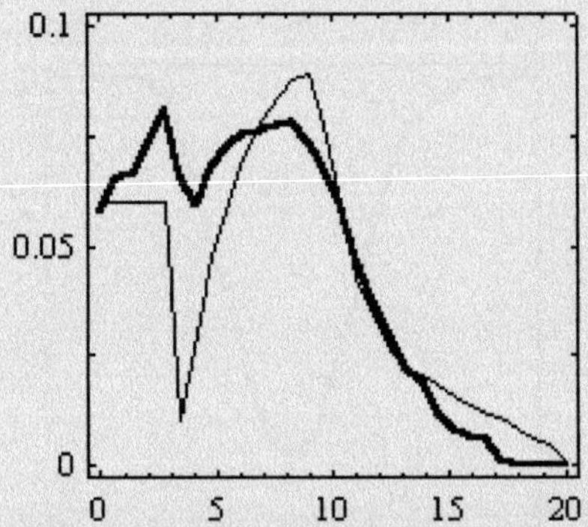	*TWI*		
	min	0.00	0.00		min	0.00	0.00
	max	4.10	5.60		max	14.60	18.85
	$\overline{x}$	0.73	0.72		$\overline{x}$	5.53	5.80
	s	0.88	1.09		*s*	4.04	4.92
	Dn (*P*)	0.361 (0.00)			Dn (*P*)	0.361 (0.00)	

[a]*Distribution density values are along* Y-*axis; morphometric variable values are along* X-*axis;* heavy and thin lines, *morphometric variables at the plot and the entire site, respectively; min,* minimum; *max,* maximum; $\overline{x}$, average; *s,* standard deviation; *Dn (P),* Kolmogorov-Smirnov statistics (significance level).

(From Florinsky, I.V., 2010. Theory and Applications of Mathematical Cartographic Modeling of Topography (D.Sc. thesis). Institute of Mathematical Problems of Biology, Russian Academy of Sciences, Pushchino, Russia (in Russian), Table 3.4.)

TABLE 12.3 The Minnedosa Site: density traces and statistics for morphometric variables at the plot and the entire site.[a]

Density trace		Plot	Site	Density trace		Plot	Site
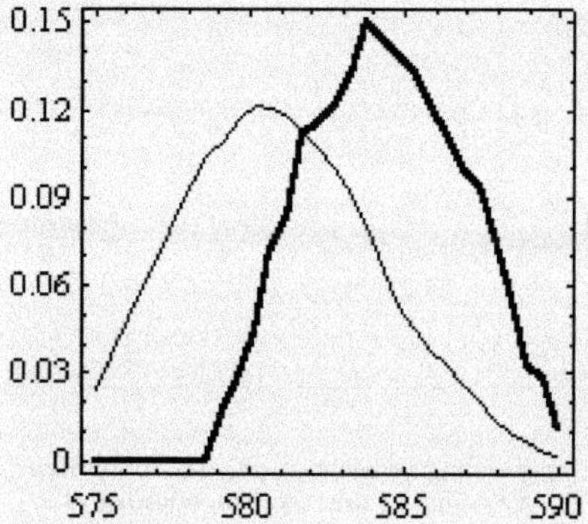	*z*, m			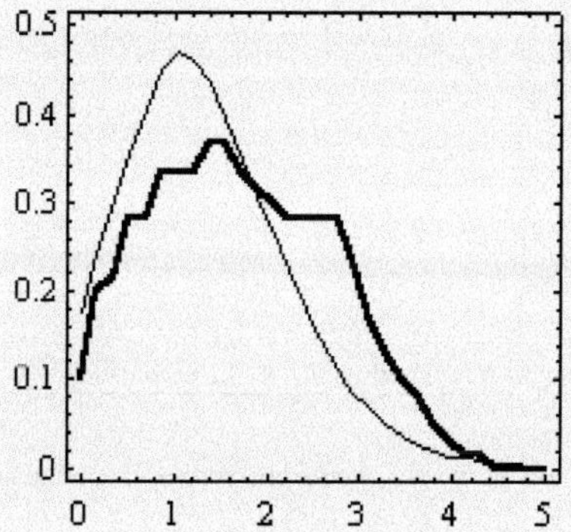	*G*, °		
	min	580.89	575.13		min	0.3	0.0
	max	588.49	588.57		max	3.7	4.8
	$\overline{x}$	584.44	580.88		$\overline{x}$	1.7	1.3
	s	2.16	2.83		*s*	0.9	0.8
	Dn (*P*)	0.566 (0.00)			Dn (*P*)	0.280 (0.00)	
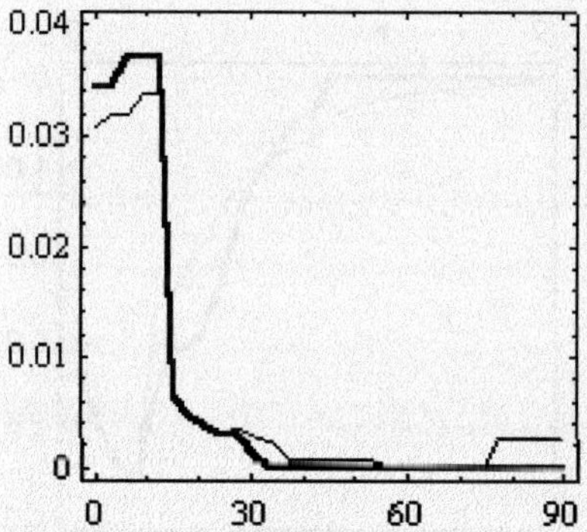	*SCA*, $\times 10^2$, m			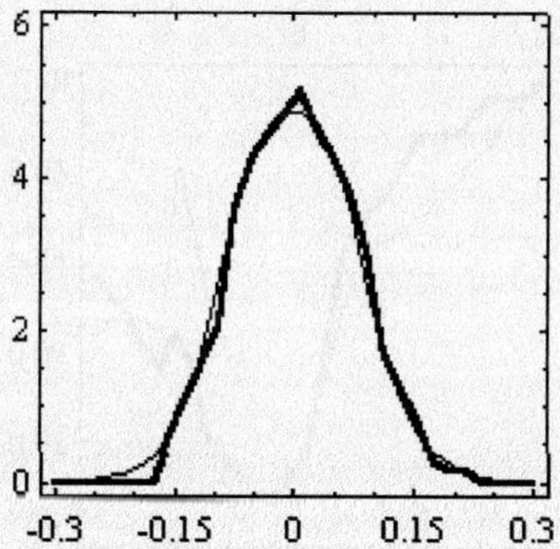	*H*, $\times 10^{-2}$, m^{-1}		
	min	0.00	0.00		min	−0.08	−0.26
	max	17.54	89.00		max	0.14	0.22
	$\overline{x}$	2.09	9.17		$\overline{x}$	0.01	0.00
	s	4.70	22.97		*s*	0.05	0.06
	Dn (*P*)	0.333 (0.00)			Dn (*P*)	0.139 (0.43)	
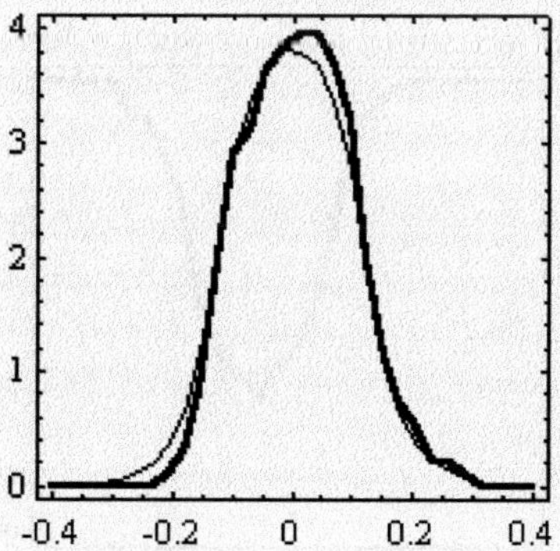	k_h, $\times 10^{-2}$, m^{-1}			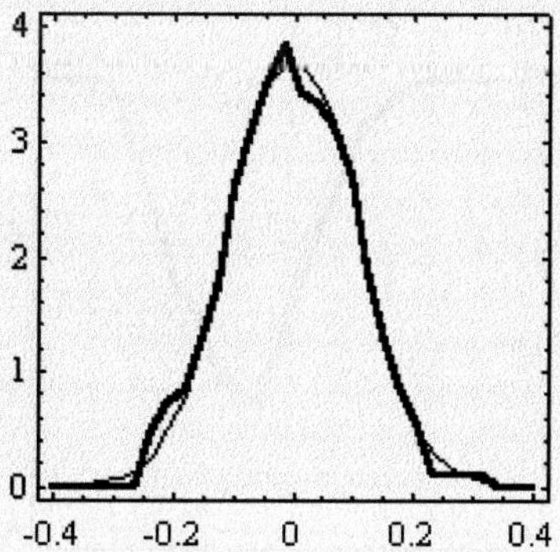	k_v, $\times 10^{-2}$, m^{-1}		
	min	−0.09	−0.32		min	−0.14	−0.34
	max	0.19	0.40		max	0.22	0.29
	$\overline{x}$	0.01	0.00		$\overline{x}$	0.00	0.00
	s	0.06	0.07		*s*	0.08	0.08
	Dn (*P*)	0.143 (0.40)			Dn (*P*)	0.084 (0.94)	

Continued

TABLE 12.3 The Minnedosa Site: density traces and statistics for morphometric variables at the plot and the entire site—cont'd

Density trace		Plot	Site	Density trace		Plot	Site
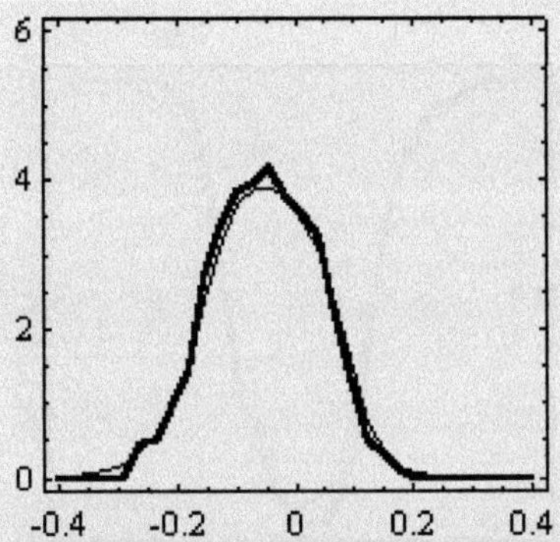	k_{min}, $\times 10^{-2}$, m^{-1}			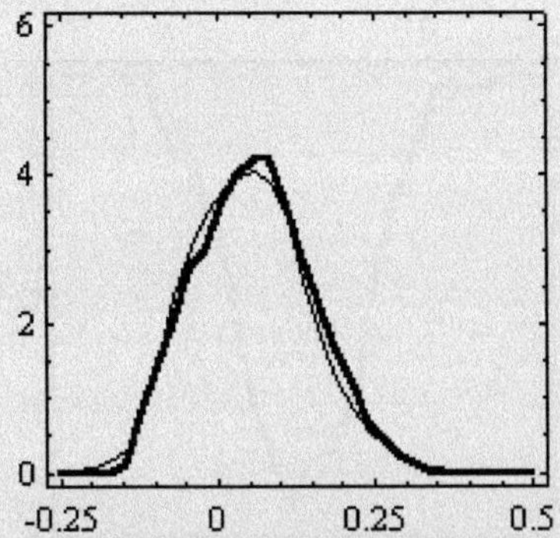	k_{max}, $\times 10^{-2}$, m^{-1}		
	min	−0.16	−0.37		min	−0.04	−0.23
	max	0.06	0.16		max	0.21	0.42
	$\bar{x}$	−0.05	−0.05		$\bar{x}$	0.06	0.05
	s	0.05	0.06		s	0.06	0.07
	Dn (P)	0.090 (0.91)			Dn (P)	0.136 (0.46)	
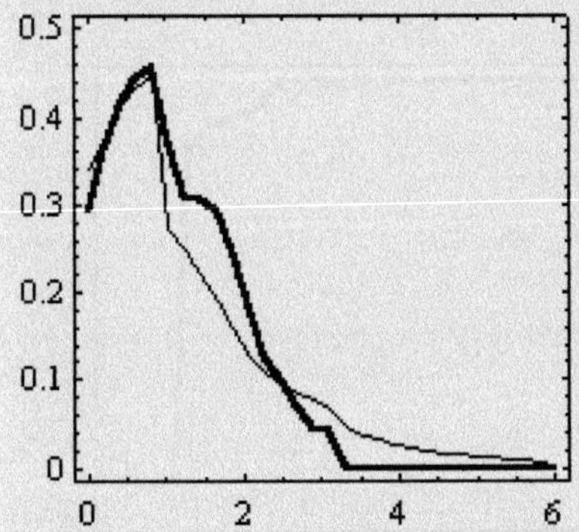	*SPI*			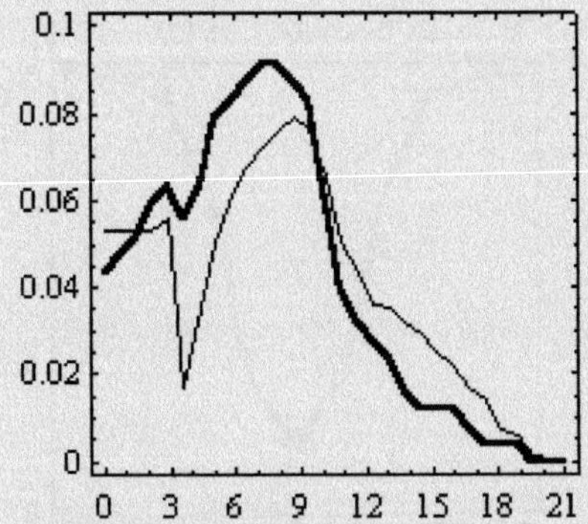	*TWI*		
	min	0.00	0.00		min	0.00	0.00
	max	2.33	5.98		max	16.13	20.39
	$\bar{x}$	0.84	0.93		$\bar{x}$	5.90	6.37
	s	0.74	1.11		s	4.19	5.15
	Dn (P)	0.333 (0.00)			Dn (P)	0.333 (0.00)	

[a]*For explanation, see footnotes to Table 12.2.*

(From Florinsky, I.V., McMahon, S., Burton, D.L., 2004. Topographic control of soil microbial activity: a case study of denitrifiers. Geoderma 119 (1–2), 33–53, Table 4, reproduced with permission from Elsevier.)

12.4 Results and discussion

12.4.1 Variability in relationships between soil and morphometric variables

For the Miniota site, the results of correlation analysis (Table 12.6) show that soil moisture at the depth 0–0.3 m was dependent on many topographic variables (except k_{he}, k_{ve}, E, M, K_r, K_a, and K) This was expected and supported by physical interpretations of topographic variables (Table 2.1) and results of previous research (see details in Chapter 9).

12.4.1.1 *Temporal variability*

However, different correlation coefficients describe statistical relationships between soil moisture and topography in different seasons (Table 12.6). For different seasons, we also obtained different regression equations of soil moisture content at depths 0–0.3 m with $R^2 \geq 0.40$ (Tables 12.8 and 12.9).[a] This is clear evidence of temporal variability in soil-topography relationships.

The existence of this phenomenon was quantitatively established for soil moisture and some dynamic soil properties that depend on soil moisture, such as residual phosphate and potassium (Florinsky et al., 1999). This was supported by results from other research groups that analyzed the spatiotemporal regularities of the distribution of soil moisture (Western et al., 1999; Sulebak et al., 2000). Topography controls many soil properties because its characteristics determine the ways of the lateral migration of water and zones of water accumulation (Chapter 9). Therefore, although temporal variability was certainly established for soil moisture only, we believe that this regularity can be extended to other dynamic soil properties associated with soil moisture.

Soil properties are the integral result of various processes having (1) different typical times (i.e., related to different temporal scales) and (2) different temporal variabilities (Stepanov et al., 1991; Targulian and Sokolova, 1996; Targulian and Krasilnikov, 2007). That is the reason why relationships between soil and topographic attributes are temporally unstable. Indeed, tectonic and erosion processes change the land surface relatively slowly, so morphometric variables can be seen as temporally stable determinants of soil development. Other factors of soil formation (e.g., characteristics of plant cover) have higher temporal variability. This leads to temporal variability in a spatially distributed soil response, which can be observed as temporal variability in relationships between soil and topographic properties. The rate of this temporal variability may be connected with the dynamic rate of a particular soil property. For example, relationships between topography and relatively static soil attributes such as soil horizon thickness may have low temporal variability; that is, they may remain unaltered over years.

12.4.1.2 *Depth variability*

Absolute values of the correlation coefficients of the soil moisture content with a topographic variable decreased and, at times, became statistically insignificant with increasing the depth of the soil layer (Table 12.6). This demonstrates a decrease in the topographic influence on soil moisture with the soil layer depth, at least for the conditions of the Miniota site.

The variability in relations between soil and topography depending on the soil layer depth (let us call this phenomenon *depth variability*) may stem from the spatial variability in the characteristic decline of hydraulic conductivity with depth. If this decline were the same at all points in the landscape, we would observe equal correlations between soil and topographic attributes at all depths. The spatial variability of the decline in hydraulic conductivity with depth can be associated with spatial variability of pedogenetic processes, the existence of relict soil patterns, and random inclusions of sand or silt lenses in glacial till. The strongest dependence of soil properties on topography occurred at 0–0.3 m depths within the Miniota site.

We suppose that in different landscapes, one may observe different depths of effective soil layers wherein relations of soil to topography are statistically significant. Moreover, more intricate situations may be found in some natural conditions and in some soil properties. In particular, topographic control of a soil property may increase with depth.

Indeed, the depth variability of topography-soil relationships was first quantitatively described by analyzing soil moisture and some dynamic soil properties depending on soil moisture, such as residual phosphate and potassium (Florinsky et al., 1999). For residual phosphate, the highest correlations with morphometric variables were observed for the depths 0.3–0.6 m (negative correlations with curvatures and positive ones with nonlocal variables). All correlations decreased and became insignificant with increases in the soil layer depth. For residual potassium, we observed a relatively

a. Such results are acceptable because $R^2 > 0.70$ is unusual and $R^2 \approx 0.50$ is common for quantitative soil spatial models (Beckett and Webster, 1971; Malone et al., 2009).

TABLE 12.4 The Miniota Site: density traces and statistics for the soil moisture content (%).[a]

Depth, m	Density trace		May		Density trace	
			1997	1998		
0–0.3	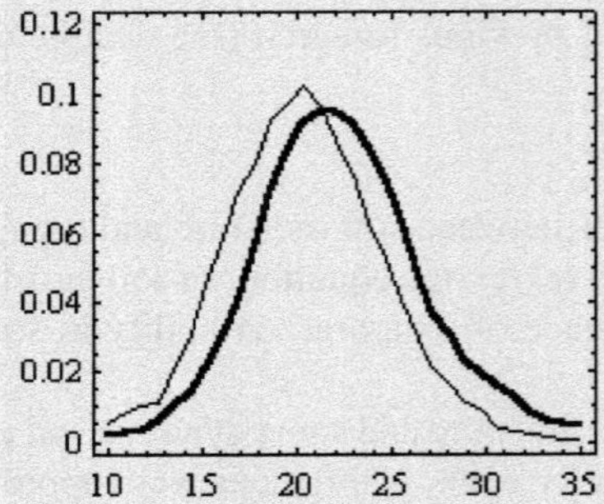	min	12.46	10.78	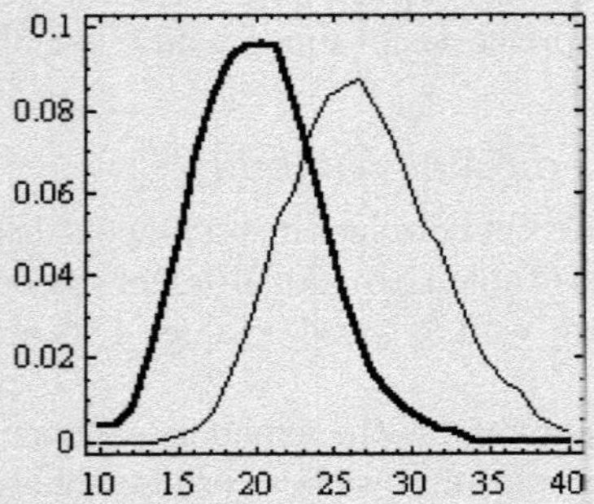	min
		max	34.09	34.05		max
		$\bar{x}$	22.34	20.34		$\bar{x}$
		s	3.88	3.59		*s*
		CV	17	18		CV
		Dn (*P*)	0.256 (0.00)			Dn (*P*)
0.3–0.6	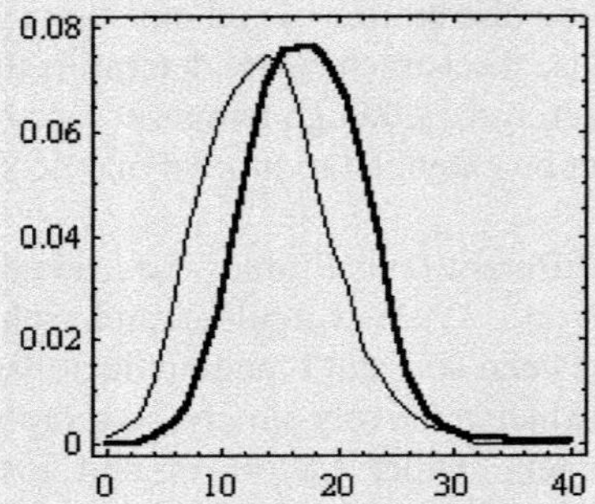	min	8.48	3.82	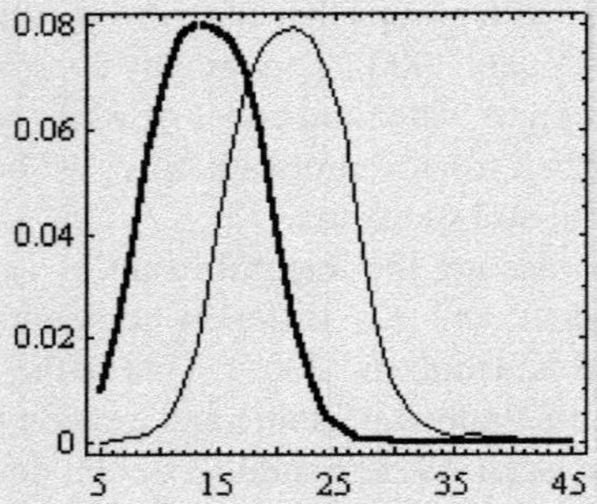	min
		max	39.86	25.20		max
		$\bar{x}$	17.29	13.64		$\bar{x}$
		s	3.60	3.85		*s*
		CV	21	28		CV
		Dn (*P*)	0.465 (0.00)			Dn (*P*)
0.6–0.9	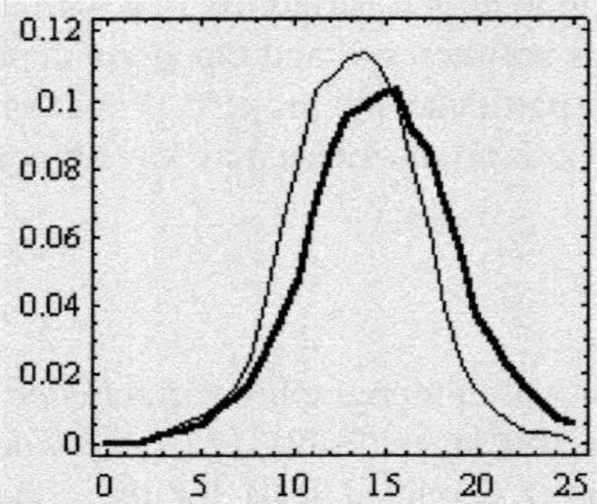	min	5.51	3.57	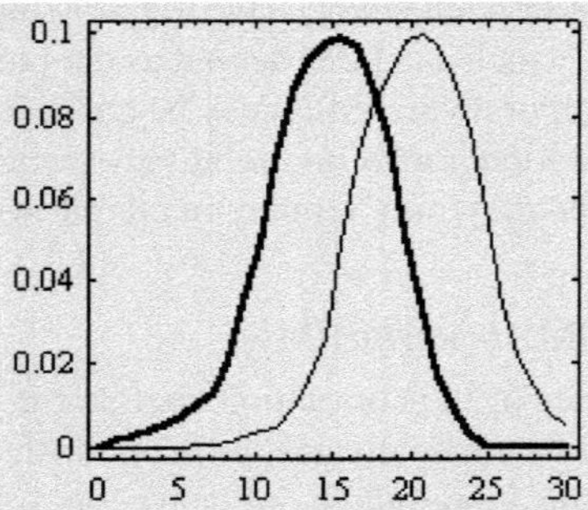	min
		max	24.97	22.58		max
		$\bar{x}$	14.85	13.21		$\bar{x}$
		s	3.34	2.71		*s*
		CV	22	21		CV
		Dn (*P*)	0.293 (0.00)			Dn (*P*)
0.9–1.2	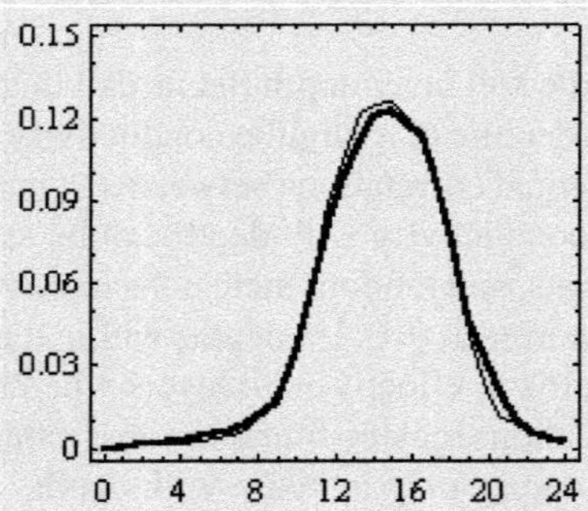	min	4.31	4.57	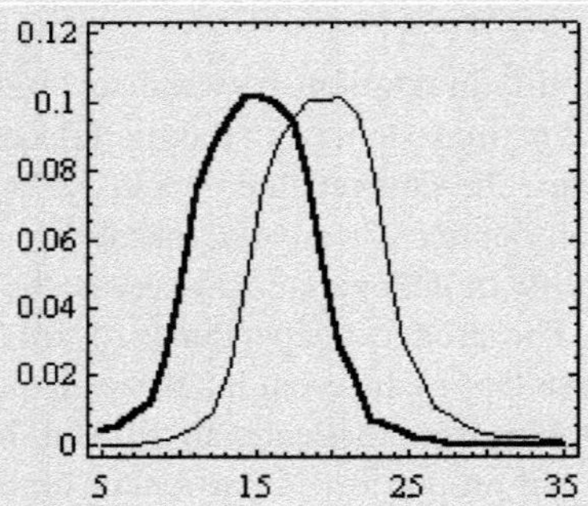	min
		max	22.93	23.86		max
		$\bar{x}$	14.66	14.57		$\bar{x}$
		s	2.58	2.40		*s*
		CV	18	16		CV
		Dn (*P*)	0.083 (0.48)			Dn (*P*)

[a]*Distribution density values are along* Y*-axis; soil moisture values are along* X*-axis;* heavy and thin lines, *soil moisture in 1997 and 1998, respectively;* CV, *coefficient of variation. For other symbols, see footnotes to Table 12.2.*
(From Florinsky, I.V., 2010. Theory and Applications of Mathematical Cartographic Modeling of Topography (D.Sc. thesis). Institute of Mathematical Problems of Biology, Russian Academy of Sciences, Pushchino, Russia (in Russian), Table 3.6.)

July		Density trace		August	
1997	1998			1997	1998
11.87	17.95	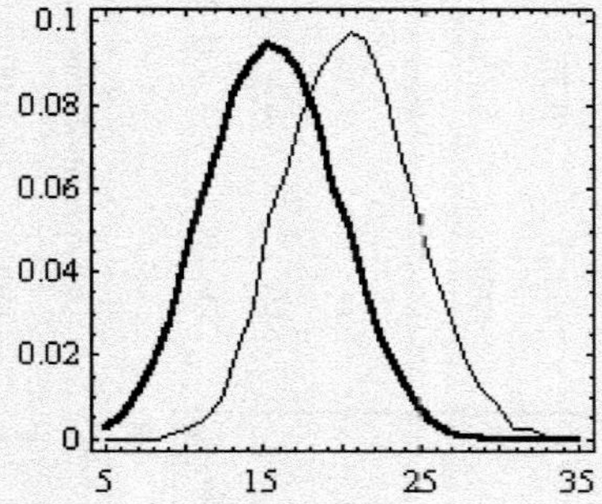	min	7.18	12.65
29.71	37.11		max	24.39	30.71
20.16	26.70		$\bar{x}$	15.59	20.49
3.00	3.76		s	3.13	3.03
15	14		CV	20	15
0.709 (0.00)			Dn (P)	0.600 (0.00)	
5.61	11.13	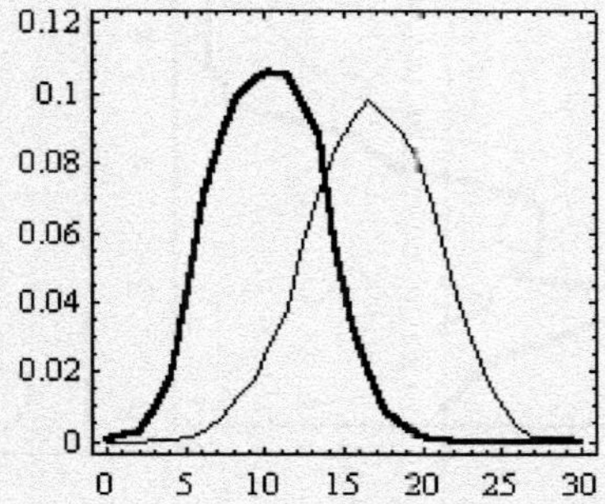	min	3.94	7.60
24.83	42.26		max	18.16	28.73
14.05	21.19		$\bar{x}$	10.23	16.75
2.71	3.26		s	2.12	3.03
19	15		CV	21	18
0.833 (0.00)			Dn (P)	0.795 (0.00)	
4.98	10.47	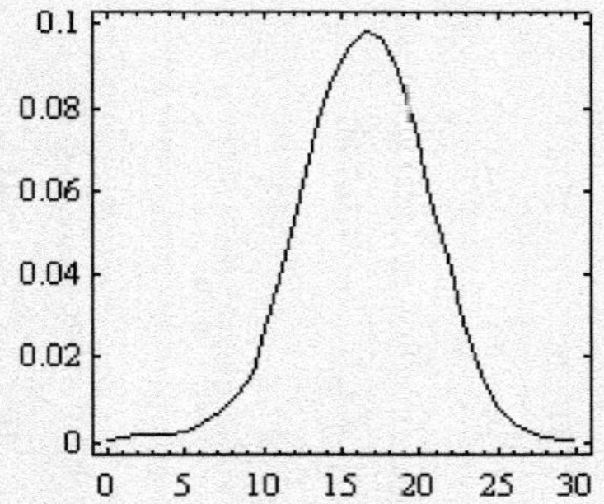	min	–	3.39
20.72	29.49		max	–	26.79
14.7[illegible]	20.45		$\bar{x}$	–	16.35
2.91	2.83		s	–	3.12
20	14		CV	–	19
0.759 (0.00)					
5.45	11.63	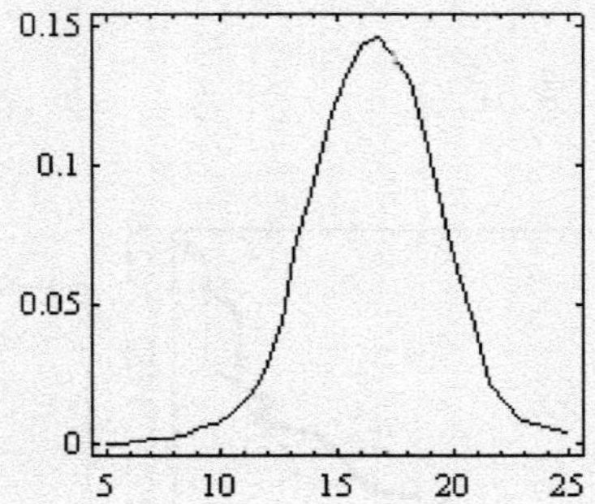	min	–	9.30
23.18	32.94		max	–	24.68
14.90	19.47		$\bar{x}$	–	16.63
2.6[illegible]	2.75		s	–	2.17
17	14		CV	–	13
0.737 (0.00)					

TABLE 12.5 The Minnedosa Site: density traces and statistics for soil attributes.[a]

Density trace		2000	2001	Density trace		2000	2001
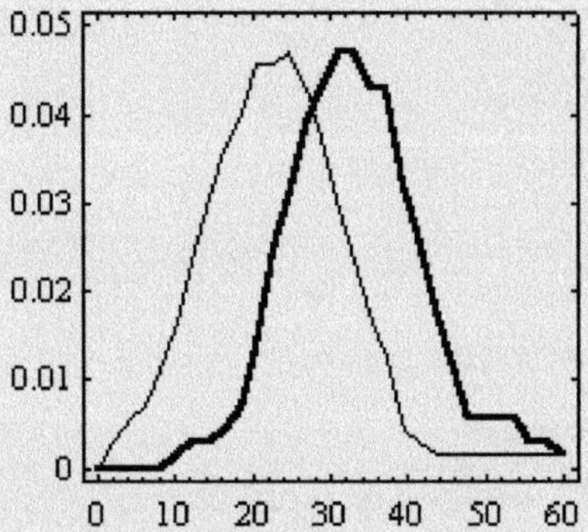	Soil moisture content, %			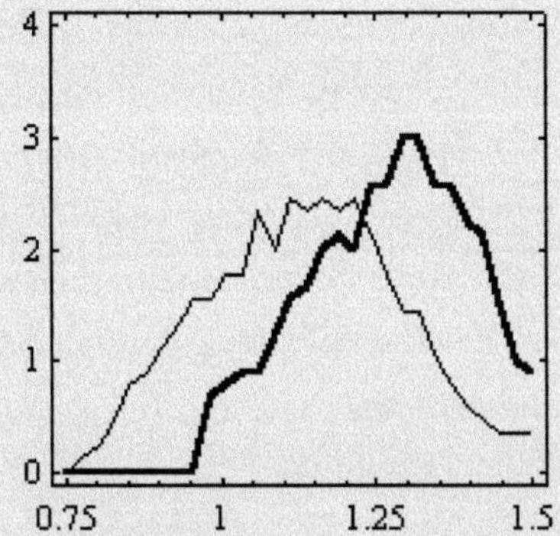	Bulk density, $g\,cm^{-3}$		
	min	19.00	9.83		min	1.07	0.87
	max	57.00	52.27		max	1.51	1.41
	$\bar{x}$	33.35	23.24		$\bar{x}$	1.27	1.13
	s	7.16	7.60		s	0.12	0.14
	CV	21	33		CV	9	13
	Dn (P)	0.650 (0.00)			Dn (P)	0.475 (0.00)	
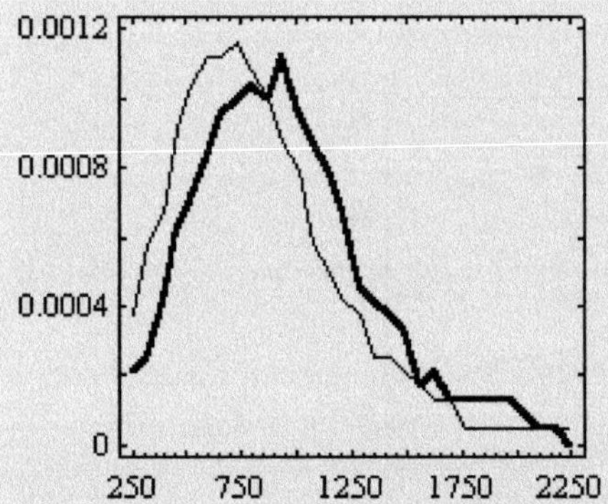	Microbial biomass carbon content, $\mu g_C g_{soil}^{-1}$			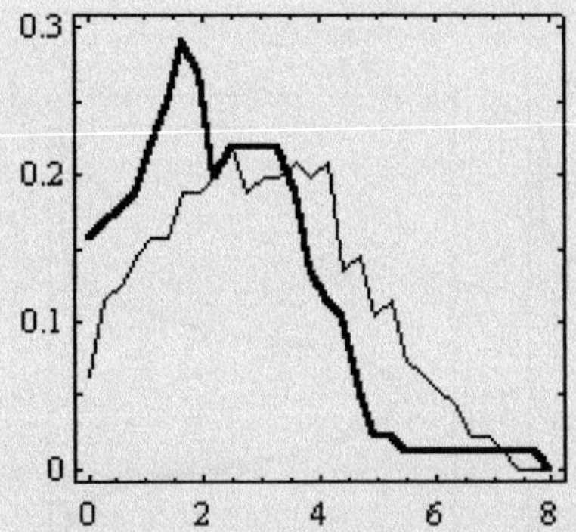	Denitrifier enzyme activity, $\mu g_N g_{soil}^{-1} h^{-1}$		
	min	316.20	257.02		min	0.58	0.60
	max	1907.20	2053.82		max	6.75	6.24
	$\bar{x}$	926.88	803.89		$\bar{x}$	2.25	2.92
	s	364.71	361.80		s	1.36	1.56
	CV	39	45		CV	60	54
	Dn (P)	0.225 (0.26)			Dn (P)	0.300 (0.05)	
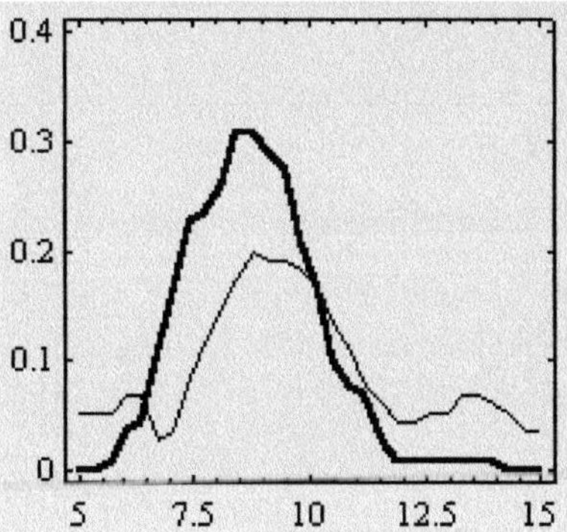	Most probable number, $\#_{organisms} g_{soil}^{-1}$			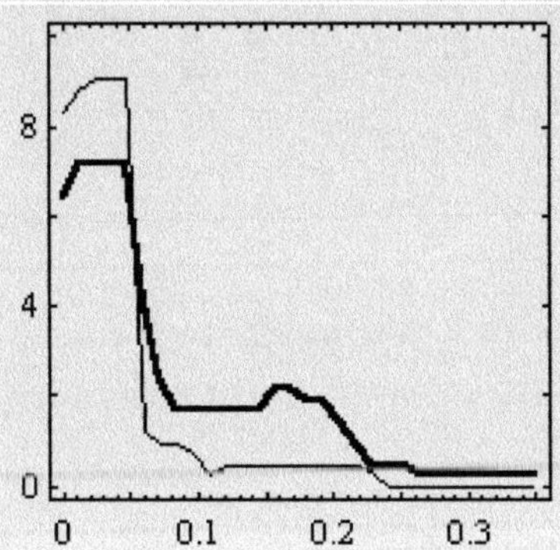	Denitrification rate, $\mu g_N g_{soil}^{-1} h^{-1}$		
	min	1056	141		min	0.00	0.00
	max	293,919	3,004,034		max	0.32	0.18
	$\bar{x}$	15,306	238,779		$\bar{x}$	0.06	0.01
	s	45,754	631,858		s	0.08	0.04
	CV	299	265		CV	136	276
	Dn (P)	0.325 (0.03)			Dn (P)	0.550 (0.00)	

TABLE 12.5 The Minnedosa Site: density traces and statistics for soil attributes[a]</ce:cross-ref—cont'd

Density trace		2000	2001	Density trace		2000	2001
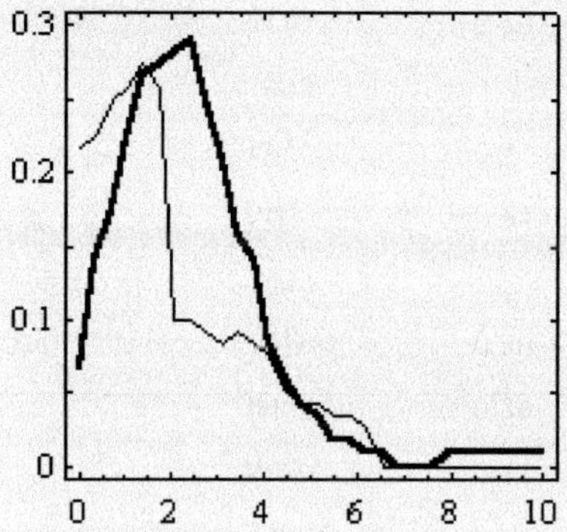	Respiration rate, $\mu g_{CO_2} g_{soil}^{-1} h^{-1}$			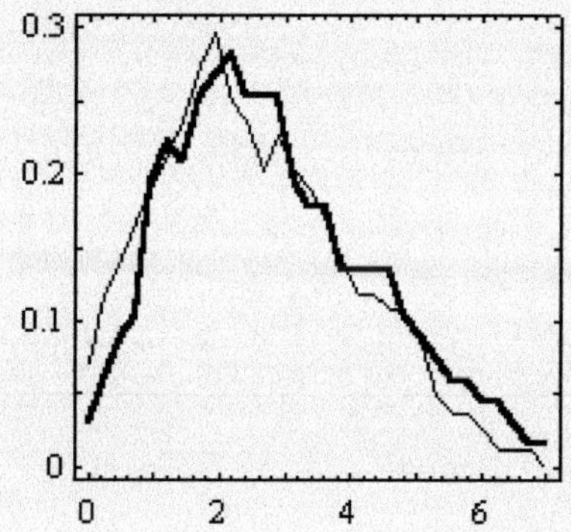	N_2O flux, $ng_{N_2O} m^{-2} s^{-1}$		
	min	0.27	0.14		min	0.00	0.92
	max	9.21	5.04		max	643.35	327.64
	$\overline{x}$	2.39	1.40		$\overline{x}$	41.99	31.51
	s	1.50	1.54		*s*	113.35	58.60
	CV	63	110		CV	270	186
	Dn (*P*)	0.575 (0.00)			Dn (*P*)	0.138 (0.89)	

[a]*Distribution density values are along* Y-*axis; soil property values are along* X-*axis;* heavy and thin lines, *samples collected in 2000 and 2001, respectively;* CV, *coefficient of variation. For other symbols, see footnotes to Table 12.2.*

(From Florinsky, I.V., 2010. Theory and Applications of Mathematical Cartographic Modeling of Topography (D.Sc. thesis). Institute of Mathematical Problems of Biology, Russian Academy of Sciences, Pushchino, Russia (in Russian), Table 3.7.)

TABLE 12.6 The Miniota Site: rank correlations[a] between the soil moisture content and morphometric variables.

Depth, m	Season	z	G	A_E	A_N	k_h	k_v	H	k_{min}	k_{max}	SCA	TWI	SPI
0–0.3	1997												
	May	−0.43	−0.38	–	0.17	−0.29	−0.41	−0.40	−0.35	−0.38	0.34	0.33	0.32
	July	−0.37	−0.39	–	0.27	−0.23	−0.37	−0.34	−0.25	−0.37	0.36	0.32	0.30
	August	−0.35	−0.25	–	0.14	−0.17	−0.24	−0.23	−0.21	−0.19	0.25	0.26	0.20
	1998												
	May	−0.44	−0.31	–	0.20	−0.29	−0.48	−0.44	−0.34	−0.46	0.46	0.46	0.42
	July	−0.29	−0.41	0.33	–	−0.23	−0.18	−0.26	−0.21	−0.24	0.29	0.30	0.22
	August	−0.36	−0.39	–	0.17	−0.23	−0.33	−0.31	−0.23	−0.36	0.33	0.30	0.24
0.3–0.6	1997												
	May	−0.23	−0.17	−0.19	–	–	–	–	–	–	–	–	–
	July	−0.26	−0.27	–	–	–	−0.21	–	–	–	0.19	–	0.14
	August	−0.19	−0.30	–	–	–	−0.20	−0.17	−0.15	−0.17	0.19	0.16	–
	1998												
	May	−0.38	−0.15	−0.25	–	–	−0.29	−0.22	−0.20	−0.19	0.21	0.17	0.21
	July	−0.23	−0.24	0.22	–	–	–	–	–	–	0.15	–	–
	August	−0.26	–	–	0.19	–	–	–	–	–	–	–	–
0.6–0.9	1997												
	May	−0.28	−0.17	–	–	–	–	–	–	–	0.14	0.13	–
	July	−0.26	–	–	–	–	–	–	–	–	–	–	–
	1998												
	May	−0.33	–	−0.26	0.15	–	−0.15	–	−0.15	–	–	–	–
	July	−0.29	−0.16	–	0.19	–	–	–	–	–	–	–	–
	August	−0.30	–	–	0.14	–	−0.14	–	–	–	–	–	–
0.9–1.2	1997												
	May	−0.24	–	–	–	–	–	–	–	–	–	–	–
	July	−0.41	–	–	–	–	−0.16	−0.18	−0.21	–	0.20	0.17	0.20
	1998												
	May	0.29	–	−0.21	–	–	−0.15	–	–	–	–	–	–
	July	−0.37	−0.15	0.17	0.19	−0.14	0.17	0.20	−0.24	–	–	–	–
	August	−0.21	–	–	0.20	–	–	–	–	–	–	–	–

[a] $P \leq 0.05$ for statistically significant correlations; dashes are statistically insignificant correlations. There were no significant correlations between soil properties and the following morphometric variables: k_{he}, k_{ve}, E, M, K_r, K_a, and K.

(From Florinsky, I.V., 2010. Theory and Applications of Mathematical Cartographic Modeling of Topography (D.Sc. thesis). Institute of Mathematical Problems of Biology, Russian Academy of Sciences, Pushchino, Russia (in Russian), Table 3.8.)

TABLE 12.7 The Minnedosa Site: rank correlations[a] between soil properties and morphometric variables in the wetter and drier conditions.

		Morphometric variable									
Soil property	**Year**	z	G	A_E	k_v	H	k_{min}	k_{max}	SCA	TWI	SPI
Soil moisture content	2000	−0.51	–	–	−0.60	−0.48	−0.33	−0.43	0.42	0.53	0.34
	2001	−0.50	–	–	−0.37	−0.31	–	−0.33	–	–	–
Bulk density	2000	0.53	0.35	–	0.65	0.57	0.36	0.58	−0.47	−0.65	−0.49
	2001	–	–	–	–	–	–	–	–	–	–
Most probable number	2000	–	−0.41	−0.40	–	–	–	–	–	–	–
	2001	–	–	–	–	–	–	–	–	–	–
Microbial biomass carbon content	2000	−0.48	−0.30	–	−0.39	−0.32	–	−0.33	0.38	0.50	–
	2001	–	–	–	–	–	–	–	–	–	–
Denitrifier enzyme activity	2000	–	–	–	–	–	–	–	–	–	–
	2001	–	–	–	–	–	–	–	–	–	–
Denitrification rate	2000	–	–	–	–	–	–	–	0.52	0.46	0.40
	2001	–	–	–	–	–	–	–	–	–	–
Microbial respiration rate	2000	–	−0.40	–	–	–	–	–	–	–	–
	2001	–	–	–	–	–	–	–	–	–	–
N_2O flux	2000	–	–	–	−0.37	–	–	–	–	–	–
	2001	–	–	–	–	–	–	–	–	–	–

[a]*$P \leq 0.05$ for statistically significant correlations; dashes are statistically insignificant correlations. There were no significant correlations between soil properties and the following variables: k_h, k_{he}, k_{ve}, E, M, K_r, K_a, K, and A_N.*

(From Florinsky, I.V., 2010. Theory and Applications of Mathematical Cartographic Modeling of Topography (D.Sc. thesis). Institute of Mathematical Problems of Biology, Russian Academy of Sciences, Pushchino, Russia (in Russian), Table 3.9.)

TABLE 12.8 The Miniota Site: parameters and statistics of the regression equation for the dependence of the soil moisture content (the depth 0–0.3 m, July 1997) on morphometric variables.

			95% Confidence interval	
Predictor	**Estimate**	**Standard error**	**Lower limit**	**Upper limit**
Constant	250.02	89.11	74.31	425.73
z	−0.45	0.17	−0.79	−0.10
G	−2.41	0.34	−3.08	−1.74
A_E	−0.79	0.24	−1.26	−0.32
A_N	1.20	0.35	0.52	1.89
k_v	−22.11	4.62	−31.22	−13.01
k_{min}	16.82	5.16	6.65	27.00
ln*SCA*	0.34	0.10	0.15	0.52

Source	**Sum of squares**	**Df**	**Mean square**	***F*-ratio**	***P*-value**
Model	854.65	7	122.09	24.08	0.00
Residual	1024.02	202	5.07		
Total (Corr.)	1878.67	209			
R^2		0.45			
Standard error of the estimate		2.25			
Durbin-Watson statistic		1.75			

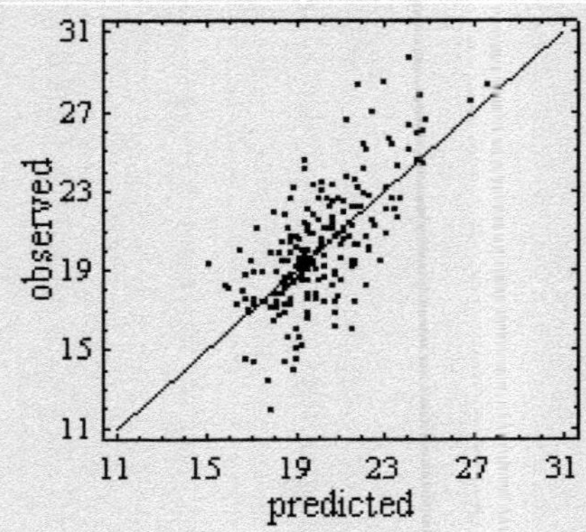

(From Florinsky, I.V., 2010. Theory and Applications of Mathematical Cartographic Modeling of Topography (D.Sc. thesis). Institute of Mathematical Problems of Biology, Russian Academy of Sciences, Pushchino, Russia (in Russian), Table 3.10.)

stable negative correlation with G for all depths within the 1.2-m soil layer. Other morphometric variables did not influence the spatial distribution of residual potassium at 0.3–0.9 m depths. For the surficial layer (0–0.3 m), we observed slight positive correlations with nonlocal topographic attributes and slight negative ones with some curvatures. However, for the depths of 0.9–1.2 m, we found the inverse situation: there were slight negative correlations with nonlocal topographic attributes and slight positive ones with curvatures. The cause of such a behavior of residual potassium is not clear.

Later Mitusov (2001) and Shary (2005) also studied depth variability. They found that the dependence of soil properties on morphometric variables may both decrease and increase as the soil layer depth increases. In particular, for a cultivated field located in the southern Moscow Region, the dependence of available phosphorus (P_2O_5) as well as calcium and magnesium exchange cations on morphometric metrics of overland flows (i.e., k_h, k_v, k_{he}, and CA) decreases as the soil layer depth increases. More complex relationships were found for the topographic control of potassium and sodium exchange cations. For potassium cations, as the soil layer depth increases, there is a decrease in correlations with local flow attributes (i.e., k_h and E) and an increase in those with nonlocal morphometric variables. In contrast, for sodium cations, as the soil layer depth increases, there is a decrease in correlations with nonlocal topographic variables and an increase in those with local variables (i.e., k_h). The cause of such a behavior of potassium and sodium cations is not clear.

For topography-NO_3^- relationships, there were no significant correlations for the depths 0–0.2 m (Mitusov, 2001; Shary, 2005). This finding may be explained by a stronger impact of nontopographic factors (biota, etc.) on this nitrogen form in

TABLE 12.9 The Miniota Site: parameters and statistics of the regression equation for the dependence of the soil moisture content (the depth 0–0.3 m, May 1998) on morphometric variables.

			95% Confidence interval	
Predictor	**Estimate**	**Standard error**	**Lower limit**	**Upper limit**
Constant	319.80	96.71	129.12	510.49
z	−0.58	0.19	−0.96	−0.21
G	−2.56	0.39	−3.33	−1.79
A_E	−0.88	0.27	−1.42	−0.35
k_v	−33.87	5.29	−44.30	−23.44
k_{min}	25.68	5.94	13.97	37.38
lnSCA	0.55	0.11	0.33	0.76

Source	**Sum of squares**	**Df**	**Mean square**	**F-ratio**	***P*-value**
Model	1326.28	6	221.05	32.91	0.00
Residual	1363.44	203	6.72		
Total (Corr.)	2689.72	209			
R^2		0.49			
Standard error of the estimate		2.59			
Durbin-Watson statistic		1.802			

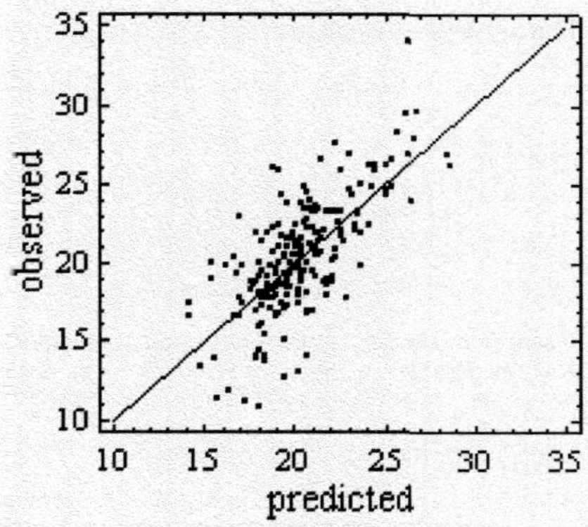

(From Florinsky, I.V., 2010. Theory and Applications of Mathematical Cartographic Modeling of Topography (D.Sc. thesis). Institute of Mathematical Problems of Biology, Russian Academy of Sciences, Pushchino, Russia (in Russian), Table 3.11.)

upper soil layers. For deeper soil layers, correlations of NO_3^- with all flow attributes decrease, while correlations with form attributes increase. Moreover, the spatial distribution of NO_3^- in the deepest soil layer is the reverse of that in the medium soil layer, which is demonstrated by the change of signs of correlation coefficients (as in the case with residual potassium; as discussed earlier). Thus, a lateral redistribution of NO_3^- takes place during its vertical movement. The causes of this phenomenon are not understood. It is clear that the depth variability of topography-soil relationships invites further investigations.

Temporal and depth variabilities in relationships between soil and topography should be considered along with regional and scale variabilities in the topographic control of soil attributes. Regional variability refers to distinctions in the topographic control of soil properties under different natural conditions. Scale variability refers to the change in the character of soil-topography relationships under changes at the hierarchical level of the biogeocoenosis and/or study scale (Chapter 10).

12.4.2 Topography and denitrification

Different rainfall conditions in July 2000 and July 2001 (Section 12.3.1) resulted in different levels of soil water content in the landscape: the averages of the soil moisture content in the plot differed by 10% in these seasons (Table 12.5). In wetter and drier soil conditions, there were strong differences in relationships between soil environmental and microbial properties (Florinsky et al., 2004), as well as between soil properties and topographic attributes (Table 12.7).

TABLE 12.10 The Minnedosa Site: parameters and statistics of the regression equation for the dependence of the microbial biomass carbon content (July 2000) on morphometric variables.

			95% Confidence interval	
Predictor	**Estimate**	**Standard error**	**Lower limit**	**Upper limit**
Constant	47,247.7	19,038.5	8309.6	86,185.8
z	−79.17	32.61	−145.85	−12.48
ln*SCA*	73.29	34.29	3.15	143.42

Source	**Sum of squares**	**Df**	**Mean square**	***F*-ratio**	***P*-value**
Model	2,285,180	2	1,142,590	12.14	0.00
Residual	2,730,100	29	94,141.3		
Total (Corr.)	5,015,280	31			
R^2		0.46			
Standard error of the estimate		306.83			
Durbin-Watson statistic		2.32			

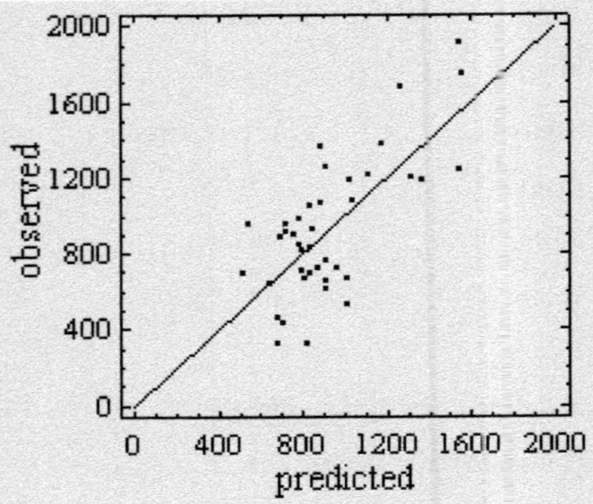

(From Florinsky, I.V., 2010. Theory and Applications of Mathematical Cartographic Modeling of Topography (D.Sc. thesis). Institute of Mathematical Problems of Biology, Russian Academy of Sciences, Pushchino, Russia (in Russian), Table 3.12.)

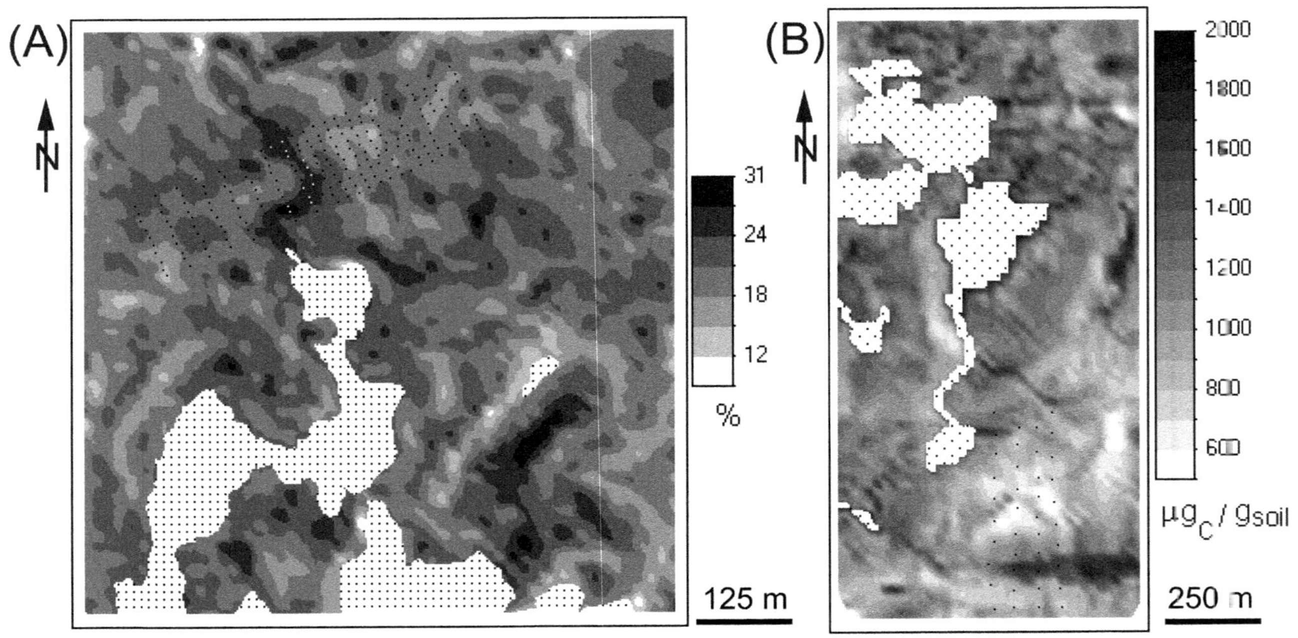

FIG. 12.4 Prediction of the spatial distribution of soil properties: (A) The Miniota site, soil moisture content (the depth 0–0.3 m, May 1998). (B) The Minnedosa site, the microbial biomass carbon content (July 2000). *Dots*, soil-sampling sites; *hatching*, depressions omitted from the prediction. *(From Florinsky, I.V., 2010. Theory and Applications of Mathematical Cartographic Modeling of Topography (D.Sc. thesis). Institute of Mathematical Problems of Biology, Russian Academy of Sciences, Pushchino, Russia (in Russian), Fig. 3.10.)*

12.4.2.1 Wetter conditions

An earlier correlation analysis (Florinsky et al., 2004) demonstrated the significant dependence of all soil microbial variables, notably the microbial biomass carbon content and denitrification enzyme activity, on the soil moisture content in the wetter soil conditions of July 2000. The microbial biomass carbon content, most probable number, and N_2O flux also depended on the bulk density. These results were consistent with previous observations (Myrold and Tiedje, 1985; Groffman and Tiedje, 1989; Webster and Hopkins, 1996). Indeed, gravimetric moisture content and bulk density are indicative of aeration status in the soil. These parameters also influence the movement of water through soil and thereby the distribution of nitrogen and organic carbon, the proximal regulators of denitrification. An increase in water content (water-filled pores) and/or bulk density (namely, decreased total porosity) will result in a lower air-filled porosity and, therefore, a greater number of anaerobic sites in the soil that increases the suitability of the environment to support denitrification.

Note that a topographic influence on the spatial distribution of the soil organic carbon content has been previously demonstrated at this site (Bergstrom et al., 2001b). This was expected and stems from the spatial differentiation of organic matter and moistening according to the land surface morphology (Ototzky, 1901; Moore et al., 1993; Arrouays et al., 1998; Florinsky et al., 1999, 2002; Bell et al., 2000; Gessler et al., 2000; Chaplot et al., 2001; Mitusov, 2001; Shary, 2005).

There was a relatively strong influence of topography on the spatial distribution of soil moisture for the wetter soil conditions (Table 12.7). This was expected and supported by interpretations of topographic variables (Table 2.1) and previous results (Chapter 9). The soil moisture content had the highest values in such locations where k_v had negative values (concave profiles) and *SCA* had high values (draining of large upslope areas).

Relatively high correlations between the bulk density and topographic attributes (Table 12.7) resulted from its dependence on soil moisture, soil texture, and soil organic matter, which are usually distributed according to the land surface morphology. Consequently, convex-profile upslopes ($k_v > 0$, low values of *SCA*) had higher values of bulk density than downslopes and depressions within the plot.

Almost all soil microbial indices, in one way or another, depended on topographic variables in the wetter soil conditions (Table 12.7). This was expected, as essential factors for denitrification—the soil moisture content, soil organic carbon content, and bulk density—depended on topographic attributes within the plot (as discussed earlier). The denitrification rate depended on nonlocal and combined topographic variables: *CA*, *TWI*, and *SPI* (Table 12.7). The microbial biomass carbon content depended on both nonlocal and combined topographic variables (*SCA* and *TWI*) and some local attributes (z, G, k_v, H, and k_{max}). Only local topographic variables influenced the N_2O flux, number of denitrifiers, and microbial respiration rate: they depended on k_v, G, and A (Table 12.7).

Thus, under wetter soil conditions, the spatial variability of the denitrification rate was mostly affected by redistribution and accumulation of soil moisture and soil organic matter due to their gain along a slope from top to bottom because of their additional amounts contributed from upslope (i.e., according to the relative position of the point in the landscape). However, the N_2O emission, number of denitrifiers, and microbial respiration rate were affected by the distribution of soil water and organic matter according to the local slope geometry. Both groups of topographic factors (nonlocal and local) of spatial redistribution and accumulation of soil moisture and organic matter influenced the microbial biomass content. These results were generally consistent with other observations that denitrification hotspots are associated with downslope positions (Pennock et al., 1992; Van Kessel et al., 1993; Corre et al., 1996; Anderson et al., 2015).

12.4.2.2 Drier conditions

Based on the Kolmogorov-Smirnov statistics (Table 12.5), in the drier July 2001, there were marked decreases in the soil moisture content and bulk density, decreases in the denitrification and microbial respiration rates, a slight increase in the number of denitrifiers, and no significant changes in the microbial biomass carbon content, denitrification enzyme activity, or N_2O flux (Table 12.5).

Comparisons between the two sampling events demonstrated significant changes in the spatial differentiation of the soil properties. In the drier soil conditions, the coefficient of variation values showed pronounced increases in variation of the soil moisture content, denitrification and microbial respiration rates and a pronounced decrease in variation of the N_2O flux (Table 12.5). Correlations between indices of soil microbial activity and selected soil environmental properties became statistically insignificant in the drier soil conditions, except for the microbial biomass carbon content (Florinsky et al., 2004).

We observed a decrease in the topographic control of the soil moisture content in the drier July 2001, while associations of other soil properties with topographic attributes became insignificant (Table 12.7). This demonstrates a phenomenon of temporal variability in topographic control of dynamic soil properties (Section 12.4.1.1) (Florinsky et al., 1999, 2002).

From the observed distribution of the denitrification rate, it may be deduced that the denitrifier activity continued to persist under the drier soil conditions, but it was reduced and ceased to depend on the spatial distribution of soil moisture and thus land surface morphology. This likely reflects a transition of some critical level of the soil moisture status and the ability of denitrifiers to be effective aerobic heterotrophs under aerobic conditions. The soil moisture status was still sufficient for the activity of these organisms but was no longer a dominant force in influencing their spatial patterns.

12.4.2.3 Interpretations

It is possible to conclude that to keep the topographic control of the spatial distribution of denitrifiers and their activity, the landscape should contain a sufficient amount of soil water (higher than some threshold value). Physically, this idea seems reasonable, as topographically controlled gravity-driven lateral transport of substances (e.g., nutrients) generally acts through the medium of gravimetric soil water. It is possible to generalize that the spatial distribution of soil dynamic properties depends on topographic variables only if soil moisture content is higher than some threshold value.

In drier and wetter soil conditions, the number of denitrifiers was different. It depended on topography in the wetter season, while the denitrifier enzyme activity was essentially the same on both sampling dates, ceasing to be affected by topography (Tables 12.5 and 12.7). This result reinforces the view that a direct relationship seldom exists between the number of denitrifiers and the amount of denitrifying enzymes in the soil (Parsons et al., 1991). This is because of the dual aerobic/anaerobic nature of the ecology and physiology of denitrifiers. The occurrence of denitrifying bacteria in any given habitat is primarily controlled by their ability to compete as heterotrophs rather than their ability to denitrify (Groffman and Tiedje, 1989). The expression of denitrifying enzyme is just a response to anaerobic conditions, reflecting soil aeration status.

Although the spatial distribution of denitrifiers was affected by topography in the wetter season, other groups of soil microbiota were probably more sensitive to the land surface morphology. This is reflected in the higher topographic effect on the microbial biomass carbon content compared with the most probable number (Table 12.7). Recall that microbial biomass carbon content measures the total microbial biomass in the soil, including both denitrifiers and other organisms (Table 12.1).

The lower topographic control of the spatial differentiation of the N_2O emission compared with other soil microbial variables in wetter soil conditions as well as the disappearance of this control in drier soil conditions (Table 12.7) may reflect the high temporal and spatial variability of this attribute (Parsons et al., 1991) as well as the nature of N_2O production. Because N_2O production results from both autotrophic, aerobic processes (nitrification) and heterotrophic, anaerobic processes (denitrification) and is merely an intermediate in denitrification, it is not surprising that N_2O production and flux are highly variable and do not always reflect the environmental prerequisites of either of the microbial groups producing this gas.

The results obtained relate to the microbial activity in the upper soil layer wherein samples were collected and may vary with depth. This is caused by two factors:

- Soil microbial communities and enzyme activity change with depth (Zvyagintsev, 1994; Lehman, 2007).
- Topographic control of soil properties may both decrease and increase as the soil layer depth increases (Section 12.4.1.2) (Florinsky et al., 1999, 2002).

For the Minnedosa site, the regression equations explained 46% of the variability of the microbial biomass carbon content in the wetter soil conditions (Table 12.10). Spatial heterogeneity of some other soil microbial attributes can be predicted using DTM-based regression, as demonstrated by the predictive maps of microbial biomass carbon content (Fig. 12.4B). However, DTM-based regression should be used with caution as different equations may be obtained in different seasons because of:

- temporal variability in the dependence of dynamic soil properties on topography (Section 12.4.1.1) (Florinsky et al., 1999, 2002);
- seasonal dynamics of soil microbiota (Golovchenko and Polyanskaya, 1996);
- soil microbial succession (Polyanskaya and Zvyagintsev, 1995);
- spatiotemporal oscillations of soil microbial populations (Semenov et al., 1999) and other oscillating processes in soils (Florinsky, 2006; Florinsky et al., 2009b).

Part III

Digital terrain modeling in geology

Chapter 13

Folds and folding

13.1 Introduction

Folds of various hierarchical levels are the most abundant and investigated geological features (Ramsay, 1967). In the context of fold studies, methods of digital terrain modeling (i.e., curvature derivation from DEMs of the land and stratigraphic surfaces) are applied to solve at least three problems (Bergbauer, 2007):

1. Analyzing the geometry of folds and their classification.
2. Predicting the degree of deformation or strain of folds as well as the fracture orientation and density of folded strata.
3. Analyzing the evolution of folded strata and estimating the degree of their plasticity.

13.2 Fold geometry and fold classification

Folds are characterized by the great diversity of geometric forms. This can be associated with regional lithological differences, the deformation degree of geological strata, stress distribution during deformation, and so on. A quantitative description of the fold geometry can provide the key information necessary to analyze, reconstruct, and model the origin and evolution of folds (Bastida et al., 1999).

Bergbauer (2007, p. 185) reasoned that for structural geology, "surface curvature is a unique descriptor of shape. Describing the geometry of a horizon quantitatively is an essential first step when attempting to compare physical and numerical models with natural surfaces." From the standpoint of terrain modeling, however, visualization of a shape by itself is a standard procedure (Fig. 13.1). So we do not discuss this issue in depth. Examples of visualization of the fold geometry with data on geological surface curvatures can be found elsewhere (Bergbauer, 2007; Hart and Sagan, 2007; Mynatt et al., 2007a).

Fold visualization combined with fold classification is a more challenging task. Hudleston (1973) proposed a quantitative classification of folds based on stereographic projections of bedding attitudes and cylindrical fold profiles. However, an assumption of cylindricity of folds was the main problem of such fold descriptions: it is obvious that a folded geological surface cannot be totally cylindrical.

Lisle (1994), Lisle and Robinson (1995), and Stewart and Podolski (1998) used the sign of the Gaussian curvature to distinguish three classes of folds: elliptical (domes or basins), hyperbolical (saddles), and cylindrical folds (Section 2.7.1). Roberts (2001) used a combination of the signs of the Gaussian and mean curvatures to distinguish six classes of folds (Section 2.7.1): domes ($K>0$ and $H>0$), bowls ($K>0$ and $H<0$), saddles ($K<0$), ridges ($K=0$ and $H>0$), planes ($K=0$ and $H=0$), and synforms ($K=0$ and $H<0$). Within the framework of this approach, Lisle and Toimil (2007) suggested a generalized definition of a fold as a patch consisting of contiguous points with K and H of consistent sign. They distinguished four classes of folds: synclastic antiforms ($K>0$ and $H>0$), anticlastic antiforms ($K<0$ and $H>0$), anticlastic synforms ($K<0$ and $H<0$), and synclastic synforms ($K>0$ and $H<0$). For the continuous classification of folded surfaces, these authors applied the shape index (Koenderink and van Doorn, 1992) (Section 2.6.2). Mynatt et al. (2007a) elaborated the approaches of Roberts (2001) and Lisle and Toimil (2007); they proposed a classification including eight theoretically possible fold types (Fig. 2.22A), four of which correspond to "rare forms" (for details, see Sections 2.7.1 and 2.7.3). See Fig. 6.7 for an example of the classification of the geological surface by the signs of K and H.

We should note, however, that the landform classification by the signs of the Gaussian and mean curvatures is not complete and includes "rare forms." The Shary classification (Section 2.7.3) holds much promise.

Digital Terrain Analysis. https://doi.org/10.1016/B978-0-443-24798-9.00020-0

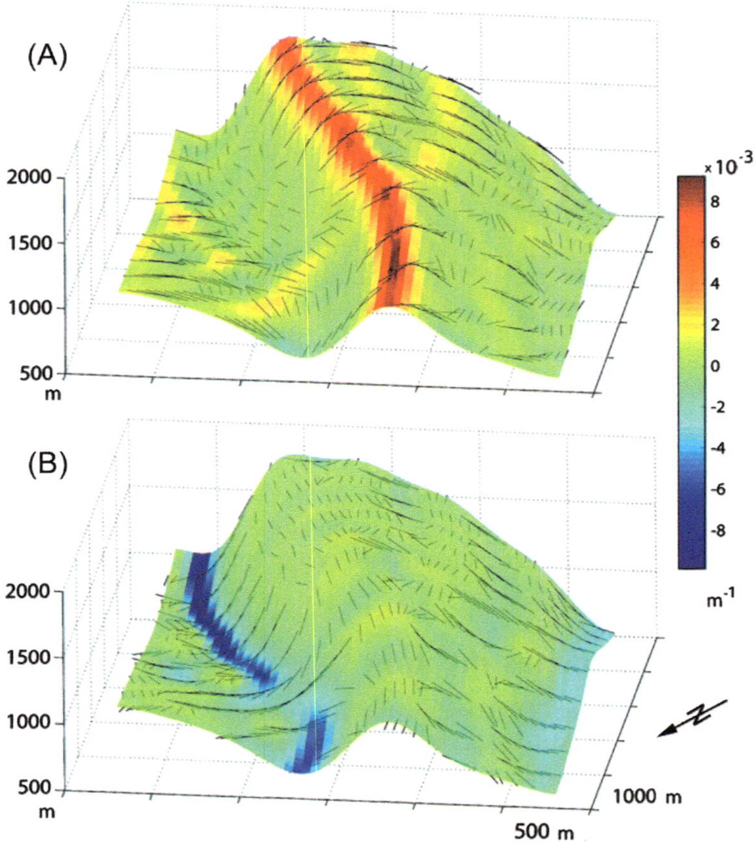

FIG. 13.1 Magnitudes (*colors*) and axis directions (*black tic marks*) of the principal curvatures draped on the 3D model of the Sheep Mountain Anticline, Wyoming, United States: (A) Maximal curvature. (B) Minimal curvature. For the DEM description, see Fig. 6.7. *(From Mynatt, I., Bergbauer, S., Pollard, D.D., 2007. Using differential geometry to describe 3-D folds. J. Struct. Geol. 29 (7), 1256–1266, Fig. 4, reproduced with permission from Elsevier.)*

13.3 Predicting the degree of fold deformation and fracturing

Assume there is an elastic, thin beam. It is known that if the beam is bent, the shearing constraints at a point of the beam are proportional to its curvature (Timoshenko and Goodier, 1951). If these constraints reach some threshold, they are responsible for the beam's rupture. Let us also consider a geological layer as an elastic beam. In this case, fractures may be typical for the layer portions where the shearing constraints are higher than the threshold (Samson and Mallet, 1997).

Murray Jr. (1968) was probably the first to employ this idea and to link mathematically the curvature of a folded surface with the degree of fracture porosity and permeability, and so with the degree of oil- and gas-bearing potential of folded geological structures. He proposed that the product of the curvature and bed thickness can be used to estimate fracture porosity, while the third power of the product can be used to estimate fracture permeability. The computerization of the Murray approach to reveal areas of the highest fracture intensity was based on the Laplacian[a] derivation from DEMs of stratigraphic surfaces, which were generated from structural contours (Muñoz-Espinoza, 1968; Pirson, 1970; Schultz-Ela and Yeh, 1992). Lisle (1994) argued that areas having high values of the Gaussian curvature correspond to zones of greater deformation and higher density of fractures. Fischer and Wilkerson (2000) proposed that joints formed by a deformation are parallel to axis directions of k_{min} normal sections of a geological surface.

However, Schultz-Ela and Yeh (1992) did not observe relationships between the high absolute values of the Laplacian and rock permeability (fracturing), which was estimated from the maximum annual gas productivity. Using results of

a. The Laplacian can be considered as an approximation of mean curvature (Section 2.2.4.13).

detailed field studies, Fiore Allwardt et al. (2007, p. 419) concluded: "Fracture intensities do not correlate directly with curvature" because "curvature is only one mechanism for fracture development. Other mechanisms, such as stretching, faulting, bedding-plane slip, the influence of preexisting fractures, and tectonic loading may contribute to fracture development and should be evaluated. … A complete understanding of the heterogeneity in fracture intensities requires analysis beyond curvature calculations." Bergbauer (2007) and Pearce et al. (2011) came to a similar conclusion.

13.4 Folding models and the *Theorema Egregium*

In the *Theorema Egregium*,[b] Gauss (1828, 1902) demonstrated that the total (Gaussian) curvature is a bending invariant for the surface if bending takes place without stretching, compression, or breaking of the surface. In other words, the Gaussian curvature does not change its values at each point of a surface during its deformations if the surface is bent and/or inclined without breaking, compressing, or stretching. Changes in the Gaussian curvature value indicate compression or stretching of the surface in the vicinity of these changes.

In a pioneer work, Belonin and Zhukov (1968) utilized the *Theorema Egregium* to study the uplift evolution of an elevated block of crust. These authors used DEMs of several conformable stratigraphic surfaces. To estimate the level, direction, and unevenness of stretching and bending of the block, they analyzed K as well as H, k_{min}, and k_{max} of these surfaces.

The *Theorema Egregium* was implemented to estimate "the degree of plasticity of the tectonics" (Samson and Mallet, 1997), that is, to investigate the geometry and properties of oceanic lithospheric plates in subduction zones, in particular, whether or not the oceanic lithosphere maintains its mechanical integrity during subduction (Bevis, 1986; Cahill and Isacks, 1992; Ansell and Bannister, 1996; Nothard et al., 1996). However, DEMs of lithospheric plates were generated from data on earthquake hypocenter locations. It is clear that the accuracy of such DEMs and, hence, the certainty of conclusions about lithospheric plate plasticity are highly conjectural.

It would be interesting to apply this property of the Gaussian curvature to reconstruct paleotopography. Such a reconstruction can be best explained by the following example (Shary et al., 1991). Let us consider a plane affected by two folding processes with different vergences (Fig. 13.2A and B). As a result, a topographic surface arises with a pronounced regular cell-like structure, which can be seen on the elevation map (Fig. 13.2C). Assume that the plane is affected by two other folding processes with different vergences (Fig. 13.2D and E). In this case, the processes are characterized by the same directions but shorter wavelengths compared with the first two processes. A topographic surface arises with a pronounced regular cell-like structure, but the cell size is less than it was in the first case (Fig. 13.2F). Now assume that all four folding processes affect the plane. As a result, a topographic surface arises with a rather complicated cell-like structure, which can, however, be seen on the elevation map (Fig. 13.2I). Finally, assume that the surface (Fig. 13.2I) was inclined (Fig. 13.2G) and bent (Fig. 13.2H). The final topographic surface (Fig. 13.2J) obviously keeps some traces of the four folding processes, but they are hidden by the inclination and bending. On the final elevation map, it is impossible to see the cell-like structure; however, a valley network is well marked.

Let us assume that (1) all surface deformations were free of breaks, stretching, and compression and (2) there was an erosion-free condition. There is a need to reveal the hidden cell-like folding structure. Let us derive the Gaussian, horizontal, and vertical curvatures from the DEM of the topographic surface before (Fig. 13.2I) and after (Fig. 13.2J) its inclination and bending. One can see that there are distinct patterns on the maps of horizontal curvature of the surface before and after the inclination and bending (Fig. 13.3A). The same is true for the maps of vertical curvature (Fig. 13.3B). Moreover, there is no clear evidence of the cell-like structure (Fig. 13.2I) on the maps of these two curvatures. However, the desired cell-like structure can be seen on the maps of the Gaussian curvature (Fig. 13.3C). Notice that the maps of the Gaussian curvature of the surface before and after the inclination and bending have almost the same patterns (minor differences are caused by discretization errors).

Modeling was performed by the software LandLord (Florinsky, 2016).

We should finally note that it is difficult to use the *Theorema Egregium* to reconstruct paleotopography from DEMs of actual terrains. This is due to surface deformations caused by erosion as well as tectonic stretching, compression, and faulting. This problem calls for further research.

b. This is Latin for "Remarkable Theorem."

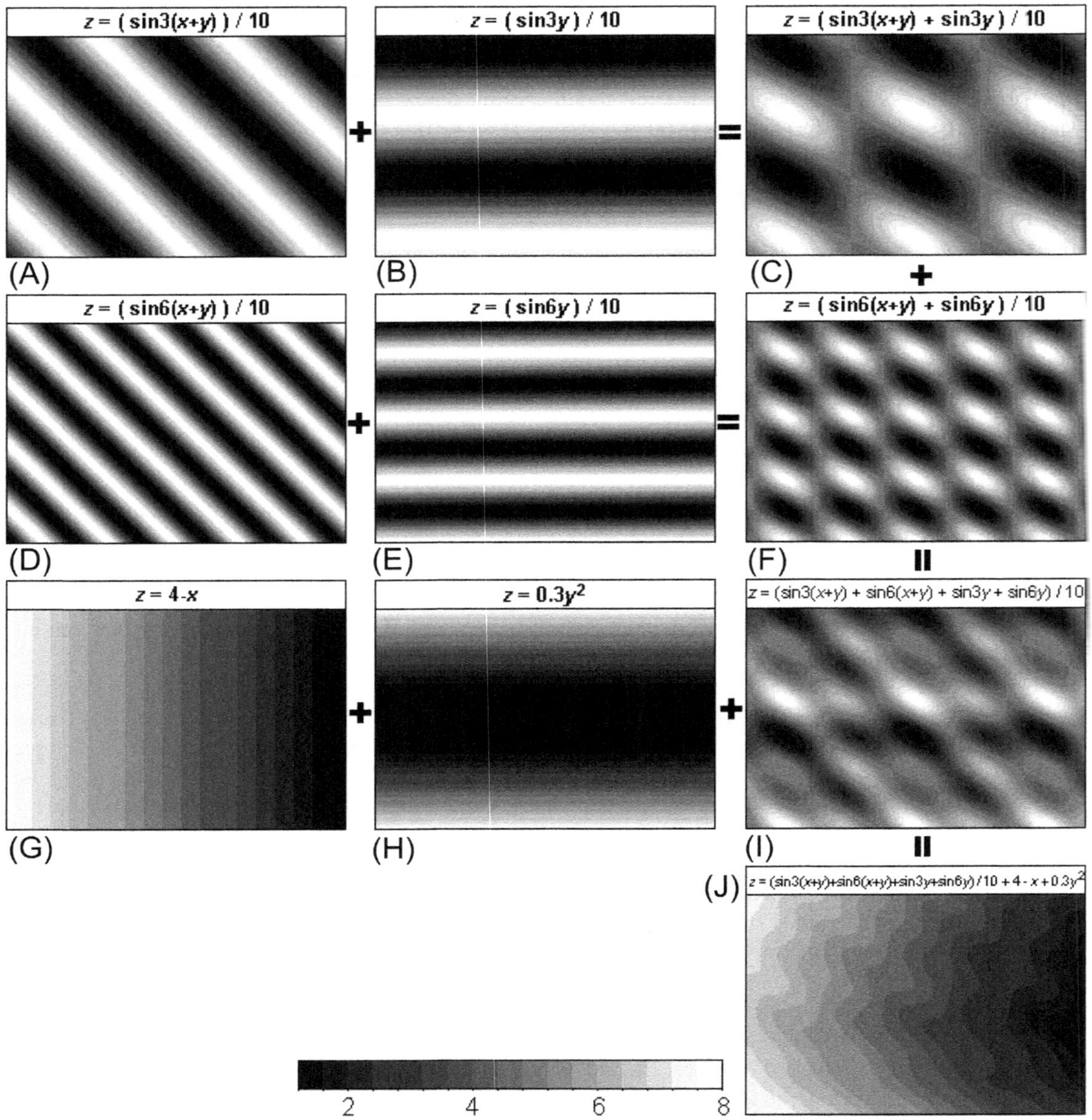

FIG. 13.2 Simulation of the formation of the hidden complex folding (Shary et al., 1991). Two folding processes with different vergences, (A) and (B), form a cell-like folding structure (C). Two other folding processes with different vergences, (D) and (E), form the other cell-like folding structure (F). All four folding processes form the complex folding structure (I). Its inclination (G) and bending (H) form the final surface (J). Elevation maps of the surfaces are presented.

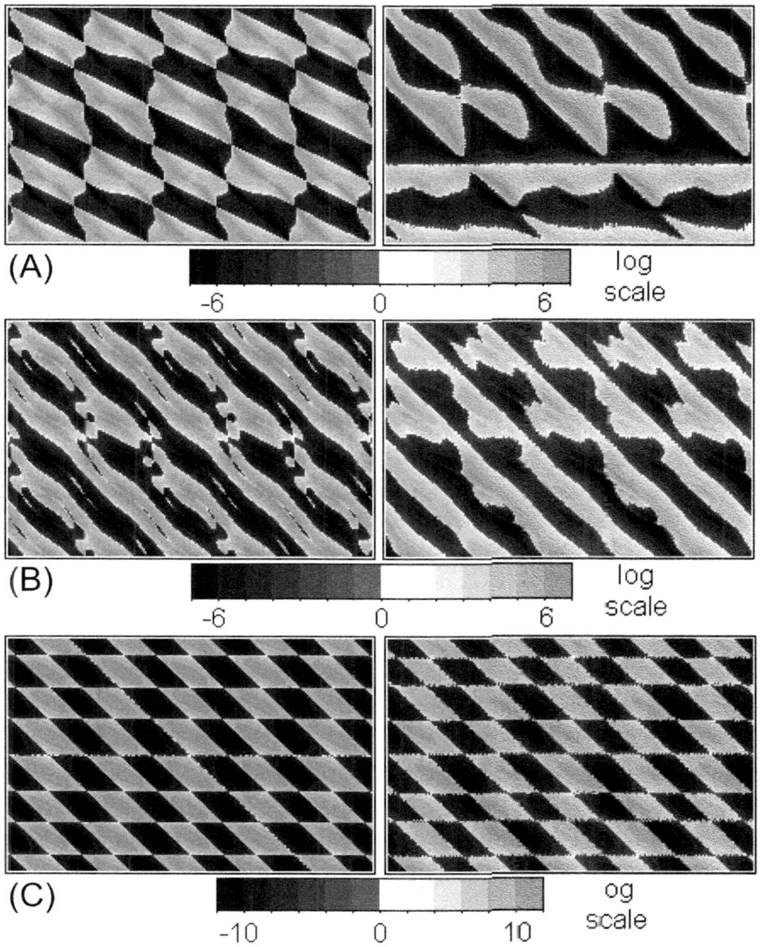

FIG. 13.3 Revelation of the hidden complex folding structure using the *Theorema Egregium* (Shary et al., 1991): (A) Horizontal curvature. (B) Vertical curvature. (C) Gaussian curvature. *Left*, curvatures of the surface with the hidden structure before the inclination and bending (Fig. 13.2I). *Right*, curvatures of the surface with the hidden structure after the inclination and bending (Fig. 13.2J).

Chapter 14

Lineaments and faults

14.1 Motivation

It is well known that data on geological faults (Peive, 1956a; Slemmons and Depolo, 1986; Trifonov, 1995; Nikonov, 1995; Keller and Pinter, 2002; Groshong Jr., 2006; Handy et al., 2007; Burbank and Anderson, 2012, chap. 4) are required in neotectonic and seismic studies, ore and petroleum exploration, and so on. Revealing and monitoring of local and regional faults are necessary in the development and operation of some potentially dangerous objects, such as nuclear power plants and hydroelectric power stations (Tabor, 1986; Chekunov et al., 1990; Poletaev et al., 1991). Indeed, the construction and operation of such objects can cause changes in regional tectonic activity, such as, tectonic movements along previously stable faults. It is obvious that this can affect the security of an object and its operation.

Faults are as a rule indicated by lineaments. The term *lineament* is commonly used in reference to any linear geological features of different origin, age, depth, and scale (Hobbs, 1904; Sonder, 1938; Hills, 1956; Katterfeld and Charushin, 1973; Hodgson et al., 1976; O'Leary et al., 1976; Makarov, 1981; Trifonov et al., 1983; Wise et al., 1985; Kats et al., 1986; Bryukhanov and Mezhelovsky, 1987; Twidale and Bourne, 2007). Lineaments are usually associated with faults, linear zones of fracturing, bending deformation, and increased permeability of the crust as well as linear chains of some geological features (i.e., laccolites, volcanoes, etc.). Lineaments are as a rule expressed in topography. They can be observed from air- and spacecraft as well as are recorded on aerial and satellite images, three-dimensional (3D) seismic models, and thematic maps at a wide range of scales.

Geometric characteristics of linear topographic dislocations formed by vertical tectonic motions differ from those formed by horizontal tectonic movements (Ollier, 1981; Trifonov, 1983; Keller, 1986; Keller and Pinter, 2002, chap. 2). Qualitative and quantitative signs of these differences can be used to recognize the origin (or type) of topographically expressed faults.

Qualitative approaches for revelation and classification of faults using topographic data have been repeatedly discussed (Trifonov, 1983; Slemmons and Depolo, 1986; Burbank and Anderson, 2012, chap. 4). One of the most popular approaches was visual analysis of topographic maps (Hobbs, 1904, 1911; Nezametdinova, 1970; Yabrova, 1981) and remotely sensed images (Wilson, 1941; Lattman, 1958; Skaryatin, 1973; Hodgson et al., 1976; Trifonov et al., 1983; Bryukhanov and Mezhelovsky, 1987). To recognize lineaments and ring structures visually, some researchers used hill-shaded photographic images obtained by taking pictures of 3D plastic terrain models illuminated at different angles (Hills, 1956; Wise, 1969; Saul, 1978; Wise et al., 1985). Analog photogrammetric procedures were utilized to reveal and classify faults as well as measure their dip and strike angles (Vinogradova and Eremin, 1971).

To reveal and classify faults, geologists employed some standard techniques of digital terrain modeling, such as 3D topographic models (Section 8.5) dropped by aerial/satellite images or geological maps (Campagna and Levandowski, 1991; Morris, 1991; Kukowski et al., 2001, 2008; Ganas et al., 2005; Palyvos et al., 2006; Nazari et al., 2009) (Fig. 14.1), thalweg mapping (Eliason and Eliason, 1987; Thienssen et al., 1994; Jordan, 2003), as well as slope gradient and aspect mapping (Onorati et al., 1992; Steen et al., 1998; Kopp, 2000; Ganas et al., 2005; Kukowski et al., 2008). Lisle (1994) argued that high values of the Gaussian curvature may indicate fault zones. To reveal faults, Steen et al. (1998) approximated the derivation of vertical curvature by double derivation of slope gradient from DEMs of stratigraphic surfaces (that is, derivation of G from the digital model of G). To reveal lineament, Wladis (1999) applied the geophysical method of second derivatives (Elkins, 1951) to DEMs. Maximal and minimal curvatures were also used to locate faults (Samson and Mallet, 1997; Bergbauer et al., 2003; Wynn and Stewart, 2003).

Hill shading (Sections 2.5 and 8.3) is the most popular technique for revealing faults (Moore and Simpson, 1983; Schowengerdt and Glass, 1983; Zeilik et al., 1989; Onorati et al., 1992; Deffontaines et al., 1997; Chorowicz et al., 1998, 1999; Dhont et al., 1998a, 1998b; Adiyaman et al., 2001; Szynkaruk et al., 2004; Ganas et al., 2005; Dhont and

Digital Terrain Analysis. https://doi.org/10.1016/B978-0-443-24798-9.00014-5

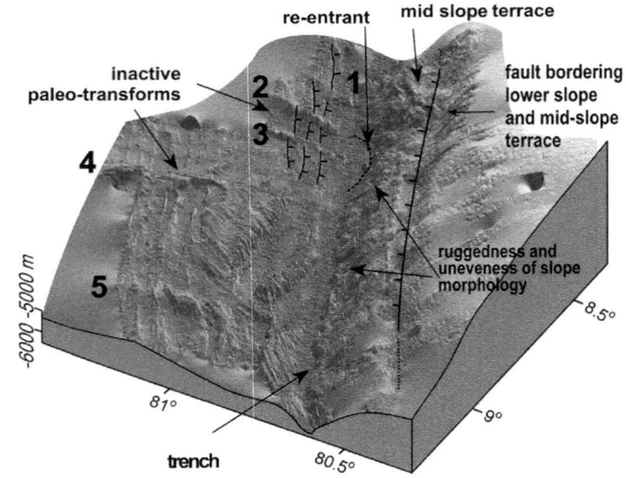

FIG. 14.1 Expression of faults on a 3D topographic model exemplified by the Nazca Plate, the trench, and the Peruvian submarine continental slope. *1–5*, ridges. View from the south. A DEM was derived from swath bathymetry data. *(From Kukowski, N., Hampel, A., Hoth, S., Bialas, J., 2008. Morphotectonic and morphometric analysis of the Nazca plate and the adjacent offshore Peruvian continental slope—implications for submarine landscape evolution. Mar. Geol. 254 (1–2), 107–120, Fig. 4C, reproduced with permission from Elsevier.)*

Chorowicz, 2006; Palyvos et al., 2006; Kukowski et al., 2008; Saadi et al., 2009) (Fig. 14.2). It has been noted that this technique is usually more effective than interpretation of remotely sensed images in terms of fault recognition (Schowengerdt and Glass, 1983). On hill-shaded maps, one can find up to 90% of faults, which may be revealed by conventional geological and geophysical methods (Onorati et al., 1992).

In all these techniques, DEM processing allows to enhance the "contrast" of existing lineaments and faults, which are poorly expressed in topography. The final revealing of lineaments is usually carried out by the human visual system.[a] Moreover, the use of these geomorphometric approaches without ancillary geological data does not allow for the classification of revealed faults. This is mainly connected with difficulties in formalizing fault diagnostic signs. Detailed qualitative descriptions of such signs can be found elsewhere (Gzovsky, 1954; Ollier, 1981; Keller, 1986; Nikonov, 1995; Keller and Pinter, 2002, chap. 2; Groshong Jr., 2006). For instance, Dhont and Chorowicz (2006, p. 36) described the following principles to reveal and classify faults using hill-shaded maps:

> *"Neotectonic faults can be identified by (1) their morphology, forming asymmetric ranges with one side corresponding to breaks in slope or scarps, (2) the displacement of late Neogene volcanic boundaries, structural or erosional surfaces, and (3) the occurrence of straight lines of several tens of kilometers in length. … Strike-slip faults have rectilinear traces and they locally bound push-up hills or extensional basins at step-over or bends of the fault trace. They can be associated with typical patterns, such as tail-crack or horse-tail structures at fault ends. Reverse faults have sinuous traces and they are associated with half-cylindrical-shaped hills of the uplifted blocks due to drag folds deforming ancient planar erosion surface in the hanging wall. Normal faults are recognized by the following geomorphic characters: (1) they generally have a widely arched trace, concave (mainly) or convex toward the footwall, in contrast to the strike-slip faults whose trace is generally straighter, (2) they bound tilted plateaus (tilted blocks), and (3) as is also the case for the strike-slip faults, they are not related to half-cylindrical-shaped hills corresponding to recent drag folds, which accompany active reverse faulting."*

Revealing topographically expressed lineaments can be done by k_h and k_v calculation and mapping; k_h and k_v values should be subdivided into two levels relative to the zero value (Florinsky, 1992). It has been observed that lineaments recorded on k_h and k_v maps have essentially distinct statistical properties (i.e., orientation and density). This can geomorphologically be explained by the fact that k_h mapping reveals predominantly valley and crest spurs, while k_v mapping reveals mainly terraces (Evans, 1980). Shary, P.A. (1992, personal communication) supposed that lineaments displayed on k_h and k_v maps may be associated with faults generated by horizontal and vertical tectonic motions, respectively.

a. There is a tendency to use techniques of image processing and pattern recognition for auto-delineation of lineaments from DEMs (Masoud and Koike, 2006, 2011a, b; Zlatopolsky, 2011; Yeomans et al., 2019; Kelka et al., 2022).

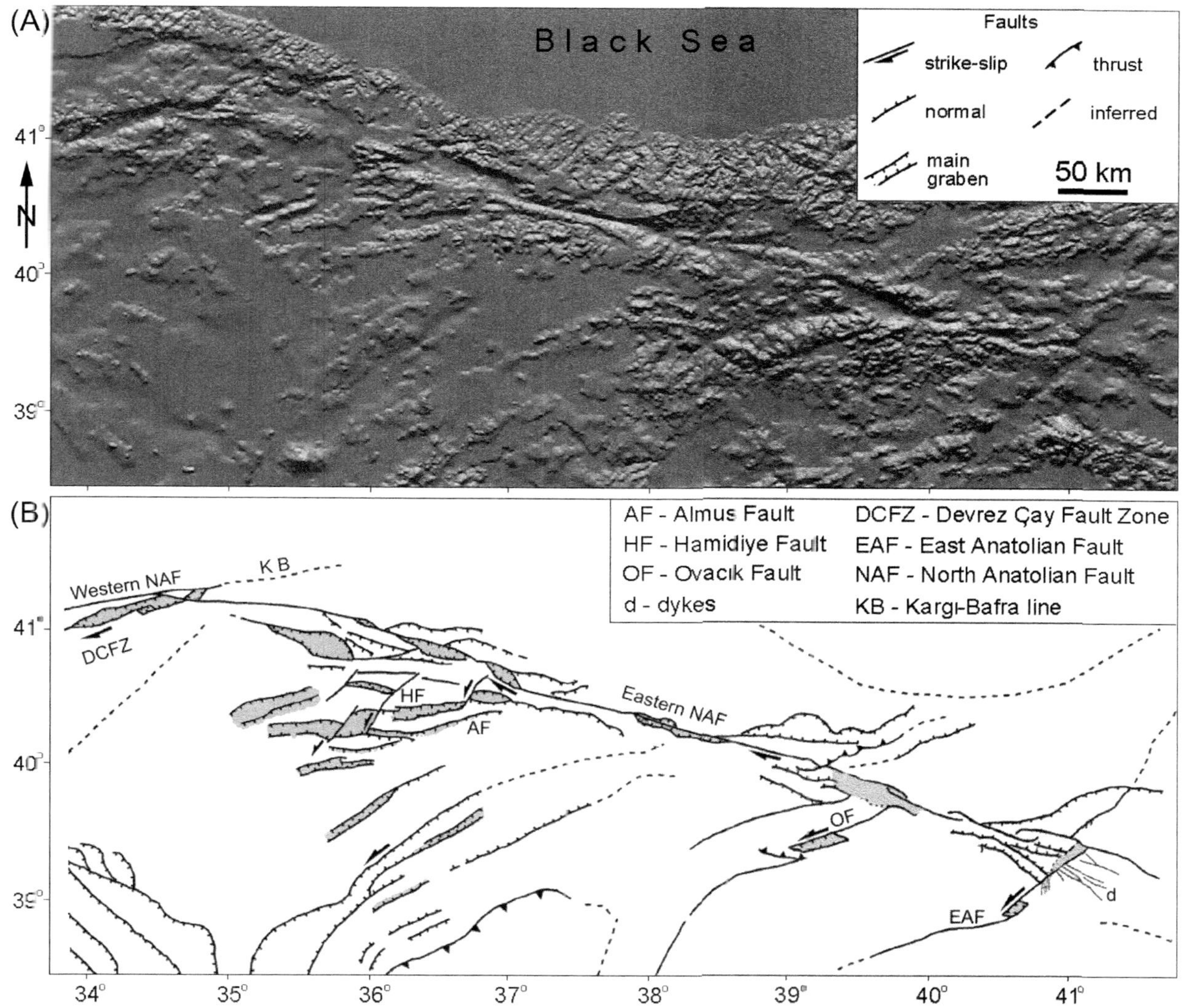

FIG. 14.2 Fault recognition from a hill-shaded map exemplified by Central and Eastern Anatolia, Turkey: (A) Hill-shaded map with illumination from the north. A DEM was derived from the Digital Chart of the World (US DMA, 1992) by contour interpolation; DEM grid spacing is 300 m. (B) Fault patterns visually recognized on the hill-shaded map; *gray areas*, Neogene-Quaternary basins. *(From Chorowicz, J., Dhont, D., Gündoğdu, N., 1999. Neotectonics in the eastern North Anatolian fault region (Turkey) advocates crustal extension: mapping from SAR ERS imagery and digital elevation model. J. Struct. Geol. 21 (5), 511–532, Fig. 1B and C, reproduced with permission from Elsevier.)*

In this chapter, we describe a method to reveal and classify topographically expressed lineaments. The method is based on the derivation of horizontal and vertical curvatures from DEMs of the land or stratigraphic surfaces (Florinsky, 1996).

14.2 Theory

Consider an arbitrary surface (Fig. 14.3A). Recall that k_h and k_v are positive for convex areas and negative for concave ones (Sections 2.2.4.2 and 2.2.4.3). Let us use k_h and k_v binary maps subdividing k_h and k_v values into two levels relative to the zero value as well as displaying areas with k_h and k_v positive values in white while displaying areas with k_h and k_v negative values in black.

Suppose a dip-slip (or reverse) fault occurs in the zone of k_v positive values (Fig. 14.3B). In the deformation zone, the k_h and k_v values will change. In particular, the k_v will take negative values along the fault line x–x. In this case, one will be able to see an indicator of the dip-slip (or reverse) fault as a black lineament on a white background on the k_v binary map. A similar lineament will be seen on the k_v binary map if the surface was plane before the vertical motion. If k_v values were

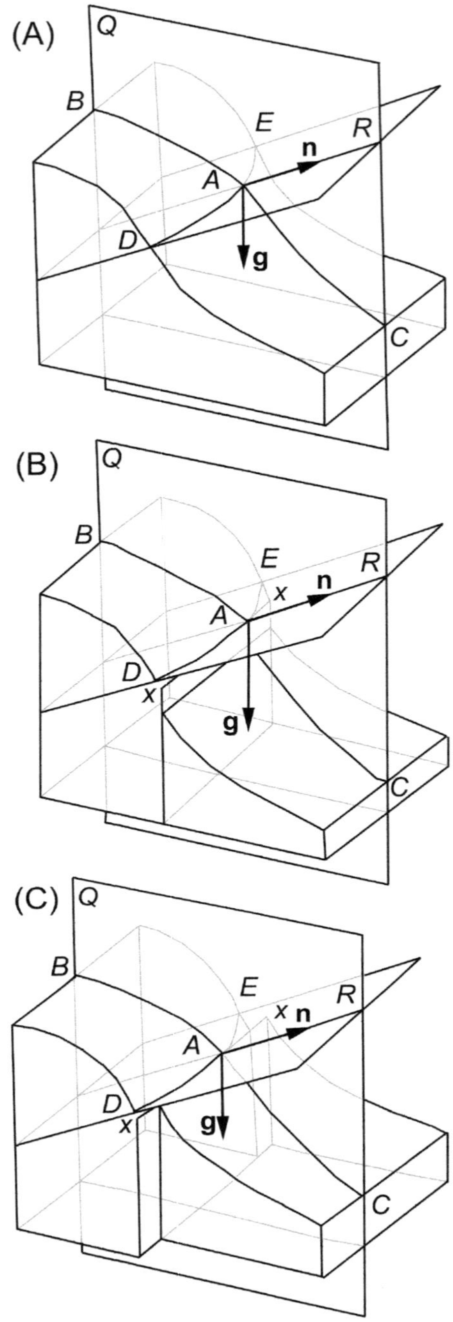

FIG. 14.3 Block diagrams illustrating sign changes in horizontal and vertical curvature values after tectonic motions: (A) Initial surface. (B) Surface with a dip-slip fault after a vertical displacement. (C) Surface with a strike-slip fault after a horizontal displacement. k_v is the curvature of the normal section *BAC* formed by the intersection of the surface with the plane *Q* containing the gravitational acceleration vector **g** at the point *A*. k_h is the curvature of the normal section *DAE* formed by the intersection of the surface with the plane *R* orthogonal to the normal vertical section *BAC* at the point *A*. **n**, external normal to the surface at the point *A*; *x*–*x*, fault lines.

negative before the movement, the k_v sign will also change along the fault line. In this case, the dip-slip fault will be indicated by a white lineament on a black background on the k_v binary map. A lineament consisting of black and white spots will be recorded on the k_v binary map after the vertical movement if the surface has a complex shape.

In like manner, a lineament indicating a thrust will be revealed on the k_v binary map because thrusting as a rule produces a scarp. However, lineaments indicating dip-slip, reverse, and thrust faults will not be recorded on the k_h binary map because changes of the k_h sign along fault lines will be random rather than systematic (some nonlinear changes in patterns will arise on the k_h map anyway).

Suppose a strike-slip fault occurs in the zone of k_h positive values (Fig. 14.3C). The k_h and k_v values will also change in the deformation zone. However, the k_h will take negative values along the length of the fault line *x–x*, while changes of the k_v sign will be random rather than systematic. Therefore, the following lineaments indicating horizontal movement traces will be recorded on the k_h binary map: (1) a black lineament on a white background for the given case and planar surface before deformation; (2) a white lineament on a black background for the surface with negative k_h values; and (3) a lineament consisting of white and black spots for the complex surface. Some nonlinear traces of horizontal movements will be recorded on the k_v binary map.

After an oblique-slip and a gaping fault formation, both curvatures ought to change signs systematically along fault lines. Therefore, we can expect that lineaments indicating these faults will be recorded on the binary maps of both k_h and k_v

There is a need to use ancillary geological, geophysical, and geomorphic data to distinguish and/or separate: (1) lineaments of nontectonic (e.g., erosion) origin; (2) lineaments associated with flexures and some folds; (3) dip-slip, reverse, and thrust faults equally revealed by k_v; and (4) oblique-slip and gaping faults revealed by k_h in a similar manner.

Notice that for platform areas the direct association between lineaments and faults may be conjectural. In such cases, the lineaments on k_h and k_v binary maps can be interpreted as linear zones of concentration of bending deformations (Trifonov, V.G., 2010, personal communication):

- Lineaments revealed on k_h binary maps are predominantly associated with zones of horizontal bending deformations.
- Lineaments recorded on k_v binary maps mostly relate to zones of vertical bending deformations (e.g., flexures).
- Lineaments revealed on both k_h and k_v binary maps can indicate zones of bending deformations formed by both vertical and horizontal movements.

14.3 Method validation

14.3.1 Materials and data processing

The method validation was performed using a simulated surface. The site (Fig. 14.4A) measured 60 m × 60 m; the maximum elevation difference is about 7.5 m. It included an east-striking valley and two adjacent crests. An irregular DEM included 129 points. The following seven simple typical faults (Gzovsky, 1954) were modeled by deformations of the initial irregular DEM:

1. A vertical dip-slip fault with 1-m displacement (Fig. 14.4B)
2. A low-angle dip-slip fault with 1-m displacement and 30° dip (Fig. 14.4C)
3. A strike-slip fault with 3.5-m displacement (Fig. 14.4D)
4. An oblique-slip fault with 3.5-m horizontal and 1-m vertical displacements (Fig. 14.4E)
5. An overthrust with 5-m displacement (Fig. 14.4F)
6. An overthrust with 15-m displacement (Fig. 14.4G)
7. A gaping fault with a trench of 1-m width and 0.2-m depth (Fig. 14.4H).

Eight irregular DEMs of the initial and deformed surfaces were converted into regular ones by the Delaunay triangulation and a piecewise quadric polynomial interpolation (Agishtein and Migdal, 1991). The grid spacing was 2 m. Digital models of k_h and k_v (Fig. 14.4) were derived from all regular DEMs by the Evans-Young method (Section 4.1). Data were processed by the LandLord software (Florinsky, 2016).

14.3.2 Results and discussion

Visual analysis of the obtained maps (Fig. 14.4) confirmed the general correctness of the method. Indeed, for the dip-slip fault models (Fig. 14.4B and C), the k_v binary maps display clear lineaments passing strictly along the fault axes. k_h binary mapping also revealed some traces of vertical motion but no lineaments along the fault axes. The lineament corresponding to the vertical dip-slip fault (Fig. 14.4B) does not essentially differ from the lineament corresponding to the low-angle dip-slip fault (Fig. 14.4C).

For the strike-slip fault model (Fig. 14.4D), the k_h binary map reveals a small lineament in the lower part of the map strictly passing along the fault axis. The k_v binary map does not contain lineaments, although strike-slip fault traces can be found: these are the break of the valley and the displacement of its eastern portion to the north.

FIG. 14.4 Modeling faults on the simulated surface: (A) Initial surface. (B) Vertical dip-slip fault. (C) Low-angle dip-slip fault. (D) Strike-slip fault. (E) Oblique-slip fault. (F) Overthrust with 5-m displacement. (G) Overthrust with 15-m displacement. (H) Gaping fault. *Upper*, elevations; *middle*, horizontal curvatures; *lower*, vertical curvatures; *x–x*, fault axes. *(From Florinsky, I.V., 1996. Quantitative topographic method of fault morphology recognition. Geomorphology 16 (2), 103–119, Fig. 2, reproduced with permission from Elsevier.)*

For the oblique-slip fault model (Fig. 14.4E), the clearly expressed lineament is revealed on the k_v binary map. Traces of vertical and horizontal motions are recorded by k_h mapping, but there are no lineaments along the fault axis. This fact contradicts the theoretical expectation and may be connected with the drawbacks of the oblique-slip fault model.

For the thrust models (Fig. 14.4F and G), clear lineaments passing along the fault axes are revealed by k_v mapping. Some overthrust traces are also recorded on the k_h binary maps, but they are not regular. For the gaping fault model (Fig. 14.4H), lineaments are recorded on the k_h and k_v binary maps.

To validate the method, we applied a very simple DEM and modeled single, simple faults. It is obvious that the effectiveness and impartiality of fault revealing and classification will be lower for DEMs of actual terrains with many faults of different strikes and nontectonic lineaments.

14.3.3 Strike, dip, and displacement estimation

DEMs are utilized to measure dip and strike of faults and other geological contacts (Chorowicz et al., 1991; Morris, 1991; Koike et al., 1998; Masoud and Koike, 2011a, b) (Fig. 14.5). For a locally planar geological contact, three points belonging to the contact define a local approximation of the contact surface between two lithologic units. If these points also belong to a DEM grid, one can estimate the following characteristics from the point coordinates: (1) the strike, the azimuth of the intersection of the contact surface with a horizontal plane; (2) the dip direction, the azimuth of local slope lines; and (3) the dip, the angle between local slope lines and a horizontal plane (Chorowicz et al., 1991).

For a simple surface with a single fault (Fig. 14.4), DTM-based quantitative estimation of a fault displacement may be done as follows: To estimate a vertical displacement, one may use an average value of elevation differences at points where flow lines intersect a lineament displayed on a k_v binary map (Fig. 14.6A). The estimation accuracy obviously depends on the erosion intensity, the DEM accuracy and resolution, as well as the number of measurements along flow lines. A horizontal displacement of a strike-slip fault may be estimated by analyzing relative displacements of valley portions displayed on a k_v binary map along a fault line (Fig. 14.6B). However, displacement estimation will be complicated in the case of an actual terrain having complex topography and tectonics.

14.4 Two case studies

The method is exemplified by a part of the Crimean Peninsula and the adjacent sea bottom (Florinsky, 1996) as well as the Kurchatov city area (Florinsky et al., 1995). The study regions are representative of seismically active mountainous terrains and platform plain territories, respectively.

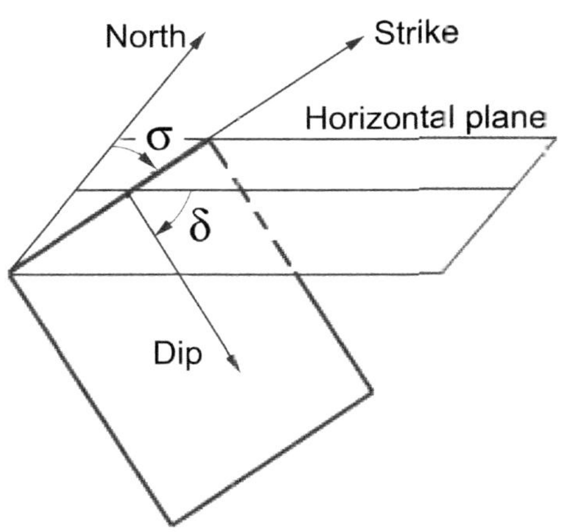

FIG. 14.5 Estimation of the dip (δ) and strike (σ) of a fault or geological contact. *(From Chorowicz, J., Bréard, J.-Y., Guillande, R., Morasse, C.-R., Prudon, D., Rudant, J.-P., 1991. Dip and strike measured systematically on digitized three-dimensional geological maps. Photogrammet. Eng. Remote Sens. 57 (4), 431–436, Fig. 2, reproduced with permission from American Society for Photogrammetry and Remote Sensing.)*

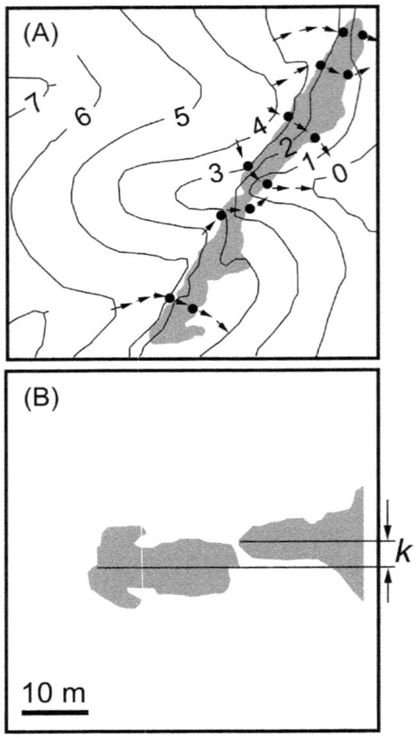

FIG. 14.6 Quantitative estimation of some properties of faults: (A) Estimation of the vertical displacement modeled in Fig. 14.4B; *arrows*, flow lines; *dots*, intersections of flow lines with the lineament; *gray area*, lineament. (B) Estimation of the horizontal displacement modeled in Fig. 14.4D; *gray area*, valley parts; *k*, horizontal displacement.

14.4.1 The Crimean peninsula

The part of the Crimean Peninsula and the adjacent sea bottom measures 210 km × 132 km (Fig. 14.7). This territory was selected for two reasons. First, it has been much studied. A lot of geological, geophysical, and satellite data are available that can be used to test results of lineament revealing and classification. Second, the topographic (Fig. 14.7) and geological diversity of the region allow us to demonstrate the possibilities of the method in different geomorphic and geological conditions.

14.4.1.1 Geological setting

Rather complicated topography and geological setting are typical for this territory due to an interaction of the Precambrian East European Platform and the Alpine-Himalayan orogenic belt. It is customary to distinguish three main parts of the territory: (1) the Epi-Paleozoic Scythian Plate correlating with the Crimean Plain; (2) the Crimean Meganticlinorium topographically expressed as the Crimean Mountains; and (3) an adjacent foredeep system topographically related to the continental shelf and slope of the Black Sea Basin. One can distinguish several elevated blocks of the Proterozoic folded basement covered by Miocene and Pliocene sediments in the Scythian Plate. The central portion of the meganticlinorium—the Main Ridge of the Crimean Mountains—is mainly comprised of the Tauric series and Middle and Late Jurassic rocks. The Main Ridge and piedmonts include several anticlinoria and synclinoria covered by Early Cretaceous, Paleogene, and Neogene limestones, clays, and marls (Muratov, 1969).

Two systems of transregional deep mantle fault zones intersect the Crimean Peninsula (Fig. 14.8A) (Chekunov et al., 1965; Kovalevsky, 1965; Pustovitenko and Trostnikov, 1977):

1. The pre-Riphean, approximately north-striking fault zones of the southern part of the East European Platform crossing the Ukrainian Shield and Scythian Plate and extending southward to the Anatolian Peninsula.
2. The Paleozoic, approximately east-striking fault zone separating the Crimean Mountains from the Crimean Plain and the Black Sea Basin.

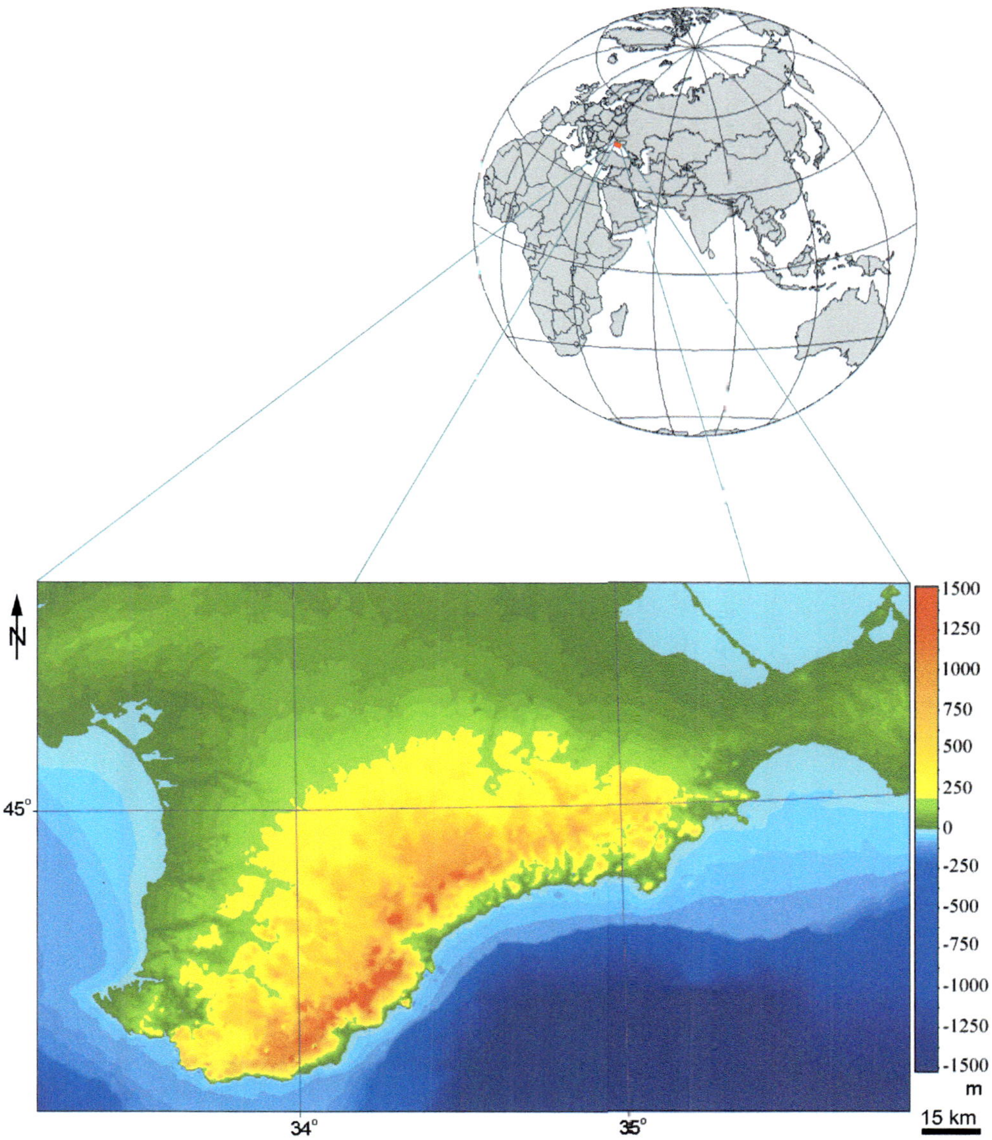

FIG. 14.7 The Crimean Peninsula and the adjacent sea bottom: geographical location and elevations. *Publisher's note: Map lines delineate the study area and do not necessarily depict accepted national boundaries. (Modified from Florinsky, I.V., 2010. Theory and Applications of Mathematical Cartographic Modeling of Topography (D.Sc. thesis). Institute of Mathematical Problems of Biology, Russian Academy of Sciences, Pushchino, Russia (in Russian), Fig. 4.5.)*

These fault zones range in width from 15 km to 20 km and include numerous smaller faults (Pustovitenko and Trostnikov, 1977). The latest reactivation of the Eupatoria-Skadovsk, Salgir-Oktyabrskoe, and Korsak-Theodosia fault zones occurred in the Mesozoic time, whereas the latest reactivation of the Orekhovo-Pavlograd, Foothill Crimean-Caucasian, and Central Crimean-Caucasian fault zones took place during the neotectonic period. A network of deep crustal faults (Fig. 14.8A) was also activated at that time (Borisenko, 1986).

The spatial orientation of regional and local faults (Fig. 14.8A and B) is generally controlled by deep crustal faults. However, different authors described different systems of regional and local lineaments and faults (Shalimov, 1966; Muratov, 1969; Rastsvetaev, 1977; Kats et al., 1981; Borisenko, 1986; Yudin and Gerasimov, 2001; Wolfman et al., 2008; Verkhovtsev, 2008). For example, Borisenko (1986) proposed that there are two diagonal systems of faults in the Crimean Mountains: (1) approximately northeast-striking normal and reverse dip-slip faults with vertical displacements of up to several hundred meters; and (2) approximately northwest-striking oblique- and strike-slip faults with horizontal displacements of up to several hundred meters. These faults probably originated in the Triassic to early Jurassic time. After being reactivated several times, they are currently active.

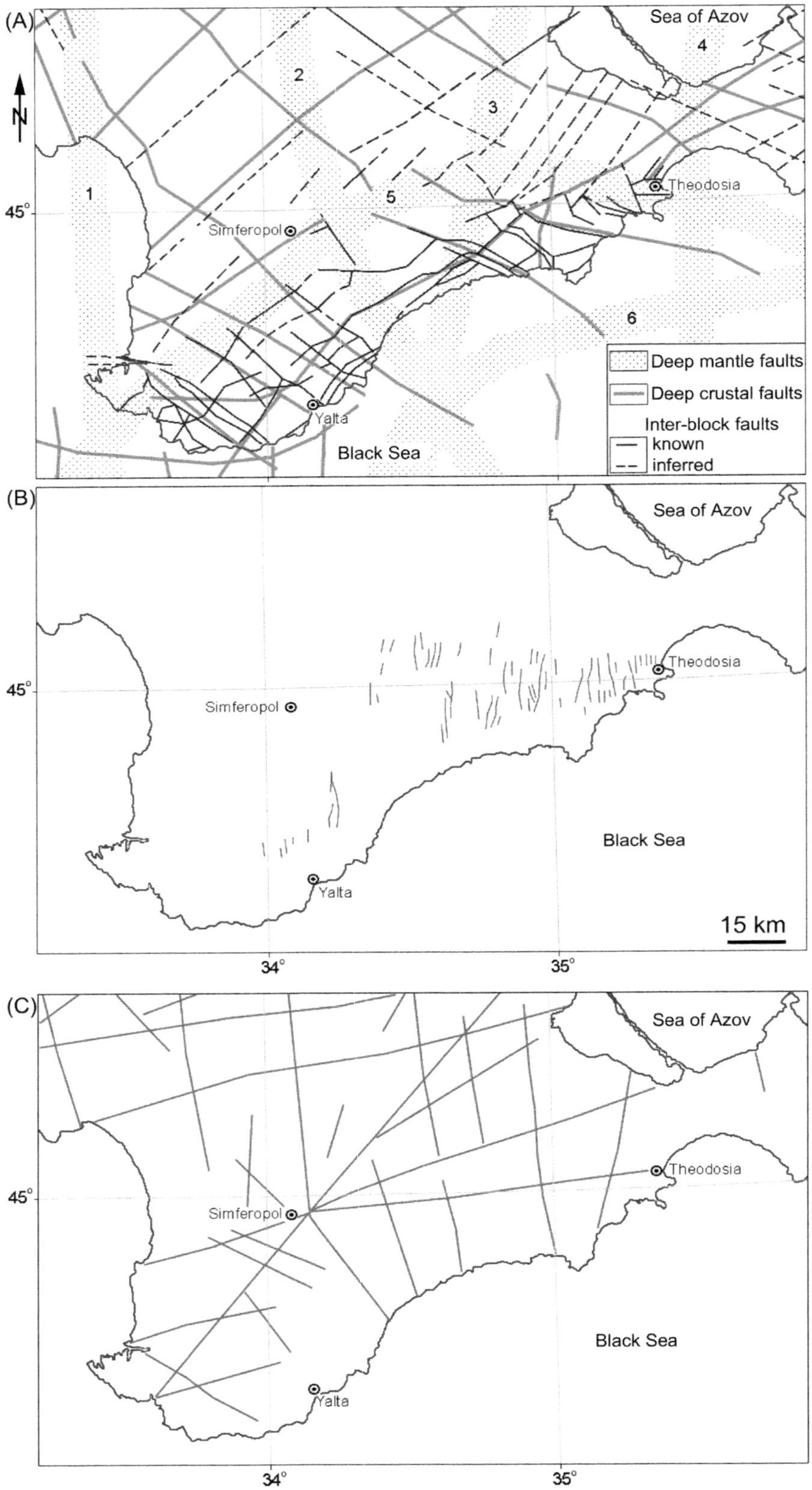

FIG. 14.8 The Crimean Peninsula and the adjacent sea bottom, known faults and lineaments: (A) Deep crustal faults (Borisenko, 1986) and deep mantle fault zones (Chekunov et al., 1965; Kovalevsky, 1965; Lebedev and Orovetsky, 1966; Pustovitenko and Trostnikov, 1977): *1*, Eupatoria–Skadovsk; *2*, Salgir–Oktyabrskoe; *3*, Orekhovo–Pavlograd; *4*, Korsak–Theodosia; *5*, Foothill Crimean–Caucasian; *6*, Central Crimean–Caucasian. (B) North-striking faults in southeastern Crimea (Muratov, 1969). (C) Main lineaments from satellite images (Kats et al., 1981). *(From Florinsky, I.V., 2010. Theory and Applications of Mathematical Cartographic Modeling of Topography (D.Sc. thesis). Institute of Mathematical Problems of Biology, Russian Academy of Sciences, Pushchino, Russia (in Russian), Fig. 4.6.)*

Rastsvetaev (1977, p. 99) suggested that

> *"approximately north-striking faults dominate in the Crimean Mountains, which are up to several tens of kilometers long. There are horizontal sinistral displacements of geological contacts and portions of large folds along most faults of this system .... Horizontal displacements are not less than 3–5 km along each fault .... Available data allow us to characterize approximately north-striking faults of the Crimean Mountains as a system of sinistral oblique-slip faults with the dominant horizontal displacement."*

Some of these structures are represented in Fig. 14.8B. That author also described two other groups of faults in the Crimean Mountains: (1) northeast-striking thrusts 10–15 km long, with up to 3-km horizontal displacements and 30°–45° dips and (2) dextral northwest-striking strike-slip faults with steep dips (Rastsvetaev, 1977).

14.4.1.2 Materials and data processing

We used an irregular DEM (Fig. 14.7) (Florinsky, 1992) generated from topographic maps (Central Board of Geodesy and Cartography, 1953; USSR General Headquarters, 1986). The irregular DEM included 11,936 points. It was converted into the regular DEM by the Delaunay triangulation and a piecewise quadric polynomial interpolation (Agishtein and Migdal, 1991). The grid spacing was 500 m. Digital models of k_h and k_v (Fig. 14.9) were derived from the regular DEM by the Evans-Young method (Section 4.1); the grid spacing was 3000 m. Data were processed by the LandLord software (Florinsky, 2016).

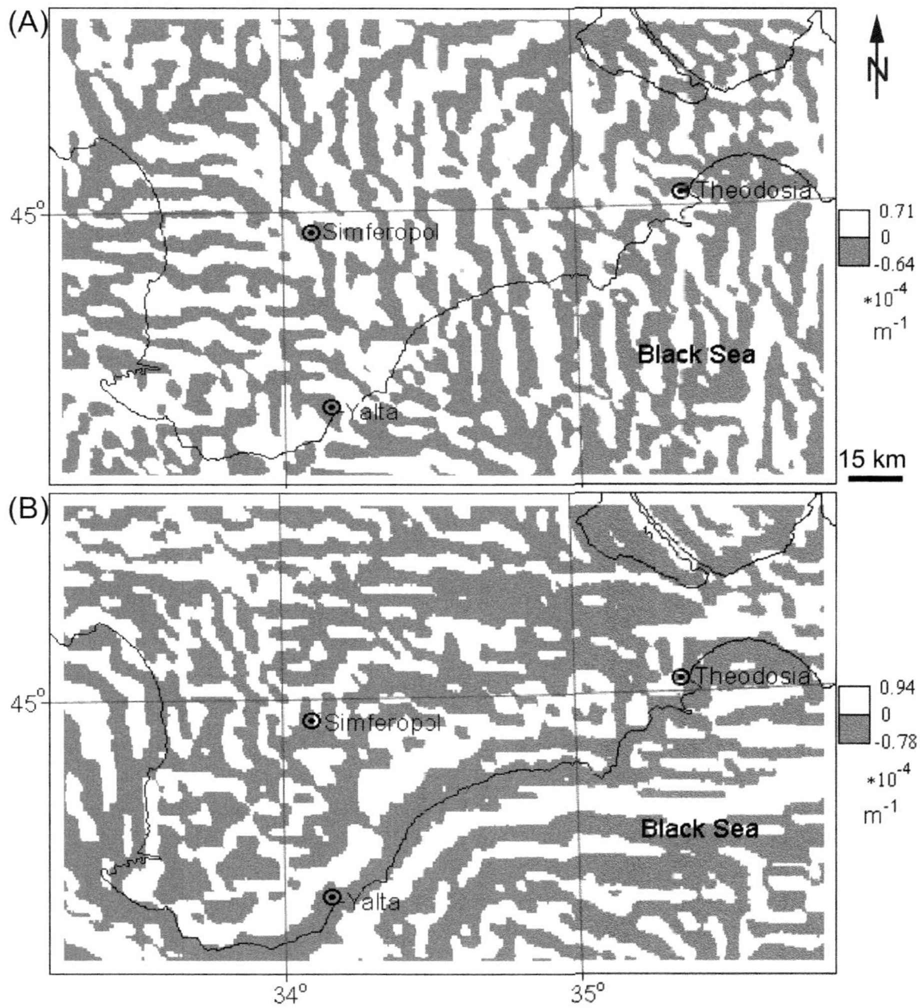

FIG. 14.9 The Crimean Peninsula and the adjacent sea bottom, binary maps: (A) Horizontal curvature. (B) Vertical curvature. *(From Florinsky, I.V., 1996. Quantitative topographic method of fault morphology recognition. Geomorphology 16 (2), 103–119, Fig. 5A and B, reproduced with permission from Elsevier.)*

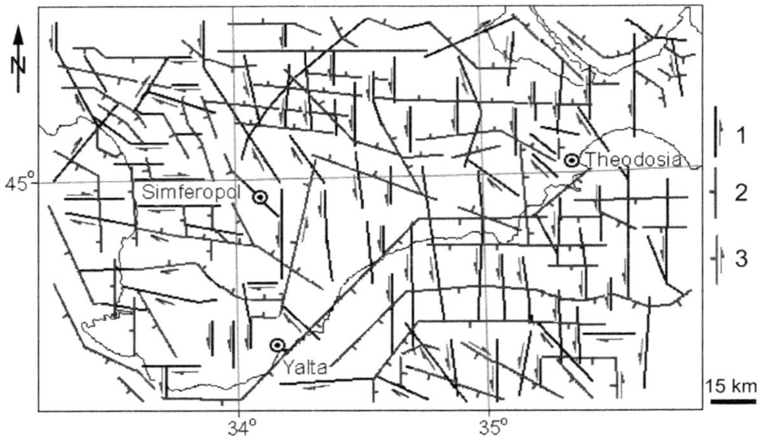

FIG. 14.10 The Crimean Peninsula and the adjacent sea bottom: classification of lineaments revealed by horizontal and vertical curvature binary mapping: *1*, strike-slip faults; *2*, dip-slip faults and thrusts; *3*, oblique-slip faults. *(From Florinsky, I.V., 1996. Quantitative topographic method of fault morphology recognition. Geomorphology 16 (2), 103–119, Fig. 5C, reproduced with permission from Elsevier.)*

The map of revealed and classified lineaments (Fig. 14.10) was produced by a visual analysis of the k_h and k_v binary maps (Fig. 14.9); the skeleton of linear patterns was constructed manually. To estimate the effectiveness of the method, we compared this map (Fig. 14.10) and available geological data (Fig. 14.8).

14.4.1.3 Results and discussion

The k_h mapping (Fig. 14.9A) allowed us to reveal (1) a system of approximately northwest-striking lineaments in the eastern and central parts of the region, (2) a system of approximately east-striking lineaments in the western part of the region, (3) some approximately northeast-striking linear structures in the northern part of the region, and (4) some approximately northwest-striking lineaments in the southern part of the region. We believe that these lineaments are associated with strike-slip faults (Fig. 14.10).

The k_v mapping (Fig. 14.9B) allowed us to reveal (1) a system of approximately west-striking lineaments in the eastern and central parts of the region, (2) a system of approximately north-striking lineaments mainly in the western part of the region, and (3) some approximately northwest-striking linear structures mostly in the southern and the northern parts of the region. We interpreted these lineaments as dip-slip faults and thrusts (Fig. 14.10).

Lineaments revealed on both k_h and k_v binary maps (Fig. 14.9) are interpreted as oblique-slip faults (Fig. 14.10).

The map of classified lineaments (Fig. 14.10) displays their complicated spatial distribution. As a rule, lineaments of the same origin unite into systems, whereas lineaments associated with dip-slip faults stretch across lineaments relating to strike-slip faults. There are some complex faults including dip-slip, strike-slip, thrust, and oblique-slip offsets. A system of approximately north-striking strike-slip faults passes through areas of different geological origins (i.e., the Scythian Plate and the Mountain Crimean meganticlinorium).

A comparison of the lineament map (Fig. 14.10) and fault maps based on geological and geophysical data (Fig. 14.8A) demonstrated that a part of revealed lineaments correlates with known faults. For example, the lineament map reflects the Korsak-Theodosia, Salgir-Oktyabrskoe, Eupatoria-Skadovsk, and Central Crimean-Caucasian deep mantle fault zones (1, 2, 4, and 6, respectively, in Fig. 14.8A). However, most of the revealed lineaments do not correlate with previously mapped faults. We expected such a result: it is well known that lineaments detected on topographic maps and satellite images do not necessarily correlate with faults established by conventional geological methods and displayed on geological maps (Makarov, 1981; Trifonov et al., 1983; Bryukhanov and Mezhelovsky, 1987). However, the k_h and k_v binary maps (Fig. 14.9) show most lineaments recognized from satellite images (Kats et al., 1981) (Fig. 14.8C). The latter authors related the approximately north-striking lineaments to the north-striking deep mantle faults (Fig. 14.8A).

Most of the revealed lineaments relate to the main groups of known regional faults (Section 14.4.1.1). In particular, the system of approximately north-striking lineaments, which are, in our view, associated with strike-slip faults (Fig. 14.10), relates to the system of sinistral oblique-slip faults with the dominant horizontal displacement described by Rastsvetaev (1977) (see Section 14.4.1.1). Northwest-striking lineaments, which, from our standpoint, indicate strike-slip, dip-slip, and oblique-slip faults (Fig. 14.10), relate to groups of such faults described by both Rastsvetaev (1977) and Borisenko (1986) (see Section 14.4.1.1).

The lineament map (Fig. 14.10) has a somewhat subjective nature. First, we generally drew fault lines as medians of linear patterns recorded on the k_h and k_v binary maps. Second, a visual analysis of these maps may result in the loss of some lineaments. Third, the lineament map corresponds to the only DEM grid spacing. One may obtain maps containing more or fewer lineaments using smaller or larger grid spacings, respectively. To reveal and classify virtually all topographically expressed lineaments as well as to grade them into transregional, regional, and local scales, it is necessary to handle a series of k_h and k_v binary maps with different resolutions.

14.4.2 The Kurchatov city area

A study area around the Kurchatov city measured 68 km × 48 km (Fig. 14.11).

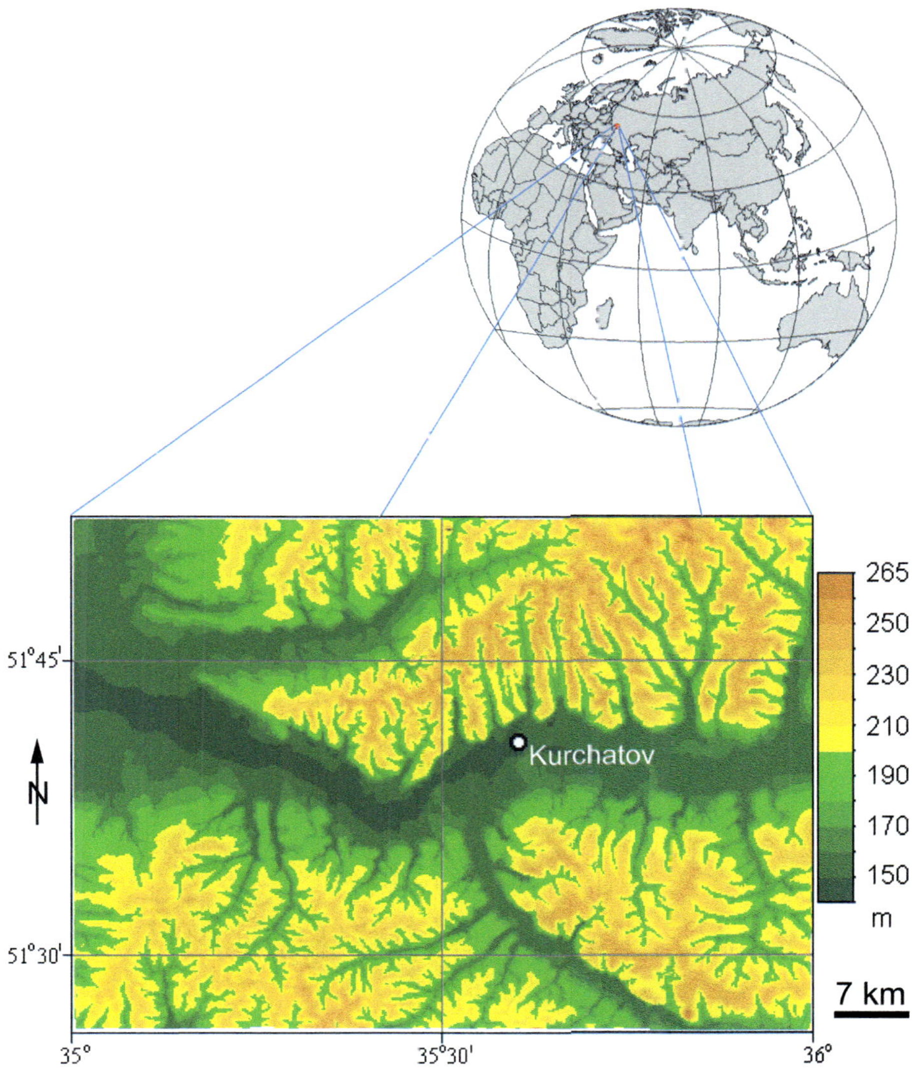

FIG. 14.11 The Kurchatov city area, geographical location, and elevations. *Publisher's note: Map lines delineate the study area and do not necessarily depict accepted national boundaries. (Modified from Florinsky, I.V., 2010. Theory and Applications of Mathematical Cartographic Modeling of Topography (D.Sc. thesis). Institute of Mathematical Problems of Biology, Russian Academy of Sciences, Pushchino, Russia (in Russian), Fig. 2.6A.)*

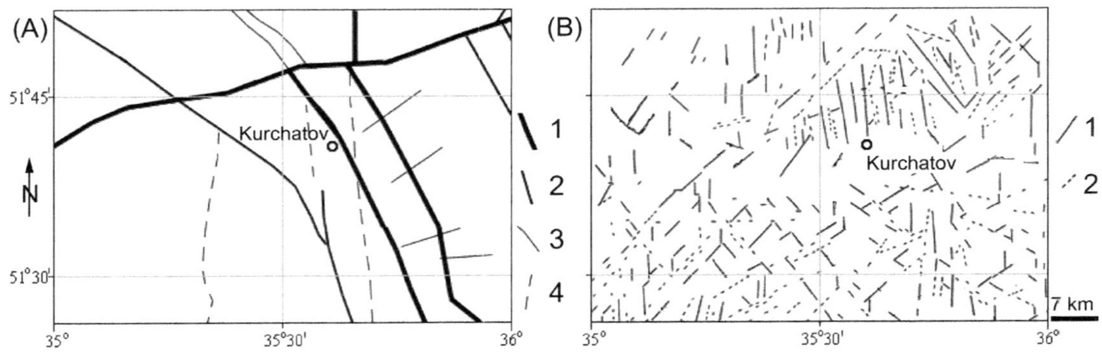

FIG. 14.12 The Kurchatov city area, known faults and lineaments: (A) Faults revealed from geophysical and satellite data: *1*, deep crustal faults; *2*, regional faults; *3*, other faults; *4*, inferred (Myrzin et al., 1988). (B) Lineaments revealed from topographic maps and satellite images (Polevaev et al., 1992): *1*, true; *2*, inferred. *(Modified from Florinsky, I.V., 2010. Theory and Applications of Mathematical Cartographic Modeling of Topography (D. Sc. thesis). Institute of Mathematical Problems of Biology, Russian Academy of Sciences, Pushchino, Russia (in Russian), Fig. 4.10.)*

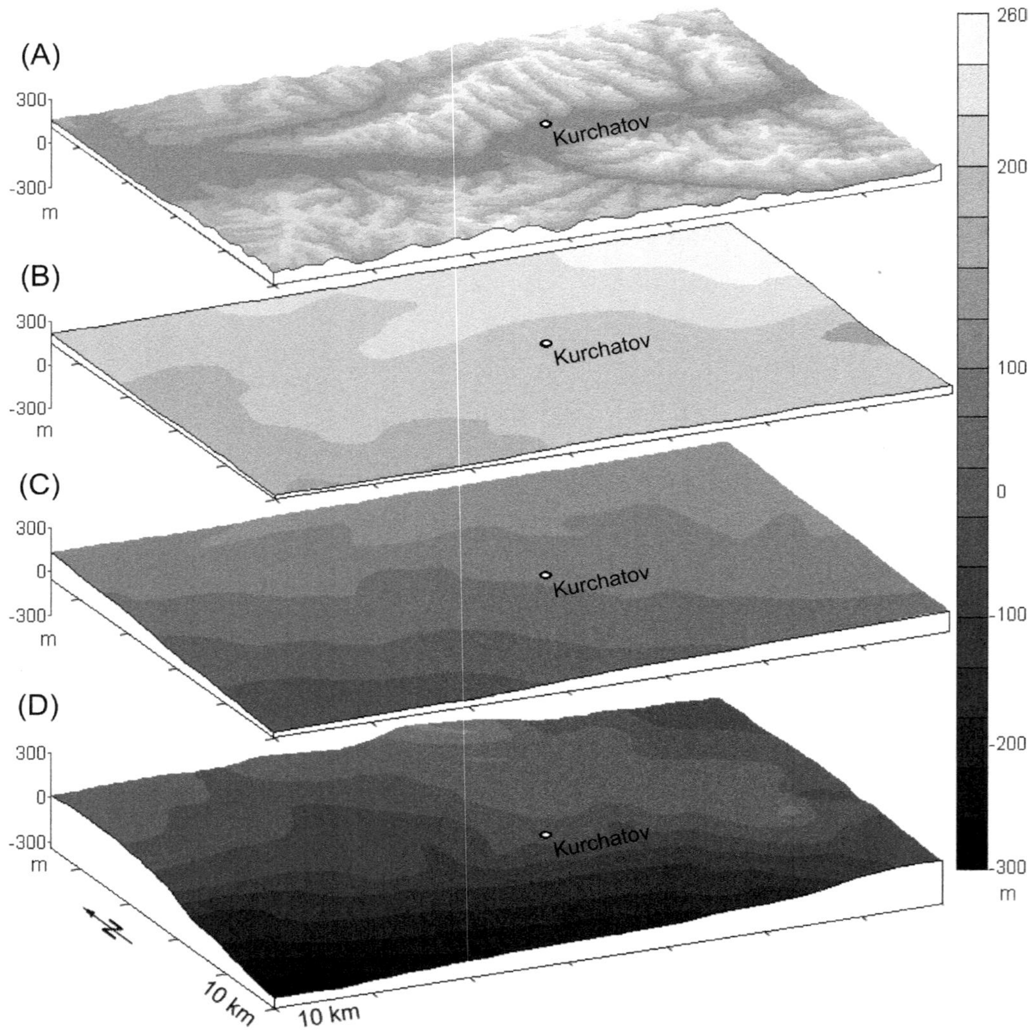

FIG. 14.13 The Kurchatov city area, elevation: (A) Land surface. (B) Cretaceous deposit surface. (C) Cenomanian deposit surface. (D) Crystalline basement surface. All elevations are above sea level; 30× exaggeration of the vertical scale. *(Modified from Florinsky, I.V., 2010. Theory and Applications of Mathematical Cartographic Modeling of Topography (D.Sc. thesis). Institute of Mathematical Problems of Biology, Russian Academy of Sciences, Pushchino, Russia (in Russian), Fig. 4.11.)*

14.4.2.1 Geological setting

The territory is composed of Archean, Proterozoic, Devonian, Carboniferous, Jurassic, Cretaceous, Paleogene, Neogene, and Quaternary rocks (Dubyansky and Khakman, 1949; Nalivkin, 1970, 1974; Myrzin et al., 1988; Poletaev et al., 1992). Precambrian rocks (crystalline and foliated schists, marbles, dolomites, quartzites, etc.) composing the crystalline basement are highly metamorphosed and dislocated (Fig. 14.12A). The basement includes Archean blocks, horst-anticlinoria, and synclinoria extended to the northwest and separated by faults of different hierarchical levels. Devonian sediments are inclined to the northwest, while Carboniferous, Mesozoic, and Cenozoic sediments are inclined to the southwest.

The depth of the crystalline basement surface ranges from −300 to 50 m above sea level. Poletaev et al. (1992) argued that the key structural elements of the basement (e.g., faults) may be expressed as lineaments on the land surface. They should reflect both the northwestern strike of the main folding structures and faults and the northeastern strike of transverse faults. In particular, those authors considered rectilinear, northwest-striking coasts of pre-Mesozoic sea bays and northeast-striking coasts of Paleogene ones as reactivated basement faults. Poletaev et al. (1992) also supposed that approximately north-striking linear elements of the modern topography may reflect both portions of the pre-Bajocian north-striking paleo-drainage network (Myrzin et al., 1988) and some elements of the deep structure of the crust also associated with approximately north-striking faults of the Precambrian basement.

14.4.2.2 Materials and data processing

As input data, we used a topographic map (USSR General Headquarters, 1981) and 1:200,000-scale geological and geophysical cartographic materials (Myrzin et al., 1988). Using a digitizer, we produced irregular DEMs of the land surface (Figs. 14.11 and 14.13A), the Cretaceous deposit surface (Fig. 14.13B), the Cenomanian deposit surface (Fig. 14.13C), and the crystalline basement surface (Fig. 14.13D). The irregular DEMs included 46,694, 941, 1002, and 1308 points, respectively.

The irregular DEMs were converted into regular ones by the Delaunay triangulation and a piecewise quadric polynomial interpolation (Agishtein and Migdal, 1991). Digital models of k_h and k_v (Fig. 14.14) were derived from the regular DEMs by the Evans-Young method (Section 4.1); the grid spacing was 1500 m. It is obvious that all land surface DTMs have higher accuracy and resolution than DTMs of stratigraphic surfaces.

Maps of revealed and classified lineaments (Fig. 14.15) were produced by visual analysis of the k_h and k_v binary maps (Fig. 14.14); skeletons of linear patterns were constructed manually.

14.4.2.3 Results and discussion

We are considering a platform territory, so it is reasonable to interpret lineaments as zones of concentration of bending deformations rather than faults (Section 14.2).

Lineaments probably associated with zones of horizontal bending deformations were detected on all surfaces. For the crystalline basement, these are predominantly north-striking structures (Fig. 14.15D). For the Cenomanian and Cretaceous deposits, there are generally northeast-striking structures, but there are also minor north- and northwest-striking ones (Fig. 14.15B and C). For the land surface, there are predominantly north-striking structures (Fig. 14.15A). On all surfaces, the lineaments associated with flexures generally stretch across the lineaments, indicating zones of horizontal bending deformations (Fig. 14.15).

As for Crimea (Section 14.4.1.3), the revealed lineaments do not relate to the basement faults recognized by geological and geophysical techniques (Fig. 14.12A). However, one can see a similarity between the two lineament maps of the land surface derived from satellite images (Fig. 14.12B) and morphometric maps (Fig. 14.15A). Moreover, our results are consistent with the conclusion by Poletaev et al. (1992, p. 34) that "landscape elements of the Kurchatov city area can reflect northwest- and northeast-striking folds and faults of the basement as well as north-striking elements of the deep structure and paleotopography (the pre-Bajocian drainage network), and coasts of pre-Mesozoic and Paleogene sea bays, which have northwestern and northeastern orientations."

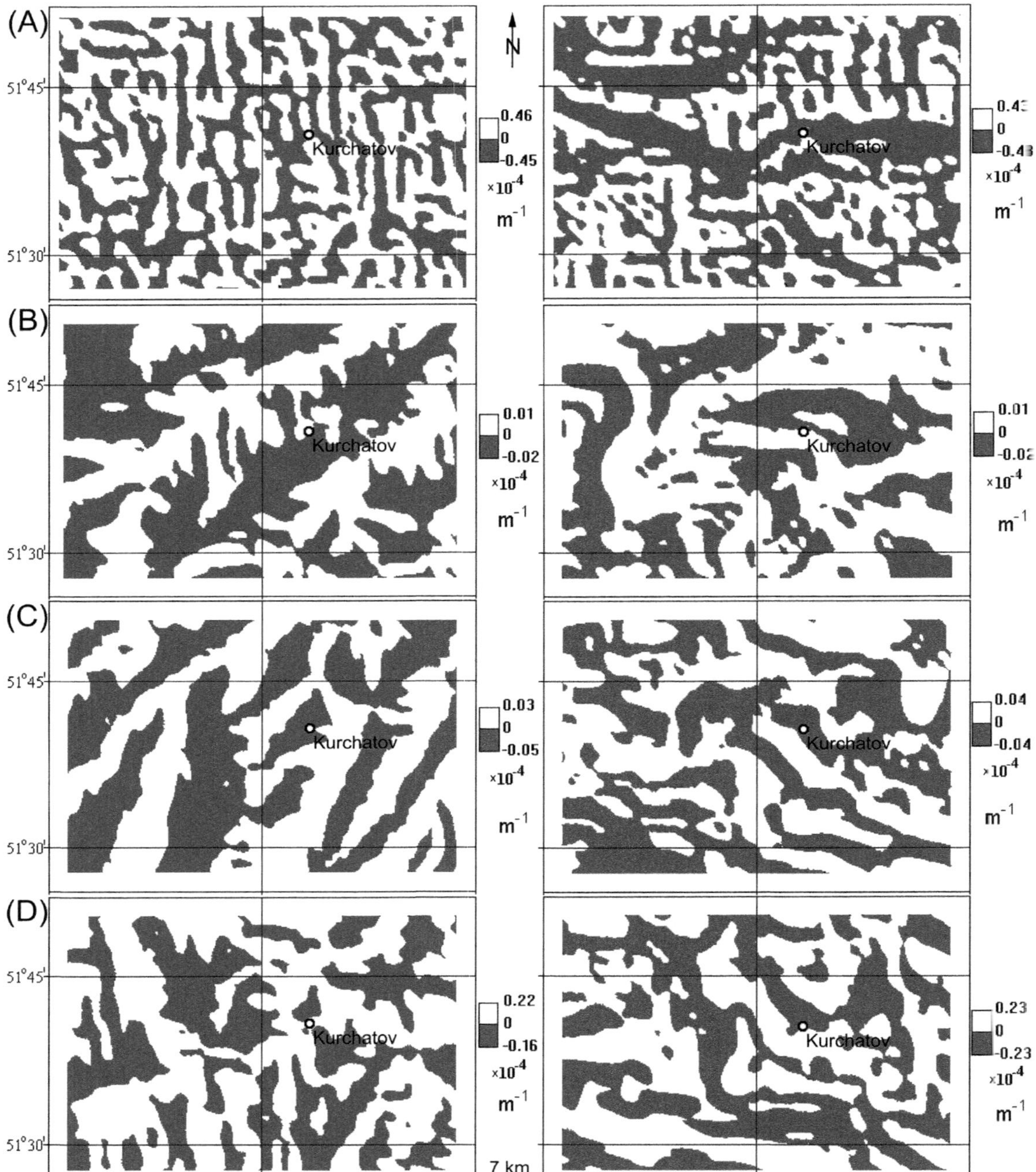

FIG. 14.14 The Kurchatov city area, binary maps of horizontal curvature (*left*) and vertical curvature (*right*): (A) Land surface. (B) Cretaceous deposit surface. (C) Cenomanian deposit surface. (D) Crystalline basement surface. *(Modified from Florinsky, I.V., 2010. Theory and Applications of Mathematical Cartographic Modeling of Topography (D.Sc. thesis). Institute of Mathematical Problems of Biology, Russian Academy of Sciences, Pushchino, Russia (in Russian), Fig. 4.12.)*

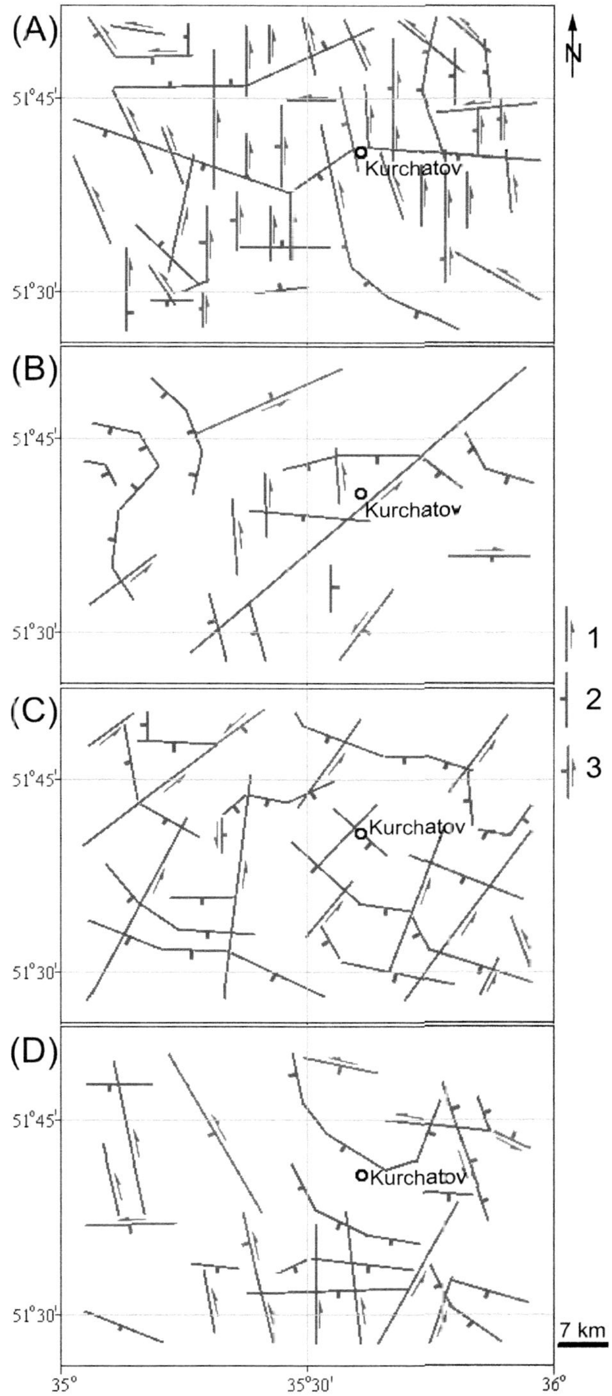

FIG. 14.15 The Kurchatov city area, classification of lineaments revealed on the binary maps of horizontal and vertical curvatures: (A) Land surface. (B) Cretaceous deposit surface. (C) Cenomanian deposit surface. (D) Crystalline basement surface: *1*, zones of horizontal bending deformations; *2*, flexures; *3*, zones of bending deformations formed by both vertical and horizontal movements. Directions of horizontal motions are arbitrary. *(Modified from Florinsky, I.V., 2010. Theory and Applications of Mathematical Cartographic Modeling of Topography (D.Sc. thesis). Institute of Mathematical Problems of Biology, Russian Academy of Sciences, Pushchino, Russia (in Russian), Fig. 4.13.)*

Chapter 15

Accumulation zones and fault intersections

15.1 Motivation

Topography in many ways controls the lateral transport of substances, namely gravity-driven migration and accumulation of water as well as mineral and organic materials over the land surface and in the soil (Huggett and Cheesman, 2002). At the same time, there are strong associations between topography and tectonics (Penck, 1924; Meshcheryakov, 1965; Ollier, 1981; Ufimtsev, 1988; Scheidegger, 2004). In particular, valley networks, determining principal routes of overland flows, can often be associated with networks of faults and fracturing zones of different scales (Zernitz, 1931; Filosofov, 1960; Gerasimov and Korzhuev, 1979; Hantke and Scheidegger, 1999; Gudmundsson, 2000; Twidale, 2004). In turn, faults and fracturing zones serve as pathways for upward transport of deep groundwaters, fluids, and gases to the Earth's surface (Peive, 1956b; Kerrich, 1986; Caine et al., 1996; Gudmundsson, 2000; Handy et al., 2007) and hence determine areas of soil geochemical anomalies (Kovda, 1973, pp. 214–221; Kasimov et al., 1978; Kasimov, 1980). Therefore, some migration routes of surface flows can coincide with some sites of fluid discharge. The study of such associations can be useful in gaining a better understanding of relationships between (1) geological structures and landforms and (2) endo- and exogenous processes influencing the evolution of the biosphere.

Gravity-driven overland and intrasoil transport can be interpreted in terms of flow divergence/convergence and flow deceleration/acceleration determined by horizontal and vertical curvatures, respectively (Sections 2.2.4.2 and 2.2.4.3). Flow convergence and deceleration result in the accumulation of substances in soil caused by the slowing down or termination of overland and intrasoil transport. At different scales, the intensity and direction of the transport of substances in many respects depend on the spatial distribution of three types of landforms (Fig. 2.22B) (Section 2.7.2):

1. *Relative accumulation zones*. There are concurrent convergence and deceleration of flows ($k_h < 0$ and $k_v < 0$).
2. *Relative dissipation zones*. There are concurrent divergence and acceleration of flows ($k_h > 0$ and $k_v > 0$).
3. *Transit zones*. There is no concurrent action of "unidirectional" processes (k_h and k_v have different signs or are zero).

Qualitatively, the role of relative accumulation zones in the formation of soil properties has been repeatedly discussed. Accumulation zones are usually closed depressions of different scales; soil moisture content can be highly increased in such sites (Sibirtsev, 1899, p. 134; Ponagaibo, 1915; Neustruev, 1927; Rode, 1953; Fedoseev, 1959). In arid and semiarid regions, there occurs buildup of salts, secondary salinization of soils, and groundwater salinization in these zones at macro- and mesotopographic scales (Kovda, 1946, 1971).

Quantitatively, it was demonstrated that the wettest soil areas and saturation zones are typical for relative accumulation zones (O'Loughlin, 1981; Feranec et al., 1991). These may initiate overflows in accumulation zones (Wood et al., 1990). Landsliding (Lanyon and Hall, 1983), soil gleying, maximum thickness of the A horizon, and maximum depth to calcium carbonate (Pennock et al., 1987) are seen in these zones due to increased water content in soils and grounds.

Fault intersections of different scales are important features of the geoenvironment (Poletaev, 1992). They are characterized by increased fracturing, porosity, and permeability of rocks. It is established that fault intersections are often marked by increased seismicity (Gelfand et al., 1972, 1976; Rantsman, 1979; Kozlov, 1991; Keilis-Borok, 2002; Gorshkov et al., 2003; Soloviev et al., 2014), high discharge of springs and boreholes (Lattman and Parizek, 1964; Morozov et al., 1988; Babiker and Gudmundsson, 2004), development of swamps (Garetsky et al., 1983), abnormal structure of soil horizons and composition of soil cover (Romashkevich et al., 1997), as well as intensification of erosion, karstification, and landsliding (Korobeynik et al., 1982; Glasko and Rantsman, 1996; Karakhanian, 1981). Fault intersections can control intensive magmatism and volcanism (Poletaev, 1992), ore fields and deposits (Kutina, 1974; Wertz, 1974; Volchanskaya, 1981; Trifonov et al., 1983; O'Driscoll, 1986; Lopatin, 2002; Tripp and Vearncombe, 2004), as well as hydrocarbon fields and leakage zones (Guberman et al., 1997; Gartrell et al., 2004). These do not imply that all the phenomena are observed at each fault intersection; this depends on the geological background and other environmental factors.

Digital Terrain Analysis. https://doi.org/10.1016/B978-0-443-24798-9.00005-4

Garetsky et al. (1983) were probably the first to establish an association of some topographic depressions with fault intersections by visual interpretation of satellite images. Analyzing geological and remotely sensed data, Poletaev (1992) found qualitatively that fault intersections are, as a rule, expressed by topographic depressions. Florinsky (1993) proved quantitatively that topographically expressed accumulation zones, as a rule, coincide with fault intersections (at least, in tectonic regions).

Indeed, k_h and k_v mapping reveal two groups of topographically expressed lineaments (Florinsky, 1992). Structures of the first group correspond to convergence areas ($k_h < 0$), whereas lineaments of the second group relate to deceleration areas ($k_v < 0$). These lineaments indicate predominantly strike-slip and dip-slip faults, respectively (Florinsky, 1996) (Chapter 14). It is clear that an intersection of two lineaments or faults of different groups relates to a topographically expressed accumulation zone by its definition (as discussed earlier). Therefore, values $k_h < 0$ with $k_v < 0$ determine fault intersections, values $k_h > 0$ with $k_v > 0$ (dissipation zones) determine blocks between faults, while other combinations of k_h and k_v values (transit zones) determine fault segments outside fault intersections.[a]

In this chapter, we analyze relationships of some natural phenomena a priori associated with fault intersections (i.e., sites of intensive rock fracturing and springs/boreholes with abnormally high discharges) with zones of flow accumulation, transit, and dissipation (Florinsky, 2000).

15.2 Study area

We studied a part of the Crimean Peninsula (Fig. 14.6). For geological setting, see Section 14.4.1.1.

15.3 Materials and methods

We employed the DEM of the part of the Crimean Peninsula and adjacent sea bottom (Fig. 14.7). For a description of the DEM and calculation of k_h and k_v digital models (Fig. 14.9), see Section 14.4.1.2. A map of accumulation, transit, and dissipation zones (Fig. 15.1) was derived by combining k_h and k_v models.

For the analysis, we selected the following natural phenomena, which can be a priori associated with fault intersections:

- springs and boreholes with abnormally high discharges (Fig. 15.2) (Morozov et al., 1988)
- intensive rock fracturing, wherein discharge of deep thermal groundwater occurs (Fig. 15.2) (Shtengelov, 1978, 1982; Korobeynik et al., 1982)

The location of springs and boreholes with abnormally high discharges (Fig. 15.2) is controlled by intersections of approximately north-striking transregional fault zones with northwest- and approximately west-striking local faults (Morozov et al., 1988). Those authors did not publish quantitative data on all discharges. An example is the Vaucluse-type

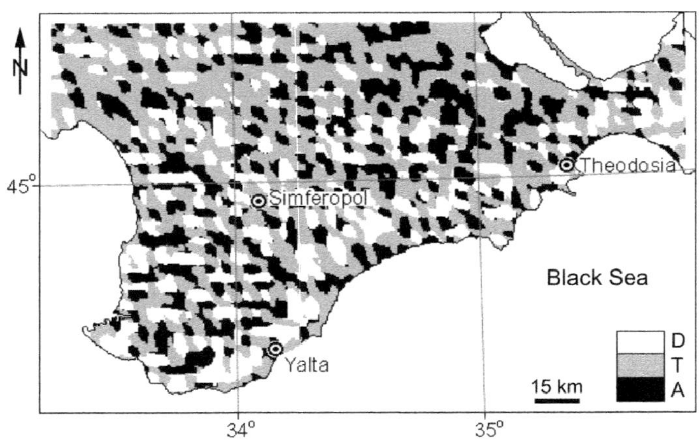

FIG. 15.1 The Crimean Peninsula: topographically expressed zones of dissipation, transit, and accumulation. *A*, accumulation zones; *D*, dissipation zones; *T*, transit zones. *(From Florinsky, I.V., 2000. Relationships between topographically expressed zones of flow accumulation and sites of fault intersection: analysis by means of digital terrain modelling. Environ. Model. Software 15 (1), 87–100, Fig. 6), reproduced with permission from Elsevier.)*

a. We do not advocate that geological structures control all accumulation zones. Some of them can be formed by exogenous processes, such as erosion.

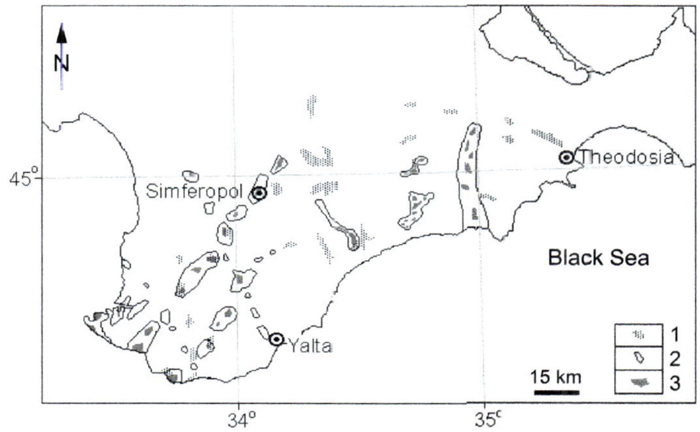

FIG. 15.2 The Crimean Peninsula, phenomena under study: *1*, sites of springs and boreholes with abnormally high discharges (Morozov et al., 1988); *2*, sites of intensive rock fracturing; *3*, sites of highly intensive rock fracturing (Shtengelov, 1978, 1982; Korobeynik et al., 1982). *(From Florinsky, I.V., 2000. Relationships between topographically expressed zones of flow accumulation and sites of fault intersection: analysis by means of digital terrain modelling. Environ. Model. Software 15 (1), 87–100, Fig. 4, reproduced with permission from Elsevier.)*

spring of Kara Sou Bashi, the greatest Crimean spring located at one of the fault intersections. Its mean annual discharge is about 1450 L/s.

Sites of intensive and highly intensive rock fracturing (Fig. 15.2) were recognized by a gamma-metric technique based on a relationship between rock radioactivity and the degree of rock fracturing (Shtengelov, 1978, 1982). For lithologically homogeneous rocks, the greater the rock fracturing, the less the gamma activity. This is caused by increased leakage of fine materials, the major contributor to gamma activity for most sedimentary and metamorphic rocks. Shtengelov (1978, 1982) and Korobeynik et al. (1982) did not publish quantitative data on the background gamma activity and gamma activity for sites of intensive rock fracturing.

To estimate the association of sites of intensive/highly intensive rock fracturing and springs/boreholes having abnormally high discharge with accumulation, transit, and dissipation zones, we estimated the following parameters (Tables 15.1 and 15.2):

1. Total areas of accumulation, transit, and dissipation zones S_a, S_t, and S_d, respectively.
2. Proportions of the total area of the region S_Σ corresponding to the total areas of accumulation, transit, and dissipation zones (E_a, E_t, and E_d, respectively): $E_i = S_i / S_\Sigma$, where i denotes a, t, and d.

TABLE 15.1 Parameters of topographically expressed accumulation, transit, and dissipation zones within the study region.

Topographic zone	Parameter	Value
Accumulation	Total area, S_a (km^2)	6588
	Proportion of the total area, E_a	0.25
	Weight, W_a	1.09
Transit	Total area, S_t (km^2)	14,751
	Proportion of the total area, E_t	0.55
	Weight, W_t	0.78
Dissipation	Total area, S_d (km^2)	5364
	Proportion of the total area, E_d	0.20
	Weight, W_d	1.13

The total area of the study area $S_\Sigma = 26{,}703\,km^2$.
(From Florinsky, I.V., 2000. Relationships between topographically expressed zones of flow accumulation and sites of fault intersection: analysis by means of digital terrain modelling. Environ. Model. Software 15 (1), 87–100, Table 1, reproduced with permission from Elsevier.)

TABLE 15.2 Parameters for the association of the phenomena under study with topographically expressed accumulation (A), transit (T), and dissipation (D) zones.

Phenomenon	Total area, U_Σ (km^2)	Parameter	Topographic zone		
			A	T	D
Sites of springs and boreholes with abnormally high discharges	301	Total area, U_i (km^2)	204	90	7
		Proportion of the total area, P_i	0.68	0.30	0.02
		Association coefficient, R_i^{as}	0.74	0.23	0.03
Sites of intensive rock fracturing	1006	Total area, U_i (km^2)	492	382	132
		Proportion of the total area, P_i	0.49	0.38	0.13
		Association coefficient, R_i^{as}	0.54	0.31	0.15
Sites of highly intensive rock fracturing	134	Total area, U_i (km^2)	95	25	14
		Proportion of the total area, P_i	0.71	0.19	0.10
		Association coefficient, R_i^{as}	0.74	0.15	0.11

(From Florinsky, I.V., 2000. Relationships between topographically expressed zones of flow accumulation and sites of fault intersection: analysis by means of digital terrain modelling. Environ. Model. Software 15 (1), 87–100, Table 2, reproduced with permission from Elsevier.)

3. Weights of accumulation, transit, and dissipation zones are W_a, W_t, and W_d, respectively: $W_i = \frac{4}{3} - E_i$
4. The total area of a particular phenomenon under study U_Σ and its total areas located in accumulation, transit, and dissipation zones U_a, U_t, and U_d, respectively.
5. Proportions of the total area of a particular phenomenon corresponding to its total areas located in accumulation, transit, and dissipation zones (P_a, P_t, and P_d, respectively): $P_i = U_i/U_\Sigma$.

Finally, we applied "association coefficients" R_a^{as}, R_t^{as}, and R_d^{as} (Table 15.2) to characterize the association of a particular phenomenon with accumulation, transit, and dissipation zones, respectively: $R_i^{as} = W_i P_i$. For each phenomenon, R_i^{as} was normalized: $R_d^{as} + R_t^{as} + R_a^{as} = 1$.

15.4 Results and discussion

The map of accumulation, transit, and dissipation zones (Fig. 15.1) shows a regular, cell-like spatial distribution of these landforms that is probably connected with intersections of north-, east-, northwest-, and northeast-striking regional faults (Florinsky, 1996) (Section 14.4.1.1).

Location of springs/boreholes with abnormally high discharge and sites of highly intensive rock fracturing are closely associated with accumulation zones; the coefficients are 0.74 (Table 15.2 and Fig. 15.3). This is no surprise: as accumulation zones are connected with fault intersections, so phenomena located at fault intersections should correlate strongly with accumulation zones. Sites of intensive rock fracturing are associated with accumulation zones to a lesser extent: the coefficient is 0.54 (Table 15.2 and Fig. 15.3).

The association of intensive rock fracturing sites with transit zones is described by the coefficient of 0.31 (Table 15.2 and Fig. 15.3). This is also not surprising because transit zones are connected with fault segments outside intersections (Section 15.1). At these segments, fracturing is less than at fault intersections, so transit zones can control lesser parts of phenomena typical for accumulation zones. Indeed, the associations of sites of springs/boreholes having abnormally high discharge and highly intensive rock fracturing with transit zones are described by the coefficients of 0.23 and 0.15, respectively (Table 15.2 and Fig. 15.3).

The smallest parts of the phenomena under study are associated with dissipation zones: the coefficients range from 0.03 for abnormally high discharge sites to 0.15 for intensive rock-fracturing sites (Table 15.2 and Fig. 15.3). We anticipate that

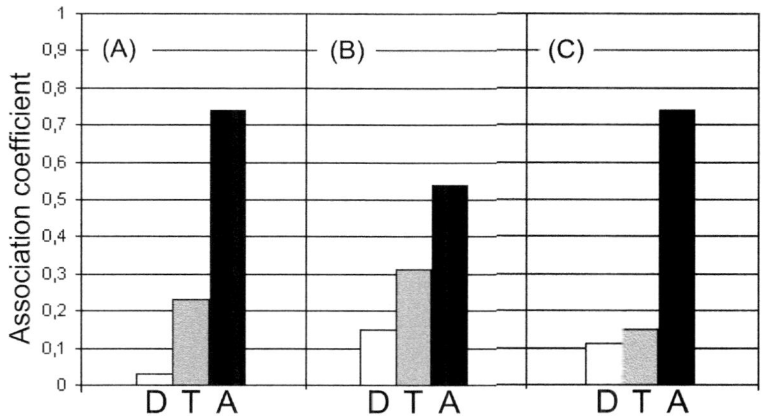

FIG. 15.3 The association of phenomena under study with topographically expressed zones of dissipation, transit, and accumulation: (A) Sites of springs and boreholes with abnormally high discharges. (B) Sites of intensive rock fracturing. (C) Sites of highly intensive rock fracturing. *A*, accumulation zones; *D*, dissipation zones; *T*, transit zones. *(From Florinsky, I.V., 2000. Relationships between topographically expressed zones of flow accumulation and sites of fault intersection: analysis by means of digital terrain modelling. Environ. Model. Software 15 (1), 87–100, Fig. 7, reproduced with permission from Elsevier.)*

these are artifacts caused by data errors because rock fracturing is generally not typical for dissipation zones (that is, blocks between faults).

It is clear that the numerical results of the analysis (Table 15.2) are area-specific. Associations concerned depend on geological and geomorphic conditions as well as the study scale.

Topographically expressed accumulation zones are closely associated with sites of abnormally high discharges and highly intensive rock fracturing (Table 15.2 and Fig. 15.3), which control upward transport of deep groundwaters. In accumulation zones, the increased soil moisture can therefore be controlled by increased upwelling of deep fluids coincident with the accumulation of water from overland and intrasoil lateral flows. This is consistent with the known facts:

- Increased discharges of springs and boreholes are typical for both accumulation zones (Lukin, 1987) and fault intersections (Lattman and Parizek, 1964; Morozov et al., 1988; Babiker and Gudmundsson, 2004).
- Increased soil and ground moisture is observed in both accumulation zones (Rode, 1953; Feranec et al., 1991) and fault intersections (Garetsky et al., 1983).
- Increased soil and ground moisture lead to landslide development in both accumulation zones (Lanyon and Hall, 1983) and fault intersections (Karakhanian, 1981).
- Soil development in depressions at fault zones is affected by the upwelling of saline and fresh groundwater as well as overland flows carrying dissolved materials. This results in the formation of salt-affected and hydromorphic soils (Kasimov et al., 1978).

Thus, accumulation zones/fault intersections are areas of contact and interaction between overland flows and deep fluids. The association of accumulation zones with fault intersections is probably caused by increased rock fracturing typical for fault intersections. They have a higher degree of fracturing as compared to faults outside intersections. Therefore, rocks at fault intersections are more prone to weathering, erosion, and other processes leading to the occurrence and development of depressions. An increase in soil and ground moisture in accumulation zones may increase erosion, which in turn may enhance the topographic expression of these areas.

Chapter 16

Global topography and tectonic structures

16.1 Motivation

Faults and fracture zones are expressed as lineaments on the Earth's surface (Section 14.1). One can geometrically consider lineaments as planar straight lines at a regional scale and spatial curves at continental and global scales.

Much attention has been paid to planetary systems of lineaments in global tectonic studies and modeling. Three groups of investigations have been presented:

1. Detection of regularities in the global distribution and orientation of lineaments by analyzing physiographic and geological maps and then developing a model for the regularities observed (Chebanenko, 1963; Moody, 1966; Katterfeld and Charushin, 1973; Besprozvanny et al., 1994).
2. Development of a physical and mathematical model of a global tectonic process causing an ideal planetary network of lineaments and then comparison of the ideal and actual lineament networks (Weinberg, 1934b; Vening Meinesz, 1947; Dolitsky and Kiyko, 1963; Chebanenko and Fedorin, 1983).
3. Laboratory simulation of a global lineament network using rotatable spheres (Knetsch, 1965; Cherednichenko et al., 1966). In this case, the origin of global lineament systems was usually associated with rotational forces.

Rance (1968) developed a physical and mathematical model of the torsional deformation of a sphere. The torsion was attributed to the action of mantle convection currents on the crust. According to the model, there are two systems of traces of torsional failure surfaces on the surface of the sphere: shear fractures and cleavage cracks (Rance, 1967). Traces of torsional deformation constitute two systems of double helices encircling the sphere from pole to pole (Fig. 16.1A). The traces vary in inclination at the equator: a system of double helices tracing shear fractures has an inclination of 15°–18° and 165°–162°, and another system of double helices tracing cleavage cracks has an inclination of 56°–62° and 124°–118°. A search for actual global helical tectonic features resulted in the detection of several relatively small lineaments referring to faults, trenches, ridges, fracture zones, and seamount chains in basins of the Pacific and Indian Oceans (Rance, 1967, 1969).

Using physiographic maps, O'Driscoll (1980) visually detected two global topographically and tectonically expressed double-helical zones. The zones have the same inclination at the equator: about 22° and 160° (Fig. 16.1B). O'Driscoll (1980) believed that these are fundamental structural belts governing the global deformation network and planetary evolution. Volkov (1995) reported six global double-helical structures that were also detected by a visual analysis of physiographic maps. At the equator, three of them are inclined at about 12° and 168°, and the other three structures are inclined at about 22° and 158° (Fig. 16.1C). Volkov (1995) presumed that these are ancient traces of tidal effects within the Earth-Moon resonance system.

Although much attention has been paid to global lineaments, their existence is still questionable. This is because of:

- the qualitative character of physiographic and geological maps analyzed in previous works;
- inaccurate presentation of seafloor bathymetry on those maps produced before reasonably accurate bathymetric data became available;
- the impossibility of considering all natural conditions in a mathematical model;
- obvious differences between the Earth's rotation and its laboratory simulation;
- a basic conflict between the mainstream plate tectonic theory and the possibility of the existence of global topographic and tectonic structures.

Technical flaws can be obviated using quantitative descriptions of global topography (a global DEM) and methods of digital terrain modeling. Indeed, resulting from the interaction of endogenous and exogenous geophysical processes on different spatial and temporal scales, topography carries information on both surface processes and tectonic features. Thus, if global helical structures really exist, there is a good chance that they are expressed on the topographic surface.

Digital Terrain Analysis. https://doi.org/10.1016/B978-0-443-24798-9.00013-3

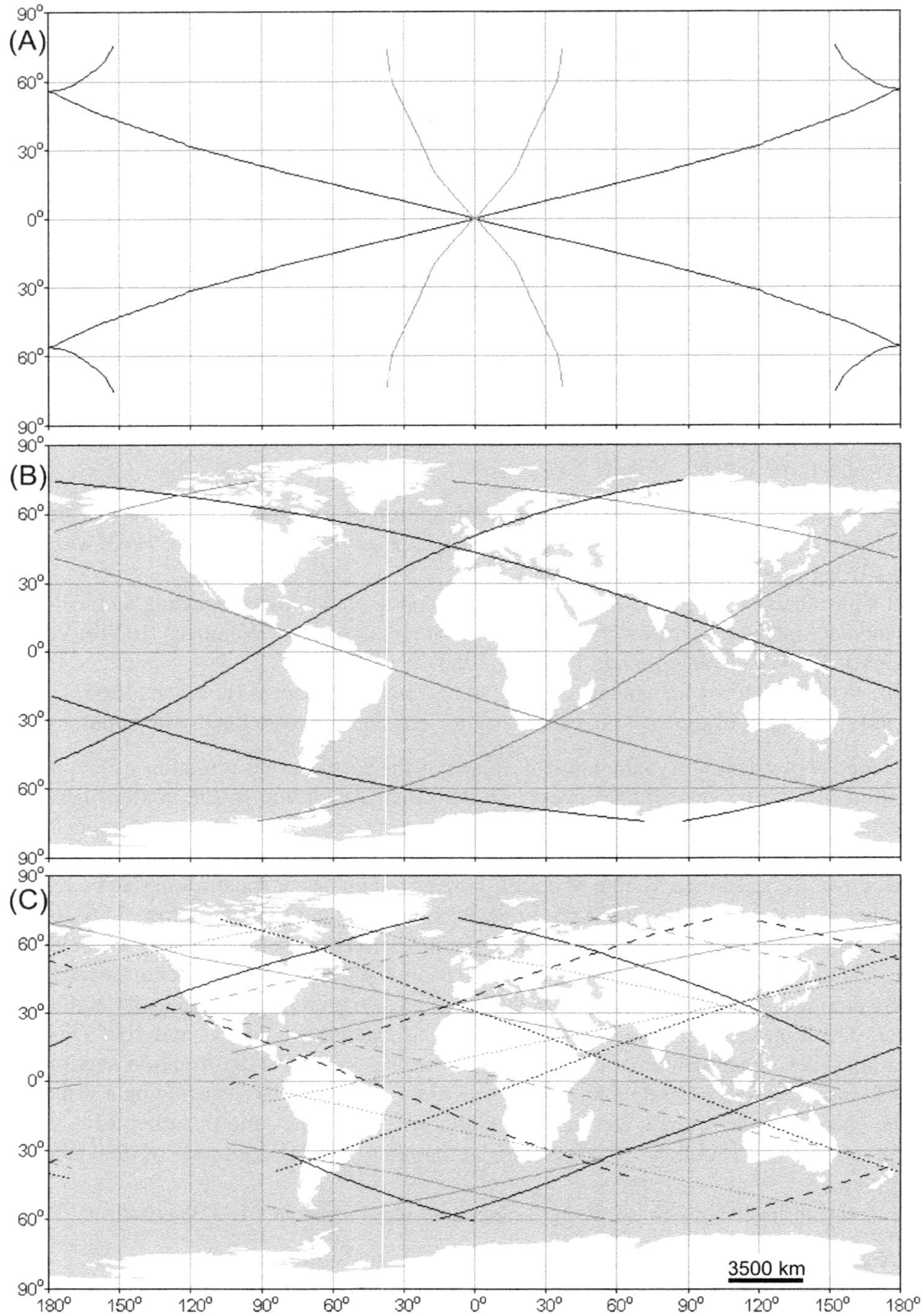

FIG. 16.1 The Earth, global helices: (A) Theoretical traces of torsional deformation (Rance, 1967); *black*, shear fractures; *gray*, cleavage cracks. (B) Axes of double-helical zones (O'Driscoll, 1980). (C) Six double helices (Volkov, 1995); *lines of different styles*, different structures. *(From Florinsky, I.V., 2008b. Global lineaments: application of digital terrain modelling. In: Zhou, Q., Lees, B., Tang, G.-A. (Eds.), Advances in Digital Terrain Analysis. Lecture Notes in Geoinformation and Cartography Book Series. Springer Berlin, Germany, 365–382, Fig. 1, reproduced with permission from Springer Nature.)*

In this chapter, we present the results of global-scale geomorphometric modeling of the Earth and some other solid planetary bodies of the Solar System to detect global helical structures to support or refute the hypotheses outlined earlier (Florinsky, 2008b,c).

16.2 Materials and data processing

The study was based on a 30′-gridded global DEM (Fig. 16.2) assembled from several sources. Elevations of the land topography were derived from GLOBE, the 30″-gridded global DEM (GLOBE Task Team and Others, 1999). Most of the seafloor topography was taken from ETOPO2, the 2′-gridded global DEM (NOAA, 2001). Bathymetry of the Antarctic Continental Shelf, the Caspian Sea, and some large lakes was digitized from topographic maps. The DEM included 260,281 points (the elevation matrix 721 × 361). For Antarctica and Greenland, GLOBE includes elevations of ice sheets rather than subglacial topography (Hastings and Dunbar, 1998). These areas were included in the DEM to retain a united configuration of data.

We selected Mars, Venus, and the Moon for a comparative analysis. A DEM of Mars (Fig. 16.3A) was extracted from the Mars Orbiter Laser Altimeter (MOLA) data archive (Smith et al., 2003) of the Mars Global Surveyor mission (Albee, 2000). A DEM of Venus (Fig. 16.3B) was compiled using data from two data archives (Ford et al., 1992; Sjogren et al., 1997) of the Magellan mission (Saunders and Pettengill, 1991). A DEM of the Moon (Fig. 16.3C) was extracted from the Lunar Orbiter Laser Altimeter (LOLA) data archive (Neumann, 2008) of the Lunar Reconnaissance Orbiter mission (Chin et al., 2007). Each DEM was based on the global spheroidal equal angular grid with the grid spacing of 30′ including 260,281 points (the elevation matrix 721 × 361).

To suppress high-frequency noise, the initial DEMs were smoothed using the 3 × 3 moving window. Our experiments demonstrated that one to three smoothing iterations should be done to obtain readable global morphometric maps.

For all four planetary bodies, we derived models of all local morphometric attributes (Table 2.1) by the author's method for spheroidal equal angular grids (Section 4.3). Models of catchment and dispersive areas (Table 2.1) were calculated by the Martz-de Jong method adapted to spheroidal equal angular grids (Section 4.4.2). All topographic variables were derived from smoothed DEMs. The global DEMs were processed as virtually closed spheroidal matrices of elevations. Each global spheroidal DTM included 260,281 points (the matrix 721 × 361); the grid spacing was 30′.

To estimate linear sizes of spheroidal trapezoidal windows in DTM calculation and smoothing (Section 4.3.3), standard values of the major and minor semiaxes of the Krasovsky ellipsoid (6,378,245 and 6,356,863 m) and the Martian ellipsoid (3,396,190 and 3,376,200 m) were used for the Earth and Mars, respectively. Venus and the Moon were considered spheres of 6,051,848 and 1,738,000 m radii, respectively.

We derived small-scale global maps of all calculated topographic attributes (Figs. 16.4–16.9). To deal with the large dynamic range of their values, we logarithmically transformed these digital models by Eq. (8.1) with $n = 9$. We also derived "binary" maps of CA and DA, subdividing their values into two levels (Figs. 16.10 and 16.11). Data were processed by the LandLord software (Florinsky, 2016).

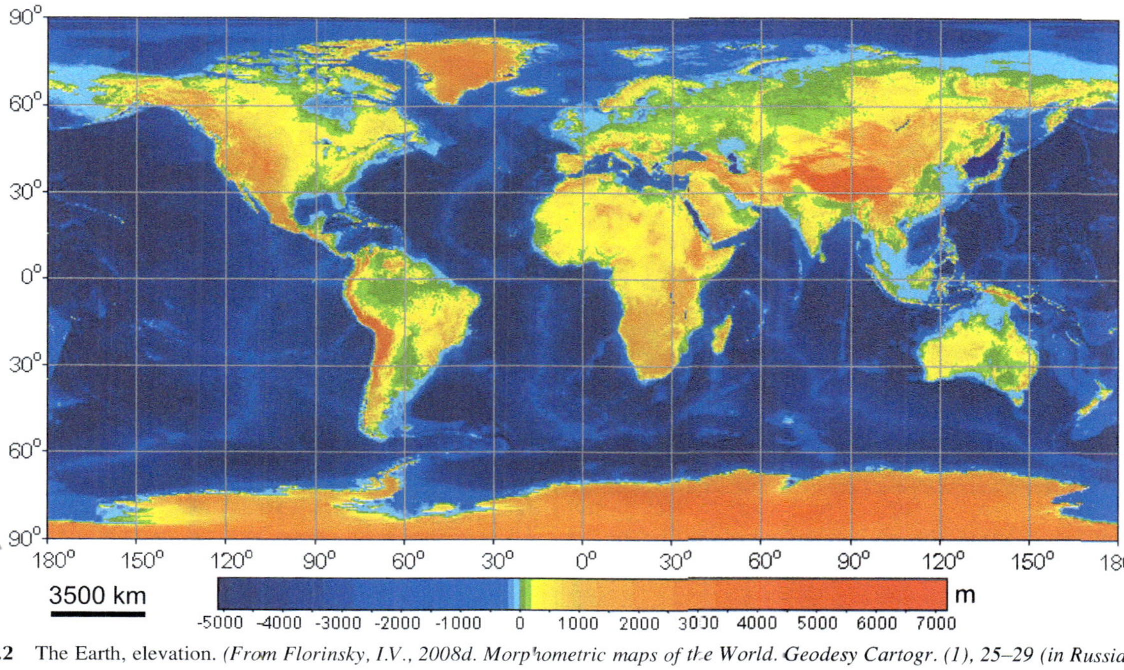

FIG. 16.2 The Earth, elevation. *(From Florinsky, I.V., 2008d. Morphometric maps of the World. Geodesy Cartogr. (1), 25–29 (in Russian), Fig. 1.)*

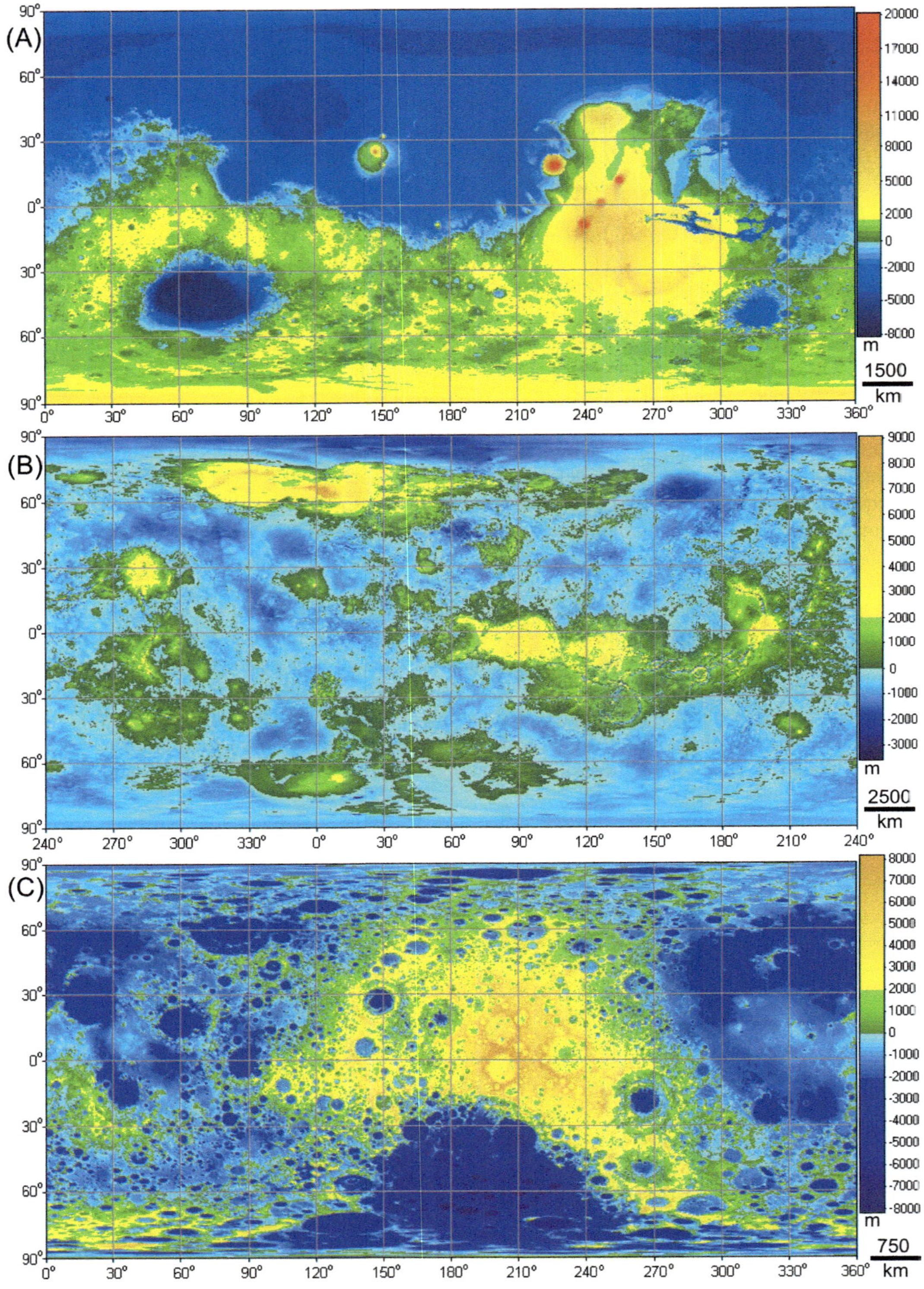

FIG. 16.3 Elevation: (A) Mars. (B) Venus. (C) The Moon. *(Modified from Florinsky, I.V., 2010. Theory and Applications of Mathematical Cartographic Modeling of Topography (D.Sc. thesis). Institute of Mathematical Problems of Biology, Russian Academy of Sciences, Pushchino, Russia (in Russian), Fig. 4.19.)*

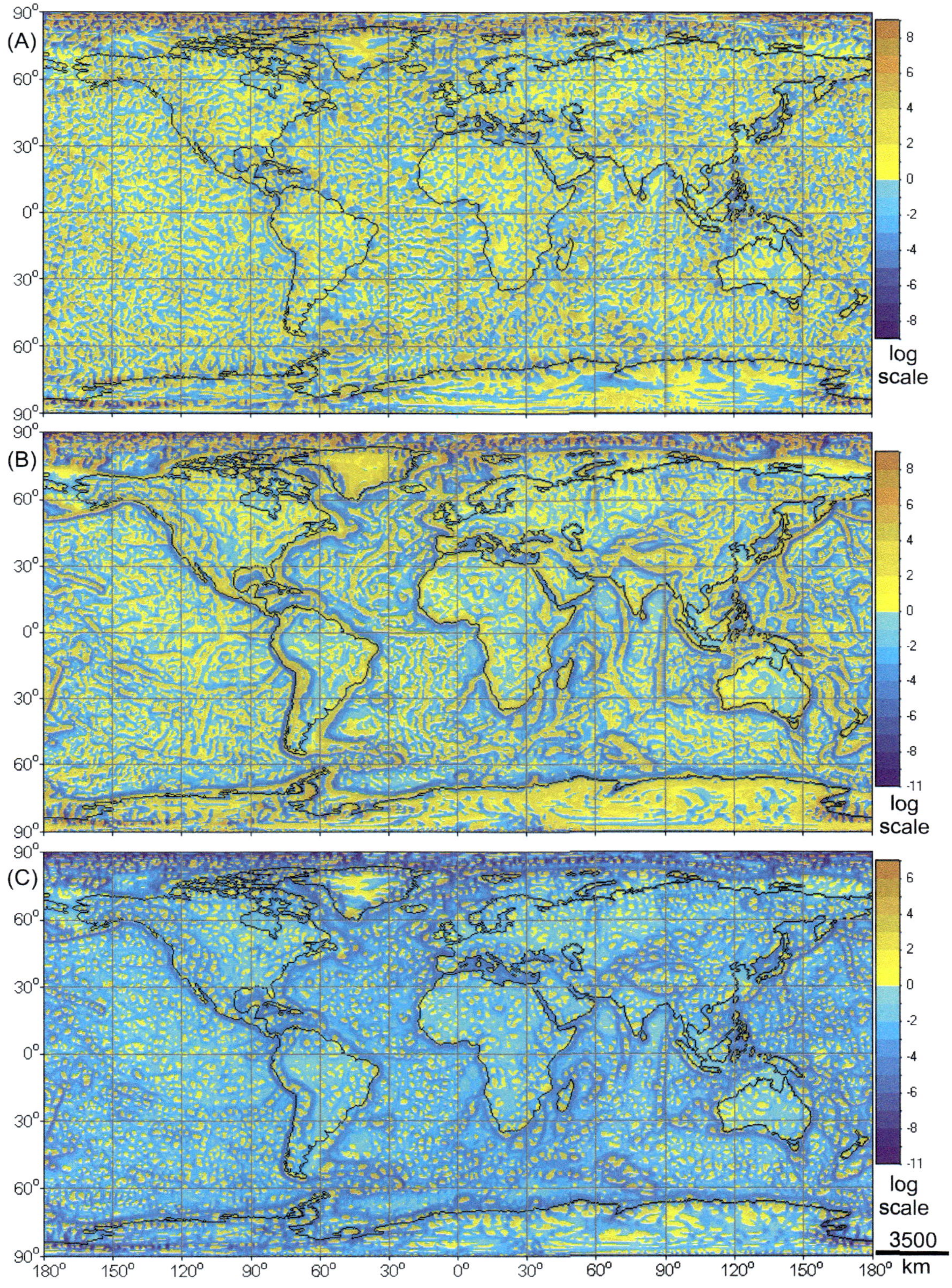

FIG. 16.4 The Earth, global morphometric maps derived from the thrice-smoothed DEM: (A) Horizontal curvature. (B) Vertical curvature. (C) Minimal curvature. *(From Florinsky, I.V., 2008d. Morphometric maps of the World. Geodesy Cartogr. (1), 25–29 (in Russian), Fig. 2A, B, and C.)*

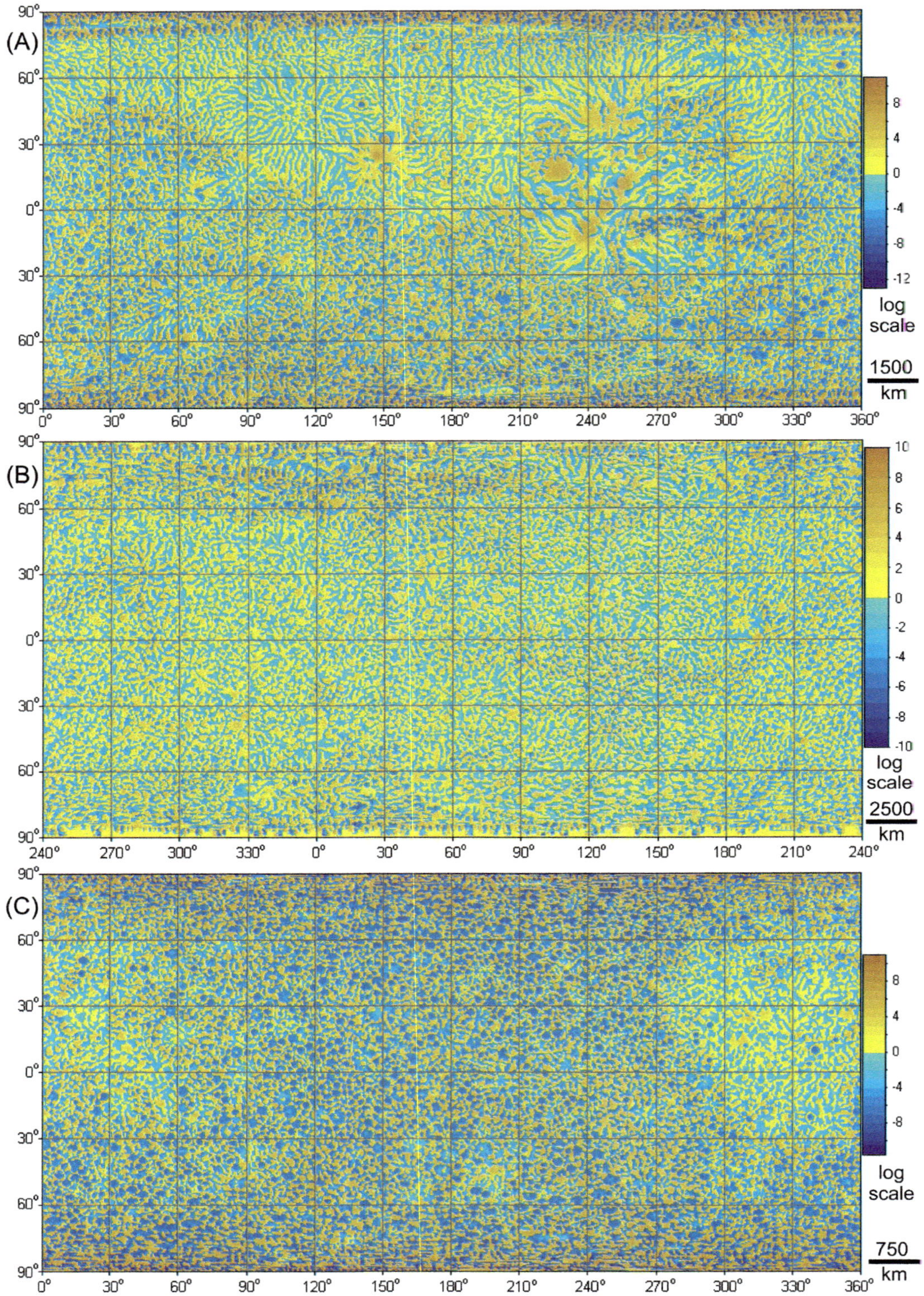

FIG. 16.5 Horizontal curvature derived from the twice-smoothed DEMs: (A) Mars. (B) Venus. (C) The Moon. *(Modified from Florinsky, I.V., 2010. Theory and Applications of Mathematical Cartographic Modeling of Topography (D.Sc. thesis). Institute of Mathematical Problems of Biology, Russian Academy of Sciences, Pushchino, Russia (in Russian), Fig. 4.21.)*

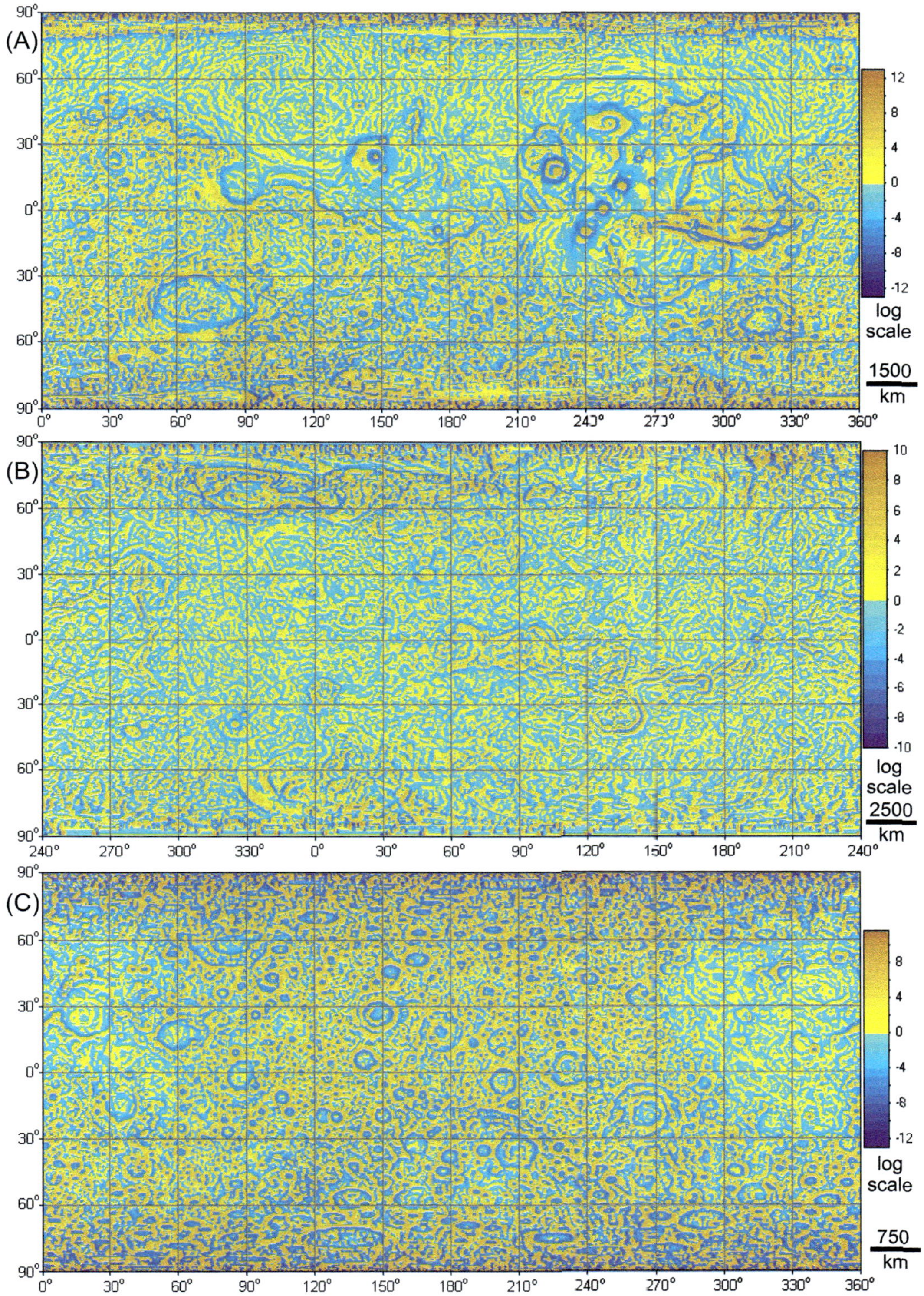

FIG. 16.6 Vertical curvature derived from the twice-smoothed DEMs: (A) Mars. (B) Venus. (C) The Moon. *(Modified from Florinsky, I.V., 2010. Theory and Applications of Mathematical Cartographic Modeling of Topography (D.Sc. thesis). Institute of Mathematical Problems of Biology, Russian Academy of Sciences, Pushchino, Russia (in Russian), Fig. 4.22.)*

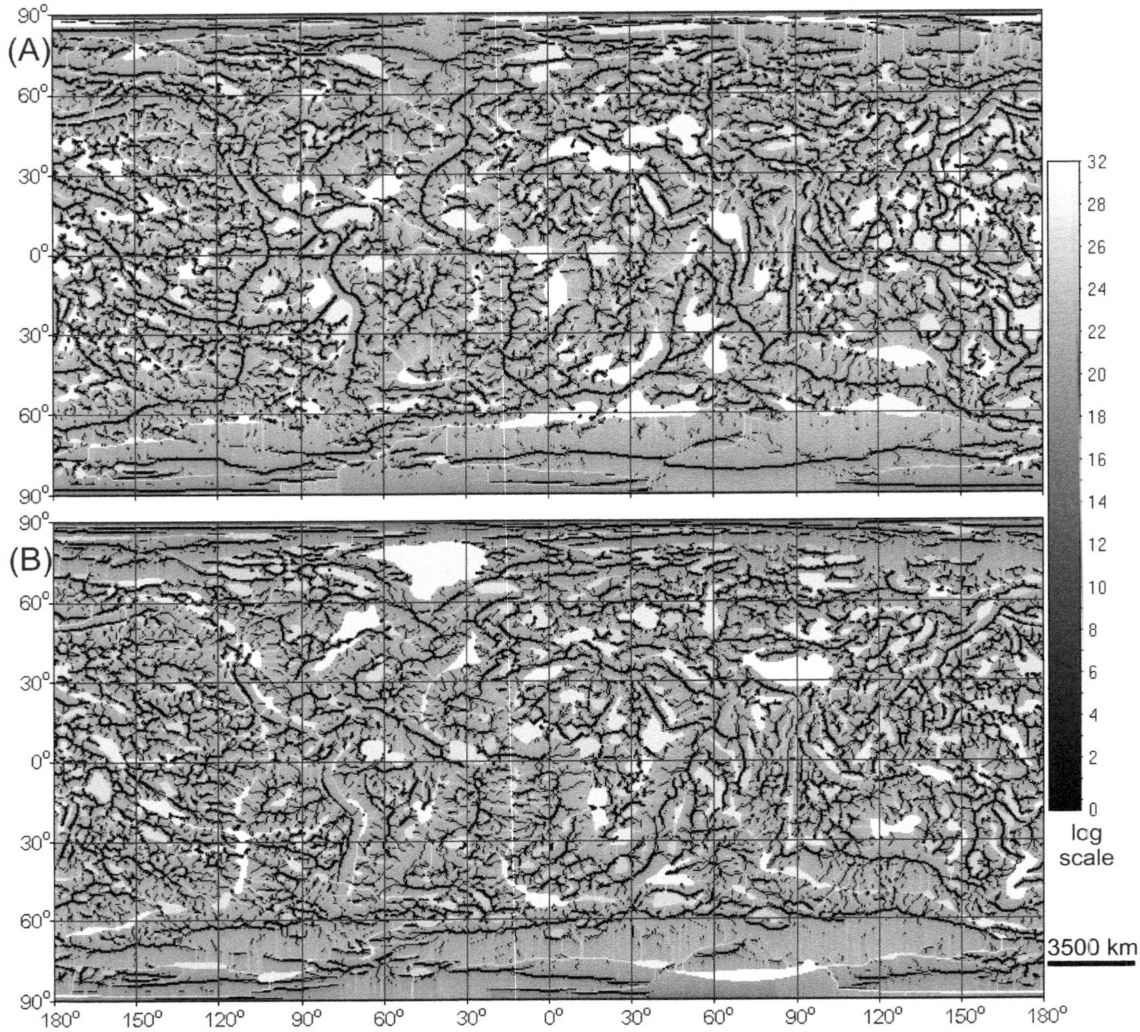

FIG. 16.7 The Earth, global morphometric maps derived from the thrice-smoothed DEM: (A) Catchment area. (B) Dispersive area. *(Modified from Florinsky, I.V., 2008d. Morphometric maps of the World. Geodesy Cartogr. (1), 25–29 (in Russian), Fig. 3.)*

Maps obtained were visually examined. Attention was paid to lineaments running over the entire globe or a hemisphere. Contrary to many lineaments of regional and continental scales, global lineaments are not expressed as uninterrupted linear image patterns or their sequences. A global lineament may be visually detected due to traits of the image texture strung out along some line running over the planet.

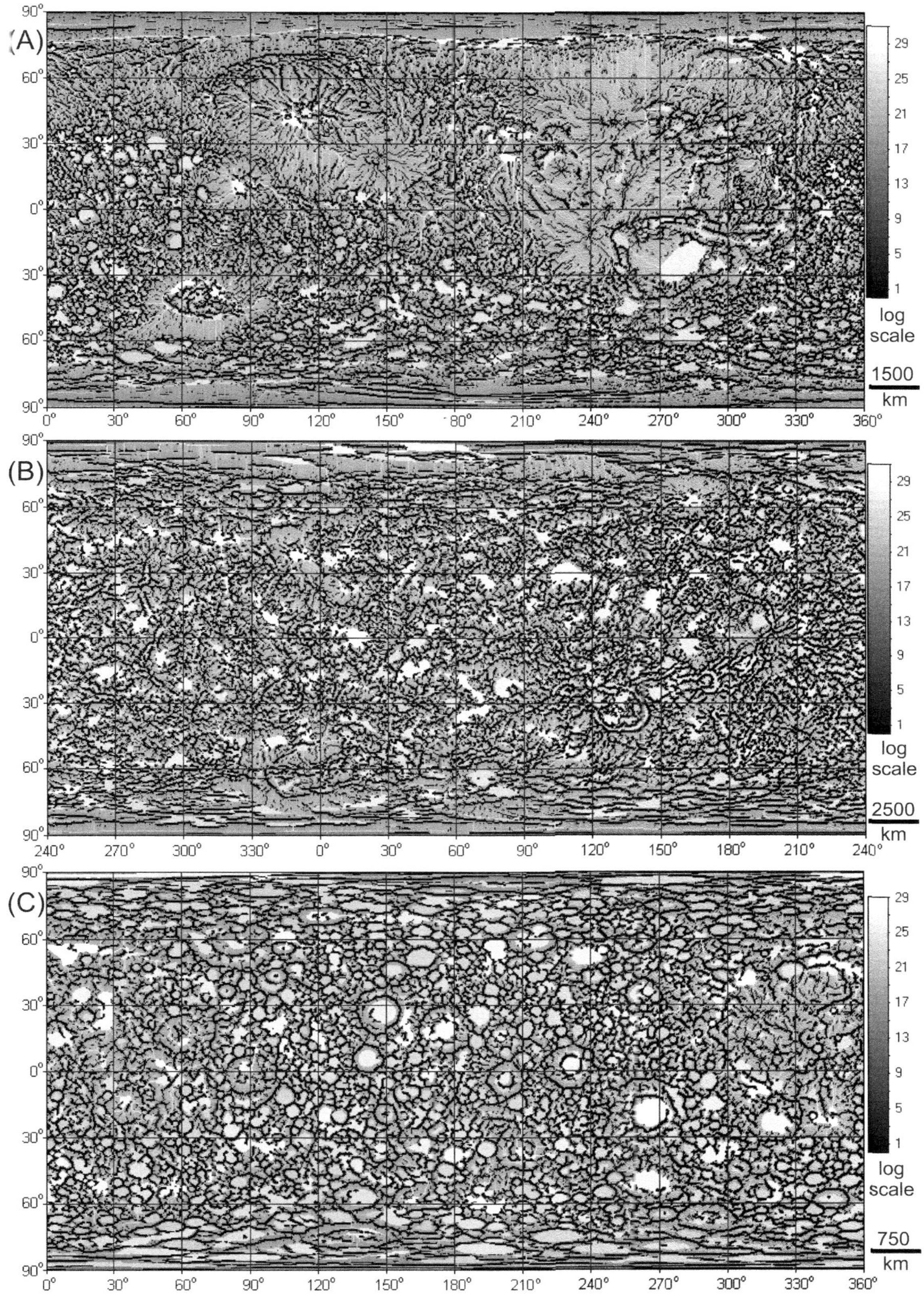

FIG. 16.8 Catchment area derived from the twice-smoothed DEMs: (A) Mars. (B) Venus. (C) The Moon. *(Modified from Florinsky, I.V., 2010. Theory and Applications of Mathematical Cartographic Modeling of Topography (D.Sc. thesis). Institute of Mathematical Problems of Biology, Russian Academy of Sciences, Pushchino, Russia (in Russian), Fig. 4.24.)*

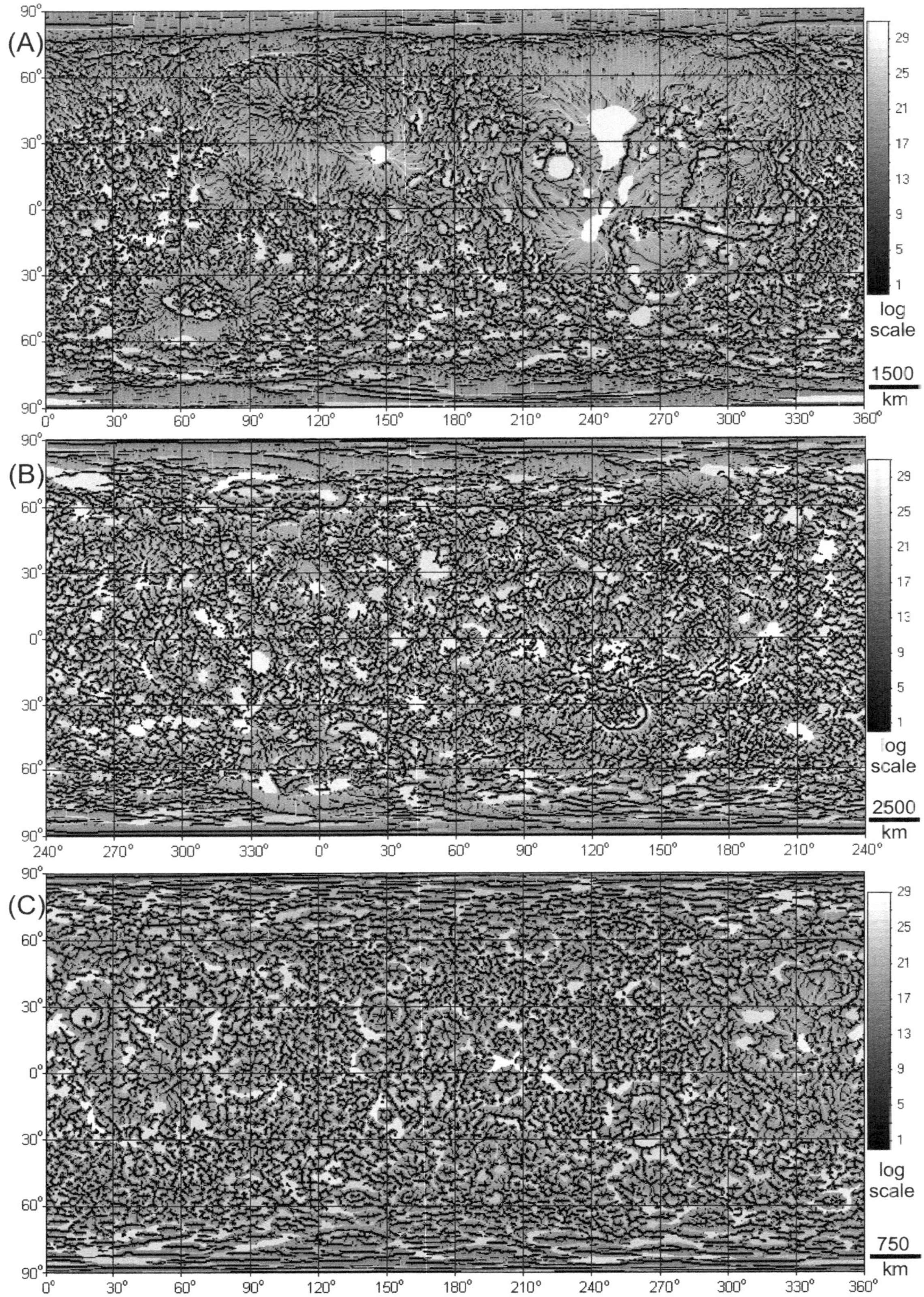

FIG. 16.9 Dispersive area derived from the twice-smoothed DEMs: (A) Mars. (B) Venus. (C) The Moon. *(Modified from Florinsky, I.V., 2010. Theory and Applications of Mathematical Cartographic Modeling of Topography (D.Sc. thesis). Institute of Mathematical Problems of Biology, Russian Academy of Sciences, Pushchino, Russia (in Russian), Fig. 4.25.)*

FIG. 16.10 The Earth, global binary maps of catchment area derived from: (A) Smoothed DEM. (B) Twice-smoothed DEM. (C) Thrice-smoothed DEM. *(Modified from Florinsky, I.V., 2010. Theory and Applications of Mathematical Cartographic Modeling of Topography (D.Sc. thesis). Institute of Mathematical Problems of Biology, Russian Academy of Sciences, Pushchino, Russia (in Russian), Fig. 4.26.)*

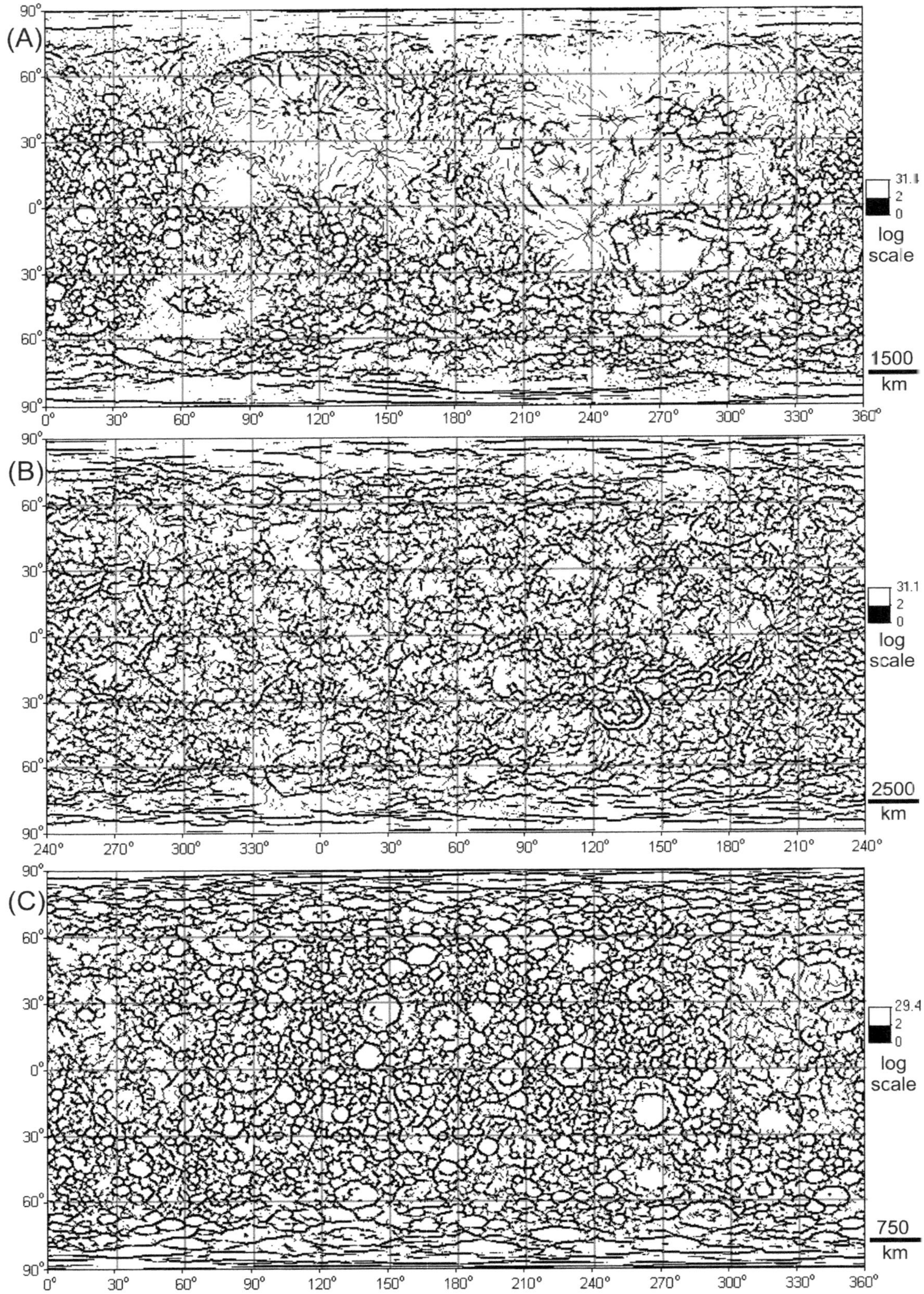

FIG. 16.11 Global binary maps of the catchment area derived from the twice-smoothed DEMs: (A) Mars. (B) Venus. (C) The Moon. *(Modified from Florinsky, I.V., 2010. Theory and Applications of Mathematical Cartographic Modeling of Topography (D.Sc. thesis). Institute of Mathematical Problems of Biology, Russian Academy of Sciences, Pushchino, Russia (in Russian), Fig. 4.27.)*

16.3 Results and discussion

16.3.1 General interpretation

For the same angular grid spacing of 30′, DTMs and morphometric maps of the Earth, Mars, Venus, and the Moon have distinct linear grid spacings (around 55.2, 29.5, 52.8, and 15.2 km on the equator, respectively) and scales. It is recommended to use the same angular resolution in comparative cartographic analysis of different-sized planetary bodies: the ratio of map scales of planetary bodies should be equal to the ratio of their sizes (Burba, 1984). For example, a Venus map at the scale of 1:1,000,000 approximately corresponds to a Mars map at the scale of 1:500,000 and a Moon map at the scale of 1:250,000.

Global maps of morphometric variables represent peculiarities of megatopography in different ways, according to the physical and mathematical sense of a particular variable. Horizontal curvature delineates areas of flow divergence and convergence (positive and negative values, respectively). These areas relate to spurs of valleys and ridges (blue and yellow patterns on the k_h maps, respectively) forming flow structures. For the Earth, they are most pronounced in ocean basins (Fig. 16.4A). On the k_h map of Mars (Fig. 16.5A), one can see flow structures, probably of lava origin, beginning on the slopes of Alba Patera and forming a huge fan in the North Polar Basin (30°–75° N, 200°–310° E). There is a system of flow structures incoming to Utopia Planitia from Nilosyrtis and Protonilus Mensae, as well as Elysium Planitia and Mons (5°–70° N, 75°–150° E). On Venus, flow structures appear slightly at the global scale (Fig. 16.5B). For example, one can see them on Beta Regio slopes (15°–45° N, 270°–300° E). For the Moon, the horizontal curvature represents cell-like patterns (Fig. 16.5C) resulting from a predominance of craters at the global scale.

Vertical curvature is a measure of relative acceleration and deceleration of flows (positive and negative values, respectively). Among other features, the k_v map of the Earth shows "mega-scarps," such as the edges of continents and mountain systems (Fig. 16.4B). On the k_v map of Mars (Fig. 16.6A), one can see boundaries of Hellas Planitia (30°–50° S, 50°–90° E), Isidis Planitia (10°–25° N, 75°–100° E), Valles Marineris (10°–20° S, 270°–335° E), foothills of Olympus Mons (15°–20° N, 220°–230° E), Alba Patera (30°–50° N, 225°–265° E), and so on. Artemis Chasma (30°–45° S, 120°–145° E) and some other surface features are pronounced on the k_v map of Venus (Fig. 16.6B). On the k_v map of the Moon (Fig. 16.6C), one can see well-marked boundaries of relatively large craters.

Catchment area measures an upslope area potentially drained through a given point on the topographic surface (Table 2.1). For the Earth (Fig. 16.7A), low values of *CA* delineate land and ocean ridges as black lines (e.g., the Andes and mid-ocean ridges). High values of *CA* show land valleys and ocean canyons as white lines as well as land depressions and ocean basins as light areas (e.g., the Mediterranean Sea, Gulf of Mexico, Angola Basin). On the *CA* map of Mars (Fig. 16.8A), one can see the planetary network of valleys and canyons as well as a large feature, Solis Planum (15°–30° S, 270°–290° E), and a plethora of smaller depressions, predominantly craters. The *CA* map of the Moon also displays craters (Fig. 16.8C).

Dispersive area measures a downslope area potentially exposed by slope lines passing through the given point on the topographic surface (Table 2.1). For the Earth, high values of *DA* delineate mountain systems and highlands as light areas (e.g., Tibet, Ethiopian Highlands, Greenland) as well as land and ocean ridges as light lines (Fig. 16.7B). On the *DA* map of Mars (Fig. 16.9A), one can see Alba Patera (30°–50° N, 230°–260° E), Tharsis Montes (15° N–15° S, 230°–260° E), and other surface features.

Global morphometric maps of the Earth are complementary to the existing global- and continental-scale geological maps (e.g., Trifonov, 1995; Ryakhovsky et al., 2003). The geological maps of Mars (Scott and Carr, 1978), Venus (Ivanov, 2008), and the Moon (USGS, 1972) can also be combined with the morphometric maps of these planetary bodies. This may be useful in solving various tasks of tectonics, geophysics, and planetary science, such as refining shapes of topographic and geological structures, describing them quantitatively, and analyzing their spatial distribution over the planetary surface.

Recognizable artifacts are typical for polar regions on all maps (Figs. 16.4–16.9). They were caused by a relatively low accuracy of the description of these areas in the databases used. The Venus DEM was compiled by combining the Global Topography Data Record and Magellan Spherical Harmonic Models and Digital Maps. These archives have different resolutions and accuracies. As a result, one can see several linear artifacts—marks of this combination (e.g., 15°–55°N 325°E, 60°–80°S 325°–355°E)—on the maps of elevation, k_v, *CA*, and *DA* of Venus (Figs. 16.3B, 16.6B, 16.8B, and 16.9B). There are also computational artifacts on the maps of *CA* and *DA* expressed as straight parallel lines, predominantly in polar regions (Figs. 16.7–16.9). These are well-known artifacts of single-flow direction algorithms common for flat slopes (Liang and Mackay, 2000).

16.3.2 Global helical structures

Subdivision of *CA* values into two levels allowed the display of ridge networks of the Earth (Fig. 16.10), Mars (Fig. 16.11A), Venus (Fig. 16.11B), and the Moon (Fig. 16.11C). The greater the number of DEM smoothings, the more generalized the picture of the network (Fig. 16.10).

It was observed that maps of catchment areas with values subdivided into two levels were best suited to detect global lineaments. Analysis of the *CA* binary maps of the Earth (Fig. 16.12) revealed five systems of double-helical-like tectonic structures encircling the planet from pole to pole (Fig. 16.13). It is clear that the structures revealed are helical zones many kilometers wide rather than simply lines. Each helical zone transgresses plate boundaries and regions dissimilar with respect to their tectonic origin, rock composition, and age. Each double helix was named after the area(s) of the intersection(s) of its arms (Table 16.1). Arms running clockwise upward and counterclockwise upward (dextral and sinistral helices) were called right and left arms, respectively.

For Mars and Venus, several global lineaments were also detected by the visual analysis of the *CA* binary maps (Fig. 16.14). The global helical structures encircle these planets from pole to pole.

The global lineaments revealed cannot be artifacts due to DEM errors, treatment, or grid geometry (Section 5.7). First, DEM noise and errors usually have a random spatial distribution. Second, smoothing and derivation of topographic variables were carried out using local filters (3 × 3 moving windows). Third, the grid geometry may amplify its own preferential directions: orthogonal (north-south and east-west) as well as diagonal (northeast-southwest and northwest-southeast). However, the structures detected have (1) a global character relative to the DEMs and (2) directions distinct from orthogonal and diagonal ones (Figs. 16.13 and 16.14).

Although visual analysis may cause artifacts, it is questionable whether an application of automated tools (e.g., Fukue et al., 1981; Takahashi, 1981; Masoud and Koike, 2011a; Zlatopolsky, 1992, 2011) may reveal radically different lineaments. Indeed, experiments demonstrated that the ideal observer (Swets, 1961) and the human visual system have almost similar capabilities to recognize geometrical patterns on noisy binary images (Krasilnikov et al., 2000). Numerous geological studies have demonstrated that visual analysis of maps and remotely sensed images can be successfully used to detect lineaments (Trifonov et al., 1983; Kats et al., 1986; Bryukhanov and Mezhelovsky, 1987; Kozlov, 1991).

The artifacts described in Section 16.3.1 did not influence the detection of global helical structures because the artifacts are located only in the polar regions.

The Caucasus-Clipperton double helix (Fig. 16.13A) coincides with one of the structures reported by Volkov (1995) (Fig. 16.1C). The left arm of the Biscay-Santa Cruz structure (Fig. 16.13B) partly agrees with the left arm of one of the helices detected by O'Driscoll (1980) (Fig. 16.1B). A comparison of inclination angles of theoretical traces of torsional deformation (Section 16.1 and Fig. 16.1A) and those of the double helices revealed (Table 16.1) shows that the Caucasus-Clipperton and Biscay-Santa Cruz structures can be assigned to traces of shear fractures, whereas the Dakar and Palawan structures can be assigned to traces of cleavage cracks. The mean deviation of the inclination angles of the structures from the theoretical values is 2.8°. There are, however, two disagreements with the Rance model: (1) the arms of each double helix meet off the equator and (2) the Marcus structure has "abnormal" inclination. These disagreements may be attributed to (1) the deviation of the Earth's shape from a sphere used in the Rance model and (2) the impossibility of considering all natural factors in mathematical models.

Thus, of the five double-helical structures detected, four have the inclination angles fitting the theoretical inclination angles for traces of torsional deformation of the sphere. This suggests that one may consider topographically expressed

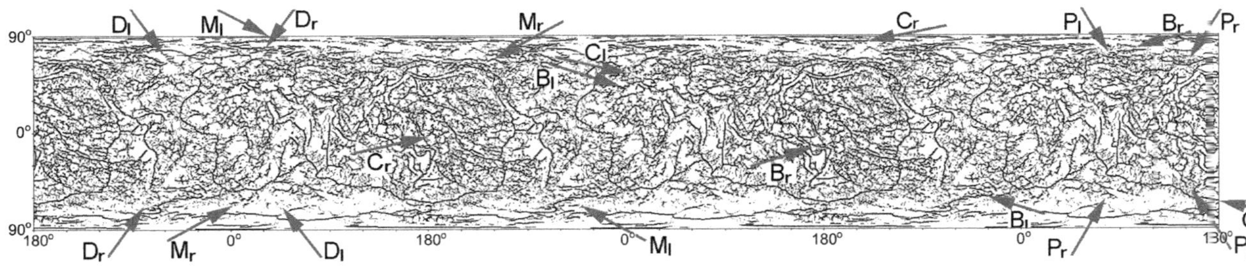

FIG. 16.12 The Earth, a scheme for visual detection of global helical structures on the threefold map of catchment area derived from the twice-smoothed DEM (Fig. 16.10B). Structures (Table 16.1): *C*, Caucasus-Clipperton; *B*, Biscay-Santa Cruz; *M*, Marcus; *D*, Dakar; *P*, Palawan. Subscripts: *r*, right arm; *l*, left arm. *Arrows*, positions of helical arms. *(Modified from Florinsky, I.V., 2008b. Global lineaments: application of digital terrain modelling. In: Zhou, Q., Lees, B., Tang, G.-A. (Eds.), Advances in Digital Terrain Analysis. Lecture Notes in Geoinformation and Cartography Book Series. Springer Berlin, Germany, 365–382, Fig. 4, reproduced with permission from Springer Nature.)*

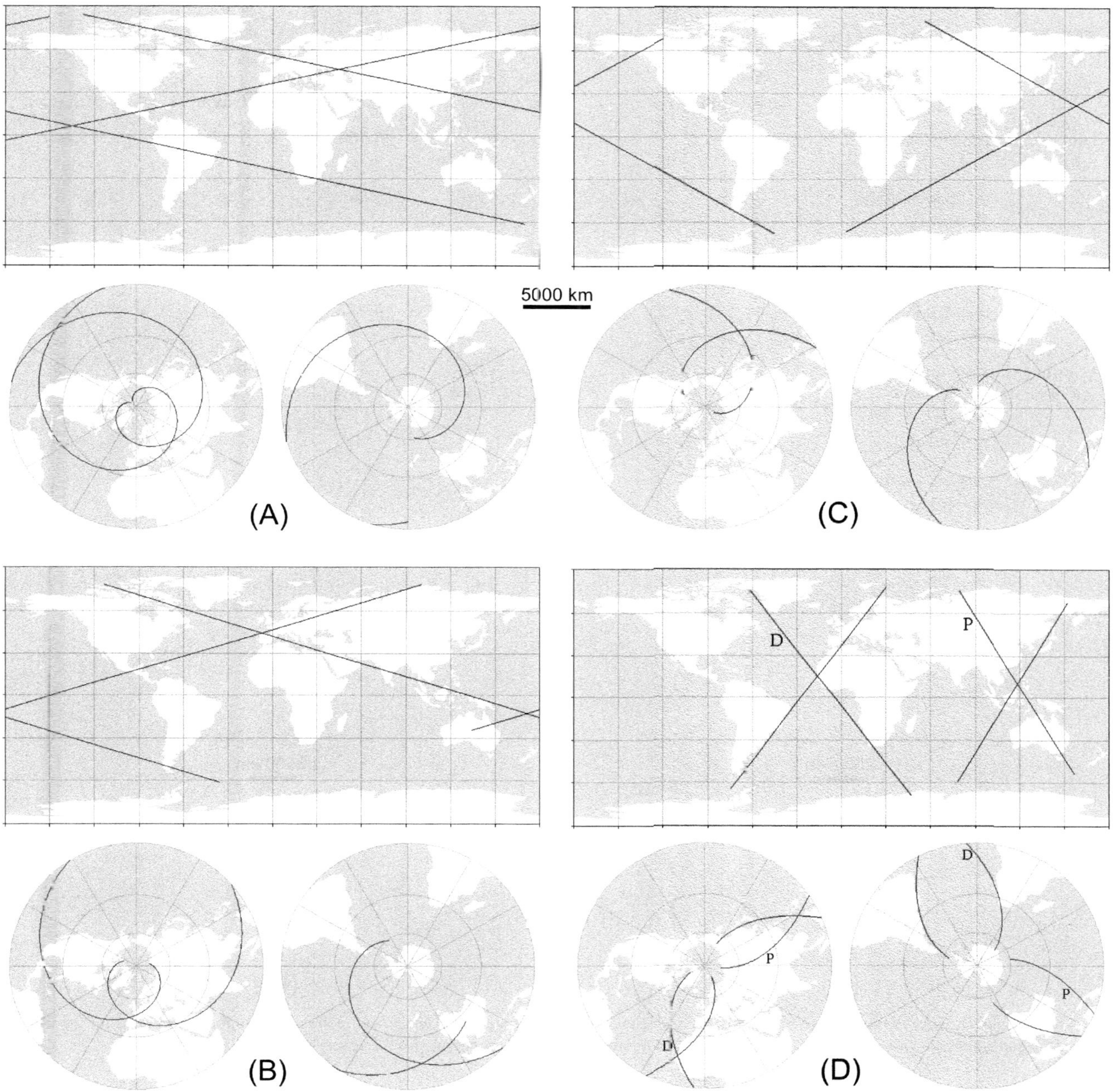

FIG. 16.13 The Earth, global double-helical structures: (A) Caucasus-Clipperton. (B) Biscay-Santa Cruz. (C) Marcus. (D) Dakar and Palawan, *D* and *P*, respectively. The Plate Carrée projection (*upper*) and polar stereographic projections for the northern (*lower left*) and southern (*lower right*) hemispheres. *Gray color*, oceans and seas.

helical structures as fracture traces of global torsion stresses and deformations. They may be caused by the rotational unevenness of the layers of the Earth's spheroid, which is a rheologically and lithologically nonuniform body.

The helical structures are expressed in the modern topography; each of the structures crosses adjacent tectonic plates. Therefore, the helical structures could have occurred in the Cenozoic only if the plate tectonic hypothesis is true. If the hypothesis is rejected, the helical structures could have arisen in earlier geological periods.

Notice that some spiral structures have earlier been documented for the Martian polar caps (Howard, 1978; Weijermars, 1985,'86; Fishbaugh and Head III, 2001) (Fig. 16.15A). These are troughs and scarps 5–30 km wide and up to several hundred kilometers long. Trough depths are up to 500–1000 m on polar caps edges and 100–200 m near poles.

TABLE 16.1 Parameters of the global topographic helices of the Earth.

Structure		Caucasus-Clipperton	Biscay-Santa Cruz	Marcus	Dakar	Palawan
Left arm	Lengths (km)	55,800	39,600	26,500	17,700	15,400
	Inclination at the equator (°)	167.5	162.2	150.6	126.9	121.3
Right arm	Lengths (km)	31,500	29,800	24,900	17,200	15,300
	Inclination at the equator (°)	12.5	17.5	29.7	53.3	59.5
Coordinates of the arm intersection(s)		46.4°N 44.81°E, 5.9°N 134.7°W	44.4°N 7.3°W, 12.9°S 171.4°E	21.4°N 157.5°E	14.9°N 16.0°W	9.9°N 119.1°E

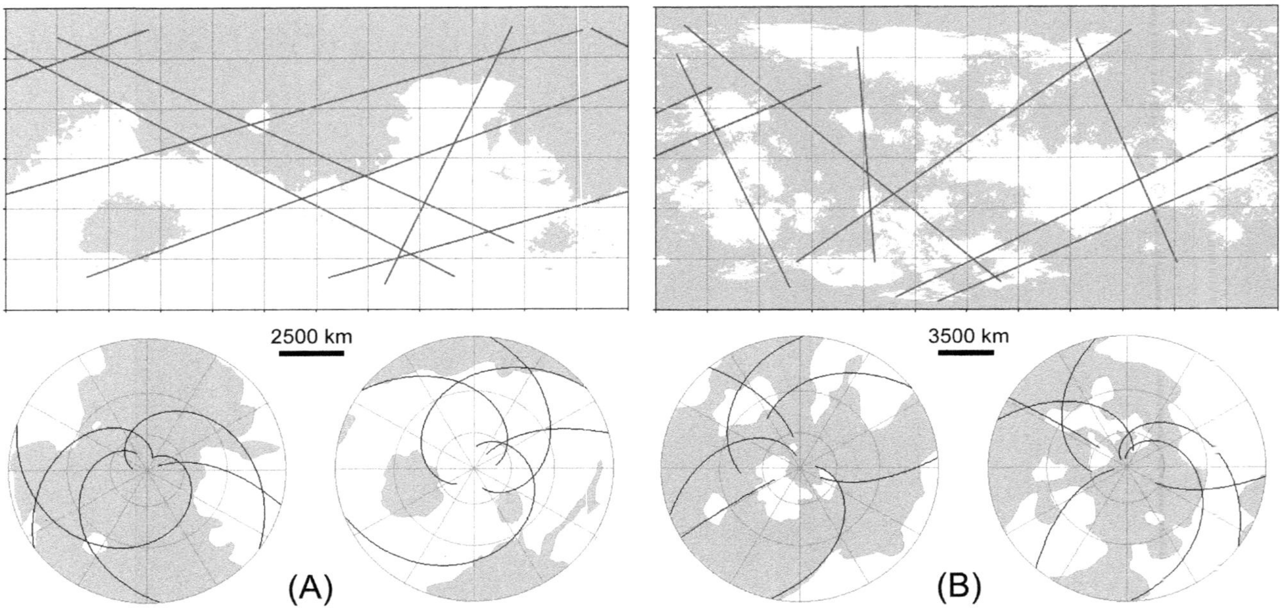

FIG. 16.14 Global helical structures: (A) Mars. (B) Venus. The Plate Carrée projection (*upper*) and polar stereographic projections for the northern (*lower left*) and southern (*lower right*) hemispheres. *Gray color*, areas located below datum (for reference, see Fig. 16.3).

There are several hypotheses for the origin of Martian polar spiral structures, including the manifestation of ice elastic deformations and sublimation as well as the influence of katabatic wind flows affected by the Coriolis forces (Howard, 1978; Weijermars, 1985/86; Fisher, 1993).

For the northern hemisphere of Venus, Slyuta et al. (1989) discovered a dense, regular network of dextral and sinistral spiral structures (Fig. 16.15B). They are topographically expressed as troughs, scarps, and depressions. These authors believed that strong rotational forces had formed the network during the deceleration of Venus's rotation. They suggested that the helical network is a relict feature, an "imprint" of ancient rotational stress fields, because the current rotational velocity of Venus is quite slow.

These facts on the Martian and Venusian spiral structures are circumstantial evidence that (1) global helical structures of the Earth are not artifacts and (2) their origin may be associated with planetary rotational (torsional) stresses.

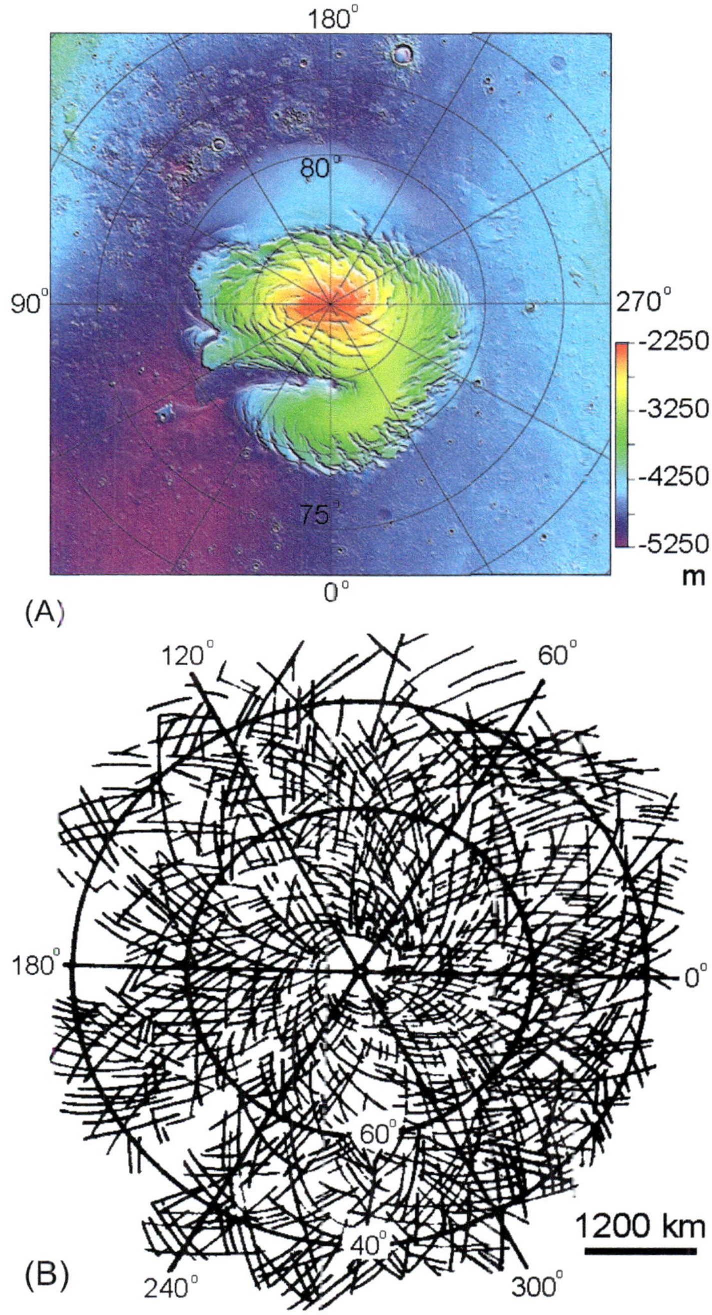

FIG. 16.15 Helical structures of Mars and Venus detected using satellite data: (A) Structures of the northern polar cap of Mars from the MOLA DEM. (B) Structures of the northern hemisphere of Venus from the Venera-15 and -16 satellite radar data. *((A) From Fishbaugh, K.E., Head, J.W., III, 2001. Comparison of the north and south polar caps of Mars: new observations from MOLA data and discussion of some outstanding questions. Icarus. 154 (1), 145–161, Fig. 3-top, reproduced with permission from Elsevier. (B) From Slyuta, E.N., Kudrin, L.V., Shilo, V.P., 1989. Preliminary data on the nature of a planetary system of lineaments observed in radar images of Venus (data from Venera-15 and -16). Cosmic Res. 27 (6) 786–797, Fig. 2, © Plenum Publishing, reproduced with permission from Springer Nature.)*

Part IV

Digital terrain modeling in glaciology and polar research

Chapter 17

Glacier motion and evolution

17.1 Introduction

Glaciers and ice sheets hold an essential key to the knowledge of the past, present, and future global environmental conditions (Menzies, 2002). In particular, Antarctica, the largest center of glaciation, is of great importance in forming the climate of the Southern Hemisphere and the Earth as a whole. To estimate the state of the Antarctic ice sheet and its response to the circulation processes in the ocean and atmosphere, one can analyze, among other information, sizes of the ice sheet and changes in the position of edges of shelf and outlet Antarctic glaciers (Zakharov, 1988; Jacobs et al., 1992; Benn et al., 2007; Rignot et al., 2011, 2013, 2019; Shepherd et al., 2018; Kusahara, 2020; Noble et al., 2020).

A glacier is a natural object with a constantly and dynamically changing surface. General characteristics of glacier changes, motion, and evolution are usually obtained by comparing multitemporal cartographic and geophysical data, aerial and satellite imagery as well as satellite altimetry measurements (Lucchitta and Ferguson, 1986; Zakharov, 1988, 1994; Rémy et al., 1999; Wingham et al., 2006; Stearns, 2011; Markov et al., 2016; Yu et al., 2019; Chen et al., 2020). The velocity and direction of movement of a glacier surface are traditionally estimated by geodetic methods using multitemporal, high-precision measurements of marked ground points (Zakharov, 2002; Zhang et al., 2008; Zhidkov and Zakharov, 2009; An et al., 2015; Sunil et al., 2017; Ai et al., 2019; Fu et al., 2021).

In the late 1960s, Knizhnikov (1967, 1973) proposed a method of pseudo-parallaxes to study the spatial distribution of ice flow velocities of mountain glaciers using images obtained by phototheodolite. The essence of the method was to determine the glacier displacement by measuring the parallaxes at the corresponding points according to the total deformed model generated from images of a moving glacier, which were captured from two basis points in a time interval. The parallaxes were stereoscopically measured. As a result, one could create maps of glacier displacements. Skrypitsyna (2022) proposed a modification of the pseudo-parallaxes method applicable to imagery obtained by manned or unmanned aerial surveys.

To determine the speed of glacier movement in Antarctica, Scambos et al. (1992) proposed a method of cross-correlation for automated finding of corresponding points on multitemporal satellite scenes. Li R. et al. (2017) adopted the pseudo-parallax method for archived, multitemporal satellite images.

With the advent of unmanned aerial systems (UASs), it became possible to expand and intensify glaciological studies (Bhardwaj et al., 2016b; Gaffey and Bhardwaj, 2020; Pina and Vieira, 2022), in particular, to produce high-accuracy, high-resolution DEMs and orthomosaics of glaciers as well as to determine characteristics of glacier movement. The main achievement of the UAS era is the ability to generate DEMs and orthomosaics using data from unmanned aerial surveys without ground control points (GCPs) applying a post-processed kinematic (PPK) direct georeferencing approach (Chiang et al., 2012; Turner et al., 2014; Bliakharskii, 2019; Bliakharskii et al., 2019a; Chudley et al., 2019; Li T. et al., 2020).

Geomorphometry-based studies of glacier motion and evolution, as a rule, include comparative analysis and processing of multitemporal glacier DEMs and morphometric models of glacier surface. Such studies are generally focused on three interrelated issues characterizing the glacier state and behavior: glacier mass balance, ice flow velocity, and glacier force balance.

17.2 Glacier mass balance

The mass balance describes how a glacier changes in size over time (Hughes, 2011).

Let there be two DEMs describing a glacier surface at the times t_1 and t_2 (two different years, seasons, or days). For these two points in time, the glacier mass balance can approximately be estimated by subtracting one glacier DEM from another. A result of the DEM subtraction is a difference matrix, each point of which describes the change in the glacier

Digital Terrain Analysis. https://doi.org/10.1016/B978-0-443-24798-9.00007-8

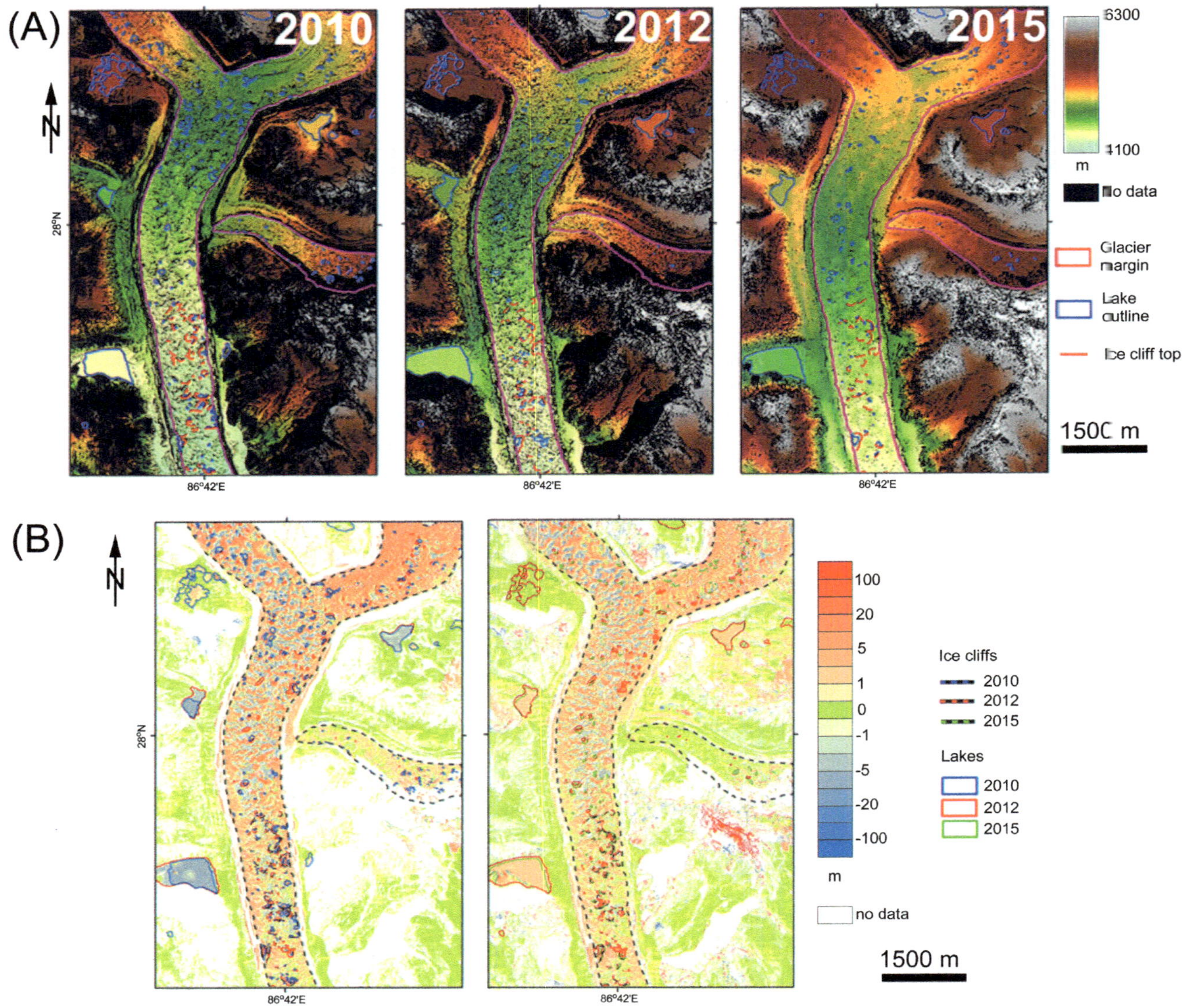

FIG. 17.1 Changes in the surface elevation of Ngozumpa Glacier, Nepal: (A) A time series of 1-m gridded DEMs constructed from three sets of high-resolution (0.5 m) satellite stereo imagery captured by GeoEye-1 and WorldView-3 in June 2010, December 2012, and January 2015. (B) DEM differences between June 2010 and December 2012 (*left*) and December 2012 and January 2015 (*right*). The scale shows the difference in elevation between the two DEMs with the positive and negative numbers indicating areas of elevation increase and decrease, respectively. *(Modified from Thompson, S., Benn, D., Mertes, J., Luckman, A., 2016. Stagnation and mass loss on a Himalayan debris-covered glacier: processes, patterns and rates. J. Glaciol. 62 (233), 467–485, Figs. 2 and 5, © Authors, distributed under the terms and conditions of the Creative Commons Attribution (CC BY) license.)*

surface elevation. If there are several multitemporal glacier DEMs (that is, a DEM time series), they can be sequentially processed in pairs (Fig. 17.1).

Variations of this approach have widely been used in glaciology for about 30 years (Etzelmüller and Sollid, 1997; Etzelmüller, 2000; Rippin et al., 2003; Ragettli et al., 2016; Thompson et al., 2016; Vincent et al., 2016; Brun et al., 2017; Wigmore and Mark, 2017; Zhou Y. et al., 2017; Bash et al., 2018; Bliakharskii et al., 2019b; Bash and Moorman, 2020; Skrypitsyna et al., 2023). DEM-derived information on vertical changes in glacier surface can be utilized for estimating glacier mass balance, glacier melting rate, and related issues.

Etzelmüller and Sollid (1997) noted:

> *"The change of the glacier surface elevation over time is given as the mass balance minus the glacier flux gradient, which describes the vertical flow velocity (emergence velocity) at a location. The surface elevation change averaged over the whole glacier surface gives the volume change during a period, which expresses the long-term net mass balance when divided by the glacier surface area. An average density of melted or accumulated snow and ice is assumed. Emergence velocity can be ignored here."*

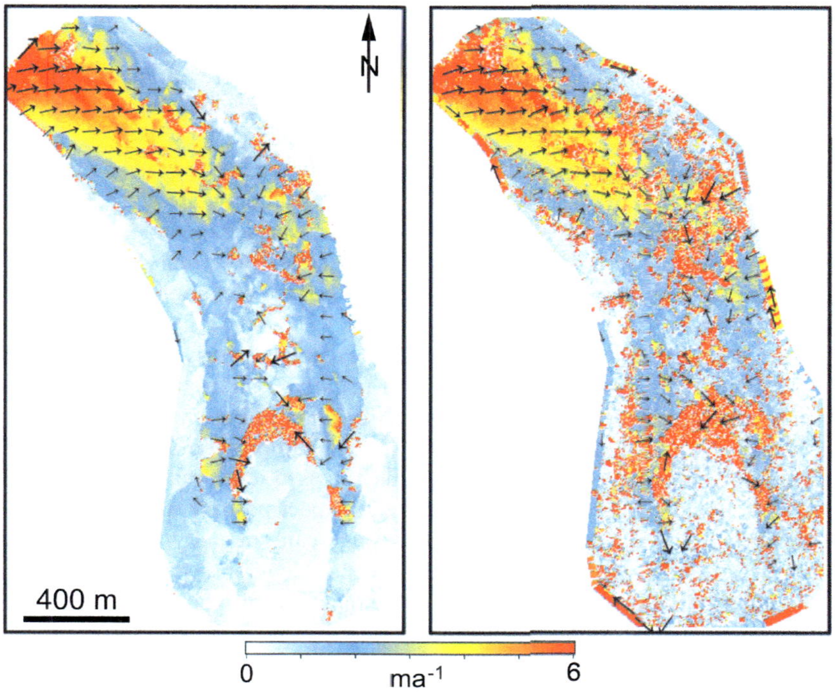

FIG. 17.2 Surface velocity fields for Lirung Glacier, Nepal (summer period) obtained by the COSI-Corr method applied to two different input data types: the UAS-based multitemporal orthomosaics (*left*) and hill-shaded maps derived from UAS-based multitemporal DEMs (*right*). Grid spacing of the input data was 0.2 m. The vectors are not linearly scaled; vectors with a magnitude less than 1.4 ma^{-1} are not displayed. *(Modified from Kraaijenbrink, P.D.A., Meijer, S.W., Shea, J.M., Pellicciotti, F., de Jong, S.M., Immerzeel, W.W. 2016. Seasonal surface velocities of a Himalayan glacier derived by automated correlation of unmanned aerial vehicle imagery. Ann. Glaciol. 57 (71), 103–113, Fig. 4, © Authors, distributed under the terms and conditions of the Creative Commons Attribution (CC BY) license.)*

17.3 Ice flow velocity

The glacier ice flow velocity can be estimated from glacier horizontal displacements. The latter can be determined by comparing the coordinates of GCPs or well-defined ground features (e.g., rocks and large crevasses) observed on multitemporal glacier orthoimagery (Scambos et al., 1992; Scherler et al., 2008; Kraaijenbrink et al., 2016a; Ragettli et al., 2016; Wigmore and Mark, 2017; Brun et al., 2018; Skrypitsyna et al., 2023) or multitemporal hill-shaded maps (Section 2.5.1.1) of a glacier (Kraaijenbrink et al., 2016a).

To process multitemporal glacier orthoimages, one can apply common photogrammetric techniques of cross-correlation (Luhmann et al., 2020), which allow for searching the coordinates of corresponding points in stereopairs. In our case, two multitemporal orthoimages are used as a pseudo-stereopair. Cross-correlation processing of such pseudo-stereopairs allows one to automatically determine the shifts of well-defined ground points on multitemporal scenes and, thus, to produce models of glacier horizontal displacements.

An alternative way to determine glacier horizontal displacements is the application of the Co-registration of Optically Sensed Images and Correlation (COSI-Corr) method (Leprince et al., 2007; Scherler et al., 2008). In COSI-Corr, glacier horizontal displacements are retrieved from the sub-pixel correlation of multitemporal orthorectified remotely sensed images or multitemporal morphometric products (e.g., multitemporal hill-shaded maps) (Kraaijenbrink et al., 2016a). Image or map correlation is carried out with an iterative, unbiased processor estimating the phase plane in the Fourier domain.

Final products—maps of flow velocity fields (Fig. 17.2)—are produced from models of glacier horizontal displacement.

17.4 Glacier force balance

The force balance describes how glaciers change in shape (surface geometry and thickness) over time and thus reflects glacier dynamics (Hughes, 2011). Etzelmüller and Sollid (1997) described how changes in glacier shape can be investigated by analyzing changes in slope gradient and curvatures of the glacier surface.

As a measure of slope gradient change, one can take a gradient of long-term change in glacier surface elevation. It can be defined as the average slope of a line determined by the average change in glacier surface elevation between some elevation intervals. By comparing this gradient between glaciers of the same region and with measured annual mass-balance gradients, one can estimate the emergence velocity and therefore the dynamics of glaciers under study (Etzelmüller and Sollid, 1997).

For example, if at a given point of the glacier surface, the average slope is negative and the values of two gradients—annual mass-balance gradient and long-term surface elevation change gradient—are close, then the emergence velocity is low. If the annual mass-balance gradient is higher than the long-term surface elevation change gradient at the negative average slope, then the emergence velocity is high. In this case, glacier flow replaces a part of the mass loss by ice. The negative values of the gradient of long-term change in glacier surface elevation may indicate a surge during a measurement period (Etzelmüller and Sollid, 1997).

Changes in both slope gradient and elevation values can indicate changes in driving stress. For instance, driving stresses increase under a steepening of the glacier surface at positive or negative minor changes in surface elevation. Such a situation is common for surge-type glaciers in their quiescent phase (Etzelmüller and Sollid, 1997).

The change in sign or values of curvatures of the glacier surface can provide information about changes in ice flow behavior. In particular, horizontal curvature (Section 2.2.4.2) may be used to estimate the magnitude and change of emergence velocity. For the ablation and accumulation areas, one can expect positive and negative values of horizontal curvature, respectively. A low-dynamic glacier in a downwasting stage can gradually acquire a surface topography closely corresponding to the subglacial topography (the glacier surface may develop a more concave shape). Thus, high rates of change in surface topography during surface lowering can indicate a mirroring of the subglacial topography on the glacier surface. In contrast, small changes in the glacier surface, despite high rates of surface lowering, suggest that such a glacier maintains high flow velocity, counteracting the loss of mass due to high ice flow (Etzelmüller and Sollid, 1997).

These examples show the need for more in-depth theoretical research on the possibilities of using geomorphometry, in particular, the complete system of curvatures (Section 2.2.4), in glacier studies.

Chapter 18

Crevasses

18.1 Motivation

Crevasses are fractures or cracks in glaciers and ice sheets (Fig. 18.1), a few meters to thousands of meters long, a few millimeters to several meters wide, and up to tens of meters deep (Van der Veen, 2011). Since the size and orientation of crevasses are closely associated with glacier dynamics, crevasses can be considered a visible manifestation of glacier stress regimes (Van der Veen, 1999; Colgan et al., 2016).

Crevasses pose a danger to polar researchers and glaciologists. For polar studies, one of the most important support tasks is rapid and safe monitoring of sledge routes, ice airfields, and other visited areas for recognition of open and hidden crevasses. Crevasses hidden by snow bridges are extremely dangerous. Monitoring and timely recognition of crevasses are important for the safety of participants in sledge and caterpillar-track caravans.

There are ground-based and remote-sensing approaches for rapid recognition of hidden crevasses. The ground-based approaches involve studying a glacier with geophysical methods. Ground penetrating radars (GPRs) are particularly applied, with antennas usually mounted in front of a manned or unmanned vehicle (Foessel-Bunting et al., 1999; Taurisano et al., 2006; Koh et al., 2010; Colgan et al., 2016). However, the safety and effectiveness of this approach are questionable because the skills and responsiveness of an operator play a significant role. Indeed, the operator must not only recognize a suspected crevasse on the radar monitor while a vehicle is moving but also have time to stop the vehicle.

Remote-sensing approaches include analysis of data from airborne GPRs (Delaney and Arcone, 1995; Thompson et al., 2020), airborne and satellite imagery (Glasser and Scambos, 2008; Holt et al., 2013; Colgan et al., 2016; Thompson et al., 2020; Chudley et al., 2021; Marsh et al., 2021), as well as satellite laser altimetry (Liu et al., 2014). However, key disadvantages of this approach are relatively low resolution of data, low promptness, dependence on weather conditions (e.g., cloudiness), and relatively high cost.

The processes of crevasse formation and propagation are reflected in the geometry of a glacier's surface. A hidden crevasse has a snow bridge, which makes it difficult to recognize. However, a snow bridge can sink under gravity and so form some sort of ditch. In fact, hidden crevasses are micro-landforms of the ice sheet topography. They should therefore be registered in high-resolution DEMs of glaciers.

However, it does not make sense to use DEMs derived from manned aerial surveys or satellite images because of their relatively low resolution. Such DEMs cannot contain information about crevasses 1–5 m wide. Such crevasses are the most dangerous because they can be covered by snow and are wide enough for a person or vehicle to fall into them.

This problem can be solved by using the results of unmanned aerial surveys (Section 3.1.4). In recent years, unmanned aerial systems (UASs) and UAS-derived products—orthomosaics and DEMs—have been increasingly used in glaciology (Section 1.3.2.5). UAS-derived DEMs can have centimeter-to-meter resolutions, centimeter-range planimetric and vertical accuracies, and contain information about glacial surface deformations associated with open and hidden crevasses.

In this chapter, we show the possibility of revealing hidden crevasses by processing UAS-derived glacier DEMs (Florinsky and Bliakharskii, 2019a).

Digital Terrain Analysis. https://doi.org/10.1016/B978-0-443-24798-9.00003-0

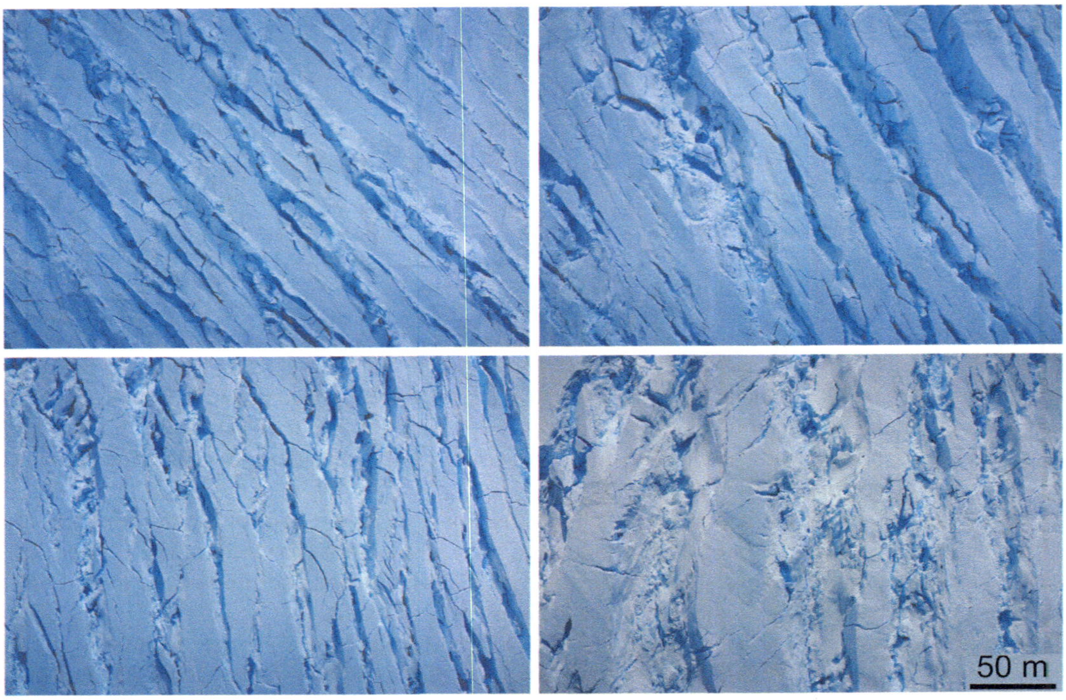

FIG. 18.1 Examples of crevasses on the Dålk Glacier (Princess Elizabeth Land, East Antarctica). Images were captured by Geoscan 201 Geodesy UAS in February 2023 during the 68th Russian Antarctic Expedition; drone pilots: D.V. Fedorov, M.A. Rybko, E.G. Seleznev, O.A. Khazov, and M.I. Fedotkin (Aerogeodezia JSC, St. Petersburg, Russia). *(Modified from Florinsky, I.V., 2023c. Study on Mathematical and Cartographic Modeling of the Topography in the Area of Progress Station. Interim Report on the Program 'Study on Mathematical and Cartographic Modeling of the Topography in the Areas of Progress Station, Molodezhnaya and Gora Vechernyaya Bases as well as Novolazarevskaya and Bellingshausen Stations'. 68th Russian Antarctic Expedition, Progress Station (in Russian), Fig. 2.2.)*

18.2 Study area

In the austral summer of 2016/2017, an unmanned aerial survey was conducted near the Larsemann Hills, East Antarctica, during the 62nd Russian Antarctic Expedition (Bliakharskii et al., 2019a). The UAS survey was performed along an initial section of a sledge route from the Progress Station to the Vostok Stations, from 69°27′24″ S, 76°19′04″ E to 69°40′40″ S, 76°32′46″ E (Fig. 18.2). The surveyed sledge route zone had a length of about 30 km and a width of around 3 km. The territory is a snow-covered ice sheet surface gradually rising from the north to the south, from about 300 to 750 m above sea level.

There are a lot of open and hidden crevasses in the sledge route zone. We selected a typical area—a crevasse testing ground—measuring 1300 m × 770 m (Fig. 18.2). It included two (long and short) snow-covered hidden crevasses. We do not discuss open crevasses there because they are well recognizable both in the field and in aerial images.

Within the crevasse testing ground, the crevasses were examined in the field by D.P. Bliakharskii. In particular, two portions of each crevasse were cleared and measured (field works were limited due to safety reasons). The long crevasse had a width of about 1.5 m and an observed depth of 2 m. The short crevasse had a width of about 1 m. The crevasses were covered by the snow layer with a thickness ranging from 0.2 to 1.2 m (Florinsky and Bliakharskii, 2019a).

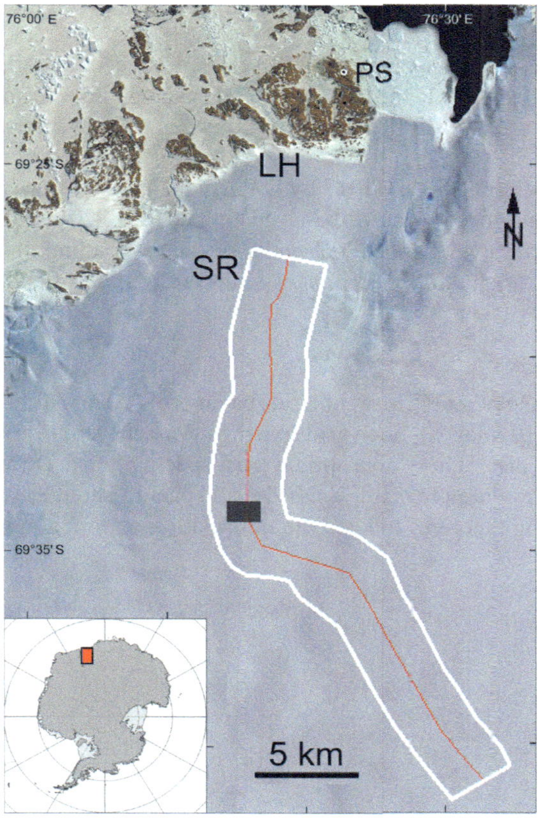

FIG. 18.2 Geographical location of the study area on the background of the Google Earth image mosaic: *LH*, Larsemann Hills; *PS*, Progress Station; *red line*, initial section of the sledge route from the Progress Station to the Vostok Station; *white frame SR*, surveyed zone along the sledge route; *gray rectangle*, crevasse testing ground. *(From Florinsky, I.V., Bliakharskii, D.P., 2019a. Detection of crevasses by geomorphometric treatment of data from unmanned aerial surveys. Remote Sens. Lett. 10 (4), 323–332, Fig. 1, reproduced with permission from Taylor & Francis Ltd.)*

18.3 Materials and methods

18.3.1 UAS characteristics

The study area was surveyed using Geoscan 201 Geodesy, a professional flying-wing UAS with a wingspan of 2.3 m. Geoscan 201 is launched by a catapult and landed with a parachute. It can be used at wind speeds up to 15 m/s and air temperature of ±40 °C. Its flight duration is up to 3 h with a cruising speed of 110 km/h. A flight altitude can range from 100 to 4000 m (Geoscan, 2016–2024).

On board, there are (1) a field-calibrated, visible-band digital camera Sony DSC-RX1 equipped with a Carl Zeiss Vario Sonnar T lens (a central leaf shutter; a 35 mm focal length) and a 35.8 mm × 23.9 mm sensor with a Bayer filter (6000 pixels × 4000 pixels matrix with a pixel size of 6 μm × 6 μm); (2) a global navigation satellite systems (GNSS) receiver Topcon; and (3) a modem for the telemetry communication with a laptop ground control station (GCS). A ground GNSS base station includes a receiver Topcon HiPer V. This combination of the camera and GNSS receiver provides planimetric and vertical accuracy of up to 2 cm and 3–4 cm, respectively, for determining image projection centers (Florinsky et al., 2018b).

The Geoscan 201 flight is performed using an autopilot. A flight mission is specified in the GCS by selecting a study area and setting one of the three interrelated parameters: ground sample distance (GSD), flight altitude, or flight strip width. The GCS calculates the flight lines and coordinates of image capture positions

18.3.2 UAS surveying

In this study, a key precondition was a grid spacing of a DEM to be produced. In earlier experiments with Geoscan UASs (e.g., Florinsky et al., 2018b), we found that (1) the best DEM grid spacing ranges from 20 to 30 cm (the use of smaller grid spacing led to the production of noisy DEMs) and (2) the best ratio "image pixel spacing:DEM grid spacing" is 1:4. Thus, a GSD of 6 cm was selected.

The sledge route was surveyed in mid-January 2017 by D.P. Bliakharskii. The weather was sunny; the air temperature ranged from −5°C to −10°C; the wind speed was 10–15 m/s gusting up to 18–20 m/s at the flight altitude, 300–400 m above ground level. Forward and side overlaps were about 70% and 50%, correspondingly. For the sledge route zone, there were nine flights consisting of 195 strips, including 9381 images with a resolution of about 6 cm. Each flight took about 2 h.

18.3.3 Data processing

For the crevasse testing ground, three glacier DEMs with grid spacings of 25 cm, 50 cm, and 1 m (Fig. 18.3B) were produced from a set of 133 images (7 strips × 19 images). Such grid spacings were chosen for four reasons: (1) Considering the 6-cm resolution of images, 24–25 cm is the highest appropriate grid spacing for a DEM of this area (for explanation, see Section 18.3.2). (2) To detect a 1-m-wide crevasse, a DEM grid spacing should also be, at least, 25 cm. This is due to the first consequence of the sampling theorem: to keep the information on topographic features with typical planar sizes

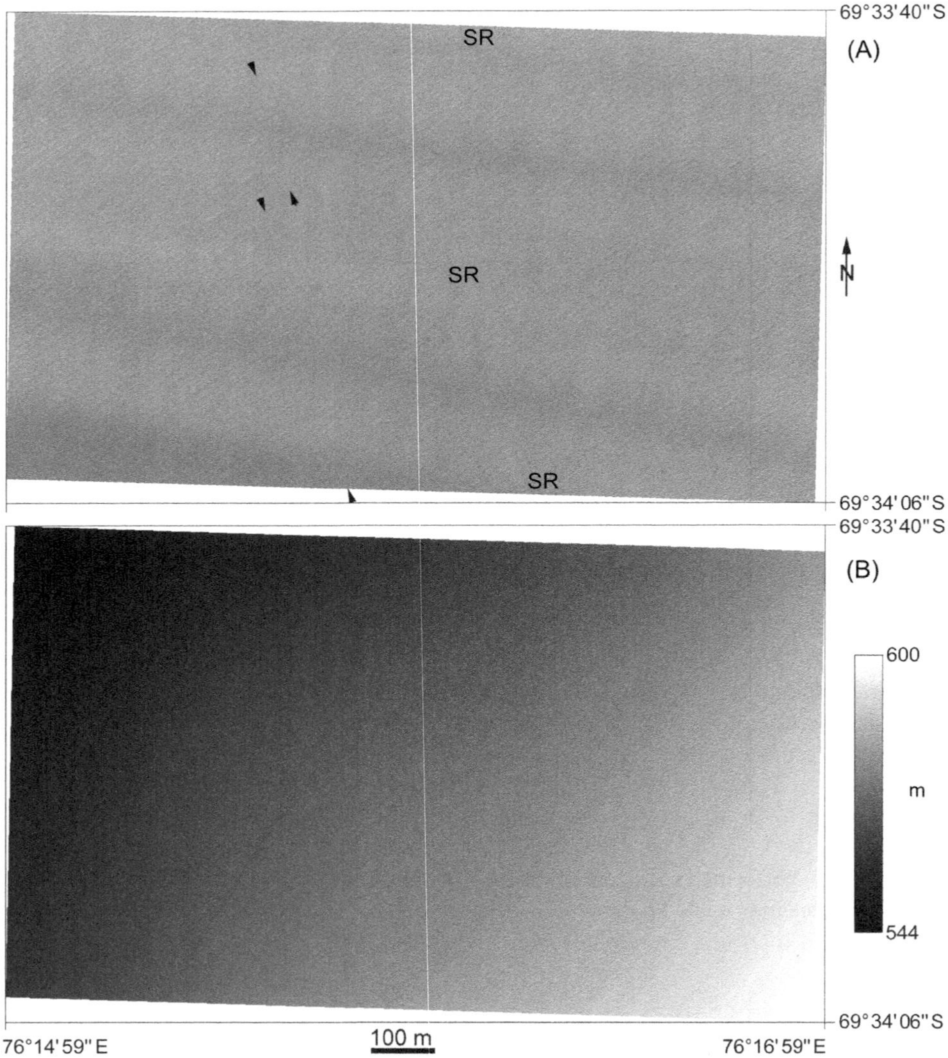

FIG. 18.3 The crevasse testing ground: (A) Orthomosaic. (B) Elevations derived from the 1-m gridded DEM. *Arrows*, snow-covered crevasses; *SR*, sledge route. *(From Florinsky, I.V., Bliakharskii, D.P., 2019a. Detection of crevasses by geomorphometric treatment of data from unmanned aerial surveys. Remote Sens. Lett. 10 (4), 323–332, Fig. 4, reproduced with permission from Taylor & Francis Ltd.)*

λ in a DEM, one should use the DEM grid spacing $w \leq \lambda/(2n)$, where $n \geq 2$ (Section 3.3). (3) Grid spacings coarser than 1 m do not make sense because narrow crevasses will be located between DEM nodes. (4) The 50-cm grid spacing completes a series of the chosen values.

The data processing workflow included five main steps (Section 3.1.4):

(1) Determining the high-precision coordinates of image projection centers. These data were then utilized for the post-processed kinematic direct georeferencing (Turner et al., 2014).
(2) Aligning aerial photos.
(3) Building a dense point cloud using a semi-global matching approach (Hirschmüller, 2008).
(4) Generating DEMs by interpolation of the dense point cloud (Fig. 18.3B).
(5) Producing orthomosaics (Fig. 18.3A).

Step 1 was performed with Magnet Office Tools software (Topcon Positioning Systems, Inc.). Next steps were carried out using PhotoScan Professional 1.3.2 software (Agisoft, 2017).

To suppress high-frequency noise, the obtained DEMs were smoothed 1–3 times by a weighted average filter using a 3×3 moving window (Section 6.2.5.1).

Finally, we derived models of 16 morphometric variables (Fig. 18.4) from the smoothed DEMs, including slope gradient, slope aspect, 12 curvatures (horizontal, vertical, minimal, maximal, etc.), as well as catchment and dispersive areas.

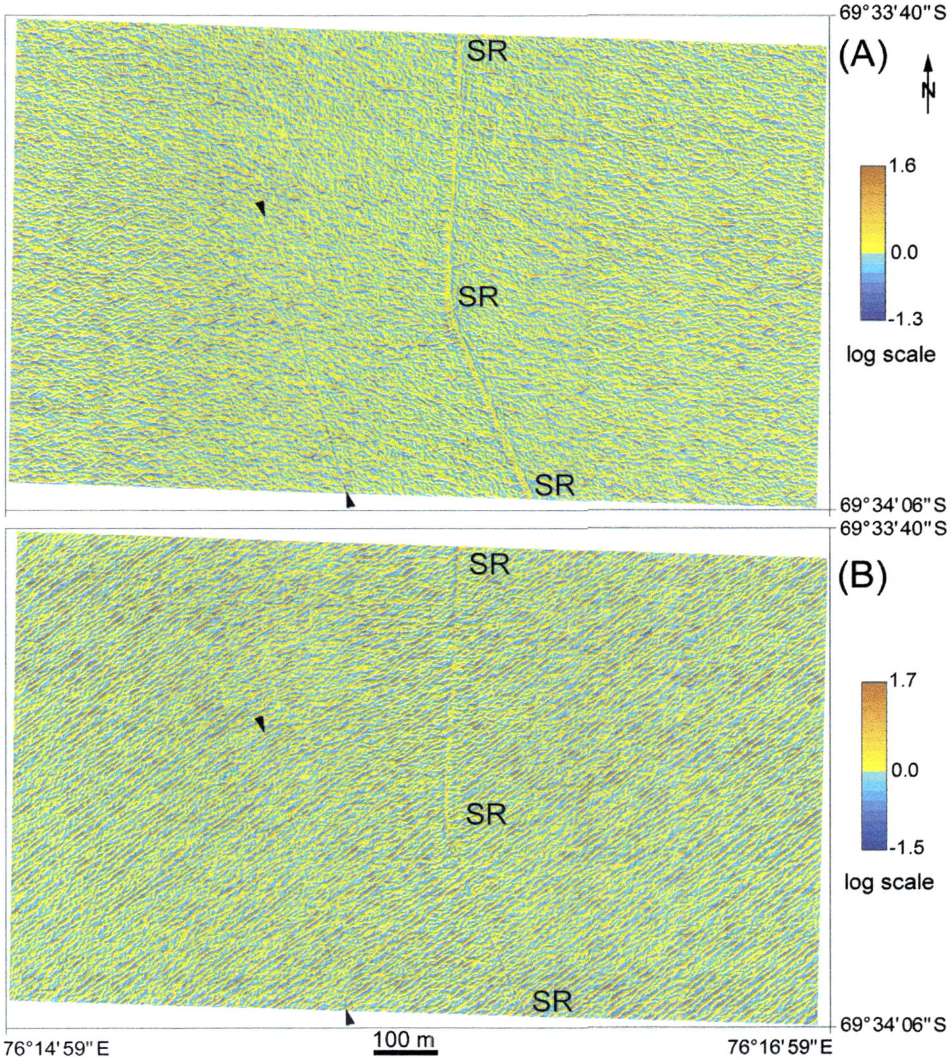

FIG. 18.4 The crevasse testing ground, morphometric maps derived from the thrice-smoothed 1-m gridded DEM: (A) Horizontal curvature. (B) Vertical curvature. (C) Minimal curvature.

(Continued)

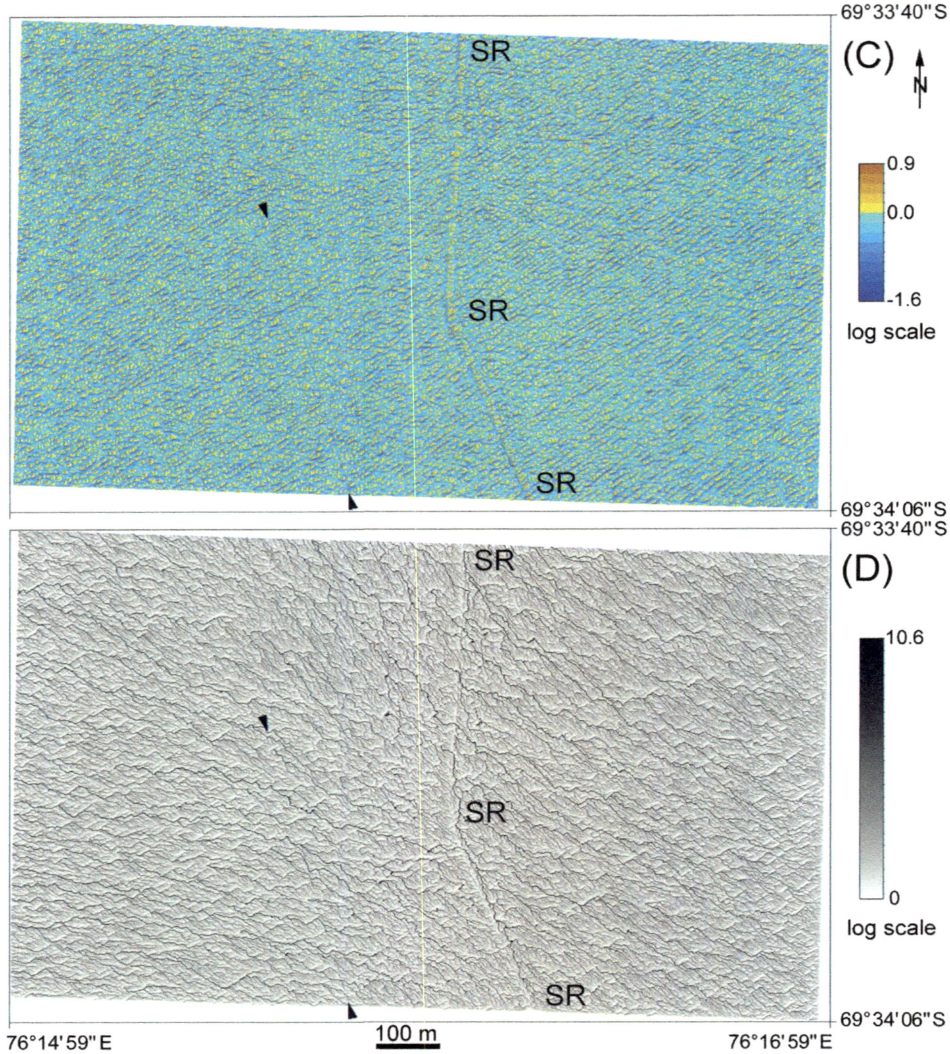

FIG. 18.4, CONT'D (D) Catchment area. *Arrows*, revealed long crevasse; *SR*, sledge route. *(From Florinsky, I.V., Bliakharskii, D.P., 2019a. Detection of crevasses by geomorphometric treatment of data from unmanned aerial surveys. Remote Sens. Lett. 10 (4), 323–332, Fig. 5, reproduced with permission from Taylor & Francis Ltd.)*

These variables have a clear physical and mathematical sense and describe a particular property of the topographic surface (Chapter 2, Table 2.1). We applied this set of variables rather than particular ones because it is a priori unclear how exactly hidden crevasses are reflected in the glacier surface geometry. Slope gradient, slope aspect, and curvatures were derived using a finite-difference method (Section 4.2); catchment and dispersive areas were calculated using a flow-routing algorithm (Section 4.4.1). Here, we utilized LandLord software (Florinsky, 2016).

18.4 Results and discussion

Visually, the both snow-covered crevasses cannot be easily found in the orthomosaic and elevation maps of the crevasse testing ground (Fig. 18.3). However, morphometric variables derived from the 1-m gridded DEM revealed the long crevasse as a lineament (Fig. 18.4). Maps of all variables somehow display this crevasse because it is reflected in the geometry of the glacier surface. However, different variables have different abilities to reveal the crevasse because a particular morphometric attribute describes topographic features according to their physical, mathematical, and geomorphic sense (Chapter 2, Table 2.1). The crevasse was best detected by k_h and k_{min} (Fig. 18.4A and C). Although the measured width of the crevasse is 1.5 m, the width of its trench manifested in the snow microtopography should be 5–6 m. Otherwise, it would not be revealed by the 1-m gridded models due to the first consequence of the sampling theorem (Section 18.3.3). The short crevasse is not detected on the 1-m gridded morphometric maps (Fig. 18.4) because it is narrower than the long crevasse and can be located between DEM nodes.

The sledge route is also detected by morphometric variables (Fig. 18.4) because it is manifested as a worn snow path surrounded by the virgin snow surface.

Processing the 25- and 50-cm-gridded DEMs did not allow us to reveal the crevasses. These DEMs included myriads of high-frequency patterns representing both microtopographic snow features and noise/artifacts. The latter was associated with specificities of the formation and photogrammetric processing of high-resolution images of the glistening snow surface. The related morphometric maps were almost unreadable (Florinsky and Bliakharskii, 2019a).

It was impossible to adequately remove these noises and artifacts. It is well known that filtering/smoothing not only suppresses noise but also changes the desired signal (Chapter 6). If signal-to-noise ratio is high enough (as in the case of the 1-m gridded DEM), this is not a serious problem, and a crevasse can be detected. However, for the 25- and 50-cm-gridded DEMs, signal and noise levels were comparable, and the applied smoothing removed both the noise and signal patterns equally.

The orthomosaic (Fig. 18.3A) displays many elongated snow patterns of eolian origin (apparently, little *sastrugi*), which are stretched from southwest to northeast, perpendicular to the local katabatic wind direction. Eolian snow microtopographic patterns can also be seen on the maps of k_h, k_v, and CA (Fig. 18.4A, B, and D). The k_v map demonstrates a system of snow microterraces stretched from southwest to northeast (Fig. 18.4B), which are also perpendicular to the local katabatic wind. The k_h and CA maps (Fig. 18.4A and D) show a wind-derived system of snow microflows, which are stretched from southeast to northwest, from uphill to downhill, parallel to the local katabatic wind. Thus, high-resolution, UAS-derived glacier DEMs can also be used in studying wind-driven distribution of snow and ice patterns.

This study and its further expansion, in which the described procedure was tested on a larger number of crevasses (Ishalina et al., 2021), demonstrated that the geomorphometric detection of hidden crevasses can be functional, at least for crevasses with a width of 1.5 m and wider (i.e., most dangerous crevasses for persons and all-terrain vehicles) covered by the snow layer with a thickness of up to 1.2 m. The approach can be used as an alternative or complement to existing methods of crevasse detection.

Chapter 19

Catastrophic glacier events

19.1 Motivation

Rapid and deep subsidences of large areas in glaciers and ice sheets are rare and poorly investigated phenomena that may be connected with surficial, englacial, and subglacial hydrological processes (Gudmundsson et al., 1997; Björnsson, 2003). Monitoring and investigation of these events are essential for a deeper understanding of processes taking place in the cryosphere as well as for ensuring the safety of glaciologists and polar researchers. However, these phenomena are objectively difficult to observe and study because of the inability to predict the time and place of a subsidence. In this regard, such events and their interpretation are almost not mentioned in the scientific literature, except for ice cauldrons in Iceland caused by subglacial geothermal and volcanic activity (Björnsson, 2003; Evatt and Fowler, 2007).

In this chapter, we describe the study of the 2017 catastrophic subsidence in the Dålk Glacier, East Antarctica, using data from unmanned aerial surveys and geomorphometric modeling (Florinsky and Bliakharskii, 2019b).

19.2 Study area

The study was conducted in and near the Larsemann Hills, East Antarctica (Fig. 19.1). The Larsemann Hills is an Antarctic coastal oasis including ice-free, low-rounded hills with an area of about 40 km^2 located on the southeastern coast of the Prydz Bay, Princess Elizabeth Land (ATCM, 2021). There are three year-round operated polar stations in the oasis: Progress (Russia), Zhongshan (China), and Bharati (India). There is the Zenit Airfield (Russia) on the glacial plateau located to the south of the Progress Station (Fig. 19.1). For a detailed description of the Larsemann Hills, see Section 20.2.

The Dålk Glacier is a typical, marine-terminating outlet glacier draining into the Dålkoy Bay east of the Larsemann Hills (Fig. 19.1). The glacier length is about 15 km; its width is about 3 km. The glacier elevation varies from 50 to 140 m above sea level. The movement direction of the Dålk Glacier is controlled by the bedrock depressions of a subglacial and submarine fjord. The surface of the Dålk Glacier is covered by crevasses of various lengths, widths, and directions (Fig. 18.1) (Zakharov, 2002; Bliakharskii et al., 2019a; Chen et al., 2020; Yuan et al., 2020; Skrypitsyna et al., 2023).

On the morning of January 30, 2017, an unexpected subsidence in the Dålk Glacier occurred. A wide and deep ice-surface depression formed abruptly in the northwestern margin of the Dålk Glacier, at the coordinates 69°23′58″ S, 76° 24′49″ E (Florinsky and Bliakharskii, 2019b; Li T. et al., 2020; Skrypitsyna et al., 2023). The subsidence destroyed a section of a sledge route from the Progress Station to the Zenit Airfield.

Digital Terrain Analysis. https://doi.org/10.1016/B978-0-443-24798-9.00026-1

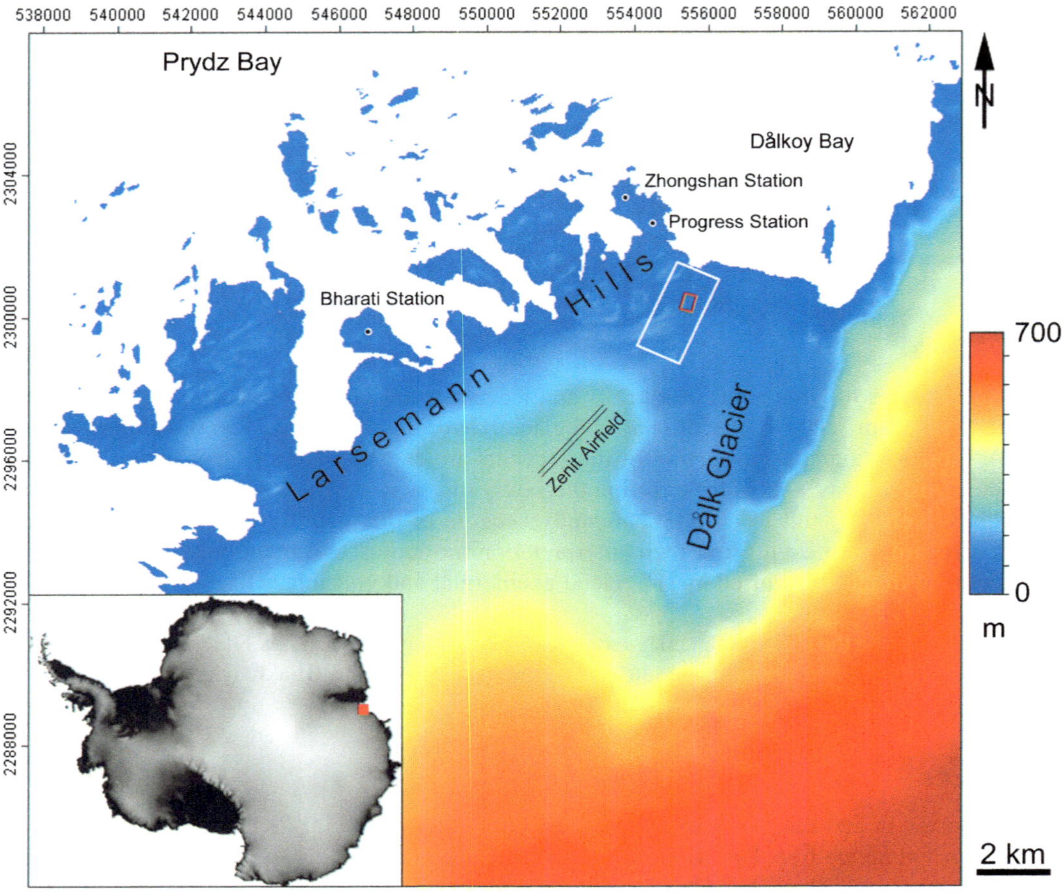

FIG. 19.1 Geographical location of the Larsemann Hills and the Dålk Glacier. *Red frame*, the collapsed zone (see Figs. 19.3 and 19.4); *white frame*, a territory between the Boulder Lake and the Dålkoy Bay (see Fig. 19.5). Elevations are extracted from the Reference Elevation Model of Antarctica (Howat et al., 2019).

19.3 Materials and methods

19.3.1 UAS surveying

In the austral summer 2016/2017, unmanned aerial surveys with Geoscan 201 Geodesy, a professional flying-wing UAS (Section 18.3.1), were carried out in the Larsemann Hills during the 62nd Russian Antarctic Expedition (Bliakharskii et al., 2019a). A territory including the subsidence zone with an area of about 7.5 km^2 was surveyed by D.P. Bliakharskii in two flights performed on 20 January and February 9, 2017, that is, 10 days before and 10 days after the event. The first flight was carried out as a part of the UAS survey of the eastern portion of the Larsemann Hills for updating a local topographic map (Fig. 19.2A). The second flight was specially organized to survey the subsidence zone (Fig. 19.2C).

During both flights, the weather was sunny. The air temperature ranged from −1°C to −5°C; the wind was northeast; the wind speed was 0–5 m/s at the flight altitude of 310–350 m above ground level.

Each flight took about 2 h. The aerial surveying consisted of 20 and 30 flight strips, including 1594 and 1011 images for the first and second flights, respectively. In the first flight, forward and side overlaps were about 70% and 50%, respectively. This provided triple image overlaps, guaranteeing no photogrammetric voids. In the second, crosswise flight, both overlaps were about 85%. During the both flights, an average ground sample distance was 6 cm.

An unmanned aerial video recording of the subsidence zone was conducted 1 h after the event (Fig. 19.2B). It was performed using the Phantom 4, a consumer copter-type UAS equipped with a simple video camera (DJI, 2018–2024).

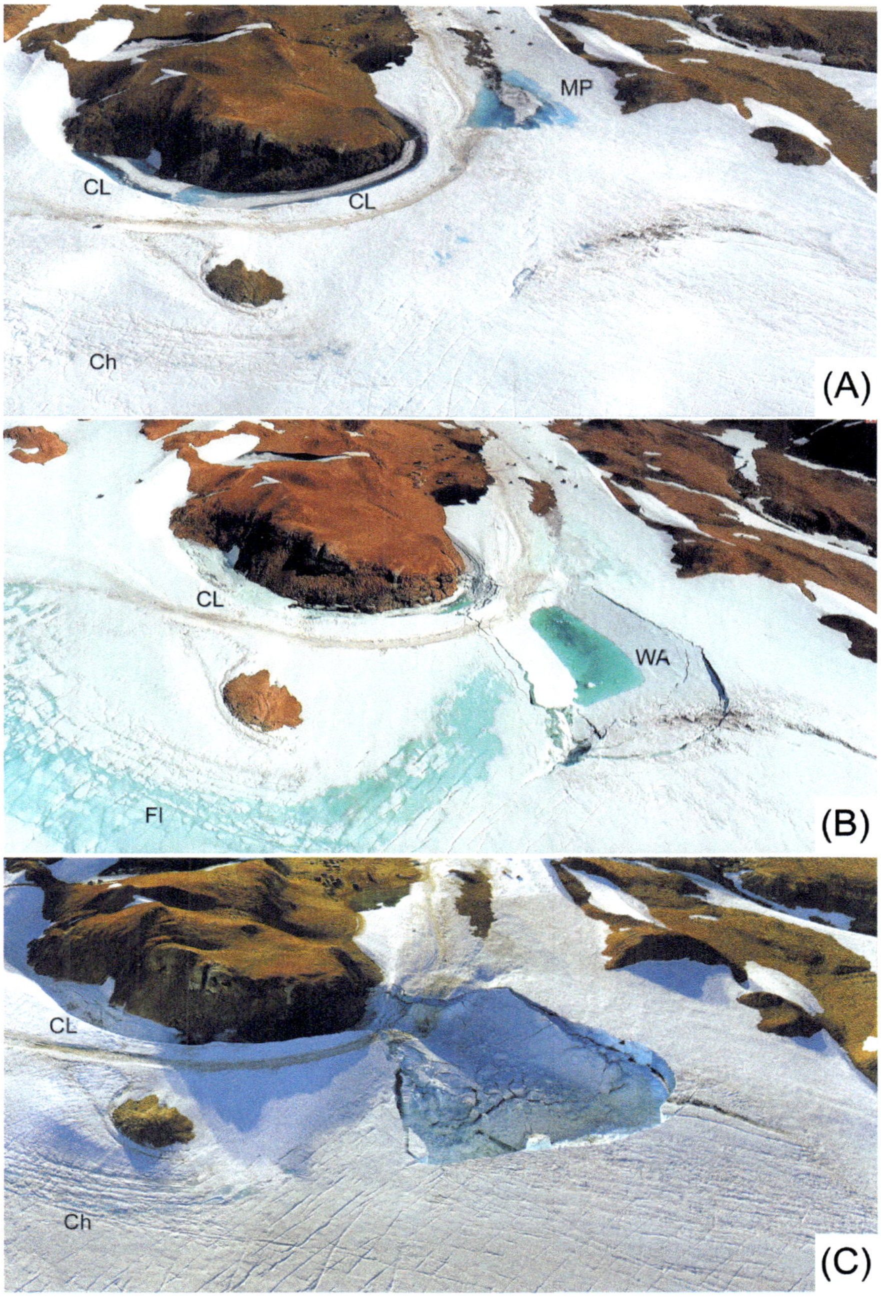

FIG. 19.2 The Dålk Glacier collapsed zone, aerial perspective views: (A) Ten days before the event (B) One hour after the event. (C) Ten days after the event. *Ch*, drainage channels; *CL*, unnamed crescent-shaped lake; *Fl*, intensive meltwater flow from the Boulder Lake; *MP*, meltwater pond; *WA*, meltwater accumulated on the cavern rooftop. The views (A) and (C) were produced from aerial photographs captured by the Geoscan 201. The view (B) is a frame from a video recorded by the Phantom 4. *(From Florinsky, I.V., Bliakharskii, D.P., 2019b. The 2017 catastrophic subsidence in the Dålk Glacier, East Antarctica: unmanned aerial survey and terrain modelling. Remote Sens. Lett. 10 (4), 333–342 Fig. 3, reproduced with permission from Taylor & Francis Ltd.)*

19.3.2 Data processing

First, we generated pre- and post-event DEMs of the collapsed zone (two 8400 × 7655 elevation matrices) with a grid spacing of 22 cm (Fig. 19.3). The DEMs were produced from Geoscan 201-captured images and describe a mixed surface of ice and bedrock. Two pre- and post-event orthomosaics of the collapsed zone were also generated (Fig. 19.4). The data processing workflow was standard; it is described in Section 18.3.3.

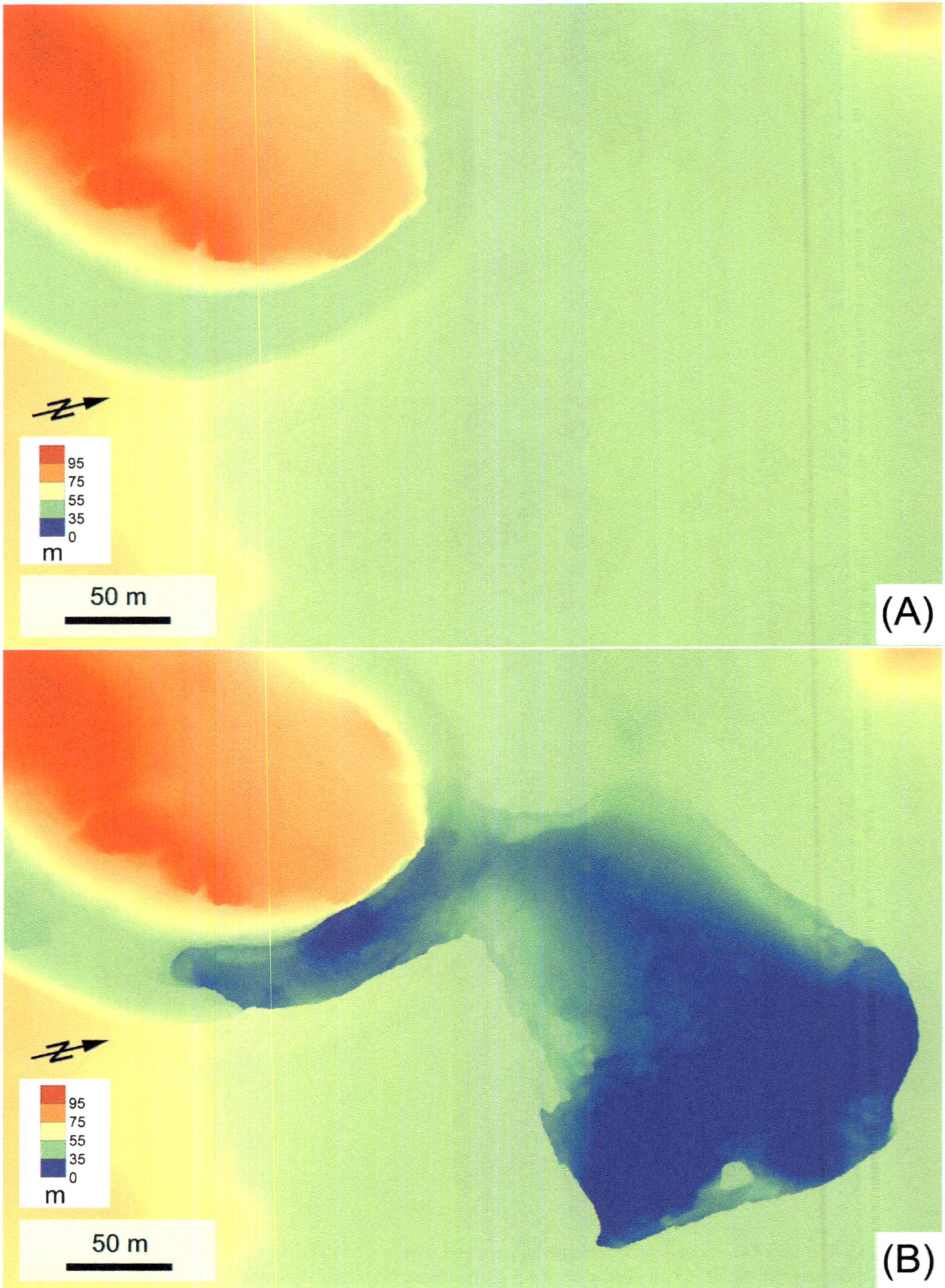

FIG. 19.3 The Dålk Glacier collapsed zone, elevations: (A) Ten days before the event. (B) Ten days after the event. *(From Florinsky, I.V., Bliakharskii, D.P., 2019b. The 2017 catastrophic subsidence in the Dålk Glacier, East Antarctica: unmanned aerial survey and terrain modelling. Remote Sens. Lett. 10 (4), 333–342, Fig. 2c and d, reproduced with permission from Taylor & Francis Ltd.)*

To estimate the volume of the pre-collapsed englacial cavern, the pre-event DEM (Fig. 19.3A) was subtracted from the post-event one (Fig. 19.3B) using QGIS software (QGIS, 2009–2023). To superpose the two DEMs, seven control points (stones on adjacent ice-free hills) were utilized. Lateral (x and y) and vertical (z) mean errors and standard deviations of the superposition were as follows: $x = 0.006 \pm 0.050$ m, $y = -0.020 \pm 0.058$ m, and $z = 0.056 \pm 0.141$ m.

Next, for a broader territory between the Boulder Lake and the Dålkoy Bay, a pre-event DEM (a 2601×5801 elevation matrix) with a grid spacing of 50 cm (Fig. 19.5A) as well as two pre- and post-event orthomosaics (Fig. 19.5B and C) were produced.

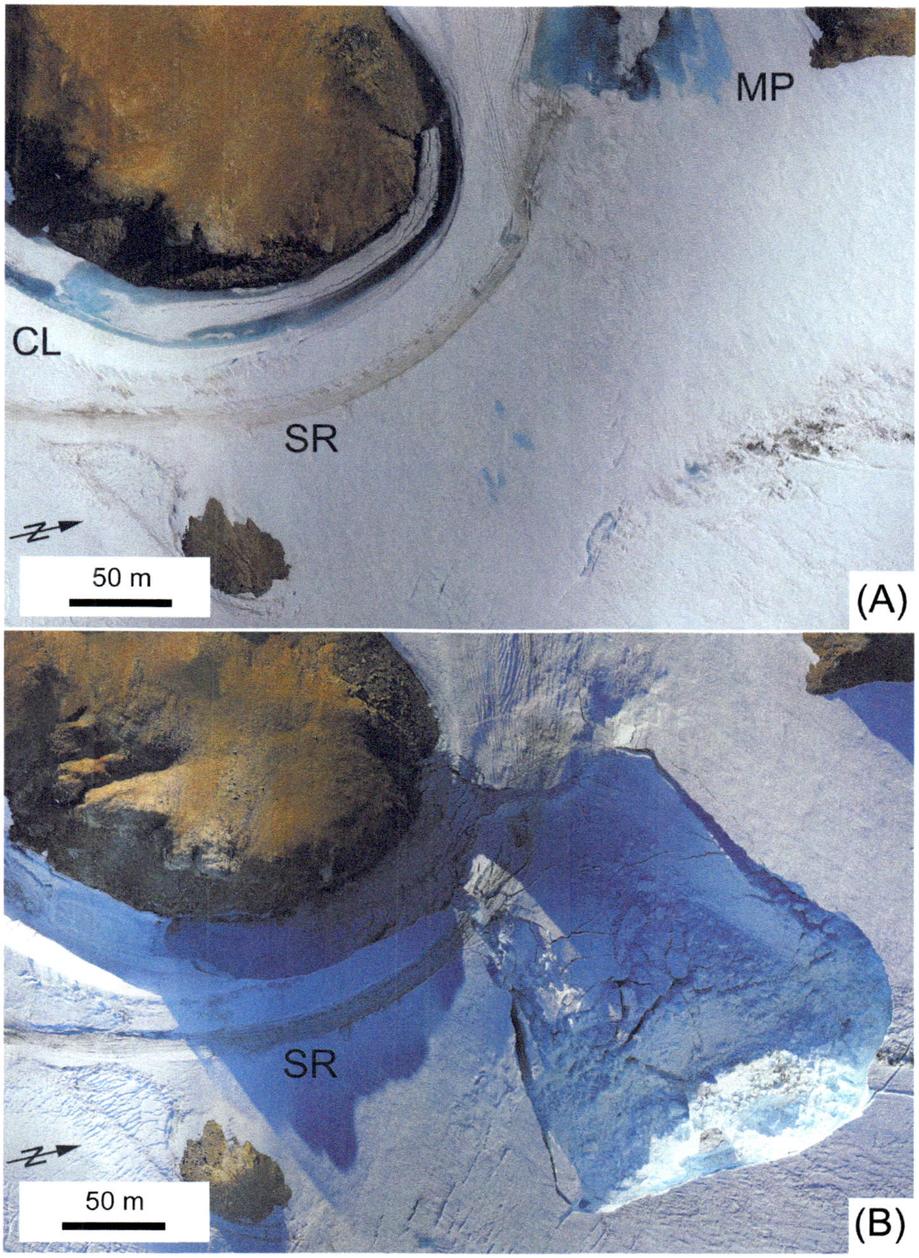

FIG. 19.4 The Dålk Glacier collapsed zone, orthomosaics: (A) Ten days before the event. (B) Ten days after the event. *CL*, unnamed crescent-shaped lake; *MP*, meltwater pond; *SR*, sledge route from the Progress Station to the Zenit Airfield. *(From Florinsky, I.V., Bliakharskii, D.P., 2019b. The 2017 catastrophic subsidence in the Dålk Glacier, East Antarctica: unmanned aerial survey and terrain modelling. Remote Sens. Lett. 10 (4), 333–342, Fig. 2a and b, reproduced with permission from Taylor & Francis Ltd.)*

Finally, for the territory between the Boulder Lake and the Dålkoy Bay, an ice-surface channel network from the pre-event DEM was derived (Fig. 19.5) by a standard geomorphometric procedure: (1) hydrological correction of the DEM by sink filling, (2) calculation of a catchment area from the corrected DEM by a flow-routing algorithm, and (3) extraction of the channels thresholding the *CA* values. Here, SAGA GIS software (Conrad et al., 2015) was used.

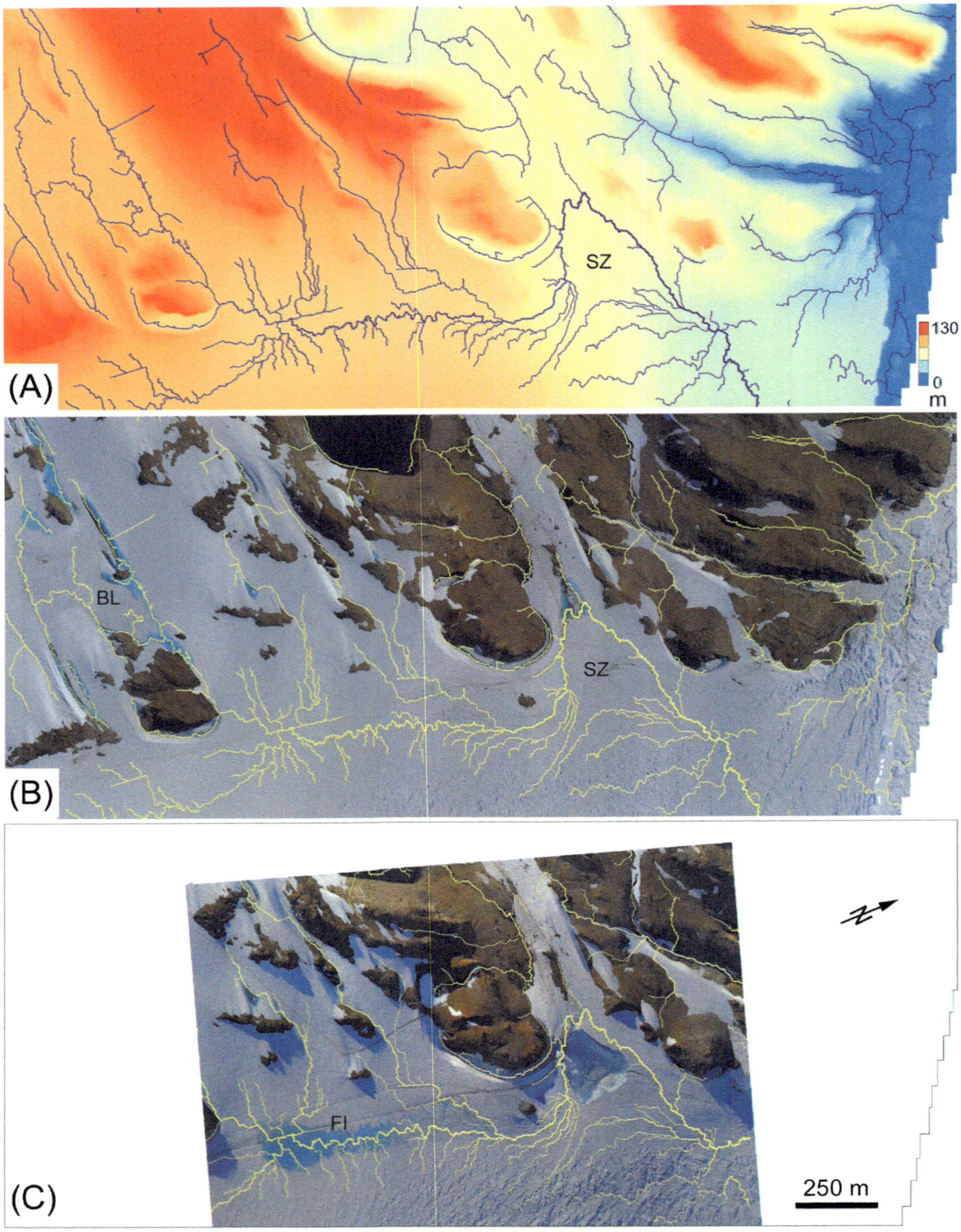

FIG. 19.5 The territory between the Boulder Lake and the Dålkoy Bay with the collapsed zone. The ice-surface channel network (*blue and yellow lines*) is superposed on the pre-event elevations (A) and orthomosaics obtained 10 days before (B) and 10 days after (C) the event. *BL*, Boulder Lake; *Fl*, remains of the meltwater flow from the Boulder Lake; *SZ*, subsidence zone. *(From Florinsky, I.V., Bliakharskii, D.P., 2019b. The 2017 catastrophic subsidence in the Dålk Glacier, East Antarctica: unmanned aerial survey and terrain modelling. Remote Sens. Lett. 10 (4), 333–342, Fig. 4, reproduced with permission from Taylor & Francis Ltd.)*

19.4 Results and interpretation

Fig. 19.2 shows the subsidence development. In the image recorded 1 h after the event (Fig. 19.2B), one can see that the englacial cavern roof collapsed under the weight of water, which is supplied through a wide ice-surface meltwater flow from the Boulder Lake (see details below). In the image captured 10 days after the event, one can find only some residues of this flow (Fig. 19.2C).

The orthomosaics (Fig. 19.4) and DEMs (Fig. 19.3) of the collapsed zone clearly demonstrate the subsidence results. According to these data, the ice-surface depression measured 183 m by 220 m, and its area reached up to 40,000 m^2. The maximum depth of the ice-surface depression was 43 m. The approximate volume of the pre-collapse englacial cavern was about 835,000 m^3.

An analysis of the obtained data allowed proposing a mechanism for formation of the glacier subsidence. The supposed chain of events is as follows (Florinsky and Bliakharskii, 2019b).

First, the englacial cavern with a relatively thin roof existed there (long) before the event (it is well-known that englacial and subglacial caves are common features for glaciers—Gulley and Fountain, 2019). The cavern has formed at the glacier margin close to and between ice-free hills (Figs. 19.2–19.5). Thus, the cavern was isolated from the continuous flow of the Dålk Glacier toward the Dålkoy Bay.

Second, there was a channel of the surface and near-subsurface, snow-covered meltwater flow, which originated in the Boulder Lake located 1.5 km south of the subsidence zone (Fig. 19.5B). It is essential that the channel from the Boulder Lake to the Dålkoy Bay round the englacial cavern (Fig. 19.5). This testifies that the channel corresponded to a system of crevasses along the cavern roof.

Ten days before the event, there were visual observations of numerous small meltwater flows between the Boulder Lake and the future subsidence zone. About a day before the event, an intensive meltwater flow began from the Boulder Lake. This water overflowed the existing channel (Fig. 19.2B). The water volume and the flow rate were too great. The water did not have time to drain toward the Dålkoy Bay, and, as a result, meltwater began to accumulate on the rooftop of the englacial cavern. This resulted in the initial collapse of the cavern roof (Fig. 19.2B). After this, meltwater flowed to the emerging ice-surface depression from a small pond located to the northwest of the depression, between ice-free hills (cf. Fig. 19.2A and B as well as Fig. 19.4A and B). In several days, the meltwater accumulated on the cavern rooftop, completing the cavern collapse (cf. Fig. 19.2B and C). Finally, a vast and deep ice depression formed.

In subsequent years the depression was gradually filled with ice. According to the author's observations during the 68th Russian Antarctic Expedition, by the austral summer of 2022/2023, the depression was completely level with the adjacent surface of the Dålk Glacier.

The 2017 Dålk Glacier catastrophic event is of scientific interest in the context of studying glacier evolution, pseudo-karst, and hydrography. UAS-derived data demonstrate that ice-surface hydrological processes can cause rapid, deep subsidences of large areas in glaciers, without involving subglacial or englacial water outbursts from subglacial lakes as well as subglacial geothermal or volcanic activity. This conclusion can be extrapolated to other regions because the event does not seem to be specific to the Dålk Glacier or the Antarctic environment.

The subsidence destroyed a section of the sledge route from the Progress Station to the Zenit Airfield (cf. Fig. 19.4A and B). Several vehicles passed this section 2 h before the event. Thus, the study of such phenomena is important for ensuring the safety of people in Antarctica as well as ice sheets and glaciers in other regions.

Chapter 20

Antarctic oases

20.1 Motivation

Antarctic oases are glacier-free areas of the Antarctic coastal zone characterized by the presence of three essential features: (1) a local climate largely determined by the adjacent or surrounding ice sheet; (2) unfrozen water in the form of seasonal streams and unfrozen lakes; and (3) primitive soils, plants, and biota (Simonov, 1971; Sokratova, 2010). Despite their relatively small sizes and total area (from 0.03% to 1% of the Antarctic area depending on different estimations), Antarctic oases are logistically and scientifically important territories. First, Antarctic oases are suitable for the construction of year-round stations and seasonal bases because of their relative accessibility. Second, Antarctic oases attract researchers by the possibility of obtaining a considerable amount of scientific material in a short period of austral summer (Sokratova, 2010).

The prospects of geomorphometric modeling and mapping of Antarctic oases are quite obvious, both for obtaining new fundamental knowledge about quantitative characteristics of the topography of these unique terrains and for further use of the obtained morphometric information for solving fundamental and applied tasks of geomorphology, geology, geophysics, glaciology, climatology, etc. Nevertheless, geomorphometric modeling and mapping of Antarctic oases have not been performed so far.

This chapter presents the first results of geomorphometric modeling and mapping of the Larsemann Hills, one of the most popular Antarctic oases, obtained during the 68th Russian Antarctic Expedition (RAE) in January-April 2023 (Florinsky, 2023a,b). The goals of this chapter are: (1) to expose the polar scientific community to geomorphometric methods (computer modeling and field interpretation) in the context of studying Antarctic oases; (2) to present a system of large-scale morphometric maps derived, for the first time, for the Larsemann Hills; and (3) to discuss possible interpretation and application of these maps.

20.2 Study area

The Larsemann Hills (Fig. 20.1) is an Antarctic coastal oasis including ice-free, low-rounded hills with a total area of about 40 km^2 located on the southeastern coast of the Prydz Bay of the Commonwealth Sea, Ingrid Christensen Coast, Princess Elizabeth Land, East Antarctica. The oasis consists of 2 large peninsulas (Stornes and Broknes Peninsulas), 3 small peninsulas (Grovnes, Brattnevet, and Stinear Peninsulas), and about 130 islands including 3 relatively large ones (Fisher, Manning, and McLeod Islands). The peninsulas and islands are separated by a system of fjords and bays, such as Thala, Nella, and Clemence Fjords, as well as Dålkoy, Wilcock, and Quilty Bays. The oasis is bounded on the east by the Dålk outlet glacier, on the south by the ice plateau, and on the west by the Polarårboken outlet glacier.

The Larsemann Hills are mainly composed of Proterozoic garnet-biotite gneisses and garnet-sillimanite schists; there are also intrusions of pegmatites and granites (Stüwe et al., 1989; Carson and Grew, 2007; ATCM, 2021). Unique borosilicate and phosphate complexes are developed in the Stornes Peninsula (ATCM, 2014).

The Larsemann Hills were deglaciated between 20,000 and 10,000 years ago (Burgess et al., 1994; Hodgson et al., 2001; Verleyen et al., 2004; Kiernan et al., 2009). The oasis topography is a structurally controlled, exarated rocky hill terrain (Bolshiyanov, 2011; Biryukov, 2022). The elongated shape of the large landform elements of the Larsemann Hills is generally related to compositional layering and the presence of folds and faults in the metamorphic bedrock (ATCM, 2021). Nivation dominates among exogenous topography-forming processes. The landscape is rugged with large, structurally controlled fjords and valleys with steep slopes that are rarely deeper than 100 m on land and have a maximum length of 3 km. Elevations range from 30 to 120 m. The maximum height above sea level is 158 m (Blundell Peak of the Stornes Peninsula).

There are about 150 lakes within the oasis (Gillieson et al., 1990), but the drainage network is not developed. Snowfields are of widespread occurrence in the oasis. There is a small passive glacial dome on the Stornes Peninsula.

Digital Terrain Analysis. https://doi.org/10.1016/B978-0-443-24798-9.00012-1

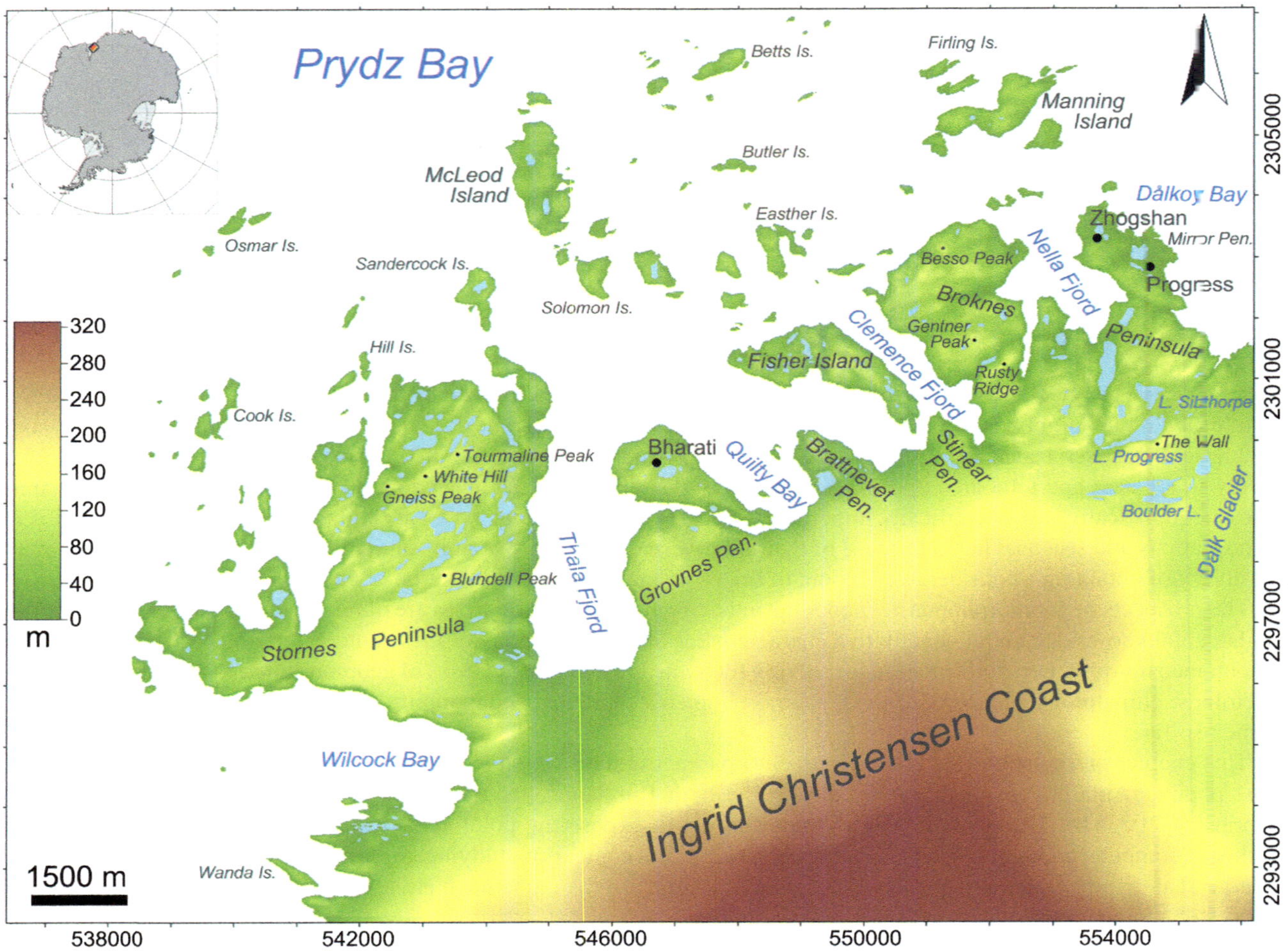

FIG. 20.1 The geographic location and elevations of the Larsemann Hills extracted from the REMA DEM. *(From Florinsky, I.V., 2023b. Larsemann Hills: geomorphometric modeling and mapping. Pol. Sci. 38, 100969, Fig. 1, reproduced with permission from Elsevier.)*

The Larsemann Hills is the Antarctic Specially Managed Area No. 6 (ATCM, 2021), while its largest portion, the Stornes Peninsula, is the Antarctic Specially Protected Area No. 174, which is a sort of mineralogical reserve (ATCM, 2014). There are three year-round operated polar stations in the oasis: Progress (Russia), Zhongshan (China), and Bharati (India) (Fig. 20.1). There is the Zenit snow-ice airfield (Russia) on the glacial plateau, to the south of the Progress Station.

20.3 Materials and methods

20.3.1 Geomorphometric modeling

We used the Reference Elevation Model of Antarctica (REMA) (Howat et al., 2019; REMA, 2018–2022) as input data for geomorphometric calculations, modeling, and mapping. To date, this DEM obtained photogrammetrically from WorldView-1, WorldView-2, and WorldView-3 satellite images is the most complete and accurate DEM of Antarctica. It describes both the glacial cover and glacier-free areas—oases, mountain ranges, nunataks, and islands (so, strictly speaking, REMA is DSM rather than DEM). Our preliminary assessment of the accuracy and quality of the REMA showed that within the Larsemann Hills, it generally meets the requirements of geomorphometric modeling, namely, low level of high-frequency noise, low number of obvious artifacts, and sufficient smoothness, which do not require additional smoothing or filtering.

For the preprocessing and editing of the DEM, geomorphometric calculation, and mapping, we used the System for Automated Geoscientific Analyses (SAGA) version 7.8.2 software (Conrad et al., 2015).

From the REMA ver. 1.1, we extracted a fragment of 18,168 m × 14,968 m, which includes the Larsemann Hills oasis and adjacent areas of the ice sheet and outlet glaciers. The extracted DEM contains 4,253,184 points (the elevation matrix of 2272 × 1872); the DEM grid spacing is 8 m. Since the REMA does not describe the submarine topography, the extracted DEM includes 2,068,674 cells with elevation values.

The REMA is presented in the polar stereographic projection and referenced to the WGS84 ellipsoid horizontal datum (Polar Geospatial Center, 2023). Before further processing, the extracted DEM was reprojected into the UTM projection, zone 43S (Fig. 20.1). The reprojected DEM also has a grid spacing of 8 m.

At the next stage, we manually edited the extracted DEM in order to remove island-like artifacts, which actually represent icebergs. As a reference, we used the available topographic map (Australian Antarctic Division, 2005).

From the edited DEM of the Larsemann Hills (Fig. 20.1), we derived digital models of 17 morphometric characteristics, namely: 14 local morphometric variables, such as G, A, k_h, k_v, E, k_{he}, k_{ve}, K_a, K_r, k_{min}, k_{max}, H, K, and M; one nonlocal morphometric variable, CA; as well as two combined morphometric variables: TWI and SPI (Tables 2.1 and 2.2).

It should be noted that the capabilities of SAGA allow the direct computation of only a part of the mentioned morphometric variables. In particular, we directly calculated digital models of six local morphometric variables using the Evans-Young method (Evans, 1972) (Section 4.1): G, A, k_h, k_v, k_{min}, and k_{max}. Digital models of the remaining eight local morphometric variables—E, k_{he}, k_{ve}, K_a, K_r, H, K, and M—were sequentially calculated using their formulas (Table 2.2) and the SAGA option allowing arithmetic operations with raster layers. Also, we derived CA, TWI, and SPI using a digital model of "total catchment area" directly computed in SAGA. In calculating TWI and SPI, we used the previously derived model of G.

The REMA does not include the lake bathymetry. However, DEM cells relating to lakes are not indicated as "no data value" like submarine topography cells. Lake cells include "elevation" values obtained by interpolation of lakeshore elevations; that is, these values are artifacts of data processing. Due to this fact, we masked all lakes on all maps. To create the mask, we utilized the available topographic map (Australian Antarctic Division, 2005).

Finally, we produced maps of the morphometric variables from the calculated models (Figs. 20.2–20.9). All maps are presented in the UTM projection, zone 43S.

20.3.2 Field geomorphometric interpretation

The main goal of field geomorphometric interpretation is to provide correct physical geographic, geological, and geomorphological interpretations of morphometric maps. The technology of field works is based on traversing a network of foot routes over the territory for which the morphometric maps are produced. For the spatial resolution of morphometric maps of the Larsemann Hills calculated from the REMA with the grid spacing of 8 m, the distance between routes can be 300–500 m depending on the type of topography, the slope steepness, the type of surface (e.g., rocks, kurums, snowfields), as well as the presence of lakes or other landscape elements affecting the possibility and speed of movement on foot.

The distance between routes was estimated as follows. According to one of the consequences of the sampling theorem (Section 3.3), a DEM with a grid spacing of w may correctly represent landforms with minimal typical sizes of $4–6 \times w$. Therefore, both the REMA with the grid spacing of 8 m and morphometric models derived from this DEM can describe digitally landforms with minimal typical sizes ranging from 32 to 48 m. Thus, the minimal distance between geomorphometric field routes in the Larsemann Hills can be taken as 30–50 m. However, a reasonable distance between routes should be estimated taking into account the following factors: (1) typical sizes of landforms within the oasis; (2) its total area; (3) terrain roughness affecting the movement on foot; (4) limited time of field works in the austral summer; and (5) a number of researchers in the field. Considering all these factors, we decided to apply a reasonable distance between routes 10 times larger than the minimal distance, that is, 300–500 m.

In conditions of very rugged terrain and Antarctic climate in the austral summer, the optimal length and duration of a foot route of geomorphometric interpretation are 6–10 km and 5–8 h, respectively. In this mode, it is possible to carry out field geomorphometric interpretation daily for a long time (a month or more) without deterioration of work quality because of physical or psychological fatigue. The planned foot route can be corrected in the course of its passage depending on real terrain and weather conditions. In particular, light snowfalls and winds up to 10 m/s at air temperatures ranging from +5°C to −5°C do not generally hamper the passage of the route (snow is blown from slopes and ridges and usually melts on the next day).

A heavy multiday snowfall at air temperatures ranging from −5°C to −15°C can significantly complicate or indefinitely suspend the field geomorphometric works because the capability of traversing the terrain becomes problematic due to the emerging loose snow cover (up to half a meter or thicker in valleys and depressions). However, wind with speeds of 30–40 m/s, as a rule, removes loose snow from slopes in a matter of hours, while several sunny days with air temperatures

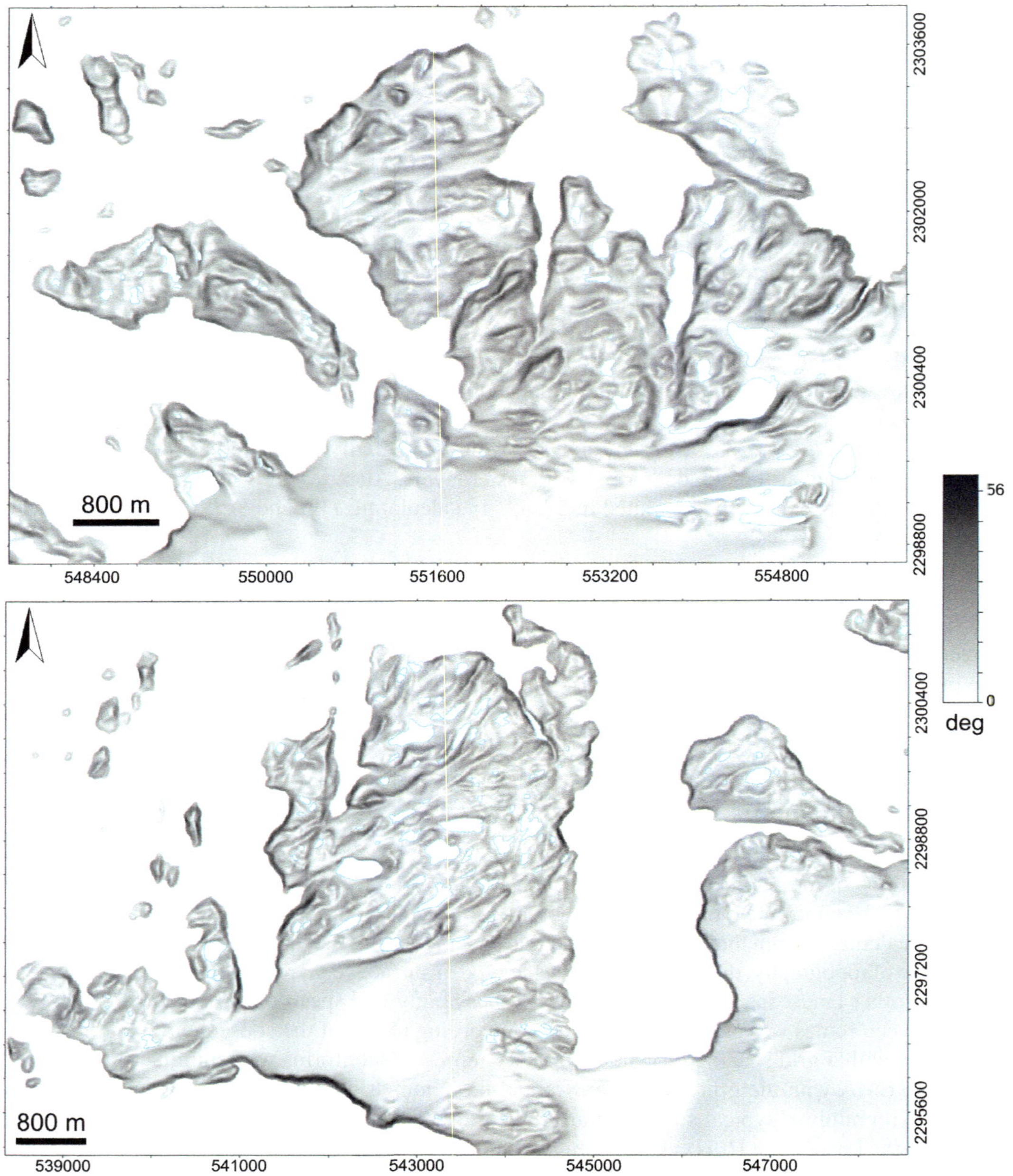

FIG. 20.2 Larsemann Hills, slope gradient. Upper: Broknes Peninsula. Lower: Stornes Peninsula. *(From Florinsky, I.V., 2023b. Larsemann Hills: geomorphometric modeling and mapping. Pol. Sci. 38, 100969, Fig. 2, reproduced with permission from Elsevier.)*

of about −20°C lead to the formation of a firm snow layer in the valleys, depressions, and snowfields. As a result, the capability of traversing the terrain resumes.

During the field geomorphometric interpretation, we made a comparative analysis of the elements of the real topography and their representation on the morphometric maps (an approach to visual observation). During the austral summer in the Larsemann Hills (from 6 January till April 22, 2023), we performed 54 geomorphometric interpretation foot routes with a total length of about 422 km (the route schemes can be found elsewhere—Florinsky, 2023b).

20.4 Results

One can see the most informative morphometric maps of the Larsemann Hills in Figs. 20.2–20.9.

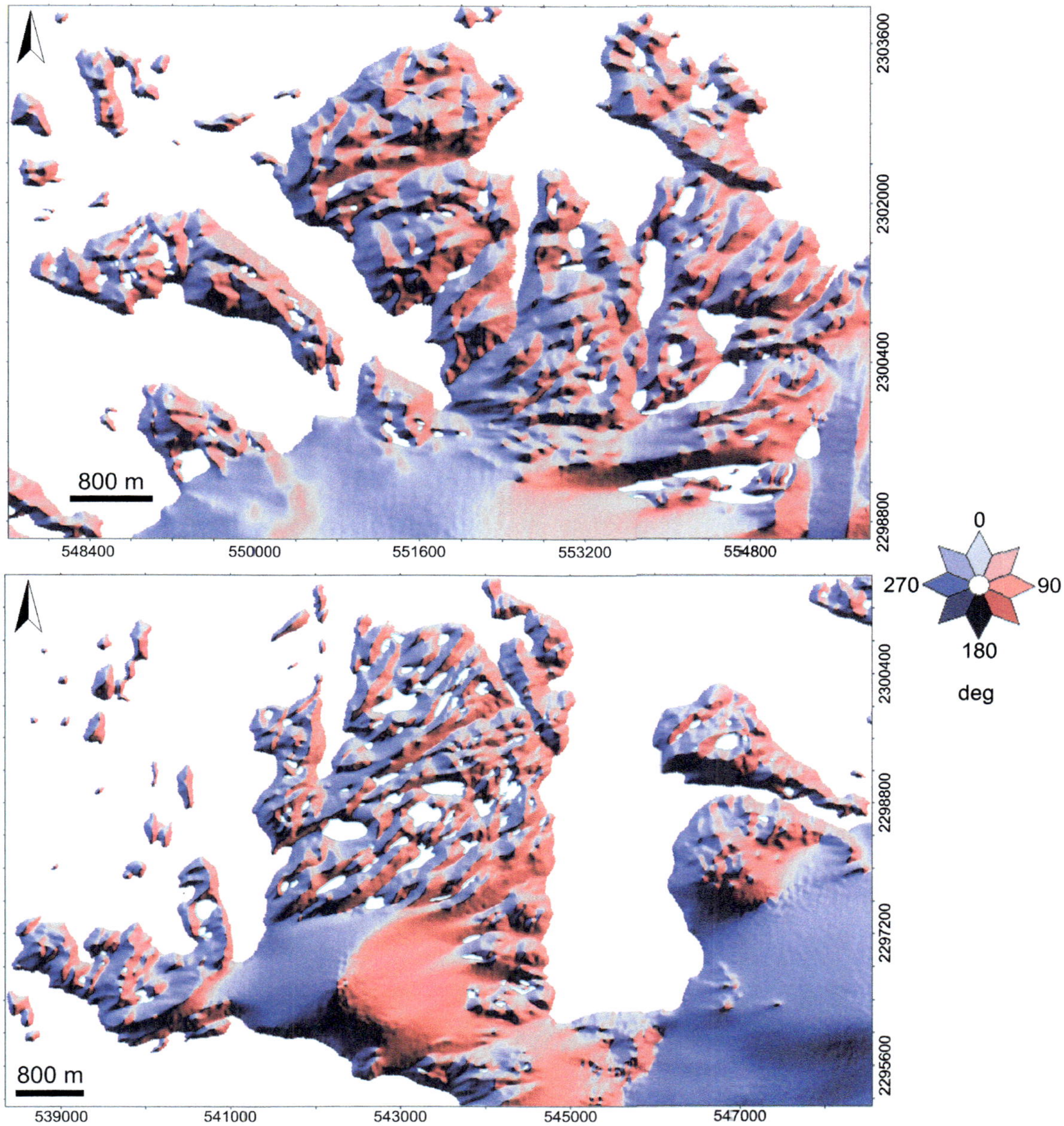

FIG. 20.3 Larsemann Hills, slope aspect. Upper: Broknes Peninsula. Lower: Stornes Peninsula. *(From Florinsky, I.V., 2023b. Larsemann Hills: geomorphometric modeling and mapping. Pol. Sci. 38, 100969, Fig. 3, reproduced with permission from Elsevier.)*

The spatial distribution of slope steepness is clearly shown in the *G* map (Fig. 20.2). The steepest slopes are typical for the Wall Mount (southern Broknes Peninsula), Rusty Ridge, and Castle Bluff (western Broknes Peninsula), the unnamed mountain south of the Canterbury Hill, and the coastal slopes of the northwestern Broknes and southern Stornes Peninsulas, where they coincide with the edge cliff of the passive glacial dome. The hill slopes in the eastern Fisher Island are also steep. It should generally be noted that in spite of the low elevations of the Larsemann Hills (the maximum height above sea level is 158 m at the Blundell Peak of the Stornes Peninsula) and the use of the term "hills" to describe the local topography, in many parts of the oasis the terrain looks like steep, rocky mountains, and some hills are called peaks.

The *A* map (Fig. 20.3) shows the spatial distribution of slope aspects. A large part of the slopes of the structurally predetermined elongated hills of the Broknes and Stornes Peninsulas are oriented to the northwest and southeast, corresponding to the general lineament network of the region. The hill slopes of the Grovnes, Brattnevet, and Stinear Peninsulas, as well as Fisher Island, which seem to wedge between the Broknes and Stornes Peninsulas, are oriented

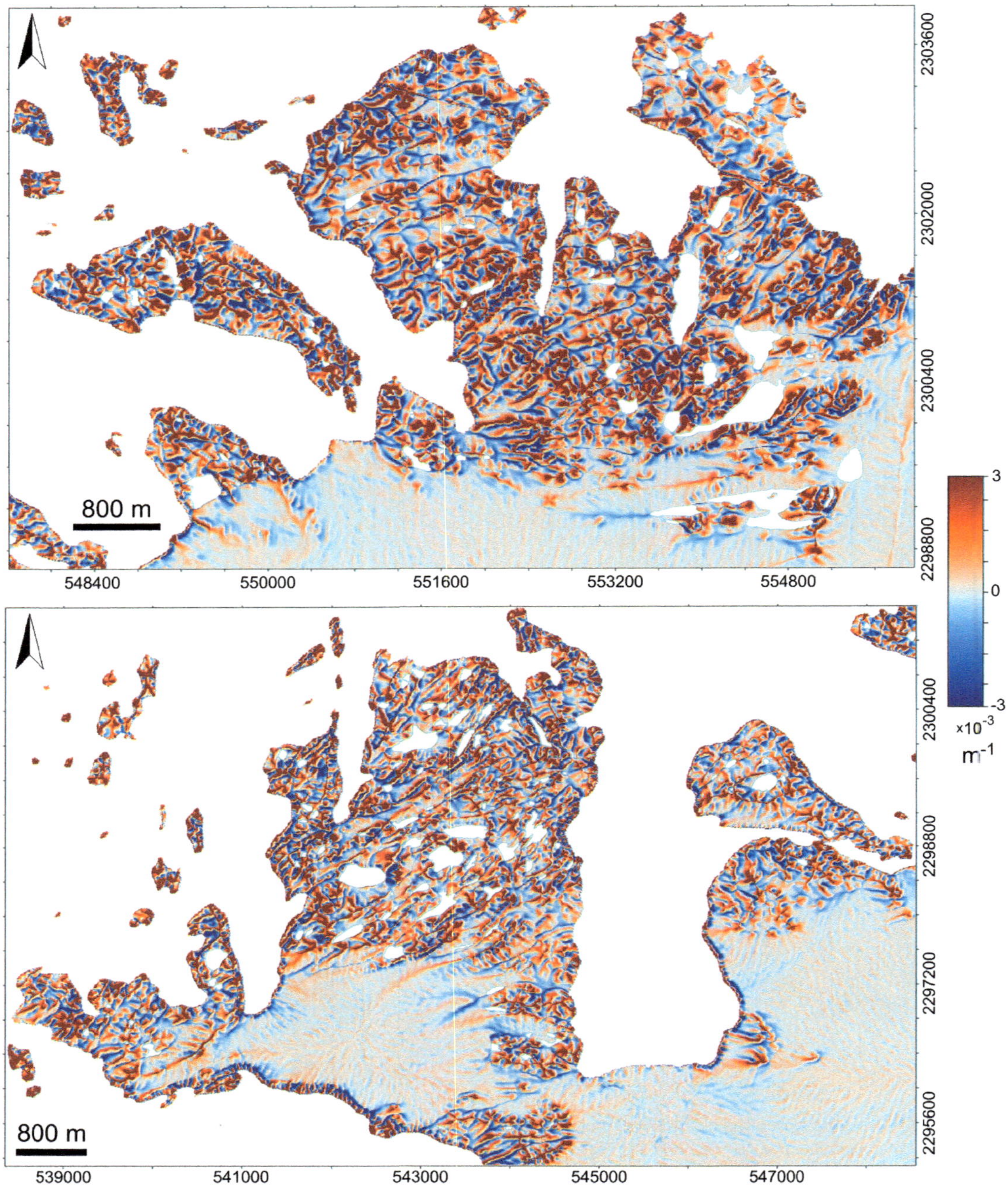

FIG. 20.4 Larsemann Hills, horizontal curvature. Upper: Broknes Peninsula. Lower: Stornes Peninsula. *(From Florinsky, I.V., 2023b. Larsemann Hills: geomorphometric modeling and mapping. Pol. Sci. 38, 100969, Fig. 4, reproduced with permission from Elsevier.)*

predominantly to the southwest and northeast. This may serve as an indirect indication of the difference in the geological origin of these two groups of territories.

The k_h map (Fig. 20.4) shows the spatial distribution of convergence and divergence zones of surface flows ($k_h < 0$ and $k_h > 0$, respectively); in the geomorphological sense, these are spurs of valleys and ridges (blue and brown colors, respectively) (Section 2.2.4.2). In combination, the convergence and divergence zones represent the flow structure of the terrain, which is structurally predetermined here. However, the drainage network of the Larsemann Hills is poorly developed, and it is hard to find classical, long and vast flow structures with different patterns (correlated with different geological properties of the landscape), which can be easily recognized in other territories (Stepanov, 1989; Kuryakova and Florinsky, 1991).

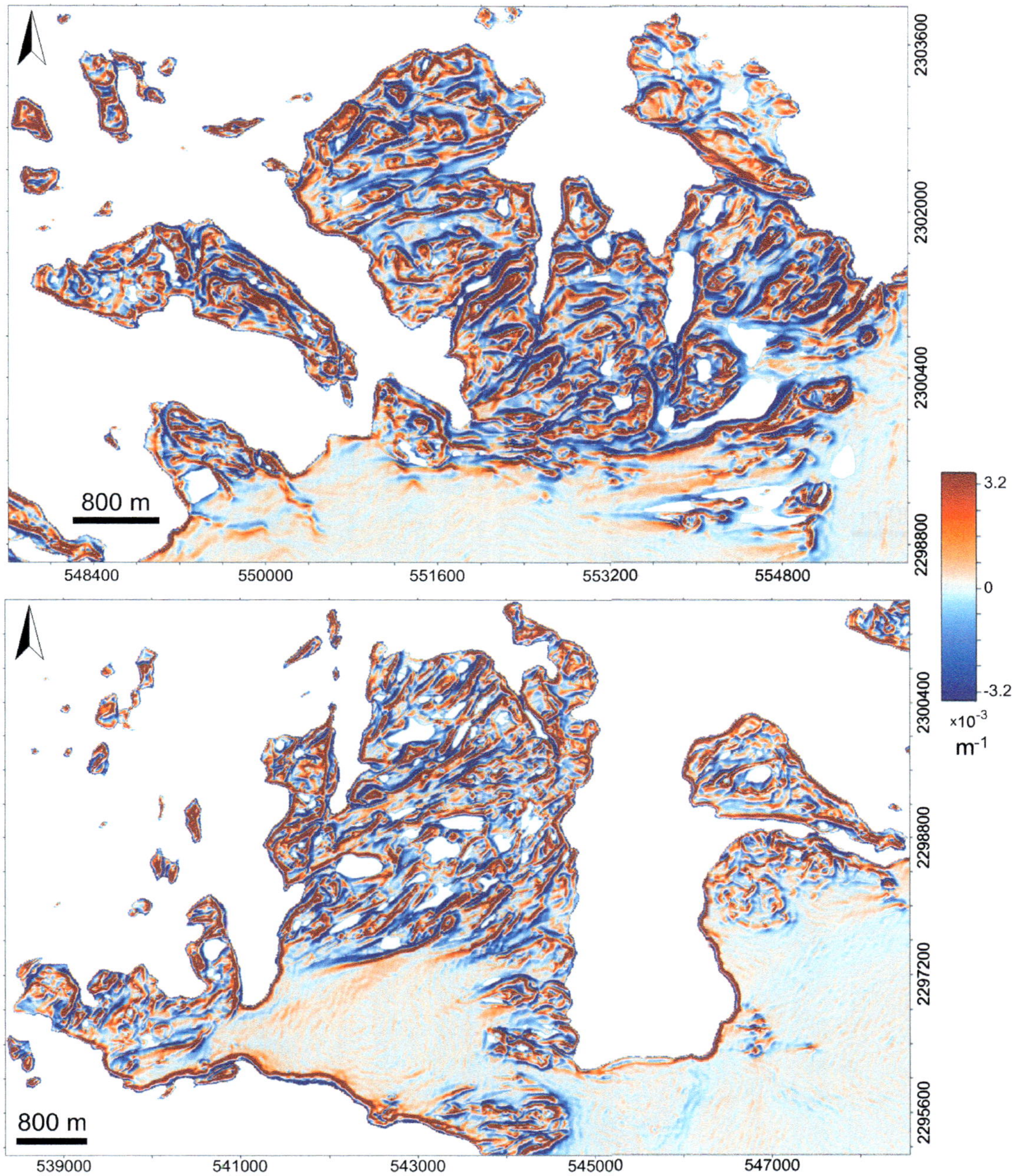

FIG. 20.5 Larsemann Hills, vertical curvature. Upper: Broknes Peninsula. Lower: Stornes Peninsula. *(From Florinsky, I.V., 2023b. Larsemann Hills: geomorphometric modeling and mapping. Pol. Sci. 38, 100969, Fig. 5, reproduced with permission from Elsevier.)*

The k_v map (Fig. 20.5) shows the spatial distribution of zones of relative deceleration and acceleration of surface flows ($k_v < 0$ and $k_v > 0$, respectively). Geomorphologically, this map shows areas of sharp change of elevation (e.g., scarps) (Section 2.2.4.3). The geological structure of the territory is also well readable on this map. One can clearly see the southern geological edge of the Broknes Peninsula (Fig. 20.5, upper), which is topographically an escarp with north, northeast, and northwest steep slopes, including the southern cliffs of the Wall Mount, their western prolongation, and the northeastern steep slopes of the Stinear Peninsula and Fisher Island, as well as a canyon between the Dålkoy Bay and a plain of the Lakes Progress and Sibthorpe.

Also, k_v map (Fig. 20.5, upper) clearly demonstrates that the Mirror Peninsula is separated from the rest part of the Eastern Broknes by a probable dip-slip fault with an escarp oriented to the southwest. The Western Broknes is separated

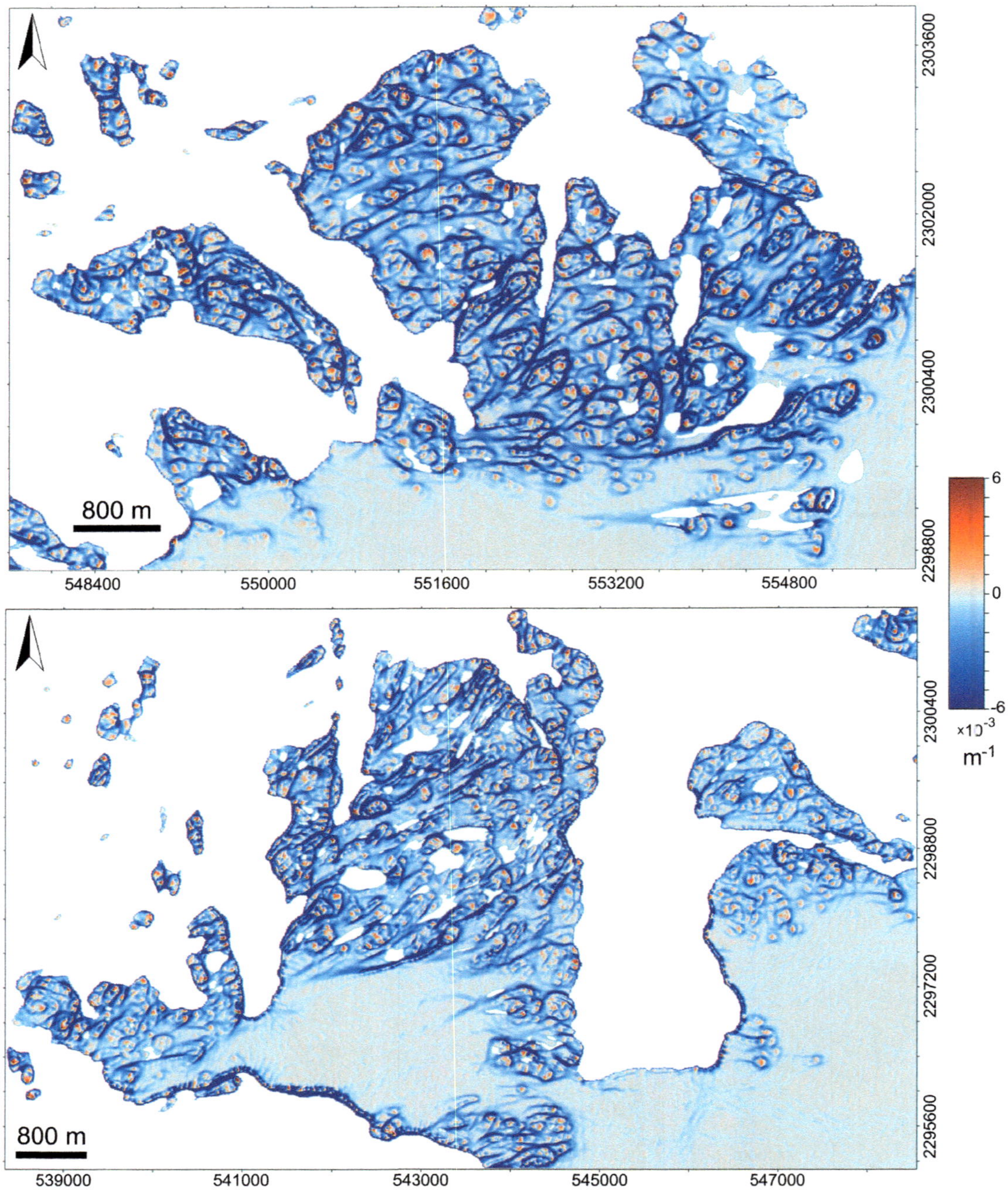

FIG. 20.6 Larsemann Hills, minimal curvature. Upper: Broknes Peninsula. Lower: Stornes Peninsula. *(From Florinsky, I.V., 2023b. Larsemann Hills: geomorphometric modeling and mapping. Pol. Sci. 38, 100969, Fig. 6, reproduced with permission from Elsevier.)*

from the rest territory of the oasis by a probable gaping fault topographically manifested as the deep intermountain valley between the Rusty Ridge and Gentner Peak.

On the Stornes Peninsula (Fig. 20.5, lower), a key escarp of the northwestern orientation, probably corresponding to a dip-slip fault, separates the northern (generally lower) and central (generally higher) portions of this peninsula. The escarp includes the northwestern steep slopes of the Gneiss Peak, White Hill, and Tourmaline Peak, that is, the northwestern edge of a 4-km southwest-to-northeast-striking belt, where rare phosphate and borosilicate minerals (e.g., prismatine) are observed (Carson and Grew, 2007).

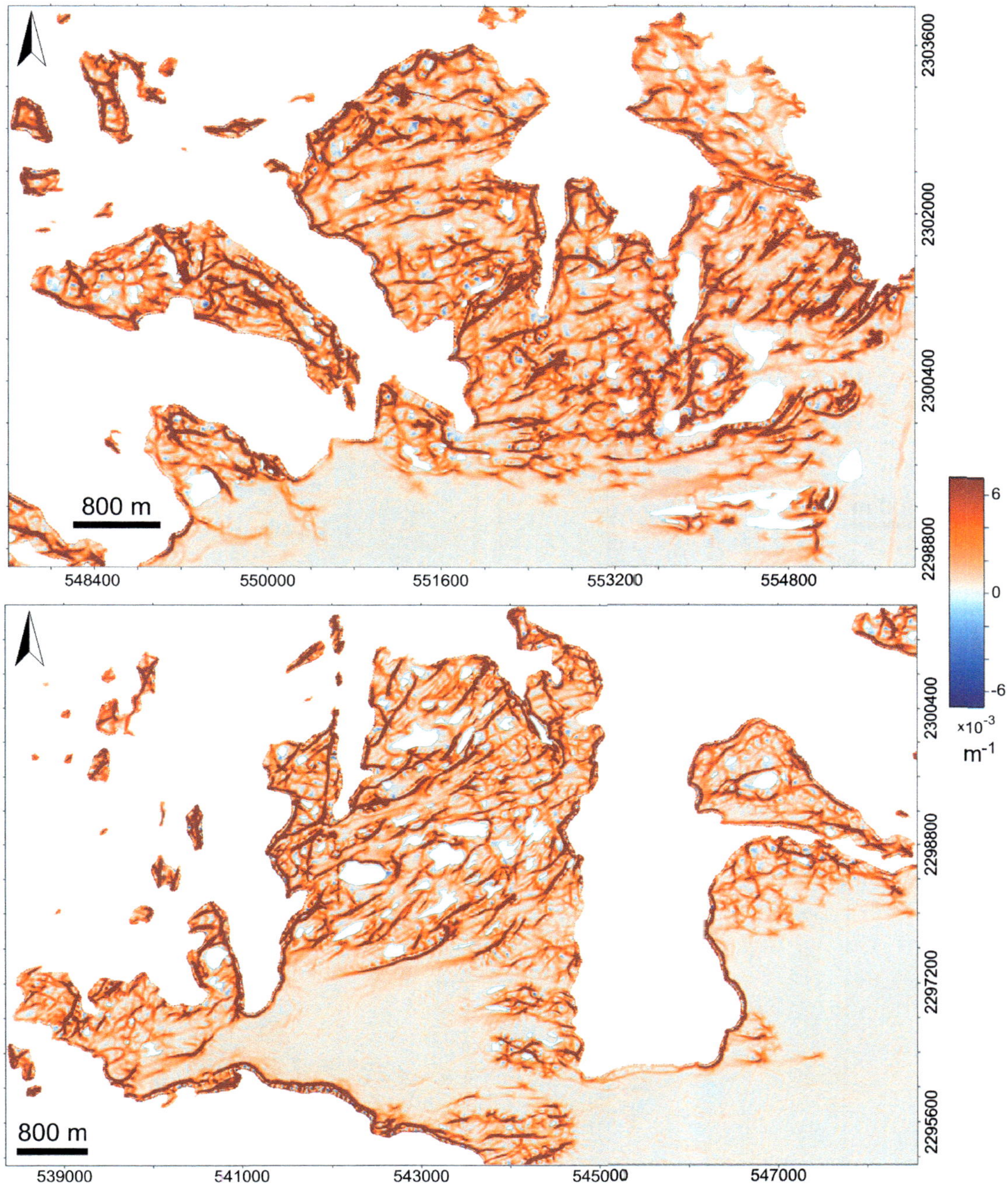

FIG. 20.7 Larsemann Hills, maximal curvature. Upper: Broknes Peninsula. Lower: Stornes Peninsula. *(From Florinsky, I.V., 2023b. Larsemann Hills: geomorphometric modeling and mapping. Pol. Sci. 38, 100969, Fig. 7, reproduced with permission from Elsevier.)*

It is reasonable to consider maps of k_{min} and k_{max} (Figs. 20.6 and 20.7) together (Section 2.2.4.5). Their analysis reveals two types of elongated linear forms: concave (e.g., valleys) represented by dark blue lines (Fig. 20.6, $k_{min} < 0$) and convex (e.g., ridges) represented by dark brown lines (Fig. 20.7, $k_{max} > 0$). In terms of revealing lineaments (faults, folds, etc.) of the oasis, the k_{min} and k_{max} maps are the most informative material because all the structures are visible on these maps with the naked eye, and their identification does not require additional data processing.

CA, the nonlocal morphometric variable, is a measure of the contributing area (Section 2.3.1). On the *CA* map (Fig. 20.8), ridges and valleys are clearly distinguished as light and dark lines (low and high *CA* values), respectively.

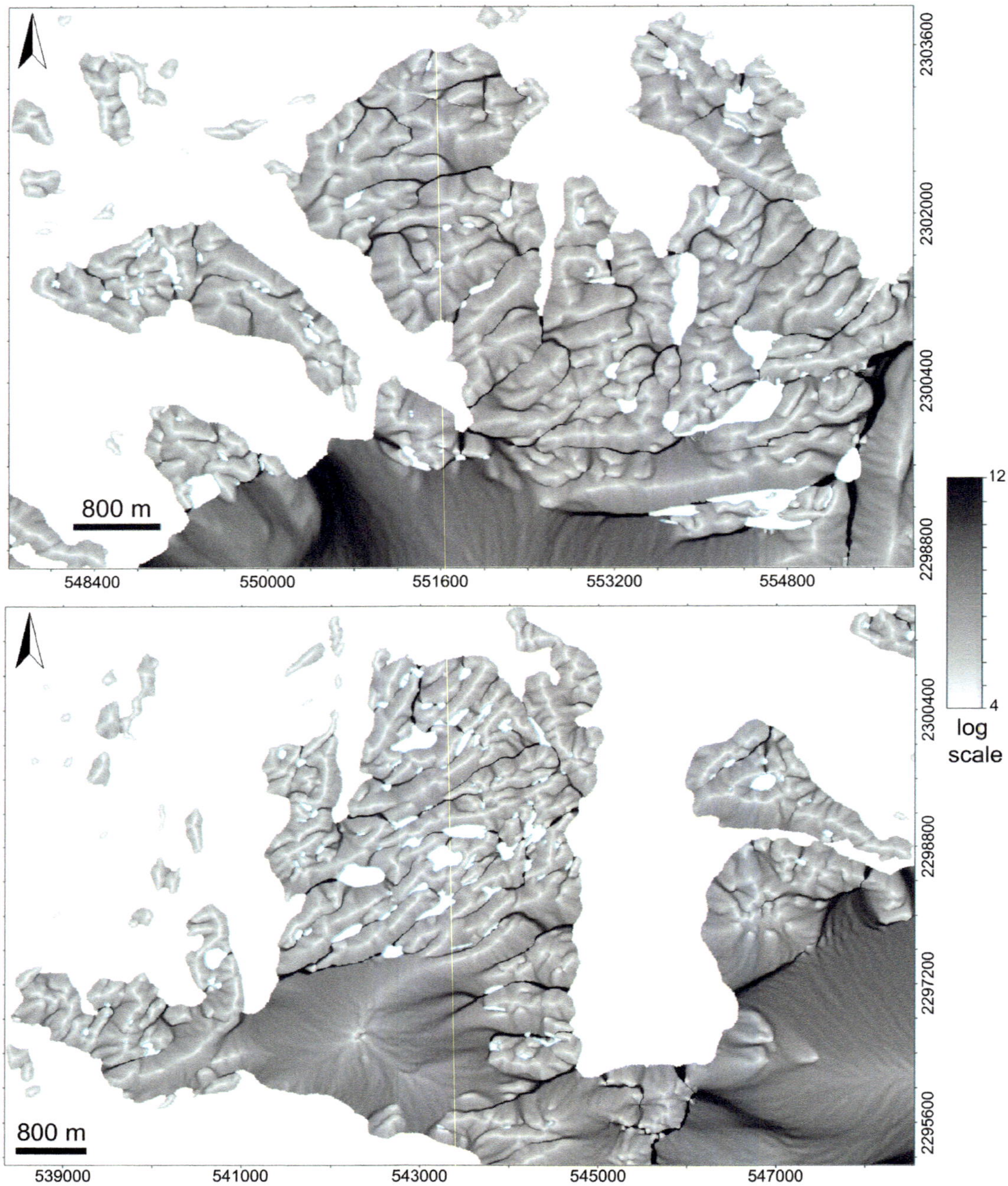

FIG. 20.8 Larsemann Hills, catchment area. Upper: Broknes Peninsula. Lower: Stornes Peninsula. *(From Florinsky, I.V., 2023b. Larsemann Hills: geomorphometric modeling and mapping. Pol. Sci. 38, 100969, Fig. 8, reproduced with permission from Elsevier.)*

TWI, the combined morphometric variable, is a measure of the potential degree of water accumulation in the landscape (Section 2.6.1). *TWI* reaches high values in areas with high values of *CA* at low values of *G* (e.g., terrain with a large upslope contributing area and flat local topography) (Fig. 20.9).

20.5 Discussion

The morphometric models and maps of the Larsemann Hills can, first of all, be useful for revealing and mapping topographically manifested faults and folds (Chapters 13 and 14), which are represented as lineaments on the curvature maps (Figs. 20.4–20.7). As noted in Section 20.4, the k_{min} and k_{max} maps (Figs. 20.6 and 20.7) are the most useful for this purpose

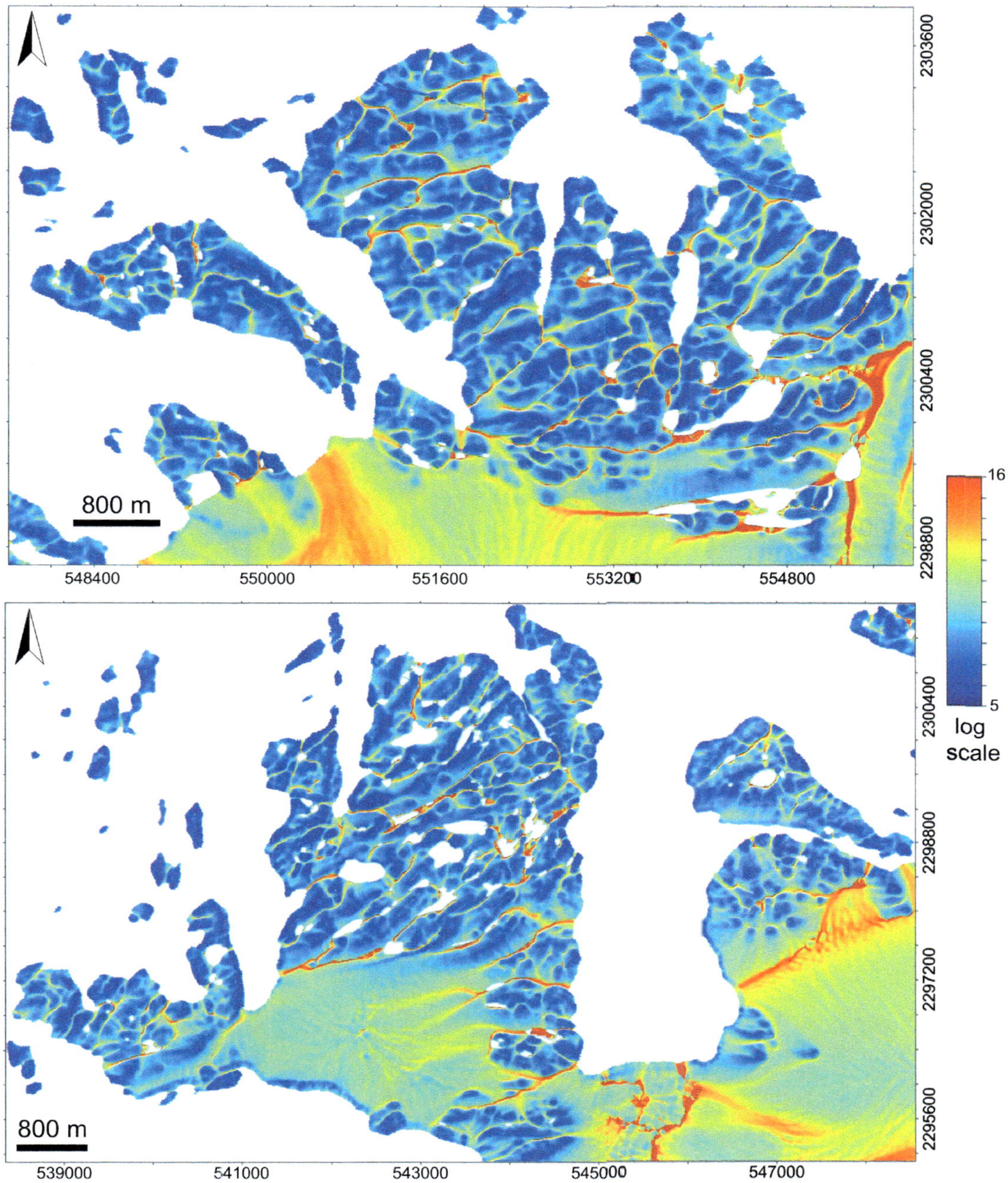

FIG. 20.9 Larsemann Hills, topographic wetness index. Upper: Broknes Peninsula. Lower: Stornes Peninsula. *(From Florinsky, I.V., 2023b. Larsemann Hills: geomorphometric modeling and mapping. Pol. Sci. 38, 100969, Fig. 9, reproduced with permission from Elsevier.)*

in the topographic conditions of this region. The maps of other curvatures can be used to clarify position, geometry, and types of faults and folds recognized on the k_{min} and k_{max} maps (Fig. 20.10). From the geological point of view, the oasis is generally open: first, there is no soil and vegetation cover; second, Quaternary deposits are found in only some valleys and depressions; and third, bedrock outcrops are observed everywhere (snowfields are the only problem for field geological studies, especially at the Stornes Peninsula). Therefore, field verification of the revealed lineaments cannot, in most cases, cause serious difficulties.

The combined use of k_h and k_v models (Figs. 20.4 and 20.5) allows identifying zones of relative accumulation of surface flows, that is, areas characterized by both $k_h < 0$ and $k_v < 0$ (Section 2.7.2). These zones coincide with the sites of fault intersection and are marked by an increased fragmentation and permeability of rocks (Chapter 15). In these zones, there

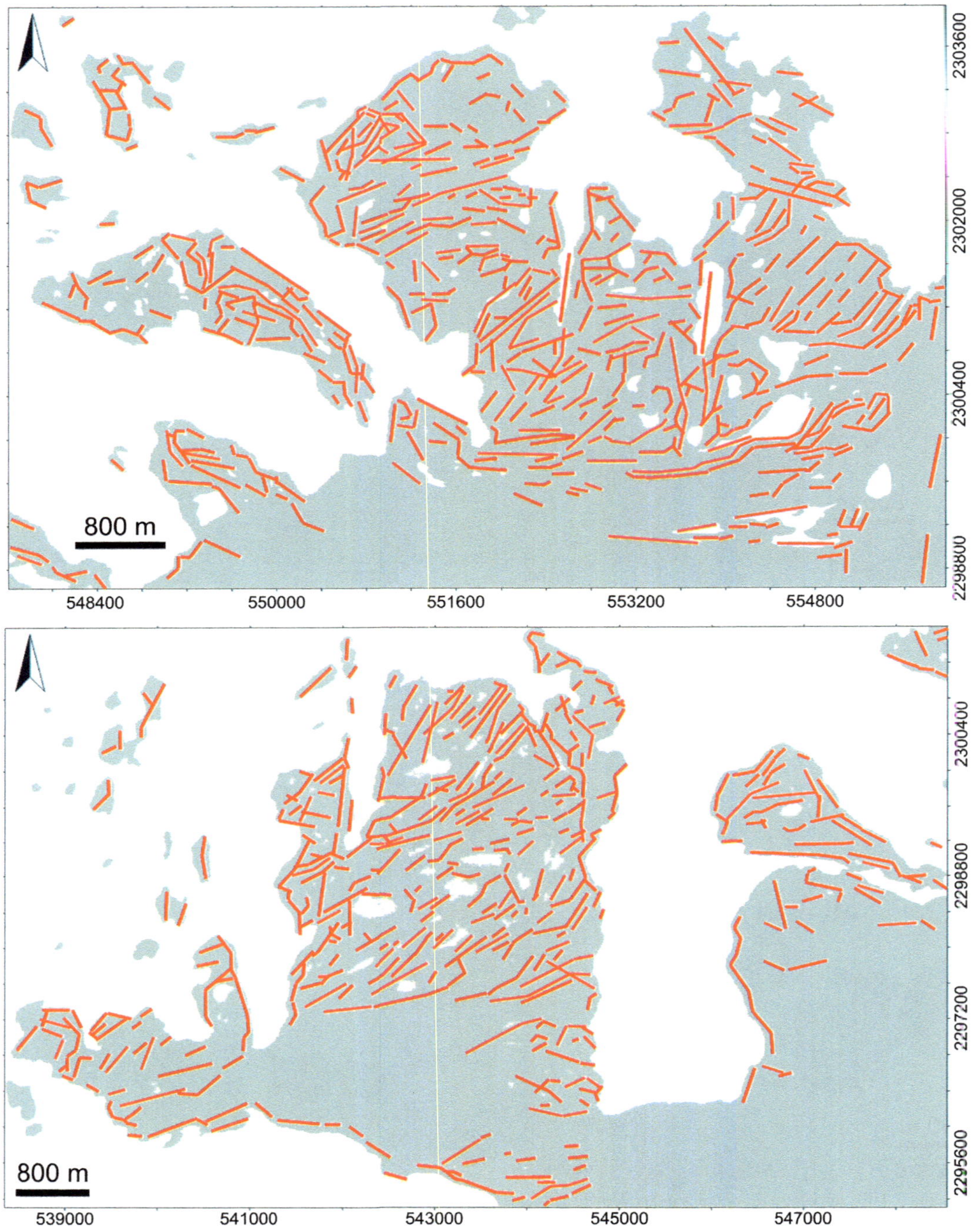

FIG. 20.10 Larsemann Hills, lineaments revealed from k_{min} and k_{max} maps (Figs. 20.6 and 20.7, respectively). Upper: Broknes Peninsula. Lower: Stornes Peninsula. *(From Florinsky, I.V., 2023b. Larsemann Hills: geomorphometric modeling and mapping. Pol. Sci. 38, 100969, Fig. 10, reproduced with permission from Elsevier.)*

is an interaction and exchange between two types of substance flows: (1) lateral, gravity-driven substance flows moved along the land surface and in the near-surface layer (e.g., water, dissolved and suspended substances) and (2) vertical, upward substance flows (e.g., fluids, groundwater of various mineralizations and temperatures). The map of accumulation zones derived from the k_h and k_v models (Fig. 20.11, lower) could be useful for geochemical and mineralogical studies of the oasis, especially the Stornes Peninsula, where one can observe borosilicate and phosphate complexes, the origin of which is not completely clear.

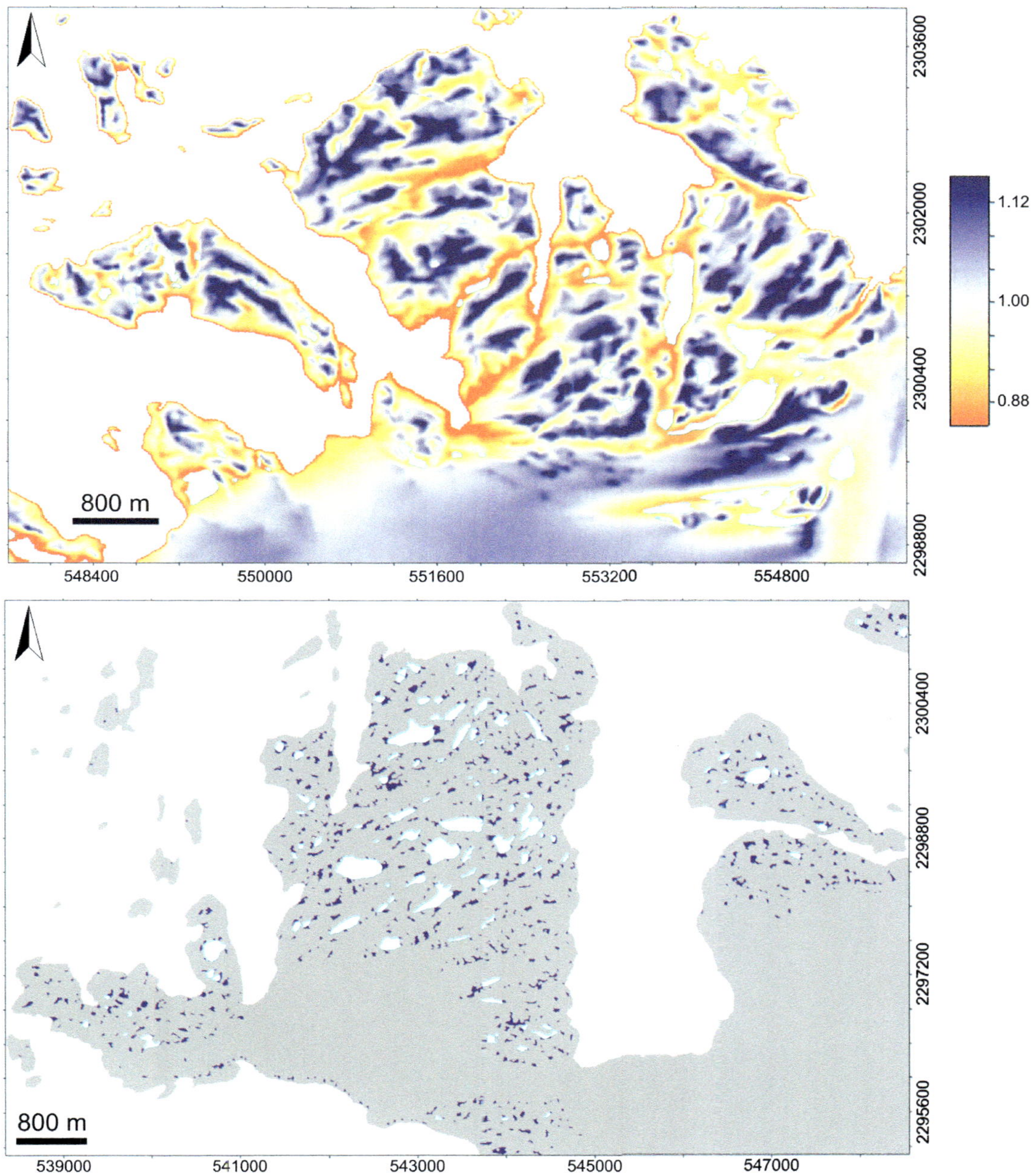

FIG. 20.11 Larsemann Hills. Upper: Broknes Peninsula, wind exposition index (values <1 indicate wind-shadowed areas; values >1 indicate areas exposed to wind). Lower: Stornes Peninsula, zones of relative flow accumulations derived from k_h and k_m models (Figs. 20.4 and 20.5, respectively). Data were processed by the SAGA software (Conrad et al., 2015). *(From Florinsky, I.V., 2023b. Larsemann Hills: geomorphometric modeling and mapping. Pol. Sci. 38, 100969, Fig. 11, reproduced with permission from Elsevier.)*

It is well known that topography largely determines the thermal, wind, and hydrological regimes of slopes, thus controlling the distribution of soils and vegetation covers and their properties (Chapter 9). In particular, solar morphometric variables (Section 2.5.1) can be used to model and predict the thermal characteristics of landscapes. Wind morphometric attributes (Section 2.5.2) can be utilized to predict spatial differentiation of snow accumulation and to identify areas affected by and protected from wind impact. This can, for example, be estimated using wind exposition index (Fig. 20.11, upper), a measure of the average exposition of slopes to wind flows in all possible wind directions at a given point on the topographic surface.

Such information is important for soil, plant, and ecological studies of this region, where one of the main meteorological factors determining the microclimate is a strong periodic katabatic wind. Information on the differentiation of slopes by insolation level and on areas protected from wind impact may be useful to clarify and predict the spatial distribution of primitive soils (Mergelov, 2014; Dolgikh et al., 2015) and plants, such as mosses, lichens, and algae (Gupta, 2015), in the oasis. To refine this prediction, one can use the *TWI* map (Fig. 20.9), which shows the topographic prerequisites of water migration pathways and areas of moisture accumulation in the landscape during the austral summer.

For the Larsemann Hills, it is urgent to monitor and predict outbursts of lakes, some of which are cascade systems (Boronina, 2022). On the Broknes Peninsula, such outbursts may particularly pose a threat to the transport connectivity of the Progress Station with the Zenit Airfield and Vostok Station. For the eastern and southeastern part of the Broknes Peninsula, the *TWI* map (Fig. 20.9, top) clearly shows potential floodwater discharge pathways (narrow red areas) from the Lake LH73 into the Lake Progress, then into the Lake Sibthorpe, and then through a canyon to the Dålkoy Bay, as well as from the Boulder Lake to the zone of the 2017 catastrophic subsidence in the Dålk Glacier (Chapter 19), and then toward the Dålkoy Bay. We suppose that the *TWI* model can be applied to reveal and analyze topographic prerequisites of floodwater discharge pathways along the lake cascades.

Chapter 21

Concluding remarks and pending problems

21.1 Geomorphometry today

By the end of the first quarter of the 21st century, geomorphometry had become a scientific discipline with a well-developed physical and mathematical theory (Chapter 2) and a powerful set of computational methods and algorithms (Chapter 4). The methods of digital terrain analysis can be applied to DEMs of any resolution produced by any technique. This allows modeling and analysis of topography at any hierarchical level in studies of any spatial scale. Reproducibility, relative simplicity, and flexibility of geomorphometric methods determine their practical potential for geosciences.

The last several years were marked by the development of technologies for unmanned aerial and underwater surveys (Section 1.3.1.1) as well as sophisticated methods for processing remote sensing data (Sections 1.3.1.3), in particular, structure-from-motion algorithms (Section 3.1.4). These achievements led to two consequences. First, hundreds of users got the opportunity to produce their own inexpensive, high-resolution DEMs of relatively small areas. Second, there was the emergence of new free, publicly available global, quasi-global, and regional DEMs derived from satellite imagery, such as Copernicus GLO-30, ALOS AW3D30, REMA, ArcticDEM, etc. (Section 1.3.1.2). These DEMs provide researchers with the data to carry out studies of different tasks at regional, continental, and global scales without additional costs.

These factors, as well as the adaptation of machine learning methods, have led to a rapid increase in the number and quality of studies implementing digital terrain analysis to solve multiscale problems in various geosciences (Section 1.3.2). It should be realized, however, that the governing factor for the evolution and current success of digital terrain analysis is the advancements in the theory of the topographic surface (Chapter 2), which is a rigorous physical and mathematical foundation for both computational algorithms and applied issues of topographic modeling.

21.2 Theory

To date, several dozens of morphometric variables have been described in literature, and many of them are known under different names (Behrens et al., 2005; Lecours et al., 2016a, 2017c; Favalli and Fornaciai, 2017; Amatulli et al., 2018, 2020; Newman et al., 2018; Wilson, 2018; Minár et al., 2020, 2024; Maxwell and Shabe, 2022; Temme et al., 2022). For the successful application of geomorphometric methods, a user should be oriented toward these variables, understand their physical and mathematical meaning, and know their interpretations. Thus, a classification of morphometric variables based on their physical and mathematical meaning (Section 2.1.3) can play an important role.

A family of roughness indices is widely used to study the topography of the Earth (Grohmann et al., 2011b; Trevisani et al., 2012, 2023a,b,c; Trevisani and Rocca, 2015; Trevisani and Cavalli, 2016; Groom et al., 2018; Korzeniowska et al., 2018) and other solid celestial bodies (Kreslavsky and Head, 2000; Kreslavsky et al., 2014; Fa et al., 2016; Ermakov et al., 2019; Deutsch et al., 2020). Roughness indices can be generally defined as statistical metrics of morphometric variables determined at each point of the topographic surface. Roughness indices remain outside of the classes of morphometric variables discussed in Chapter 2. This is because of the unclear physical meaning of roughness indices and the empirical nature of their interpretations.

A similar problem exists with the Haralick image textural characteristics (Haralick et al., 1973; Haralick, 1979), including contrast, correlation, variance, inverse difference moment (homogeneity), sum average, sum variance, sum entropy, entropy, difference variance, and difference entropy. An approach based on the use of these variables in image analysis was formally adapted to DEM analysis (Sathyamoorthy, 2013; Zhao et al., 2017; Ding et al., 2018; Ishalina et al., 2021). However, interpretations as well as physical and mathematical meanings of the Haralick textural characteristics still remain unclear in the topographic context.

Digital Terrain Analysis. https://doi.org/10.1016/B978-0-443-24798-9.00008-X

For the geomorphometric theory, it would be important to clarify the situation with the roughness indices and Haralick textural characteristics as well as generalize the classification of morphometric variables, taking into account these two groups of variables.

21.3 Data processing

Traditionally, DEM interpolation (Section 3.4), DEM denoising, and generalization (Sections 6.1.2 and 6.1.3), as well as computation of local morphometric variables (Sections 4.1–4.3) are separately solved using procedures implemented in different algorithms and even software. The progress in the theory and computational practice of polynomial approximation made it possible to combine these procedures within the universal spectral analytical method, which is based on high-order expansions of the Chebyshev polynomials of the first kind with the subsequent Fejér summation (Chapter 7). The universal spectral analytical method is intended for the processing of regularly gridded DEMs within a single framework, including DEM global approximation, smoothing, denoising, and generalization, as well as calculating the partial derivatives of elevation and local morphometric variables. The approximation is characterized by monotonic convergence and the possibility of fast, deep expansion of the elevation function (up to 50 million expansion coefficients). We believe that this method can become an effective alternative to common techniques of digital terrain analysis. Further progress in the theory of polynomial approximation (Pankratov and Tetuev, 2023) can lead to the generalization of the universal spectral analytical method to other orthogonal polynomial bases, such as the Chebyshev of the second kind, Legendre, Fourier, Jacobi, and Hermite polynomials.

Another alternative to the traditional scalar calculations of local morphometric variables is the vector calculation of these functions (Hu G. et al., 2020, 2021). It can allow obtaining smoother morphometric models than traditional finite-difference approaches described in Sections 4.1–4.3, and its calculation results depend less on the DEM errors and grid spacing.

Achievements in the development of calculation methods for local morphometric variables became possible largely due to the fact that a rigorous and clear physical and mathematical theory of topographic surface—an extension of the section of differential geometry devoted to surface curvature—was created 30 years ago (Shary, 1991, 1995). The development of the theory of nonlocal morphometry succeeded 20 years later (Gallant and Hutchinson, 2011). Various logical flow routing algorithms (Section 4.4) have been developed and applied to derive nonlocal morphometric variables from regular DEMs (Grimm et al., 2018; Wilson, 2018; Zhang et al., 2018) and point clouds (Rheinwalt et al., 2019). Despite an attempt to substantiate multiple flow direction algorithms mathematically using Manning-Strickler models for water flow (Coatléven, 2020), such algorithms are formal and therefore often lead to poorly reproducible and implausible results. The use of different algorithms for the same area leads to very different flow models (Huang and Lee, 2016; Clarke and Romero, 2017; Li Z. et al., 2020, 2021; Riihimäki et al., 2021). Analytical methods for calculating nonlocal topographic attributes (Qin et al., 2017; Bonetti et al., 2018; Li Z. et al., 2021), which implement and extend solutions of the Gallant-Hutchinson differential equation (Gallant and Hutchinson, 2011), should replace those obsolete algorithms.

A close situation has arisen with the extraction of thalweg networks (Section 4.5.1). The usual sequence of this procedure is as follows: (1) minor depressions are filled in a DEM (Barnes, 2016; Lindsay, 2016a; Wei et al., 2018; Xiong et al., 2019); (2) a model of catchment area is derived from the DEM using a flow routing algorithm; and (3) a threshold value is selected and applied to the catchment area model, providing visualization of thalwegs. The threshold choice formalization (Qin et al., 2016) and alternative ways of thalweg network derivation (Freitas et al., 2016; Hooshyar et al., 2016; Zhang et al., 2017a; Romero and Clarke, 2018; Zhou et al., 2019; Xu et al., 2020; Lyu et al., 2021) are not theoretically substantiated. Until a rigorous mathematical theory of structural morphometric variables (Section 2.4) is created, a correct and reproducible determination of the thalweg and ridge networks will be impossible.

Digital terrain analysis can correctly be applied to DEMs based on plain square grids (Sections 4.1, 4.2, and 4.4.1) as well as spheroidal equal angular grids on a sphere and an ellipsoid of revolution (Sections 4.3 and 4.4.2). Geomorphometric calculations are trivial when modeling the Earth, Mars, the Moon, Venus, and Mercury because the shape of these solid celestial bodies can be described by either an ellipsoid of revolution or a sphere. For these two surfaces, there are algorithms for solving the inverse geodesic problem and determining the area of a spheroidal trapezoid. The solution to the inverse geodesic problem is required to determine (1) the sizes of the moving window depending on latitude when calculating locale morphometric variables and (2) the trapezoidal area when calculating nonlocal topographic attributes. In the case of small celestial bodies, such as asteroids and some moons like Phobos, one may apply a general analytical and computational solution for geomorphometric modeling on the surface of a triaxial ellipsoid (Section 4.9).

Nyrtsov et al. (2007) proposed to use multiaxial ellipsoids as reference surfaces for mapping small celestial bodies with complex irregular shapes. For geomorphometric modeling, the use of multiaxial ellipsoids is a nontrivial mathematical task. This is because geodesics, the inverse geodetic problem, and related computational issues have, as far as we know, not been adequately studied for such complex surfaces. Furthermore, for the case of a very irregular body—e.g., comet Churyumov–Gerasimenko (Preusker et al., 2015)—it seems that it would be appropriate to derive models of morphometric variables analytically from its 3D model without using a single reference ellipsoid (Nyrtsov et al., 2018) or a combination of several reference ellipsoids.

Choosing DEM grid spacing is a very important stage of any DTM-based research (Chapter 10). There is no universal solution, and the competent DEM grid spacing depends on the typical sizes of the studied objects. A minimal value of the DEM grid spacing is determined by the fundamental sampling theorem and its three consequences (Section 3.3). Ignoring these rules inevitably leads to the production of artifacts reflecting the properties of an interpolator rather than the topographic surface. The DEM grid spacing influences the results of both geomorphometric calculations and further DTM-based modeling of characteristics of natural objects and phenomena (Pentížek et al., 2016; Gruber et al., 2017; Mondal et al., 2017; Goyal et al., 2018; Grimm et al., 2018; Groom et al., 2018; Dai et al., 2019b; Guo et al., 2019; Sîrbu et al., 2019; Depicker et al., 2020; Li Z. et al., 2021; Riihimäki et al., 2021; Zhang and Shang, 2021; Piedallu et al., 2022). The availability of lidar and drone data causes a temptation to use the most detailed DEMs. Sometimes this allows for obtaining more plausible modeling results (Woodrow et al., 2016; Kasprak et al , 2019; Žabota et al., 2019; Rocha et al., 2020; Li A. et al., 2021). However, the race for maximum resolution is often unjustified because modeling with less resolution gives similar or even better results (Han et al., 2018; Jacobs et al., 2018; Chang et al., 2019; Maxwell and Warner, 2019; Larson et al., 2022). Besides, ultra-high-resolution DEMs with a grid spacing of 1–25 cm can be saturated with high-frequency noise, which is difficult to remove without seriously damaging the signal. Such DEMs are useless for digital terrain analysis (Florinsky and Bliakharskii, 2019a).

At DEM preprocessing stages, one can try to suppress high-frequency noise, errors, and artifacts, which can be found in any DEM, by applying various types of filtering and smoothing (Chapter 6). It is, however, obvious that it is impossible to remove the noise without damaging the signal. A user solves empirically the problem of choosing the optimal degree of DEM smoothing or filtering determined by their parameters because it is impossible to develop mathematically rigorous, universal criteria for these procedures for any type of topography. In practice, one tries to produce a filtered or smoothed DEM retaining the highest possible formal statistical accuracy and non-formal visual plausibility. As a result, a user can obtain several DEMs with different degrees of smoothness, and each of them may be utilized for subsequent geomorphometric calculations. Further using these DEMs, a researcher will, however, derive different models of the same morphometric variable as well as different results in DTM-based prediction or modeling of objects, processes, or phenomena under study (Van Nieuwenhuizen et al., 2021b; Ciglič et al., 2022). The choice of a particular filtered or smoothed DEM is a nontrivial task depending on the formulation of the problem to be solved as well as the experience of the user.

A similar problem arises in DEM generalization (Duan et al., 2017; Behrens et al., 2018b; Wu et al., 2019; Feciskanin and Minár, 2021; Samsonov, 2022). Strictly speaking, it is often difficult to draw a line between DEM smoothing for noise suppression and DEM generalization for adequate visualization of morphometric models, generation of a multiscale model of a morphometric variable, or landform classification at several hierarchical levels (Bandura et al., 2021; Popov et al., 2021).

21.4 Applications

In geomorphology, soil science, hydrology, and some other sciences, it is already difficult to find studies involving spatial modeling, which would be carried out without one or another geomorphometric approach (Section 1.3.2).

The wide use of digital terrain analysis in soil research stems from the fact that topography is one of the soil-forming factors (Chapter 9). Dozens of methods can be applied to analyze, model, and map the spatial distribution of soil properties using DTMs (Chapters 11 and 12). Although spatial prediction of soil properties can be performed by many statistical and mathematical approaches, the success of such modeling depends primarily on the correct selection of predictors for a particular soil property. Morphometric attributes can be considered the most useful covariates.

Geology-oriented digital terrain modeling is based on the fact that topography is the most observable indicator of the geological composition of the crust. One can use this feature of topography at any scale: to describe the geometry of folds and to predict the degree of fold deformation and strain (Chapter 13); to reveal hidden lineaments and faults and to classify these structures (Chapter 14); to study the interaction between overland and deep substance flows (Chapter 15); to carry out comparative planetological studies at the global scale (Chapter 16), etc. Achievements of 3D seismic survey (Section 3.1.11) offer a wide field of activity for the application of geomorphometric analysis to contact surfaces between stratigraphic horizons and geological bodies.

Over the past several years, seven new DEMs of Antarctica have been created: REMA, BedMachine Antarctica, IBCSO version 4, TanDEM-X PolarDEM 90 m, CryoSat-2 DEM, Gapless-REMA100, and ICESat-2 DEM (Section 1.3.1.2). This fact testifies to a sharp increase in interest in the study and exploration of the ice continent. In this regard, we expect an increase in the number and quality of DTM-based glaciological, geomorphological, geological, permafrost, soil, biological, oceanological, and geophysical research studies of Antarctica. In particular, there will be further development in studying the motion and evolution of glaciers (Chapter 17), detecting crevasses (Chapter 18), researching catastrophic glacier events (Chapter 19), mapping topographic peculiarities of Antarctic oases (Chapter 20), revealing relic pre-glacial landscapes beneath the Antarctic Ice Sheet (Jamieson et al., 2023), etc.

The scientific community extensively uses common geoinformation systems and specialized geomorphometric software to produce maps for conducting particular studies. However, the development of thematic atlases remains the fundamental task of cartography. An atlas, as a large cartographic product, always solves several interrelated scientific problems that cannot be resolved through deriving individual maps for particular research tasks. Each atlas is neither just a set of various maps, nor their mechanical combination in the form of a book, compact disk, or website. An atlas includes a system of maps, which are interrelated to one another and complement each other (Salichtchev, 1990, p. 195). Of particular interest and importance is the development of a geomorphometric atlas of Antarctic oases and other ice-free territories of Antarctica (Chapter 20), which are remote, hard-to-reach, and poorly explored territories with limited possibility of fieldwork.

Geomorphometry significantly expands the possibilities of geosciences. I hope that the third edition of this book provides more insight into the theory, methods, and applications of digital terrain analysis, as well as further understanding of the relationships between topography and other components of geosystems.

References

Aandahl, A.R., 1948. The characterization of slope positions and their influence on the total nitrogen content of a few virgin soils of western Iowa. Soil Sci. Soc. Am. Proc. 13, 449–454.

Abdel-Fattah, M., Saber, M., Kantoush, S.A., Khalil, M.F., Sumi, T., Sefelnasr, A.M., 2017. A hydrological and geomorphometric approach to understanding the generation of wadi flash floods. Water 9 (7), 553.

Abdelguerfi, M., Wynne, C., Cooper, E., Roy, L., Shaw, K., 1998. Representation of 3-D elevation in terrain databases using hierarchical triangulated irregular networks: a comparative analysis. Int. J. Geogr. Inf. Sci. 12 (8), 853–873.

Abelsky, A.M., Lastochkin, A.N., 1969. Revealing and analysis of relief-forming wave-like deformations by a method of directional stacking with simultaneous frequency filtering (exemplified by the northeast of the Siberian Platform). Sci. Notes Arc. Geol. Res. Inst. Region. Geol. Ser. 14, 101–108 (in Russian).

Aber, J.S., Marzolff, I., Ries, J.B., Aber, S.E.W., 2019. Small-Format Aerial Photography and UAS Imagery: Principles, Techniques, and Geoscience Applications, second ed. Elsevier, Amsterdam, the Netherlands.

Abramova, A.S., 2020a. Method of assessment of magnitude of artifacts in digital bathymetric models. Izvestia Vuzov. Geodesy Aerophotosurveying 64 (5), 522–531 (in Russian).

Abramova, A.S., 2020b. Types of artifacts in digital bathymetric models (exemplified by the Arctic Basin). Izvestia Vuzov. Geodesy Aerophotosurveying 64 (6), 642–652 (in Russian).

Abramova, A.S., 2021a. Accuracy assessment of publicly available digital elevation models of the ocean floor at the polygons of multibeam data coverage in the Norwegian and Barents Seas. Geodesy Cartogr. (1), 13–22 (in Russian).

Abramova, A.S., 2021b. Method to assess accuracy of interpolation in digital bathymetric models. Izvestia Vuzov. Geodesy Aerophotosurveying 65 (1), 52–60 (in Russian).

Abrams, M., Crippen, R., Fujisada, H., 2020. ASTER Global Digital Elevation Model (GDEM) and ASTER Global Water Body Dataset (ASTWBD). Remote Sens. (Basel) 12 (7), 1156.

ACE2 Team, 2008. ACE2. EAPRS Laboratory, De Montfort University, Leicester, UK. https://www.tech.dmu.ac.uk/EAPRS/projects_ace2.html. (Accessed 27 June 2024).

Acker III, W.L., 1974. Basic Procedures for Soil Sampling and Core Drilling. Acker Drill Co., Scranton, PA.

Ackermann, F., 1978. Experimental investigation into the accuracy of contouring from DTM. Photogramm. Eng. Remote Sens. 44 (12), 1537–1548.

Adhikari, K., Mishra, U., Owens, P.R., Libohova, Z., Wills, S.A., Riley, W.J., et al., 2020. Importance and strength of environmental controllers of soil organic carbon changes with scale. Geoderma 375, 114472.

Adiyaman, Ö., Chorowicz, J., Arnaud, O.N., Gündogdu, M.N., Gourgaud, A., 2001. Late cenozoic tectonics and volcanism along the North Anatolian Fault: new structural and geochemical data. Tectonophysics 338 (2), 135–165.

Afanasiev, J.N., 1927. The Classification Problem in Russian Soil Science. Academy of Sciences of the USSR, Leningrad, USSR.

Agishtein, M.E., Migdal, A.A., 1991. Smooth surface reconstruction from scattered data points. Comput. Graph. 15 (1), 29–39.

Agisoft, 2017. Agisoft PhotoScan User Manual: Professional Edition, Version 1.3. Agisoft LLC, St. Petersburg, Russia.

Agrafiotis, P., Skarlatos, D., Georgopoulos, A., Karantzalos, K., 2019. DepthLearn: learning to correct the refraction on point clouds derived from aerial imagery for accurate dense shallow water bathymetry based on SVMs-Fusion with LiDAR point clouds. Remote Sens. (Basel) 11 (19), 2225.

Agterberg, F., 2014. Geomathematics: Theoretical Foundations, Applications and Future Developments. Quantitative Geology and Geostatistics Book Series. Springer, Cham, Switzerland.

Aguilar, F.J., Mills, J.P., 2008. Accuracy assessment of lidar-derived digital elevation models. Photogramm. Rec. 23 (122), 148–169.

Aguilar, F.J., Agüera, F., Aguilar, M.A., Carvajal, F., 2005. Effects of terrain morphology, sampling density, and interpolation methods on grid DEM accuracy. Photogramm. Eng. Remote Sens. 71 (7), 805–816.

Ai, S., Wang, S., Li, Y., Moholdt, G., Zhou, C., Liu, L., Yang, Y., 2019. High-precision ice-flow velocities from ground observations on Dalk Glacier, Antarctica. Pol. Sci. 19, 13–23.

Ai, T., Li, J., 2010. A DEM generalization by minor valley branch detection and grid filling. ISPRS J. Photogrammet. Remote Sens. 65 (2), 198–207.

Airbus DS, 2021. WorldDEM™ Neo. Technical Product Description. Version 1.0. Airbus Defence and Space, Taufkirchen, Germany.

Aitkenhead, M., Coull, M., 2020. Mapping soil profile depth, bulk density and carbon stock in Scotland using remote sensing and spatial covariates. Eur. J. Soil Sci. 71 (4), 553–567.

Akeno, K., 1996. DEM generation from multisensor stereopairs—AVHRR and MSS. In: Kraus, K., Waldhäusl, P. (Eds.), XVIII ISPRS Congress, Technical Commission IV: Mapping and Geographic Information Systems, Vienna, Austria, 9–19 July 1996. Int. Arch. Photogramm. Remote Sens. XXXI (B4), pp. 36–40.

Akima, H., 1974. A method of bivariate interpolation and smooth surface fitting based on local procedures. Commun. ACM 17 (1), 18–20.

Akovetsky, V.G., 1994. Efficiency improvement of stereoscopic measurements. Geodesy Cartogr. (1), 29–33 (in Russian).

Akpa, S.I.C., Odeh, I.O.A., Bishop, T.F.A., Hartemink, A.E., Amapu, I.Y., 2016. Total soil organic carbon and carbon sequestration potential in Nigeria. Geoderma 271, 202–215.

Albani, M., Klinkenberg, B., 2003. A spatial filter for the removal of striping artifacts in digital elevation models. Photogrammet. Eng. Remote Sens. 69 (7), 755–765.

Albani, M., Klinkenberg, B., Andison, D.W., Kimmins, J.P., 2004. The choice of window size in approximating topographic surfaces from digital elevation models. Int. J. Geogr. Inf. Sci. 18 (6), 577–593.

Albee, A.L., 2000. Mars 2000. Annu. Rev. Earth Planet. Sci. 28, 281–304.

Alexander, C., Deák, B., Heilmeier, H., 2016. Micro-topography driven vegetation patterns in open mosaic landscapes. Ecol. Indic. 60, 906–920.

Alifu, H., Johnson, B.A., Tateishi, R., 2016. Delineation of debris-covered glaciers based on a combination of geomorphometric parameters and a TIR/NIR/SWIR band ratio. IEEE J. Sel. Topics Appl. Earth Obs. Remote Sens. 9 (2), 781–792.

Alifu, H., Vuillaume, J.-F., Johnson, B.A., Hirabayashi, Y., 2020. Machine-learning classification of debris-covered glaciers using a combination of Sentinel-1/-2 (SAR/optical), Landsat 8 (thermal) and digital elevation data. Geomorphology 369, 107365.

Alonso-Sarría, F., Gomariz-Castillo, F., Cánovas-García, F., 2018. A new approach to the openness index for landform characterisation. Comput. Geosci. 119, 68–79.

Al-Rousan, N., Cheng, P., Petrie, G., Toutin, T., Valadan Zoej, M.J., 1997. Automated DEM extraction and orthoimage generation from SPOT level 1B imagery. Photogrammet. Eng. Remote Sens. 63 (8), 965–974.

Alvarez, L., Moreno, H., Segales, A., Pham, T., Pillar-Little, E., Chilson, P., 2018. Merging unmanned aerial systems (UAS) imagery and echo soundings with an adaptive sampling technique for bathymetric surveys. Remote Sens. (Basel) 10 (9), 1362.

Alvioli, M., Guzzetti, F., Marchesini, I., 2020. Parameter-free delineation of slope units and terrain subdivision of Italy. Geomorphology 358, 107124.

Amante, C., Eakins, B.W., 2009. ETOPO1 1 arc-minute global relief model: procedures, data sources and analysis. NOAA Technical Memorandum NESDIS NGDC-24. National Geophysical Data Center, National Oceanic and Atmospheric Administration, Boulder, CO.

Amatulli, G., Domisch, S., Tuanmu, M.N., Parmentier, B., Ranipeta, A., Malczyk, J., et al., 2018. A suite of global, cross-scale topographic variables for environmental and biodiversity modeling. Sci. Data 5, 180040.

Amatulli, G., McInerney, D., Sethi, T., Strobl, P., Domisch, S., 2020. Geomorpho90m, empirical evaluation and accuracy assessment of global high-resolution geomorphometric layers. Sci. Data 7, 162.

An, M., Wiens, D., An, C., Shi, G., Zhao, Y., Li, Y., 2015. Antarctic ice velocities from GPS locations logged by seismic stations. Antarct. Sci. 27 (Suppl. 2), 210–222.

Anderson, K., Westoby, M.J., James, M.R., 2019. Low-budget topographic surveying comes of age: structure from motion photogrammetry in geography and the geosciences. Progr. Phys. Geogr. 43 (2), 163–173.

Anderson, M.G., Burt, T.P., 1978. The role of topography in controlling throughflow generation. Earth Surf. Process. 3 (4), 331–344.

Anderson, M.G., Burt, T.P., 1980. The role of topography in controlling throughflow generation: a reply. Earth Surf. Process. 5 (2), 193–195.

Anderson, T.R., Groffman, P.M., Walter, M.T., 2015. Using a soil topographic index to distribute denitrification fluxes across a northeastern headwater catchment. J. Hydrol. 522, 123–134.

Andreani, L., Gloaguen, R., 2016. Geomorphic analysis of transient landscapes in the Sierra Madre de Chiapas and Maya Mountains (northern Central America), implications for the North American–Caribbean–Cocos plate boundary. Earth Surf. Dyn. 4 (2), 71–102.

Angelini, M.E., Heuvelink, G.B.M., Kempen, B., Morrás, H.J.M., 2016. Mapping the soils of an Argentine Pampas region using structural equation modelling. Geoderma 281, 102–118.

Angelini, M.E., Heuvelink, G.B.M., Kempen, B., 2017. Multivariate mapping of soil with structural equation modelling. Eur. J. Soil Sci. 68 (5), 575–591.

Anisimov, I.G., Bordunov, A.A., Deeva, N.F., Kamalov, L.F., Lopachev, N.A., Muratova, V.S., et al., 1977. A technique for compilation of the series of medium-scale thematic maps "Natural and reclamation estimation of the Mid-Region of the USSR". In: Stepanov, I.N. (Ed.), Estimation of Natural and Reclamation Conditions and Prediction of Their Changes (Exemplified by Central Asia). Biological Research Centre Press, Pushchino, USSR, pp. 23–93 (in Russian).

Anokhin, V.M., Odesskii, I.A., 2001. Characteristics of the global pattern of planetary fracturing. Geotectonics 35 (5), 335–340.

Ansell, J.H., Bannister, S.C., 1996. Shallow morphology of the subducted Pacific plate along the Hikurangi margin, New Zealand. Phys. Earth Planet. In. 93 (1–2), 3–20.

Antonić, O., Legović, T., 1999. Estimating the direction of an unknown air pollution source using a digital elevation model and a sample of deposition. Ecol. Model. 124 (1), 85–95.

Arabelos, D., 2000. Intercomparisons of the global DTMs ETOPO5, TerrainBase and JGP95E. Phys. Chem. Earth, Pt A Solid Earth Geod. 25 (1), 89–93.

Argyriou, A.V., Sarris, A., Teeuw, R.M., 2016a. Using geoinformatics and geomorphometrics to quantify the geodiversity of Crete, Greece. Int. J. Appl. Earth Obs. Geoinf. 51, 47–59.

Argyriou, A.V., Teeuw, R.M., Rust, D., Sarris, A., 2016b. GIS multi-criteria decision analysis for assessment and mapping of neotectonic landscape deformation: a case study from Crete. Geomorphology 253, 262–274.

Argyriou, A.V., Teeuw, R.M., Sarris, A., 2017a. GIS-based landform classification of Bronze Age archaeological sites on Crete Island. PLoS One 12 (2), e0170727.

Argyriou, A.V., Teeuw, R.M., Soupios, P., Sarris, A., 2017b. Neotectonic control on drainage systems: GIS-based geomorphometric and morphotectonic assessment for Crete, Greece. J. Struct. Geol. 104, 93–111.

Ariza-López, F.J., Reinoso-Gordo, J.F., 2021. Comparison of gridded DEMs by buffering. Remote Sens. (Basel) 13 (15), 3002.

Armstrong, R.N., Martz, L.W., 2003. Topographic parameterization in continental hydrology: a study in scale. Hydrol. Process. 17 (18), 3763–3781.

Arndt, J.E., Schenke, H.W., Jakobsson, M., Nitsche, F.O., Buys, G., Goleby, B., et al., 2013. The International Bathymetric Chart of the Southern Ocean (IBCSO) Version 1.0—a new bathymetric compilation covering circum-Antarctic waters. Geophys. Res. Lett. 40 (12), 3111–3117.

Arnold, N., 2010. A new approach for dealing with depressions in digital elevation models when calculating flow accumulation values. Progr. Phys. Geogr. 34 (6), 781–809.

Aronov, V.I., 1977. Methods of Mathematical Processing of Geological Data by a Computer. Nedra, Moscow, USSR (in Russian).

Arrell, K., Wise, S., Wood, J., Donoghue, D., 2008. Spectral filtering as a method of visualising and removing striped artefacts in digital elevation data. Earth Surf. Process. Landf. 33 (6), 943–961.

Arrouays, D., Daroussin, J., Kicin, J.L., Hassika, P., 1998. Improving topsoil carbon storage prediction using a digital elevation model in temperate forest soils of France. Soil Sci. 163 (2), 103–108.

Arundel, S.T., Sinha, G., 2019. Validating the use of object-based image analysis to map commonly recognized landform features in the United States. Cartogr. Geogr. Inf. Sci. 46 (5), 441–455.

Arundel, S.T., Thiem, P.T., Constance, E.W., 2018. Automated extraction of hydrographically corrected contours for the conterminous United States: the US Geological Survey US Topo product. Cartogr. Geogr. Inf. Sci. 45 (1), 31–55.

Ashphaq, M., Srivastava, P.K., Mitra, D., 2021. Review of near-shore satellite derived bathymetry: classification and account of five decades of coastal bathymetry research. J. Ocean Eng. Sci. 6 (4), 340–359.

ATCM, 2014. Management plan for Antarctic Specially Protected Area No. 174 Stornes, Larsemann Hills, Princess Elizabeth Land. In: Antarctic Treaty Consultative Meeting (ATCM) XXXVII Final Report, Measure 12 Annex. Secretariat of the Antarctic Treaty, Buenos Aires, Argentina.

ATCM, 2021. Larsemann Hills, East Antarctica Antarctic Specially Managed Area No. 6 management plan. In: Antarctic Treaty Consultative Meeting (ATCM) XLIII Final Report. Measure 1. Secretariat of the Antarctic Treaty, Buenos Aires, Argentina.

Atkinson, J., de Clercq, W., Rozanov, A., 2020. Multi-resolution soil-landscape characterisation in KwaZulu Natal: using geomorphons to classify local soilscapes for improved digital geomorphological modelling. Geoderma Reg. 22, e00291.

Audet, P., 2011. Directional wavelet analysis on the sphere: application to gravity and topography of the terrestrial planets. J. Geophys. Res. Planet. 116 (E1), E01003.

Australian Antarctic Division, 2005. Larsemann Hills, Princess Elizabeth Land, Antarctica. Environmental Management Map, 1:25,000 Scale. Australian Antarctic Division, Kingston, TS.

Australian Drilling Industry Training Committee, 1997. Drilling: The Manual of Methods, Applications, and Management, fourth ed. Lewis, Boca Raton, FL.

Avanzi, F., Bianchi, A., Cina, A., de Michele, C., Maschio, P., Pagliari, D., et al., 2018. Centimetric accuracy in snow depth using unmanned aerial system photogrammetry and a MultiStation. Remote Sens. (Basel) 10 (5), 765.

Awange, J.L., 2012. Environmental Monitoring Using GNSS: Global Navigation Satellite Systems. Environmental Science and Engineering Book Series. Springer, Berlin, Germany.

Axelsson, P., 1999. Processing of laser scanner data—algorithms and applications. ISPRS J. Photogrammet. Remote Sens. 54 (2–3), 138–147.

Babiker, M., Gudmundsson, A., 2004. The effects of dykes and faults on groundwater flow in an arid land: the Red Sea Hills, Sudan. J. Hydrol. 297 (1–4), 256–273.

Badaro-Saliba, N., Adjizian-Gerard, J., Zaarour, R., Najjar, G., 2021. LCZ scheme for assessing Urban Heat Island intensity in a complex urban area (Beirut, Lebanon). Urban Clim. 37, 100846.

Badger, J., Lundtang Petersen, E., Mann, J., Hahmann, A.N., 2021. Wind atlas trends. In: Holst Jørgensen, B., Hauge Madsen, P., Giebel, G., Martí, I., Thomsen, K. (Eds.), DTU International Energy Report 2021: Perspectives on Wind Energy. DTU Wind Energy, Kongens Lyngby, Denmark, pp. 25–33.

Bagheri, H., Sadjadi, S.Y., Sadeghian, S., 2014. Exploring the role of genetic algorithms and artificial neural networks for interpolation of elevation in geoinformation models. In: Isikdag, U. (Ed.), Innovations in 3D Geo-Information Sciences. Lecture Notes in Geoinformation and Cartography Book Series. Springer, Cham, Switzerland, pp. 107–121.

Bagratuni, G.V., 1967. Course in Spheroidal Geodesy. Foreign Technology Division, Wright-Patterson, OH.

Bailey, J.J., Boyd, D.S., Hjort, J., Lavers, C.P., Field, R., 2017. Modelling native and alien vascular plant species richness: at which scales is geodiversity most relevant? Glob. Ecol. Biogeogr. 26 (7), 763–776.

Bailey, J.J., Boyd, D.S., Field, R., 2018. Models of upland species' distributions are improved by accounting for geodiversity. Landsc. Ecol. 33 (12), 2071–2087.

Baillard, J.-M., 2013. Geodesics on a triaxial ellipsoid for the HP-41. HP41Programs. http://hp41programs.yolasite.com/geod3axial.php. (Accessed 27 June 2024).

Bakanova, V.V., 1983. Large Scale Topographic Surveys. Nedra, Moscow, USSR (in Russian).

Baker, K.D., 1982. Basic image processing concepts. In: Jones, N.B. (Ed.), Digital Signal Processing. Peter Peregrinus, Stevenage, UK, pp. 287–318.

Bakhvalov, N.S., 1977. Numerical Methods: Analysis, Algebra, Ordinary Differential Equations. Mir, Moscow, USSR.

Bakker, M., Lane, S.N., 2017. Archival photogrammetric analysis of river-floodplain systems using Structure from Motion (SfM) methods. Earth Surf. Process. Landf. 42 (8), 1274–1286.

Balce, A.E., 1987. Determination of optimum sampling interval in grid digital elevation models (DEM) data acquisition. Photogrammet. Eng. Remote Sens. 53 (3), 323–330.

Baldi, P., Bonvalot, S., Briole, P., Marsella, M., 2000. Digital photogrammetry and kinematic GPS applied to the monitoring of Vulcano Island, Aeolian Arc, Italy. Geophys. J. Int. 142 (3), 801–811.

Baldwin, D., Naithani, K.J., Lin, H., 2017. Combined soil-terrain stratification for characterizing catchment-scale soil moisture variation. Geoderma 285, 260–269.

Ballabio, C., 2009. Spatial prediction of soil properties in temperate mountain regions using support vector regression. Geoderma 151 (3–4), 338–350.

Ballabio, C., Comolli, R., 2010. Mapping heavy metal content in soils with multi-kernel SVR and LiDAR derived data. In: Boettinger, J.L., Howell, D.W., Moore, A.C., Hartemink, A.E., Kienast-Brown, S. (Eds.), Digital Soil Mapping: Bridging Research, Environmental Application, and Operation. Progress in Soil Science Book Series. Springer, Dordrecht, the Netherlands, pp. 205–216.

Ballabio, C., Panagos, P., Monatanarella, L., 2016. Mapping topsoil physical properties at European scale using the LUCAS database. Geoderma 261, 110–123.

Baltensweiler, A., Heuvelink, G.B.M., Hanewinkel, M., Walthert, L., 2020. Microtopography shapes soil pH in flysch regions across Switzerland. Geoderma 380, 114663.

Baltsavias, E.P., 1999a. A comparison between photogrammetry and laser scanning. ISPRS J. Photogrammet. Remote Sens. 54 (2–3), 83–94.

Baltsavias, E.P., 1999b. Airborne laser scanning: basic relations and formulas. ISPRS J. Photogrammet. Remote Sens. 54 (2–3), 199–214.

Bamber, J.L., Ekholm, S., Krabill, W.B., 2001a. A new, high-resolution digital elevation model of Greenland fully validated with airborne laser altimeter data. J. Geophys. Res. Solid Earth 106 (B4), 6733–6745.

Bamber, J.L., Layberry, R.L., Gogineni, S.P., 2001b. A new ice thickness and bed data set for the Greenland ice sheet. 1. Measurement, data reduction, and errors. J. Geophys. Res. Atmos. 106 (D24), 33773–33780.

Bamber, J.L., Gomez-Dans, J.L., Griggs, J.A., 2009. A new 1 km digital elevation model of the Antarctic derived from combined satellite radar and laser data—Part 1: data and methods. Cryosphere 3 (1), 101–111.

Bamber, J.L., Griggs, J.A., Hurkmans, R.T.W.L., Dowdeswell, J.A., Gogineni, S.P., Howat, I., et al., 2013. A new bed elevation data-set for Greenland. Cryosphere 7 (2), 499–510.

Band, L.E., 1986. Topographic partition of watersheds with digital elevation models. Water Resour. Res. 22 (11), 15–24.

Band, L.E., 1993. Extraction of channel networks and topographic parameters from digital elevation data. In: Beven, K., Kirkby, M.J. (Eds.), Channel Network Hydrology. Wiley, Chichester, UK, pp. 13–42.

Band, L.E., Moore, I.D., 1995. Scale: landscape attributes and geographical information systems. Hydrol. Process. 9 (3–4), 401–422.

Band, L.E., Patterson, P., Nemani, R., Running, S.W., 1993. Forest ecosystem processes at the watershed scale: incorporating hillslope hydrology. Agric. For. Meteorol. 63 (1–2), 93–126.

Bandura, P., Minár, J., Bielik, M., 2021. Physically based morphostructural land surface segmentation: case of the Alps and Western Carpathians. Trans. GIS 25 (5), 2394–2418.

Baransky, N.N., 1946. Generalization in cartography and geographical textual description. Sci. Notes Lomonosov Moscow State Univ. 119 (2), 180–205 (in Russian).

Barbarella, M., di Benedetto, A., Fiani, M., 2022. A method for obtaining a DEM with curved abscissa from MLS Data for linear infrastructure survey design. Remote Sens. (Basel) 14 (4), 889.

Bargain, A., Marchese, F., Savini, A., Taviani, M., Fabri, M.-C., 2017. Santa Maria di Leuca Province (Mediterranean Sea), identification of suitable mounds for cold-water coral settlement using geomorphometric proxies and Maxent methods. Front. Mar. Sci. 4, 338.

Barnes, R., 2016. Parallel Priority-Flood depression filling for trillion cell digital elevation models on desktops or clusters. Comput. Geosci. 96, 56–68.

Barringer, J.R.F., Hewitt, A.E., Lynn, I.H., Schmidt, J., 2008. National mapping of landform elements in support of S-Map, a New Zealand soils database. In: Zhou, Q., Lees, B., Tang, G.-A. (Eds.), Advances in Digital Terrain Analysis. Lecture Notes in Geoinformation and Cartography Book Series. Springer, Berlin, Germany, pp. 443–458.

Barton, K.E., Howell, D.G., Vigil, J.F., Reed, J.C., Wheeler, J.O., 2003. The North America Tapestry of Time and Terrain. Geologic Investigations Series Map I-2781, Scale 1:8,000,000. US Geological Survey, Denver, CO.

Bash, E.A., Moorman, B.J., 2020. Surface melt and the importance of water flow—an analysis based on high-resolution unmanned aerial vehicle (UAV) data for an Arctic glacier. Cryosphere 14 (2), 549–563.

Bash, E.A., Moorman, B.J., Gunther, A., 2018. Detecting short-term surface melt on an Arctic glacier using UAV surveys. Remote Sens. (Basel) 10 (10), 1547.

Bastida, F., Aller, J., Bobillo-Ares, N.C., 1999. Geometrical analysis of folded surfaces using simple functions. J. Struct. Geol. 21 (7), 729–742.

Bater, C.W., Coops, N.C., 2009. Evaluating error associated with lidar-derived DEM interpolation. Comput. Geosci. 35 (2), 289–300.

Batschelet, E., 1981. Circular Statistics in Biology. Academic Press, London, UK.

Batson, R.M., Edwards, K., Eliason, E.M., 1975. Computer-generated shaded-relief images. J. Res. US Geol. Survey 3 (4), 401–408.

Baxter, S.J., Oliver, M.A., 2005. The spatial prediction of soil mineral N and potentially available N using elevation. Geoderma 128 (3–4), 325–339.

Bear, J., 1972. Dynamics of Fluids in Porous Media. Elsevier, New York, NY.

Bear, J., Braester, C., 1972. On the flow of two immiscible fluids in fractured porous media. In: International Association for Hydraulic Research (Ed.), Fundamentals of Transport Phenomena in Porous Media. Developments in Soil Science, vol. 2. Elsevier, Amsterdam, the Netherlands, pp. 177–202.

Beauchamp, E.G., Bergstrom, D.W., 1993. Denitrification. In: Carter, M.R. (Ed.), Soil Sampling and Methods of Analysis. Lewis, Boca Raton, FL, pp. 351–357.

Becker, J.J., Sandwell, D.T., Smith, W.H.F., Braud, J., Binder, B., Depner, J., et al., 2009. Global bathymetry and elevation data at 30 arc seconds resolution: SRTM30_PLUS. Mar. Geod. 32 (4), 355–371.

Becker, K.J., Robinson, M.S., Becker, T.L., Weller, L.A., Edmundson, K.L., Neumann, G.A., et al., 2016. First global digital elevation model of Mercury. In: 47th Lunar and Planetary Science Conference, 21–25 March 2016, Woodlands, TX. Universities Space Research Association, Washington, DC. 2959.

Beckett, P.H.T., Webster, R., 1971. Soil variability: a review. Soils Fert. 34 (1), 1–15.

Bedard-Haughn, A.K., Pennock, D.J., 2002. Terrain controls on depressional soil distribution in a hummocky morainal landscape. Geoderma 110 (3–4), 169–190.

Behrens, T., Scholten, T., 2007. A comparison of data-mining techniques in predictive soil mapping. In: Lagacherie, P., McBratney, A.B., Voltz, M. (Eds.), Digital Soil Mapping: An Introductory Perspective. Developments in Soil Science, vol. 31. Elsevier, Amsterdam, the Netherlands, pp. 353–364.

Behrens, T., Förster, H., Scholten, T., Steinrücken, U., Spies, E.-D., Goldschmitt, M., 2005. Digital soil mapping using artificial neural networks. J. Plant Nutr. Soil Sci. 168 (1), 21–33.

Behrens, T., Zhu, A.-X., Schmidt, K., Scholten, T., 2010. Multi-scale digital terrain analysis and feature selection for digital soil mapping. Geoderma 155 (3–4), 175–185.

Behrens, T., Schmidt, K., MacMillan, R.A., Viscarra Rossel, R.A., 2018a. Multiscale contextual spatial modelling with the Gaussian scale space. Geoderma 310, 128–137.

Behrens, T., Schmidt, K., MacMillan, R.A., Viscarra Rossel, R.A., 2018b. Multi-scale digital soil mapping with deep learning. Sci. Rep. 8, 15244.

Behrens, T., MacMillan, R.A., Rossel, R.A.V., Schmidt, K., Lee, J., 2019. Teleconnections in spatial modelling. Geoderma 354, 113854.

Bektaş, S., 2015. Geodetic computations on triaxial ellipsoid. Int. J. Min. Sci. 1 (1), 25–34.

Belgiu, M., Drăguţ, L., 2016. Random forest in remote sensing: a review of applications and future directions. ISPRS J. Photogrammet. Remote Sens. 114, 24–31.

Bell, J.C., Cunningham, R.L., Havens, M.W., 1992. Calibration and validation of a soil-landscape model for predicting soil drainage class. Soil Sci. Soc. Am. J. 56 (6), 1860–1866.

Bell, J.C., Cunningham, R.L., Havens, M.W., 1994a. Soil drainage class probability mapping using a soil-landscape model. Soil Sci. Soc. Am. J. 58 (2), 464–470.

Bell, J.C., Thompson, J.A., Butler, C.A., McSweeney, K., 1994b. Modeling soil genesis from a landscape perspective. In: Etchevers, B.J.D. (Ed.), Transactions of the 15th World Congress of Soil Science, Acapulco, Mexico, July 1994. International Society of Soil Science, vol. 6a. Acapulco, Mexico, pp. 179–195.

Bell, J.C., Grigal, D.F., Bates, P.C., 2000. A soil–terrain model for estimating spatial patterns of soil organic carbon. In: Wilson, J., Gallant, J. (Eds.), Terrain Analysis: Principles and Applications. Wiley, New York, NY, pp. 295–310.

Bellian, J.A., Kerans, C., Jennette, D.D., 2005. Digital outcrop models: applications of terrestrial scanning lidar technology in stratigraphic modeling. J. Sediment. Res. 75 (2), 166–176.

Belonin, M.D., Zhukov, I.M., 1968. Geometrical properties of surfaces of the Alexeevskoye elevated block, Kuibyshev Region. In: Romanova, M.A. (Ed.), Problems of Mathematical Geology. Nauka, Leningrad, USSR, pp. 194–207 (in Russian).

Bemis, S.P., Micklethwaite, S., Turner, D., James, M.R., Akciz, S., Thiele, S.T., et al., 2014. Ground-based and UAV-based photogrammetry: a multi-scale, high-resolution mapping tool for structural geology and paleoseismology. J. Struct. Geol. 59 (A), 163–178.

Benn, D.I., Warren, C.R., Mottram, R.H., 2007. Calving processes and the dynamics of calving glaciers. Earth Sci. Rev. 82 (3–4), 143–179.

Bergbauer, S., 2007. Testing the predictive capability of curvature analyses. In: Jolley, S.J., Barr, D., Walsh, J.J., Knipe, R.J. (Eds.), Structurally Complex Reservoirs. Geol. Soc. London Spec. Publ., vol. 292, pp. 185–202.

Bergbauer, S., Pollard, D.D., 2003. How to calculate normal curvatures of sampled geological surfaces. J. Struct. Geol. 25 (2), 277–289.

Bergbauer, S., Mukerji, T., Hennings, P., 2003. Improving curvature analyses of deformed horizons using scale-dependent filtering techniques. Am. Assoc. Pet. Geol. Bull. 87 (8), 1255–1272.

Bergonse, R., Reis, E., 2016. Controlling factors of the size and location of large gully systems: a regression-based exploration using reconstructed pre-erosion topography. Catena 147, 621–631.

Bergstrom, D.W., Monreal, C.M., St. Jacques, E., 2001a. Influence of tillage practice on carbon sequestration is scale-dependent. Can. J. Soil Sci. 81 (1), 63–70.

Bergstrom, D.W., Monreal, C.M., St. Jacques, E., 2001b. Spatial dependence of soil organic carbon mass and its relationship to soil series and topography. Can. J. Soil Sci. 81 (1), 53–62.

Berlyant, A.M., 1966. Cartographic methods to study contemporary tectonics and their classification. Izvestiya Academii Nauk SSSR: Geogr. Ser. (2), 71–80 (in Russian).

Berlyant, A.M., 1986. Image of the Space: Map and Information. Mysl, Moscow, USSR (in Russian).

Bernhardson, M., Alexanderson, H., 2017. Early Holocene dune field development in Dalarna, central Sweden: a geomorphological and geophysical case study. Earth Surf. Process. Landf. 42 (12), 1847–1859.

Berry, P.A.M., 2000. Topography from land radar altimeter data: possibilities and restrictions. Phys. Chem. Earth, Pt A Solid Earth Geod. 25 (1), 81–88.

Berry, P.A.M., Hilton, R.D., Johnson, C.P.D., 2002. ACE: first full release of the new 30″ GDEM incorporating satellite altimeter derived heights. In: XXVII General Assembly of the European Geophysical Society, Nice, France, 21–26 April 2002. European Geophysical Society, Nice, France, p. 4405.

Berry, P.A.M., Smith, R.G., Benveniste, J., 2010. ACE2: the new global digital elevation model. In: Mertikas, S.P. (Ed.), Gravity, Geoid and Earth Observation. IAG Commission 2: Gravity Field, Chania, Crete, Greece, 23–27 June 2008. International Association of Geodesy Symposia, vol. 135. Springer, Berlin, Germany, pp. 231–237.

Bespalov, N.A., 1980. Methods for Solving Problems of Spheroidal Geodesy. Nedra, Moscow, USSR (in Russian).

Besprozvanny, P.A., Borodzich, E.V., Bush, V.A., 1994. Numerical analysis of ordering relations in the global network of lineaments. Phys. Solid Earth 30 (2), 150–159.

Beven, K., 1997. TOPMODEL: a critique. Hydrol. Process. 11 (9), 1069–1085.

Beven, K., Freer, J., 2001. A dynamic TOPMODEL. Hydrol. Process. 15 (10), 1993–2011.

Beven, K.J., Kirkby, M.J., 1979. A physically-based variable contributing area model of basin hydrology. Hydrol. Sci. Bull. 24 (1), 43–69.

Beven, K., Kirkby, M.J. (Eds.), 1993. Channel Network Hydrology. Wiley, Chichester, UK.

Beven, K.J., Moore, I.D. (Eds.), 1993. Terrain Analysis and Distributed Modelling in Hydrology. Wiley, Chichester, UK.

Beven, K.J., Kirkby, M.J., Schoffield, N., Tagg, A.F., 1984. Testing a physically-based flood forecasting model (TOPMODEL) for three UK catchments. J. Hydrol. 69 (1–4), 119–143.

Beven, K.J., Lamb, R., Quinn, P., Romanowicz, R., Freer, J., 1995. TOPMODEL. In: Singh, V.P. (Ed.), Computer Models of Watershed Hydrology. Water Resources Publications, Englewood, CO, pp. 627–668.

Beven, K.J., Kirkby, M.J., Freer, J.E., Lamb, R., 2021. A history of TOPMODEL. Hydrol. Earth Syst. Sci. 25 (2), 527–549.

Bevis, M., 1986. The curvature of Wadati–Benioff Zones and the torsional rigidity of subducting plates. Nature 323 (6083), 52–53.

Bhadra, S., Li, R., Wu, D., Wang, G., Rekabdar, B., 2021. Assessing the impacts of anthropogenic drainage structures on hydrologic connectivity using high-resolution digital elevation models. Trans. GIS 25 (5), 2596–2611.

Bhardwaj, A., Sam, L., Bhardwaj, A., Martín-Torres, F.J., 2016a. LiDAR remote sensing of the cryosphere: present applications and future prospects. Remote Sens. Environ. 177, 125–143.

Bhardwaj, A., Sam, L., Akanksha, Martín-Torres, F.J., Kumar, R., 2016b. UAVs as remote sensing platform in glaciology: present applications and future prospects. Remote Sens. Environ. 175, 196–204.

Bhardwaj, A., Sam, L., Martin-Torres, F.J., Zorzano, M.-P., 2019a. Distribution and morphologies of transverse aeolian ridges in ExoMars 2020 Rover landing site. Remote Sens. (Basel) 11 (8), 912.

Bhardwaj, A., Sam, L., Martín-Torres, F.J., Zorzano, M.-P., Ramírez Luque, J.A., 2019b. UAV imaging of a Martian brine analogue environment in a fluvio-aeolian setting. Remote Sens. (Basel) 11 (18), 2104.

Biau, G., Scornet, E., 2016. A random forest guided tour. Test 25, 197–227.

Bielski, C., López-Vázquez, C., Grohmann, C.H., Guth, P.L., Hawker, L., Gesch, D., et al., 2024. Novel approach for ranking DEMs: Copernicus DEM improves one arc second open global topography. IEEE Trans. Geosci. Remote Sens. 62, 4503922.

Bierkens, M.F.P., Finke, P.A., de Willigen, P. (Eds.), 2000. Upscaling and Downscaling Methods for Environmental Research. Kluwer, Dordrecht, the Netherlands.

Bigelow, P., Benda, L., Pearce, S., 2016. Delineating incised stream sediment sources within a San Francisco Bay tributary basin. Earth Surf. Dyn. 4 (2), 531–547.

Bindlish, R., Barros, A.P., 1996. Aggregation of digital terrain data using a modified fractal interpolation scheme. Comput. Geosci. 22 (8), 907–917.

Biryukov, A.S., 2022. Geological and geophysical study of East Antarctica: the eastern part of the Larsemann Hills (Princess Elizabeth Land). In: Egorov, M.S. (Ed.), Performing Geological and Geophysical Field and Laboratory Works During the 66th Russian Antarctic Expedition: Report on the Government Contract No. 1.1.4/2020/PMGRE, 17.11.2020, vol. 2. VNIIOkeangeologia, Rosgeologia, Polar Marine Geological Prospecting Expedition. Lomonosov, St. Petersburg, Russia, pp. 17–47 (in Russian).

Bishop, T.F.A., McBratney, A.B., 2001. A comparison of prediction methods for the creation of field-extent soil property maps. Geoderma 103 (1–2), 149–160.

Bishop, T.F.A., Minasny, B., McBratney, A.B., 2006. Uncertainty analysis for soil–terrain models. Int. J. Geogr. Inf. Sci. 20 (2), 117–134.

Bispo, P.d.C., dos Santos, J.R., Valeriano, M.d.M., Graça, P.M.L.d.A., Balzter, H., França, H., et al., 2016. Predictive models of primary tropical forest structure from geomorphometric variables based on SRTM in the Tapajós Region, Brazilian Amazon. PLoS One 11 (6), e0152009.

Bispo, P.d.C., Balzter, H., Malhi, Y., Slik, J.W.F., dos Santos, J.R., Rennó, C.D., et al., 2017. Drivers of metacommunity structure diverge for common and rare Amazonian tree species. PLoS One 12 (11), e0188300.

Bjerhammar, A., 1973. Theory of Errors and Generalized Matrix Inverses. Elsevier, Amsterdam, the Netherlands.

Björk, G., Jakobsson, M., Assmann, K., Andersson, L.G., Nilsson, J., Stranne, C., et al., 2018. Bathymetry and oceanic flow structure at two deep passages crossing the Lomonosov Ridge. Ocean Sci. 14 (1), 1–13.

Bjørke, J.T., Nilsen, S., 2003. Wavelets applied to simplification of digital terrain models. Int. J. Geogr. Inf. Sci. 17 (7), 601–621.

Bjørke, J.T., Nilsen, S., 2004. Examination of a constant-area quadrilateral grid in representation of global digital elevation models. Int. J. Geogr. Inf. Sci. 18 (7), 653–664.

Björnsson, H., 2003. Subglacial lakes and jökulhlaups in Iceland. Global Planet. Change 35 (3–4), 255–271.

Bladin, K., Axelsson, E., Broberg, E., Emmart, C., Ljung, P., Bock, A., et al., 2018. Globe browsing: contextualized spatio-temporal planetary surface visualization. IEEE Trans. Vis. Comput. Graph. 24 (1), 802–811.

Blain, J.M., 2012. The Complete Guide to Blender Graphics: Computer Modeling and Animation. CRC Press, Boca Raton, FL.

Blaschke, T., Donert, K., Gossette, F., Kienberger, S., Marani, M., Qureshi, S., et al., 2012. Virtual globes: serving science and society. Information 3 (3), 372–390.

Blazkova, S., Beven, K.J., Kulasova, A., 2002. On constraining TOPMODEL hydrograph simulations using partial saturated area information. Hydrol. Process. 16 (2), 441–458.

Blender Foundation, 2003–2023. Blender. Stichting Blender Foundation, Amsterdam, the Netherlands. https://www.blender.org. (Accessed 27 June 2024).

Blender Foundation, 2018. Blender 2.79 Reference Manual. Stichting Blender Foundation, Amsterdam, the Netherlands.

Bliakharskii, D.P., 2019. Specifics of aerial survey of mountain and sheet glaciers using unmanned aerial vehicles. Izvestia Vuzov. Geodesy Aerophotosurveying 63 (6), 650–661 (in Russian).

Bliakharskii, D.P., Florinsky, I.V., Skrypitsyna, T.N., 2019a. Modelling glacier topography in Antarctica using unmanned aerial survey: assessment of opportunities. Int. J. Remote Sens. 40 (7), 2517–2541.

Bliakharskii, D.P., Volgusheva, N.E., Kazakov, E.E., 2019b. UAS photosurveying method for glacier monitoring during ablation season: Potanin and Alexandra Glaciers, Mongolian Altai Mountains. Izvestia Vuzov. Geodesy Aerophotosurveying 63 (2), 168–179 (in Russian).

Blinn, J.F., 1977. Models of light reflection for computer synthesized pictures. ACM SIGGRAPH Comput. Graph. 11 (2), 192–198.

Blomdin, R., Stroeven, A.P., Harbor, J.M., Gribenski, N., Caffee, M.W., Heyman, J., et al., 2018. Timing and dynamics of glaciation in the Ikh Turgen Mountains, Altai region, High Asia. Quat. Geochronol. 47, 54–71.

Blondel, P., 2009. The Handbook of Sidescan Sonar. Springer Praxis Books Geophysical Sciences Series. Springer, Berlin, Germany.

Bockheim, J.G., Gennadiyev, A.N., 2010. Soil-factorial models and earth-system science: a review. Geoderma 159 (3–4), 243–251.

Bogorodsky, V.V., Bentley, C.R., Gudmandsen, P.E., 1985. Radioglaciology. D Reidel, Dordrecht, the Netherlands.

Bogunovic, I., Trevisani, S., Pereira, P., Vukadinovic, V., 2018. Mapping soil organic matter in the Baranja region (Croatia): geological and anthropic forcing parameters. Sci. Total Environ. 643, 335–345.

Böhner, J., 2004. Regionalisierung bodenrelevanter Klimaparameter für das Niedersächsische Landesamt für Bodenforschung (NLfB) und die Bundesanstalt für Geowissenschaften und Rohstoffe (BGR). Arbeitshefte Boden 4, 17–66.

Böhner, J., 2005. Advancements and new approaches in climate spatial prediction and environmental modelling. Arbeitsberichte des Geographischen Instituts der HU zu Berlin 109, 49–90.

Böhner, J., Antonić, O., 2009. Land-surface parameters specific to topo-climatology. In: Hengl, T., Reuter, H.I. (Eds.), Geomorphometry: Concepts, Software, Applications. Developments in Soil Science, vol. 33. Elsevier, Amsterdam, the Netherlands, pp. 195–226.

Bohoyo, F., Larter, R.D., Galindo-Zaldívar, J., Leat, P.T., Maldonado, A., Tate, A.J., et al., 2019. Morphological and geological features of Drake Passage, Antarctica, from a new digital bathymetric model. J. Maps 15 (2), 49–59.

Bollati, I.M., Cavalli, M., 2021. Unraveling the relationship between geomorphodiversity and sediment connectivity in a small alpine catchment. Trans. GIS 25 (5), 2481–2500.

Bolshakov, V.D., Gaidaev, P.A., 1977. Theory of Mathematical Processing of Geodetic Measurements, second rev. enl. ed. Nedra, Moscow, USSR (in Russian).

Bolshakov, V.D., Markuze, Y.I., Golubev, V.V., 1989. Adjustment of Geodetic Measurements. Nedra, Moscow, USSR (in Russian).

Bolshiyanov, D.Y., 2011. Geomorphic structure of the Larsemann Hills. In: Lastochkin, A.N. (Ed.), Antarctica. Geomorphological Atlas. Karta, St. Petersburg, Russia, pp. 222–225 (in Russian).

Bolstad, P.V., Stowe, T., 1994. An evaluation of DEM accuracy: elevation, slope, and aspect. Photogrammet. Eng. Remote Sens. 60 (11), 1327–1332.

Bonetti, S., Bragg, A.D., Porporato, A., 2018. On the theory of drainage area for regular and non-regular points. Proc. R. Soc. A Math. Phys. Eng. Sci. 474 (2211), 20170693.

Booysen, R., Zimmerman, R., Lorenz, S., Gloaguen, R., Nex, P.A.M., Andreani, L., et al., 2019. Towards multiscale and multisource remote sensing mineral exploration using RPAS: a case study in the Lofdal carbonatite-hosted REE deposit, Namibia. Remote Sens. (Basel) 11 (21), 2500.

Borisenko, L.S., 1986. Geological criteria of seismic activity in Crimea. Seismol. Res. 9, 38–48 (in Russian).

Boronina, A.S., 2022. Large-scale outbursts of lakes in the Antarctic oases: current knowledge. Ice Snow 62 (1), 141–160 (in Russian).

Boulton, S.J., Stokes, M., 2018. Which DEM is best for analyzing fluvial landscape development in mountainous terrains? Geomorphology 310, 168–187.

Bourennane, H., King, D., Couturier, A., 2000. Comparison of kriging with external drift and simple linear regression for predicting soil horizon thickness with different sample densities. Geoderma 97 (3–4), 255–271.

Bourillet, J.F., Edy, C., Rambert, F., Satra, C., Loubrieu, B., 1996. Swath mapping system processing: bathymetry and cartography. Mar. Geophys. Res. 18 (2–4), 487–506.

Boussinesq, M.J., 1872. Sur les lignes de faîte et de thalweg. C. R. Hebd. Séances Acad. Sci. 75 (15), 835–837.

Bowman, W.J., 1968. Graphic Communication. Wiley, New York, NY

Bowring, B.R., 1996. Total inverse solutions for the geodesic and great elliptic. Survey Rev. 33 (261), 461–476.

Boyko, A.V., 1980. Methods and Tools for Automatization of Topographic Surveys. Nedra, Moscow, USSR (in Russian).

Boyko, A.V., Limontov, L.Y., 1975. Digital terrain models and topographic data collection data in large-scale mapping. Geodesy Cartogr. (10), 46–53 (in Russian).

Brassel, K., 1974. A model for automatic hill-shading. Am. Cartograph. 1 (1), 15–27.

Brassel, K.E., Weibel, R., 1988. A review and conceptual framework of automated map generalization. Int. J. Geogr. Inf. Syst. 2 (3), 229–244.

Brecheisen, Z.S., Cook, C.W., Heine, P.R., Richter, D.d.B., 2019. Micro-topographic roughness analysis (MTRA) highlights minimally eroded terrain in a landscape severely impacted by historic agriculture. Remote Sens. Environ. 222, 78–89.

Brenner, A.C., DiMarzio, J.P., Zwally, H.J., 2007. Precision and accuracy of satellite radar and laser altimeter data over the continental ice sheets. IEEE Trans. Geosci. Remote Sens. 45 (2), 321–331.

Bretar, F., Chehata, N., 2010. Terrain modeling from lidar range data in natural landscapes: a predictive and Bayesian framework. IEEE Trans. Geosci. Remote Sens. 48 (3 Pt 2), 1568–1578.

Breton de Champ, M., 1877. Mémoire sur les lignes de faîte et de thalweg que l'on est conduit à considérer en topographie. J. Math. Pure Appl. 3 Sér. 3, 99–114.

Breunig, M., 1996. Integration of Spatial Information for Geo-Information Systems. Lect. Notes Earth Sci., vol. 61. Springer, Berlin, Germany.

Břežný, M., Pánek, T., 2017. Deep-seated landslides affecting monoclinal flysch morphostructure: evaluation of LiDAR-derived topography of the highest range of the Czech Carpathians. Geomorphology 285, 44–57.

Brocklehurst, S.H., 2010. Tectonics and geomorphology. Progr. Phys. Geogr. 34 (3), 357–383.

Brovelli, M.A., Cannata, M., Longoni, U.M., 2004. LIDAR data filtering and DTM interpolation within GRASS. Trans. GIS 8 (2), 155–174.

Brown, A.R., 2011. Interpretation of Three-Dimensional Seismic Data. seventh ed. AAPG Memoir 42, SEG Investigations in Geophysics, no. 9 Society of Exploration Geophysicists, American Association of Petroleum Geologists, Tulsa, OK.

Brown, D.G., Bara, T.J., 1994. Recognition and reduction of systematic error in elevation and derivative surfaces from 7.5-minute DEMs. Photogrammet. Eng. Remote Sens. 60 (2), 189–194.

Brun, F., Buri, P., Miles, E.S., Wagnon, P., Steiner, J., Berthier, E., et al., 2016. Quantifying volume loss from ice cliffs on debris-covered glaciers using high-resolution terrestrial and aerial photogrammetry. J. Glaciol. 62 (234), 684–695.

Brun, F., Berthier, E., Wagnon, P., Kääb, A., Treichler, D., 2017. A spatially resolved estimate of High Mountain Asia glacier mass balances from 2000 to 2016. Nat. Geosci. 10 (9), 668–673.

Brun, F., Wagnon, P., Berthier, E., Shea, J.M., Immerzeel, W.W., Kraaijenbrink, P.D.A., et al., 2018. Ice cliffs contribution to the tongue-wide ablation of Changri Nup Glacier, Nepal, central Himalaya. Cryosphere 12 (11), 3439–3457.

Brunier, G., Fleury, J., Anthony, E.J., Gardel, A., Dussouillez, P., 2016. Close-range airborne structure-from-motion photogrammetry for high-resolution beach morphometric surveys: examples from an embayed rotating beach. Geomorphology 261, 76–88.

Brunsdon, C., Fotheringham, S., Charlton, M., 1998. Geographically weighted regression. J. R. Stat. Soc. Ser. D Stat. 47 (3), 431–443.

Brusnikin, E.S., Kreslavsky, M.A., Zubarev, A.E., Patratiy, V.D., Krasilnikov, S.S., Head, J.W., et al., 2016. Topographic measurements of slope streaks on Mars. Icarus 278, 52–61.

Bryukhanov, V.N., Mezhelovsky, N.V. (Eds.), 1987. Geology of the USSR from Space. Nedra, Moscow, USSR (in Russian).

Buckley, S.J., Howell, J.A., Enge, H.D., Kurz, T.H., 2008. Terrestrial laser scanning in geology: data acquisition, processing and accuracy considerations. J. Geol. Soc. London 165 (3), 625–638.

Bugaevsky, L.M., 1999. Theory of Cartographic Projections of Regular Surfaces. Zlatoust, Moscow, Russia (in Russian).

Bugnicourt, P., Guitet, S., Santos, V.F., Blanc, L., Sotta, E.D., Barbier, N., et al., 2018. Using textural analysis for regional landform and landscape mapping, Eastern Guiana Shield. Geomorphology 317, 23–44.

Bühler, Y., Adams, M.S., Bösch, R., Stoffel, A., 2016. Mapping snow depth in alpine terrain with unmanned aerial systems (UASs), potential and limitations. Cryosphere 10 (3), 1075–1088.

Bühler, Y., Adams, M., Stoffel, A., Boesch, R., 2017. Photogrammetric reconstruction of homogenous snow surfaces in alpine terrain applying near-infrared UAS imagery. Int. J. Remote Sens. 38 (8–10), 3135–3158.

Bühler, Y., von Rickenbach, D., Stoffel, A., Margreth, S., Stoffel, L., Christen, M., 2018. Automated snow avalanche release area delineation—validation of existing algorithms and proposition of a new object-based approach for large scale hazard indication mapping. Nat. Haz. Earth Syst. Sci. 18 (12), 3235–3251.

Bui, E.N., Moran, C.J., 2001. Disaggregation of polygons of surficial geology and soil maps using spatial modelling and legacy data. Geoderma 103 (1–2), 79–94.

Bui, E.N., Moran, C.J., 2003. A strategy to fill gaps in soil survey over large spatial extents: an example from the Murray–Darling basin of Australia. Geoderma 111 (1–2), 21–44.

Buivydaite, V.V., Mozgeris, G., 2007. Comparison of automated landform classification and soil mapping units at a farm level. Int. J. Ecol. Dev. 8 (F07), 26–38.

Bulatović, V., Sušić, Z., Ninkov, T., 2012. Estimate of the ASTER GDEM regional systematic errors and their removal. Int. J. Remote Sens. 33 (18), 5915–5926.

Burba, G.A., 1984. Application of Cartographic Method to Study Topography of the Terrestrial Planets (Ph.D. thesis). Lomonosov Moscow State University, Moscow, USSR (in Russian).

Burbank, D.W., Anderson, R.S., 2012. Tectonic Geomorphology, second ed. Wiley–Blackwell, Chichester, UK.

Burgess, J.S., Spate, A.P., Shevlin, J., 1994. The onset of deglaciation in the Larsemann Hills, Eastern Antarctica. Antarct. Sci. 6 (4), 491–495.

Bürgmann, R., Rosen, P.A., Fielding, E.J., 2000. Synthetic aperture radar interferometry to measure Earth's surface topography and its deformation. Annu. Rev. Earth Planet. Sci. 28, 169–209.

Buri, P., Pellicciotti, F., Steiner, J.F., Miles, E.S., Immerzeel, W.W., 2016. A grid-based model of backwasting of supraglacial ice cliffs on debris covered glaciers. Ann. Glaciol. 57 (71), 199–211.

Burrough, P.A., 1986. Principles of Geographical Information Systems for Land Resources Assessment. Clarendon Press, Oxford, UK.

Burrough, P.A., 1993. Soil variability: a late 20th century view. Soils Fert. 56 (5), 529–562.

Burrough, P.A., Bouma, J., Yates, S.R., 1994. The state of the art in pedometrics. Geoderma 62 (1–3), 311–326.

Burt, T.P., Butcher, D.P., 1985. Topographic controls of soil moisture distributions. J. Soil Sci. 36 (3), 469–486.

Burton, D.L., McMahon, S.K., Chen, Y., 2000. Influence of method of manure application on greenhouse gas emissions. In: Proceedings of the 43rd Annual Meeting of the Manitoba Society of Soil Science, Winnipeg, MB, 25–26 Jan. 2000. Manitoba Society of Soil Science, Winnipeg, MB, pp. 22–30.

Butler, J., Lane, S., Chandler, J., Porfiri, E., 2002. Through-water close range digital photogrammetry in flume and field environments. Photogrammet. Rec. 17 (99), 419–439.

Byrne, C.J., 2007. A large basin on the near side of the moon. Earth Moon Planet. 101 (3–4), 153–188.

Byzov, L., Sankov, V., 2024. The evolution of continental rifts revealed from the numerical modeling: a case study of the Baikal rift system, Siberia. J. Asian Earth Sci. 259, 105871.

Caflisch, R.E., 1998. Monte Carlo and quasi-Monte Carlo methods. Acta Num. 7, 1–49.

Caglar, B., Becek, K., Mekik, C., Ozendi, M., 2018. On the vertical accuracy of the ALOS World 3D-30m digital elevation model. Remote Sens. Lett. 9 (6), 607–615.

Cahill, T. Isacks, B.L., 1992. Seismicity and shape of the subducted Nazca Plate. J. Geophys. Res. Solid Earth 97 (B12), 17503–17529.

Cai, Y., Wan, F., Lang, S., Cui, X., Yao, Z., 2023. Multi-branch deep neural network for bed topography of Antarctica super-resolution: reasonable integration of multiple remote sensing data. Remote Sens. (Basel) 15 (5), 1359.

Caine, J.S., Evans, J.P., Forster, C.B., 1996. Fault zone architecture and permeability structure. Geology 24 (11), 1025–1028.

Callow, J.N., van Niel, K.P., Boggs, G.S., 2007. How does modifying a DEM to reflect known hydrology affect subsequent terrain analysis? J. Hydrol. 332 (1–2), 30–39.

Camiz, S., Poscolieri, M., 2018. An approach to DEM analysis for landform classification based on local gradients. Earth Sci. Inform. 11 (2), 287–305.

Campagna, D.J., Levandowski, D.W., 1991. The recognition of strike-slip fault systems using imagery, gravity, and topographic data sets. Photogrammet. Eng. Remote Sens. 57 (9), 1195–1201.

Campbell, J.B., 1979. Spatial variability of soils. Ann. Assoc. Am. Geogr. 69 (4), 544–556.

Canuto, M.A., Estrada-Belli, F., Garrison, T.G., Houston, S.D., Acuña, M.J., Kováč, M., et al., 2018. Ancient lowland Maya complexity as revealed by airborne laser scanning of northern Guatemala. Science 361 (6409), eaau0137.

Cao, W., Sofia, G., Tarolli, P., 2020. Geomorphometric characterisation of natural and anthropogenic land covers. Prog Earth Planet Sci 7, 2.

Carabajal, C.C., Harding, D.J., 2006. SRTM C-band and ICESat laser altimetry elevation comparisons as a function of tree cover and relief. Photogrammet. Eng. Remote Sens. 72 (3), 287–298.

Cardinale, M., Silvestro, S., Vaz, D.A., Michaels, T., Bourke, M.C., Komatsu, G., et al., 2016. Present-day aeolian activity in Herschel Crater, Mars. Icarus 265, 139–148.

Carey, S W., 1988. Theories of the Earth and Universe. A History of Dogma in the Earth Sciences. Stanford University Press, Stanford, CA.

Carless, D., Kulessa, B., Booth, A.D., Drocourt, Y., Sinnadurai, P., Street-Perrott, F.A., et al , 2021. An integrated geophysical and GIS based approach improves estimation of peatland carbon stocks. Geoderma 402, 115176.

Carlisle, B.H., 2005. Modelling the spatial distribution of DEM error. Trans. GIS 9 (4), 521–540.

Carrara, A., Guzzetti, F. (Eds.), 1995. Geographical Information Systems in Assessing Natural Hazards. Advances in Natural and Technological Hazards Research, vol. 5. Kluwer Academic Publishers, Dordrecht, the Netherlands.

Carrara, A., Cardinali, M., Detti, R., Guzzetti, F., Pasqui, V., Reichenbach, P., 1991. GIS techniques and statistical models in evaluating landslide hazard. Earth Surf. Process. Landf. 16 (5), 427–445.

Carrara, A., Bitelli, G., Carla, R., 1997. Comparison of techniques for generating digital terrain models from contour lines. Int. J. Geogr. Inf. Sci. 11 (5), 451–473.

Carré, F , Girard, M.C., 2002. Quantitative mapping of soil types based on regression kriging of taxonomic distances with landform and land cover attributes. Geoderma 110 (3–4), 241–263.

Carré, F., McBratney, A.B., 2005. Digital terron mapping. Geoderma 128 (3–4), 340–353.

Carré, F , Reuter, H.I., Daroussin, J., Scheurer, O., 2008. From a large to a small scale soil map: top-down against bottom-up approaches. Application to the Aisne soil map (France). In: Hartemink, A.E., McBratney, A., Mendonça-Santos, M.L. (Eds.), Digital Soil Mapping with Limited Data. Springer, Dordrecht, the Netherlands, pp. 203–212.

Carrera-Hernández, J.J., 2021. Not all DEMs are equal: an evaluation of six globally available 30 m resolution DEMs with geodetic benchmarks and LiDAR in Mexico. Remote Sens. Environ. 261, 112474.

Carrick Utsi, E., 2017. Ground Penetrating Radar: Theory and Practice. Butterworth-Heinemann, Oxford, UK.

Carrivick, J.L., Smith, M.W., Quincey, D.J., 2016. Structure from Motion in the Geosciences. Wiley, Chichester, UK.

Carson, C.J., Grew, E.S., 2007. Geology of the Larsemann Hills, Princess Elizabeth Land, Antarctica, 1:25,000 Scale Map. Geoscience Australia, Canberra, ACT.

Carson, M.A., Kirkby, M.J., 1972. Hillslope Form and Process. Cambridge University Press, Cambridge, MA.

Carter, J.R., 1988. Digital representations of topographic surfaces. Photogrammet. Eng. Remote Sens. 54 (11), 1577–1580.

Carter, J.R., 1992. The effect of data precision on the calculation of slope and aspect using gridded DEMs. Cartographica 29 (1), 22–34.

Carter, M.R. (Ed.), 1993. Soil Sampling and Methods of Analysis. Lewis, Boca Raton, FL.

Carvalho, R.C., Allan, B., Kennedy, D.M., Leach, C., O'Brien, S., Ierodiaconou, D., 2021. Quantifying decadal volumetric changes along sandy beaches using improved historical aerial photographic models and contemporary data. Earth Surf. Process. Landf. 46 (10), 1882–1897.

Catmull, E., Clark, J., 1978. Recursively generated B-spline surfaces on arbitrary topological meshes. Comput. Aided Des. 10 (6), 350–355.

Caubet, M., Dobarco, M.R., Arrouays, D., Minasny, B., Saby, N.P., 2019. Merging country, continental and global predictions of soil texture: lessons from ensemble modelling in France. Geoderma 337, 99–110.

Caumon, G., Collon-Drouaillet, P., Le Carlier de Veslud, C., Sausse, J., 2009. Surface-based 3D modeling of geological structures. Math. Geosci. 41 (8), 927–945.

Cavalli, M., Goldin, B., Comiti, F., Brardinoni, F., Marchi, L., 2017. Assessment of erosion and deposition in steep mountain basins by differencing sequential digital terrain models. Geomorphology 291, 4–16.

Cavazzi, S., Corstanje, R., Mayr, T., Hannam, J., Fealy, R., 2013. Are fine resolution digital elevation models always the best choice in digital soil mapping? Geoderma 195–196, 111–121.

Cayley, A., 1859. On contour and slope lines. London, Edinburgh, Dublin Philos. Mag. J. Sci. Ser. 4 (18), 264–268.

Central Board of Geodesy and Cartography, 1953. Topographic Map, Scale 1:300,000, Sheets VI-L-36 (Dzhankoi), VIII-L-36 (Sevastopol), IX-L-36 (Simferopol), V-L-36 (Kherson). Central Board of Geodesy and Cartography, Moscow, USSR (in Russian).

Central Board of Geodesy and Cartography, 1968. Topographic Map, Scale 1:1,000,000, Sheet L-38 (Pyatigorsk). Central Board of Geodesy and Cartography, Moscow, USSR (in Russian).

Čeru, T., Šegina, E., Gosar, A., 2017. Geomorphological dating of pleistocene conglomerates in central Slovenia based on spatial analyses of dolines using LiDAR and ground penetrating radar. Remote Sens. (Basel) 9 (12), 1213.

Cesium GS, 2012–2023. Cesium: The Platform for 3D Geospatial. Cesium GS, Philadelphia, PA. https://cesium.com. (Accessed 27 June 2024).

Chai, X., Shen, C., Yuan, X., Huang, Y., 2008. Spatial prediction of soil organic matter in the presence of different external trends with REML-EBLUP. Geoderma 148 (2), 159–166.

Chaney, N.W., Wood, E.F., McBratney, A.B., Hempel, J.W., Nauman, T.W., Brungard, C.W., et al., 2016. POLARIS: a 30-meter probabilistic soil series map of the contiguous United States. Geoderma 274, 54–67.

Chang, J.M.-H., Lam, Y.F., Lau, S.P.-W., Wong, W.-K., 2021. Development of fine-scale spatiotemporal temperature forecast model with urban climatology and geomorphometry in Hong Kong. Urban Clim. 37, 100816.

Chang, K.-T., Tsai, B.-W., 1991. The effect of DEM resolution on slope and aspect mapping. Cartogr. Geogr. Inf. Syst. 18 (1), 69–77.

Chang, K.-T., Merghadi, A., Yunus, A.P., Pham, B.T., Dou, J., 2019. Evaluating scale effects of topographic variables in landslide susceptibility models using GIS-based machine learning techniques. Sci. Rep. 9, 12296.

Chaplot, V., Walter, C., 2003. Subsurface topography to enhance the prediction of the spatial distribution of soil wetness. Hydrol. Process. 17 (13), 2567–2580.

Chaplot, V., Walter, C., Curmi, P., 2000. Improving soil hydromorphy prediction according to DEM resolution and available pedological data. Geoderma 97 (3–4), 405–422.

Chaplot, V., Bernoux, M., Walter, C., Curmi, P., Herpin, U., 2001. Soil carbon storage prediction in temperate hydromorphic soils using a morphologic index and digital elevation model. Soil Sci. 166 (1), 48–60.

Chaplot, V., Walter, C., Curmi, P., Lagacherie, P., King, D., 2004. Using the topography of the saprolite upper boundary to improve the spatial prediction of the soil hydromorphic index. Geoderma 123, 343–354.

Chase, D.Z., Chase, A.F., 2017. Caracol, Belize, and changing perceptions of Ancient Maya society. J. Archaeol. Res. 25 (3), 185–249.

Chebanenko, I.I., 1963. Principal Regularities of Fault Tectonics of the Earth's Crust and Its Problems. Ukrainian Academic Press, Kiev, USSR (in Russian).

Chebanenko, I.I., Fedorin, Y.V., 1983. On a new type of rotational tectonic lines in the Earth's lithosphere. Doklady Acad. Nauk SSSR 270 (2), 406–409 (in Russian).

Chekina, A.A., Khromykh, V.V., 2023. Visualization of floods in lowland rivers on a map: digital modelling of floodplain relief as a basis for mapping flood zones (exemplified by the Ob River in the southern part of the Tomsk Region). Geosph. Res. (2), 123–141 (in Russian).

Chekunov, A.V., Garkalenko, I.A., Kharchenko, G.E., 1965. Ancient Precambrian faults of the southern Russian Platform and their extension in the northern Black Sea coast area. Geophys. Commun. 14, 24–34 (in Russian).

Chekunov, A.V., Gavrilenko, N.M., Shnyukov, E.F., Tikhonenkov, E.P., Baisarovich, M.N., Yakovlev, E.V., et al., 1990. Geological structure and geodynamics of the region of the Crimean Nuclear Power Plant. Geophys. J. 12 (3), 3–27 (in Russian).

Chelton, D.B., Ries, J.C., Haines, B.J., Fu, L.-L., Callahan, P.S., 2001. Satellite altimetry. In: Fu, L.-L., Cazenave, A. (Eds.), Satellite Altimetry and Earth Sciences: A Handbook of Techniques and Applications. International Geophysics, vol. 69. Academic Press, San Diego, CA, pp. 1–131.

Chen, A., Bailey, J. (Eds.), 2011. Virtual globes in science. Comput. Geosci. 37 (1), 1–110.

Chen, C., Gao, Y., Li, Y., Bei, Y., 2022. Structure tensor-based interpolation for the derivation of accurate digital elevation models. Catena 208, 105733.

Chen, L.C., Rau, J.Y., 1993. A unified solution for digital terrain model and orthoimage generation from SPOT stereopairs. IEEE Trans. Geosci Remote Sens. 31 (6), 1243–1252.

Chen, Q., Liu, G., Ma, X., Mariethoz, G., He, Z., Tian, Y., et al., 2018. Local curvature entropy-based 3D terrain representation using a comprehensive quadtree. ISPRS J. Photogrammet. Remote Sens. 139, 30–45.

Chen, W., Pourghasemi, H.R., Kornejady, A., Zhang, N., 2017a. Landslide spatial modeling: introducing new ensembles of ANN, MaxEnt, and SVM machine learning techniques. Geoderma 305, 314–327.

Chen, W., Xie, X., Wang, J., Pradhan, B., Hong, H., Tien Bui, D., et al., 2017b. A comparative study of logistic model tree, random forest, and classification and regression tree models for spatial prediction of landslide susceptibility. Catena 151, 147–160.

Chen, Y., Zhou, C., Ai, S., Liang, Q., Zheng, L., Liu, R., et al., 2020. Dynamics of Dalk Glacier in East Antarctica derived from multisource satellite observations since 2000. Remote Sens. (Basel) 12 (11), 1809.

Chentsov, V.N., 1940. Morphometric indices of topography as applied to geomorphological maps. Trans. Inst. Geogr. 36, 69–71 (in Russian).

Cherednichenko, A.I., Burmistenko, V.M., Tokovenko, V.S., Chebanenko, I.I., 1966. An attempt of laboratory simulation of planetary faults (lineaments) of the Earth. Dopovidi Acad. Nauk Ukrainy (10), 1333–1336 (in Ukrainian).

Chiang, K.-W., Tsai, M.-L., Chu, C.-H., 2012. The development of an UAV borne direct georeferenced photogrammetric platform for ground control point free applications. Sensors 12 (7), 9161–9180.

Chibunichev, A.G., 2022. Photogrammetry. MIIGAiK Press, Moscow. 327 p. (in Russian).

Chibunichev, A.G., Mikhailov, A.P., Starshov, V.V., 2015. Automatic generation of a dense point cloud using multiple images, based on semi-global method of matching corresponding points. Izvestia Vuzov. Geodesy Aerophotosurveying 59 (2), 14–18 (in Russian).

Chigirev, A.A. (Ed.), 1976. Digital Filtering of Aerial Photos, Stereomodels, and Maps. Nedra, Leningrad, USSR (in Russian).

Chin, G., Brylow, S., Foote, M., Garvin, J., Kasper, J., Keller, J., et al., 2007. Lunar Reconnaissance Orbiter overview: the instrument suite and mission. Space Sci. Rev. 129 (4), 391–419.

Chopra, S., Marfurt, K.J., 2005. Seismic attributes—a historical perspective. Geophysics 70 (5), 3SO–28SO.

Chopra, S., Marfurt, K.J., 2007a. Seismic Attributes for Prospect Identification and Reservoir Characterization. Geophysical Developments, no. 11. Society of Exploration Geophysicists, Tulsa, OK.

Chopra, S., Marfurt, K.J., 2007b. Volumetric curvature attributes add value to 3D seismic data interpretation. Lead. Edge 26 (7), 856–867.

Chorley, R.J., Haggett, P., 1965. Trend-surface mapping in geographical research. Trans. Inst. Br. Geogr. 37, 47–67.

Chorowicz, J., 2005. The East African rift system. J. Afr. Earth Sci. 43 (1–3), 379–410.

Chorowicz, J., Bréard, J.-Y., Guillande, R., Morasse, C.-R., Prudon, D., Rudant, J.-P., 1991. Dip and strike measured systematically on digitized three-dimensional geological maps. Photogrammet. Eng. Remote Sens. 57 (4), 431–436.

Chorowicz, J., Collet, B., Bonavia, F.F., Mohr, P., Parrot, J.F., Korme, T., 1998. The Tana basin, Ethiopia: intra-plateau uplift, rifting and subsidence. Tectonophysics 295 (3–4), 351–367.

Chorowicz, J., Dhont, D., Gündoğdu, N., 1999. Neotectonics in the eastern North Anatolian fault region (Turkey) advocates crustal extension: mapping from SAR ERS imagery and digital elevation model. J. Struct. Geol. 21 (5), 511–532.

Choubin, B., Borji, M., Mosavi, A., Sajedi-Hosseini, F., Singh, V.P., Shamshirband, S., 2019. Snow avalanche hazard prediction using machine learning methods. J. Hydrol. 577, 123929.

Chu, H.-J., Chen, Y.-C., Ali, M.Z., Höfle, B., 2019. Multi-parameter relief map from high-resolution DEMs: a case study of mudstone badland. Int. J. Environ. Res. Public Health 16 (7), 1109.

Chudley, T.R., Christoffersen, P., Doyle, S.H., Abellan, A., Snooke, N., 2019. High accuracy UAV photogrammetry of ice sheet dynamics with no ground control. Cryosphere 13 (3), 955–968.

Chudley, T.R., Christoffersen, P., Doyle, S.H., Dowling, T.P.F., Law, R., Schoonman, C.M., et al., 2021. Controls on water storage and drainage in crevasses on the Greenland ice sheet. J. Geophys. Res. Earth Surf. 126 (9), e2021JF006287.

Chui, C.K., 1992. An Introduction to Wavelets. Academic Press, San Diego, CA.

Cialella, A.T., Dubayah, R., Lawrence, W., Levine, E., 1997. Predicting soil drainage class using remotely sensed and digital elevation data. Photogrammet. Eng. Remote Sens. 63 (2), 171–178.

Ciglič, R., Čonč, Š., Breg Valjavec, M., 2022. The impact of digital elevation model preprocessing and detection methods on karst depression mapping in densely forested Dinaric Mountains. Remote Sens. (Basel) 14 (10), 2416.

Cimoli, E., Marcer, M., Vandecrux, B., Bøggild, C.E., Williams, G., Simonsen, S.B., 2017. Application of low-cost UASs and digital photogrammetry for high-resolution snow depth mapping in the Arctic. Remote Sens. (Basel) 9 (11), 1144.

Civera, J., Davison, A.J., Martínez Montiel, J.M., 2012. Structure from Motion Using the Extended Kalman Filter. Springer Tracts in Advanced Robotics, vol. 75. Springer, Berlin, Germany.

Clapuyt, F., Vanacker, V., van Oost, K., 2016. Reproducibility of UAV-based earth topography reconstructions based on structure-from-motion algorithms. Geomorphology 260, 4–15.

Clark, L.A., Pregibon, D., 1997. Tree-based models. In: Chambers, J.M., Hastie, T.J. (Eds.), Statistical Models in S. Chapman & Hall Computer Science Series. Chapman & Hall, London, UK, pp. 377–419.

Clark, R.L., Lee, R., 1998. Development of topographic maps for precision farming with kinematic GPS. Trans. Am. Soc. Agric. Eng. 41 (4), 909–916.

Clarke, D.D., Phillips, C.C., 2003. Three-dimensional geologic modeling and horizontal drilling bring more oil out of the Wilmington oil field of southern California. In: Carr, T.R., Mason, E.P., Feazel, C.T. (Eds.), Horizontal Wells: Focus on the Reservoir. AAPG Methods in Exploration, vol. 14. American Association of Petroleum Geologists, Tulsa, OK, pp. 27–47.

Clarke, J.I., 1966. Morphometry from maps. In: Dury, G.H. (Ed.), Essays in Geomorphology. Heinemann, London, UK, pp. 235–274.

Clarke, K.C., 1988. Scale-based simulation of topographic relief. Am. Cartograph. 15 (2), 173–181.

Clarke, K.C., Romero, B.E., 2017. On the topology of topography: a review. Cartogr. Geogr. Inf. Sci. 44 (3), 271–282.

Claus, M., 1984. Digital terrain models through digital stereo correlation. Photogrammetria 39 (4–6), 183–192.

Clayton, J.S., Ehrlich, W.A., Cann, D.B., Day, J.H., Marshall, I.B., 1977. Soils of Canada, vol. 1: Soil Report. Research Branch, Canada Department of Agriculture, Ottawa, ON.

Clenshaw, C.W., Hayes, J.G., 1965. Curve and surface fitting. IMA J. Appl. Math. 1 (2), 164–183.

Clubb, F.J., Mudd, S.M., Attal, M., Milodowski, D.T., Grieve, S.W.D., 2016. The relationship between drainage density, erosion rate, and hilltop curvature: implications for sediment transport processes. J. Geophys. Res. Earth Surf. 121 (10), 1724–1745.

Clubb, F.J., Mudd, S.M., Milodowski, D.T., Valters, D.A., Slater, L.J., Hurst, M.D., et al., 2017. Geomorphometric delineation of floodplains and terraces from objectively defined topographic thresholds. Earth Surf. Dyn. 5, 369–385.

Coatléven, J., 2020. Some multiple flow direction algorithms for overland flow on general meshes. ESAIM. Math. Model. Num. Anal. 54 (6), 1917–1949.

Codilean, A.T., Bishop, P., Hoey, T.B., 2006. Surface process models and the links between tectonics and topography. Progr. Phys. Geogr. 30 (3), 307–333.

Cole, N.J., Boettinger, J.L., 2007. Pedogenic understanding raster classification methodology for mapping soils, Powder River Basin, Wyoming, USA. In: Lagacherie, P., McBratney, A.B., Voltz, M. (Eds.), Digital Soil Mapping: An Introductory Perspective. Developments in Soil Science, vol. 31. Elsevier, Amsterdam, the Netherlands, pp. 377–388.

Colgan, W., Rajaram, H., Abdalati, W., McCutchan, C., Mottram, R., Moussavi, M.S., et al., 2016. Glacier crevasses: observations, models, and mass balance implications. Rev. Geophys. 54 (1), 119–161.

Collet, B., Parrot, J.F., Taud, H., 2000a. Orientation of absolute African plate motion revealed by tomomorphometric analysis of the Ethiopian dome. Geology 28 (12), 1147–1149.

Collet, B., Taud, H., Parrot, J.F., Bonavia, F., Chorowicz, J., 2000b. A new kinematic approach for the Danakil block using a digital elevation model representation. Tectonophysics 316 (3–4), 343–357.

Collin, A., Hench, J.L., Pastol, Y., Planes, S., Thiault, L., Schmitt, R.J., et al., 2018. High resolution topobathymetry using a Pleiades-1 triplet: Moorea Island in 3D. Remote Sens. Environ. 208, 109–119.

Colomina, I., Molina, P., 2014. Unmanned aerial systems for photogrammetry and remote sensing: a review. ISPRS J. Photogrammet. Remote Sens. 92, 79–97.

Colucci, R.R., Boccali, C., Žebre, M., Guglielmin, M., 2016. Rock glaciers, protalus ramparts and pronival ramparts in the south-eastern Alps. Geomorphology 269, 112–121.

Conrad, O., Bechtel, B., Bock, M., Dietrich, H., Fischer, E., Gerlitz, L., et al., 2015. System for automated geoscientific analyses (SAGA) v. 2.1 4. Geosci. Model Dev. 8 (7), 1991–2007.

Conrad, R., 1996. Soil microorganisms as controllers of atmospheric trace gases (H_2, CO, CH_4, OCS, N_2O, and NO). Microbiol. Rev. 60 (4), 609–640.

Conway, S.J., Balme, M.R., 2016. A novel topographic parameterization scheme indicates that martian gullies display the signature of liquid water. Earth Planet. Sci. Lett. 454, 36–45.

Cook, R.L., Torrance, K.E., 1982. A reflectance model for computer graphics. ACM Trans. Graph. 1 (1), 7–24.

Cook, S.E., Corner, R.J., Grealish, G., Gessler, P.E., Chartres, C.J., 1996. A rule-based system to map soil properties. Soil Sci. Soc. Am. J. 60 (6), 1893–1900.

Cooley, J.W., Tukey, J.W., 1965. An algorithm for the machine calculation of complex Fourier series. Math. Comput. 19 (90), 297–301.

Cooper, M.A.R., 1998. Datums, coordinates and differences. In: Lane, S.N., Richards, K.S., Chandler, J.H. (Eds.), Landform Monitoring, Modelling and Analysis. Wiley, Chichester, UK, pp. 21–35.

Copping, J.P., Stewart, B.D., McClean, C.J., Hancock, J., Rees, R., 2018. Does bathymetry drive coastal whale shark (*Rhincodon typus*) aggregations? PeerJ 6, e4904.

Corre, M.D., van Kessel, C., Pennock, D.J., 1996. Landscape and seasonal patterns of nitrous oxide emissions in a semiarid region. Soil Sci. Soc. Am. J. 60 (6), 1806–1815.

Courant, R., Hilbert, D., 1989. Methods of Mathematical Physics, vol. 1. Wiley, New York, NY.

Cozzi, P., Ring, K., 2011. 3D Engine Design for Virtual Globes. A K Peters/CRC Press, Boca Raton, FL.

Crippen, R., Buckley, S., Agram, P., Belz, E., Gurrola, E., Hensley, S., et al., 2016. NASADEM global elevation model: methods and progress. In: Halounova, L., Šafář, V., Jiang, J., Olešovská, H., Dvořáček, P., Holland, D., et al. (Eds.), XXIII ISPRS Congress, Commission IV, 12–19 July 2016, Prague, Czech Republic. Int. Arch. Photogramm. Remote Sens. Spatial Inf. Sci. XLI (B4), pp. 125–128.

Cucchiaro, S., Cazorzi, F., Marchi, L., Crema, S., Beinat, A., Cavalli, M., 2019. Multi-temporal analysis of the role of check dams in a debris-flow channel: linking structural and functional connectivity. Geomorphology 345, 106844.

Cucchiaro, S., Maset, E., Cavalli, M., Crema, S., Marchi, L., Beinat, A., 2020. How does co-registration affect geomorphic change estimates in multi-temporal surveys? GISci. Remote Sens. 57 (5), 611–632.

Cunha, N.S., Magalhães, M.R., Domingos, T., Abreu, M.M., Küpfer, C., 2017. The land morphology approach to flood risk mapping: an application to Portugal. J. Environ. Manage. 193, 172–187.

Curcio, A.C., Peralta, G., Aranda, M., Barbero, L., 2022. Evaluating the performance of high spatial resolution UAV-photogrammetry and UAV-LiDAR for salt marshes: the Cádiz Bay study case. Remote Sens. (Basel) 14 (15), 3582.

D'Oleire-Oltmanns, S., Marzolff, I., Peter, K.D., Ries, J.B., 2012. Unmanned aerial vehicle (UAV) for monitoring soil erosion in Morocco. Remote Sens. (Basel) 4 (11), 3390–3416.

Da Ros, D., Borga, M., 1997. Use of digital elevation model data for the derivation of the geomorphological instantaneous unit hydrograph. Hydrol. Process. 11 (1), 13–33.

Dąbski, M., Zmarz, A., Pabjanek, P., Korczak-Abshire, M., Karsznia, I., Chwedorzewska, K.J., 2017. UAV-based detection and spatial analyses of periglacial landforms on Demay Point (King George Island, South Shetland Islands, Antarctica). Geomorphology 290, 29–38.

Dai, C., Howat, I.M., 2017. Measuring lava flows with ArcticDEM: application to the 2012–2013 eruption of Tolbachik, Kamchatka. Geophys. Res. Lett. 44 (24), 12133–12140.

Dai, C., Durand, M., Howat, I.M., Altenau, E.H., Pavelsky, T.M., 2018. Estimating river surface elevation from ArcticDEM. Geophys. Res. Lett. 45 (7), 3107–3114.

Dai, C., Howat, I.M., Freymueller, J.T., Vijay, S., Jia, Y., 2020. Characterization of the 2008 phreatomagmatic eruption of Okmok from ArcticDEM and InSAR: deposition, erosion, and deformation. J. Geophys. Res. Solid Earth 125 (6), e2019JB018977.

Dai, W., Hu, G., Huang, N., Zhang, P., Yang, X., Tang, G., 2019a. A contour-directional detection for deriving terrace ridge from open source images and digital elevation models. IEEE Access 7, 129215–129224.

Dai, W., Yang, X., Na, J., Li, J., Brus, D., Xiong, L., et al., 2019b. Effects of DEM resolution on the accuracy of gully maps in loess hilly areas. Catena 177, 114–125.

Dai, W., Na, J., Huang, N., Hu, G., Yang, X., Tang, G., et al., 2020. Integrated edge detection and terrain analysis for agricultural terrace delineation from remote sensing images. Int. J. Geogr. Inf. Sci. 34 (3), 484–503.

Dall'Asta, E., Forlani, G., Roncella, R., Santise, M., Diotri, F., Morra di Cella, U., 2017. Unmanned aerial systems and DSM matching for rock glacier monitoring. ISPRS J. Photogrammet. Remote Sens. 127, 102–114.

Danchenkov, A., Belov, N., Stont, Z., 2019. Using the terrestrial laser scanning technique for aeolian sediment transport assessment in the coastal zone in seasonal scale. Estuar. Coast. Shelf Sci. 223, 105–114.

Daniel, W.W., 2000. Applied Nonparametric Statistics, second ed. Cengage Learning, Belmont, CA.

Danielson, J.J., Gesch, D.B., 2011. Global Multi-Resolution Terrain Elevation Data 2010 (GMTED2010). USGS Open-File Report 2011-1073. US Geological Survey, Reston, VA.

Danilov, D.L., Zhigljavsky, A.A. (Eds.), 1997. Principal Components of Time Series: The "Caterpillar" Method. St. Petersburg University Press, St. Petersburg, Russia (in Russian).

Danko, D.M., 1992. The digital chart of the World project. Photogrammet. Eng. Remote Sens. 58 (8), 1125–1128.

Darabi, H., Choubin, B., Rahmati, O., Haghighi, A.T., Pradhan, B., Kløve, B., 2019. Urban flood risk mapping using the GARP and QUEST models: a comparative study of machine learning techniques. J. Hydrol. 569, 142–154.

Daubechies, I., 1992. Ten Lectures on Wavelets. Society for Industrial and Applied Mathematics, Philadelphia, PA.

Davis, F.W., Dozier, J., 1990. Information analysis of a spatial database for ecological land classification. Photogrammet. Eng. Remote Sens. 56 (8), 605–613.

Davis, J.C., 2002. Statistics and Data Analysis in Geology, third ed. Wiley, New York, NY.

Davis, P.J., 1975. Interpolation and Approximation. Dover, New York, NY.

De Boer, D.H., 1992. Hierarchies and spatial scale in process geomorphology: a review. Geomorphology 4 (5), 303–318.

De Bruin, S., Stein, A., 1998. Soil-landscape modelling using fuzzy *c*-means clustering of attribute data derived from a Digital Elevation Model (DEM). Geoderma 83 (1–2), 17–33.

De Donatis, M., Borraccini, F., Susini, S., 2009. Sheet 280–Fossombrone 3D: a study project for a new geological map of Italy in three dimensions. Comput. Geosci. 35 (1), 19–32.

De Floriani, L., Mirra, D., Puppo, E., 1993. Extracting contour lines from a hierarchical surface model. Comput. Graph. Forum 12 (3), 249–260.

De Floriani, L., Marzano, P., Puppo, E., 1994. Line-of-sight communication on terrain models. Int. J. Geogr. Inf. Syst. 8 (4), 329–342.

De Floriani, L., Magillo, P., Puppo, E., 1997. Multiresolution representation and reconstruction of triangulated surfaces. In: Arcelli, C., Cordella, L., Sanniti di Baja, G. (Eds.), Advances in Visual Form Analysis. World Scientific, Singapore, pp. 140–149.

De Floriani, L., Magillo, P., Puppo, E., 2000. VARIANT: a system for terrain modeling at variable resolution. GeoInformatica 4 (3), 287–315.

De Michele, C., Avanzi, F., Passoni, D., Barzaghi, R., Pinto, L., Dosso, P., et al., 2016. Using a fixed-wing UAS to map snow depth distribution: an evaluation at peak accumulation. Cryosphere 10 (2), 511–522.

De Moustier, C., 1988. State of the art in swath bathymetry survey systems. Int. Hydrogr. Rev. 65 (2), 25–54.

De Reu, J., Bourgeois, J., Bats, M., Zwertvaegher, A., Gelorini, V., de Smedt, P., et al., 2013. Application of the topographic position index to heterogeneous landscapes. Geomorphology 186, 39–49.

De Saint-Venant, M., 1852. Surfaces à plus grande pente constitutées sur des lignes courbes. Extraits Procès-Verbaux Séances Soc. Philomatique Paris: 5 Sér. 17, 24–30.

De Waele, J., Fabbri, S., Santagata, T., Chiarini, V., Columbu, A., Pisani, L., 2018. Geomorphological and speleogenetical observations using terrestrial laser scanning and 3D photogrammetry in a gypsum cave (Emilia Romagna, N. Italy). Geomorphology 319, 47–61.

Debella-Gilo, M., Etzelmüller, B., 2009. Spatial prediction of soil classes using digital terrain analysis and multinomial logistic regression modeling integrated in GIS: examples from Vestfold County, Norway. Catena 77 (1), 8–18.

Declercq, F.A.N., 1996. Interpolation methods for scattered sample data: accuracy, spatial patterns, processing time. Cartogr. Geogr. Inf. Syst. 23 (3), 128–144.

Dedkova, A.A., Florinsky, I.V., 2023. Geomorphometry and microelectronic metrology: converged realms. Trans. GIS 27 (6), 1642–1661.

Dedus, A.F., Dedus, F.F., Makhortykh, S.A., Ustinin, M.N., 1995. Generalized spectral-analytic method. Part I. Theoretical foundations. In: Kuznetsov, N. A., Soifer, V.A. (Eds.), 5th International Workshop on Image Processing and Computer Optics (DIP–94), Samara, Russia, 22–26 Aug. 1994. Proc. SPIE 2363, pp. 109–112.

Dedus, F.F., Makhortykh, S.A., Ustinin, M.N., 1995. Generalized spectral-analytic method. Part II. Applications. In: Kuznetsov, N.A., Soifer, V.A. (Eds.), 5th International Workshop on Image Processing and Computer Optics (DIP–94), Samara, Russia, 22–26 Aug. 1994. Proc. SPIE 2363, pp. 113–118.

Dedus, F.F., Makhortykh, S.A., Ustinin, M.N., Dedus, A.F., 1999. Generalized Spectral Analytical Method for Data Set Processing: Problems of Image Analysis and Pattern Recognition. Mashinostroenie, Moscow, Russia (in Russian).

Dedus, F.F., Makhortykh, S.A., Ustinin, M.N., 2002. Application of the generalized spectral-analytic method in information problems. Pattern Recogn. Image Anal. 12 (4), 429–437.

Dedus, F.F., Kulikova, L.I., Pankratov, A.N., Tetuev, R.K., 2004. Classical Orthogonal Bases in Problems of Analytical Description and Processing of Information Signals. Moscow University Press, Moscow, Russia (in Russian).

Deffontaines, B., Lacombe, O., Angelier, J., Chu, H.T., Mouthereau, F., Lee, C.T., et al., 1997. Quaternary transfer faulting in the Taiwan Foothills: evidence from a multisource approach. Tectonophysics 274 (1–3), 61–82.

Dehecq, A., Millan, R., Berthier, E., Gourmelen, N., Trouve, E., 2016. Elevation changes inferred from TanDEM-X data over the Mont-Blanc area: impact of the X-band interferometric bias. IEEE J. Sel. Topics Appl. Earth Obs. Remote Sens. 9 (8), 3870–3882.

Dekavalla, M., Argialas, D., 2017a. Evaluation of a spatially adaptive approach for land surface classification from digital elevation models. Int. J. Geogr. Inf. Sci. 31 (10), 1978–2000.

Dekavalla, M., Argialas, D., 2017b. Object-based classification of global undersea topography and geomorphological features from the SRTM30_PLUS data. Geomorphology 288, 66–82.

Delaney, A.J., Arcone, S.A., 1995. Detection of Crevasses Near McMurdo Station, Antarctica with Airborne Short-Pulse Radar. U.S. Army Corps of Engineers, Cold Regions Research & Engineering Laboratory Special Report 95-7.

Delarue, F., Cornu, S., Daroussin, J., Salvador-Blanes, S., Bourennane, H., Albéric, P., et al., 2009. 3D representation of soil distribution: an approach for understanding pedogenesis. C. R. Geosci. 341 (6), 486–494.

Delaunay, B., 1934. Sur la sphère vide (a la memoire de Georges Voronoi). Bull. Acad. Sci. USSR, Classes Sci. Math. Natur. 7 (6), 793–800.

Delchiaro, M., Fioramonti, V., Della Seta, M., Cavinato, G.P., Mattei, M., 2021. Fluvial inverse modeling for inferring the timing of Quaternary uplift in the Simbruini range (Central Apennines, Italy). Trans. GIS 25 (5), 2455–2480.

Deng, Y., 2007. New trends in digital terrain analysis: landform definition, representation, and classification. Progr. Phys. Geogr. 31 (4), 405–419.

Depicker, A., Jacobs, L., Delvaux, D., Havenith, H.-B., Mateso, J.-C.M., Govers, G., et al., 2020. The added value of a regional landslide susceptibility assessment: the western branch of the East African Rift. Geomorphology 353, 106886.

Desmet, P.J.J., 1997. Effects of interpolation errors on the analysis of DEMs. Earth Surf. Process. Landf. 22 (6), 563–580.

Desmet, P.J.J., Govers, G., 1995. GIS-based simulation of erosion and deposition patterns in an agricultural landscape: a comparison of model results with soil map information. Catena 25 (1–4), 389–401.

Desmet, P.J.J., Govers, G., 1996. Comparison of routing algorithms for digital elevation models and their implications for predicting ephemeral gullies. Int. J. Geogr. Inf. Sci. 10 (3), 311–332.

Desmet, P.J.J., Govers, G., 1997. Two-dimensional modelling of the within-field variation in rill and gully geometry and location related to topography. Catena 29 (3–4), 283–306.

Deutsch, A.N., Head, J.W., Neumann, G.A., Kreslavsky, M.A., Barker, M.K., 2020. Assessing the roughness properties of circumpolar Lunar craters: implications for the timing of water-ice delivery to the Moon. Geophys. Res. Lett. 47 (15), e2020GL087782.

Devdariani, A.S., 1967. Mathematical Analysis in Geomorphology. Nedra, Moscow, USSR (in Russian).

DeWalle, D.R., Rango, A., 2008. Principles of Snow Hydrology. Cambridge University Press, Cambridge, UK.

Dharumarajan, S., Hegde, R., Singh, S.K., 2017. Spatial prediction of major soil properties using Random Forest techniques—a case study in semi-arid tropics of South India. Geoderma Reg. 10, 154–162.

Dhingra, R.D., Barnes, J.W., Yanites, B.J., Kirk, R.L., 2018. Large catchment area recharges Titan's Ontario Lacus. Icarus 299, 331–338.

Dhont, D., Chorowicz, J., 2006. Review of the neotectonics of the eastern Turkish–Armenian Plateau by geomorphic analysis of digital elevation model imagery. Int. J. Earth Sci. 95 (1), 34–49.

Dhont, D., Chorowicz, J., Yürür, T., Froger, J.-L., Köse, O., Gündogdu, N., 1998a. Emplacement of volcanic vents and geodynamics of Central Anatolia, Turkey. J. Volcanol. Geotherm. Res. 85 (1–4), 33–54.

Dhont, D., Chorowicz, J., Yürür, T., Köse, O., 1998b. Polyphased block tectonics along the North Anatolian Fault in the Tosya Basin area (Turkey). Tectonophysics 299 (1–3), 213–227.

Dhont, D., Luxey, P., Chorowicz, J., 2005. 3-D modeling of geologic maps from surface data. Am. Assoc. Pet. Geol. Bull. 89 (11), 1465–1474.

Dietrich, J.T., 2016. Bathymetric structure-from-motion: extracting shallow stream bathymetry from multi-view stereo photogrammetry. Earth Surf. Process. Landf. 42 (2), 355–364.

Dikau, R., 1988. Case studies in the development of derived geomorphic maps. Geol. Jahr. A104, 329–338.

Dikau, R., 1989. The application of a digital relief model to landform analysis in geomorphology. In: Raper, J. (Ed.), Three Dimensional Applications in GIS. Taylor and Francis, London, UK, pp. 51–77.

Dikau, R., 1990. Derivatives from detailed geoscientific maps using computer methods. Zeit. Geomorph. Suppl. 80, 45–55.

Dikau, R., Schmidt, J., 1999. Georeliefklassifikation. In: Schneider-Sliwa, R., Schaub, D., Gerold, G. (Eds.), Angewandte Landschaftsökologie: Grundlagen und Methoden. Springer, Berlin, Germany, pp. 217–244.

Dikau, R., Brabb, E.E., Mark, R.K., Pike, R.J., 1995. Morphometric landform analysis of New Mexico. Zeit. Geomorph. Suppl. 101, 109–126.

Dimarzio, J.P., Brenner, A.C., Fricker, H.A., Schutz, B.E., Shuman, C.A., Zwally, H.J., 2005. Digital elevation models of the Antarctic and Greenland ice sheets from ICESat. In: AGU Fall Meeting 2005, San Francisco, CA, 5–9 Dec. 2005. American Geophysical Union, San Francisco, CA. C51B-0276.

Ding, H., Na, J., Huang, X., Tang, G., Liu, K., 2018. Stability analysis unit and spatial distribution pattern of the terrain texture in the northern Shaanxi Loess Plateau. J. Mountain Sci. 15 (3), 577–589.

Dirscherl, M., Rossi, C., 2018. Geomorphometric analysis of the 2014–2015 Bárðarbunga volcanic eruption, Iceland. Remote Sens. Environ. 204, 244–259.

Dixon, T.H., Naraghi, M., McNutt, M.K., Smith, S.M., 1983. Bathymetric prediction from SEASAT altimeter data. J. Geophys. Res. Oceans 88 (C3), 1563–1571.

DJI, 2018–2024. Support Phantom 4. DJI, Shenzhen, China. https://www.dji.com/support/product/phantom-4. (Accessed 27 June 2024).

Dmitriev, E.A., 1998. The concept of relief flexure in relation to soil science. Eurasian Soil Sci. 31 (3), 338–346.

Dobos, E., Micheli, E., Baumgardner, M.F., Biehl, L., Helt, T., 2000. Use of combined digital elevation model and satellite radiometric data for regional soil mapping. Geoderma 97 (3–4), 367–391.

Dobos, E., Micheli, E., Montanarella, L., 2007. The population of a 500-m resolution soil organic matter spatial information system for Hungary. In: Lagacherie, P., McBratney, A.B., Voltz, M. (Eds.), Digital Soil Mapping: An Introductory Perspective. Developments in Soil Science, vol. 31. Elsevier, Amsterdam, the Netherlands, pp. 487–495.

Dokuchaev, P.M., 2017. Constructing a Digital Soil Map and Soil Carbon Map by Digital Soil Mapping Methods (Exemplified by the Vyatka–Kama Province of Soddy-Podzolic Soils of the Southern Taiga) (Ph.D. thesis). Lomonosov Moscow State University, Moscow, Russia (in Russian).

Dokuchaev, V.V., 1883. Russian Chernozem: Report to the Imperial Free Economic Society. Decleron and Evdokimov Press, St. Petersburg, Russia (in Russian).

Dokuchaev, V.V., 1886. Materials for Land Evaluation of the Nizhny Novgorod Governorate. Natural and Historical Part: Report to the Nizhny Novgorod Governorate Zemstvo, vol. 1: Key Points in the History of Land Evaluation in the European Russia, with Classification of Russian Soils. Evdokimov Press, St. Petersburg, Russia (in Russian).

Dokuchaev, V.V., 1891. On the relationships between, on the one hand, age and elevation of the terrain, and, on the other hand, character and distribution of chernozems, forest soils, and solonets. Rev. Sci. Natur. (1), 1–16; (2), 57–67; (3), 112–123 (in Russian).

Dokuchaev, V.V., 1899. Report to the Transcaucasian Statistical Committee on Soil Taxonomy in General and Especially for the Transcaucasia: Horizontal and Vertical Soil Zones. Office Press of the Civilian Affairs Commander-in-Chief in the Caucasus, Tiflis, Russia (in Russian).

Dolgikh, A.V., Mergelov, N.S., Abramov, A.A., Lupachev, A.V., Goryachkin, S.V., 2015. Soils of Enderby Land. In: Bockheim, G. (Ed.), The Soils of Antarctica. Springer, Cham, Switzerland, pp. 45–63.

Dolitsky, A.V., Kiyko, I.A., 1963. On the causes of deformation of the Earth's crust. In: Nalivkin, D.V., Tupitsin, N.V. (Eds.), Problems of Planetary Geology. Gosgeoltekhizdat, Moscow, USSR, pp. 291–312 (in Russian).

Dong, Y., Zhao, J., Li, C., Liao, M., 2022. Gapless-REMA100: a gapless 100-m reference elevation model of Antarctica with voids filled by multi-source DEMs. ISPRS J. Photogrammet. Remote Sens. 186, 70–82.

Dornik, A., Drăguţ, L., Urdea, P., 2018. Classification of soil types using geographic object-based image analysis and random forests. Pedosphere 28 (6), 913–925.

Dornik, A., Cheţan, M.A., Drăguţ, L., Dicu, D.D., Iliuţă, A., 2022. Optimal scaling of predictors for digital mapping of soil properties. Geoderma 405, 115453.

Dorschel, B., Hehemann, L., Viquerat, S., Warnke, F., Dreutter, S., Schulze Tenberge, Y., et al., 2022. The International Bathymetric Chart of the Southern Ocean Version 2. Sci. Data 9, 275.

Dos Reis, A.A., Franklin, S.E., de Mello, J.M., Acerbi Jr., F.W., 2019. Volume estimation in a *Eucalyptus* plantation using multi-source remote sensing and digital terrain data: a case study in Minas Gerais State, Brazil. Int. J. Remote Sens. 40 (7), 2683–2702.

Dos Reis, A.A., Franklin, S.E., Acerbi Jr., F.W., Ferraz Filho, A.C., de Mello, J.M., 2022. Classification of *Eucalyptus* plantation site index (SI) and mean annual increment (MAI) prediction using DEM-based geomorphometric and climatic variables in Brazil. Geocarto Int. 37 (5), 1256–1273.

Douglas, D.H., 1986. Experiments to locate ridges and channels to create a new type of digital elevation model. Cartographica 23 (4), 29–61.

Doumit, J.A., 2018. Multiscale landforms classification based on UAV datasets. Sustain. Environ. 3 (2), 128–131.

Doumit, J.A., Sakr, S.C., 2021. Urban topo climatic factors effects on heating islands based on high-resolution digital surface models. Afr. J. Land Policy Geospat. Sci. 4 (1), 1–14.

Doyle, F.J., 1978. Digital terrain models: an overview. Photogrammet. Eng. Remote Sens. 44 (12), 1481–1485.

Dozier, J., Frew, J., 1990. Rapid calculation of terrain parameters for radiation modeling from digital elevation data. IEEE Trans. Geosci. Remote Sens. 28 (5), 963–969.

Drăguţ, L., Dornik, A., 2016. Land-surface segmentation as a method to create strata for spatial sampling and its potential for digital soil mapping. Int. J. Geogr. Inf. Sci. 30 (7), 1359–1376.

Drummond, J., Christou, J., 2008. Triaxial ellipsoid dimensions and rotational poles of seven asteroids from Lick Observatory adaptive optics images, and of Ceres. Icarus 197 (2), 480–496.

Du Carla, M., 1782. Expression des Nivellemens, ou Méthode Nouvelle pour Marquer Rigoureusement sur les Cartes Terrestres et Marines les Hauteurs et les Configurations du Terrain. Cellot, Paris, France.

Du, L., You, X., Li, K., Meng, L., Cheng, G., Xiong, L., et al., 2019. Multi-modal deep learning for landform recognition. ISPRS J. Photogrammet. Remote Sens. 158, 63–75.

Duan, X., Li, L., Zhu, H., Shen, Y., 2017. A high-fidelity multiresolution digital elevation model for Earth systems. Geosci. Model Dev. 10 (1), 239–253.

Dubayah, R., Rich, P.M., 1995. Topographic solar radiation models for GIS. Int. J. Geogr. Inf. Syst. 9 (4), 405–419.

Dubyansky, A.A., Khakman, S.A. (Eds.), 1949. Geology of the USSR, vol. 6: Braynsk, Orel, Kursk, Voronezh, and Tambov Regions, pt. 1: Geological Description. Geological Literature State Press, Moscow, USSR (in Russian).

Durand, P., Robson, A., Neal, C., 1992. Modelling the hydrology of submediterranean mountain catchments (Mont Lozere, France), using TOPMODEL: initial results. J. Hydrol. 139 (1–4), 1–14.

Durham, L.S., 1999. 3-D models are making cents. AAPG Explorer (11), 8–9, 18.

Dury, G.H., 1962. Map Interpretation, second ed. Pitman and Sons, London, UK.

Duszyński, F., Jancewicz, K., Kasprzak, M., Migoń, P., 2017. The role of landslides in downslope transport of caprock-derived boulders in sedimentary tablelands, Stołowe Mts, SW Poland. Geomorphology 295, 84–101.

Dutton, G., 1999. A Hierarchical Coordinate System for Geoprocessing and Cartography. Lect. Notes Earth Sci., vol. 79. Springer, Berlin, Germany.

Duxbury, T.C., 1974. Phobos: control network analysis. Icarus 23 (2), 290–299.

Dvornikov, Y., Slukovskaya, M., Yaroslavtsev, A., Meshalkina, J., Ryazanov, A., Sarzhanov, D., et al., 2022. High-resolution mapping of soil pollution by Cu and Ni at a polar industrial barren area using proximal and remote sensing. Land Degrad. Dev. 33 (10), 1731–1744.

Dyba, K., Jasiewicz, J., 2022. Toward geomorphometry of plains—country-level unsupervised classification of low-relief areas (Poland). Geomorphology 413, 108373.

EAPRS Lab, 2002. ACE GDEM. EAPRS Laboratory, De Montfort University, Leicester, UK. http://www.cse.dmu.ac.uk/EAPRS/products_ace_overview.html. (Accessed 27 June 2024).

Eberly, D., Gardner, R., Morse, B., Pizer, S., Scharlach, C., 1994. Ridges for image analysis. J. Math. Imag. Vis. 4 (4), 351–371.

Eckerstorfer, M., Bühler, Y., Frauenfelder, R., Malnes, E., 2016. Remote sensing of snow avalanches: recent advances, potential, and limitations. Cold Reg. Sci. Technol. 121, 126–140.

Eckert, M., 1908. On the nature of maps and map logic. Bull. Am. Geogr. Soc. 40 (6), 344–351.

Edwards, K., Davis, P.A., 1994. The use of intensity-hue-saturation transformation for producing color shaded-relief images. Photogrammet. Eng. Remote Sens. 60 (11), 1369–1374.

Efremov, Y.K., 1949. An experience on morphological classification of elements and simple forms of topography. Prob. Geogr. 11, 109–136 (in Russian).

Efron, B., 1982. The Jackknife, the Bootstrap, and Other Resampling Plans. Society for Industrial Mathematics, Philadelphia, PA.

Egels, Y., Kasser, M. (Eds.), 2002. Digital Photogrammetry. Taylor and Francis, London, UK.

Eguez, A., Alvarado, A., Yepes, H., Machette, M.N., Costa, C., Dart, R.L., 2003. Database and Map of Quaternary Faults and Folds of Ecuador and Its Offshore Regions. USGS Open-File Report 03-289, US Geological Survey, Denver, CO.

Ehsani, A.H., Quiel, F., 2008. Application of self organizing maps and SRTM data to characterize yardangs in the Lut Desert, Iran. Remote Sens. Environ. 112 (7), 3284–3294.

Eisenbeiß, H., 2009. UAV Photogrammetry. Institut für Geodäsie und Photogrammetrie, Eidgenössische Technische Hochschule, Zurich, Switzerland.

Eklundh, L., Mårtensson, U., 1995. Rapid generation of digital elevation models from topographic maps. Int. J. Geogr. Inf. Sci. 9 (3), 329–340.

Ekman, M., 1988. Gaussian curvature of postglacial rebound and the discovery of caves created by major earthquakes in Fennoscandia. Geophysica 24 (1–2), 47–56.

El Koundi, M., Dhaoui, M., El Ghali, A., 2024. Lineament depth deduction using geomorphometry and 3D Euler solution. Acta Geophys. 72, 1327–1342.

Elez, J., Silva, P.G., Huerta, P., Perucha, M.Á., Civis, J., Roquero, E., et al., 2016. Quantitative paleotopography and paleogeography around the Gibraltar Arc (South Spain) during the Messinian Salinity Crisis. Geomorphology 275, 26–45.

Eliason, J.R., Eliason, V.L.C., 1987. Process for structural geologic analysis of topography and point data. US Patent No. 4698759, 107 p.

Elkins, T.A., 1951. The second derivative method of gravity interpretation. Geophysics 16 (1), 29–50.

Elmahdy, S.I., Mohamed, M.M., Ali, T.A., 2021. Automated detection of lineaments express geological linear features of a tropical region using topographic fabric grain algorithm and the SRTM DEM. Geocarto Int. 36 (1), 76–95.

Elmahdy, S.I., Ali, T.A., Mohamed, M.M., Yahia, M., 2022. Topographically and hydrologically signatures express subsurface geological structures in an arid region: a modified integrated approach using remote sensing and GIS. Geocarto Int. 37 (7), 2095–2115.

El-Sheimy, N., Valeo, C., Habib, A., 2005. Digital Terrain Modeling: Acquisition, Manipulation, and Applications. Artech House, Boston, MA.

Elsner, J.B., Tsonis, A.A., 1996. Singular Spectrum Analysis: A New Tool in Time Series Analysis. Plenum Press, New York, NY.

Elsner, P., Bonnici, M., 2007. Vertical accuracy of Shuttle Radar Topography Mission (SRTM) elevation and void-filled data in the Libyan desert. Int. J. Ecol. Dev. 8 (F07), 66–80.

Eltner, A., Kaiser, A., Castillo, C., Rock, G., Neugirg, F., Abellán, A., 2016. Image-based surface reconstruction in geomorphometry—merits, limits and developments. Earth Surf. Dyn. 4 (2), 359–389.

Ely, J.C., Graham, C., Barr, I.D., Rea, B.R., Spagnolo, M., Evans, J., 2017. Using UAV acquired photography and structure from motion techniques for studying glacier landforms: application to the glacial flutes at Isfallsglaciären. Earth Surf. Process. Landf. 42 (6), 877–888.

Emeis, S., Knoche, H.R., 2009. Applications in meteorology. In: Hengl, T., Reuter, H.I. (Eds.), Geomorphometry: Concepts, Software, Applications. Developments in Soil Science, vol. 33. Elsevier, Amsterdam, the Netherlands, pp. 603–622.

Endreny, T.A., Wood, E.F., Lettenmaier, D.P., 2000. Satellite-derived digital elevation model accuracy: hydrogeomorphological analysis requirements. Hydrol. Process. 14 (1), 1–20.

Ermakov, A.I., Kreslavsky, M.A., Scully, J.E.C., Hughson, K.H.G., Park, R.S., 2019. Surface roughness and gravitational slope distributions of Vesta and Ceres. J. Geophys. Res. Planet. 124 (1), 14–30.

Erskine, R.H., Green, T.R., Ramirez, J.A., MacDonald, L.H., 2007. Digital elevation accuracy and grid cell size: effects on estimated terrain attributes. Soil Sci. Soc. Am. J. 71 (4), 1371–1380.

ESA, 2022. Copernicus DEM – Global and European Digital Elevation Model (COP-DEM). European Space Agency, https://doi.org/10.5270/ESA-c5d3d65.

Etzelmüller, B., 2000. On the quantification of surface changes using grid-based digital elevation models (DEMs). Trans. GIS 4 (2), 129–143.

Etzelmüller, B., Sollid, J.L., 1997. Glacier geomorphometry—an approach for analysing long-term glacier surface changes using grid-based digital elevation models. Ann. Glaciol. 24, 135–141.

Evans, D., 2016. Airborne laser scanning as a method for exploring long-term socioecological dynamics in Cambodia. J. Archaeol. Sci. 74, 164–175.

Evans, I.S., 1972. General geomorphometry, derivatives of altitude, and descriptive statistics. In: Chorley, R.J. (Ed.), Spatial Analysis in Geomorphology. Methuen, London, UK, pp. 17–90.

Evans, I.S., 1979. Statistical Characterization of Altitude Matrices by Computer. An Integrated System of Terrain Analysis and Slope Mapping. The Final Report on Grant DA-ERO-591-73-G0040, Department of Geography, University of Durham, Durham, UK.

Evans, I.S., 1980. An integrated system of terrain analysis and slope mapping. Zeit. Geomorph. Suppl. 36, 274–295.

Evans, I.S., Minár, J., 2011. A classification of geomorphometric variables. In: Hengl, T., Evans, I.S., Wilson, J.P., Gould, M. (Eds.), Proceedings of Geomorphometry 2011, Redlands, CA, 7–11 Sept. 2011. International Society for Geomorphometry, pp. 105–108.

Evans, J.S., Hudak, A.T., 2007. A multiscale curvature algorithm for classifying discrete return LiDAR in forested environments. IEEE Trans. Geosci. Remote Sens. 45 (4), 1029–1038.

Evatt, G.W., Fowler, A.C., 2007. Cauldron subsidence and subglacial floods. Ann. Glaciol. 45, 163–168.

Ewertowski, M.W., Tomczyk, A.M., Evans, D.J.A., Roberts, D.H., Ewertowski, W., 2019. Operational framework for rapid, very-high resolution mapping of glacial geomorphology using low-cost unmanned aerial vehicles and structure-from-motion approach. Remote Sens. (Basel) 11 (1), 65.

Eyton, J.R., 1986. Digital elevation model perspective plot overlays. Ann. Assoc. Am. Geogr. 76 (4), 570–576.

Fa, W., Cai, Y., Xiao, Z., Tian, W., 2016. Topographic roughness of the northern high latitudes of Mercury from MESSENGER Laser Altimeter data. Geophys. Res. Lett. 43 (7), 3078–3087.

Fabbri, S., Sauro, F., Santagata, T., Rossi, G., de Waele, J., 2017. High-resolution 3-D mapping using terrestrial laser scanning as a tool for geomorphological and speleogenetical studies in caves: an example from the Lessini mountains (North Italy). Geomorphology 280, 16–29.

Fairbanks, D., Shepard, C., Murphy, M., Rasmussen, C., Chorover, J., Rich, V., et al., 2020. Depth and topographic controls on microbial activity in a recently burned sub-alpine catchment. Soil Biol. Biochem. 148, 107844.

Fallati, L., Savini, A., Argentino, C., Bünz, S., Panieri, G., 2022. Using ROV video photogrammetry to reconstruct seafloor landforms of Arctic cold seep environments. In: EGU22, the 24th EGU General Assembly, Vienna, Austria & Online, 23–27 May 2022. European Geophysical Union, Vienna, Austria. EGU22-13267.

Falorni, G., Teles, V., Vivoni, E.R., Bras, R.L., Amaratunga, K.S., 2005. Analysis and characterization of the vertical accuracy of digital elevation models from the Shuttle Radar Topography Mission. J. Geophys. Res. Earth Surf. 110 (F2), F02005.

Famiglietti, J.S., Wood, E.F., 1995. Effects of spatial variability and scale on areally averaged evapotranspiration. Water Resour. Res. 31 (3), 699–712.

Fan, B., Tao, W., Qin, G., Hopkins, I., Zhang, Y., Wang, Q., et al., 2020. Soil micro-climate variation in relation to slope aspect, position, and curvature in a forested catchment. Agric. For. Meteorol. 290, 107999.

Farenhorst, A., Florinsky, I.V., Monreal, C.M., Muc, D., 2003. Evaluating the use of digital terrain modelling for quantifying the spatial variability of 2,4-D sorption by soil within agricultural landscapes. Can. J. Soil Sci. 83 (5), 557–564.

Farr, T.G., Rosen, P.A., Caro, E., Crippen, R., Duren, R., Hensley, S., et al., 2007. The Shuttle Radar Topography Mission. Rev. Geophys. 45 (2), RG2004.

Fasshauer, G.E., Schumaker, L.L., 1998. Scattered data fitting on the sphere. In: Dæhlen, M., Lyche, T., Schumaker, L.L. (Eds.), Mathematical Methods for Curves and Surface II. Vanderbilt University Press, Nashville, TN, pp. 117–166.

Favalli, M., Fornaciai, A., 2017. Visualization and comparison of DEM-derived parameters: application to volcanic areas. Geomorphology 290, 69–84.

Feciskanin, R., Minár, J., 2021. Polygonal simplification and its use in DEM generalization for land surface segmentation. Trans. GIS 25 (5), 2361–2375.

Fedorov, A.E., 1991. Hexagonal Networks of Linear Heterogeneities of the Earth. Nedra, Moscow, USSR (in Russian).

Fedoseev, A.P., 1959. Soil moisture and terrain topography. In: Konyukhov, N.A. (Ed.), Agricultural Meteorology. Hydrometeoizdat, Moscow, USSR, pp. 66–88 (in Russian).

Feizizadeh, B., Blaschke, T., Tiede, D., Moghaddam, M.H.R., 2017. Evaluating fuzzy operators of an object-based image analysis for detecting landslides and their changes. Geomorphology 293 (A), 240–254.

Felicísimo, A.M., 1994a. Modelos Digitales del Terreno: Introducción y Aplicaciones en las Ciencias Ambientales. Pentalfa Ediciones, Oviedo, Spain.

Felicísimo, A.M., 1994b. Parametric statistical method for error detection in digital elevation models. ISPRS J. Photogrammet. Remote Sens. 49 (4), 29–33.

Felicísimo, A.M., 1995. Error propagation analysis in slope estimation by means of digital elevation models. In: Cartography Crossing Borders: Proceedings 1 of the 17th International Cartographic Conference and 10th General Assembly of International Cartographic Association, Barcelona, Spain, 3–9 Sept. 1995. Institut Cartogràfic de Catalunya, Barcelona, Spain, pp. 94–98.

Feltens, J., 2009. Vector method to compute the Cartesian (X, Y, Z) to geodetic (φ, λ, h) transformation on a triaxial ellipsoid. J. Geodesy 83 (2), 129–137.

Feranec, J., Kolár, J., Krcho, J., 1991. Mapping of the surface water logging intensity of the soils by applying Landsat TM data and complex digital terrain model. Bull. Comité Franç. Cartogr. 127–128, 154–157.

Fernandez, O., Jones, S., Armstrong, N., Johnson, G., Ravaglia, A., Muñoz, J.A., 2009. Automated tools within workflows for 3D structural construction from surface and subsurface data. Geoinformatica 13 (3), 291–304.

Fichtenholz, G.M., 1966a. A Course in Differential and Integral Calculus, vol. 1, sixth ed. Nauka, Moscow, USSR (in Russian).

Fichtenholz, G.M., 1966b. A Course in Differential and Integral Calculus, vol. 3, fourth ed. Nauka, Moscow, USSR (in Russian).

Filatov, M.M., 1927. Diagrammatic Meridional Soil Profile of the European Part of USSR. Mospoligraf, Moscow, USSR.

Filippov, S.V., 2018. Blender software platform as an environment for modeling objects and processes of science disciplines. KIAM Prepr. 230, 1–42 (in Russian).

Filippov, Y.V. (Ed.), 1955. Principles of Generalization of Small-Scale Geographical Maps. Geodezizdat, Moscow, USSR (in Russian).

Filosofov, V.P., 1960. A Brief Guide of the Morphometric Method to Search Tectonic Structures. Saratov University Press, Saratov, USSR (in Russian).

Filosofov, V.P., 1975. Principles of the Morphometric Method to Search Tectonic Structures. Saratov University Press, Saratov, USSR (in Russian).

Finkl, C.W., Benedet, L., Andrews, J.L., 2005. Interpretation of seabed geomorphology based on spatial analysis of high-density airborne laser bathymetry. J. Coast. Res. 21 (3), 501–514.

Fiore Allwardt, P.F., Bellahsen, N., Pollard, D.D., 2007. Curvature and fracturing based on global positioning system data collected at Sheep Mountain anticline, Wyoming. Geosphere 3 (6), 408–421.

Fischer, M.P., Keating, D.P., 2005. Photogrammetric techniques for analyzing displacement, strain, and structural geometry in physical models: application to the growth of monoclinal basement uplifts. Geol. Soc. Am. Bull. 117 (3–4), 369–382.

Fischer, M.P., Wilkerson, M.S., 2000. Predicting the orientation of joints from fold shape: results of pseudo-three-dimensional modeling and curvature analysis. Geology 28 (1), 15–18.

Fishbaugh, K.E., Head III, J.W., 2001. Comparison of the north and south polar caps of Mars: new observations from MOLA data and discussion of some outstanding questions. Icarus 154 (1), 145–161.

Fisher, D., 1993. If Martian ice caps flow: ablation mechanisms and appearance. Icarus 105 (2), 501–511.

Fisher, P.F., 1995. An exploration of probable viewsheds in landscape planning. Environ. Plan. B:Plan. Des. 22 (5), 527–546.

Fisher, P.F., Tate, N.J., 2006. Causes and consequences of error in digital elevation models. Progr. Phys. Geogr. 30 (4), 467–489.

Fissore, C., Dalzell, B.J., Berhe, A.A., Voegtle, M., Evans, M., Wu, A., 2017. Influence of topography on soil organic carbon dynamics in a Southern California grassland. Catena 149 (1), 140–149.

Fitzmaurice, J., Eilers, R.G., St. Jacques, E., Waddell, A., 1999. Soils of SE 32-14-25W—Miniota Precision Agriculture Research Site. Special Report Series 99-1, Land Resource Unit, Brandon Research Centre, Agriculture and Agri-Food Canada, Winnipeg, MB.

Florinsky, I.V., 1991. Map Generalization—A Brief Review of the Problem. Pushchino Research Centre Press, Pushchino, USSR (in Russian).

Florinsky, I.V., 1992. Recognition of Lineaments and Ring Structures: Quantitative Topographic Techniques. Pushchino Research Centre Press, Pushchino, Russia (in Russian).

Florinsky, I.V., 1993. Analyzing Digital Elevation Models to Reveal Linear Structures of the Land Surface (Ph.D. thesis). Institute of Soil Science and Photosynthesis, Russian Academy of Sciences, Pushchino, Russia (in Russian).

Florinsky, I.V., 1996. Quantitative topographic method of fault morphology recognition. Geomorphology 16 (2), 103–119.

Florinsky, I.V., 1998a. Accuracy of local topographic variables derived from digital elevation models. Int. J. Geogr. Inf. Sci. 12 (1), 47–61.

Florinsky, I.V., 1998b. Combined analysis of digital terrain models and remotely sensed data in landscape investigations. Progr. Phys. Geogr. 22 (1), 33–60.

Florinsky, I.V., 1998c. Derivation of topographic variables from a digital elevation model given by a spheroidal trapezoidal grid. Int. J. Geogr. Inf. Sci. 12 (8), 829–852.

Florinsky, I.V., 2000. Relationships between topographically expressed zones of flow accumulation and sites of fault intersection: analysis by means of digital terrain modelling. Environ. Model. Software 15 (1), 87–100.

Florinsky, I.V., 2002. Errors of signal processing in digital terrain modelling. Int. J. Geogr. Inf. Sci. 16 (5), 475–501.

Florinsky, I.V., 2005. Artificial lineaments in digital terrain modelling: can operators of topographic variables cause them? Mathemat. Geol. 37 (4), 357–372.

Florinsky, I.V., 2006. Oscillating processes in soils. In: Lakhno, V.D. (Ed.), Mathematical Biology and Bioinformatics: Proceedings of the 1st International Conference, Pushchino, Russia, 9–15 Oct. 2006. Max Press, Moscow, Russia, pp. 146–147.

Florinsky, I.V., 2007a. Small-scale morphometric maps of the Northern Eurasia. Geodesy Cartogr. (2), 15–21 (in Russian).

Florinsky, I.V., 2007b. Solving three problems of exploration and engineering geology by digital terrain analysis. Int. J. Ecol. Dev. 8 (F07), 52–65.

Florinsky, I.V., 2008a. Estimating the derivating function of elevation to detect structural lines using satellite data and topographic maps. Issledovanie Zemli iz Cosmosa (6), 43–51 (in Russian).

Florinsky, I.V., 2008b. Global lineaments: application of digital terrain modelling. In: Zhou, Q., Lees, B., Tang, G.-A. (Eds.), Advances in Digital Terrain Analysis. Lecture Notes in Geoinformation and Cartography Book Series. Springer, Berlin, Germany, pp. 365–382.

Florinsky, I.V., 2008c. Global morphometric maps of Mars, Venus, and the Moon. In: Moore, A., Drecki, I. (Eds.), Geospatial Vision: New Dimensions in Cartography. Lecture Notes in Geoinformation and Cartography Book Series. Springer, Berlin, Germany, pp. 171–192.

Florinsky, I.V., 2008d. Morphometric maps of the World. Geodesy Cartogr. (1), 25–29 (in Russian).

Florinsky, I.V., 2008e. On calculation accuracy in digital terrain modeling. Geodesy Cartogr. (6), 28–32 (in Russian).

Florinsky, I.V., 2008f. 100 years anniversary of the Tunguska Event: digital terrain modeling of the epicenter area. Geodesy Cartogr. (8), 20–22 (in Russian).

Florinsky, I.V., 2009a. Accurate method for derivation of local topographic variables. Geodesy Cartogr. (4), 19–23 (in Russian).

Florinsky, I.V., 2009b. Computation of the third-order partial derivatives from a digital elevation model. Int. J. Geogr. Inf. Sci. 23 (2), 213–231.

Florinsky, I.V., 2009c. Soil mapping based on digital terrain modeling (using data from kinematic GPS and soil surveys). Issledovanie Zemli iz Cosmosa (6), 56–65 (in Russian).

Florinsky, I.V., 2010. Theory and Applications of Mathematical Cartographic Modeling of Topography (D.Sc. thesis). Institute of Mathematical Problems of Biology, Russian Academy of Sciences, Pushchino, Russia (in Russian).

Florinsky, I.V., 2012. The Dokuchaev hypothesis as a basis for predictive digital soil mapping (on the 125th anniversary of its publication). Eurasian Soil Sci. 45 (4), 445–451.

Florinsky, I.V., 2016. LandLord—a brief description of the software. In: Florinsky, I.V., Digital Terrain Analysis in Soil Science and Geology, second ed. Academic Press, Amsterdam, the Netherlands, pp. 413–414.

Florinsky, I.V., 2017a. An illustrated introduction to general geomorphometry. Progr. Phys. Geogr. 41 (6), 723–752.

Florinsky, I.V., 2017b. Spheroidal equal angular DEMs: the specificity of morphometric treatment. Trans. GIS 21 (6), 1115–1129.

Florinsky, I.V., 2018a. Geomorphometry on the surface of a triaxial ellipsoid: towards the solution of the problem. Int. J. Geogr. Inf. Sci. 32 (8), 1558–1571.

Florinsky, I.V., 2018b. Multiscale geomorphometric modeling of Mercury. Planet. Space Sci. 151, 56–70.

Florinsky, I.V., 2021. Geomorphometry today. InterCarto InterGIS 27 (2), 394–448 (in Russian).

Florinsky, I.V., 2022. Unmanned aerial survey in the summer season of the 67th Russian Antarctic Expedition. InterCarto InterGIS 28 (1), 284–304 (in Russian).

Florinsky, I.V., 2023a. Geomorphometric modeling and mapping of Antarctic oases. arXiv Preprint. 2305.07523.

Florinsky, I.V., 2023b. Larsemann Hills: geomorphometric modeling and mapping. Pol. Sci. 38, 100969.

Florinsky, I.V., 2023c. Study on Mathematical and Cartographic Modeling of the Topography in the Area of Progress Station. Interim Report on the Program 'Study on Mathematical and Cartographic Modeling of the Topography in the Areas of Progress Station, Molodezhnaya and Gora Vechernyaya Bases as well as Novolazarevskaya and Bellingshausen Stations'. 68th Russian Antarctic Expedition, Progress Station (in Russian).

Florinsky, I.V., Arlashina, H.A., 1998. Quantitative topographic analysis of gilgai soil morphology. Geoderma 82 (4), 359–380.

Florinsky, I.V., Bliakharskii, D.P., 2019a. Detection of crevasses by geomorphometric treatment of data from unmanned aerial surveys. Remote Sens. Lett. 10 (4), 323–332.

Florinsky, I.V., Bliakharskii, D.P., 2019b. The 2017 catastrophic subsidence in the Dålk Glacier, East Antarctica: unmanned aerial survey and terrain modelling. Remote Sens. Lett. 10 (4), 333–342.

Florinsky, I.V., Eilers, R.G., 2002. Prediction of the soil organic carbon content at micro-, meso- and macroscales by digital terrain modelling. In: Transactions of the 17th World Congress of Soil Science, Bangkok, Thailand, 14–21 Aug. 2002, Symposium 52. International Society of Soil Science, Bangkok, Thailand, 24.

Florinsky, I.V., Filippov, S.V., 2015. Development of virtual morphometric globes using Blender. arXiv Preprint. 1512.08511.

Florinsky, I.V., Filippov, S.V., 2017. A desktop system of virtual morphometric globes for Mars and the Moon. Planet. Space Sci. 137, 32–39.

Florinsky, I.V., Filippov, S.V., 2019a. Three-dimensional desktop morphometric models for the Arctic Ocean floor. Proc. Int. Cartogr. Assoc. 2, 32.

Florinsky, I.V., Filippov, S.V., 2019b. Three-dimensional geomorphometric modelling of the Arctic Ocean submarine topography (a testing, low-resolution desktop system). InterCarto InterGIS 25 (1), 275–289 (in Russian).

Florinsky, I.V., Filippov, S.V., 2019c. Three-dimensional terrain modeling with multiple-source illumination. Trans. GIS 23 (5), 937–959.

Florinsky, I.V., Filippov, S.V., 2021. Three-dimensional geomorphometric modeling of the Arctic Ocean submarine topography: a low-resolution desktop application. IEEE J. Ocean. Eng. 46 (1), 88–101.

Florinsky, I.V., Kuryakova, G.A., 1996. Influence of topography on some vegetation cover properties. Catena 27 (2), 123–141.

Florinsky, I.V., Kuryakova, G.A., 2000. Determination of grid size for digital terrain modelling in landscape investigations—exemplified by soil moisture distribution at a micro-scale. Int. J. Geogr. Inf. Sci. 14 (8), 815–832.

Florinsky, I.V., Pankratov, A.N., 2015. Digital terrain modeling with orthogonal polynomials. Mach. Learn. Data Anal. 1 (12), 1647–1659.

Florinsky, I.V., Pankratov, A.N., 2016. A universal spectral analytical method for digital terrain modeling. Int. J. Geogr. Inf. Sci. 30 (12), 2506–2528.

Florinsky, I.V., Grokhlina, T.I., Mikhailova, N.L., 1995. LANDLORD 2.0: the software for analysis and mapping of geometric characteristics of topography. Geodesy Cartogr. (5), 46–51 (in Russian).

Florinsky, I.V., Eilers, R.G., Manning, G., Fuller, L.G., 1999. Application of digital terrain modelling to prediction of soil properties in the Prairie Ecozone. In: Papers of 42nd Annual Manitoba Society of Soil Science Meeting, Winnipeg, MB, 2–3 Feb. 1999. Manitoba Society of Soil Science, Winnipeg, MB, pp. 140–154.

Florinsky, I.V., Eilers, R.G., Lelyk, G.W., 2000. Prediction of soil salinity risk by digital terrain modelling in the Canadian prairies. Can. J. Soil Sci. 80 (3), 455–463.

Florinsky, I.V., Eilers, R.G., Manning, G., Fuller, L.G., 2002. Prediction of soil properties by digital terrain modelling. Environ. Model. Software 17 (3), 295–311.

Florinsky, I.V., McMahon, S., Burton, D.L., 2004. Topographic control of soil microbial activity: a case study of denitrifiers. Geoderma 119 (1–2), 33–53.

Florinsky, I.V., Eilers, R.G., Burton, D.L., McMahon, S.K., Monreal, C.M., Farenhorst, A., 2009a. Predictive soil mapping based on digital terrain modeling. Geoinformatika (1), 22–32 (in Russian).

Florinsky, I.V., Eilers, R.G., Wiebe, B.H., Fitzgerald, M.M., 2009b. Dynamics of soil salinity in the Canadian prairies: application of singular spectrum analysis. Environ. Model. Software 24 (10), 1182–1195.

Florinsky, I.V., McMahon, S., Burton, D.L., 2009c. Topographic factors of nitrous oxide emission. In: Sheldon, A.I., Barnhart, E.P. (Eds.), Nitrous Oxide Emissions Research Progress. Nova Science, New York, NY, pp. 105–126.

Florinsky, I.V., Selezneva, E.V., Kulikova, A.I., 2015. GIS-based support for the complex botanical studies at the Molnieboi Spur, Altai. arXiv Preprint. 1512.05143.

Florinsky, I.V., Garov, A.S., Karachevtseva, I.P., 2018a. A web system of virtual morphometric globes for Mars and the Moon. Planet. Space Sci. 159, 105–109.

Florinsky, I.V., Kurkov, V.M., Bliakharskii, D.P., 2018b. Geomorphometry from unmanned aerial surveys. Trans. GIS 22 (1), 58–81.

Florinsky, I.V., Skrypitsyna, T.N., Luschikova, O.S., 2018c. Comparative accuracy of the AW3D30 DSM, ASTER GDEM, and SRTM1 DEM: a case study on the Zaoksky testing ground, Central European Russia. Remote Sens. Lett. 9 (7), 706–714.

Florinsky, I.V., Skrypitsyna, T.N., Trevisani, S., Romaikin, S.V., 2019. Statistical and visual quality assessment of nearly-global and continental digital elevation models of Trentino, Italy. Remote Sens. Lett. 10 (8), 726–735.

Foessel-Bunting, A., Apostolopoulos, D., Whittaker, W.L., 1999. Radar sensor for an autonomous Antarctic explorer. In: Choset, H.M., Gage, D.W., Kachroo, P., Kourjanski, M.A., de Vries, M.J. (Eds.), Mobile Robots XIII and Intelligent Transportation Systems. Proc. SPIE 3525, pp. 117–124.

Fonstad, M.A., Dietrich, J.T., Courville, B.C., Jensen, J.L., Carbonneau, P.E., 2013. Topographic structure from motion: a new development in photogrammetric measurement. Earth Surf. Process. Landf. 38 (4), 421–430.

Ford, P.G., Pettengill, G., Liu, F., Quigley, J., 1992. Magellan Global Topography, Emissivity, Reflectivity, and Slope Data Record (GXDR). NASA Planetary Data System Geosciences Node, Washington University, St, Louis, MO. http://pds-geosciences.wustl.edu/missions/magellan/gxdr/index.htm. (Accessed 27 June 2024).

Forkuor, G., Hounkpatin, O.K.L., Welp, G., Thiel, M., 2017. High resolution mapping of soil properties using remote sensing variables in south-western Burkina Faso: a comparison of machine learning and multiple linear regression models. PLoS One 12 (1), e0170478.

Foroutan, M., Marshall, S.J., Menounos, B., 2019. Automatic mapping and geomorphometry extraction technique for crevasses in geodetic mass-balance calculations at Haig Glacier, Canadian Rockies. J. Glaciol. 65 (254), 971–982.

Found, W.C., 1970. The electronic map-coordinate digitizer. Cartographica 7 (2), 131–136.

Fox, A.J., Cziferszky, A., 2008. Unlocking the time-capsule of historic aerial photography to measure changes in Antarctic Peninsula glaciers. Photogrammet. Rec. 23 (121), 51–68.

Fox, A.J., Gooch, M.J., 2001. Automatic DEM generation for Antarctic terrain. Photogrammet. Rec. 17 (98), 275–290.

Fox, L., Parker, I.B., 1968. Chebyshev Polynomials in Numerical Analysis. Oxford University Press, London, UK.

Franke, R., 1982. Scattered data interpolation: tests of some methods. Math. Comput. 38 (157), 181–200.

Franke, R., 1985. Thin plate spline with tension. Comput. Aided Geomet. Des. 2 (1–3), 87–95.

Franklin, J., 1995. Predictive vegetation mapping: geographic modelling of biospatial patterns in relation to environmental gradients. Progr. Phys. Geogr. 19 (4), 474–499.

Franklin, S.E., 1989. Ancillary data input to satellite remote sensing of complex terrain phenomena. Comput. Geosci. 15 (5), 799–808.

Franklin, S.E., 2020. Interpretation and use of geomorphometry in remote sensing: a guide and review of integrated applications. Int. J. Remote Sens. 41 (19), 7700–7733.

Frederiksen, P., 1981. Terrain analysis and accuracy prediction by means of the Fourier transformation. Photogrammetria 36 (4), 145–157.

Freeland, T., Heung, B., Burley, D.V., Clark, G., Knudby, A., 2016. Automated feature extraction for prospection and analysis of monumental earthworks from aerial LiDAR in the Kingdom of Tonga. J. Archaeol. Sci. 69, 64–74.

Freeman, T.G., 1991. Calculating catchment area with divergent flow based on a regular grid. Comput. Geosci. 17 (3), 413–422.

Freer, J., McDonnell, J.J., Beven, K.J., Peters, N.E., Burns, D.A., Hooper, R.P., et al., 2002. The role of bedrock topography on subsurface storm flow. Water Resour. Res. 38 (12), 1269.

Freitas, H.R.A., Freitas, C.D.C., Rosim, S., Oliveira, J.R.F., 2016. Drainage networks and watersheds delineation derived from TIN-based digital elevation models. Comput. Geosci. 92, 21–37.

French, J.R., 2003. Airborne LiDAR in support of geomorphological and hydraulic modelling. Earth Surf. Process. Landf. 28 (3), 321–335.

Fretwell, P., Pritchard, H.D., Vaughan, D.G., Bamber, J.L., Barrand, N.E., Bell, R., et al., 2013. Bedmap2: improved ice bed, surface and thickness datasets for Antarctica. Cryosphere 7 (1), 375–393.

Fridland, V.M., 1976. Pattern of the Soil Cover. Israel Program for Scientific Translations, Jerusalem, Israel.

Fridland, V.M. (Ed.), 1988. Soil Map of the Russian Soviet Federative Socialist Republic, Scale 1:2,500,000. Dokuchaev Soil Institute, Moscow USSR (in Russian).

Friedman, J.H., 2002. Stochastic gradient boosting. Comput. Stat. Data Anal. 38 (4), 367–378.

Fryer, J.G., Kniest, H.T., 1985. Errors in depth determination caused by waves in throughwater photogrammetry. Photogrammet. Rec. 11 (66) 745–753.

Fu, L., Guo, J., Chen, X., 2021. Measurement of ice flow velocities from GPS positions logged by short-period seismographs in East Antarctica. Sci. China Earth Sci. 64 (8), 1381–1389.

Fugazza, D., Scaioni, M., Corti, M., d'Agata, C., Azzoni, R.S., Cernuschi, M., et al., 2018. Combination of UAV and terrestrial photogrammetry to assess rapid glacier evolution and conditions of glacier hazards. Nat. Haz. Earth Syst. Sci. 18 (4), 1055–1071.

Fujisada, H., Urai, M., Iwasaki, A., 2012. Technical methodology for ASTER global DEM. IEEE Trans. Geosci. Remote Sens. 50 (10), 3725–3736.

Fukue, K., Shimoda, H., Sakata, T., 1981. Complete lineament extraction with the aid of shadow-free Landsat image. In: Machine Processing of Remotely Sensed Data with Special Emphasis on Range, Forest, and Wetlands Assessment: Proceedings of the 7th International Symposium, West Lafayette, IN, 23–26 June 1981. Purdue University, West Lafayette, IN, pp. 94–102.

Fuss, C.E., Berg, A.A., Lindsay, J.B., 2016. DEM fusion using a modified *k*-means clustering algorithm. Int. J. Dig. Earth 9 (12), 1242–1255.

Gabrlik, P., La Cour-Harbo, A., Zalud, L., Janata, P., 2018. Calibration and accuracy assessment in a direct georeferencing system in UAS photogrammetry. Int. J. Remote Sens. 39 (15–16), 4931–4959.

Gaffey, C., Bhardwaj, A., 2020. Applications of unmanned aerial vehicles in cryosphere: latest advances and prospects. Remote Sens. (Basel) 12 (6), 948.

Galin, E., Guérin, E., Peytavie, A., Cordonnier, G., Cani, M.P., Benes, B., et al., 2019. A review of digital terrain modeling. Comput. Graph. Forum 38 (2), 553–577.

Gallant, J., 2011. Adaptive smoothing for noisy DEMs. In: Hengl, T., Evans, I.S., Wilson, J.P., Gould, M. (Eds.), Proceedings of Geomorphometry 2011, Redlands, CA, 7–11 Sept. 2011. International Society of Geomorphometry, pp. 37–40.

Gallant, J.C., Hutchinson, M.F., 1996. Towards an understanding of landscape scale and structure. In: Proceedings of the 3rd International Conference/Workshop on Integrating GIS and Environmental Modeling, Santa Fe, CA, 21–25 Jan. 1996. National Center for Geographic Information and Analysis, Santa Barbara, CA.

Gallant, J.C., Hutchinson, M.F., 1997. Scale dependence in terrain analysis. Math. Comput. Simul. 43 (3–6), 313–321.

Gallant, J.C., Hutchinson, M.F., 2011. A differential equation for specific catchment area. Water Resour. Res. 47 (5), W05535.

Gallant, J.C., Read, A., 2009. Enhancing the SRTM data for Australia. In: Purves, R., Gruber, S., Hengl, T., Straumann, R. (Eds.), Proceedings of Geomorphometry 2009, Zurich, Switzerland, 31 Aug. – 2 Sept. 2009. University of Zurich, Zurich, Switzerland, pp. 149–154.

Gallay, M., Hochmuth, Z., Kaňuk, J., Hofierka, J., 2016. Geomorphometric analysis of cave ceiling channels mapped with 3D terrestrial laser scanning. Hydrol. Earth Syst. Sci. 20 (5), 1827–1849.

Gallwey, J., Eyre, M., Tonkins, T., Coggan, J., 2019. Bringing lunar LiDAR back down to Earth: mapping our industrial heritage through deep transfer learning. Remote Sens. (Basel) 11 (17), 1994.

Ganas, A., Pavlides, S., Karastathis, V., 2005. DEM-based morphometry of range-front escarpments in Attica, central Greece, and its relation to fault slip rates. Geomorphology 65 (3–4), 301–319.

Gao, J., 2009. Bathymetric mapping by means of remote sensing: methods, accuracy and limitations. Progr. Phys. Geogr. 33 (1), 103–116.

Gao, Y., Wang, W., Yao, T., Lu, N., Lu, A., 2018. Hydrological network and classification of lakes on the Third Pole. J. Hydrol. 560, 582–594.

García-Gil, A., Baquedano, C., Marazuela, M.Á., Martínez-León, J., Cruz-Pérez, N., Hernández-Gutiérrez, L.E., et al., 2023. A 3D geological model of El Hierro volcanic island reflecting intraplate volcanism cycles. Groundw. Sustain. Dev. 21, 100936.

García-Romero, L., Delgado-Fernández, I., Hesp, P.A., Hernández-Calvento, L., Viera-Pérez, M., Hernández-Cordero, A.I., et al., 2019. Airflow dynamics, vegetation and aeolian erosive processes in a shadow zone leeward of a resort in an arid transgressive dune system. Aeolian Res. 38, 48–59.

Gardiner, V., Park, C.C., 1978. Drainage basin morphometry: review and assessment. Progr. Phys. Geogr. 2 (1), 1–35.

Garetsky, R.G., Karasev, O.I., Levkov, E.V., Svyatogorov, A.A., 1983. The western part of the East European Platform. In: Trifonov, V.G., Makarov, V.I., Saforov, Y.G., Florensky, P.V. (Eds.), Space Information in Geology. Nauka, Moscow, USSR, pp. 185–189 (in Russian).

Garosi, Y., Sheklabadi, M., Besalatpour, A.A., Pourghasemi, H.R., Conoscenti, C., van Oost, K., 2018. Comparison of the different resolution and source of controlling factors for gully erosion susceptibility mapping. Geoderma 330, 65–78.

Garosi, Y., Sheklabadi, M., Conoscenti, C., Pourghasemi, H.R., van Oost, K., 2019. Assessing the performance of GIS-based machine learning models with different accuracy measures for determining susceptibility to gully erosion. Sci. Total Environ. 664, 1117–1132.

Garov, A.S., Karachevtseva, I.P., Matveev, E.V., Zubarev, A.E., Florinsky, I.V., 2016. Development of a heterogenic distributed environment for spatial data processing using cloud technologie. In: Halounova, L., Šafář, V., Jiang, J., Olešovská, H., Dvořáček, P., Holland, D., et al. (Eds.), XXIII ISPRS Congress, Commission IV, 12–19 July 2016, Prague, Czech Republic. Int. Arch. Photogramm. Remote Sens. Spatial Inf. Sci. XLI (B4), pp. 385–390.

Garrison, T.G., Houston, S., Alcover Firpi, O., 2019. Recentering the rural: lidar and articulated landscapes among the Maya. J. Anthropol. Archaeol. 53, 133–146.

Garstki, K., 2017. Virtual representation: the production of 3D digital artifacts. J. Archaeol. Method Theory 24 (3), 726–750.

Gartrell, A., Zhang, Y., Lisk, M., Dewhurst, D., 2004. Fault intersections as critical hydrocarbon leakage zones: integrated field study and numerical modelling of an example from the Timor Sea, Australia. Mar. Petrol. Geol. 21 (9), 1165–1179.

Gasperini, L., Alvisi, F., Biasini, G., Bonatti, E., Longo, G., Pipan, M., et al., 2007. A possible impact crater for the 1908 Tunguska Event. Terra Nova 19 (4), 245–251.

Gatter, R., Cavalli, M., Crema, S., Bossi, G., 2018. Modelling the dynamics of a large rock landslide in the Dolomites (eastern Italian Alps) using multi-temporal DEMs. PeerJ 6, e5903.

Gauch, J.M., Pizer, S.M., 1993. Multiresolution analysis of ridges and valleys in gray-scale images. IEEE Trans. Pattern Anal. Mach. Intell. 15 (6), 635–646.

Gaucherel, C., Frelat, R., Salomon, L., Rouy, B., Pandey, N., Cudennec, C., 2017. Regional watershed characterization and classification with river network analyses. Earth Surf. Process. Landf. 42 (13), 2068–2081.

Gauss, C.F., 1828. Disquisitiones generales circa superficies curvas. Commentationes Societatis Regiae Scientiarum Gottingensis Recentiores 6, 99–146.

Gauss, K.F., 1902. General investigations of curved surfaces. In: Gauss, K.F., General Investigations of Curved Surfaces of 1827 and 1825. Princeton University Library, Princeton, NJ, pp. 1–44.

Gautschi, W., 2004. Orthogonal Polynomials: Computation and Approximation. Oxford University Press, Oxford, UK.

Gawrysiak, L., Kociuba, W., 2023. LiDAR-derived relief typology of loess patches (East Poland). Remote Sens. (Basel) 15 (7), 1875.

Gayen, A., Pourghasemi, H.R., Saha, S., Keesstra, S., Bai, S., 2019. Gully erosion susceptibility assessment and management of hazard-prone areas in India using different machine learning algorithms. Sci. Total Environ. 668, 124–138.

Gdulová, K., Marešová, J., Moudrý, V., 2020. Accuracy assessment of the global TanDEM-X digital elevation model in a mountain environment. Remote Sens. Environ. 241, 111724.

GEBCO, 2014. GEBCO 30 Arc-Second Grid. General Bathymetric Chart of the Oceans, British Oceanographic Data Centre, Liverpool, UK. http://www.gebco.net/data_and_products/gridded_bathymetry_data/gebco_30_second_grid. (Accessed 27 June 2024).

GEBCO, 2023a. Historical GEBCO Data Sets. General Bathymetric Chart of the Oceans, British Oceanographic Data Centre, Liverpool, UK. https://www.gebco.net/data_and_products/historical_data_sets/. (Accessed 27 June 2024).

GEBCO, 2023b. The GEBCO_2023 Grid. General Bathymetric Chart of the Oceans, British Oceanographic Data Centre, Liverpool, UK. https://www.gebco.net/data_and_products/gridded_bathymetry_data/gebco_2023/. (Accessed 27 June 2024).

Geiger, R., 1927. Das Klima der bodennahen Luftschicht. Vieweg, Braunschweig, Germany.

Geisthövel, R., Hurni, L., 2018. Automated Swiss-style relief shading and rock hachuring. Cartogr. J. 55 (4), 341–361.

Gelautz, M., Paillou, P., Chen, C.W., Zebker, H.A., 2003. Radar stereo- and interferometry-derived digital elevation models: comparison and combination using Radarsat and ERS-2 imagery. Int. J. Remote Sens. 24 (24), 5243–5264.

Gelfand, I.M., Guberman, S.I., Izvekova, M.L., Keilis-Borok, V.I., Rantsman, E.Y., 1972. Criteria of high seismicity, determined by pattern recognition. Tectonophysics 13 (1–4), 415–422.

Gelfand, I.M., Guberman, S.I., Keilis-Borok, V.I., Knopoff, L., Press, F., Ranzman, E.Y., et al., 1976. Pattern recognition applied to earthquake epicenters in California. Phys. Earth Planet. In. 11 (3), 227–283.

Geoscan, 2016–2024. Geoscan 201 Geodesy. Geoscan Ltd, St. Petersburg, Russia. https://www.geoscan.aero/en/products/geoscan201/geo/. (Accessed 27 June 2024).

Gerasimov, I.P., 1959. Structural Features of the Land Surface Topography of the USSR and Their Origin. Soviet Academic Press, Moscow, USSR (in Russian).

Gerasimov, I.P., Korzhuev, S.S. (Eds.), 1979. Morphostructural Analysis of the Drainage Network of the USSR. Nauka, Moscow, USSR (in Russian).

Gerke, H.H., Vogel, H.-J., Weber, T.K.D., van der Meij, W.M., Scholten, T., 2022. 3–4D soil model as challenge for future soil research: quantitative soil modeling based on the solid phase. J. Plant Nutr. Soil Sci. 185 (6), 720–744.

Gerlitz, L., Conrad, O., Böhner, J., 2015. Large scale atmospheric forcing and topographic modification of precipitation rates over High Asia—a neural network based approach. Earth Syst. Dynam. 6 (1), 1–21.

Gerrard, A.J., 1981. Soils and Landforms: An Integration of Geomorphology and Pedology. George Allen and Unwin, London, UK.

Gesch, D.B., 2012. Global digital elevation model development from satellite remote-sensing data. In: Yang, X., Li, J. (Eds.), Advances in Mapping from Remote Sensor Imagery: Techniques and Applications. CRC Press, Boca Raton, FL, pp. 91–117.

Gesch, D.B., 2018. Best practices for elevation-based assessments of sea-level rise and coastal flooding exposure. Front. Earth Sci. 6, 230.

Gesch, D.B., Verdin, K.L., Greenlee, S.K., 1999. New land surface digital elevation model covers the Earth. Eos 80 (6), 69–70.

Gessler, P.E., Moore, I.D., McKenzie, N.J., Ryan, P.J., 1995. Soil–landscape modelling and spatial prediction of soil attributes. Int. J. Geogr. Inf. Syst. 9 (4), 421–432.

Gessler, P.E., Chadwick, O.A., Chamran, F., Althouse, L., Holmes, K., 2000. Modeling soil–landscape and ecosystem properties using terrain attributes. Soil Sci. Soc. Am. J. 64 (6), 2046–2056.

Gevaert, C., Persello, C., Nex, F., Vosselman, G., 2018. A deep learning approach to DTM extraction from imagery using rule-based training labels. ISPRS J. Photogrammet. Remote Sens. 142, 106–123.

Gheyle, W., Stichelbaut, B., Saey, T., Note, N., van den Berghe, H., van Eetvelde, V., et al., 2018. Scratching the surface of war. Airborne laser scans of the Great War conflict landscape in Flanders (Belgium). Appl. Geogr. 90, 55–68.

Ghilani, C.D., 2022. Elementary Surveying: An Introduction to Geomatics, sixteenth ed. Pearson, London, UK.

Ghorbanzadeh, O., Blaschke, T., Gholamnia, K., Meena, S., Tiede, D., Aryal, J., 2019. Evaluation of different machine learning methods and deep-learning convolutional neural networks for landslide detection. Remote Sens. (Basel) 11 (2), 196.

Ghosh, A., Bera, B., 2023. Landform classification and geomorphological mapping of the Chota Nagpur Plateau, India. Quat. Sci. Adv. 10, 100082.

Ghosh, J.K., Delampady, M., Samanta, T., 2006. An Introduction to Bayesian Analysis: Theory and Methods. Springer Texts in Statistics Book Series. Springer, New York, NY.

Giasson, E., Figueiredo, S.R., Tornquist, C.G., Clarke, R.T., 2008. Digital soil mapping using logistic regression on terrain parameters for several ecological regions in southern Brazil. In: Hartemink, A.E., McBratney, A., Mendonça-Santos, M.L. (Eds.), Digital Soil Mapping with Limited Data. Springer, Dordrecht, the Netherlands, pp. 225–232.

Gibson, M.J., Glasser, N.F., Quincey, D.J., Mayer, C., Rowan, A.V., Irvine-Fynn, T.D., 2017. Temporal variations in supraglacial debris distribution on Baltoro Glacier, Karakoram between 2001 and 2012. Geomorphology 295, 572–585.

Giles, P.T., Franklin, S.E., 1996. Comparison of derivative topographic surfaces of a DEM generated from stereoscopic SPOT images with field measurements. Photogrammet. Eng. Remote Sens. 62 (10), 1165–1171.

Gillieson, D., Burgess, J., Spate, A., Cochrane, A., 1990. An Atlas of the Lakes of the Larsemann Hills, Princess Elizabeth Land, Antarctica. ANARE Research Notes, No. 74. Antarctic Division Australia, Kingston, TS.

Gillings, M., Hacıgüzeller, P., Lock, G. (Eds.), 2020. Archaeological Spatial Analysis: A Methodological Guide. Routledge, London, UK.

Gindraux, S., Boesch, R., Farinotti, D., 2017. Accuracy assessment of digital surface models from unmanned aerial vehicles' imagery on glaciers. Remote Sens. (Basel) 9 (2), 186.

Gioia, D., Danese, M., Corrado, G., di Leo, P., Minervino Amodio, A., Schiattarella, M., 2021. Assessing the prediction accuracy of geomorphon-based automated landform classification: an example from the Ionian Coastal Belt of Southern Italy. ISPRS Int. J. Geo Inf. 10 (11), 725.

Glasko, M.P., Rantsman, E.Y., 1996. Morphostructural groups as activization sites of natural processes. Trans. (Doklady) Russ. Acad. Sci., Earth Sci. Sect. 350 (7), 1171–1174.

Glasser, N., Scambos, T., 2008. A structural glaciological analysis of the 2002 Larsen B ice-shelf collapse. J. Glaciol. 54 (184), 3–16.

GLOBE Task Team and Others (Hastings, D.A., Dunbar, P.K., Elphingstone, G.M., Bootz, M., Murakami, H., Maruyama, H., et al.) (Eds.), 1999. The Global Land One-Kilometer Base Elevation (GLOBE) Digital Elevation Model, Version 1.0. National Oceanic and Atmospheric Administration, National Geophysical Data Center, Boulder, CO. http://www.ngdc.noaa.gov/mgg/topo/globe.html. (Accessed 27 June 2024).

Gogineni, S., Tammana, D., Braaten, D., Leuschen, C., Akins, T., Legarsky, J., et al., 2001. Coherent radar ice thickness measurements over the Greenland ice sheet. J. Geophys. Res. Atmos. 106 (D24), 33761–33772.

Golbreich, I.G., Zabaluev, V.V., Lastochkin, A.N., Mirkin, G.R., Reinin, I.V., 1968. Morphostructural Methods for Study of Tectonics of Oil- and Gas-Bearing Covered Platform Areas. Nedra, Leningrad, USSR (in Russian).

Gold, C., Mostafavi, M.A., 2000. Towards the global GIS. ISPRS J. Photogrammet. Remote Sens. 55 (3), 150–163.

Golovchenko, A.V., Polyanskaya, L.M., 1996. Seasonal dynamics of population and biomass of microorganisms in the soil profile. Eurasian Soil Sci. 29 (10), 1145–1150.

Golts, S., Rosenthal, E., 1993. A morphotectonic map of the northern Arava in Israel, derived from isobase lines. Geomorphology 7 (4), 305–315.

Golyandina, N., Stepanov, D., 2005. SSA-based approaches to analysis and forecast of multidimensional time series. In: Ermakov, S.M., Melas, V.B., Pepelyshev, A.N. (Eds.), Proceedings of the 5th St. Petersburg Workshop on Simulation, St. Petersburg, Russia, 26 June –2 July 2005. St. Petersburg State University, St. Petersburg, Russia, pp. 293–298.

Golyandina, N.E., Usevich, K.D., 2010. 2D-extension of singular spectrum analysis: algorithm and elements of theory. In: Olshevsky, V., Tyrtyshnikov, E. (Eds.), Matrix Methods: Theory, Algorithms, Applications. World Scientific, Singapore, pp. 449–473.

Golyandina, N., Zhigljavsky, A., 2013. Singular Spectrum Analysis for Time Series. Springer Briefs in Statistics Book Series. Springer, Heidelberg, Germany.

Golyandina, N.E., Nekrutkin, V.V., Zhigljavsky, A.A., 2001. Analysis of Time Series Structure: SSA and Related Techniques. Monographs on Statistics and Applied Probability, vol. 90. Chapman and Hall/CRC, London, UK.

Golyandina, N.E., Usevich, K.D., Florinsky, I.V., 2007. Filtering of digital terrain models by two-dimensional singular spectrum analysis. Int. J. Ecol. Dev. 8 (F07), 81–94.

Golyandina, N.E., Usevich, K.D., Florinsky, I.V., 2008. Singular spectrum analysis for filtering of digital terrain models. Geodesy Cartogr. (5), 21–28 (in Russian).

Gomez, C., 2018. Understanding volcanic geomorphology from derivatives and wavelet analysis: a case study at Miyakejima Volcano, Izu Islands, Japan. J. Volcanol. Geotherm. Res. 354, 57–66.

Gomez, C., 2022. Point Cloud Technologies for Geomorphologists. From Data Acquisition to Processing. Springer Textbooks in Earth Sciences, Geography and Environment Book Series. Springer, Cham, Switzerland.

Gong, J., Li, Z., Zhu, Q., Sui, H., Zhou, Y., 2000. Effects of various factors on the accuracy of DEMs: an intensive experimental investigation. Photogramm. Eng. Remote Sens. 66 (9), 1113–1117.

Gonzalez, C., Rizzoli, P., 2018. Landcover-dependent assessment of the relative height accuracy in TanDEM-X DEM products. IEEE Geosci. Remote Sens. Lett. 15 (12), 1892–1896.

González-García, J., Gómez-Espinosa, A., Cuan-Urquizo, E., García-Valdovinos, L.G., Salgado-Jiménez, T., Cabello, J.A.E., 2020. Autonomous underwater vehicles: localization, navigation, and communication for collaborative missions. Appl. Sci. 10 (4), 1256.

González-Moradas, M.d.R., Viveen, W., 2020. Evaluation of ASTER GDEM2, SRTMv3.0, ALOS AW3D30 and TanDEM-X DEMs for the Peruvian Andes against highly accurate GNSS ground control points and geomorphological-hydrological metrics. Remote Sens. Environ. 237, 111509.

Goodchild, M., Kimerling, A.J. (Eds.), 2002. Discrete Global Grids. National Center for Geographic Information and Analysis, Santa Barbara, CA.

Google, 2005–2023. Google Earth. Google. https://earth.google.com/web/. (Accessed 27 June 2024).

Goovaerts, P., 1997. Geostatistics for Natural Resources Evaluation. Oxford University Press, New York, NY.

Gorini, M.A.V., Mota, G.L.A., 2016. Dealing with double vagueness in DEM morphometric analysis. Int. J. Geogr. Inf. Sci. 30 (8), 1644–1666.

Gorshkov, A., Kossobokov, V., Soloviev, A., 2003. Recognition of earthquake-prone areas. In: Keilis-Borok, V.I., Soloviev, A.A. (Eds.), Nonlinear Dynamics of the Lithosphere and Earthquake Prediction. Springer Series in Synergetics. Springer, Berlin, Germany, pp. 239–310.

Gosteva, T.S., Patrakova, V.S., Abramkina, V.A., 1983. Establishing the principles of spatial distribution of ring structures based on trend analysis of relief. Soviet Geol. Geophys. 24 (8), 62–67.

Goudge, T.A., Milliken, R.E., Head, J.W., Mustard, J.F., Fassett, C.I., 2017. Sedimentological evidence for a deltaic origin of the western fan deposit in Jezero Crater, Mars and implications for future exploration. Earth Planet. Sci. Lett. 458, 357–365.

Goyal, M.K., Panchariya, V.K., Sharma, A., Singh, V., 2018. Comparative assessment of SWAT model performance in two distinct catchments under various DEM scenarios of varying resolution, sources and resampling methods. Water Resour. Manage. 32 (2), 805–825.

Grachev, A.F., Magnitsky, V.A., Mukhamediev, S.A., Nikolaev, V.A., 2001. The effect of neotectonic movements on the gradients and curvatures of the Northern Eurasian lithosphere surface. Phys. Solid Earth 37 (2), 89–106.

Graciano, A., Rueda, A.J., Feito, F.R., 2018. A formal framework for the representation of stack-based terrains. Int. J. Geogr. Inf. Sci. 32 (10), 1999–2022.

Graf, L., Moreno-de-las-Heras, M., Ruiz, M., Calsamiglia, A., García-Comendador, J., Fortesa, J., et al., 2018. Accuracy assessment of digital terrain model dataset sources for hydrogeomorphological modelling in small Mediterranean catchments. Remote Sens. (Basel) 10 (12), 2014.

Graupe, D., 2007. Principles of Artificial Neural Networks, second ed. World Scientific, Singapore.

Greve, M.H., Greve, M.B., Bou Kheir, R., Bøcher, P.K., Larsen, R., McCloy, K., 2010. Comparing decision tree modeling and indicator kriging for mapping the extent of organic soils in Denmark. In: Boettinger, J.L., Howell, D.W., Moore, A.C., Hartemink, A.E., Kienast-Brown, S. (Eds.), Digital Soil Mapping: Bridging Research, Environmental Application, and Operation. Progress in Soil Science Book Series, Springer, Dordrecht, the Netherlands, pp. 267–280.

Grieve, S.W.D., Mudd, S.M., Hurst, M.D., Milodowski, D.T., 2016. A nondimensional framework for exploring the relief structure of landscapes. Earth Surf. Dyn. 4 (2), 309–325.

Grigorenko, A.M., 1998. Some Problems of Technical Information Theory. Ubex Press, Moscow, Russia (in Russian).

Grimm, K., Tahmasebi Nasab, M., Chu, X., 2018. TWI computations and topographic analysis of depression-dominated surfaces. Water 10 (5), 663.

Grinand, C., Arrouays, D., Laroche, B., Martin, M.P., 2008. Extrapolating regional soil landscapes from an existing soil map: sampling intensity, validation procedures, and integration of spatial context. Geoderma 143 (1–2), 180–190.

Groffman, P.M., Hanson, G.C., 1997. Wetland denitrification: influence of site quality and relationships with wetland delineation protocols. Soil Sci. Soc. Am. J. 61 (1), 323–329.

Groffman, P.M., Tiedje, J.M., 1989. Denitrification in north temperate forest soils: spatial and temporal patterns at the landscape and seasonal scales. Soil Biol. Biochem. 21 (5), 613–620.

Grohmann, C.H., 2004a. Morphometric analysis in geographic information systems: applications of free software GRASS and R. Comput. Geosci. 30 (9–10), 1055–1067.

Grohmann, C.H., 2004b. Técnicas de Geoprocessamento Aplicadas a Análise Morfométrica (M.Sc. thesis). Institute of Geosciences, University of São Paulo, São Paulo, Brazil.

Grohmann, C.H., 2005. Trend-surface analysis of morphometric parameters: a case study in southeastern Brazil. Comput. Geosci. 31 (8), 1007–1014.

Grohmann, C.H., 2014. Comparative analysis of global digital elevation models. In: 2014 Annual Meeting of the Geological Society of America, Vancouver, BC, 19–22 Oct. 2014. Geological Society of America, Vancouver, BC, pp. 34–35.

Grohmann, C.H., 2018. Evaluation of TanDEM-X DEMs on selected Brazilian sites: comparison with SRTM, ASTER GDEM and ALOS AW3D30. Remote Sens. Environ. 212, 121–133.

Grohmann, C.H., Riccomini, C., Alves, F.M., 2007. SRTM-based morphotectonic analysis of the Poços de Caldas Alkaline Massif, southeastern Brazil. Comput. Geosci. 33 (1), 10–19.

Grohmann, C.H., Riccomini, C., Chamani, M.A.C., 2011a. Regional scale analysis of landform configuration with base-level (isobase) maps. Hydrol. Earth Syst. Sci. 15 (5), 1493–1504.

Grohmann, C.H., Smith, M.J., Riccomini, C., 2011b. Multiscale analysis of topographic surface roughness in the Midland Valley, Scotland. IEEE Trans. Geosci. Remote Sens. 49 (4), 1200–1213.

Grohmann, C.H., Garcia, G.P.B., Affonso, A.A., Albuquerque, R.W., 2020. Dune migration and volume change from airborne LiDAR, terrestrial LiDAR and structure from motion–multi view stereo. Comput. Geosci. 143, 104569.

Groom, J., Bertin, S., Friedrich, H., 2018. Evaluation of DEM size and grid spacing for fluvial patch-scale roughness parameterisation. Geomorphology 320, 98–110.

Groshong Jr., R.H., 2006. 3-D Structural Geology: A Practical Guide to Quantitative Surface and Subsurface Map Interpretation, second ed. Springer, Berlin, Germany.

Gruber, F.E., Baruck, J., Geitner, C., 2017. Algorithms vs. surveyors: a comparison of automated landform delineations and surveyed topographic positions from soil mapping in an Alpine environment. Geoderma 308, 9–25.

Gruen, A., Baltsavias, E.P., Henricsson, O. (Eds.), 1997. Automatic Extraction of Man-Made Objects from Aerial and Space Images (II). Birkhäuser, Basel, Switzerland.

Grunwald, S., 2009. Multi-criteria characterization of recent digital soil mapping and modeling approaches. Geoderma 152 (3–4), 195–207.

Grunwald, S., Barak, P., McSweeney, K., Lowery, B., 2000. Soil landscape models at different scales portrayed in virtual reality modeling language. Soil Sci. 165 (8), 598–615.

Grunwald, S., Ramasundaram, V., Comerford, N.B., Bliss, C.M., 2007. Are current scientific visualisation and virtual reality techniques capable to represent real soil-landscapes? In: Lagacherie, P., McBratney, A.B., Voltz, M. (Eds.), Digital Soil Mapping: An Introductory Perspective. Developments in Soil Science, vol. 31. Elsevier, Amsterdam, the Netherlands, pp. 571–580.

Guan, L., Hu, J., Pan, H., Wu, W., Sun, Q., Chen, S., et al., 2020. Fusion of public DEMs based on sparse representation and adaptive regularization variation model. ISPRS J. Photogrammet. Remote Sens. 169, 125–134.

Guardiano, F.B., Srivastava, R.M., 1993. Multivariate geostatistics: beyond bivariate moments. In: Soares, A. (Ed.), Geostatistics-Tróia'92, vol. 1. Quantitative Geology and Geostatistics, vol. 5. Kluwer, Dordrecht, the Netherlands, pp. 133–144.

Guberman, S., Pikovski, Y., Rantsman, E., 1997. Methodology for prediction of the locations of giant oil and gas reservoirs: field results. In: Proceedings of the SPE 67th Annual Western Regional Meeting, Long Beach, CA, 25–27 June 1997. Society of Petroleum Engineers, Richardson, TX, pp. 321–330.

Gudmundsson, A., 2000. Active fault zones and groundwater flow. Geophys. Res. Lett. 27 (18), 2993–2996.

Gudmundsson, M.T., Sigmundsson, F., Björnsson, H., 1997. Ice–volcano interaction of the 1996 Gjálp subglacial eruption, Vatnajökull, Iceland. Nature 389 (6654), 954–957.

Guevara, M., Vargas, R., 2019. Downscaling satellite soil moisture using geomorphometry and machine learning. PLoS One 14 (9), e0219639.

Guevara, M., Arroyo, C., Brunsell, N., Cruz, C.O., Domke, G., Equihua, J., et al., 2020. Soil organic carbon across Mexico and the conterminous United States (1991–2010). Global Biogeochem. Cycles 34 (3), e2019GB006219.

Gugan, D.J., Dowman, I.J., 1988. Topographic mapping from SPOT imagery. Photogrammet. Eng. Remote Sens. 54 (10), 1409–1414.

Guida, D., Cuomo, A., Palmieri, V., 2016. Using object-based geomorphometry for hydro-geomorphological analysis in a Mediterranean research catchment. Hydrol. Earth Syst. Sci. 20 (9), 3493–3509.

Guillaume, B., Dhont, D., Brusset, S., 2008. Three-dimensional geologic imaging and tectonic control on stratigraphic architecture: upper Cretaceous of the Tremp Basin (south-central Pyrenees, Spain). Am. Assoc. Pet. Geol. Bull. 92 (2), 249–269.

Guillon, H., Byrne, C.F., Lane, B.A., Solis, S.S., Pasternack, G.B., 2020. Machine learning predicts reach-scale channel types from coarse-scale geospatial data in a large river basin. Water Resour. Res. 56 (3), e2019WR026691.

Guirado, E., Alcaraz-Segura, D., Rigol-Sánchez, J.P., Gisbert, J., Martínez-Moreno, F.J., Galindo-Zaldívar, J., et al., 2018. Remote-sensing-derived fractures and shrub patterns to identify groundwater dependence. Ecohydrology 11 (6), e1933.

Guisado-Pintado, E., Jackson, D.W.T., Rogers, D., 2019. 3D mapping efficacy of a drone and terrestrial laser scanner over a temperate beach-dune zone. Geomorphology 328, 157–172.

Guisan, A., Weiss, S.B., Weiss, A.D., 1999. GLM versus CCA spatial modeling of plant species distribution. Plant Ecol. 143 (1), 107–122.

Gulley, J.D., Fountain, A.G., 2019. Glacier caves. In: White, W.B., Culver, D.C., Pipan, T. (Eds.), Encyclopedia of Caves, third ed. Academic Press, London, UK, pp. 468–473.

Gunia, D.A., 2004–2018. Earth3D. http://www.earth3d.org (Accessed 27 June 2024).

Guo, Y., Liu, Y., Oerlemans, A., Lao, S., Wu, S., Lew, M.S., 2016. Deep learning for visual understanding: a review. Neurocomputing 187, 27–48.

Guo, Z., Adhikari, K., Chellasamy, M., Greve, M.B., Owens, P.R., Greve, M.H., 2019. Selection of terrain attributes and its scale dependency on soil organic carbon prediction. Geoderma 340, 303–312.

Gupta, P., 2015. Biodiversity of Larsemann Hills, Antarctica. Clim. Change 1 (3), 174–183.

Guth, P.L., 1999. Contour line "ghosts" in USGS Level 2 DEMs. Photogrammet. Eng. Remote Sens. 65 (3), 289–296.

Guth, P.L., 2012. Automated export of GIS maps to Google Earth: tool for research and teaching. Geol. Soc. Am. Spec. Paper 492, 165–182.

Guth, P.L., Geoffroy, T.M., 2021. LiDAR point cloud and ICESat-2 evaluation of 1 second global digital elevation models: Copernicus wins. Trans. GIS 25 (5), 2245–2261.

Guth, P.L., Van Niekerk, A., Grohmann, C.H., Muller, J.-P., Hawker, L., Florinsky, I.V., et al., 2021. Digital elevation models: terminology and definitions. Remote Sens. (Basel) 13 (18), 3581.

Guyot, A., Hubert-Moy, L., Lorho, T., 2018. Detecting neolithic burial mounds from LiDAR-derived elevation data using a multi-scale approach and machine learning techniques. Remote Sens. (Basel) 10 (2), 225.

Gvin, V.Y., 1963. The use of morphometry in structural studies of the Upper and Mid Volga and Kama Regions. Prob. Geogr. 63, 64–80 (in Russian).

Gzovsky, M.V., 1954. The main problems of fault classification. Sov. Geol. 41, 131–169 (in Russian).

Haggett, P., Chorley, R.J., Stoddart, D.R., 1965. Scale standards in geographical research: a new measure of areal magnitude. Nature 205 (4974), 844–847.

Hall, J.K., 1996. Digital topography and bathymetry of the area of the Dead Sea depression. Tectonophysics 266 (1–4), 177–185.

Han, X., Liu, J., Mitra, S., Li, X., Srivastava, P., Guzman, S.M., et al., 2018. Selection of optimal scales for soil depth prediction on headwater hillslopes: a modeling approach. Catena 163, 257–275.

Hancock, G.R., 2005. The use of digital elevation models in the identification and characterization of catchments over different grid scales. Hydrol. Process. 19 (9), 1727–1749.

Hancock, G., Willgoose, G., 2001. The production of digital elevation models for experimental model landscapes. Earth Surf. Process. Landf. 26 (5), 475–490.

Handy, M.R., Hirth, G., Hovius, N. (Eds.), 2007. Tectonic Faults: Agents of Change on a Dynamic Earth. MIT Press, Cambridge, MA.

Hannah, M.J., 1981. Error detection and correction in digital terrain models. Photogrammet. Eng. Remote Sens. 47 (1), 63–69.

Hantke, R., Scheidegger, A.E., 1999. Tectonic predesign in geomorphology. In: Hergarten, S., Neugebauer, H.J. (Eds.), Process Modelling and Landform Evolution. Lect. Notes Earth Sci., vol. 78, pp. 251–266.

Haralick, R.M., 1979. Statistical and structural approaches to texture. Proc. IEEE 67 (5), 786–804.

Haralick, R., 1983. Ridges and valleys on digital images. Comput. Vis. Graph. Image Process. 22 (10), 28–38.

Haralick, R.M., Shanmugam, K., Dinstein, I., 1973. Textural features for image classification. IEEE Trans. Syst. Man Cybern. 3 (6), 610–621.

Harding, C., Hasiuk, F., Wood, A., 2021. TouchTerrain—3D printable terrain models. ISPRS Int. J. Geo Inf. 10 (3), 108.

Harper, H., Sandwell, D.T., 2024. Global predicted bathymetry using neural networks. Earth Space Sci. 11 (3), e2023EA003199.

Harrison, J.M., Lo, C.-P., 1996. PC-based two-dimensional discrete Fourier transform programs for terrain analysis. Comput. Geosci. 22 (4), 419–424.

Hart, B.S., Sagan, J.A., 2007. Curvature for visualization of seismic geomorphology. In: Davies, R.J., Posamentier, H.W., Wood, L.J., Cartwright, J.A. (Eds.), Seismic Geomorphology: Applications to Hydrocarbon Exploration and Production. Geol. Soc. London Spec. Publ., vol. 277, pp. 139–149.

Hartemink, A.E., Zhang, Y., Bockheim, J.G., Curi, N., Silva, S.H.G., Grauer-Gray, J., et al., 2020. Soil horizon variation: a review. Adv. Agron. 160 (1), 125–185.

Hasiuk, F.J., Harding, C., Renner, A.R., Winer, E., 2017. TouchTerrain: a simple web-tool for creating 3D-printable topographic models. Comput. Geosci. 109, 25–31.

Hastie, T., Tibshirani, R., 1990. Generalized Additive Models. Chapman and Hall, London, UK.

Hastings, D.A., Dunbar, P.K., 1998. Development and assessment of the global land one-km base elevation digital elevation model (GLOBE). In: Fritsch, D., Englich, M., Sester, M. (Eds.), ISPRS Commission IV Symposium "GIS Between Visions and Applications", Stuttgart, Germany, 7–10 Sept. 1998. ISPRS Arch. XXXII (4), pp. 218–221.

Hastings, D.A., Dunbar, P.K. (Eds.), 1999. Global Land One-Kilometer Base Elevation (GLOBE). Digital Elevation Model, Version 1.0. Documentation Version 1.0. National Geophysical Data Center, National Oceanic and Atmospheric Administration, Boulder, CO.

Hato, M., Tsu, H., Tachikawa, T., Abrams, M., Bailey, B., 2009. The ASTER Global Digital Elevation Model (GDEM)—for societal benefit. In: AGU Fall Meeting 2009, San Francisco, CA, 14–18 Dec. 2009. American Geophysical Union, San Francisco, CA. U33B-0065.

Haugerud, R.A., Harding, D.J., Johnson, S.Y., Harless, J.L., Weaver, C.S., Sherrod, B.L., 2003. High-resolution lidar topography of the Puget Lowland, Washington—a bonanza for earth science. GSA Today 13 (6), 4–10.

Hawker, L., Uhe, P., Paulo, L., Sosa, J., Savage, J., Sampson, C., Neal, J., 2022. A 30 m global map of elevation with forests and buildings removed. Environ. Res. Lett. 17 (2), 024016.

Hayes, A.G., Wolf, A.S., Aharonso, O., Zebker, H., Lorenz, R., Kirk, R.L., et al., 2010. Bathymetry and absorptivity of Titan's Ontario Lacus. J. Geophys. Res. Planet. 115 (E9), E09009.

Hayes, A.G., Birch, S.P.D., Dietrich, W.E., Howard, A.D., Kirk, R.L., Poggiali, V., et al., 2017. Topographic constraints on the evolution and connectivity of Titan's lacustrine basins. Geophys. Res. Lett. 44 (23), 11745–11753.

Heckmann, T., Cavalli, M., Cerdan, O., Foerster, S., Javaux, M., Lode, E., et al., 2018. Indices of sediment connectivity: opportunities, challenges and limitations. Earth Sci. Rev. 187, 77–108.

Heddadj, D., Gascuel-Odoux, C., 1999. Topographic and seasonal variations of unsaturated hydraulic conductivity as measured by tension disc infiltrometers at the field scale. Eur. J. Soil Sci. 50 (2), 275–283.

Heifetz, B.S., 1964. Approximation of the topographic surface by the Chebyshev orthogonal polynomials. Izvestia Vuzov. Geodesy Aerophotosurveying (2), 78–86 (in Russian).

Heller, M., 1990. Triangulation algorithms for adaptive terrain modeling. In: Brassel, K., Kisomoto, H. (Eds.), Proceedings of the 4th International Symposium on Spatial Data Handling, Zurich, Switzerland, July 1990, vol. 1. University of Zurich, Zurich, Switzerland, pp. 163–174.

Helm, V., Humbert, A., Miller, H., 2014. Elevation and elevation change of Greenland and Antarctica derived from CryoSat-2. Cryosphere 8 (4), 1539–1559.

Helmert, F.R., 1880. Die mathematischen und physikalischen Theorieen der höheren Geodäsie, vol. 1: Die mathematischen Theorieen. Teubner, Leipzig, Germany.

Helvaci, C., Bayram, B., 2004. Semi automatic digitizing of contours from 1:25,000 scaled maps. In: Altan, O. (Ed.), XX ISPRS Congress, Technical Commission III, Istanbul, Turkey, 12–23 July 2004. ISPRS Arch. XXXV (B3), pp. 461–464.

Hendrickx, H., Vivero, S., de Cock, L., de Wit, B., de Maeyer, P., Lambiel, C., et al., 2019. The reproducibility of SfM algorithms to produce detailed digital surface models: the example of PhotoScan applied to a high-alpine rock glacier. Remote Sens. Lett. 10 (1), 11–20.

Hengl, T., 2006. Finding the right pixel size. Comput. Geosci. 32 (9), 1283–1298.

Hengl, T., Reuter, H.I., 2009. Geomorphometry: Concepts, Software, Applications. Developments in Soil Science, vol. 33 Elsevier, Amsterdam, the Netherlands.

Hengl, T., Heuvelink, G.B.M., Stein, A., 2004. A generic framework for spatial prediction of soil variables based on regression-kriging. Geoderma 120 (1–2), 75–93.

Hengl, T., Heuvelink, G.B.M., Rossiter, D.G., 2007a. About regression kriging: from equations to case studies. Comput. Geosci. 33 (10), 1301–1315.

Hengl, T., Toomanian, N., Reuter, H.I., Malakouti, M.J., 2007b. Methods to interpolate soil categorical variables from profile observations: lessons from Iran. Geoderma 140 (4), 417–427.

Hengl, T., Heuvelink, G.B.M., Kempen, B., Leenaars, J.G.B., Walsh, M.G., Shepherd, K.D., et al., 2015. Mapping soil properties of Africa at 250 m resolution: random forests significantly improve current predictions. PLoS One 10 (6), e0125814.

Hengl, T., Leenaars, J.G.B., Shepherd, K.D., Walsh, M.G., Heuvelink, G.B.M., Mamo, T., et al., 2017a. Soil nutrient maps of Sub-Saharan Africa: assessment of soil nutrient content at 250 m spatial resolution using machine learning. Nutr. Cycl. Agroecosyst. 109 (1), 77–102.

Hengl, T., Mendes de Jesus, J., Heuvelink, G.B.M., Ruiperez Gonzalez, M., Kilibarda, M., Blagotić, A., et al., 2017b. SoilGrids250m: global gridded soil information based on machine learning. PLoS One 12 (2), e0169748.

Hengl, T., Nussbaum, M., Wright, M.N., Heuvelink, G.B.M., Gräler, B., 2018. Random forest as a generic framework for predictive modeling of spatial and spatio-temporal variables. PeerJ 6, e5518.

Hengl, T., Miller, M.A.E., Križan, J., Shepherd, K.D., Sila, A., Kilibarda, M., et al., 2021. African soil properties and nutrients mapped at 30 m spatial resolution using two-scale ensemble machine learning. Sci. Rep. 11, 6130.

Herbst, M., Diekkrüger, B., Vereecken, H., 2006. Geostatistical co-regionalization of soil hydraulic properties in a micro-scale catchment using terrain attributes. Geoderma 132 (1–2), 206–221.

Hergarten, S., Robl, J., Stüwe, K., 2016. Tectonic geomorphology at small catchment sizes—extensions of the stream-power approach and the χ method. Earth Surf. Dyn. 4 (2), 1–9.

Hess, R., 2010. Blender Foundations: The Essential Guide to Learning Blender 2.6. Focal Press, Amsterdam, the Netherlands.

Heung, B., Ho, H.C., Zhang, J., Knudby, A., Bulmer, C.E., Schmidt, M.G., 2016. An overview and comparison of machine-learning techniques for classification purposes in digital soil mapping. Geoderma 265, 62–77.

Heuvelink, G.B.M., Bierkens, M.F.P., 1992. Combining soil maps with interpolations from point observations to predict quantitative soil properties. Geoderma 55 (1–2), 1–15.

Heuvelink, G.B.M., Webster, R., 2001. Modelling soil variation: past, present, and future. Geoderma 100 (3–4), 269–301.

Heuvelink, G.B.M., Burrough, P.A., Stein, A., 1989. Propagation of errors in spatial modelling with GIS. Int. J. Geogr. Inf. Syst. 3 (4), 303–322.

Hewitt, E., Hewitt, R.E., 1980. The Gibbs–Wilbraham phenomenon: an episode in Fourier analysis. Arch. Hist. Exact Sci. 21 (1), 129–160.

Hilley, G.E., Mynatt, I., Pollard, D.D., 2010. Structural geometry of Raplee Ridge monocline and thrust fault imaged using inverse Boundary Element Modeling and ALSM data. J. Struct. Geol. 32 (1), 45–58.

Hills, E.S., 1956. A contribution to the morphotectonics of Australia. J. Geol. Soc. Aust. 3 (1), 1–15.

Hirano, A., Welch, R., Lang, H., 2003. Mapping from ASTER stereo image data: DEM validation and accuracy assessment. ISPRS J. Photogrammet. Remote Sens. 57 (5–6), 356–370.

Hirschmüller, H., 2008. Stereo processing by semiglobal matching and mutual information. IEEE Trans. Pattern Anal. Mach. Intell. 30 (2), 328–341.

Hirt, C., Filmer, M.S., Featherstone, W.E., 2010. Comparison and validation of the recent freely available ASTER-GDEM ver1, SRTM ver4.1 and GEODATA DEM-9S ver3 digital elevation models over Australia. Aust. J. Earth Sci. 57 (3), 337–347.

Hjelle, Ø., Dæhlen, M., 2006. Triangulations and Applications. Mathematics and Visualization Book Series. Springer, Berlin, Germany.

Hobbs, K.F., 1999. An investigation of RGB multi-band shading for relief visualisation. Int. J. Appl. Earth Obs. Geoinf. 1 (3–4), 181–186.

Hobbs, W.H., 1904. Lineaments of Atlantic Border region. Geol. Soc. Am. Bull. 15, 483–506.

Hobbs, W.H., 1911. Repeating patterns in the relief and in the structure of the land. Geol. Soc. Am. Bull. 22 (2), 123–176.

Hodgetts, D., 2013. Laser scanning and digital outcrop geology in the petroleum industry: a review. Mar. Petrol. Geol. 46, 335–354.

Hodgson, D.A., Noon, P.E., Vyvermann, W., Bryant, C.L., Gore, D.B., Appleby, P., et al., 2001. Were the Larsemann Hills ice-free through the last glacial maximum? Antarct. Sci. 13 (4), 440–454.

Hodgson, M.E., 1995. What cell size does the computed slope/aspect angle represent? Photogrammet. Eng. Remote Sens. 61 (5), 513–517.

Hodgson, R.A., Gay, S.P., Benjamins, J.Y. (Eds.), 1976. Proceedings of the 1st International Conference on the New Basement Tectonics, 3–7 June 1974, Salt Lake City, UT. Utah Geological Association, Salt Lake City, UT.

Hodúl, M., Bird, S., Knudby, A., Chénier, R., 2018. Satellite derived photogrammetric bathymetry. ISPRS J. Photogrammet. Remote Sens. 142, 268–277.

Hofierka, J., Gallay, M., Bandura, P., Šašak, J., 2018. Identification of karst sinkholes in a forested karst landscape using airborne laser scanning data and water flow analysis. Geomorphology 308, 265–277.

Hofierka, J., Gallay, M., Onačillová, K., Hofierka Jr., J., 2020. Physically-based land surface temperature modeling in urban areas using a 3-D city model and multispectral satellite data. Urban Clim. 31, 100566.

Hollingsworth, I.D., Bui, E.N., Odeh, I.O.A., McLeod, P., 2007. Rule-based land unit mapping of the Tiwi Islands, Northern Territory, Australia. In: Lagacherie, P., McBratney, A.B., Voltz, M. (Eds.), Digital Soil Mapping: An Introductory Perspective. Developments in Soil Science, vol. 31. Elsevier, Amsterdam, the Netherlands, pp. 401–414.

Holmes, K.W., Chadwick, O.A., Kyriakidis, P.C., 2000. Error in a USGS 30-meter digital elevation model and its impact on terrain modeling. J. Hydrol. 233 (1–4), 154–173.

Holt, T., Glasser, N., Quincey, D., 2013. The structural glaciology of southwest Antarctic Peninsula ice shelves (ca. 2010). J. Maps 9 (4), 523–531.

Hong, D., Gao, L., Yokoya, N., Yao, J., Chanussot, J., Du, Q., et al., 2021. More diverse means better: multimodal deep learning meets remote-sensing imagery classification. IEEE Trans. Geosci. Remote Sens. 59 (5), 4340–4354.

Hong, H., Llia, L., Tsangaratos, P., Chen, W., Xu, C., 2017. A hybrid fuzzy weight of evidence method in landslide susceptibility analysis on the Wuyuan area, China. Geomorphology 290, 1–16.

Hooshyar, M., Wang, D., Kim, S., Medeiros, S.C., Hagen, S.C., 2016. Valley and channel networks extraction based on local topographic curvature and *k*-means clustering of contours. Water Resour. Res. 52 (10), 8081–8102.

Horáček, M., Samec, P., Minár, J., 2018. The mapping of soil taxonomic units via fuzzy clustering—a case study from the Outer Carpathians, Czechia. Geoderma 326, 111–122.

Horn, B.K.P., 1981. Hill shading and the reflectance map. Proc. IEEE 69 (1), 14–47.

Horn, C., Ivarsson, O., Lindhé, C., Potter, R., Green, A., Ling, J., 2022. Artificial intelligence, 3D documentation, and rock art—approaching and reflecting on the automation of identification and classification of rock art images. J. Archaeol. Method Theory 29 (1), 188–213.

Horton, R.E., 1945. Erosional development of streams and their drainage basins, hydrophysical approach to quantitative morphology. Geol. Soc. Am. Bull. 56 (3), 275–370.

Hounkpatin, O.K.L., de Hipt, F.O., Bossa, A.Y., Welp, G., Amelung, W., 2018. Soil organic carbon stocks and their determining factors in the Dano catchment (Southwest Burkina Faso). Catena 166, 298–309.

Houser, C., Bishop, M., Wernette, P., 2017. Multi-scale topographic anisotropy patterns on a barrier island. Geomorphology 297, 153–158.

Howard, A.D., 1978. Origin of the stepped topography of the Martian poles. Icarus 34 (3), 581–599.

Howat, I.M., Negrete, A., Smith, B.E., 2014. The Greenland Ice Mapping Project (GIMP) land classification and surface elevation data sets. Cryosphere 8 (4), 1509–1518.

Howat, I.M., Porter, C., Smith, B.E., Noh, M.-J., Morin, P., 2019. The Reference Elevation Model of Antarctica. Cryosphere 13 (2), 665–674.

Hu, G., Dai, W., Li, S., Xiong, L., Tang, G., 2020. A vector operation to extract second-order terrain derivatives from digital elevation models. Remote Sens. (Basel) 12 (19), 3134.

Hu, G., Dai, W., Li, S., Xiong, L., Tang, G., Strobl, J., 2021. Quantification of terrain plan concavity and convexity using aspect vectors from digital elevation models. Geomorphology 375, 107553.

Hu, Q., Zhou, Y., Wang, S., Wang, F., 2020. Machine learning and fractal theory models for landslide susceptibility mapping: case study from the Jinsha River Basin. Geomorphology 351, 106975.

Huang, H., Chen, X., Wang, X., Wang, X., Liu, L., 2019. A depression-based index to represent topographic control in urban pluvial flooding. Water 11 (10), 2115.

Huang, P.-C., Lee, K.T., 2016. Distinctions of geomorphological properties caused by different flow-direction predictions from digital elevation models. Int. J. Geogr. Inf. Sci. 30 (2), 168–185.

Huang, X., Senthilkumar, S., Kravchenko, A., Thelen, K., Qi, J., 2007. Total carbon mapping in glacial till soils using near infrared spectroscopy, Landsat imagery, and topographical information. Geoderma 141 (1–2), 34–42.

Hudleston, P.J., 1973. Fold morphology and some geometrical implications of theories of fold development. Tectonophysics 16 (1–2), 1–46.

Huffman, D.P., 2014. Shaded relief in Blender. In: Banff2014: Papers and Slide Presentations, 9th ICA Mountain Cartography Workshop, Banff, Canada, April 22–26, 2014. ICA Commission on Mountain Cartography. https://carto.geogr.msu.ru/cmc/activities/workshops/banff_canada/papers/huffman.pdf. (Accessed 27 June 2024).

Huffman, D.P., 2017. Creating Shaded Relief in Blender. Somethingaboutmaps. https://somethingaboutmaps.wordpress.com/2017/11/16/creating-shaded-relief-in-blender/. (Accessed 27 June 2024).

Huffman, D., Patterson, T., 2013. The design of Gray Earth: a monochrome terrain dataset of the world. Cartogr. Perspect. 74, 61–70.

Hugenholtz, C., Brown, O., Walker, J., Barchyn, T., Nesbit, P., Kucharczyk, M., et al., 2016. Spatial accuracy of UAV-derived orthoimagery and topography: comparing photogrammetric models processed with direct geo-referencing and ground control points. Geomatica 70 (1), 21–30.

Huggett, R.J., 1975. Soil landscape systems: a model of soil genesis. Geoderma 13 (1), 1–22.

Huggett, R., 2023. Soil as part of the Earth system. Progr. Phys. Geogr. 47 (3), 454–466.

Huggett, R.J., Cheesman, J., 2002. Topography and the Environment. Pearson Education, Harlow, UK.

Hughes, T., 2011. Glacier motion/ice velocity. In: Singh, V.P., Singh, P., Haritashya, U.K. (Eds.), Encyclopedia of Snow, Ice and Glaciers. Encyclopedia of Earth Sciences Series. Springer, Dordrecht, the Netherlands, pp. 408–414.

Huising, E.J., Gomes-Pereira, L.M., 1998. Errors and accuracy estimates of laser data acquired by various laser scanning systems for topographic applications. ISPRS J. Photogrammet. Remote Sens. 53 (5), 245–261.

Hunter, G.J., Goodchild, M.F., 1995. Dealing with error in spatial databases: a simple case study. Photogrammet. Eng. Remote Sens. 61 (5), 529–537.

Hurni, L., 2010. Cartographic relief presentation revisited—forty years after Eduard Imhof. In: Otto, J.-C., Dikau, R. (Eds.), Landform—Structure, Evolution, Process Control: Proceedings of the International Symposium on Landform Organised by the Research Training Group 437. Lect. Notes Earth Sci., vol. 115. Springer, Berlin, Germany, pp. 1–20.

Hurskainen, P., Adhikari, H., Siljander, M., Pellikka, P., Hemp, A., 2019. Auxiliary datasets improve accuracy of object-based land use/land cover classification in heterogeneous savanna landscapes. Remote Sens. Environ. 233, 111354.

Hutchinson, G.L., Mosier, A.R., 1981. Improved soil cover method for field measurement of nitrous oxide fluxes. Soil Sci. Soc. Am. J. 45 (2), 311–316.

Hutchinson, M.F., 1989. New procedure for gridding elevation and stream line data with automatic removal of spurious pits. J. Hydrol. 106 (3–4), 211–232.

Hutchinson, M.F., 2008. Adding the Z dimension. In: Wilson, J.P., Fotheringham, A.S. (Eds.), The Handbook of Geographic Information Science. Blackwell, Oxford, UK, pp. 144–168.

Ibanez, D.M., Almeida-Filho, R., Miranda, F.P., 2016. Analysis of SRTM data as an aid to hydrocarbon exploration in a frontier area of the Amazonas Sedimentary Basin, northern Brazil. Mar. Petrol. Geol. 73, 528–538.

IBCAO Compilation Group, 2012. The International Bathymetric Chart of the Arctic Ocean (IBCAO) Version 3.0. NOAA, National Geophysical Data Center, Marine Geology and Geophysics. http://www.ngdc.noaa.gov/mgg/bathymetry/arctic/ibcaoversion3.html. (Accessed 27 June 2024).

IBCAO Compilation Group, 2020. The International Bathymetric Chart of the Arctic Ocean (IBCAO) Version 4.0. British Oceanographic Data Centre, National Oceanography Centre, NERC, UK, https://doi.org/10.5285/a01d292f-b4a0-1ef7-e053-6c86abc0a4b2.

IBCSO Compilation Group, 2022. The International Bathymetric Chart of the Southern Ocean (IBCSO) Version 2.0. https://ibcso.org/current_version/. (Accessed 27 June 2024).

Ichoku, C., Chorowicz, J., Parrot, J.F., 1994. Computerized construction of geological cross sections from digital maps. Comput. Geosci. 20 (9), 1321–1327.

Idrees, M.O., Pradhan, B., 2016. A decade of modern cave surveying with terrestrial laser scanning: a review of sensors, method and application development. Int. J. Speleol. 45 (1), 71–88.

Imhof, E., 1965. Kartographische Geländedarstellung. De Gruyter, Berlin, Germany.

Imhof, E., 1982. Cartographic Relief Presentation. De Gruyter, Berlin, Germany.

Inomata, T., Triadan, D., Pinzón, F., Burham, M., Ranchos, J.L., Aoyama, K., et al., 2018. Archaeological application of airborne LiDAR to examine social changes in the Ceibal region of the Maya lowlands. PLoS One 13 (2), e0191619.

Ioffe, A.I., Kozhurin, A.I., 1997. Active tectonics and geoenvironmental zoning of the Moscow Region. Bull. Moscow Soc. Nat. Geol. Ser. 72 (5), 31–35 (in Russian).

Isaacson, D.L., Ripple, W.J., 1990. Comparison of 7.5-minute and 1-degree digital elevation models. Photogrammet. Eng. Remote Sens 56 (11), 1523–1527.

Ishalina, O.T., Bliakharskii, D.P., Florinsky, I.V., 2021. Recognition of crevasses with high-resolution digital elevation models: application of geomorphometric modeling and texture analysis. Trans. GIS 25 (5), 2529–2552.

Ivanov, M.A., 2008. Global geological map of Venus: preliminary results. In: Abstracts of the 39th Lunar and Planetary Science Conference, League City, TX, 10–14 March 2008. 1017.

Ivanov, S.S., 1994. Estimation of fractal dimension of the global relief. Oceanology 34 (1), 94–98.

Ivanov, V.I., Kruzhkov, V.A., 1992. Evaluation of the optimal discretization step for a digital elevation model. Geodesy Cartogr. (5), 47–50 (in Russian).

Iwahashi, J., Pike, R.J., 2007. Automated classifications of topography from DEMs by an unsupervised nested-means algorithm and a three-part geometric signature. Geomorphology 86 (3–4), 409–440.

Iwahashi, J., Kamiy, I., Matsuoka, M., Yamazaki, D., 2018. Global terrain classification using 280m DEMs: segmentation, clustering and reclassification. Prog Earth Planet Sci 5, 1.

Jacobi, C.G.J., 1839. Note von der geodätischen Linie auf einem Ellipsoid und den verschiedenen Anwendungen einer merkwürdigen analytischen Substitution. Crelle J. reine angewandte Math. 19, 309–313.

Jacobs, L., Dewitte, O., Poesen, J., Sekajugo, J., Nobile, A., Rossi, M., et al., 2018. Field-based landslide susceptibility assessment in a data-scarce environment: the populated areas of the Rwenzori Mountains. Nat. Haz. Earth Syst. Sci. 18 (1), 105–124.

Jacobs, L., Kervyn, M., Reichenbach, P., Rossi, M., Marchesini, I., Alvioli, M., et al., 2020. Regional susceptibility assessments with heterogeneous landslide information: slope unit- vs. pixel-based approach. Geomorphology 356, 107084.

Jacobs, S.S., Helmer, H.H., Doake, C.S.M., Jenkins, A., Frolich, R.M., 1992. Melting of ice shelves and the mass balance of Antarctica. J. Glaciol. 38 (130), 375–387.

Jain, A.O., Thaker, T., Chaurasia, A., Patel, P., Singh, A.K., 2018. Vertical accuracy evaluation of SRTM-GL1, GDEM-V2, AW3D30 and CartoDEM-V3.1 of 30-m resolution with dual frequency GNSS for lower Tapi Basin India. Geocarto Int. 33 (11), 1237–1256.

Jakobsson, M., Macnab, R., Mayer, L., Anderson, R., Edwards, M., Hatzky, J., et al., 2008. An improved bathymetric portrayal of the Arctic Ocean: implications for ocean modeling and geological, geophysical and oceanographic analyses. Geophys. Res. Lett. 35 (7), L07602.

Jakobsson, M., Mayer, L., Coakley, B., Dowdeswell, J.A., Forbes, S., Fridman, B., et al., 2012. The International Bathymetric Chart of the Arctic Ocean (IBCAO) Version 3.0. Geophys. Res. Lett. 39 (12), L12609.

Jakobsson, M., Mayer, L.A., Bringensparr, C., Castro, C.F., Mohammad, R., Johnson, P., et al., 2020. The International Bathymetric Chart of the Arctic Ocean Version 4.0. Sci. Data 7 (176), 14.

James, L.A., Monohan, C., Ertis, B., 2019. Long-term hydraulic mining sediment budgets: connectivity as a management tool. Sci. Total Environ. 651 (2), 2024–2035.

James, M.R., Robson, S., 2012. Straightforward reconstruction of 3D surfaces and topography with a camera: accuracy and geoscience application. J. Geophys. Res. Earth Surf. 117 (F3), F03017.

James, M.R., Robson, S., 2014. Mitigating systematic error in topographic models derived from UAV and ground-based image networks. Earth Surf. Process. Landf. 39 (10), 1413–1420.

James, M.R., Robson, S., d'Oleire-Oltmanns, S., Niethammer, U., 2017a. Optimising UAV topographic surveys processed with structure-from-motion: ground control quality, quantity and bundle adjustment. Geomorphology 280, 51–66.

James, M.R., Robson, S., Smith, M.W., 2017b. 3-D uncertainty-based topographic change detection with structure-from-motion photogrammetry: precision maps for ground control and directly georeferenced surveys. Earth Surf. Process. Landf. 42 (12), 1769–1788.

James, M.R., Antoniazza, G., Robson, S., Lane, S.N., 2020. Mitigating systematic error in topographic models for geomorphic change detection: accuracy, precision and considerations beyond off-nadir imagery. Earth Surf. Process. Landf. 45, 2251–2271.

Jamieson, S.S.R., Ross, N., Paxman, G.J.G., Clubb, F.J., Young, D.A., Yan, S., et al., 2023. An ancient river landscape preserved beneath the East Antarctic Ice Sheet. Nat. Commun. 14, 6507.

Jancaitis, J.R., 1978. Elevation Data Compaction by Polynomial Modeling. Report No. ETL-0140, US Army Engineer Topographic Laboratories, Fort Belvoir, VA.

Jancaitis, J.R., Junkins, J.L., 1973. Modeling irregular surfaces. Photogrammet. Eng. 39 (4), 413–420.

Jancewicz, K., Porębna, W., 2022. Point cloud does matter. Selected issues of using airborne LiDAR elevation data in geomorphometric studies of rugged sandstone terrain under forest—case study from Central Europe. Geomorphology 412, 108316.

Jancewicz, K., Migoń, P., Kasprzak, M., 2019. Connectivity patterns in contrasting types of tableland sandstone relief revealed by Topographic Wetness Index. Sci. Total Environ. 656, 1046–1062.

Janowski, L., Tylmann, K., Trzcinska, K., Rudowski, S., Tegowski, J., 2022. Exploration of glacial landforms by object-based image analysis and spectral parameters of digital elevation model. IEEE Trans. Geosci. Remote Sens. 60, 4502817.

Japan Space Systems, 2009–2019. ASTER Global Digital Elevation Model (GDEM). Japan Space Systems, Tokyo, Japan. https://www.jspacesystems.or.jp/ersdac/GDEM/E/. (Accessed 27 June 2024).

Jarvis, A., Rubiano, J., Nelson, A., Farrow, A., Mulligan, M., 2004. Practical Use of SRTM Data in the Tropics—Comparisons with Digital Elevation Models Generated from Cartographic Data. Centro Internacional de Agricultura Tropical. Cali, Columbia.

Jasiewicz, J., Stepinski, T.F., 2013. Geomorphons—a pattern recognition approach to classification and mapping of landforms. Geomorphology 182, 147–156.

JAXA EORC, 2015–2023. ALOS Global Digital Surface Model "ALOS World 3D – 30m (AW3D30)". Japan Aerospace Exploration Agency, Earth Observation Research Center, Tokyo, Japan. https://www.eorc.jaxa.jp/ALOS/en/dataset/aw3d30/aw3d30_e.htm. (Accessed 27 June 2024).

Jenčo, M., 1992. The morphometric analysis of georelief in terms of a theoretical conception of the complex digital model of georelief. Acta Fac. Rerum. Nat. Univ. Comenianae Geographica 33, 133–151.

Jenčo, M., Pacina, J., Shary, P.A., 2009. Terrain skeleton and local morphometric variables geosciences and computer vision technique. In: Horák, J., Halounová, L., Kusendová, D., Rapant, P., Voženílek, V. (Eds.), Advances in Geoinformation Technologies 2009. VSB-Technical University of Ostrava, Ostrava, Czechia, pp. 57–76.

Jenčo, M., Fulajtár, E., Bobáľová, H., Matečný, I., Saksa, M., Kožuch, M., et al., 2020. Mapping soil degradation on arable land with aerial photography and erosion models, case study from Danube Lowland, Slovakia. Remote Sens. (Basel) 12 (24), 4047.

Jenks, G.F., 1963. Generalization in statistical mapping. Ann. Assoc. Am. Geogr. 53 (1), 15–26.

Jenness, J., 2006. Topographic position index documentation. Jenness Enterprises, Flagstaff, AZ. https://www.jennessent.com/arcview/tpi.htm. (Accessed 27 June 2024).

Jenny, B., 2001. An interactive approach to analytical relief shading. Cartographica 38 (1–2), 67–75

Jenny, B., 2020. Terrain generalization with line integral convolution. Cartogr. Geogr. Inf. Sci. 48 (1), 78–92.

Jenny, B., Hurni, L., 2006. Swiss-style colour relief shading modulated by elevation and by exposure to illumination. Cartogr. J. 43 (3), 198–207.

Jenny, B., Patterson, T., 2007. Introducing plan oblique relief. Cartogr. Perspect. 57 (21–40), 88–90.

Jenny, B., Heitzler, M., Singh, D., Farmakis-Serebryakova, M., Liu, J.C., Hurni, L., 2020. Cartographic relief shading with neural networks. IEEE Trans. Vis. Comput. Graph. 27 (2), 1225–1235.

Jenny, H., 1941. Factors of Soil Formation. A System of Quantitative Pedology. McGraw Hill, New York, NY.

Jenny, H., Jenny, B., Hurni, L., 2010. Interactive design of 3D maps with progressive projection. Cartogr. J. 47 (3), 211–221.

Jensen, J.R., 1995. Issues involving the creation of digital elevation models and terrain corrected orthoimagery using soft-copy photogrammetry. Geocarto Int. 10 (1), 5–21.

Jenson, S.K., Domingue, J.Q., 1988. Extracting topographic structure from digital elevation data for geographic information system analysis. Photogrammet. Eng. Remote Sens. 54 (11), 1593–1600.

Jerri, A.J., 1998. The Gibbs Phenomenon in Fourier Analysis, Splines and Wavelet Approximations Kluwer, Boston, MA.

Jeziorska, J., Mitasova, H., Petrasova, A., Petras, V., Divakaran, D., Zajkowski, T., 2016. Overland flow analysis using time series of sUAS-derived elevation models. In: Halounova, L., Šafář, V., Raju, P.L.N., Plánka, L., Ždímal, V., Kumar, T.S., et al. (Eds.), XXIII ISPRS Congress, Commission VIII, Prague, Czech Republic, 12–19 July 2016. ISPRS Annals Photogramm. Remote Sens. Spatial Inf. Sci. III (8), pp. 159–166.

Ji, W., Civco, D.L., Kennard, W.C., 1992. Satellite remote bathymetry—a new mechanisms for modeling. Photogrammet. Eng. Remote Sens. 58 (5), 545–549.

Jing, Y., Zhao, Q., Lu, M., Wang, A., Yu, J., Liu, Y., et al., 2022. Effects of road and river networks on sediment connectivity in mountainous watersheds. Sci. Total Environ. 826, 154189.

Johansson, M., 1999. Analysis of digital elevation data for palaeosurfaces in south-western Sweden. Geomorphology 26 (4), 279–295.

Johnson, K., Nissen, E., Saripalli, S., Arrowsmith, J.R., McGarey, P., Scharer, K., et al., 2014. Rapid mapping of ultrafine fault zone topography with structure from motion. Geosphere 10 (5), 969–986.

Jol, H.M. (Ed.), 2009. Ground Penetrating Radar Theory and Applications. Elsevier, Amsterdam, the Netherlands.

Jones, R., 2002. Algorithms for using a DEM for mapping catchment areas of stream sediment samples. Comput. Geosci. 28 (9), 1051–1060.

Jones, R.R., McCaffrey, K.J.W., Clegg, P., Wilson, R.W., Holliman, N.S., Holdsworth, R.E., et al., 2009. Integration of regional to outcrop digital data: 3D visualisation of multi-scale geological models. Comput. Geosci. 35 (1), 4–18.

Jones, T.A., Johnson, C.R., 1983. Stratigraphic relationships and geologic history depicted by computer mapping. Am. Assoc. Pet. Geol. Bull. 67 (9), 1415–1421.

Jordan, G., 2003. Morphometric analysis and tectonic interpretation of digital terrain data: a case study. Earth Surf. Process. Landf. 28 (8), 807–822.

Jordan, G., 2007. Adaptive smoothing of valleys in DEMs using TIN interpolation from ridgeline elevations: an application to morphotectonic aspect analysis. Comput. Geosci. 33 (4), 573–585.

Jordan, M.C., 1872. Sur les lignes de faîte et de thalweg. C. R. Hebd. Séances Acad. Sci. 74 (23), 1457–1459.

Joughin, I., Winebrenner, D., Fahnestock, M., Kwok, R., Krabill, W., 1996. Measurement of ice-sheet topography using satellite-radar interferometry. J. Glaciol. 42 (140), 10–22.

JPL, 2009. ASTER Global Digital Elevation Model (GDEM). Jet Propulsion Laboratory, California Institute of Technology, National Aeronautics and Space Administration, Pasadena, CA. http://asterweb.jpl.nasa.gov/gdem.asp. (Accessed 27 June 2024).

Kaartinen, H., Hyyppä, J., Vastaranta, M., Kukko, A., Jaakkola, A., Yu, X., et al., 2015. Accuracy of kinematic positioning using global satellite navigation systems under forest canopies. Forests 6 (9), 3218–3236.

Kachanoski, R.G., de Jong, E., Rolston, D.E., 1985a. Spatial and spectral relationships of soil properties and microtopography: II. Density and thickness of B horizon. Soil Sci. Soc. Am. J. 49 (4), 812–816.

Kachanoski, R.G., Rolston, D.E., de Jong, E., 1985b. Spatial and spectral relationships of soil properties and microtopography: I. Density and thickness of A horizon. Soil Sci. Soc. Am. J. 49 (4), 804–812.

Kagan, V.F., 1947. Fundamentals of the Theory of Surfaces in Tensor Presentation, pt. 1: Apparatus of Research. General Foundations of the Theory and the Intrinsic Geometry of a Surface. Gostekhizdat, Moscow, USSR (in Russian).

Kaiser, K.E., McGlynn, B.L., Dore, J.E., 2018. Landscape analysis of soil methane flux across complex terrain. Biogeosciences 15 (10), 3143–3167.

Kaleita, A.L., Schott, L.R., Hargreaves, S.K., Hofmockel, K.S., 2017. Differences in soil biological activity by terrain types at the sub-field scale in central Iowa US. PLoS One 12 (7), e0180596.

Kampes, B.M., 2006. Radar Interferometry: Persistent Scatterer Technique. Remote Sensing and Digital Image Processing, vol. 12. Springer, Dordrecht, the Netherlands.

Karachevtseva, I.P., Kozlova, N.A., Kokhanov, A.A., Zubarev, A.E., Nadezhdina, I.E., Patratiy, V.D., et al., 2017. Cartography of the Luna-21 landing site and Lunokhod-2 traverse area based on Lunar Reconnaissance Orbiter Camera images and surface archive TV-panoramas. Icarus 283, 104–121.

Karachevtseva, I.P., Garov, A.S., Zubarev, A.E., Matveev, E.V., Kokhanov, A.A., Zharkova, A.Y., 2019. Geoportal of planetary data: concept, methods and implementations. In: Wu, B., Di, K., Oberst, J., Karachevtseva, I. (Eds.), Planetary Remote Sensing and Mapping. Li, Z. (Ed.), International Society for Photogrammetry and Remote Sensing (ISPRS) Book Series. CRC Press, London, UK, pp. 315–329.

Karakhanian, A.S., 1981. Recognition of large landslides by satellite image interpretation. Izvestia Vuzov. Geol. Explor. (3), 130–131 (in Russian).

Karátson, D., Yepes, J., Favalli, M., Rodríguez-Peces, M.J., Fornaciai, A., 2016. Reconstructing eroded paleovolcanoes on Gran Canaria, Canary Islands, using advanced geomorphometry. Geomorphology 253, 123–134.

Karkee, M., Steward, B.L., Aziz, S.A., 2008. Improving quality of public domain digital elevation models through data fusion. Biosyst. Eng. 101 (3), 293–305.

Karlson, M., Bastviken, D., Reese, H., 2021. Error characteristics of pan-Arctic digital elevation models and elevation derivatives in northern Sweden. Remote Sens. (Basel) 13 (22), 4653.

Karney, C.F.F., 2013. Algorithms for geodesics. J. Geodesy 87 (1), 43–55.

Karney, C.F.F., 2023. Geodesics on a triaxial ellipsoid. GeographicLib 2, 3. https://geographiclib.sourceforge.io/C++/doc/triaxial.html. (Accessed 27 June 2024).

Kasimov, N.S., 1980. Landscape Geochemistry of Fault Zones (Exemplified by Kazakhstan). Moscow University Press, Moscow, USSR (in Russian).

Kasimov, N.S., Kovin, M.I., Proskuryakov, Y.V., Shmelkova, N.A., 1978. Geochemistry of the soils of fault zones (exemplified by Kazakhstan). Soviet Soil Science 10 (4), 397–406.

Kasprak, A., Caster, J., Bangen, S.G., Sankey, J.B., 2017. Geomorphic process from topographic form: automating the interpretation of repeat survey data in river valleys. Earth Surf. Process. Landf. 42 (12), 1872–1883.

Kasprak, A., Bransky, N.D., Sankey, J.B., Caster, J., Sankey, T.T., 2019. The effects of topographic surveying technique and data resolution on the detection and interpretation of geomorphic change. Geomorphology 333, 1–15.

Kasvi, E., Salmela, J., Lotsari, E., Kumpula, T., Lane, S.N., 2019. Comparison of remote sensing based approaches for mapping bathymetry of shallow, clear water rivers. Geomorphology 333, 180–197.

Kats, Y.G., Makarova, N.V., Kozlov, V.V., Trofimov, D.M., 1981. Structural and geomorphological analysis of Crimea by satellite image interpretation. Izvestia Vuzov. Geol. Explor. (3), 8–20 (in Russian).

Kats, Y.G., Poletaev, A.I., Rumyantseva, E.F., 1986. Principles of Lineament Tectonics. Nedra, Moscow, USSR (in Russian).

Katterfeld, G.N., Charushin, G.V., 1973. General grid systems of planets. Modern Geol. 4 (4), 253–287.

Kaufmann, O., Martin, T., 2008. 3D geological modelling from boreholes, cross-sections and geological maps, application over former natural gas storages in coal mines. Comput. Geosci. 34 (3), 278–290.

KDE, 2007–2023. Marble. K Desktop Environment, Berlin, Germany. https://marble.kde.org. (Accessed 27 June 2024).

Keating, D.P., Fischer, M.P., 2008. An experimental evaluation of the curvature–strain relation in fault-related folds. Am. Assoc. Pet. Geol. Bull. 92 (7), 869–884.

Keilis-Borok, V., 2002. Earthquake prediction: state-of-the-art and emerging possibilities. Annu. Rev. Earth Planet. Sci. 30, 1–33.

Kelka, U., Martinez, C., Krapf, C., Westerlund, S., Gonzalez-Alvarez, I., Pawley, M., et al., 2022. Establishing an integrated workflow identifying and linking surface and subsurface lineaments for mineral exploration under cover: example from the Gawler Craton, South Australia. Solid Earth 13 (4), 827–847.

Keller, E.A., 1986. Investigation of active tectonics: use of surficial earth processes. In: Geophysics Study Committee, Geophysics Research Forum, and National Research Council (Eds.), Active Tectonics. Studies in Geophysics Series. National Academy Press, Washington, DC, pp. 136–147.

Keller, E.A., Pinter, N., 2002. Active Tectonics: Earthquakes, Uplift, and Landscape, second ed. Prentice Hall Earth Science Series. Prentice Hall, Upper Saddle River, NJ.

Kennedy, K.H., 2009. Introduction to 3D Data: Modeling with ArcGIS® 3D Analyst™ and Google Earth™. Wiley, Hoboken, NJ.

Kennelly, P.J., Stewart, A.J., 2014. General sky models for illuminating terrains. Int. J. Geogr. Inf. Sci. 28 (2), 383–406.

Kent, B.R., 2015. 3D Scientific Visualization with Blender®. Morgan & Claypool, San Rafael, CA.

Kerrich, R., 1986. Fluid transport in lineaments. Philos. Trans. R. Soc. Lond. A Math. Phys. Eng. Sci. 317 (1539), 219–251.

Kershaw, K.A., Looney, J.H.H., 1985. Quantitative and Dynamic Plant Ecology, third ed. Arnold, London, UK.

Keylock, C.J., Singh, A., Passalacqua, P., Foufoula-Georgiou, E., 2021. Evaluating landscape complexity and the contribution of non-locality to geomorphometry. J. Geophys. Res. Earth Surf. 126 (4), e2020JF005765.

Khalil, M.A.K., Rasmussen, R.A., 1992. The global sources of nitrous oxide. J. Geophys. Res. Atmos. 97 (D13), 14651–14660.

Khalugin, Y.I., Zhalkovsky, Y.A., Zhdanov, N.D., 1992. Digital Maps. Nedra, Moscow, Russia (in Russian).

Khanifar, J., Khademalrasoul, A., 2022. Utilization of multi-scale geomorphometric algorithms and third-order morphometric variables for soil thickness modeling. Trans. GIS 26 (2), 887–901.

Kharchenko, S.V., 2016. The experience of quantitative assessment of relief influence on the urban building aeration. Geomorfologiya (3), 45–51 (in Russian).

Kharchenko, S.V., 2020. New challenges of geomorphometry and automatic morphological classifications in geomorphology. Geomorfologiya (1), 3–21 (in Russian).

Kharchenko, S.V., 2023. Pattern recognition in the tasks of landform mapping. Izvestiya Academii Nauk Geogr. Ser. 87 (1), 192–206 (in Russian).

Kharchenko, S.V., Mukhametshin, A.R., Anikina, N.V., Bykanova, M.A., 2023. Effect of urban terrain on traffic noise propagation (a case study of Moscow). Geogr. Nat. Res. 44 (2), 133–141.

Khosravi, K., Pham, B.T., Chapi, K., Shirzadi, A., Shahabi, H., Revhaug, I., et al., 2018. A comparative assessment of decision trees algorithms for flash flood susceptibility modeling at Haraz watershed, northern Iran. Sci. Total Environ. 627, 744–755.

Khronos Group, 2008–2021. The OpenGL ES Shading Language Specification. https://www.khronos.org/registry/OpenGL/specs/es. (Accessed 27 June 2024).

Kiernan, K., Gore, D.B., Fink, D., White, D.A., McConnell, A., Sigurdsson, I.A., 2009. Deglaciation and weathering of Larsemann Hills, East Antarctica. Antarct. Sci. 21 (4), 373–382.

Kim, S., 2009. Characterization of soil moisture responses on a hillslope to sequential rainfall events during late autumn and spring. Water Resour. Res. 45 (9), W09425.

Kim, Y.-H., Rana, S., Wise, S., 2004. Exploring multiple viewshed analysis using terrain features and optimisation techniques. Comput. Geosci. 30 (9–10), 1019–1032.

Kim, Y.J., Nam, B.H., Youn, H., 2019. Sinkhole detection and characterization using LiDAR-derived DEM with logistic regression. Remote Sens. (Basel) 11 (13), 1592.

King, D., Bourennane, H., Isambert, M., Macaire, J.J., 1999. Relationship of the presence of a non-calcareous clay-loam horizon to DEM attributes in a gently sloping area. Geoderma 89 (1–2), 95–111.

King, E.C., Hindmarsh, R.C.A., Stokes, C.R., 2009. Formation of mega-scale glacial lineations observed beneath a West Antarctic ice stream. Nat. Geosci. 2 (8), 585–588.

King, P.B., Beikman, H.M., Edmonston, G.J., 1974. Geologic Map of the United States (Exclusive of Alaska and Hawaii), Scale 1:2,500,000. US Geological Survey, Reston, VA.

Kirkby, M.J., 1997. TOPMODEL: a personal view. Hydrol. Process. 11 (9), 1087–1097.

Kirkby, M.J., Chorley, R.J., 1967. Throughflow, overland flow and erosion. Bull. Int. Assoc. Sci. Hydrol. 12 (3), 5–21.

Kizyakov, A., Zimin, M., Sonyushkin, A., Dvornikov, Y., Khomutov, A., Leibman, M., 2017. Comparison of gas emission crater geomorphodynamics on Yamal and Gydan Peninsulas (Russia), based on repeat very-high-resolution stereopairs. Remote Sens. (Basel) 9 (10), 1023.

Kizyakov, A., Khomutov, A., Zimin, M., Khairullin, R., Babkina, E., Dvornikov, Y., et al., 2018. Microrelief associated with gas emission craters: remote-sensing and field-based study. Remote Sens. (Basel) 10 (5), 677.

Kleinbaum, D.G., Kupper, L.L., Nizam, A., Rosenberg, E.S., 2014. Applied Regression Analysis and Other Multivariable Methods, fifth ed. Cengage Learning, Boston, MA.

Klemeš, V., 1983. Conceptualization and scale in hydrology. J. Hydrol. 65 (1–3), 1–23.

Knetsch, G., 1965. Über ein Structur-Experiment an einer Kugel und Beziehungen zwischen Gross-Lineamenten und Pol-Lagen in der Erdeschichte. Geol. Rund. 54 (1), 523–548.

Knitter, D., Braun, R., Clare, L., Nykamp, M., Schütt, B., 2019. Göbekli Tepe: a brief description of the environmental development in the surroundings of the UNESCO World Heritage Site. Land 8 (4), 72.

Knizhnikov, Y.F., 1967. A Stereophotogrammetric Method for Studying Ice Movement of Mountain Glaciers (Development of a Pseudo-Parallax Approach) (Ph.D. thesis). Moscow Institute of Engineers for Geodesy, Aerophotography, and Cartography, Moscow, Russia (in Russian).

Knizhnikov, Y.F., 1973. Studies of Ice Movement of Mountain Glaciers Using a Stereophotogrammetric Method. Nauka, Moscow, Russia (in Russian).

Koenderink, J.J., van Doorn, A.J., 1992. Surface shape and curvature scales. Image Vis. Comput. 10 (8), 557–565.

Koenderink, J.J., van Doorn, A.J., 1993. Local features of smooth shapes: ridges and courses. In: Vemuri, B.C. (Ed.), Geometric Methods in Computer Vision II. Proc. SPIE 2013, pp. 2–13.

Koenderink, J.J., van Doorn, A.J., 1994. Two-plus-one-dimensional differential geometry. Pattern Recogn. Lett. 15 (5), 439–443.

Koh, G., Lever, J.H., Arcone, S.A., Marshall, H.-P., Ray, L.E., 2010. Autonomous FMCW radar survey of Antarctic shear zone. In: Proceedings of the XIII Internarional Conference on Ground Penetrating Radar, 21–25 June 2010, Lecce, Italy. IEEE, Curran Associates, Inc, https://doi.org/10.1109/ICGPR.2010.5550174.

Koike, K., Nagano, S., Kawaba, K., 1998. Construction and analysis of interpreted fracture planes through combination of satellite-image derived lineaments and digital elevation model data. Comput. Geosci. 24 (6), 573–583.

Koldaev, P.K., 1956. Plastic Representation of Relief on Maps. Geodezizdat, Moscow, USSR (in Russian).

Kondratyev, K.Y., Pivovarova, Z.I., Fedorova, M.P., 1978. Radiation Regime of Inclined Surfaces. Hydrometeoizdat, Leningrad, USSR (in Russian).

Konovalov, N.E., 1960. A Project of a New Railway Line, with Development of Ideas on Computers Application to Solve Particular Tasks (M.Sc. thesis). Kharkov Institute of Engineers for Railway Transport, Kharkov, USSR (in Russian).

Konvitz, J., 1987. Cartography in France, 1660–1848: Science, Engineering, and Statecraft. University of Chicago Press, Chicago, IL.

Kopp, M.L., 2000. The recent deformations of the Scythian and Southern East European Platforms as a result of pressure from the Arabian Plate. Geotectonics 34 (2), 106–120.

Koriche, S.A., Rientjes, T.H.M., 2016. Application of satellite products and hydrological modelling for flood early warning. Phys. Chem. Earth Pt A/B/C 93, 12–23.

Korobeynik, V.M., Komarova, M.V., Shtengelov, Y.S., 1982. Permeable faults of the Earth crust in Crimea and the north-western Black Sea area. Proc. Ukrain. SSR Acad. Sci. Ser. B (2), 13–16 (in Russian).

Korona, J., Berthier, E., Bernard, M., Rémy, F., Thouvenot, E., 2009. SPIRIT. SPOT 5 stereoscopic survey of Polar Ice: reference images and topographies during the fourth International Polar Year (2007–2009). ISPRS J. Photogrammet. Remote Sens. 64 (2), 204–212.

Korsakova, O.P., 2002. Morphological analysis of the topography of the north-eastern part of the Baltic Shield. Geomorfologiya (4), 87–95 (in Russian).

Korzeniowska, K., Pfeifer, N., Landtwing, S., 2018. Mapping gullies, dunes, lava fields, and landslides via surface roughness. Geomorphology 301, 53–67.

Koshkarev, A.V., 1982. Topography as an input parameter for mathematical cartographic models of geosystems. In: Evteev, O.A. (Ed.), Geographical Cartography in Scientific Research and National Economic Practices. Moscow Branch, Soviet Geographical Society, Moscow, USSR, pp. 117–131 (in Russian).

Kotelnikov, V.A., 1933. On the transmission capacity of the "ether" and wire in electrocommunications. In: Proceedings of the 1st Pan-Union Conference on the Technological Reconstruction of Communications Sector and the Development of Low-Current Engineering, Moscow, USSR, Jan. 1933. Military Communication Department Press, Moscow, USSR, pp. 1–33 (in Russian).

Kotelnikov, V.A., 2001. On the transmission capacity of the "ether" and wire in electrocommunications. In: Benedetto, J.J., Ferreira, P.J.S.G. (Eds.), Modern Sampling Theory: Mathematics and Applications. Birkhäuser, Boston, MA, pp. 27–45.

Kovaleva, O.V., 2012. Perfecting the Representation of Topography on Small-Scale Maps (Ph.D. thesis). Moscow State University of Geodesy and Cartography, Moscow, Russia (in Russian).

Kovaleva, O.V., 2014. Perfecting the representation of topography by hypsometric tints. New classification of hypsometric scales. Geodesy Cartogr. (11), 21–29 (in Russian).

Kovalevsky, S.A., 1965. The medial deep fault of the Crimean Peninsula. Doklady Acad. Nauk SSSR 162 (4), 887–890 (in Russian).

Kovda, V.A., 1946. Origin of Saline Soils and Their Regime, vol. 1. Soviet Academic Press, Moscow, USSR (in Russian).

Kovda, V.A., 1971. Origin of Saline Soils and Their Regime, vol. 1. Israel Program for Scientific Translations, Jerusalem, Israel.

Kovda, V.A., 1973. The Principles of Pedology. General Theory of Soil Formation, vol. 1. Nauka, Moscow, USSR (in Russian).

Kozlov, D.N., 2009. Digital Landscape Analysis for Large-Scale Mapping of Soil Cover Structures (Ph.D. thesis). Lomonosov Moscow State University, Moscow, Russia (in Russian).

Kozlov, V.V., 1991. Application of Remotely Sensed Data in Seismic Studies. Pan-Union Scientific Research Institute of Economics of Mineral Raw Material and Prospecting, Moscow, USSR (in Russian).

Kraaijenbrink, P.D.A., Meijer, S.W., Shea, J.M., Pellicciotti, F., de Jong, S.M., Immerzeel, W.W., 2016a. Seasonal surface velocities of a Himalayan glacier derived by automated correlation of unmanned aerial vehicle imagery. Ann. Glaciol. 57 (71), 103–113.

Kraaijenbrink, P.D.A., Shea, J.M., Pellicciotti, F., de Jong, S.M., Immerzeel, W.W., 2016b. Object-based analysis of unmanned aerial vehicle imagery to map and characterise surface features on a debris-covered glacier. Remote Sens. Environ. 186, 581–595.

Kraaijenbrink, P.D.A., Shea, J.M., Litt, M., Steiner, J.F., Treichler, D., Koch, I., et al., 2018. Mapping surface temperatures on a debris-covered glacier with an unmanned aerial vehicle. Front. Earth Sci. 6, 64.

Kramm, T., Hoffmeister, D., 2022. Comprehensive vertical accuracy analysis of freely available DEMs for different landscape types of the Rur Catchment, Germany. Geocarto Int. 37 (25), 7774–7799.

Krasilnikov, N.N., Krasilnikova, O.I., Shelepin, Y.E., 2000. Perception of achromatic, monochromatic, pure chromatic, and chromatic noisy images by real human-observer under threshold conditions. In: Krupinski, E.A. (Ed.), Medical Imaging 2000: Image Perception and Performance. Proc. SPIE 3981, pp. 78–85.

Krasovsky, F.N., 1902. Determination of the size of the Earth triaxial ellipsoid from the results of the Russian arc measurements. In: Memorial Book of the Konstantinovsky Surveying Institute for the 1900–1901 Years. Russian Association of Printing and Publishing, Moscow, Russia, pp. 19–54 (in Russian).

Krasovsky, F.N., 1936. Review and results of modern arc measurements. § 5: On the triaxiality of the Earth ellipsoid. Geodezist (7), 10–19 (in Russian).

Kraus, K., 1994. Visualization of the quality of surfaces and their derivatives. Photogrammet. Eng. Remote Sens. 60 (4), 457–462.

Kraus, K., 2007. Photogrammetry: Geometry From Images and Laser Scans, second ed. De Gruyter Textbook Series. De Gruyter, Berlin, Germany.

Kravchenko, A.N., Bollero, G.A., Omonode, R.A., Bullock, D.G., 2002. Quantitative mapping of soil drainage classes using topographical data and soil electrical conductivity. Soil Sci. Soc. Am. J. 66 (1), 235–243.

Krcho, J., 1973. Morphometric analysis of relief on the basis of geometric aspect of field theory. Acta Geogr. Univ. Comenianae Geogr. Phys. 1, 7–233.

Krcho, J., 1983. Teoretická koncepcia a interdisciplinárne aplikácie komplexného digitálneho modelu reliéfu pri modelovaní dvojdimenzionálnych polí. Geogr. Čas. 35 (3), 265–291.

Kreslavsky, M.A., Head, J.W., 2000. Kilometer-scale roughness of Mars results from MOLA data analysis. J. Geophys. Res. Planet. 105 (E11), 26695–26711.

Kreslavsky, M.A., Head, J.W., Neumann, G.A., Zuber, M.T., Smith, D.E., 2014. Kilometer-scale topographic roughness of Mercury: correlation with geologic features and units. Geophys. Res. Lett. 41 (23), 8245–8251.

Krieger, G., Moreira, A., Fiedler, H., Hajnsek, I., Werner, M., Younis, M., et al., 2007. TanDEM-X: a satellite formation for high-resolution SAR interferometry. IEEE Trans. Geosci. Remote Sens. 45 (11), 3317–3341.

Krumbein, W.C., 1959. Trend-surface analysis of contour-type maps with irregular control-point spacing. J. Geophys. Res. 64 (7), 823–834.

Kuhn, K., Fedorko, E., 2006. Development of a 1/9th arc-second elevation dataset for West Virginia. In: Proceedings of MAPPS/ASPRS 2006 Fall Conference, San Antonio, TX, 6–10 Nov. 2006. American Society for Photogrammetry and Remote Sensing, Bethesda, MD.

Kühni, A., Pfiffner, O.A., 2001a. Drainage patterns and tectonic forcing: a model study for the Swiss Alps. Basin Res. 13 (2), 169–197.

Kühni, A., Pfiffner, O.A., 2001b. The relief of the Swiss Alps and adjacent areas and its relation to lithology and structure: topographic analysis from a 250-m DEM. Geomorphology 41 (4), 285–307.

Kukowski, N., Schillhorn, T., Huhn, K., von Rad, U., Husen, S., Flueh, E.R., 2001. Morphotectonics and mechanics of the central Makran accretionary wedge off Pakistan. Mar. Geol. 173 (1–4), 1–19.

Kukowski, N., Hampel, A., Hoth, S., Bialas, J., 2008. Morphotectonic and morphometric analysis of the Nazca plate and the adjacent offshore Peruvian continental slope—implications for submarine landscape evolution. Mar. Geol. 254 (1–2), 107–120.

Kumar, A., Negi, H.S., Kumar, K., Shekhar, C., 2020. Accuracy validation and bias assessment for various multi-sensor open-source DEMs in part of the Karakoram Region. Remote Sens. Lett. 11 (10), 893–902.

Kumar, K., Pant, M.C., Satyal, G.S., Dumka, R.K., 2008. Comparison of digital surface modelling techniques for sloping hill terrain using GPS data. Int. J. Model. Simul. 28 (4), 439–447.

Kumar, S., Snehmani, Srivastava, P.K., Gore, A., Singh, M.K., 2016. Fuzzy-frequency ratio model for avalanche susceptibility mapping. Int. J. Dig. Earth 9 (12), 1168–1184.

Kumar, S., Srivastava, P.K., Snehmani, 2017. GIS-based MCDA-AHP modelling for avalanche susceptibility mapping of Nubra valley region, Indian Himalaya. Geocarto Int. 32 (11), 1254–1267.

Kumar, S., Srivastava, P.K., Bhatiya, S., 2019. Geospatial probabilistic modelling for release area mapping of snow avalanches. Cold Reg. Sci. Technol. 165, 102813.

Kumler, M.P., 1994. An intensive comparison of triangulated irregular networks (TINs) and digital elevation models (DEMs). Cartographica 31 (2), 1–99.

Kuriakose, S.L., Devkota, S., Rossiter, D.G., Jetten, V.G., 2009. Prediction of soil depth using environmental variables in an anthropogenic landscape, a case study in the Western Ghats of Kerala, India. Catena 79 (1), 27–38.

Kurkov, V.M., Kiseleva, A.S., 2020. DEM accuracy research based on unmanned aerial survey data. In: Paparoditis, N., Mallet, C., Lafarge, F., Jiang, J., Shaker, A., Zhang, H., et al. (Eds.), XXIV ISPRS Congress, Commission III. Int. Arch. Photogramm. Remote Sens. Spatial Inf. Sci. XLIII (B3-2020), pp. 1347–1352.

Kurkov, V., Bliakharskii, D., Florinsky, I., 2016. Application of unmanned aerial surveying for geomorphometric modelling. Izvestia Vuzov. Geodesy Aerophotosurveying 60 (6), 69–77 (in Russian).

Kurkov, M.V., Klestov, D.A., Brusilo, V.A., Kurkov, V.M., Kiseleva, A.S., 2021. Experience of using the Geoscan 401 LiDAR complex as an unmanned topographic system for airborne laser scanning and aerial survey. Geoprofi (6), 17–23 (in Russian).

Kuryakova, G.A., 1996. A Methodology for Studying and Preparation of Initial Data for DTM-Based Mapping of Biogeocoenosis (Ph.D. thesis). Moscow State University of Geodesy and Cartography, Moscow, Russia (in Russian).

Kuryakova, G.A., Florinsky, I.V., 1991. The Analysis of Spatial Relationships between Ring Structures, Topography, and Soil Cover. Pushchino Research Centre Press, Pushchino, USSR (in Russian).

Kuryakova, G.A., Florinsky, I.V., Shary, P.A., 1992. On the correlation between soil moisture and some topographic variables. In: Andronikov, V.L. (Ed.), Proceedings of the Conference "The Modern Problems of Soil Geography and Mapping", Moscow, USSR, 24–26 Sept. 1991. Dokuchaev Soil Institute, Moscow, Russia, pp. 70–71 (in Russian).

Kusahara, K., 2020. Interannual-to-multidecadal responses of Antarctic ice shelf–ocean interaction and coastal water masses during the twentieth century and the early twenty-first century to dynamic and thermodynamic forcing. J. Climate 33 (12), 4941–4973.

Kutina, J., 1974. Structural control of volcanic ore deposits in the context of global tectonics. Bull. Volcanol. 38 (3), 1037–1069.

Kuz'min, P.P., 1963. Snowcover and Snow Reserves. Israel Program for Scientific Translations. Jerusalem, Israel.

Kvietkauskas, V., 1963–1964. Keturspalvis morfografinis zemelapis. In: Tarvydas, S. (Ed.), Geografinis Metrastis VI–VII. Ledyninio Reljefo Morfogeneze ir Dabartiniai Egzogeniniai Procesai. Lietuvos TSR Geografine Draugija, Vilnius, USSR, pp. 87–107.

Kweon, I.S., Kanade, T., 1994. Extracting topographic terrain features from elevation maps. CVGIP Image Understand. 59 (2), 171–182.

Lacroix, S.F., 1811. Introduction à la Géographie Mathématique et Critique, et a la Géographie Physique, deuxième éd. Dentu, Paris, France.

Lagacherie, P., 2008. Digital soil mapping: a state of the art. In: Hartemink, A.E., McBratney, A., Mendonça-Santos, M.L. (Eds.), Digital Soil Mapping with Limited Data. Springer, Dordrecht, the Netherlands, pp. 3–14.

Lagacherie, P., Holmes, S., 1997. Addressing geographical data errors in a classification tree for soil unit prediction. Int. J. Geogr. Inf. Sci. 11 (2), 183–198.

Lagacherie, P., McBratney, A.B., 2007. Spatial soil information systems and spatial soil inference systems: perspectives for digital soil mapping. In: Lagacherie, P., McBratney, A.B., Voltz, M. (Eds.), Digital Soil Mapping: An Introductory Perspective. Developments in Soil Science, vol. 31. Elsevier, Amsterdam, the Netherlands, pp. 3–22.

Lagacherie, P., Robbez-Masson, J.M., Nguyen-The, N., Barthès, J.P., 2001. Mapping of reference area representativity using a mathematical soilscape distance. Geoderma 101 (3–4), 105–118.

Łajczak, A., Zarychta, R., Wałek, G., 2021. Changes in the topography of Krakow city centre, Poland, during the last millennium. J. Maps 17 (4), 186–193.

Lam, N.S., 1983. Spatial interpolation methods: a review. Am. Cartograph. 10 (2), 129–149.

Lamandé, P., 2009. Lacroix: les lignes de niveau à l'appui d'une nouvelle vision de la géographie physique. Le Monde des Cartes 199, 23–34.

Lamichhane, S., Kumar, L., Wilson, B., 2019. Digital soil mapping algorithms and covariates for soil organic carbon mapping and their implications: a review. Geoderma 352, 395–413.

Lane, B.A., Pasternack, G.B., Dahlke, H.E., Sandoval-Solis, S., 2017. The role of topographic variability in river channel classification. Progr. Phys. Geogr. 41 (5), 570–600.

Lanyon, L.E., Hall, G.F., 1983. Land-surface morphology: 2. Predicting potential landscape instability in eastern Ohio. Soil Sci. 136 (6), 382–386.

Lapen, D.R., Martz, L.W., 1993. The measurement of two simple topographic indices of wind sheltering-exposure from raster digital elevation models. Comput. Geosci. 19 (6), 769–779.

Lapen, D.R., Martz, L.W., 1996. An investigation of the spatial association between snow depth and topography in a Prairie agricultural landscape using digital terrain analysis. J. Hydrol. 184 (3–4), 277–298.

Lapen, D.R., Moorman, B.J., Price, J.S., 1996. Using ground penetrating radar to delineate subsurface features along a wetland catena. Soil Sci. Soc. Am. J. 60 (3), 923–931.

Lark, R.M., 1999. Soil–landform relationships at within-field scales: an investigation using continuous classification. Geoderma 92 (3–4), 141–165.

Lark, R.M., 2007. Decomposing digital soil information by spatial scale. In: Lagacherie, P., McBratney, A.B., Voltz, M. (Eds.), Digital Soil Mapping: An Introductory Perspective. Developments in Soil Science, vol. 31. Elsevier, Amsterdam, the Netherlands, pp. 301–326.

Lark, R.M., Bishop, T.F.A., Webster, R., 2007. Using expert knowledge with control of false discovery rate to select regressors for prediction of soil properties. Geoderma 138 (1–2), 65–78.

Larsen, H., Budka, M., Bennett, M.R., 2021. Technological innovation in the recovery and analysis of 3D forensic footwear evidence: structure from motion (SfM) photogrammetry. Sci. Justice 61 (4), 356–368.

Larson, J., Lidberg, W., Ågren, A.M., Laudon, H., 2022. Predicting soil moisture conditions across a heterogeneous boreal catchment using terrain indices. Hydrol. Earth Syst. Sci. 26 (19), 4837–4851.

Lascelles, B., Favis-Mortlock, D., Parsons, T., Boardman, J., 2002. Automated digital photogrammetry: a valuable tool for small-scale geomorphological research for the non-photogrammetrist? Trans. GIS 6 (1), 5–15.

Lastochkin, A.N., 1987. Morphodynamical Analysis. Nedra, Leningrad, USSR (in Russian).

Lattman, L.H., 1958. Technique of mapping geologic fracture traces and lineaments on aerial photographs. Photogrammet. Eng. 24 (4), 568–576.

Lattman, L.H., Parizek, R.R., 1964. Relationships between fracture traces and occurrence of ground water in carbonate rocks. J. Hydrol. 2 (2), 73–91.

Lavagnino, A.C., Bastos, A.C., Amado Filho, G.M., de Moraes, F.C., Araujo, L.S., de Moura, R.L., 2020. Geomorphometric seabed classification and potential megahabitat distribution in the Amazon continental margin. Front. Mar. Sci. 7, 190.

Lawrence, G.R.P., 1971. Cartographic Methods. Methuen, London, UK.

Lea, N.L., 1992. An aspect driven kinematic routing algorithm. In: Parsons, A.J., Abrahams, A.D. (Eds.), Overland Flow: Hydraulics and Erosion Mechanics. Chapman and Hall, New York, NY, pp. 393–407.

Lebedev, T.S., Orovetsky, Y.P., 1966. Tectonic peculiarities of the Mountain Crimea (in the context of new geological and geophysical data). Geophys. Commun. 18, 34–41 (in Russian).

Leberl, F., 1973. Interpolation in square grid DTM. ITC J. (5), 756–807.

Leberl, F.W., 1990. Radargrammetric Image Processing. Artech House, Norwood, MA.

Lecours, V., Brown, C.J., Devillers, R., Lucieer, V.L., Edinger, E.N., 2016a. Comparing selections of environmental variables for ecological studies: a focus on terrain attributes. PLoS One 11 (12), e0167128.

Lecours, V., Dolan, M.F.J., Micallef, A., Lucieer, V.L., 2016b. A review of marine geomorphometry, the quantitative study of the seafloor. Hydrol. Earth Syst. Sci. 20 (8), 3207–3244.

Lecours, V., Devillers, R., Edinger, E.N., Brown, C.J., Lucieer, V.L., 2017a. Influence of artefacts in marine digital terrain models on habitat maps and species distribution models: a multiscale assessment. Remote Sens. Ecol. Conserv. 3 (4), 232–246.

Lecours, V., Devillers, R., Lucieer, V.L., Brown, C.J., 2017b. Artefacts in marine digital terrain models: a multiscale analysis of their impact on the derivation of terrain attributes. IEEE Trans. Geosci. Remote Sens. 55 (9), 5391–5406.

Lecours, V., Devillers, R., Simms, A.E., Lucieer, V.L., Brown, C.J., 2017c. Towards a framework for terrain attribute selection in environmental studies. Environ. Model. Software 89, 19–30.

Lee, J., 1991. Analyses of visibility sites on topographic surfaces. Int. J. Geogr. Inf. Syst. 5 (4), 413–429.

Leenaers, H., Okx, J.P., Burrough, P.A., 1990. Employing elevation data for efficient mapping of soil pollution on floodplains. Soil Use Manage. 6 (3), 105–114.

Legates, D.R., Mahmood, R., Levia, D.F., DeLiberty, T.L., Quiring, S.M., Houser, C., et al., 2011. Soil moisture: a central and unifying theme in physical geography. Progr. Phys. Geogr. 35 (1), 65–86.

Legleiter, C.J., Harrison, L.R., 2019. Remote sensing of river bathymetry: evaluating a range of sensors, platforms, and algorithms on the Upper Sacramento River, California, USA. Water Resour. Res. 55 (3), 2142–2169.

Lehman, R.M., 2007. Microbial distributions and their potential controlling factors in terrestrial subsurface environments. In: Franklin, R.B., Mills, A.L. (Eds.), The Spatial Distribution of Microbes in the Environment. Springer, Dordrecht, the Netherlands, pp. 135–178.

Lehmann, J.G., 1799. Darstellung einer neuen Theorie der Bezeichnung der schiefen Flächen im Grundriß oder der Situationszeichnung der Berge. Fleischer, Leipzig, Germany.

Lehner, B., Verdin, K., Jarvis, A., 2008. New global hydrography derived from spaceborne elevation data. Eos 89 (10), 93–94.

Lei, S., Chen, H., Bian, Z., Liu, Z., 2016. Evaluation of integrating topographic wetness index with backscattering coefficient of TerraSAR-X image for soil moisture estimation in a mountainous region. Ecol. Indic. 61 (2), 624–633.

Leigh, C.L., Kidner, D.B., Thomas, M.C., 2009. Use of LiDAR in digital surface modelling: issues and errors. Trans. GIS 13 (4), 345–361.

Leitão, J.P., Moy de Vitry, M., Scheidegger, A., Rieckermann, J., 2016. Assessing the quality of digital elevation models obtained from mini unmanned aerial vehicles for overland flow modelling in urban areas. Hydrol. Earth Syst. Sci. 20 (4), 1637–1653.

Lemmens, M.J.P.M., 1999. Quality description problems of blindly sampled DEMs. In: Shi, W., Goodchild, M.F., Fisher, P.F. (Eds.), Proceedings of the International Symposium on Spatial Data Quality'99, Hong Kong, China, 18–20 July 1999. Hong Kong Polytechnic University, Hong Kong, China, pp. 210–218.

Lemon, A.M., Jones, N.L., 2003. Building solid models from boreholes and user-defined cross-sections. Comput. Geosci. 29 (5), 547–555.

Leong, W.J., Horgan, H.J., 2020. DeepBedMap: a deep neural network for resolving the bed topography of Antarctica. Cryosphere 14 (11), 3687–3705.

Leonowicz, A.M., Jenny, B., Hurni, L., 2009. Automatic generation of hypsometric layers for small-scale maps. Comput. Geosci. 35 (10), 2074–2083.

Leprince, S., Barbot, S., Ayoub, F., Avouac, J.-P., 2007. Automatic and precise orthorectification, coregistration, and subpixel correlation of satellite images, application to ground deformation measurements. IEEE Trans. Geosci. Remote Sens. 45 (6), 1529–1558.

Levorsen, A.I., 1927. Convergence studies in the Mid-Continent region. Am. Assoc. Pet. Geol. Bull. 11 (7), 657–682.

Li, A., Tan, X., Wu, W., Liu, H., Zhu, J., 2017. Predicting active-layer soil thickness using topographic variables at a small watershed scale. PLoS ONE 12 (9), e0183742.

Li, A., Zhang, X.C.J., Liu, B., 2021. Effects of DEM resolutions on soil erosion prediction using Chinese Soil Loss Equation. Geomorphology 384, 107706.

Li, H., Zhao, J., 2018. Evaluation of the newly released worldwide AW3D30 DEM over typical landforms of China using two global DEMs and ICESat/GLAS data. IEEE J. Sel. Topics Appl. Earth Obs. Remote Sens. 11 (11), 4430–4440.

Li, H., Zhao, J., Yan, B., Yue, L., Wang, L., 2022. Global DEMs vary from one to another: an evaluation of newly released Copernicus, NASA and AW3D30 DEM on selected terrains of China using ICESat-2 altimetry data. Int. J. Dig. Earth 15 (1), 1149–1168.

Li, M., Stefanakis, E., 2020. Geospatial operations of Discrete Global Grid Systems—a comparison with traditional GIS. J. Geovis. Spat. Anal. 4, 26.

Li, M., McGrath, H., Stefanakis, E., 2022. Multi-resolution topographic analysis in hexagonal Discrete Global Grid Systems. Int. J. Appl. Earth Obs. Geoinf. 113, 102985.

Li, P., Li, D., Hu, J., Fassnacht, F.E., Latifi, H., Yao, W., et al., 2024. Improving the application of UAV-LiDAR for erosion monitoring through accounting for uncertainty in DEM of difference. Catena 234, 107534.

Li, R., Ye, W., Qiao, G., Tong, X., Liu, S., Kong, F., et al., 2017. A new analytical method for estimating Antarctic ice flow in the 1960s from historical optical satellite imagery. IEEE Trans. Geosci. Remote Sens. 55 (5), 2771–2785.

Li, S., Xiong, L., Tang, G., Strobl, J., 2020. Deep learning-based approach for landform classification from integrated data sources of digital elevation model and imagery. Geomorphology 354, 107045.

Li, S., Hu, G., Cheng, X., Xiong, L., Tang, G., Strobl, J., 2022. Integrating topographic knowledge into deep learning for the void-filling of digital elevation models. Remote Sens. Environ. 269, 112818.

Li, T., Zhang, B., Xiao, W., Cheng, X., Li, Z., Zhao, J., 2020. UAV-based photogrammetry and LiDAR for the characterization of ice morphology evolution. IEEE J. Sel. Topics Appl. Earth Obs. Remote Sens. 13, 4188–4199.

Li, X., Chang, S.X., Liu, J., Zheng, Z., Wang, X., 2017. Topography–soil relationships in a hilly evergreen broadleaf forest in subtropical China. J. Soil. Sediment. 17 (4), 1101–1115.

Li, X., McCarty, G.W., Karlen, D.L., Cambardella, C.A., 2018a. Topographic metric predictions of soil redistribution and organic carbon in Iowa cropland fields. Catena 160, 222–232.

Li, X., McCarty, G.W., Karlen, D.L., Cambardella, C.A., Effland, W., 2018b. Soil organic carbon and isotope composition response to topography and erosion in Iowa. J. Geophys. Res. Biogeo. 123 (12), 3649–3667.

Li, X., McCarty, G.W., Lang, M., Ducey, T., Hunt, P., Miller, J., 2018c. Topographic and physicochemical controls on soil denitrification in prior converted croplands located on the Delmarva Peninsula, USA. Geoderma 309, 41–49.

Li, Y., Li, Y., Lu, X., Harbor, J., 2017. Geomorphometric controls on mountain glacier changes since the Little Ice Age in the Eastern Tien Shan, Central Asia. Ann. Am. Assoc. Geogr. 107 (2), 284–298.

Li, Y., Qiao, G., Popov, S., Cui, X., Florinsky, I.V., Yuan, X., et al., 2023. Unmanned aerial vehicle remote sensing for Antarctic research: a review of progress, current applications, and future use cases. IEEE Geosci. Remote Sens. Mag. 11 (1), 73–93.

Li, Z., 1994. A comparative study of the accuracy of digital terrain models (DTMs) based on various data models. ISPRS J. Photogrammet. Remote Sens. 49 (1), 2–11.

Li, Z., Zhu, Q., Gold, C., 2005. Digital Terrain Modeling: Principles and Methodology. CRC Press, New York, NY.

Li, Z., Yang, T., Xu, C.-Y., Shi, P., Yong, B., Huang, C.-S., et al., 2020. Evaluating the area and position accuracy of surface water paths obtained by flow direction algorithms. J. Hydrol. 583, 124619.

Li, Z., Yang, T., Wang, C., Shi, P., Yong, B., Song, Y., 2021. Assessing the precision of total contributing area (TCA) position accuracy of surface water paths obtained on the analytical solution of theoretical TCA on synthetic surfaces. Water Resour. Res. 57 (4), e2020WR028546.

Liang, C., Mackay, D.S., 2000. A general model of watershed extraction and representation using globally optimal flow paths and up-slope contributing areas. Int. J. Geogr. Inf. Sci. 14 (4), 337–358.

Liang, J., Shen, S., Gong, J., Liu, J., Zhang, J., 2017. Embedding user-generated content into oblique airborne photogrammetry-based 3D city model. Int. J. Geogr. Inf. Sci. 31 (1), 1–16.

Libohova, Z., Winzeler, H.E., Lee, B., Schoeneberger, J., Datta, J., Owens, P.R., 2016. Geomorphons: landform and property predictions in a glacial moraine in Indiana landscapes. Catena 142, 66–76.

Lidov, V.P., 1949. From the experience in landscape mapping of the Prioksko-Terrasny State Reserve. Prob. Geogr. 16, 179–190 (in Russian).

Ligas, M., 2012. Cartesian to geodetic coordinates conversion on a triaxial ellipsoid. J. Geodesy 86 (4), 249–256.

Lin, H. (Ed.), 2012. Hydropedology: Synergistic Integration of Soil Science and Hydrology. Academic Press, Amsterdam, the Netherlands.

Lin, H.-C., Wang, L.-L., Yang, S.-N., 1996. Automatic determination of the spread parameter in Gaussian smoothing. Pattern Recogn. Lett. 17 (12), 1247–1252.

Lin, Y., Prentice, S.E., Tran, T., Bingham, N.L., King, J.Y., Chadwick, O.A., 2016. Modeling deep soil properties on California grassland hillslopes using LiDAR digital elevation models. Geoderma Reg. 7, 67–75.

Linder, W., 2016. Digital Photogrammetry: A Practical Course, fourth ed. Springer, Berlin, Germany.

Lindsay, J.B., 2016a. Efficient hybrid breaching-filling sink removal methods for flow path enforcement in digital elevation models. Hydrol. Process. 30 (6), 846–857.

Lindsay, J.B., 2016b. The practice of DEM stream burning revisited. Earth Surf. Process. Landf. 41 (5), 658–668.

Lindsay, J.B., 2016c. Whitebox GAT: a case study in geomorphometric analysis. Comput. Geosci. 95, 75–84.

Lindsay, J.B., Creed, I.F., 2005. Removal of artifact depressions from digital elevation models: towards a minimum impact approach. Hydrol. Process. 19 (16), 3113–3126.

Lindsay, J.B., Francioni, A., Cockburn, J.M.H., 2019. LiDAR DEM smoothing and the preservation of drainage features. Remote Sens. (Basel) 11 (16), 1926.

Liner, C.L., 2016. Elements of 3D Seismology, third ed. Investigations in Geophysics, no. 19. Society of Exploration Geophysicists, Tulsa, OK.

Lisle, R.J., 1994. Detection of zones of abnormal strains in structures using Gaussian curvature analysis. Am. Assoc. Pet. Geol. Bull. 78 (12), 1811–1819.

Lisle, R.J., Robinson, J.M., 1995. The Mohr circle for curvature and its application to fold description. J. Struct. Geol. 17 (5), 739–750.

Lisle, R.J., Toimil, N.C., 2007. Defining folds on three-dimensional surfaces. Geology 35 (6), 519–522.

Lisle, R.J., Toimil, N., Aller, J., Bobillo-Ares, N., Bastida, F., 2010. The hinge lines of non-cylindrical folds. J. Struct. Geol. 32 (2), 166–171.

Liu, H., Jezek, K.C., Li, B., 1999. Development of an Antarctic digital elevation model by integrating cartographic and remotely sensed data: a GIS based approach. J. Geophys. Res. Solid Earth 104 (B10), 23,199–23,213.

Liu, H., Jezek, K., Li, B., Zhao, Z., 2001. Radarsat Antarctic Mapping Project Digital Elevation Model, Version 2. National Snow and Ice Data Center, Boulder, CO, https://doi.org/10.5067/PXKC81A7WAXD.

Liu, K., Ding, H., Tang, G., Song, C., Liu, Y., Jiang, L., et al., 2018. Large-scale mapping of gully-affected areas: an approach integrating Google Earth images and terrain skeleton information. Geomorphology 314, 13–26.

Liu, K., Song, C., Ke, L., Jiang, L., Pan, Y., Ma, R., 2019. Global open-access DEM performances in Earth's most rugged region High Mountain Asia: a multi-level assessment. Geomorphology 338, 16–26.

Liu, K., Song, C., Ke, L., Jiang, L., Ma, R., 2020a. Automatic watershed delineation in the Tibetan endorheic basin: a lake-oriented approach based on digital elevation models. Geomorphology 358, 107127.

Liu, K., Song, C., Wang, J., Ke, L., Zhu, Y., Zhu, J., et al., 2020b. Remote sensing-based modeling of the bathymetry and water storage for channel-type reservoirs worldwide. Water Resour. Res. 56 (11), e2020WR027147.

Liu, X., 2008. Airborne LiDAR for DEM generation: some critical issues. Progr. Phys. Geogr. 32 (1), 31–49.

Liu, Y., Cheng, X., Hui, F., Wang, X., Wang, F., Cheng, C., 2014. Detection of crevasses over polar ice shelves using satellite laser altimeter. Sci. China Earth Sci. 57 (6), 1267–1277.

Liu, Y., Gong, W., Hu, X., Gong, J., 2018. Forest type identification with random forest using Sentinel-1A, Sentinel 2A, multi-temporal Landsat-8 and DEM data. Remote Sens. (Basel) 10 (6), 946.

Liu, Y., Ma, X., Zhang, X., Guo, W., Kang, L., Yu, R., et al., 2022. 3D geological model-based hydraulic fracturing parameters optimization using geology–engineering integration of a shale gas reservoir: a case study. Energy Rep. 8, 10,048–10,060.

Liu, Z., Zhu, J., Fu, H., Zhou, C., Zuo, T., 2020. Evaluation of the vertical accuracy of open global DEMs over steep terrain regions using ICESat data: a case study over Hunan Province, China. Sensors 20 (17), 4865.

Lloyd, C.D., Atkinson, P.M., 2006. Deriving ground surface digital elevation models from LiDAR data with geostatistics. Int. J. Geogr. Inf. Sci. 20 (5), 535–563.

Lobanov, A.N., 1984. Photogrammetry, second enl. rev. ed. Nedra, Moscow, USSR (in Russian).

Lochner, J., Gain, J., Perche, S., Peytavie, A., Galin, E., Guérin, E., 2023. Interactive authoring of terrain using diffusion models. Comput. Graph. Forum 42 (7), e14941.

Lopatin, D.V., 2002. Lineament tectonics and giant ore deposits of the Northern Eurasia. Issledovanie Zemli iz Cosmosa (2), 77–91 (in Russian).

Lopatin, D.V., 2008. Cryptomorphic structures of the lithosphere: their reflection on space images and the Earth's surface. Doklady Earth Sci. 421 (2), 983–986.

López, A.M., Serrat, J., 1996. Tracing crease curves by solving a system of differential equations. In: Buxton, B., Cipolla, R. (Eds.), Computer Vision – ECCV '96. Proceedings of the 4th European Conference on Computer Vision, Cambridge, UK, 15–18 April 1996. Lect. Notes Comput. Sci., vol. 1064, pt. 1, pp. 241–250.

López, A.M., Lumbreras, F., Serrat, J., 1998. Creaseness from level set extrinsic curvature. In: Burkhardt, H., Neumann, B. (Eds.), Computer Vision – ECCV'98. Proceedings of the 5th European Conference on Computer Vision, Freiburg, Germany, 2–6 June 1998. Lect. Notes Comput. Sci., vol. 1407, pt. 2, pp. 156–169.

López, A.M., Lumbreras, F., Serrat, J., Villanueva, J.J., 1999. Evaluation of methods for ridge and valley detection. IEEE Trans. Pattern Anal. Mach. Intell. 21 (4), 327–335.

López, C., 1997. Locating some types of random errors in digital terrain models. Int. J. Geogr. Inf. Sci. 11 (7), 677–698.

Loro, M., Arce, R.M., Ortega, E., 2017. Identification of optimal landforms to reduce impacts on the landscape using LiDAR for hosting a new highway. Environ. Impact Assess. Rev. 66, 99–114.

Loye, A., Jaboyedoff, M., Theule, J.I., Liébault, F., 2016. Headwater sediment dynamics in a debris flow catchment constrained by high-resolution topographic surveys. Earth Surf. Dyn. 4 (2), 489–513.

Lu, S., Liu, B., Hu, Y., Fu, S., Cao, Q., Shi, Y., et al., 2020a. Soil erosion topographic factor (LS): accuracy calculated from different data sources. Catena 187, 104334.

Lu, X., Yang, K., Lu, Y., Gleason, C.J., Smith, L.C., Li, M., 2020b. Small Arctic rivers mapped from Sentinel-2 satellite imagery and ArcticDEM. J. Hydrol. 584, 124689.

Lucas, L.C., Ten Brink, U.S., Flores, C.H., Chaytor, J.D., 2022. Geological mapping in the deep ocean using structure-from-motion photogrammetry. In: AGU Fall Meeting 2022, Chicago, IL, 12–16 Dec. 2022. American Geophysical Union, Chicago, IL. T15C-0146.

Lucchitta, B.K., Ferguson, H.M., 1986. Antarctica: measuring glacier velocity from satellite images. Science 234 (4780), 1105–1108.

Luhmann, T., Robson, S., Kyle, S., Boehm, J., 2020. Close-Range Photogrammetry and 3D Imaging, third rev. exp. ed. De Gruyter STEM Book Series. De Gruyter, Berlin, Germany.

Lukatela, H., 2002. A seamless global terrain model in the Hipparchus System. In: Goodchild, M., Kimerling, A.J. (Eds.), Discrete Global Grids: A Web Book. National Center for Geographic Information and Analysis, Santa Barbara, CA, pp. 127–139.

Lukin, A.A., 1987. Development of a Technique for Morphostructural Hydrogeological Analysis. Nauka, Novosibirsk, USSR (in Russian).

Luo, L., Wang, X., Guo, H., Lasaponara, R., Zong, X., Masini, N., et al., 2019. Airborne and spaceborne remote sensing for archaeological and cultural heritage applications: a review of the century (1907–2017). Remote Sens. Environ. 232, 111280.

Luo, W., Liu, C.-C., 2018. Innovative landslide susceptibility mapping supported by geomorphon and geographical detector methods. Landslides 15 (3), 465–474.

Luo, W., Jasiewicz, J., Stepinski, T., Wang, J., Xu, C., Cang, X., 2016. Spatial association between dissection density and environmental factors over the entire conterminous United States. Geophys. Res. Lett. 43 (2), 692–700.

Luo, W., Xu, X., Liu, W., Liu, M., Li, Z., Peng, T., et al., 2019. UAV based soil moisture remote sensing in a karst mountainous catchment. Catena 174, 478–489.

Lv, G., Xiong, L., Chen, M., Tang, G., Sheng, Y., Liu, X., et al., 2017. Chinese progress in geomorphometry. J. Geogr. Sci. 27 (11), 1389–1412.

Lythe, M.B., Vaughan, D.G., The BEDMAP Consortium, 2001. BEDMAP: a new ice thickness and subglacial topographic model of Antarctica. J. Geophys. Res. Solid Earth 106 (B6), 11335–11351.

Lyu, F., Xu, Z., Ma, X., Wang, S., Li, Z., Wang, S., 2021. A vector-based method for drainage network analysis based on LiDAR data. Comput. Geosci. 156, 104892.

Lyubashin, V.N., Lisitsin, V.V., 1981. A Program of the Engineering Survey for Justification of the Technical Project of Anti-Landslide Actions and Development at the Slope of the Oka River Valley in the City of Pushchino, Serpukhov District, Moscow Region. Progress Report No. 3603. Central Trust of Engineering Surveys, Moscow, USSR (in Russian).

Lyubkov, A.N., Martynenko, A.I., 1963. Producing topographic models by computer controlled milling machines. Coll. Sci. Tech. Prod. Art. Milit. Topogr. Serv. Res. Inst. 12, 102–109 (in Russian).

Ma, C., Rokne, J., 2004. 3D seismic volume visualization. In: Zhang, D.D., Kamel, M., Baciu, G. (Eds.), Integrated Image and Graphics Technologies. Kluwer, New York, NY, pp. 241–262.

Ma, H., Zhou, W., Zhang, L., 2018. DEM refinement by low vegetation removal based on the combination of full waveform data and progressive TIN densification. ISPRS J. Photogrammet. Remote Sens. 146, 260–271.

Ma, L., Liu, Y., Zhang, X., Ye, Y., Yin, G., Johnson, B.A., 2019. Deep learning in remote sensing applications: a meta-analysis and review. ISPRS J. Photogrammet. Remote Sens. 152, 166–177.

Ma, Y., Minasny, B., Malone, B.P., McBratney, A.B., 2019. Pedology and digital soil mapping (DSM). Eur. J. Soil Sci. 70 (2), 216–235.

Macelloni, L., Simonetti, A., Knapp, J.H., Knapp, C.C., Lutken, C.B., Lapham, L.L., 2012. Multiple resolution seismic imaging of a shallow hydrocarbon plumbing system, Woolsey Mound, Northern Gulf of Mexico. Mar. Petrol. Geol. 38 (1), 128–142.

MacFerrin, M., Amante, C., Stroker, K., Carignan, K., Love, M., Lim, E., 2021. ETOPO 2022: an updated NOAA global relief model. In: AGU Fall Meeting 2021, New Orleans, LA, 13–17 Dec. 2021. American Geophysical Union, New Orleans, LA. NH25B-0560.

MacFerrin, M.J., Amante, C., Carignan, K., Love, M.R., Lim, E., Arcos, N.P., et al., 2022. ETOPO 2022: NOAA's new seamless topography-and-bathymetry bare earth surface elevation dataset. In: AGU Fall Meeting 2022, Chicago, IL, 12–16 Dec. 2022. American Geophysical Union Chicago, IL. NH22C-0433.

Mach, R., Petschek, P., 2007. Visualization of Digital Terrain and Landscape Data: A Manual. Springer, Berlin, Germany.

Mackaness, W.A., Ruas, A., Sarjakoski, L.T. (Eds.), 2007. Generalisation of Geographic Information: Cartographic Modelling and Applications. Elsevier, Amsterdam, the Netherlands.

MacMillan, R.A., 2008. Experiences with applied DSM: protocol, availability, quality and capacity building. In: Hartemink, A.E., McBratney, A., Mendonça-Santos, M.L. (Eds.), Digital Soil Mapping with Limited Data. Springer, Dordrecht, the Netherlands, pp. 113–135.

MacMillan, R.A., Marciak, L.C., 2001. Procedures Manual for Watershed-Based Salinity Management: A Comprehensive Manual for Assessing and Addressing Salinity on a Watershed Basis. Alberta Agriculture, Food and Rural Development, Conservation and Development Branch, Edmonton, AB.

MacMillan, R.A., Pettapiece, W.W., 2000. Alberta Landforms: Quantitative Morphometric Descriptions and Classification of Typical Alberta Landforms. Technical Bulletin No. 2000-2E. Semiarid Prairie Agricultural Research Centre, Agriculture and Agri-Food Canada, Swift Current, SK.

MacMillan, R.A., Shary, P.A., 2009. Landforms and landform elements in geomorphometry. In: Hengl, T., Reuter, H.I. (Eds.), Geomorphometry: Concepts, Software, Applications. Developments in Soil Science, vol. 33. Elsevier, Amsterdam, the Netherlands, pp. 227–254.

MacMillan, R.A., Pettapiece, W.W., Nolan, S.C., Goddard, T.W., 2000. A generic procedure for automatically segmenting landforms into landform elements using DEMs, heuristic rules and fuzzy logic. Fuzzy Set. Syst. 113 (1), 81–109.

MacMillan, R.A., Pettapiece, W.W., Brierley, J.A., 2005. An expert system for allocating soils to landforms through the application of soil survey tacit knowledge. Can. J. Soil Sci. 85 (1), 103–112.

Madry, S., 2015. Global Navigation Satellite Systems and Their Applications. Springer Briefs in Space Development Book Series. Springer, New York, NY.

Maerten, L., Pollard, D.D., Maerten, F., 2001. Digital mapping of three-dimensional structures of the Chimney Rock fault system, central Utah. J. Struct. Geol. 23 (4), 585–592.

Magnus, J.R., Neudecker, H., 2007. Matrix Differential Calculus with Applications in Statistics and Econometrics, third ed. Wiley, Chichester, UK.

Magruder, L., Neuenschwander, A., Klotz, B., 2021. Digital terrain model elevation corrections using space-based imagery and ICESat-2 laser altimetry. Remote Sens. Environ. 264, 112621.

Mahdavi-Amiri, A., Alderson, T., Samavati, F., 2015. A survey of Digital Earth. Comput. Graph. 53 (B), 95–117.

Makarov, V.I., 1981. Lineaments: problems and trends in airborne and spaceborne studies. Sov. J. Remote Sens. 4, 646–658.

Makarovič, B., 1972. Information transfer in reconstruction of data from sampled points. Photogrammetria 28 (4), 111–130.

Makarovič, B., 1973. Progressive sampling for digital terrain models. ITC J. (3), 397–416.

Makarovič, B., 1977. Composite sampling for digital terrain models. ITC J. (3), 406–433.

Makarovic, B., 1984a. Automatic production of DTM data by digital off-line technique. In: XV ISPRS Congress, Technical Commission II: Instrumentation for Data Reduction and Analysis, Rio de Janeiro, Brasil, 17–29 June 1984. ISPRS Arch. XXV (A2), pp. 332–343.

Makarovic, B., 1984b. On performance of digital off-line systems for automatic production of DTM data. In: XV ISPRS Congress, Technical Commission II: Instrumentation for Data Reduction and Analysis, Rio de Janeiro, Brazil, 17–29 June 1984. ISPRS Arch. XXV (A2), pp. 344–353.

Mallalieu, J., Carrivick, J., Quincey, D., Smith, M., James, W., 2017. An integrated structure-from-motion and time-lapse technique for quantifying ice-margin dynamics. J. Glaciol. 63 (242), 937–949.

Mallavan, B.P., Minasny, B., McBratney, A.B., 2010. Homosoil, a methodology for quantitative extrapolation of soil information across the globe. In: Boettinger, J.L., Howell, D.W., Moore, A.C., Hartemink, A.E., Kienast-Brown, S. (Eds.), Digital Soil Mapping: Bridging Research, Environmental Application, and Operation. Progress in Soil Science Book Series. Springer, Dordrecht, the Netherlands, pp. 137–150.

Mallet, C., Bretar, F., 2009. Full-waveform topographic lidar: state-of-the-art. ASPRS J. Photogrammet. Remote Sens. 64 (1), 1–16.

Mallet, F., Marc, V., Douvinet, J., Rossello, P., Joly, D., Ruy, S., 2020. Assessing soil water content variation in a small mountainous catchment over different time scales and land covers using geographical variables. J. Hydrol. 591, 125593.

Mallick, J., Ahmed, M., Alqadhi, S.D., Falqi, I.I., Parayangat, M., Singh, C.K., et al., 2022. Spatial stochastic model for predicting soil organic matter using remote sensing data. Geocarto Int. 37 (2), 413–444.

Malone, B.P., McBratney, A.B., Minasny, B., Laslett, G.M., 2009. Mapping continuous depth functions of soil carbon storage and available water capacity. Geoderma 154 (1–2), 138–152.

Malone, B.P., Minasny, B., Odgers, N.P., McBratney, A.B., 2014. Using model averaging to combine soil property rasters from legacy soil maps and from point data. Geoderma 232–234, 34–44.

Malone, B.P., Jha, S.K., Minasny, B., McBratney, A.B., 2016. Comparing regression-based digital soil mapping and multiple-point geostatistics for the spatial extrapolation of soil data. Geoderma 262, 243–253.

Malone, B.P., Styc, Q., Minasny, B., McBratney, A.B., 2017. Digital soil mapping of soil carbon at the farm scale: a spatial downscaling approach in consideration of measured and uncertain data. Geoderma 290, 91–99.

Malone, B.P., Odgers, N.P., Stockmann, U., Minasny, B., McBratney, A.B., 2018. Digital mapping of soil classes and continuous soil properties. In: McBratney, A.B., Minasny, B., Stockmann, U. (Eds.), Pedometrics. Progress in Soil Science Book Series. Springer, Cham, Switzerland, pp. 373–413.

Malyavsky, B.K., Zharnovsky, A.A., 1974. On digital terrain modelling. Geodesy Cartogr. (6), 31–38 (in Russian).

Mancini, F., Dubbini, M., Gattelli, M., Stecchi, F., Fabbri, S., Gabbianelli, G., 2013. Using unmanned aerial vehicles (UAV) for high-resolution reconstruction of topography: the structure from motion approach on coastal environments. Remote Sens. (Basel) 5 (12), 6880–6898.

Mann, J., Angelou, N., Arnqvist, J., Callies, D., Cantero, E., Chávez Arroyo, R., et al., 2017. Complex terrain experiments in the New European Wind Atlas. Philos. Trans. R. Soc. Ser. A Math. Phys. Sci. 375 (2091), 20160101.

Manning, G.R., 1999. Relations Between Spatial Variability of Soil Properties and Grain Yield Response to Nitrogen Fertilizer in a Variable Manitoba Soil-Landscape (M.Sc. thesis). University of Manitoba, Winnipeg, MB.

Marchese, F., Bracchi, V.A., Lisi, G., Basso, D., Corselli, C., Savini, A., 2020. Assessing fine-scale distribution and volume of Mediterranean algal reefs through terrain analysis of multibeam bathymetric data. A case study in the Southern Adriatic continental shelf. Water 12 (1), 157.

Marchetti, A., Piccini, C., Francaviglia, R., Santucci, S., Chiuchiarelli, I., 2010. Estimating soil organic matter content by regression kriging. In: Boettinger, J.L., Howell, D.W., Moore, A.C., Hartemink, A.E., Kienast-Brown, S. (Eds.), Digital Soil Mapping: Bridging Research, Environmental Application, and Operation. Progress in Soil Science Book Series. Springer, Dordrecht, the Netherlands, pp. 241–254.

Marchi, L., Comiti, F., Crema, S., Cavalli, M., 2019. Channel control works and sediment connectivity in the European Alps. Sci. Total Environ. 668, 389–399.

Mardia, K.V., 1972. Statistics of Directional Data. Academic Press, London, UK.

Mark, D.M., 1975a. Computer analysis of topography: a comparison of terrain storage methods. Geogr. Ann. Ser. A Phys. Geogr. 57 (3–4), 179–188.

Mark, D.M., 1975b. Geomorphometric parameters: a review and evaluation. Geogr. Ann. Ser. A Phys. Geogr. 57 (3–4), 165–177.

Mark, D.M., 1979. Phenomenon-based data-structuring and digital terrain modelling. Geo-Processing 1 (1), 27–36.

Mark, D.M., 1984. Automated detection of drainage networks from digital elevation models. Cartographica 21 (2–3), 168–178.

Mark, R.K., 1992. A Multidirectional, Oblique-Weighted, Shaded Relief Image of the Island of Hawaii. US Department of the Interior, Open-File Report 92-422, US Geological Survey, Reston, VA.

Markov, A.N., Dahl-Jensen, D., Kotlyakov, V.M., Golubev, V.N., Leonov, M.G., Lukin, V.V., 2016. Dynamics of Antarctic and Greenland ice sheets using the borehole, radio sounding and space observations. Ice Snow 56 (3), 309–332 (in Russian).

Marks, K.M., Smith, W.H.F., Sandwell, D.T., 2010. Evolution of errors in the altimetric bathymetry model used by Google Earth and GEBCO. Mar. Geophys. Res. 31 (3), 223–238.

Marques, K., Demattê, J., Miller, B.A., Lepsch, I., 2018. Geomorphometric segmentation of complex slope elements for detailed digital soil mapping in Southeast Brazil. Geoderma Reg. 14, e00175.

Marsh, O.J., Price, D., Courville, Z.R., Floricioiu, D., 2021. Crevasse and rift detection in Antarctica from TerraSAR-X satellite imagery. Cold Reg. Sci. Technol. 187, 103284.

Marston, B.E., Jenny, B., 2015. Improving the representation of major landforms in analytical relief shading. Int. J. Geogr. Inf. Sci. 29 (7), 1144–1165.

Marthews, T.R., Dadson, S.J., Lehner, B., Abele, S., Gedney, N., 2015. High-resolution global topographic index values for use in large-scale hydrological modelling. Hydrol. Earth Syst. Sci. 19 (1), 91–104.

Martínez-Murillo, J.F., Hueso-González, P., Ruiz-Sinoga, J.D., 2017. Topsoil moisture mapping using geostatistical techniques under different Mediterranean climatic conditions. Sci. Total Environ. 595, 400–412.

Martini, L., Picco, L., Iroumé, A., Cavalli, M., 2019. Sediment connectivity changes in an Andean catchment affected by volcanic eruption. Sci. Total Environ. 692, 1209–1222.

Martz, L.W., de Jong, E., 1987. Using Cesium-137 to assess the variability of net soil erosion and its association with topography in a Canadian Prairie landscape. Catena 14 (5), 439–451.

Martz, L.W., de Jong, E., 1988. CATCH: a Fortran program for measuring catchment area from digital elevation models. Comput. Geosci. 14 (5), 627–640.

Martz, L.W., de Jong, E., 1990. Natural radionuclides in the soils in a small agricultural basin in the Canadian prairies and their associations with topography, properties and erosion. Catena 17 (1), 85–96.

Martz, L.W., de Jong, E., 1991. Using Cesium-137 and landform classification to develop a net soil erosion budget for a small Canadian prairie watershed. Catena 18 (3–4), 289–308.

Maruyama, G., 1997. Basics of Structural Equation Modeling. Sage, Thousand Oaks, CA.

Mashimbye, Z.E., Loggenberg, K.A., 2023. Scoping review of landform classification using geospatial methods. Geomatics 3 (1), 93–114.

Masini, N., Gizzi, F.T., Biscione, M., Fundone, V., Sedile, M., Sileo, M., et al., 2018. Medieval archaeology under the canopy with LiDAR. The (re)discovery of a medieval fortified settlement in Southern Italy. Remote Sens. (Basel) 10 (10), 1598.

Maslov, L.A., Anokhin, V.M., 2006. The Earth's decelerated rotation and regularities in orientation of its surface lineaments and faults. Planet. Space Sci. 54 (22), 216–218.

Masoud, A., Koike, K., 2006. Tectonic architecture through Landsat-7 ETM+/SRTM DEM-derived lineaments and relationship to the hydrogeologic setting in Siwa region, NW Egypt. J. Afr. Earth Sci. 45 (4–5), 467–477.

Masoud, A.A., Koike, K., 2011a. Auto-detection and integration of tectonically significant lineaments from SRTM DEM and remotely-sensed geophysical data. ISPRS J. Photogrammet. Remote Sens. 66 (6), 818–832.

Masoud, A.A., Koike, K., 2011b. Morphotectonics inferred from the analysis of topographic lineaments auto-detected from DEMs: application and validation for the Sinai Peninsula, Egypt. Tectonophysics 510 (3–4), 291–308.

Masumoto, S., Raghavan, V., Yonezawa, G., Nemoto, T., Shiono, K., 2004. Construction and visualization of a three dimensional geologic model using GRASS GIS. Trans. GIS 8 (2), 211–223.

Mateo Lázaro, J., Sánchez Navarro, J.Á., García Gil, A., Edo Romero, V., 2014. 3D-geological structures with digital elevation models using GPU programming. Comput. Geosci. 70, 138–146.

Mather, A.E., Fyfe, R.M., Clason, C.C., Stokes, M., Mills, S., Barrows, T.T., 2019. Automated mapping of relict patterned ground: an approach to evaluate morphologically subdued landforms using unmanned-aerial-vehicle and structure-from-motion technologies. Progr. Phys. Geogr. 43 (2), 174–192.

Maxwell, A.E., Shabe, C., 2022. Land-surface parameters for spatial predictive mapping and modeling. Earth Sci. Rev. 226, 103944.

Maxwell, A.E., Warner, T.A., 2019. Is high spatial resolution DEM data necessary for mapping palustrine wetlands? Int. J. Remote Sens. 40 (1), 118–137.

Maxwell, A.E., Sharma, M., Kite, J.S., Donaldson, K.A., Thompson, J.A., Bell, M.L., et al., 2020. Slope failure prediction using random forest machine learning and LiDAR in an eroded folded mountain belt. Remote Sens. (Basel) 12 (3), 486.

Maxwell, J.C., 1870. On hills and dales. The London, Edinburgh Dublin Philos. Mag. J. Sci. Ser. 4 40 (269), 421–427.

Mayer, L.A., 2006. Frontiers in seafloor mapping and visualization. Mar. Geophys. Res. 27 (1), 7–17.

Mayer, L., Jakobsson, M., Allen, G., Dorschel, B., Falconer, R., Ferrini, V., et al., 2018. The Nippon Foundation—GEBCO Seabed 2030 project: the quest to see the world's oceans completely mapped by 2030. Geosciences 8 (2), 63.

McArthur, D.E., Fuentes, R.W., Devarajan, V., 2000. Generation of hierarchical multiresolution terrain databases using wavelet filtering. Photogrammet. Eng. Remote Sens. 66 (3), 287–295.

McBratney, A.B., Odeh, I.O.A., 1997. Application of fuzzy sets in soil science: fuzzy logic, fuzzy measurement, and fuzzy decisions. Geoderma 77 (2–4), 85–113.

McBratney, A.B., Odeh, I.O.A., Bishop, T.F.A., Dunbar, M.S., Shatar, T.M., 2000. An overview of pedometric techniques for use in soil survey. Geoderma 97 (3–4), 293–327.

McBratney, A.B., Mendonça Santos, M.L., Minasny, B., 2003. On digital soil mapping. Geoderma 117 (1–2), 3–52.

McBratney, A.B., Minasny, B., Stockmann, U. (Eds.), 2018. Pedometrics. Progress in Soil Science Book Series. Springer, Cham, Switzerland.

McBratney, A.B., de Gruijter, J., Bryce, A., 2019. Pedometrics timeline. Geoderma 338, 568–575.

McClean, C.J., Evans, I.S., 2000. Apparent fractal dimensions from continental scale digital elevation models using variogram methods. Trans. GIS 4 (4), 361–378.

McCullagh, M.J., 1981. Creation of smooth contours over irregularly distributed data using local surface patches. Geogr. Anal. 13 (1), 51–63

McCullagh, M.J., 1988. Terrain and surface modelling systems: theory and practice. Photogrammet. Rec. 12 (72), 747–779.

McCullagh, P., Nelder, J., 1999. Generalized Linear Models, second ed. Chapman and Hall/CRC, Boca Raton, FL.

McIntyre, M.L., Naar, D.F., Carder, K.L., Donahue, B.T., Mallison, D.J., 2006. Coastal bathymetry from hyperspectral remote sensing data: comparisons with high resolution multibeam bathymetry. Mar. Geophys. Res. 27 (2), 128–136.

McKenzie, N.J., Ryan, P.J., 1999. Spatial prediction of soil properties using environmental correlation. Geoderma 89 (1–2), 67–94.

McMahon, M.J., North, C.P., 1993. Three-dimensional integration of remotely sensed geological data: a methodology for petroleum exploration. Photogrammet. Eng. Remote Sensing 59 (8), 1251–1256.

McMahon, S., 2001. Influence of Long-Term Hydrologic Regime, as Indicated by Landscape Position, on Soil Denitrifier Populations (B.Sc thesis). University of Manitoba, Winnipeg, MB.

McMaster, R.B., Shea, K.S., 1992. Generalization in Digital Cartography. Association of American Geographers, Washington, DC.

Mead, R., 1974. A test for spatial pattern at several scales using data from a grid of contiguous quadrates. Biometrics 30 (2), 295–307.

Meddens, A.J.H., Vierling, L.A., Eitel, J.U.H., Jennewein, J.S., White, J.C., Wulder, M.A., 2018. Developing 5 m resolution canopy height and digital terrain models from WorldView and ArcticDEM data. Remote Sens. Environ. 218, 174–188.

Medvedev, A., Telnova, N., Alekseenko, N., Koshkarev, A., Kuznetchenko, P., Asmaryan, S., et al., 2020. UAV-derived data application for environmental monitoring of the coastal area of Lake Sevan, Armenia with a changing water level. Remote Sens. (Basel) 12 (22), 3821.

Meerschman, E., van Meirvenne, M., van de Vijver, E., de Smedt, P., Islam, M.M., Saey, T., 2013. Mapping complex soil patterns with multiple-point geostatistics. Eur. J. Soil Sci. 64 (2), 183–191.

Meijering, E., 2002. A chronology of interpolation: from ancient astronomy to modern signal and image processing. Proc. IEEE 90 (3), 319–342.

Meixner, F.X., Eugster, W., 1999. Effects of landscape pattern and topography on emissions and transport. In: Tenhunen, J.D., Kabat, P. (Eds.), Integrating Hydrology, Ecosystem Dynamics, and Biogeochemistry in Complex Landscapes. Wiley, Chichester, UK, pp. 147–175.

Meles, M.B., Younger, S.E., Jackson, C.R., Du, E., Drover, D., 2020. Wetness index based on landscape position and topography (WILT), modifying TWI to reflect landscape position. J. Environ. Manage. 255, 109863.

Mello, F.A.O., Demattê, J.A.M., Rizzo, R., Dotto, A.C., Poppiel, R.R., Mendes, W.d.S., et al., 2021. Expert-based maps and highly detailed surface drainage models to support digital soil mapping. Geoderma 384, 114779.

Mello, F.A.O., Melo Demattê, J.A., Rizzo, R., de Mello, D.C., Poppiel, R.R., Silvero, N.E.Q., et al., 2022. Complex hydrological knowledge to support digital soil mapping. Geoderma 409, 115638.

Mendonça-Santos, M.L., Guenat, C., Bouzelboudjen, M., Golay, F., 2000. Three-dimensional GIS cartography applied to the study of the spatial variation of soil horizons in a Swiss floodplain. Geoderma 97 (3–4), 351–366.

Mendonça-Santos, M.L., Santos, H.G., Dart, R.O., Pares, J.G., 2008. Digital mapping of soil classes in Rio de Janeiro State, Brazil: data, modelling and prediction. In: Hartemink, A.E., McBratney, A., Mendonça-Santos, M.L. (Eds.), Digital Soil Mapping with Limited Data. Springer, Dordrecht, the Netherlands, pp. 381–396.

Mendonça-Santos, M.L., Dart, R.O., Santos, H.G., Coelho, M.R., Berbara, R.L.L., Lumbreras, J.F., 2010. Digital soil mapping of topsoil organic carbon content of Rio de Janeiro State, Brazil. In: Boettinger, J.L., Howell, D.W., Moore, A.C., Hartemink, A.E., Kienast-Brown, S. (Eds.), Digital Soil Mapping: Bridging Research, Environmental Application, and Operation. Progress in Soil Science Book Series. Springer, Dordrecht, the Netherlands, pp. 255–266.

Menna, F., Agrafiotis, P., Georgopoulos, A., 2018. State of the art and applications in archaeological underwater 3D recording and mapping. J. Cult. Herit. 33, 231–248.

Menz, G., 1988. Deduction of human-bioclimatological maps by means of remote sensing data and a digital terrain model using a correlation approach. Geol. Jahr. A104, 383–391.

Menzies, J. (Ed.), 2002. Modern and Past Glacial Environments, rev. student ed. Butterworth-Heinemann, Oxford, UK.

Mergelov, N.S., 2014. Soils of wet valleys in the Larsemann Hills and Vestfold Hills Oases (Princess Elizabeth Land, East Antarctica). Eurasian Soil Sci. 47 (9), 845–862.

Merot, P., Ezzahar, B., Walter, C., Aurousseau, P., 1995. Mapping waterlogging of soils using digital terrain models. Hydrol. Process. 9 (1), 27–34.

Merriman, M., 1899. Elements of Precise Surveying and Geodesy. Wiley, New York, NY.

Mertes, J.R., Gulley, J.D., Benn, D.I., Thompson, S.S., Nicholson, L.I., 2017. Using structure-from-motion to create glacier DEMs and orthoimagery from historical terrestrial and oblique aerial imagery. Earth Surf. Process. Landf. 42 (14), 2350–2364.

Mesa-Mingorance, J.L., Ariza-López, F.J., 2020. Accuracy assessment of digital elevation models (DEMs): a critical review of practices of the past three decades. Remote Sens. (Basel) 12 (16), 2630.

Meshcheryakov, Y.A., 1965. Structural Geomorphology of Plain Lands. Nauka, Moscow, USSR (in Russian).

Micheletti, N., Tonini, M., Lane, S.N., 2017. Geomorphological activity at a rock glacier front detected with a 3D density-based clustering algorithm. Geomorphology 278, 287–297.

Middleton, M., Heikkonen, J., Nevalainen, P., Hyvönen, E., Sutinen, R., 2020. Machine learning-based mapping of micro-topographic earthquake-induced paleo-Pulju moraines and liquefaction spreads from a digital elevation model acquired through laser scanning. Geomorphology 358, 107099.

Midgley, N.G., Tonkin, T.N., Graham, D.J., Cook, S.J., 2018. Evolution of high-Arctic glacial landforms during deglaciation. Geomorphology 311, 63–75.

Migoń, P., Kasprzak, M., 2016. Pathways of geomorphic evolution of sandstone escarpments in the Góry Stołowe tableland (SW Poland)—insights from LiDAR-based high-resolution DEM. Geomorphology 260, 51–63.

Migoń, P., Jancewicz, K., Różycka, M., Duszyński, F., Kasprzak, M., 2017. Large-scale slope remodelling by landslides—geomorphic diversity and geological controls, Kamienne Mts, Central Europe. Geomorphology 289, 134–151.

Milledge, D.G., Lane, S.N., Warburton, J., 2009. The potential of digital filtering of generic topographic data for geomorphological research. Earth Surf. Process. Landf. 34 (1), 63–74.

Miller, C.L., Leflamme, R.A., 1958. The digital terrain model—theory and application. Photogrammet. Eng. 24 (3), 433–442.

Minár, J., Evans, I.S., 2008. Elementary forms for land surface segmentation: the theoretical basis of terrain analysis and geomorphological mapping. Geomorphology 95 (3–4), 236–259.

Minár, J., Jenčo, M., Evans, I.S., Minár Jr., J., Kadlec, M., Krcho, J., et al., 2013. Third-order geomorphometric variables (derivatives): definition, computation and utilization of changes of curvatures. Int. J. Geogr. Inf. Sci. 27 (7), 1381–1402.

Minár, J., Krcho, J., Evans, I.S., 2016. Geomorphometry: quantitative land-surface analysis. In: Elias, S.A. (Ed.), Reference Module in Earth Systems and Environmental Sciences. Elsevier, Amsterdam, the Netherlands, https://doi.org/10.1016/B978-0-12-409548-9.10260-X.

Minár, J., Evans, I.S., Jenčo, M., 2020. A comprehensive system of definitions of land surface (topographic) curvatures, with implications for their application in geoscience modelling and prediction. Earth Sci. Rev. 211, 103414.

Minár, J., Drăguţ, L., Evans, I.S., Feciskanin, R., Gallay, M., Jenčo, M., et al., 2024. Physical geomorphometry for elementary land surface segmentation and digital geomorphological mapping. Earth Sci. Rev. 248, 104631.

Minasny, B., McBratney, A.B., 2007. Spatial prediction of soil properties using EBLUP with the Matérn covariance function. Geoderma 140 (4), 324–336.

Minasny, B., McBratney, A.B., 2016. Digital soil mapping: a brief history and some lessons. Geoderma 264 (B), 301–311.

Mishra, D., Narumalani, S., Lawson, M., Rundquist, D., 2004. Bathymetric mapping using IKONOS multispectral data. GISci. Remote Sens. 41 (4), 301–321.

Mishustin, E.N., Shilnikova, V.K., 1971. Biological Fixation of Atmospheric Nitrogen. Macmillan, London, UK.

Misiuk, B., Lecours, V., Bell, T., 2018. A multiscale approach to mapping seabed sediments. PLoS One 13 (2), e0193647.

Mitas, L., Mitasova, H., 2005. Spatial interpolation. In: Longley, P., Goodchild, M.F., Maguire, D.J., Rhind, D.W. (Eds.), Geographical Information Systems: Principles, Techniques, Management and Applications, second abr. ed. Wiley, Hoboken, NJ, pp. 481–492.

Mitášová, H., Hofierka, J., 1993. Interpolation by regularized spline with tension: II. Application to terrain modeling and surface geometry analysis. Mathemat. Geol. 25 (6), 657–669.

Mitášová, H., Mitáš, L., 1993. Interpolation by regularized spline with tension: I. Theory and implementation. Mathemat. Geol. 25 (6), 641–655.

Mitasova, H., Harmon, R.S., Weaver, K.J., Lyons, N.J., Overton, M.F., 2012. Scientific visualization of landscapes and landforms. Geomorphology 137 (1), 122–137.

Mithan, H.T., Hales, T.C., Cleall, P.J., 2019. Supervised classification of landforms in Arctic mountains. Permafr. Periglac. Process. 30 (3), 131–145.

Mitusov, A.V., 2001. The Role of Topography in Soil Fertility Development in Biogeocoenosis of the Forest Steppe Zone of the European Part of Russia (Ph.D. thesis). Orel State Agricultural University, Orel, Russia (in Russian).

Mitusov, A.V., Burian, L., Khrisanov, V.R., 2017. Distribution of local landforms at head and end points of gullies on different grid spacing. Catena 159, 159–170.

Mohamed, M.A., 2020. Classification of landforms for digital soil mapping in urban areas using LiDAR data derived terrain attributes: a case study from Berlin, Germany. Land 9 (9), 319.

Mölg, N., Bolch, T., 2017. Structure-from-motion using historical aerial images to analyse changes in glacier surface elevation. Remote Sens. (Basel) 9 (10), 1021.

Moličová, H., Grimaldi, M., Bonell, M., Hubert, P., 1997. Using TOPMODEL towards identifying and modelling the hydrological patterns within a head-water, humid, tropical catchment. Hydrol. Process. 11 (9), 1169–1196.

Mondal, A., Khare, D., Kundu, S., Mukherjee, S., Mukhopadhyay, A., Mondal, S., 2017. Uncertainty of soil erosion modelling using open source high resolution and aggregated DEMs. Geosci. Front. 8 (3), 425–436.

Mongus, D., Žalik, B., 2012. Parameter-free ground filtering of LiDAR data for automatic DTM generation. ISPRS J. Photogrammet. Remote Sensing 67, 1–12.

Moody, J.D., 1966. Crustal shear patterns and orogenesis. Tectonophysics 3 (6), 479–522.

Moon, S., Perron, J.T., Martel, S.J., Holbrook, W.S., St. Clair, J., 2017. A model of three-dimensional topographic stresses with implications for bed-rock fractures, surface processes, and landscape evolution. J. Geophys. Res. Earth Surf. 122 (4), 823–846.

Mooney, W., Laske, G., Master, T., 1998. CRUST 5.1: a global crustal model at 5×5. J. Geophys. Res. Solid Earth 103 (B1), 727–747.

Moonjun, R., Farshad, A., Shrestha, D.P., Vaiphasa, C., 2010. Artificial neural network and decision tree in predictive soil mapping of Hoi Num Rin sub-watershed, Thailand. In: Boettinger, J.L., Howell, D.W., Moore, A.C., Hartemink, A.E., Kienast-Brown, S. (Eds.), Digital Soil Mapping: Bridging Research, Environmental Application, and Operation. Progress in Soil Science Book Series. Springer, Dordrecht, the Netherlands, pp. 151–164.

Moore, I.D., Mackay, S.M., Wallbrink, P.J., Burch, G.J., O'Loughlin, E.M., 1986. Hydrologic characteristics and modelling of a small forested catchment in southeastern New South Wales. Pre-logging condition. J. Hydrol. 83 (3–4), 307–335.

Moore, I.D., Grayson, R.B., Ladson, A.R., 1991. Digital terrain modelling: a review of hydrological, geomorphological and biological applications. Hydrol. Process. 5 (1), 3–30.

Moore, I.D., Gessler, P.E., Nielsen, G.A., Peterson, G.A., 1993. Soil attribute prediction using terrain analysis. Soil Sci. Soc. Am. J. 57 (2), 443–452.

Moore, R.F., Simpson, C.J., 1983. Image analysis—a new aid in morphotectonic studies. In: Proceedings of the 17th International Symposium on Remote Sensing of Environment, Ann Arbor, MI, 9–13 May 1983, vol. 3. Environmental Research Institute of Michigan, Ann Arbor, MI, pp. 99[illegible]–1002.

Mor, M., Lamdan, T., 1972. A new approach to automatic scanning of contour maps. Commun. ACM 15 (9), 809–812.

Moran, C.J., Bui, E., 2002. Spatial data mining for enhanced soil map modelling. Int. J. Geogr. Inf. Sci. 16 (6), 533–542.

Mora-Vallejo, A., Claessens, L., Stoorvogel, J., Heuvelink, G.B.M., 2008. Small scale digital soil mapping in Southeastern Kenya. Catena 76 (1), 44–53.

Moreno-Gómez, M., Liedl, R., Stefan, C., Moreno-Gómez, M., Liedl, R., Stefan, C., 2019. A new GIS-based model for karst dolines mapping using LiDAR; application of a multidepth threshold approach in the Yucatan Karst, Mexico. Remote Sens. (Basel) 11 (10), 1147.

Morlighem, M., Rignot, E., Seroussi, H., Larour, E., Ben Dhia, H., Aubry, D., 2011. A mass conservation approach for mapping glacier ice thickness. Geophys. Res. Lett. 38 (19), L19503.

Morlighem, M., Williams, C.N., Rignot, E., An, L., Arndt, J.E., Bamber, J.L., et al., 2017. BedMachine v3: complete bed topography and ocean bathymetry mapping of Greenland from multibeam echo sounding combined with mass conservation. Geophys. Res. Lett. 44, 11051–11061.

Morlighem, M., Rignot, E., Binder, T., Blankenship, D., Drews, R., Eagles, G., et al., 2020. Deep glacial troughs and stabilizing ridges unveiled beneath the margins of the Antarctic ice sheet. Nat. Geosci. 13, 132–137.

Morozov, V.I., Kovalenko, A.P., Pasynkov, A.A., 1988. Sites of abnormally high discharges of the Crimean Mountains. Geologichesky J. (2), 65–69 (in Russian).

Morozov, V.P., 1979. A Course in Spheroidal Geodesy, second enl. rev. ed. Nedra, Moscow, USSR (in Russian).

Morris, K., 1991. Using knowledge-base rules to map the three-dimensional nature of geological features. Photogrammet. Eng. Remote Sens. 57 (9), 1209–1216.

Moudrý, V., Lecours, V., Gdulová, K., Gábor, L., Moudrá, L., Kropáček, J., et al., 2018. On the use of global DEMs in ecological modelling and the accuracy of new bare-earth DEMs. Ecol. Model. 383, 3–9.

Mourier, B., Walter, C., Merot, P., 2008. Soil distribution in valleys according to stream order. Catena 72 (3), 395–404.

Mudd, S.M., 2020. Topographic data from satellites. Dev. Earth Surf. Process. 23, 91–128.

Mueller-Dombois, D., Ellenberg, H., 1974. Aims and Methods of Vegetation Ecology. Wiley, New York, NY.

Mulder, V.L., de Bruin, S., Schaepman, M.E., Mayr, T.R., 2011. The use of remote sensing in soil and terrain mapping—a review. Geoderma 162 (1–2), 1–19.

Mulder, V.L., Lacoste, M., Richer-de-Forges, A.C., Arrouays, D., 2016a. GlobalSoilMap France: high-resolution spatial modelling the soils of France up to two meter depth. Sci. Total Environ. 573, 1352–1369.

Mulder, V.L., Lacoste, M., Richer-de-Forges, A.C., Martin, M.P., Arrouays, D., 2016b. National versus global modelling the 3D distribution of soil organic carbon in mainland France. Geoderma 263, 16–34.

Müller, R.D., Qin, X., Sandwell, D.T., Dutkiewicz, A., Williams, S.E., Flament, N., et al., 2016. The GPlates portal: cloud-based interactive 3D visualization of global geophysical and geological data in a web browser. PLoS One 11 (3), e0150883.

Muñoz-Espinoza, R.E., 1968. Fracture Finding by Structural Curvature Mapping (M.Sc. thesis). University of Texas at Austin, Austin, TX.

Murase, T., Tanaka, M., Tani, T., Miyashita, Y., Ohkawa, N., Ishiguro, S., et al., 2008. A photogrammetric correction procedure for light refraction effects at a two-medium boundary. Photogrammet. Eng. Remote Sens. 74 (9), 1129–1136.

Muratov, M.V. (Ed.), 1969. Geology of the USSR, vol. 8: Crimea, pt. 1: Geological Description. Nedra, Moscow, USSR (in Russian).

Murray Jr., G.H., 1968. Quantitative fracture study—Sanish Pool, McKenzie County, North Dakota. Am. Assoc. Pet. Geol. Bull. 52 (1), 57–65.

Murray, R.E., Parsons, L.L., Smith, M.S., 1989. Kinetics of nitrate utilization by mixed populations of denitrifying bacteria. Appl. Environ. Microbiol. 55 (3), 717–721.

Muthusamy, M., Rivas Casado, M., Butler, D., Leinster, P., 2021. Understanding the effects of digital elevation model resolution in urban fluvial flood modelling. J. Hydrol. 596, 126088.

Mynatt, I., Bergbauer, S., Pollard, D.D., 2007a. Using differential geometry to describe 3-D folds. J. Struct. Geol. 29 (7), 1256–1266.

Mynatt, I., Hilley, G.E., Pollard, D.D., 2007b. Inferring fault characteristics using fold geometry constrained by Airborne Laser Swath Mapping at Raplee Ridge, Utah. Geophys. Res. Lett. 34 (16), L16315.

Myrold, D.D., Tiedje, J.M., 1985. Establishment of denitrification capacity in soil: effects of carbon, nitrate and moisture. Soil Biol. Biochem. 17 (6), 819–822.

Myrzin, Y.N., Furinevich, O.S., Artemyeva, E.S., 1988. Report on Hydrogeological and Engineering Geological Surveys (Scale 1:200,000) for Areas of Topographic Sheets M-36-V, -VI, and -XII, 1984–1988. Centregeologia, Moscow, USSR (in Russian).

Na, J., Ding, H., Zhao, W., Liu, K., Tang, G., Pfeifer, N., 2021. Object-based large-scale terrain classification combined with segmentation optimization and terrain features: a case study in China. Trans. GIS 25 (6), 2939–2962.

Nachappa, T.G., Kienberger, S., Meena, S.R., Hölbling, D., Blaschke, T., 2020. Comparison and validation of per-pixel and object-based approaches for landslide susceptibility mapping. Geomat. Nat. Haz. Risk 11 (1), 572–600.

Nagy, G., 1994. Terrain visibility. Comput. Graph. 18 (6), 763–773.

Nagy-Reis, M.B., Estevo, C.A., Setz, E.Z.F., Ribeiro, M.C., Chiarello, A.G., Nichols, J.D., 2017a. Relative importance of anthropogenic landscape characteristics for Neotropical frugivores at multiple scales. Anim. Conserv. 20 (6), 520–531.

Nagy-Reis, M.B., Nichols, J.D., Chiarello, A.G., Ribeiro, M.C., Setz, E.Z.F., 2017b. Landscape use and co-occurrence patterns of neotropical spotted cats. PLoS One 12 (1), e0168441.

Nalivkin, D.V. (Ed.), 1970. Geological Map of the Russian Platform and Adjacent Areas, Scale 1:1,500,000. Pan-Union Aerial Geological Trust, Moscow, USSR (in Russian).

Nalivkin, D.V. (Ed.), 1974. Basement Structure of Platform Areas of the USSR. Nedra, Moscow, USSR (in Russian).

NASA, 2003–2023. NASA World Wind. National Aeronautics and Space Administration. http://worldwind.arc.nasa.gov. (Accessed 27 June 2024).

NASA, 2010. SRTM30 Documentation. National Aeronautics and Space Administration. Goddard Space Flight Center, Greenbelt, MD. https://icesat.gsfc.nasa.gov/icesat/tools/SRTM30_Documentation.html. (Accessed 27 June 2024).

NASA JPL, 2021. NASADEM Global Digital Elevation Model. OpenTopography, La Jolla, CA, https://doi.org/10.5069/G93T9FD9.

Natural Resources Canada, 1997. Canadian Digital Elevation Data: Standards and Specifications. Centre for Topographic Information, Natural Resources Canada, Sherbrooke, QC.

Nazari, H., Ritz, J.-F., Salamati, R., Shafei, A., Ghassemi, A., Michelot, J.-L., et al., 2009. Morphological and palaeoseismological analysis along the Taleghan Fault (Central Alborz, Iran). Geophys. J. Int. 178 (2), 1028–1041.

Neal, J., Hawker, L., 2023. FABDEM V1-2. University of Bristol Research Data Service, Bristol, UK, https://doi.org/10.5523/bris.s5hqmjcdj8yo2ibzi9b4ew3sn.

Neumann, G.A., 2008. Lunar Reconnaissance Orbiter LOLA Instrument Science Data Archive. NASA Planetary Data System Geosciences Node, Washington University, St. Louis, MO. http://pds-geosciences.wustl.edu/missions/lro/lola.htm (Accessed 27 June 2024).

Neustruev, S.S., 1915. On soil combinations in plain and mountainous terrains. Pochvovedenie 17 (1), 62–73 (in Russian).

Neustruev, S.S., 1927. Genesis of Soils. Academy of Sciences of the USSR, Leningrad, USSR.

Neustruev, S.S., 1930. Elements of Soil Geography. Selkhozgiz, Moscow, USSR (in Russian).

Nevalainen, P., Middleton, M., Sutinen, R., Heikkonen, J., Pahikkala, T., 2016. Detecting terrain stoniness from airborne laser scanning data. Remote Sens. (Basel) 8 (9), 720.

Newman, D.R., Lindsay, J.B., Cockburn, J.M.H., 2018. Evaluating metrics of local topographic position for multiscale geomorphometric analysis. Geomorphology 312, 40–50.

Nex, F., Remondino, F., 2013. UAV for 3D mapping applications: a review. Appl. Geomat. 6 (1), 1–15.

Nezametdinova, S.S., 1970. Strike Analysis of Regional Faults within Oil and Gas Bearing Areas (Exemplified by the Precaucasus) (Ph.D. thesis). Leningrad Mining Institute, Leningrad, USSR (in Russian).

Ngo, P.-T.T., Pham, T.D., Nhu, V.-H., Le, T.T., Tran, D.A., Phan, D.C., et al., 2021. A novel hybrid quantum-PSO and credal decision tree ensemble for tropical cyclone induced flash flood susceptibility mapping with geospatial data. J. Hydrol. 596, 125682.

Nhu, V.-H., Ngo, P.-T.T., Pham, T.D., Dou, J., Song, X., Hoang, N.-D., et al., 2020. A new hybrid firefly–PSO optimized random subspace tree intelligence for torrential rainfall-induced flash flood susceptible mapping. Remote Sens. (Basel) 12 (17), 2688.

Nico, G., Leva, D., Fortuny-Guasch, J., Antonello, G., Tarchi, D., 2005a. Generation of digital terrain models with a ground-based SAR system. IEEE Trans. Geosci. Remote Sens. 43 (1), 45–49.

Nico, G., Rutigliano, P., Benedetto, C., Vespe, R., 2005b. Terrain modelling by kinematical GPS survey. Nat. Haz. Earth Syst. Sci. 5 (2), 293–299.

Nicoll, T., Brierley, G., 2017. Within-catchment variability in landscape connectivity measures in the Garang catchment, upper Yellow River. Geomorphology 277, 197–209.

Nicu, I.C., Asăndulesei, A., 2018. GIS-based evaluation of diagnostic areas in landslide susceptibility analysis of Bahluieţ River Basin (Moldavian Plateau, NE Romania), are Neolithic sites in danger? Geomorphology 314, 27–41.

Niculiţă, M., 2016. Automatic landslide length and width estimation based on the geometric processing of the bounding box and the geomorphometric analysis of DEMs. Nat. Haz. Earth Syst. Sci. 16 (8), 2021–2030.

Niculiţă, M., 2020. Geomorphometric methods for burial mound recognition and extraction from high-resolution LiDAR DEMs. Sensors 20 (4), 1192.

Niculiţă, M., Mărgărint, M.C., Cristea, A.I., 2019. Using archaeological and geomorphological evidence for the establishment of a relative chronology and evolution pattern for Holocene landslides. PLoS One 14 (12), e0227335.

Nielson, G., Franke, R., 1984. A method for construction of surfaces under tension. Rocky Mount. J. Mathemat. 14 (1), 203–221.

Nikolova, V., Kamburov, A., Rizova, R., 2021. Morphometric analysis of debris flows basins in the Eastern Rhodopes (Bulgaria) using geospatial technologies. Nat. Hazards 105 (1), 159–175.

Nikolsky, S.M., 1977a. A Course of Mathematical Analysis, vol. 1. Mir, Moscow, USSR.

Nikolsky, S.M., 1977b. A Course of Mathematical Analysis, vol. 2. Mir, Moscow, USSR.

Nikonov, A.A., 1995. Active faults: definition and recognition problems. Geoecologiya (4), 16–27 (in Russian).

NIMA, 1997. World Geodetic System 1984, Its Definition and Relationships with Local Geodetic Systems. Technical Report TR8350.2, third ed. National Imagery and Mapping Agency, Bethesda, MD.

NIMA, 2000. Performance Specification: Digital Terrain Elevation Data (DTED). MIL-PRF-89020B, National Imagery and Mapping Agency, Fairfax, VA.

NOAA, 1988. Digital Relief of the Surface of the Earth. Data Announcement 88-MGG-02, NOAA, National Geophysical Data Center, Boulder, CO.

NOAA, 1993. 5-Minute Gridded Global Relief Data Collection (ETOPO5). National Oceanic and Atmospheric Administration, World Data Center for Geophysics & Marine Geology, Boulder, CO, https://doi.org/10.7289/V5D798BF.

NOAA, 1995. TerrainBase, Global 5 Arc-Minute Ocean Depth and Land Elevation. National Oceanic and Atmospheric Administration, National Geophysical Data Center, Boulder, CO, https://doi.org/10.5065/E08M-4482.

NOAA, 2001. ETOPO2, 2-Minute Gridded Global Relief Data. National Oceanic and Atmospheric Administration, National Centers for Environmental Information, Boulder, CO.

NOAA, 2006. 2-Minute Gridded Global Relief Data (ETOPO2) v2. National Oceanic and Atmospheric Administration, National Centers for Environmental Information, Boulder, CO, https://doi.org/10.7289/V5J1012Q.

NOAA, 2009. ETOPO1, 1 Arc-Minute Global Relief Model. National Oceanic and Atmospheric Administration, National Centers for Environmental Information, Boulder, CO, https://doi.org/10.7289/V5C8276M.

NOAA, 2022. ETOPO 2022 15 Arc-Second Global Relief Model. National Oceanic and Atmospheric Administration, National Centers for Environmental Information, Boulder, CO, https://doi.org/10.25921/fd45-gt74.

Noble, T.L., Rohling, E.J., Aitken, A.R.A., Bostock, H.C., Chase, Z., Gomez, N., et al., 2020. The sensitivity of the Antarctic Ice Sheet to a changing climate: past, present, and future. Rev. Geophys. 58 (4), e2019RG000663.

Noriega-Londoño, S., Restrepo-Moreno, S.A., Vinasco, C., Bermúdez, M.A., Min, K., 2020. Thermochronologic and geomorphometric constraints on the Cenozoic landscape evolution of the Northern Andes: northwestern Central Cordillera, Colombia. Geomorphology 351, 106890.

Nothard, S., McKenzie, D., Haines, J., Jackson, J., 1996. Gaussian curvature and the relationship between the shape and the deformation of the Tonga slab. Geophys. J. Int. 127 (2), 311–327.

Novaczek, E., Devillers, R., Edinger, E., 2019. Generating higher resolution regional seafloor maps from crowd-sourced bathymetry. PLoS One 14 (6), e0216792.

Nowosad, J., Stepinski, T.F., 2021. Pattern-based identification and mapping of landscape types using multi-thematic data. Int. J. Geogr. Inf. Sci. 35 (8), 1634–1649.

Nozette, S., Rustan, P., Pleasance, L.P., Horan, D.M., Regeon, P., Shoemaker, E.M., et al., 1994. The Clementine mission to the Moon: scientific overview. Science 266 (5192), 1835–1839.

Nussbaum, M., Spiess, K., Baltensweiler, A., Grob, U., Keller, A., Greiner, L., et al., 2018. Evaluation of digital soil mapping approaches with large sets of environmental covariates. Soil 4 (1), 1–22.

Nyrtsov, M.V., Bugaevsky, L.M., Stooke, P.J., 2007. The multiple axis ellipsoids as reference surfaces for mapping of small celestial bodies. In: Proceedings of the 23rd International Cartographic Conference, Moscow, Russia, 4–10 Aug. 2007. ICA, Moscow, Russia. 2.1.5.

Nyrtsov, M.V., Fleis, M.E., Borisov, M.M., Stooke, P.J., 2015. Equal-area projections of the triaxial ellipsoid: first time derivation and implementation of cylindrical and azimuthal projections for small solar system bodies. Cartogr. J. 52 (2), 114–124.

Nyrtsov, M.V., Fleis, M.E., Nikolaeva, D.D., Orendarchuk, A.V., 2018. Mapping of celestial bodies with bi-lobate shape in triaxial ellipsoid projections on the example of binary nucleus of the 67P/Churyumov–Gerasimenko comet. Izvestia Vuzov. Geodesy Aerophotosurveying 62 (5), 514–523 (in Russian).

O'Callaghan, J.F., Mark, D.M., 1984. The extraction of drainage networks from digital elevation data. Comput. Vis. Graph. Image Process. 28 (3), 323–344.

O'Driscoll, E.S.T., 1980. The double helix in global tectonics. Tectonophysics 63 (1–4), 397–417.

O'Driscoll, E.S.T., 1986. Observations of the lineament–ore relation. Philos. Trans. R. Soc. Lond. A Math. Phys. Eng. Sci. 317 (1539), 195–218.

O'Leary, D.W., Friedman, J.D., Pohn, H.A., 1976. Lineament, linear, lineation: some proposed new standards for old terms. Geol. Soc. Am. Bull. 87 (10), 1463–1469.

O'Loughlin, E.M., 1981. Saturation regions in catchments and their relations to soil and topographic properties. J. Hydrol. 53 (3–4), 229–246.

O'Loughlin, F.E., Paiv, R.C.D., Durand, M., Alsdorf, D.E., Bates, P.D.A., 2016. A multi-sensor approach towards a global vegetation corrected SRTM DEM product. Remote Sens. Environ. 182, 49–59.

O'Neil, G.L., Goodall, J.L., Watson, L.T., 2018. Evaluating the potential for site-specific modification of LiDAR DEM derivatives to improve environmental planning-scale wetland identification using Random Forest classification. J. Hydrol. 559, 192–208.

O'Neil, G.L., Saby, L., Band, L.E., Goodall, J.L., 2019. Effects of LiDAR DEM smoothing and conditioning techniques on a topography-based wetland identification model. Water Resour. Res. 55 (5), 4343–4363.

O'Reilly, D., Evans, D., Shewan, L., 2017. Airborne LiDAR prospection at Lovea, an Iron Age moated settlement in central Cambodia. Antiquity 91 (358), 947–965.

Odeh, I.O.A., Chittleborough, D.J., McBratney, A.B., 1991. Elucidation of soil–landform interrelationships by canonical ordination analysis. Geoderma 49 (1–2), 1–32.

Odeh, I.O.A., Chittleborough, D.J., McBratney, A.B., 1992a. Soil pattern recognition with fuzzy-*c*-means: application to classification and soil–landform interrelationships. Soil Sci. Soc. Am. J. 56 (2), 505–516.

Odeh, I.O.A., McBratney, A.B., Chittleborough, D.J., 1992b. Fuzzy-*c*-means and kriging for mapping soil as a continuous system. Soil Sci. Soc. Am. J. 56 (6), 1848–1854.

Odeh, I.O.A., McBratney, A.B., Chittleborough, D.J., 1994. Spatial prediction of soil properties from landform attributes derived from a digital elevation model. Geoderma 63 (3–4), 197–214.

Odeh, I.O.A., McBratney, A.B., Chittleborough, D.J., 1995. Further results on prediction of soil properties from terrain attributes: heterotopic cokriging and regression-kriging. Geoderma 67 (3–4), 215–226.

Odeh, I.A., Crawford, M., McBratney, A.B., 2007. Digital mapping of soil attributes for regional and catchment modelling, using ancillary covariates, statistical and geostatistical techniques. In: Lagacherie, P., McBratney, A.B., Voltz, M. (Eds.), Digital Soil Mapping: An Introductory Perspective. Developments in Soil Science, vol. 31. Elsevier, Amsterdam, the Netherlands, pp. 437–453.

OGC, 2010. OpenGIS Web Map Tile Service Implementation Standard OGC 07-057r7. Open Geospatial Consortium, Arlington, VA. https://www.ogc.org/standard/wmts/. (Accessed 27 June 2024).

Oke, T.R., 2002. Boundary Layer Climates, second ed. Routledge, New York, NY.

Oksanen, J., Sarjakoski, T., 2005. Error propagation of DEM-based surface derivatives. Comput. Geosci. 31 (8), 1015–1027.

Oliver, M.A., Webster, R., 1986. Combining nested and linear sampling for determining the scale and form of spatial variation of regionalized variables. Geogr. Anal. 18 (3), 227–242.

Ollier, C., 1981. Tectonics and Landforms. Longman, London, UK.

Olson, C.J., Becker, J.J., Sandwell, D.T., 2014. A new global bathymetry map at 15 arcsecond resolution for resolving seafloor fabric: SRTM15_PLUS. In: AGU Fall Meeting 2014, San Francisco, CA, 15–19 Dec. 2014. American Geophysical Union, San Francisco, CA. OS34A-03.

Onorati, G., Poscolieri, M., Salvi, S., Trigila, R., 1987. Use of TM Landsat data as a support to classical ground-based methodologies in the investigation of a volcanic site in Central Italy the Caldera of Latera. In: IGARSS'87. Remote Sensing: Understanding the Earth as a System: Proceedings of the IEEE International Geoscience and Remote Sensing Symposium, Ann Arbor, MI, 18–21 May 1987, vol. 2. Institute of Electrical and Electronics Engineers, New York, NY, pp. 1173–1178.

Onorati, G., Poscolieri, M., Ventura, R., Chiarini, V., Crucilla, U., 1992. The digital elevation model of Italy for geomorphology and structural geology. Catena 19 (2), 147–178.

Oren, M., Nayar, S.K., 1994. Generalization of Lambert's reflectance model. In: SIGGRAPH'94: Proceedings of the 21st Annual Conference on Computer Graphics and Interactive Techniques, Orlando, FL, 24–29 July 1994. ACM, New York, NY, pp. 239–246.

Orlandini, S., Moretti, G., 2009. Determination of surface flow paths from gridded elevation data. Water Resour. Res. 45 (3), W03417.

Orlandini, S., Moretti, G., Corticelli, M.A., Santangelo, P.E., Capra, A., Rivola, R., et al., 2012. Evaluation of flow direction methods against field observations of overland flow dispersion. Water Resour. Res. 48 (10), W10523.

Orlandini, S., Moretti, G., Gavioli, A., 2014. Analytical basis for determining slope lines in grid digital elevation models. Water Resour. Res. 50 (1), 526–539.

Orlova, A.V., 1975. Block Structures and Topography. Nedra, Moscow, USSR (in Russian).

Orti, M.V., Negussie, K., Corral-Pazos-de-Provens, E., Höfle, B., Bubenzer, O., 2019. Comparison of three algorithms for the evaluation of TanDEM-X data for gully detection in Krumhuk Farm (Namibia). Remote Sens. (Basel) 11 (11), 1327.

Ototzky, P.V., 1901. On the relationship between elevation and the character of chernozems in the Poltava Governorate. Pochvovedenie 3 (2), 197–206 (in Russian).

Owono, F.M., Ntamak-Nida, M.-J., Dauteuil, O., Guillocheau, F., Njom, B., 2016. Morphology and long-term landscape evolution of the South African Plateau in South Namibia. Catena 142, 47–65.

Özyeşil, O., Voroninski, V., Basri, R., Singer, A., 2017. A survey of structure from motion. Acta Num. 26, 305–364.

Pacheco-Ruiz, R., Adams, J., Pedrotti, F., 2018. 4D modelling of low visibility underwater archaeological excavations using multi-source photogrammetry in the Bulgarian Black Sea. J. Archaeol. Sci. 100, 120–129.

Pachepsky, Y., Hill, R.L., 2017. Scale and scaling in soils. Geoderma 287, 4–30.

Pachepsky, Y.A., Timlin, D.J., Rawls, W.J., 2001. Soil water retention as related to topographic variables. Soil Sci. Soc. Am. J. 65 (6), 1787–1795.

Pacina, J., 2010. Comparison of approximation methods for partial derivatives of surfaces on regular grids. Geomorph. Slovaca Bohemica 1 (10), 25–32.

Padarian, J., Minasny, B., McBratney, A.B., 2019. Using deep learning for digital soil mapping. Soil 5 (1), 79–89.

Pajares, G., 2015. Overview and current status of remote sensing applications based on unmanned aerial vehicles (UAVs). Photogrammet. Eng. Remote Sens. 81 (4), 281–329.

Pajarola, R., Gobbetti, E., 2007. Survey of semi-regular multiresolution models for interactive terrain rendering. Vis. Comput. 23 (8), 583–605.

Palyvos, N., Bantekas, I., Kranis, H., 2006. Transverse fault zones of subtle geomorphic signature in northern Evia Island (central Greece extensional province): an introduction to the Quaternary Nileas graben. Geomorphology 76 (3–4), 363–374.

Pánek, T., Břežný, M., Kapustová, V., Lenart, J., Chalupa, V., 2019. Large landslides and deep-seated gravitational slope deformations in the Czech Flysch Carpathians: new LiDAR-based inventory. Geomorphology 346, 106852.

Panin, A.V., Gelman, R.N., 1997. Experience of the GPS technique application to derive detailed digital terrain models. Geodesy Cartogr. (10), 22–27 (in Russian).

Pankratov, A.N., 2004a. Algebraic Operations Over Orthogonal Series in Problems of Data Processing (Ph.D. thesis). Institute of Mathematical Problems of Biology, Russian Academy of Sciences, Pushchino, Russia (in Russian).

Pankratov, A.N., 2004b. On the implementation of algebraic operations on orthogonal function series. Comput. Math. Math. Phys. 44 (12), 2017–2023.

Pankratov, A.N., Britenkov, A.K., 2004. A generalized spectral analytical method: problems of description of digital data by families of orthogonal polynomials. Vestnik Lobachevsky State Univ. Nizhny Novgorod: Ser. Radiophys. (2), 5–14 (in Russian).

Pankratov, A.N., Kulikova, L.I., 2006. On the derivatives calculation under uniform and mean square approximation of signals. In: Computer Applications in Scientific Researches. Proceedings of the IVTN-2006 Session. NC Group/NVK VIST, Moscow, Russia. 0606 (in Russian).

Pankratov, A.N., Tetuev, R.K., 2023. Algebra of orthogonal series. Pattern Recogn. Image Anal. 33 (4), 1309–1314.

Pankratov, A.N., Pyatkov, M.I., Tetuev, R.K., Kulikova, L.I., 2011. Algorithms of Spectral Analysis Using Intel IPP and MKL Libraries. Intel – Moscow State University, Moscow, Russia (in Russian).

Pankratov, A.N., Tetuev, R.K., Pyatkov, M.I., Toigildin, V.P., Popova, N.N., 2015. Spectral analytical method of recognition of inexact repeats in character sequences. Proc. Inst. Syst. Program. Russ. Acad. Sci. 27 (6), 335–344.

Pannekoek, A.J., 1967. Generalized contour maps, summit level maps, and streamline surface maps as geomorphological tools. Zeit. Geomorph. 11 (2), 169–182.

Panou, G., 2013. The geodesic boundary value problem and its solution on a triaxial ellipsoid. J. Geodet. Sci. 3 (3), 240–249.

Panou, G., Korakitis, R., Delikaraoglou, D., 2016. Triaxial coordinate systems and their geometrical interpretation. In: Fotiou, A., Paraschakis, I., Rossikopoulos, D. (Eds.), Measuring and Mapping the Earth. Ziti, Thessaloniki, Greece, pp. 126–135.

Papageorgaki, I., Nalbantis, I., 2016. Classification of drainage basins based on readily available information. Water Resour. Manage. 30 (15), 5559–5574.

Papo, H.B., Gelbman, E., 1984. Digital terrain models for slopes and curvatures. Photogrammet. Eng. Remote Sens. 50 (6), 695–701.

Papworth, H., Ford, A., Welham, K., Thackray, D., 2016. Assessing 3D metric data of digital surface models for extracting archaeological data from archive stereo-aerial photographs. J. Archaeol. Sci. 72, 85–104.

Paraskevas, T., 2011. Virtual globes and geological modeling. Int. J. Geosci. 2 (4), 648–656.

Pardo-Igúzquiza, E., Dowd, P.A., 2021. Fractal analysis of the Martian landscape: a study of kilometre-scale topographic roughness. Icarus 372, 114727.

Park, S.J., van de Giesen, N., 2004. Soil–landscape delineation to define spatial sampling domains for hillslope hydrology. J. Hydrol. 295 (1–4), 28–46.

Park, S.J., McSweeney, K., Lowery, B., 2001. Identification of the spatial distribution of soils using a process-based terrain characterization. Geoderma 103 (3–4), 249–272.

Parsons, L.L., Murray, R.E., Smith, M.S., 1991. Soil denitrification dynamics: spatial and temporal variations of enzyme activity, populations, and nitrogen gas loss. Soil Sci. Soc. Am. J. 55 (1), 90–95.

Patterson, T., 2001. DEM manipulation and 3-D terrain visualization: techniques used by the US National Park Service. Cartographica 38 (1–2), 89–101.

Patterson, T., Jenny, B., 2011. The development and rationale of cross-blended hypsometric tints. Cartogr. Perspect. 69, 31–45.

Patton, N.R., Lohse, K.A., Godsey, S.E., Crosby, B.T., Seyfried, M.S., 2018. Predicting soil thickness on soil mantled hillslopes. Nat. Commun. 9, 3329.

Patton, N.R., Ellerton, D., Shulmeister, J., 2019. High-resolution remapping of the coastal dune fields of south East Queensland, Australia: a morphometric approach. J. Maps 15 (2), 578–589.

Pavlis, N.K., Holmes, S.A., Kenyon, S.C., Factor, J.K., 2012. The development and evaluation of the Earth Gravitational Model 2008 (EGM2008). J. Geophys. Res. Solid Earth 117 (B4), B04406.

Payne, W.J., 1981. Denitrification. Wiley, New York, NY.

Pe'eri, S., Gardner, J.V., Ward, L.G., Morrison, J.R., 2011. The seafloor: a key factor in lidar bottom detection. IEEE Trans. Geosci. Remote Sens. 49 (3), 1150–1157.

Pearce, M.A., Jones, R.R., Smith, S.A.F., McCaffrey, K.J.W., Clegg, P., 2006. Numerical analysis of fold curvature using data acquired by high-precision GPS. J. Struct. Geol. 28 (9), 1640–1646.

Pearce, M.A., Jones, R.R., Smith, S.A.F., McCaffrey, K.J.W., 2011. Quantification of fold curvature and fracturing using terrestrial laser scanning. Am. Assoc. Pet. Geol. Bull. 95 (5), 771–794.

Peckham, S.D., Stoica, M., Jafarov, E., Endalamaw, A., Bolton, W.R., 2017. Reproducible, component-based modeling with TopoFlow, a spatial hydrologic modeling toolkit. Earth Space Sci. 4 (6), 377–394.

Pedersen, G.B.M., 2016. Semi-automatic classification of glaciovolcanic landforms: an object-based mapping approach based on geomorphometry. J. Volcanol. Geotherm. Res. 311, 29–40.

Pedersen, G.B.M., Grosse, P., Gudmundsson, M.T., 2020. Morphometry of glaciovolcanic edifices from Iceland: types and evolution. Geomorphology 370, 107334.

Pei, T., Qin, C.-Z., Zhu, A.-X., Yang, L., Luo, M., Li, B., Zhou, C., 2010. Mapping soil organic matter using the topographic wetness index: a comparative study based on different flow-direction algorithms and kriging methods. Ecol. Indic. 10 (3), 610–619.

Peive, A.V., 1956a. The main types of deep faults. 1. General characteristics, classification, and spatial distribution of deep faults. Izvestiya Acad. Nauk SSSR Geol. Ser. (1), 90–105 (in Russian).

Peive, A.V., 1956b. The main types of deep faults. 2. Relationships of sedimentation, folding, magmatism, and ore deposits with deep faults. Izvestiya Acad. Nauk SSSR Geol. Ser. (3), 57–71 (in Russian).

Pejović, M., Nikolić, M., Heuvelink, G.B.M., Hengl, T., Kilibarda, M., Bajat, B., 2018. Sparse regression interaction models for spatial prediction of soil properties in 3D. Comput. Geosci. 118, 1–13.

Penck, W., 1924. Die Morphologische Analyse. Ein Kapital der Physikalischen Geologie. J. Engelhorns Nachfolger, Stuttgart, Germany.

Penizek, V., Boruvka, L., 2008. The digital terrain model as a tool for improved delineation of alluvial soils. In: Hartemink, A.E., McBratney, A., Mendonça-Santos, M.L. (Eds.), Digital Soil Mapping with Limited Data. Springer, Dordrecht, the Netherlands, pp. 319–326.

Penížek, V., Zádorová, T., Kodešová, R., Vaněk, A., 2016. Influence of elevation data resolution on spatial prediction of colluvial soils in a Luvisol region. PLoS One 11 (11), e0165699.

Pennock, D.J., Corre, M.D., 2001. Development and application of landform segmentation procedures. Soil Tillage Res. 58 (3–4), 151–162.

Pennock, D.J., Zebarth, B.J., de Jong, E., 1987. Landform classification and soil distribution in hummocky terrain, Saskatchewan, Canada. Geoderma 40 (3–4), 297–315.

Pennock, D.J., van Kessel, C., Farrel, R.E., Sutherland, R.A., 1992. Landscape-scale variations in denitrification. Soil Sci. Soc. Am. J. 56 (3), 770–776.

Pereberin, A.V., 2001. Systematization of wavelet transforms. Num. Methods Program. 2 (3), 15–40 (in Russian).

Pereira, V., FitzPatrick, E.A., 1998. Three-dimensional representation of tubular horizons in sandy soils. Geoderma 81 (3–4), 295–303.

Petrasova, A., Harmon, B., Petras, V., Tabrizian, P., Mitasova, H., 2018. Tangible Modeling with Open Source GIS, second ed. Springer, Cham, Switzerland.

Petsch, C., Perondi, C., da Rosa, K.K., Vieira, R., de Souza Robaina, L.E., Trentin, R., 2022. Método de geomorphon aplicado à Antártica: identificação de formas de relevo paraglaciais de macro e mesoescala. Ilha Rei George, Ilhas Shetlands do Sul. Cuad. Geogr. Rev. Colomb. Geogr. 31 (1), 108–127.

Pettengill, G.H., Ford, P.G., Johnson, W.T.K., Raney, R.K., Soderblom, L.A., 1991. Magellan: radar performance and data products. Science 252 (5003), 260–265.

Pettinari, M.L., Chuvieco, E., 2020. Fire danger observed from space. Surv. Geophys. 41 (6), 1437–1459.

Peucker, K., 1898. Schattenplastik und Farbenplastik. Beiträge zur Geschichte und Theorie der Geländedarstellung. Artaria, Wien, Austria.

Peucker, T.K., 1980a. The impact of different mathematical approaches to contouring. Cartographica 17 (2), 73–95.

Peucker, T.K., 1980b. The use of computer graphics for displaying data in three dimensions. Cartographica 17 (2), 59–72.

Peucker, T.K., Chrisman, N., 1975. Cartographic data structures. Am. Cartograph. 2 (1), 55–69.

Peucker, T.K., Fowler, R.J., Little, J.J., Mark, D.M., 1976. Digital Representation of Three-Dimensional Surfaces by Triangulated Irregular Networks (TIN). Technical Report No. 10, USA Contract N00014-75-C-0886, Department of Geography, Simon Fraser University, Burnaby, BC.

Pfeifer, N., Glira, P., Briese, C., 2012. Direct georeferencing with on board navigation components of light weight UAV platforms. In: Shortis, M., Wagner, W., Hyyppä, J. (Eds.), XXII ISPRS Congress, Technical Commission VII, Melbourne, Australia, 25 Aug. – 1 Sept. 2012. Int. Arch. Photogramm. Remote Sens. Spatial Inf. Sci. XXXIX (B7), pp. 487–492.

Pflug, R., Harbaugh, J.W. (Eds.), 1992. Computer Graphics in Geology: Three-Dimensional Computer Graphics in Modeling Geologic Structures and Simulating Geologic Processes. Lect. Notes Earth Sci., vol. 41. Springer, Berlin, Germany.

Phillips, G.M., 2003. Interpolation and Approximation by Polynomials. CMS Books in Mathematics Book Series. Springer, New York, NY.

Phillips, J.D., 1988. The role of spatial scale in geomorphic systems. Geogr. Anal. 20 (4), 308–317.

Phillips, J.D., 1995. Biogeomorphology and landscape evolution: the problem of scale. Geomorphology 13 (1–4), 337–347.

Phong, B.T., 1975. Illumination for computer generated pictures. Commun. ACM 18 (6), 311–317.

Piccini, C., Marchetti, A., Rivieccio, R., Napoli, R., 2019. Multinomial logistic regression with soil diagnostic features and land surface parameters for soil mapping of Latium (Central Italy). Geoderma 352, 385–394.

Piedallu, C., Pedersoli, E., Chaste, E., Morneau, F., Seynave, I., Gégout, J.-C., 2022. Optimal resolution of soil properties maps varies according to their geographical extent and location. Geoderma 412, 115723.

Piermattei, L., Carturan, L., de Blasi, F., Tarolli, P., Dalla Fontana, G., Vettore, A., et al., 2016. Suitability of ground-based SfM–MVS for monitoring glacial and periglacial processes. Earth Surf. Dyn. 4 (2), 425–443.

Pijl, A., Bailly, J.-S., Feurer, D., El Maaoui, M.A., Boussema, M.R., Tarolli, P., 2020. TERRA: Terrain Extraction from elevation Rasters through Repetitive Anisotropic filtering. Int. J. Appl. Earth Obs. Geoinf. 84, 101977.

Pike, R.J., 1995. Geomorphometry—progress, practice, and prospect. Zeit. Geomorph. Suppl. 101, 221–238.

Pike, R.J., 2000. Geomorphometry—diversity in quantitative surface analysis. Progr. Phys. Geogr. 24 (1), 1–20.

Pike, R.J., 2001. Digital terrain modeling and industrial surface metrology: converging realms. Profess. Geogr. 53 (2), 263–274.

Pike, R.J., Evans, I.S., Hengl, T., 2009. Geomorphometry: a brief guide. In: Hengl, T., Reuter, H.I. (Eds.), Geomorphometry: Concepts, Software, Applications. Developments in Soil Science, vol. 33. Elsevier, Amsterdam, the Netherlands, pp. 3–30.

Pina, P., Vieira, G., 2022. UAVs for science in Antarctica. Remote Sens. (Basel) 14 (7), 1610.

Piñol, J., Beven, K., Freer, J., 1997. Modelling the hydrological response of Mediterranean catchments, Prades, Catalonia. The use of distributed models as aids to hypothesis formulation. Hydrol. Process. 11 (9), 1287–1306.

Pipaud, I., Lehmkuhl, F., 2017. Object-based delineation and classification of alluvial fans by application of mean-shift segmentation and support vector machines. Geomorphology 293 (A), 178–200.

Pirson, S.J., 1970. Geologic Well Log Analysis. Gulf, Houston, TX.

Pirti, A., Yucel, M.A., Gumus, K., 2013. Testing real time kinematic GNSS (GPS and GPS/GLONASS) methods in obstructed and unobstructed sites. Geod. Vest. 57 (3), 498–512.

Plattner, C., Braun, L.N., Brenning, A., 2004. Spatial variability of snow accumulation on Vernagtferner, Austrian Alps, in winter 2003/2004. Zeit. Gletscherkunde Glazialgeol. 39, 43–57.

Plewes, L.A., Hubbard, B., 2001. A review of the use of radio-echo sounding in glaciology. Progr. Phys. Geogr. 25 (2), 203–236.

Podda, S., Melis, M.T., Collu, C., Demurtas, V., Perseu, F.O., 2020. New morphometric data of lunar sinuous rilles. IEEE J. Sel. Topics Appl. Earth Obs. Remote Sens. 13, 3304–3316.

Podgórski, J., Kinnard, C., Pętlicki, M., Urrutia, R., 2019. Performance assessment of TanDEM-X DEM for mountain glacier elevation change detection. Remote Sens. (Basel) 11 (2), 187.

Poggio, L., Gimona, A., 2017. Assimilation of optical and radar remote sensing data in 3D mapping of soil properties over large areas. Sci. Total Environ. 579, 1094–1110.

Pogorelov, A.V., 1957. Differential Geometry. Noordhoff, Groningen, the Netherlands.

Pogorelov, A.V., Boyko, E.S., Petrakov, D.A., Kiselev, E.N., 2017. Fluctuations of the Fisht Glacier (West Caucasus) over 1909–2015. Ice Snow 57 (4), 498–506 (in Rissian).

Polar Geospatial Center, 2023. Guide: PGC's DEM Products—ArcticDEM, REMA, and EarthDEM. Polar Geospatial Center, University of Minnesota, Saint Paul, MN. https://www.pgc.umn.edu/guides/stereo-derived-elevation-models/pgcs-dem-products-arcticdem-rema-and-earthdem/. (Accessed 27 June 2024).

Poletaev, A.I., 1992. Fault Intersections of the Earth's Crust. Geoinformmark, Moscow, Russia (in Russian).

Poletaev, A.I., Kats, Y.G., Leonov, N.N., 1991. Revealing of active fault and lineament structures within the area of the Smolensk Nuclear Power Plant by visual and automated analysis. In: Mozhaev, B.N. (Ed.), Digital Processing of Video Information in Structural Geological and Seismotectonic Studies. Aerogeologia, Leningrad, USSR, pp. 42–55 (in Russian).

Poletaev, A.I., Tevelev, A.V., Bryantseva, G.V., Blyumkina, N.V., 1992. Study of Fault and Lineament Structure of the Area of the Kursk Nuclear Power Plant, Report No. 271, Pt. 1. Lomonosov Moscow State University, Moscow, Russia (in Russian).

Polidori, L., 1997. Cartographie Radar. Gordon & Breach, Amsterdam, the Netherlands.

Polidori, L., El Hage, M., 2020. Digital elevation model quality assessment methods: a critical review. Remote Sens. (Basel) 12 (21), 3522.

Polidori, L., Chorowicz, J., Guillande, R., 1991. Description of terrain as a fractal surface, and application to digital elevation model quality assessment. Photogrammet. Eng. Remote Sens. 57, 1329–1332.

Pollard, D.D., Fletcher, R.C., 2005. Fundamentals of Structural Geology. Cambridge University Press, Cambridge, MA.

Pollard, D.D., Bergbauer, S., Mynatt, I., 2004. Using differential geometry to characterize and analyse the morphology of joints. In: Cosgrove, J.W. (Ed.), The Initiation, Propagation, and Arrest of Joints and Other Fractures. Geol. Soc. London Spec. Publ., vol. 147, pp. 153–182.

Polyanskaya, L.M., Zvyagintsev, D.G., 1995. Microbial succession in soil. Physiol. Gen. Biol. Rev. 9 (1), 1–68.

Ponagaibo, N.D., 1915. Influence of Microtopography on Soil Character, Its Temperature, Moisture, and Productivity. Frishberg Press, Poltava, Russia (in Russian).

Popa, C.N., Knitter, D., 2016. From environment to landscape. Reconstructing environment perception using numerical data. J. Archaeol. Method Theory 23 (4), 1285–1306.

Popov, A.B., Minár, J., Gallay, M., 2021. Multiresolution land surface segmentation and generalization of DEM: searching for optimal settings. Trans. GIS 25 (5), 2376–2393.

Poppiel, R.R., Lacerda, M.P.C., Safanelli, J.L., Rizzo, R., Oliveira Jr., M.P., Novais, J., et al., 2019. Mapping at 30 m resolution of soil attributes at multiple depths in midwest Brazil. Remote Sens. (Basel) 11 (24), 2905.

Poppiel, R.R., Melo Demattê, J.A., Rosin, N.A., Campos, L.R., Tayebi, M., Bonfatti, B.R., et al., 2021. High resolution Middle Eastern soil attributes mapping via open data and cloud computing. Geoderma 385, 114890.

Porter, C., Morin, P., Howat, I., Noh, M.-J., Bates, B., Peterman, K., et al., 2018. ArcticDEM, Version 3. Harvard Dataverse, V1., https://doi.org/10.7910/DVN/OHHUKH.

Potuckova, M., 2004. Image Matching and Its Applications in Photogrammetry (Ph.D. thesis). Institut for Samfundsudvikling og Planlægning, Aalborg Universitet, Aalborg, Denmark.

Pourghasemi, H.R., Rahmati, O., 2018. Prediction of the landslide susceptibility: which algorithm, which precision? Catena 162, 177–192.

Pourghasemi, H.R., Yousefi, S., Kornejady, A., Cerdà, A., 2017. Performance assessment of individual and ensemble data-mining techniques for gully erosion modeling. Sci. Total Environ. 609, 764–775.

Pourtaghi, Z.S., Pourghasemi, H.R., Aretano, R., Semeraro, T., 2016. Investigation of general indicators influencing on forest fire and its susceptibility modelling using different data mining techniques. Ecol. Indic. 64, 72–84.

Pradervand, J.-N., Dubuis, A., Pellissier, L., Guisan, A., Randin, C., 2014. Very high resolution environmental predictors in species distribution models: moving beyond topography? Progr. Phys. Geogr. 38 (1), 79–96.

Pratt, W.K., 2001. Digital Image Processing, third ed. Wiley, New York, NY.

Press, W.H., Teukolsky, S.A., Vetterling, W.T., Flannery, B.P., 1992. Numerical Recipes in C: The Art of Scientific Computing, second ed. Cambridge University Press, Cambridge, UK.

Preusker, F., Scholten, F., Matz, K.-D., Roatsch, T., Willner, K., Hviid, S.F., et al., 2015. Shape model, reference system definition, and cartographic mapping standards for comet 67P/Churyumov-Gerasimenko—stereo-photogrammetric analysis of Rosetta/OSIRIS image data. Astron. Astrophys. 583, A33.

Prima, O.D.A., Echigo, A., Yokoyama, R., Yoshida, T., 2006. Supervised landform classification of northeast Honshu from DEM-derived thematic maps. Geomorphology 78 (3–4), 373–386.

Puliti, S., Hauglin, M., Breidenbach, J., Montesano, P., Neigh, C.S.R., Rahlf, J., et al., 2020. Modelling above-ground biomass stock over Norway using national forest inventory data with ArcticDEM and Sentinel-2 data. Remote Sens. Environ. 236, 111501.

Purinton, B., Bookhagen, B., 2017. Validation of digital elevation models (DEMs) and comparison of geomorphic metrics on the southern Central Andean Plateau. Earth Surf. Dyn. 5 (2), 211–237.

Pustovitenko, B.G., Trostnikov, V.N., 1977. Towards a relationship between seismic processes and tectonics in Crimea. Geophys. Commun. 77, 13–23 (in Russian).

Puzachenko, Y.G., Onufrenya, I.A., Aleshchenko, G.M., 2002. Analysis of hierarchical organization of topography. Izvestiya Acad. Nauk Geogr. Ser. (4), 29–38 (in Russian).

Puzachenko, M.Y., Puzachenko, Y.G., Kozlov, D.N., Fedyaeva, M.V., 2006. Mapping of the thickness of organogenic and humic horizons of forest soils and bogs in the southern taiga (south-west of the Valdai Hills) using three-dimensional topographic model and remotely sensed data (Landsat 7). Issledovanie Zemli iz Cosmosa (4), 1–9 (in Russian).

QGIS, 2009–2023. QGIS: A Free and Open Source Geographic Information System. QGIS Project. http://qgis.org. (Accessed 27 June 2024).

Qi, F., Zhu, A., 2006. Fuzzy soil mapping based on prototype category theory. Geoderma 136 (3–4), 774–787.

Qiao, G., Yuan, X., Florinsky, I., He, Y., Li, H., 2023. Topography reconstruction and evolution analysis of outlet glacier using data from unmanned aerial vehicles in Antarctica. Int. J. Appl. Earth Obs. Geoinf. 117, 103186.

Qin, C.-Z., Bao, L.-L., Zhu, A.-X., Hu, X.-M., Qin, B., 2013. Artificial surfaces simulating complex terrain types for evaluating grid-based flow direction algorithms. Int. J. Geogr. Inf. Sci. 27 (6), 1055–1072.

Qin, C.-Z., Wu, X.-W., Jiang, J.-C., Zhu, A.-X., 2016. Case-based formalization and reasoning method for knowledge in digital terrain analysis: application to extracting drainage networks. Hydrol. Earth Syst. Sci. 20 (8), 3379–3392.

Qin, C.-Z., Ai, B.-B., Zhu, A.-X., Liu, J.-Z., 2017. An efficient method for applying a differential equation to deriving the spatial distribution of specific catchment area from gridded digital elevation models. Comput. Geosci. 100, 94–102.

Qin, L., Xu, W., Tian, Y., Chen, B., Wang, S., 2018. A river channel extraction method for urban environments based on terrain transition lines. Water Resour. Res. 54 (7), 4887–4900.

Qin, N., Tan, W., Guan, H., Wang, L., Ma, L., Tao, P., et al., 2023. Towards intelligent ground filtering of large-scale topographic point clouds: a comprehensive survey. Int. J. Appl. Earth Obs. Geoinf. 125, 103566.

Queen, C.W., Nelson, F.E., Gunn, G.E., Nyland, K.E., 2021. A characteristic periglacial landform: automated recognition and delineation of cryoplanation terraces in eastern Beringia. Permafr. Periglac. Process. 32 (1), 35–46.

Quijano, L., Beguería, S., Gaspar, L., Navas, A., 2016. Estimating erosion rates using ^{137}Cs measurements and WATEM/SEDEM in a Mediterranean cultivated field. Catena 138, 38–51.

Quinn, P.F., Beven, K.J., 1993. Spatial and temporal prediction of soil moisture dynamics, runoff, variable source areas and evapotranspiration for Plynlimon, Mid-Wales. Hydrol. Process. 7 (4), 425–448.

Quinn, P.F., Beven, K.J., Chevallier, P., Planchon, O., 1991. The prediction of hillslope flowpaths for distributed modelling using digital terrain models. Hydrol. Process. 5 (1), 59–80.

Quinn, P., Beven, K., Lamb, R., 1995. The ln(a/tanβ) index: how to calculate it and how to use it within the TOPMODEL framework. Hydrol. Process. 9 (2), 161–182.

Quiroga, E., Bertoni, C., van Goethem, M., Blazevic, L.A., Ruden, F., 2022. A 3D geological model of the Horn of Africa: new insights for hydrogeological simulations of deep groundwater systems. J. Hydrol. Reg. Stud. 42, 101166.

Rabus, B., Eineder, M., Roth, A., Bamler, R., 2003. The Shuttle Radar Topography Mission—a new class of digital elevation models acquired by spaceborne radar. ISPRS J. Photogrammet. Remote Sens. 57 (4), 241–262

Ragettli, S., Bolch, T., Pellicciotti, F., 2016. Heterogeneous glacier thinning patterns over the last 40 years in Langtang Himal, Nepal. Cryosphere 10 (5), 2075–2097.

Rahmati, O., Pourghasemi, H.R., Melesse, A.M., 2016. Application of GIS-based data driven random forest and maximum entropy models for groundwater potential mapping: a case study at Mehran Region, Iran. Catena 137, 360–372.

Rahmati, O., Ghorbanzadeh, O., Teimurian, T., Mohammadi, F., Tiefenbacher, J.P., Falah, F., et al., 2019. Spatial modeling of snow avalanche using machine learning models and geo-environmental factors: comparison of effectiveness in two mountain regions. Remote Sens. (Basel) 11 (24), 2995.

Ramasundaram, V., Grunwald, S., Mangeot, A., Comerford, N.B., Bliss, C.M., 2005. Development of an environmental virtual field laboratory. Comput. Educ. 45 (1), 21–34.

Ramcharan, A., Hengl, T., Nauman, T., Brungard, C., Waltman, S., Wills, S., et al., 2018. Soil property and class maps of the conterminous United States at 100-meter spatial resolution. Soil Sci. Soc. Am. J. 82 (1), 186–201.

Ramillien, G., Cazenave, A., 1997. Global bathymetry derived from altimeter data of the ERS-1 geodetic mission. J. Geodyn. 23 (2), 129–149.

Ramsay, J.G., 1967. Folding and Fracturing of Rocks. McGraw-Hill, New York, NY.

Rance, H., 1967. Major lineaments and torsional deformation of the Earth. J. Geophys. Res. 72 (8), 2213–2217.

Rance, H., 1968. Plastic flow and fracture in a torsionally stressed planetary sphere. J. Math. Mech. 17 (10), 953–974.

Rance, H., 1969. Lineaments and torsional deformation of the Earth: Indian Ocean. J. Geophys. Res. 74 (12), 3271–3272.

Rantsman, E.Y., 1979. Earthquake Areas and Morphostructure of Mountain Regions. Nauka, Moscow, USSR (in Russian).

Raper, J.F., 1989. The 3-dimensional geoscientific mapping and modelling system: a conceptual design. In: Raper, J. (Ed.), Three Dimensional Applications in GIS. Taylor and Francis, London, UK, pp. 11–19.

Rastsvetaev, L.M., 1977. The Crimean Mountains and the Northern Black Sea area. In: Suvorov, A.I. (Ed.), Faults and Horizontal Movements in Mountain Regions of the USSR. Nauka, Moscow, USSR, pp. 95–113 (in Russian).

Raupach, M.R., Finnigan, J.J., 1997. The influence of topography on meteorological variables and surface–atmosphere interactions. J. Hydrol. 190 (3–4), 182–213.

REMA, 2018–2022. Reference Elevation Model of Antarctica (REMA). Polar Geospatial Center, University of Minnesota, Saint Paul, MN. https://www.pgc.umn.edu/data/rema/. (Accessed 27 June 2024).

Remondino, F., El-Hakim, S., 2006. Image-based 3D modelling: a review. Photogrammet. Rec. 21 (115), 269–291.

Rémy, F., Shaeffer, P., Legrésy, B., 1999. Ice flow physical processes derived from the ERS-1 high-resolution map of the Antarctica and Greenland ice sheets. Geophys. J. Int. 139 (3), 645–656.

Ren, Z., Zielke, O., Yu, J., 2018. Active tectonics in 4D high-resolution. J. Struct. Geol. 117, 264–271.

Reuter, H.I., Hengl, T., Gessler, P., Soille, P., 2009. Preparation of DEMs for geomorphometric analysis. In: Hengl, T., Reuter, H.I. (Eds.), Geomorphometry: Concepts, Software, Applications. Developments in Soil Science, vol. 33. Elsevier, Amsterdam, the Netherlands, pp. 87–120.

Rheinwalt, A., Goswami, B., Bookhagen, B., 2019. A network-based flow accumulation algorithm for point clouds: facet-flow networks (FFNs). J. Geophys. Res. Earth Surf. 124 (7), 2013–2033.

Rhind, D.W., 1971. Automated contouring—an empirical evaluation of some differing techniques. Cartogr. J. 8 (2), 145–158.

Richards, F.B., 1991. A Gibbs phenomenon for spline functions. J. Approx. Theory 66 (3), 334–351.

Richards-Rissetto, H., 2017. What can GIS + 3D mean for landscape archaeology? J. Archaeol. Sci. 84, 10–21.

Rieger, J., 1997. Topographical properties of generic images. Int. J. Comput. Vis. 23 (1), 79–92.

Rieger, W., 1996. Accuracy of slope information derived from DEM-data. In: Kraus, K., Waldhäusl, P. (Eds.), XVIII ISPRS Congress, Technical Commission IV: Mapping and Geographic Information Systems, Vienna, Austria, 9–19 July 1996. ISPRS Arch. XXXI (B4), pp. 690–695.

Rieger, W., 1998. A phenomenon-based approach to upslope contributing area and depressions in DEMs. Hydrol. Process. 12 (6), 857–872

Rignot, E., Mouginot, J., Scheuchl, B., 2011. Ice flow of the Antarctic ice sheet. Science 333 (6048), 1427–1430.

Rignot, E., Jacobs, S., Mouginot, J., Scheuchl, B., 2013. Ice-shelf melting around Antarctica. Science 341 (6143), 266–270.

Rignot, E., Mouginot, J., Scheuch, B., van den Broeke, M., van Wessem, M.J., Morlighem, M., 2019. Four decades of Antarctic Ice Sheet mass balance from 1979–2017. Proc. Natl. Acad. Sci. 116 (4), 1095–1103.

Riihimäki, H., Kemppinen, J., Kopecký, M., Luoto, M., 2021. Topographic Wetness Index as a proxy for soil moisture: the importance of flow-routing algorithm and grid resolution. Water Resour. Res. 57 (10), e2021WR029871.

Rinner, K., 1969. Problems of two-medium photogrammetry. Photogrammet. Eng. 35 (2), 275–282.

Rippin, D.M., Willis, I., Arnold, N., Hodson, A., Moore, J., Kohler, J., et al., 2003. Changes in geometry and subglacial drainage of Midre Lovénbreen, Svalbard, determined from digital elevation models. Earth Surf. Process. Landf. 28 (3), 273–298.

Rivlin, T.J., 1974. The Chebyshev Polynomials. Wiley, New York, NY.

Riza, S., Sekine, M., Kanno, A., Yamamoto, K., Imai, T., Higuchi, T., 2021. Modeling soil landscapes and soil textures using hyperscale terrain attributes. Geoderma 402, 115177.

Rizzoli, P., Martone, M., Gonzalez, C., Wecklich, C., Borla Tridon, D., Bräutigam, B., et al., 2017. Generation and performance assessment of the global TanDEM-X digital elevation model. ISPRS J. Photogrammet. Remote Sens. 132, 119–139.

Robbins, S.J., Watters, W.A., Chappelow, J.E., Bray, V.J., Daubar, I.J., Craddock, R.A., et al., 2017. Measuring impact crater depth throughout the solar system. Meteorit. Planet. Sci. 53 (4), 583–637.

Roberts, A., 2001. Curvature attributes and their application to 3D interpreted horizons. First Break 19 (2), 85–100.

Roberts, A.M., Corfield, R.I., Kusznir, N.J., Matthews, S.J., Hansen, E.-K., Hooper, R.J., 2009. Mapping palaeostructure and palaeobathymetry along the Norwegian Atlantic continental margin: Møre and Vøring basins. Petrol. Geosci. 15 (1), 27–43.

Robertson, C., Chaudhuri, C., Hojati, M., Roberts, S.A., 2020. An integrated environmental analytics system (IDEAS) based on a DGGS. ISPRS J. Photogrammet. Remote Sens. 162, 214–228.

Robeson, S.M., 1997. Spherical methods for spatial interpolation: review and evaluation. Cartogr. Geogr. Inf. Syst. 24 (1), 3–20.

Robin, G.d.Q., Evans, S., Bailey, J.T., 1969. Interpretation of radio echo sounding in polar ice sheets. Philos. Trans. R. Soc. Lond A Math. Phys. Sci. 265 (1166), 437–505.

Robinson, J.E., Charlesworth, H.A.K., Ellis, M.J., 1969. Structural analysis using spatial filtering in Interior Plains of south-central Alberta. Am. Assoc. Pet. Geol. Bull. 53 (11), 2341–2367.

Robinson, N., NCEAS ENO Working Group, 2014. EarthEnv-DEM90. National Center for Ecological Analysis and Synthesis, Santa Barbara, CA. http://www.earthenv.org/DEM.html. (Accessed 27 June 2024).

Robinson, N., Regetz, J., Guralnick, R.P., 2014. EarthEnv-DEM90: a nearly-global, void-free, multi-scale smoothed, 90m digital elevation model from fused ASTER and SRTM data. ISPRS J. Photogrammet. Remote Sens. 87, 57–67.

Rocha, J., Duarte, A., Silva, M., Fabres, S., Vasques, J., Revilla-Romero, B., et al., 2020. The importance of high resolution digital elevation models for improved hydrological simulations of a Mediterranean forested catchment. Remote Sens. (Basel) 12 (20), 3287.

Rode, A.A., 1953. Origin of microtopography of the Northern Caspian Lowland. Prob. Geogr. 33, 249–260 (in Russian).

Rodríguez, E., Morris, C.S., Belz, J.E., 2006. A global assessment of the SRTM performance. Photogrammet. Eng. Remote Sens. 72 (3), 249–260.

Rodriguez, F., Maire, E., Courjault-Radé, P., Darrozes, J., 2002. The Black Top Hat function applied to a DEM: a tool to estimate recent incision in a mountainous watershed (Estibère Watershed, Central Pyrenees). Geophys. Res. Lett. 29 (6), 1085.

Rodriguez-Perez, D., Sanchez-Carnero, N., 2022. Multigrid/multiresolution interpolation: reducing oversmoothing and other sampling effects. Geomatics 2 (3), 236–253.

Roecker, S.M., Thompson, J.A., 2010. Scale effects on terrain attribute calculation and their use as environmental covariates for digital soil mapping. In: Boettinger, J.L., Howell, D.W., Moore, A.C., Hartemink, A.E., Kienast-Brown, S. (Eds.) Digital Soil Mapping: Bridging Research, Environmental Application, and Operation. Progress in Soil Science Book Series. Springer, Dordrecht, the Netherlands, pp. 55–66.

Román Dobarco, M., Arrouays, D., Lagacherie, P., Ciampalini, R., Saby, N.P.A., 2017. Prediction of topsoil texture for Region Centre (France) applying model ensemble methods. Geoderma 298, 67–77.

Romanova, E.N., 1963. Some regularities of water redistribution on slopes. In: Goltsberg, I.A. (Ed.), Problems of Microclimatology. Hydrometeoizdat, Leningrad, USSR, pp. 66–82 (in Russian).

Romanova, E.N., 1970. Seasonal humidification of soils in contrast geomorphic conditions. In: Goltsberg, I.A., Romanova, E.N. (Eds.), Microclimatology. Hydrometeoizdat, Leningrad, USSR, pp. 23–43 (in Russian).

Romanova, E.N., 1971. An approach of measurement and mapping of soil moisture using morphometric data. In: Goltsberg, I.A., Davitaya, F.F. (Eds.), Soil Climate. Hydrometeoizdat, Leningrad, USSR, pp. 39–51 (in Russian).

Romanova, E.N., 1977. Microclimatic Variability of the Main Elements of Climate. Hydrometeoizdat. Leningrad, USSR (in Russian).

Romashkevich, A.I., Rantsman, E.Y., Mikheev, G.A., 1997. The anomalies in soil properties and soil mantle composition in mountainous regions and seismotectonics. Eurasian Soil Sci. 30 (5), 472–482.

Romero, B.E., Clarke, K.C., 2018. Exploring uncertainties in terrain feature extraction across multi-scale, multi-feature, and multi-method approaches for variable terrain. Cartogr. Geogr. Inf. Sci. 45 (5), 381–399.

Rong, Y., Zhang, T., Zheng, Y., Hu, C., Peng, L., Feng, P., 2020. Three-dimensional urban flood inundation simulation based on digital aerial photogrammetry. J. Hydrol. 584, 124308.

Rosenberg, P., 1955. Information theory and electronic photogrammetry. Photogrammet. Eng. 21 (4), 543–555.

Rosenfeld, A., Kak, A.C., 1982. Digital Picture Processing, second ed., vol. 1. Academic Press, New York, NY.

Rossi, C., Rodriguez Gonzalez, F., Fritz, T., Yague-Martinez, N., Eineder, M., 2012. TanDEM-X calibrated raw DEM generation. ISPRS J. Photogrammet. Remote Sens. 73, 12–20.

Rossini, M., di Mauro, B., Garzonio, R., Baccolo, G., Cavallini, G., Mattavelli, M., et al., 2018. Rapid melting dynamics of an alpine glacier with repeated UAV photogrammetry. Geomorphology 304, 159–172.

Rothe, R., 1915. Zum Problem des Talwegs. Sitzber. Berlin. Math. Ges. 14, 51–68.

Rotnicka, J., Dłużewski, M., Dąbski, M., Rodzewicz, M., Włodarski, W., Zmarz, A., 2020. Accuracy of the UAV-based DEM of beach–foredune topography in relation to selected morphometric variables, land cover, and multitemporal sediment budget. Estuaries Coast 43 (8), 1939–1955.

Roudier, P., Malone, B.P., Hedley, C.B., Minasny, B., McBratney, A.B., 2017. Comparison of regression methods for spatial downscaling of soil organic carbon stocks maps. Comput. Electron. Agricult. 142 (A), 91–100.

Row, L.W., Hastings, D.A., Dunbar, P.K., 1995. TerrainBase Worldwide Digital Terrain Data—Documentation Manual. National Geophysical Data Center, National Oceanic and Atmospheric Administration, Boulder, CO.

Różycka, M., Migoń, P., 2022. Using geomorphometric approach to investigate spatial pattern and intensity of erosional dissection in a block-faulted topography (Orlickie-Bystrzyckie Mountains, Central Europe). Catena 211, 105937.

Różycka, M., Jancewicz, K., Migoń, P., Szymanowski, M., 2021. Tectonic versus rock-controlled mountain fronts—geomorphometric and geostatistical approach (Sowie Mts, Central Europe). Geomorphology 373, 107485

Rudy, R.M., 1999. Methods to Study Relief of the Land Surface (D.Sc. thesis). Ivano-Frankovsk State Technical University of Oil and Gas, Ivano-Frankovsk, Ukraine (in Ukrainian).

Runyon, K.D., Bridges, N.T., Ayoub, F., Newman, C.E., Quade, J.J., 2017. An integrated model for dune morphology and sand fluxes on Mars. Earth Planet. Sci. Lett. 457, 204–212.

Ryakhovsky, V., Rundquist, D., Gatinsky, Y., Chesalova, E., 2003. GIS-project: geodynamic globe for global monitoring of geological processes. Geophys. Res. Abs. 5, 11645.

Ryan, C., Boyd, M., 2003. CatchmentSIM: a new GIS tool for topographic geo-computation and hydrologic modelling. In: Boyd, M., Ball, J., Babister, M., Green, J. (Eds.), Proceedings of the 28th International Hydrology and Water Resources Symposium, Wollongong, NSW, 10–14 Nov. 2003, vol. 1. Institution of Engineers, Barton, ACT, pp. 35–42.

Saadi, N.M., Aboud, E., Watanabe, K., 2009. Integration of DEM, ETM+, geologic, and magnetic data for geological investigations in the Jifara Plain, Libya. IEEE Trans. Geosci. Remote Sens. 47 (10), 3389–3398.

Sahr, K., White, D., Kimerling, A.J., 2003. Geodesic discrete global grid systems. Cartogr. Geogr. Inf. Sci. 30 (2), 121–134.

Salichtchev, K.A., 1955. On the cartographic method of research. Moscow Univ. Bull. Phys. Mathemat. Nat. Sci. Ser. (10), 161–170 (in Russian).

Salichtchev, K.A., 1990. Map Science, third ed. Moscow University Press, Moscow, USSR (in Russian).

Sam, L., Bhardwaj, A., Kumar, R., Buchroithner, M.F., Martín-Torres, F.J., 2018. Heterogeneity in topographic control on velocities of Western Himalayan glaciers. Sci. Rep. 8, 12843.

Samarskii, A.A., 2001. The Theory of Difference Schemes. Dekker, New York, NY.

Samson, P.P., Mallet, J.L., 1997. Curvature analysis of triangulated surfaces in structural geology. Mathemat. Geol. 29 (3), 391–412.

Samsonov, T.E., 2010. Multiscale Mapping of Topography by Creation of Geodatabases (Ph.D. thesis). Lomonosov Moscow State University, Moscow, Russia (in Russian).

Samsonov, T., 2011. Multiscale hypsometric mapping. In: Ruas, A. (Ed.), Advances in Cartography and GIScience, vol. 1: Selection From ICC 2011, Paris. Lecture Notes in Geoinformation and Cartography Book Series. Springer, Berlin, Germany, pp. 497–520.

Samsonov, T., 2014. Morphometric mapping of topography by flowline hachures. Cartogr. J. 51 (1), 63–74.

Samsonov, T., 2022. Granularity of digital elevation model and optimal level of detail in small-scale cartographic relief presentation. Remote Sens. (Basel) 14 (5), 1270.

Samsonov, T., Koshel, S., Walther, D., Jenny, B., 2019. Automated placement of supplementary contour lines. Int. J. Geogr. Inf. Sci. 33 (10), 2072–2093.

Samsonova, V.P., Pozdnyakov, A.I., Meshalkina, J.L., 2007. Study of disturbed soil cover using soil electrical resistivity and topographic data. Int. J. Ecol. Dev. 8 (F07), 39–51.

Samuel-Rosa, A., Heuvelink, G.B.M., Vasques, G.M., Anjos, L.H.C., 2015. Do more detailed environmental covariates deliver more accurate soil maps? Geoderma 243–244, 214–227.

Sanchez-Fernandez, A.J., Romero, L.F., Bandera, G., Tabik, S., 2021. A data relocation approach for terrain surface analysis on multi-GPU systems: a case study on the total viewshed problem. Int. J. Geogr. Inf. Sci. 35 (8), 1500–1520.

Sandwell, D.T., Smith, W.H.F., 1997. Marine gravity anomaly from Geosat and ERS-1 satellite altimetry. J. Geophys. Res. Solid Earth 102 (B5), 10039–10054.

Sandwell, D.T., Smith, W.H.F., 2001. Bathymetric estimation. In: Fu, L.-L., Cazenave, A. (Eds.), Satellite Altimetry and Earth Sciences: A Handbook of Techniques and Applications. Academic Press, San Diego, CA, pp. 441–457.

Sandwell, D.T., Smith, W.H.F., Gille, S., Kappel, E., Jayne, S., Soofi, K., et al., 2006. Bathymetry from Space: rationale and requirements for a new, high-resolution altimetric mission. C. R. Acad. Sci. 338 (14–15), 1049–1062.

Sandwell, D.T., Smith, W.H.F., Becker, J.J., 2008. SRTM30_PLUS V11. Scripps Institution of Oceanography, University of California, San Diego, CA. ftp://topex.ucsd.edu/pub/srtm30_plus/.

Sandwell, D.T., Becker, J.J., Olson, C., Jackson, A., 2014. SRTM15_PLUS, V1. Scripps Institution of Oceanography, University of California, San Diego, CA. ftp://topex.ucsd.edu/pub/srtm15_plus/.

Saraf, A.K., Mishra, B.P., Choudhury, S., Ghosh, P., 2005. Digital elevation model (DEM) generation from NOAA–AVHRR night-time data and its comparison with USGS-DEM. Int. J. Remote Sens. 26 (18), 3879–3887.

Sărăşan, A., Józsa, E., Ardelean, A.C., Drăguţ, L., 2019. Sensitivity of geomorphons to mapping specific landforms from a digital elevation model: a case study of drumlins. Area 51 (2), 257–267.

Sarmento, E.C., Giasson, E., Webster, E.J., Flores, C.A., Hasenack, H., 2017. Regional disaggregating conventional soil maps with limited descriptive data: a knowledge-based approach in Serra Gaúcha, Brazil. Geoderma Reg. 8, 12–23.

Šašak, J., Gallay, M., Kaňuk, J., Hofierka, J., Minár, J., 2019. Combined use of terrestrial laser scanning and UAV photogrammetry in mapping alpine terrain. Remote Sens. (Basel) 11 (18), 2154.

Sasowsky, K.C., Petersen, G.W., Evans, B.M., 1992. Accuracy of SPOT digital elevation model and derivatives: utility for Alaska's North Slope. Photogrammet. Eng. Remote Sens. 58 (6), 815–824.

Satalkin, A.I. (Ed.), 1988. Soil Map of the Moscow Region, Scale 1:300,000. Central Board of Geodesy and Cartography, Moscow, USSR (in Russian).

Sathyamoorthy, D., 2013. Analysis of surface textures of physiographic features extracted from multiscale digital elevation models via grey level co-occurrence matrix. J. Geod. Geoinf. 2 (1), 19–28.

Saul, J.M., 1978. Circular structures of large scale and great age on the Earth's surface. Nature 271 (5643), 345–349.

Saulnier, G.M., Obled, C., Beven, K., 1997. Analytical compensation between DTM grid resolution and effective values of saturated hydraulic conductivity within the TOPMODEL framework. Hydrol. Process. 11 (9), 1331–1346.

Saunders, A.M., Boettinger, J.L., 2007. Incorporating classification trees into a pedogenic understanding raster classification methodology, Green River Basin, Wyoming, USA. In: Lagacherie, P., McBratney, A.B., Voltz, M. (Eds.), Digital Soil Mapping: An Introductory Perspective. Developments in Soil Science, vol. 31. Elsevier, Amsterdam, the Netherlands, pp. 389–399.

Saunders, R.S., Pettengill, G.H., 1991. Magellan: mission summary. Science 252 (5003), 247–249.

Scambos, T.A., Dutkiewicz, M.J., Wilson, J.C., Bindschadler, R.A., 1992. Application of image cross-correlation to the measurement of glacier velocity using satellite image data. Remote Sens. Environ. 42 (3), 177–186.

Schaetzl, R.J., Anderson, S., 2005. Soils: Genesis and Geomorphology. Cambridge University Press, Cambridge, MA.

Scheffers, A.M., May, S.M., Kelletat, D.H., 2015. Landforms of the World with Google Earth: Understanding Our Environment. Springer, Dordrecht, the Netherlands.

Scheidegger, A.E., 2004. Morphotectonics. Springer, Berlin, Germany.

Scherler, D., Leprince, S., Strecker, M.R., 2008. Glacier-surface velocities in alpine terrain from optical satellite imagery—accuracy improvement and quality assessment. Remote Sens. Environ. 112 (10), 3806–3819.

Schmid-McGibbon, G., 1995. Generalization of digital terrain models for use in landform mapping. Cartographica 32 (3), 26–38.

Schmidt, J., Andrew, R., 2005. Multi-scale landform characterization. Area 37 (3), 341–350.

Schmidt, J., Hewitt, A., 2004. Fuzzy land element classification from DTMs based on geometry and terrain position. Geoderma 121 (3–4), 243–256.

Schmidt, J., Evans, I.S., Brinkmann, J., 2003. Comparison of polynomial models for land surface curvature calculation. Int. J. Geogr. Inf. Sci. 17 (8), 797–814.

Schmidt, J., Tonkin, P., Hewitt, A., 2005. Quantitative soil–landscape models for the Haldon and Hurunui soil sets, New Zealand. Aust. J. Soil. Res. 43 (2), 127–137.

Schmidt, J.P., Taylor, R.K., Gehl, R.J., 2003. Developing topographic maps using a sub-meter accuracy global positioning receiver. Appl. Eng. Agric. 19 (3), 291–302.

Scholten, T., Goebes, P., Kühn, P., Seitz, S., Assmann, T., Bauhus, J., et al. 2017. On the combined effect of soil fertility and topography on tree growth in subtropical forest ecosystems—a study from SE China. J. Plant Ecol. 10 (1), 111–127.

Schowengerdt, R.A., Glass, C.E., 1983. Digitally processed topographic data for regional tectonic evaluations. Geol. Soc. Am. Bull. 94 (4), 549–556.

Schröder, P., Sweldens, W., 2000. Spherical wavelets: efficiently representing functions on a sphere. In: Klees, R., Haagmans, R. (Eds.), Wavelets in the Geosciences. Lect. Notes Earth Sci., vol. 90, pp. 158–188.

Schuler, D.L., Lee, J.S., de Grandi, G., 1996. Measurement of topography using polarimetric SAR images. IEEE Trans. Geosci. Remote Sens. 34 (5), 1266–1277.

Schuler, D.L., Ainsworth, T.L., Lee, J.S., de Grandi, G.F., 1998. Topographic mapping using polarimetric SAR data. Int. J. Remote Sens. 19 (1), 141–160.

Schultz-Ela, D.D., Yeh, J., 1992. Predicting fracture permeability from bed curvature. In: Tillerson, J.R., Wawersik, W.R. (Eds.), Proceedings of the 33rd US Symposium on Rock Mechanics, Santa Fe, NM, 3–5 June 1992. Balkema, Rotterdam, the Netherlands, pp. 579–589.

Schumann, G.J.-P., Bates, P.D., 2018. The need for a high-accuracy, open-access global DEM. Front. Earth Sci. 6, 225.

Schumm, S.A., Lichty, R.W., 1965. Time, space, and causality in geomorphology. Am. J. Sci. 263 (2), 110–119.

Schut, G.H., 1976. Review of interpolation methods for digital terrain models. Can. Survey. 30 (5), 389–412.

Schutz, B.E., Zwally, H.J., Shuman, C.A., Hancock, D., DiMarzio, J.P., 2005. Overview of the ICESat mission. Geophys. Res. Lett. 32 (21), L21S01.

Scott, D.H., Carr, M.H., 1978. Geologic Map of Mars, I-1083, Scale 1:25,000,000. US Geological Survey, Reston, VA.

Scott, M.L., Needelman, B.A., 2007. Utilizing water well logs for soil parent material mapping in the Mid-Atlantic coastal plain. Soil Sci. 172 (9), 701–720.

Scull, P., Franklin, J., Chadwick, O.A., McArthur, D., 2003. Predictive soil mapping: a review. Progr. Phys. Geogr. 27 (2), 171–197.

Scull, P., Franklin, J., Chadwick, O.A., 2005. The application of classification tree analysis to soil type prediction in a desert landscape. Ecol. Model. 181 (1), 1–15.

Segu, W.P., 1984. An investigation of the scope of high order polynomial digital elevation models. In: XV ISPRS Congress, Technical Commission III: Mathematical Analysis of Data, Rio de Janeiro, Brazil, 17–29 June 1984. ISPRS Arch. XXV (A3), pp. 996–1005.

Seibert, J., McGlynn, B.L., 2007. A new triangular multiple flow direction algorithm for computing upslope areas from gridded digital elevation models. Water Resour. Res. 43 (4), W04501.

Seier, G., Kellerer-Pirklbauer, A., Wecht, M., Hirschmann, S., Kaufmann, V., Lieb, G.K., et al., 2017. UAS-based change detection of the glacial and proglacial transition zone at Pasterze Glacier, Austria. Remote Sens. (Basel) 9 (6), 549.

Seifert, T., Tauber, F., Kayser, B., 2001. A high-resolution spherical grid topography of the Baltic Sea—2nd edition. In: Brenner, U. (Ed.), Abstracts, Baltic Sea Science Congress 2001: Past, Present and Future—A Joint Venture, Stockholm, Sweden, 25–29 Nov. 2001. Stockholm Marine Research Centre, Stockholm University, Stockholm, Sweden, p. 298.

Selezneva, E.V., 2013. Application of laser scanning in geomorphologic studies. Moscow Univ. Geogr. Bull. (2), 47–53 (in Russian).

Semenov, A.M., van Bruggen, A.H.C., Zelenev, V.V., 1999. Moving waves of bacterial populations and total organic carbon along roots of wheat. Microb. Ecol. 37 (2), 116–128.

Senthilkumar, S., Kravchenko, A.N., Robertson, G.P., 2009. Topography influences management system effects on total soil carbon and nitrogen. Soil Sci. Soc. Am. J. 73 (6), 2059–2067.

Serbenyuk, S.N., 1990. Cartography and Geoinformatics—Their Interaction. Moscow University Press, Moscow, USSR (in Russian).

Sevara, C., Verhoeven, G., Doneus, M., Draganits, E., 2018. Surfaces from the visual past: recovering high-resolution terrain data from historic aerial imagery for multitemporal landscape analysis. J. Archaeol. Method Theory 25 (2), 611–642.

Shahbazi, M., Théau, J., Ménard, P., 2014. Recent applications of unmanned aerial imagery in natural resource management. GISci. Remote Sens. 51 (4), 339–365.

Shalimov, A.I., 1966. A new tectonic scheme of Crimea and the connection between fold structures in the Mountain Crimea and the Northwest Caucasus. In: Magnitsky, V.A., Bulange, Y.D., Meshcheryakov, Y.A. (Eds.), The Structure of the Black Sea Depression. Nedra, Moscow, USSR, pp. 49–58 (in Russian).

Shan, J., Toth, C.K. (Eds.), 2008. Topographic Laser Ranging and Scanning: Principles and Processing. CRC Press, Parkway, NW.

Shannon, C.E., 1949. Communication in the presence of noise. Proc. IRE 37 (1), 10–21.

Sharaya, L.S., Shary, P.A., 2011. Geomorphometric study of the spatial organization of forest ecosystems. Russ. J. Ecol. 42 (1), 1–8.

Sharma, A., Tiwari, K.N., Bhadoria, P.B.S., 2010. Vertical accuracy of digital elevation model from Shuttle Radar Topographic Mission—a case study. Geocarto Int. 25 (4), 257–267.

Shary, P.A., 1991. The second derivative topographic method. In: Stepanov, I.N. (Ed.), The Geometry of the Earth Surface Structures. Pushchino Research Centre Press, Pushchino, USSR, pp. 30–60 (in Russian).

Shary, P.A., 1995. Land surface in gravity points classification by a complete system of curvatures. Mathemat. Geol. 27 (3), 373–390.

Shary, P.A., 2005. Estimating Interrelations Between Topography, Soil, and Plants with New Approaches of Geomorphometry (Exemplified by an Agro-landscape and a Forest Ecosystem in the Southern Moscow Region) (Ph.D. thesis). Institute of Physical, Chemical, and Biological Problems of Soil Science, Russian Academy of Sciences, Pushchino, Russia (in Russian).

Shary, P.A., 2006. Variables of geomorphometry: the current state-of-art. In: Liu, X., Wang, Y. (Eds.), Full Papers, Proceedings of International Symposium on Terrain Analysis and Digital Terrain Modelling (TADTM 2006), Nanjing, China, 23–25 Nov. 2006. Nanjing Normal University, Nanjing, China.

Shary, P.A., 2008. Models of topography. In: Zhou, Q., Lees, B., Tang, G.-A. (Eds.), Advances in Digital Terrain Analysis. Lecture Notes in Geoinformation and Cartography Book Series. Springer, Berlin, Germany, pp. 29–57.

Shary, P.A., 2012. The mathematical basis of local morphometric variables. In: Florinsky, I.V., Digital Terrain Analysis in Soil Science and Geology, first ed. Academic Press, Amsterdam, the Netherlands, pp. 289–313.

Shary, P.A., 2023. Environmental variables in predictive soil mapping: a review. Eurasian Soil Sci. 56 (3), 247–259.

Shary, P.A., Pinskii, D.L., 2013. Statistical evaluation of the relationships between spatial variability in the organic carbon content in gray forest soils, soil density, concentrations of heavy metals, and topography. Eurasian Soil Sci. 46 (11), 1076–1087.

Shary, P.A., Stepanov, I.N., 1991. Application of the method of second derivatives in geology. Trans. (Doklady) USSR Acad. Sci., Earth Sci. Sect. 320 (7), 87–92.

Shary, P.A., Kuryakova, G.A., Florinsky, I.V., 1991. On the international experience of the use of topographic techniques in landscape investigations (a brief review). In: Stepanov, I.N. (Ed.), The Geometry of the Earth Surface Structures. Pushchino Research Centre Press, Pushchino, USSR, pp. 15–29 (in Russian).

Shary, P.A., Rukhovich, O.V., Sharaya, L.S., Mitusov, A.V., 2002a. Soils and topography: accumulation zones and non-local approaches. In: Transactions of the 17th World Congress of Soil Science, Bangkok, Thailand, 14–21 Aug. 2002, Symposium 48. International Society of Soil Science, Bangkok, Thailand. 2310.

Shary, P.A., Sharaya, L.S., Mitusov, A.V., 2002b. Fundamental quantitative methods of land surface analysis. Geoderma 107 (1–2), 1–32.

Shary, P.A., Sharaya, L.S., Mitusov, A.V., 2005. The problem of scale-specific and scale-free approaches in geomorphometry. Geogr. Fis. Din. Quat. 28 (1), 81–101.

Shary, P.A., Sharaya, L.S., Mitusov, A.V., 2017a. Predictive modeling of slope deposits and comparisons of two small areas in Northern Germany. Geomorphology 290, 222–235.

Shary, P.A., Sharaya, L.S., Sidyakina, L.V., Saksonov, S.V., 2017b. Influence of solar energy and tree-crown closure on the species richness of grasses at the south of forest steppe. Contemp. Probl. Ecol. 10 (5), 464–475.

Shcherbakov, V.M., Ukolova, A.V., Skrypitsyna, T.N., Falkov, D.D., 2023. Three-dimentional modeling of the Syanov underground quarry by photogrammetric methods. Speleol. Spelestol. (1), 47–52 (in Russian).

Shean, D., 2017. High Mountain Asia 8-Meter DEM Mosaics Derived From Optical Imagery, Version 1. National Snow and Ice Data Center, Boulder, CO, https://doi.org/10.5067/kxovq9l172s2.

Shean, D.E., Alexandrov, O., Moratto, Z.M., Smith, B.E., Joughin, I.R., Porter, C., et al., 2016. An automated, open-source pipeline for mass production of digital elevation models (DEMs) from very-high-resolution commercial stereo satellite imagery. ISPRS J. Photogrammet. Remote Sens. 116, 101–117.

Shebuev, G.N., 1896. Geometric bases of geodesy on a triaxial ellipsoid, very little different from a spheroid. Proc. Topogr. Geodet. Comm. 5 70–97 (in Russian).

Shebuev, G.N., 1898. Distances, azimuths, and triangles on a triaxial ellipsoid, little different from a sphere. Proc. Topogr. Geodet. Comm. [illegible], 1–72 (in Russian).

Shen, Q., Wang, Y., Wang, X., Liu, X., Zhang, X., Zhang, S., 2019. Comparing interpolation methods to predict soil total phosphorus in the Mollisol area of Northeast China. Catena 174, 59–72.

Shen, X., Ke, C.-Q., Fan, Y., Drolma, L., 2022. A new digital elevation model (DEM) dataset of the entire Antarctic continent derived from ICESat-2. Earth Syst. Sci. Data 14 (7), 3075–3089.

Shepard, C., Schaap, M.G., Chorover, J., Rasmussen, C., 2018. Understanding critical zone evolution through predicting the three-dimensional soil chemical properties of a small forested catchment. Soil Sci. Soc. Am. J. 82 (6), 1538–1550.

Shepherd, A., Fricker, H.A., Farrell, S.L., 2018. Trends and connections across the Antarctic cryosphere. Nature 558, 223–232.

Sherstyankin, P.P., Alekseev, S.P., Abramov, A.M., Stavrov, K.G., de Batist, M., Hus, R., et al., 2006. Computer-based bathymetric map of Lake Baikal. Doklady Earth Sci. 408 (4), 564–569.

Shi, W., Deng, S., Xu, W., 2018. Extraction of multi-scale landslide morphological features based on local G_i^* using airborne LiDAR-derived DEM. Geomorphology 303, 229–242.

Shi, X., Zhu, A.-X., Burt, J.E., Qi, F., Simonson, D., 2004. A case-based approach to fuzzy soil mapping. Soil Sci. Soc. Am. J. 68 (3), 885–894.

Shi, Y., Lau, K.K.-L., Ng, E., 2017. Incorporating wind availability into land use regression modelling of air quality in mountainous high-density urban environment. Environ. Res. 157, 17–29.

Shi, Y., Katzschner, L., Ng, E., 2018. Modelling the fine-scale spatiotemporal pattern of urban heat island effect using land use regression approach in a megacity. Sci. Total Environ. 618, 891–904.

Shirani, K., Solhi, S., Pasandi, M., 2023. Automatic landform recognition, extraction, and classification using kernel pattern modeling. J. Geovis. Spat. Anal. 7, 2.

Shiryaev, E.E., 1977. New Methods of Cartographic Representation and Analysis of Geoinformation by Computers. Nedra, Moscow, USSR (in Russian).

Shtengelov, Y.S., 1978. On the fan-like recent spreading of the Earth crust, and on the nature of the Benioff zones. Doklady Acad. Nauk SSSR 240 (4), 922–925 (in Russian).

Shtengelov, Y.S., 1982. Zones of recent and modern spreading of continental crust. Int. Geol. Rev. 24 (7), 759–770.

Shults, S.S., 1971. Planetary fractures and tectonic deformations. Geotectonics 5 (4), 203–207.

Shumack, S., Hesse, P., Farebrother, W., 2020. Deep learning for dune pattern mapping with the AW3D30 global surface model. Earth Surf. Process. Landf. 45 (11), 2417–2431.

Sibirtsev, N.M., 1899. Soil Science, Pts. II and III. Skorokhodov Press, St. Petersburg, Russia (in Russian).

Siegert, M.J., 1999. On the origin, nature and uses of Antarctic ice-sheet radio-echo layering. Progr. Phys. Geogr. 23 (2), 78–98.

Siervo, V., Pescatore, E., Giano, S.I., 2023. Geomorphic analysis and semi-automated landforms extraction in different natural landscapes. Environ. Earth Sci. 82, 128.

Silva, O.L., Bezerra, F.H.R., Maia, R.P., Cazarin, C.L., 2017. Karst landforms revealed at various scales using LiDAR and UAV in semi-arid Brazil: consideration on karstification processes and methodological constraints. Geomorphology 295, 611–630.

Silva, S.H.G., de Menezes, M.D., Owens, P.R., Curi, N., 2016. Retrieving pedologist's mental model from existing soil map and comparing data mining tools for refining a larger area map under similar environmental conditions in Southeastern Brazil. Geoderma 267, 65–77.

Simonds, B., 2013. Blender Master Class: A Hands-On Guide to Modeling, Sculpting, Materials, and Rendering. No Starch Press, San Francisco, CA.

Simonov, I.A., 1971. Oases of East Antarctica. Hydrometeoizdat, Leningrad, USSR (in Russian).

Sinai, G., Zaslavsky, D., Golany, P., 1981. The effect of soil surface curvature on moisture and yield—Beer Sheba observation. Soil Sci. 132 (5), 367–375.

Singh, K.K., Frazier, A.E., 2018. A meta-analysis and review of unmanned aircraft system (UAS) imagery for terrestrial applications. Int. J. Remote Sens. 39 (15–16), 5078–5098.

Siqueira, R.G., Veloso, G.V., Fernandes-Filho, E.I., Francelino, M.R., Schaefer, C.E.G.R., Corrêa, G.R., 2022. Evaluation of machine learning algorithms to classify and map landforms in Antarctica. Earth Surf. Process. Landf. 47 (2), 367–382.

Siqueira, R.G., Moquedace, C.M., Francelino, M.R., Schaefer, C.E.G.R., Fernandes-Filho, E.I., 2023. Machine learning applied for Antarctic soil mapping: spatial prediction of soil texture for Maritime Antarctica and Northern Antarctic Peninsula. Geoderma 432, 116405.

Sîrbu, F., Drăguţ, L., Oguchi, T., Hayakawa, Y., Micu, M., 2019. Scaling land-surface variables for landslide detection. Prog Earth Planet Sci 6, 44.

Sithole, G., Vosselman, G., 2004. Experimental comparison of filter algorithms for bare-Earth extraction from airborne laser scanning point clouds. ISPRS J. Photogrammet. Remote Sens. 59 (1–2), 85–101.

Sitnikov, A.B., 1978. Water Dynamics in Unsaturated and Saturated Grounds of the Aeration Zone. Naukova Dumka, Kiev, USSR (in Russian).

Sitnikov, A.B., 1980. A concept of competent volumes. In: Babinets, A.E. (Ed.), Introduction to Modeling of Hydrogeological Processes. Naukova Dumka, Kiev, USSR, pp. 20–25 (in Russian).

Sjöberg, L.E., 2006a. Determination of areas on the plane, sphere and ellipsoid. Survey Rev. 38 (301), 583–593.

Sjöberg, L.E., 2006b. New solutions to the direct and indirect geodetic problems on the ellipsoid. Zeit. Vermessungswesen 131, 35–39.

Sjogren, W.L., Konopliv, A.S., Rappaport, N.J., 1997. Magellan Spherical Harmonics, Topography, and Gravity Data. NASA Planetary Data System Geosciences Node. Washington University, St. Louis, MO. http://pds-geosciences.wustl.edu/missions/magellan/shadr_topo_grav/index.htm. (Accessed 27 June 2024).

Skaryatin, V.D., 1973. Study of fault tectonics using a set of multiscale satellite images of the Earth (a method of multistep generalization). Izvestia Vuzov. Geol. Explor. (7), 34–50 (in Russian).

Skidmore, A.K., 1989. A comparison of techniques for calculation gradient and aspect from a gridded digital elevation model. Int. J. Geogr. Inf. Syst. 3 (4), 323–334.

Skidmore, A.K., 1990. Terrain position as mapped from a gridded digital elevation model. Int. J. Geogr. Inf. Syst. 4 (1), 33–49.

Skidmore, A.K., Ryan, P.J., Dawes, W., Short, D., O'Loughlin, E., 1991. Use of an expert system to map forest soils from a geographical information system. Int. J. Geogr. Inf. Syst. 5 (4), 431–445.

Skrypitsyna, T.N., 2022. Determination of glacier surface displacements from multi-temporal UAV data. Geodesy Cartogr. 83 (4), 39–50 (in Russian).

Skrypitsyna, T.N., Florinsky, I.V., Beloborodov, D.E., Gaydalenok, O.V., 2020. Mud volcanism at the Taman Peninsula: multiscale analysis of remote sensing and morphometric data. Remote Sens. (Basel) 12 (22), 3763.

Skrypitsyna, T.N., Florinsky, I.V., Qiao, G., 2023. Motion of the Dålk Glacier (Prydz Bay, East Antarctica): application of unmanned aerial survey. Pol. Sci. 38, 100949.

Skvortsov, A.V., 2002. Delaunay Triangulation and Its Application. Tomsk University Press, Tomsk, Russia (in Russian).

Skvortsov, P.A., 1950. On the application of the principles of art in cartography. Prob. Geogr. 22, 60–81 (in Russian).

Skvortsov, P.A., 1976. Color Design of Relief on Maps. Moscow Institute of Engineers for Geodesy, Aerophotography, and Cartography, Moscow, USSR (in Russian).

Slater, J.A., Heady, B., Kroenung, G., Curtis, W., Haase, J., Hoegemann, D., et al., 2011. Global assessment of the new ASTER global digital elevation model. Photogrammet. Eng. Remote Sens. 77 (4), 335–349.

Slater, T., Shepherd, A., McMillan, M., Muir, A., Gilbert, L., Hogg, A.E., et al., 2018. A new digital elevation model of Antarctica derived from Cryosat-2 altimetry. Cryosphere 12 (4), 1551–1562.

Slemmons, D.B., Depolo, C.M., 1986. Evaluation of active faulting and associated hazards. In: Geophysics Study Committee, Geophysics Research Forum, and National Research Council (Eds.), Active Tectonics. Studies in Geophysical Series. National Academy Press, Washington, DC, pp. 45–62.

Slyuta, E.N., Kudrin, L.V., Sinilo, V.P., 1989. Preliminary data on the nature of a planetary system of lineaments observed in radar images of Venus (data from Venera-15 and -16). Cosmic Res. 27 (6), 786–797.

Smith, B., Sandwell, D., 2003. Accuracy and resolution of Shuttle Radar Topography Mission data. Geophys. Res. Lett. 30 (9), 1467.

Smith, B., Kessler, H., Scheib, A.J., Brown, S.E., Palmer, R.C., Kuras, O., et al., 2008. 3D modelling of geology and soils—a case study from the UK. In: Hartemink, A.E., McBratney, A., Mendonça-Santos, M.L. (Eds.), Digital Soil Mapping with Limited Data. Springer, Dordrecht, the Netherlands, pp. 183–191.

Smith, D.E., Zuber, M.T., Neumann, G.A., Lemoine, F.G., 1997. Topography of the Moon from the Clementine lidar. J. Geophys. Res. Planet. 102 (E1), 1591–1611.

Smith, D.E., Zuber, M.T., Solomon, S.C., Phillips, R.J., Head, J.W., Garvin, J.B., et al., 1999. The global topography of Mars and implications for surface evolution. Science 284 (5419), 1495–1503.

Smith, D.E., Neumann, G., Arvidson, R.E., Guinness, E.A., Slavney, S., 2003. Mars Global Surveyor: MOLA MEGDRs. NASA Planetary Data System Geosciences Node, Washington University, St. Louis, MO. http://pds-geosciences.wustl.edu/missions/mgs/megdr.html (Accessed 27 June 2024).

Smith, D.E., Zuber, M.T., Jackson, G.B., Cavanaugh, J.F., Neumann, G.A., Riris, H., et al., 2010. The Lunar Orbiter Laser Altimeter investigation on the Lunar Reconnaissance Orbiter mission. Space Sci. Rev. 150 (1–4), 209–241.

Smith, M.J., Clark, C.D., 2005. Methods for the visualization of digital elevation models for landform mapping. Earth Surf. Process. Landf. 30 (7), 885–900.

Smith, M.P., Zhu, A.-X., Burt, J.E., Stiles, C., 2006. The effects of DEM resolution and neighborhood size on digital soil survey. Geoderma 137 (1–2), 58–69.

Smith, M.W., Carrivick, J.L., Quincey, D.J., 2016. Structure from motion photogrammetry in physical geography. Progr. Phys. Geogr. 40 (2), 247–275.

Smith, S., Bulmer, C., Flager, E., Frank, G., Filatow, D., 2010. Digital soil mapping at multiple scales in British Columbia, Canada. In: Proceedings of the 4th International Workshop on Digital Soil Mapping "From Digital Soil Mapping to Digital Soil Assessment: Identifying Key Gaps from Fields to Continents, Rome, Italy, 24–26 May 2010.

Smith, W.H.F., 1993. On the accuracy of digital bathymetric data. J. Geophys. Res. Solid Earth 98 (B6), 9591–9603.

Smith, W.H.F., 1998. Seafloor tectonic fabric from satellite altimetry. Annu. Rev. Earth Planet. Sci. 26, 697–747.

Smith, W.H.F., Sandwell, D.T., 1994. Bathymetric prediction from dense satellite altimetry and sparse shipboard bathymetry. J. Geophys. Res. Solid Earth 99 (B11), 21803–21824.

Smith, W.H.F., Sandwell, D.T., 1997. Global sea floor topography from satellite altimetry and ship depth soundings. Science 277 (5334), 1956–1962.

Smith, W.H.F., Sandwell, D.T., 2004. Conventional bathymetry, bathymetry from space, and geodetic altimetry. Oceanography 17 (1), 8–23.

Snyder, J.P., 1985. Conformal mapping of the triaxial ellipsoid. Survey Rev. 28 (217), 130–148.

Sobolevsky, P.K., 1932. The modern mining geometry. Soc. Reconstr. Sci. (7), 42–78 (in Russian).

Sodano, E.M., 1965. General non-iterative solution of the inverse and direct geodetic problems. Bull. Géod. 75 (1), 69–89.

Sofia, G., 2020. Combining geomorphometry, feature extraction techniques and earth-surface processes research: the way forward. Geomorphology 355, 107055.

Soil Classification Working Group, 1998. The Canadian System of Soil Classification, third ed. NRC Research Press, Ottawa, ON.

Sokratova, I.N., 2010. Antarctic Oases: History and Results of Investigations. Arctic and Antarctic Research Institute, St. Petersburg, Russia (in Russian).

Solla, M., Casqueiro, C., del Cuvillo, I., 2020. Approach to generate 3D-printed terrain models using free software and open data sources: application to military planning. Comput. Appl. Eng. Educ. 28 (3), 477–489.

Solomon, S.C., Head, J.W., Kaula, W.M., McKenzie, D., Parsons, B., Phillips, R.J., et al., 1991. Venus tectonics: initial analysis from Magellan. Science 252 (5003), 297–312.

Soloviev, A.A., Gvishiani, A.D., Gorshkov, A.I., Dobrovolsky, M.N., Novikova, O.V., 2014. Recognition of earthquake-prone areas: methodology and analysis of the results. Izvestiya Phys. Solid Earth 50 (2), 151–168.

Sonder, R.A., 1938. Die Lineamenttektonik und ihre Probleme. Eclogae Geol. Helv. 31 (1), 199–238.

Song, X.-D., Brus, D.J., Liu, F., Li, D.-C., Zhao, Y.-G., Yang, J.-L., et al., 2016. Mapping soil organic carbon content by geographically weighted regression: a case study in the Heihe River Basin, China. Geoderma 261, 11–22.

Song, X., Liu, F., Zhang, G., Li, D., Zhao, Y., Yang, J., 2017. Mapping soil organic carbon using local terrain attributes: a comparison of different polynomial models. Pedosphere 27 (4), 681–693.

Sørensen, R., Seibert, J., 2007. Effects of DEM resolution on the calculation of topographical indices: TWI and its components. J. Hydrol. 347 (1–2), 79–89.

Sorokina, N.P., Kozlov, D.N., 2009. Experience in digital mapping of soil cover patterns. Eurasian Soil Sci. 42 (2), 182–193.

Soter, S., Harris, A., 1977. The equilibrium figures of Phobos and other small bodies. Icarus 30 (1), 192–199.

Sowers, D.C., Masetti, G., Mayer, L.A., Johnson, P., Gardner, J.V., Armstrong, A.A., 2020. Standardized geomorphic classification of seafloor within the United States Atlantic canyons and continental margin. Front. Mar. Sci. 7, 9.

Speight, J.G., 1974. A parametric approach to landform regions. In: Brown, E.H., Waters, R.S. (Eds.), Progress in Geomorphology: Papers in Honour of D. L. Linton. Institute of British Geographers, London, UK, pp. 213–230.

Speight, J.G., 1980. The role of topography in controlling throughflow generation: a discussion. Earth Surf. Process. 5 (2), 187–191.

Spooner Jr., C.S., Dossi, S.W., Misulia, M.G., 1957. Let's go over the hill—potential benefits of profile scanning the stereo-model. Photogrammet. Eng. 23 (5), 909–920.

Stearns, L.A., 2011. Dynamics and mass balance of four large East Antarctic outlet glaciers. Ann. Glaciol. 52 (59), 116–126.

Steen, Ø., Sverdrup, E., Hanssen, T.H., 1998. Predicting the distribution of small faults in a hydrocarbon reservoir by combining outcrop, seismic and well data. In: Jones, G., Fisher, Q.J., Knipe, R.J. (Eds.), Faulting, Fault Sealing and Fluid Flow in Hydrocarbon Reservoirs. Geol. Soc. London Spec. Publ., vol. 147, pp. 299–312.

Stefanovic, P., Radwan, M.M., Tempfli, K., 1977. Digital terrain models: data acquisition, processing and applications. ITC J. 1, 61–76.

Steger, C., 1999. Subpixel-precise extraction of watersheds. In: Proceedings of the 7th IEEE International Conference on Computer Vision, 1999, Kerkyra, Greece, 20–27 Sept. 1999, vol. 2. Institute of Electrical and Electronics Engineers, New York, NY, pp. 884–890.

Stepanov, I.N. (Ed.), 1984. Map of Soil Cover Systems of the Turkmen Soviet Socialist Republic, Scale 1:1,500,000. Central Board of Geodesy and Cartography, Moscow, USSR (in Russian).

Stepanov, I.N. (Ed.), 1989. Map of Land Surface Systems and Soil Cover of the Part of Central Asia, Scale 1:1,500,000. Central Board of Geodesy and Cartography, Moscow, USSR (in Russian).

Stepanov, I.N., 1996. Introduction of soil geological forms of flow structures to cartography. Eurasian Soil Sci. 28 (9), 17–33.

Stepanov, I.N., Loshakova, N.A., 1998. On three types of soil delineations on the soil maps. Eurasian Soil Sci. 31 (3), 328–337.

Stepanov, I.N., Sabitova, N.I., 1983. Detection of Natural Soil Geomorphic Bodies on Topographic Maps and Aerial Photographs. Classification of Structures of the Land Surface. Biological Research Centre Press, Pushchino, USSR (in Russian).

Stepanov, I.N., Abdunazarov, U.K., Brynskikh, M.N., Deeva, N.F., Ilyina, A.A., Peido, L.P., et al., 1984. Temporal Guide for Compilation of Large- and Medium-Scale Maps of Relief Plasticity. Biological Research Centre Press, Pushchino, USSR (in Russian).

Stepanov, I.N., Loshakova, N.A., Satalkin, A.I., Andronova, M.I., 1987. Compilation of soil maps using the Relief Plasticity cartographic method. In: Kovda, V.A. (Ed.), The Method of Relief Plasticity in Thematic Mapping. Biological Research Centre Press, Pushchino, USSR, pp. 7–22 (in Russian).

Stepanov, I.N., Florinsky, I.V., Shary, P.A., 1991. On the conceptual scheme of landscape investigations. In: Stepanov, I.N. (Ed.), The Geometry of the Earth Surface Structures. Pushchino Research Centre Press, Pushchino, USSR, pp. 9–15 (in Russian).

Stepanov, I.N., Loshakova, N.A., Kovaleva, A.E., Stepanova, V.I., 1998. On the publication of regional soil maps of Russia: the problem of borders. Geodesy Cartogr. (12), 29–36 (in Russian).

Sterner, R., 1995. Color Landform Atlas of the United States. The Johns Hopkins University Applied Physics Laboratory, Laurel, MD. http://fermi.jhuapl.edu/states/. (Accessed 27 June 2024).

Stevenson, J.A., Sun, X., Mitchell, N.C., 2010. Despeckling SRTM and other topographic data with a denoising algorithm. Geomorphology 114, 238–252.

Steward, H.J., 1974. Cartographic generalisation: some concepts and explanation. Cartographica 11 (1), 1–78.

Stewart, S.A., Podolski, R., 1998. Curvature analysis of gridded geological surfaces. In: Coward, M.P., Daltaban, T.S., Johnson, H. (Eds.), Structural Geology in Reservoir Characterization. Geol. Soc. London Spec. Publ., vol. 127, pp. 133–147.

Stockmann, U., Minasny, B., McBratney, A.B., 2011. Quantifying processes of pedogenesis. Adv. Agron. 113, 1–71.

Stooke, P.J., 1998. Mapping worlds with irregular shapes. Can. Geogr. 42 (1), 61–78.

Stooke, P.J., Keller, C.P., 1990. Map projections for non-spherical worlds: the variable-radius map projections. Cartographica 27 (2), 82–100.

Strahler, A.N., 1952. Hypsometric (area-altitude) analysis of erosional topography. Geol. Soc. Am. Bull. 63 (11), 1117–1141.

Strahler, A.N., 1957. Quantitative analysis of watershed geomorphology. Trans. Am. Geophys. Union 38 (6), 913–920.

Strakhov, V.N., 2007. Change of epochs in Earth sciences. Russ. J. Earth Sci. 9, ES1001.

Strobl, P., 2020. The new Copernicus digital elevation model. GSICS Quart. 14 (1), 17–18.

Strobl, P.A., Bielski, C., Guth, P.L., Grohmann, C.H., Muller, J.-P., López-Vázquez, C., et al., 2021. The Digital Elevation Model Intercomparison eXperiment DEMIX, a community-based approach at global DEM benchmarking. In: Paparoditis, N., Mallet, C., Lafarge, F., Yang, M.Y., Zlatanova, S., Dragicevic, S., et al. (Eds.), XXIV ISPRS Congress “Imaging Today, Foreseeing Tomorrow”, Commission IV, 2021 ed., 5–9 July 2021. Int. Arch. Photogramm. Remote Sens. Spatial Inf. Sci. XLIII (B4), pp. 395–400.

Štular, B., Lozić, E., Eichert, S., 2021. Airborne LiDAR-derived digital elevation model for archaeology. Remote Sens. (Basel) 13 (9), 1855.

Stumpf, F., Schmidt, K., Goebes, P., Behrens, T., Schönbrodt-Stittc, S., Wadouxd, A., et al., 2017. Uncertainty-guided sampling to improve digital soil maps. Catena 153, 30–38.

Stüwe, K., Braun, H.-M., Peer, H., 1989. Geology and structure of the Larsemann Hills area, Prydz Bay, East Antarctica. Aust. J. Earth Sci. 36 (2), 219–241.

Suetova, I.A., Chistov, S.V., 1993. Study of ice dynamics in the Antarctic by computer modelling. Geodesy Cartogr. (3), 45–50 (in Russian).

Sulebak, J.R., Hjelle, Ø., 2003. Multiresolution spline models and their applications in geomorphology. In: Evans, I.S., Dikau, R., Tokunaga, E., Ohmori, H., Hirano, M. (Eds.), Concepts and Modelling in Geomorphology: International Perspectives. Terrapub, Tokyo, Japan, pp. 221–237.

Sulebak, J.R., Tallaksen, L.M., Erichsen, B., 2000. Estimation of areal soil moisture by use of terrain data. Geogr. Ann. Ser. B 82 (1), 89–105.

Suleymanov, A., Nizamutdinov, T., Mavlyudov, B., Abakumov, E., 2024. Predictive mapping of glacial sediment properties (Bellingshausen Dome, King George Island, Antarctica). Environ. Earth Sci. 83, 134.

Sumfleth, K., Duttmann, R., 2008. Prediction of soil property distribution in paddy soil landscapes using terrain data and satellite information as indicators. Ecol. Indic. 8 (5), 485–501.

Sun, X., Rosin, P., Martin, R., Langbein, F., 2007. Fast and effective feature-preserving mesh denoising. IEEE Trans. Vis. Comput. Graph. 13 (5), 925–938.

Sun, X.-L., Wang, H.-L., Zhao, Y.-G., Zhang, C., Zhang, G.-L., 2017. Digital soil mapping based on wavelet decomposed components of environmental covariates. Geoderma 303, 118–132.

Sundararajan, D., 2001. The Discrete Fourier Transform: Theory, Algorithms and Applications. World Scientific, Singapore.

Sunil, P.S., Reddy, C.D., Ponraj, M., Dhar, A., Jayapaul, D., 2017. GPS determination of the velocity and strain-rate fields on Schirmacher Glacier, central Dronning Maud Land, Antarctica. J. Glaciol. 53 (183), 558–564.

Svetlitchnyi, A.A., Plotnitskiy, S.V., Stepovaya, O.Y., 2003. Spatial distribution of soil moisture content within catchments and its modelling on the basis of topographic data. J. Hydrol. 277 (1–2), 50–60.

Swets, J.A., 1961. Detection theory and psychophysics: a review. Psychometrika 26 (1), 49–63.

Szeliski, R., 2022. Structure from motion and SLAM. In: Szeliski, R. (Ed.), Computer Vision: Algorithms and Applications, second ed. Texts in Computer Science Book Series. Springer, Cham, Switzerland, pp. 543–594.

Szymanowski, M., Jancewicz, K., Różycka, M., Migoń, P., 2019. Geomorphometry-based detection of enhanced erosional signal in polygenetic medium-altitude mountain relief and its tectonic interpretation, the Sudetes (Central Europe). Geomorphology 341, 115–129.

Szynkaruk, E., Graduno-Monroy, V.H., Bocco, G., 2004. Active fault systems and tectono-topographic configuration of the central Trans-Mexican Volcanic Belt. Geomorphology 61 (1–2), 111–126.

Tabik, S., Zapata, E.L., Romero, L.F., 2013. Simultaneous computation of total viewshed on large high resolution grids. Int. J. Geogr. Inf. Sci. 27 (4), 804–814.

Tabor, R.W., 1986. Changing concepts of geologic structure and the problem of siting nuclear reactors: examples from Washington State. Geology 14 (9), 738–742.

Tabrizian, P., Baran, P.K., van Berkel, D., Mitasova, H., Meentemeyer, R., 2020. Modeling restorative potential of urban environments by coupling viewscape analysis of lidar data with experiments in immersive virtual environments. Landsc. Urban Plan. 195, 103704.

Tadono, T., Ishida, H., Oda, F., Naito, S., Minakawa, K., Iwamoto, H., 2014. Precise global DEM generation by ALOS PRISM. In: Jiang, J., Zhang, H. (Eds.), ISPRS Technical Commission IV Symposium, Suzhou, China, 14–16 May 2014. ISPRS Annals Photogramm. Remote Sens. Spatial Inf. Sci. II (4), pp. 71–76.

Tadono, T., Nagai, H., Ishida, H., Oda, F., Naito, S., Minakawa, K., et al., 2016. Generation of the 30 m-mesh global digital surface model by ALOS PRISM. In: Halounova, L., Šafář, V., Jiang, J., Olešovská, H., Dvořáček, P., Holland, D., et al. (Eds.), XXIII ISPRS Congress, Commission IV, 12–19 July 2016, Prague, Czech Republic. Int. Arch. Photogramm. Remote Sens. Spatial Inf. Sci. XLI (B4), pp. 157–162.

Taghizadeh-Mehrjardi, R., Schmidt, K., Toomanian, N., Heung, B., Behrens, T., Mosavi, A., et al., 2021. Improving the spatial prediction of soil salinity in arid regions using wavelet transformation and support vector regression models. Geoderma 383, 114793.

Tagil, S., Jenness, J., 2008. GIS-based automated landform classification and topographic, landcover and geologic attributes of landforms around the Yazoren Polje, Turkey. J. Appl. Sci. 8 (6), 910–921.

Takahashi, H., 1981. A lineament enhancement technique for active fault analysis. In: Machine Processing of Remotely Sensed Data with Special Emphasis on Range, Forest, and Wetlands Assessment: Proceedings of the 7th International Symposium, West Lafayette, IN, 23–26 June 1981. Purdue University, West Lafayette, IN, pp. 103–112.

Tan, Y., Wang, S., Xu, B., Zhang, J., 2018. An improved progressive morphological filter for UAV-based photogrammetric point clouds in river bank monitoring. ISPRS J. Photogrammet. Remote Sens. 146, 421–429.

Tanyas, H., Rossi, M., Alvioli, M., van Westena, C.J., Marchesini, I., 2019. A global slope unit-based method for the near real-time prediction of earthquake-induced landslides. Geomorphology 327, 126–146.

Tapete, D., Banks, V., Jones, L., Kirkham, M., Garton, D., 2017. Contextualising archaeological models with geological, airborne and terrestrial LiDAR data: the Ice Age landscape in Farndon Fields, Nottinghamshire, UK. J. Archaeol. Sci. 81, 31–48.

Tarboton, D.G., 1997. A new method for the determination of flow directions and upslope areas in grid digital elevation models. Water Resour. Res. 33 (2), 309–319.

Targulian, V.O., Krasilnikov, P.V., 2007. Soil system and pedogenic processes: self-organization, time scales, and environmental significance. Catena 71 (3), 373–381.

Targulian, V.O., Sokolova, T.A., 1996. Soil as a biotic/abiotic natural system: a reactor, memory, and regulator of biospheric interactions. Eurasian Soil Sci. 2 (1), 30–41.

Tarolli, P., 2014. High-resolution topography for understanding Earth surface processes: opportunities and challenges. Geomorphology 216, 295–312.

Tarolli, P., Sofia, G., 2016. Human topographic signatures and derived geomorphic processes across landscapes. Geomorphology 255, 140–161.

Tarolli, P., Cao, W., Sofia, G., Evans, D., Ellis, E.C., 2019a. From features to fingerprints: a general diagnostic framework for anthropogenic geomorphology. Progr. Phys. Geogr. 43 (1), 95–128.

Tarolli, P., Cavalli, M., Masin, R., 2019b. High-resolution morphologic characterization of conservation agriculture. Catena 172, 846–856.

Taurisano, A., Tronstad, S., Brandt, O., Kohler, J., 2006. On the use of ground penetrating radar for detecting and reducing crevasse-hazard in Dronning Maud Land, Antarctica. Cold Reg. Sci. Technol. 45 (3), 166–177.

Taychinov, S.N., Fayzullin, M.M., 1958. Dynamics of soil moisture in relation to topography. Soviet Soil Science 10, 1121–1126.

Taylor, J.A., Odeh, I.O.A., 2007. Comparing discriminant analysis with binomial logistic regression, regression kriging and multi-indicator kriging for mapping salinity risk in Northwest New South Wales, Australia. In: Lagacherie, P., McBratney, A.B., Voltz, M. (Eds.), Digital Soil Mapping: An Introductory Perspective. Developments in Soil Science, vol. 31. Elsevier, Amsterdam, the Netherlands, pp. 455–464.

Taylor, L.S., Quincey, D.J., Smith, M.W., Baumhoer, C.A., McMillan, M., Mansell, D.T., 2021. Remote sensing of the mountain cryosphere current capabilities and future opportunities for research. Progr. Phys. Geogr. 45 (6), 931–964.

Temme, A.J., Schoorl, J.M., van der Meij, W.M., 2022. New geomorphometric variables for non-continuous hillslopes—assessing the value for digital soil mapping. Geoderma 418, 115848.

Tempfli, K., 1980. Spectral analysis of terrain relief for the accuracy estimation of digital terrain models. ITC J. (3), 478–510.

Terra, J.A., Shaw, J.N., Reeves, D.W., Raper, R.L., van Santen, E., Mask, P.L., 2004. Soil carbon relationships with terrain attributes, electrical conductivity, and a soil survey in a coastal plain landscape. Soil Sci. 169 (12), 819–831.

Tesch, P., Reece, R.S., Pope, M.C., Markello, J.R., 2018. Quantification of architectural variability and controls in an Upper Oligocene to Lower Miocene carbonate ramp, Browse Basin, Australia. Mar. Petrol. Geol. 91, 432–454.

Tesfa, T.K., Tarboton, D.G., Chandler, D.G., McNamara, J.P., 2010. A generalized additive soil depth model for a mountainous semi-arid watershed based upon topographic and land cover attributes. In: Boettinger, J.L., Howell, D.W., Moore, A.C., Hartemink, A.E., Kienast-Brown, S. (Eds.), Digital Soil Mapping: Bridging Research, Environmental Application, and Operation. Progress in Soil Science Book Series. Springer, Dordrecht, the Netherlands, pp. 29–41.

Tetuev, R.K., Dedus, F.F., 2007. Classical Orthogonal Polynomials: Application to Data Processing Problems. Institute of Mathematical Problems of Biology, Russian Academy of Sciences, Pushchino, Russia (in Russian).

Thelin, G.P., Pike, R.J., 1991a. Landforms of the Conterminous United States—a Digital Shaded-Relief Portrayal. Miscellaneous Investigations Map I-2206, Scale 1:3,500,000. US Geological Survey, Denver, CO.

Thelin, G.P., Pike, R.J., 1991b. Landforms of the Conterminous United States—a Digital Shaded-Relief Portrayal. Text to accompany Map I-2206. US Geological Survey, Denver, CO.

Theodoratos, N., Seybold, H., Kirchner, J.W., 2018. Scaling and similarity of a stream-power incision and linear diffusion landscape evolution model. Earth Surf. Dyn. 6 (3), 79–808.

Thienssen, R.L., Soofi, K., Sheline, H., 1994. A new expandable detector applied to digital topography and TM image data in support of petroleum exploration. Photogrammet. Eng. Remote Sens. 60 (1), 77–85.

Thistlewood, H.M.A., Gill, P., Beers, E.H., Shearer, P.W., Walsh, D.B., Rozema, B.M., et al., 2018. Spatial analysis of seasonal dynamics and overwintering of *Drosophila suzukii* (Diptera: Drosophilidae) in the Okanagan–Columbia Basin, 2010–2014. Environ. Entomol. 47 (2), 221–232.

Thomas, A.L., King, D., Dambrine, E., Couturier, A., Roque, J., 1999. Predicting soil classes with parameters derived from relief and geologic materials in a sandstone region of the Vosges mountains (Northeastern France). Geoderma 90 (3–4), 291–305.

Thomas, P.C., 1989. The shapes of small satellites. Icarus 77 (2), 248–274.

Thompson, J.A., Kolka, R.K., 2005. Soil carbon storage estimation in a forested watershed using quantitative soil–landscape modeling. Soil Sci. Soc. Am. J. 69 (4), 1086–1093.

Thompson, J.A., Bell, J.C., Butler, C.A., 2001. Digital elevation model resolution: effects on terrain attribute calculation and quantitative soil–landscape modeling. Geoderma 100 (1–2), 67–89.

Thompson, J.C., Moore, R.D., 1996. Relations between topography and water table depth in a shallow forest soil. Hydrol. Process. 10 (11), 1513–1525.

Thompson, S., Benn, D., Mertes, J., Luckman, A., 2016. Stagnation and mass loss on a Himalayan debris-covered glacier: processes, patterns and rates. J. Glaciol. 62 (233), 467–485.

Thompson, S.S., Cook, S., Kulessa, B., Winberry, J.P., Fraser, A.D., Galton-Fenzi, B.K., 2020. Comparing satellite and helicopter-based methods for observing crevasses, application in East Antarctica. Cold Reg. Sci. Technol. 178, 103128.

Thornton, J.M., Mariethoz, G., Brunner, P., 2018. A 3D geological model of a structurally complex Alpine region as a basis for interdisciplinary research. Sci. Data 5, 180238.

Tian, Y., Lei, S., Bian, Z., Lu, J., Zhang, S., Fang, J., 2018. Improving the accuracy of open source digital elevation models with multi-scale fusion and a slope position-based linear regression method. Remote Sens. (Basel) 10 (12), 1861.

Tiedje, J.M., 1994. Denitrifiers. In: Weaver, R.W., Angle, S., Bottomley, P., Bezdicek, D., Smith, S., Tabatabai, A., et al. (Eds.), Methods of Soil Analysis. Pt. 2: Microbiological and Biochemical Properties. Soil Science Society of America, Madison, WI, pp. 245–267.

Tien Bui, D., Ngo, P.-T.T., Pham, T.D., Jaafari, A., Minh, N.Q., Hoa, P.V. et al., 2019a. A novel hybrid approach based on a swarm intelligence optimized extreme learning machine for flash flood susceptibility mapping. Catena 179, 184–196.

Tien Bui, D., Shirzadi, A., Shahabi, H., Chapi, K., Omidavr, E., Pham, B.T., et al., 2019b. A novel ensemble artificial intelligence approach for gully erosion mapping in a semi-arid watershed (Iran). Sensors 19 (11), 2444.

Tillo, A.A., 1890. Orography of the European Russia from the hypsometric map. Proc. Imperial Russ. Geograph. Soc. 26 (1), 8–32 (in Russian).

Timoshenko, S., Goodier, J.N., 1951. Theory of Elasticity, second ed. McGraw-Hill, New York, NY.

Tobler, W.R., 1966. Numerical map generalization. Michigan Inter Univ. Comm. Math. Geogr. Disc Paper 8, 1–27.

Tobler, W.R., 1969. Geographical filters and their inverses. Geogr. Anal. 1 (3), 234–253.

Tomczyk, A.M., Ewertowski, M.W., 2017. Surface morphological types and spatial distribution of fan-shaped landforms in the periglacial high-Arctic environment of Central Spitsbergen, Svalbard. J. Maps 13 (2), 239–251.

Tomer, M.D., Anderson, J.L., 1995. Variation of soil water storage across a sand plain hillslope. Soil Sci. Soc. Am. J. 59 (4), 1091–1100.

Tong, R., Purser, A., Guinan, J., Unnithan, V., Yu, J., Zhang, C., 2016. Quantifying relationships between abundances of cold-water coral *Lophelia pertusa* and terrain features: a case study on the Norwegian margin. Cont. Shelf Res. 116, 13–26.

Tonkin, T.N., Midgley, N.G., Cook, S.J., Graham, D.J., 2016. Ice-cored moraine degradation mapped and quantified using an unmanned aerial vehicle: a case study from a polythermal glacier in Svalbard. Geomorphology 258, 1–10.

Tooth, S., 2006. Virtual globes: a catalyst for the re-enchantment of geomorphology? Earth Surf. Process. Landf. 31 (9), 1192–1194.

Toso, C., Madricardo, F., Molinaroli, E., Fogarin, S., Kruss, A., Petrizzo, A., et al., 2019. Tidal inlet seafloor changes induced by recently built hard structures. PLoS One 14 (10), e0223240.

Toth, C. Jóźków, G., 2016. Remote sensing platforms and sensors: a survey. ISPRS J. Photogrammet. Remote Sens. 115, 22–36.

Toutin, T., 2000. Elevation modelling from satellite data. In: Meyers, R.A. (Ed.), Encyclopedia of Analytical Chemistry: Applications, Theory and Instrumentation, vol. 10. Wiley, Chichester, UK, pp. 8543–8572.

Toutin, T., 2004a. DSM generation and evaluation from QuickBird stereo imagery with 3D physical modelling. Int. J. Remote Sens. 25 (22), 5181–5193.

Toutin, T., 2004b. DTM generation from Ikonos in-track stereo images using a 3D physical model. Photogrammet. Eng. Remote Sens. 70 (6), 695–702.

Toutin, T., 2006. Generation of DSMs from SPOT-5 in-track HRS and across-track HRG stereo data using spatiotriangulation and autocalibration. ISPRS J. Photogrammet. Remote Sens. 60 (3), 170–181.

Toutin, T., 2008. ASTER DEMs for geomatic and geoscientific applications: a review. Int. J. Remote Sens. 29 (7), 1855–1875.

Toutin, T., 2010. Impact of Radarsat-2 SAR ultrafine-mode parameters on stereo-radargrammetric DEMs. IEEE Trans. Geosci. Remote Sens. 48 (10), 3816–3823.

Toutin, T., Gray, A.L., 2000. State-of-the-art of extraction of elevation data using satellite SAR data. ISPRS J. Photogrammet. Remote Sens. 5 (1), 13–33.

Tozer, B., Sandwell, D.T., Smith, W.H.F., Olson, C., Beale, J.R., Wessel, P., 2019. Global bathymetry and topography at 15 arc seconds: SRTM15+. Earth Space Sci. 6 (10), 1847–1864.

Trangmar, B.B., Yost, R.S., Wade, M.K., Uehara, G., Sudjadi, M., 1987. Spatial variation of soil properties and rice yield on recently cleared land. Soil Sci. Soc. Am. J. 51 (3), 668–674.

Trevisani, S., Cavalli, M., 2016. Topography-based flow-directional roughness: potential and challenges. Earth Surf. Dyn. 4 (2), 343–358.

Trevisani, S., Florinsky, I.V., 2023. Morphometry. In: Sagar, B.S.D., Cheng, Q., McKinley, J., Agterberg, F. (Eds.), Encyclopedia of Mathematical Geosciences. Encyclopedia of Earth Sciences Series. Springer, Cham, Switzerland, pp. 928–934.

Trevisani, S., Rocca, M., 2015. MAD: robust image texture analysis for applications in high resolution geomorphometry. Comput. Geosci. 81, 78–92.

Trevisani, S., Cavalli, M., Marchi, L., 2012. Surface texture analysis of a high-resolution DTM: interpreting an alpine basin. Geomorphology 161–162, 26–39.

Trevisani, S., Skrypitsyna, T.N., Florinsky, I.V., 2023a. Global digital elevation models for terrain morphology analysis in mountain environments: insights on Copernicus GLO-30 and ALOS AW3D30 for a large Alpine area. Environ. Earth Sci. 82 (9), 198.

Trevisani, S., Teza, G., Guth, P., 2023b. A simplified geostatistical approach for characterizing key aspects of short-range roughness. Catena 22[illegible], 106927.

Trevisani, S., Teza, G., Guth, P., 2023c. Hacking the topographic ruggedness index. Geomorphology 439, 108838.

Tribe, A., 1992. Automated recognition of valley lines and drainage networks from grid digital elevation models: a review and a new method. J. Hydrol. 139 (1–2), 263–293.

Trifonov, V.G., 1983. Late Quaternary Tectonics. Nauka, Moscow, USSR (in Russian).

Trifonov, V.G., 1995. World map of active faults (preliminary results of studies). Quat. Int. 25 (1), 3–12.

Trifonov, V.G., Makarov, V.I., Safonov, Y.G., Florensky, P.V. (Eds.), 1983. Space Information in Geology. Nauka, Moscow, USSR (in Russian).

Tripp, G.I., Vearncombe, J.R., 2004. Fault/fracture density and mineralization: a contouring method for targeting gold exploration. J. Struct. Geol. 26 (6–7), 1087–1108.

Troeh, F.R., 1964. Landform parameters correlated to soil drainage. Soil Sci. Soc. Am. Proc. 28 (6), 808–812.

Troen, I., Lundtang Petersen, E., 1989. European Wind Atlas. Risø National Laboratory, Roskilde, Denmark.

Turner, D., Lucieer, A., Watson, C., 2012. An automated technique for generating georectified mosaics from ultra-high resolution unmanned aerial vehicle (UAV) imagery, based on structure from motion (SfM) point clouds. Remote Sens. (Basel) 4 (5), 1392–1410.

Turner, D., Lucieer, A., Wallace, L., 2014. Direct georeferencing of ultrahigh-resolution UAV imagery. IEEE Trans. Geosci. Remote Sens. 52 (5), 2738–2745.

Tuttle, B.T., Anderson, S., Huff, R., 2008. Virtual globes: an overview of their history, uses, and future challenges. Geogr. Compass 2 (5), 1478–1505.

Twidale, C.R., 2004. River patterns and their meaning. Earth Sci. Rev. 67 (3–4), 159–218.

Twidale, C.R., Bourne, J.A., 2007. Lineaments and cryptostructural effects. In: Bourne, J.A., Twidale, C.R. (Eds.), Crustal Structures and Mineral Deposits: E.S.T. O'Driscoll's Contribution to Mineral Exploration. Rosenberg, Kenthurst, NSW, pp. 153–164.

Tyuflin, Y.S., 1996. Conversion of coordinates in photogrammetric works with digital elevation models. Geodesy Cartogr. (2), 21–25 (in Russian).

Tzvetkov, J., 2018. Relief visualization techniques using free and open source GIS tools. Polish Cartogr. Rev. 50 (2), 61–71.

Ufimtsev, G.F., 1988. Topography and geological structure. In: Logachev, N.A., Timofeev, D.A., Ufimtsev, G.F. (Eds.), Problems of Theoretical Geomorphology. Nauka, Moscow, USSR, pp. 59–75 (in Russian).

US DMA, 1992. Digital Chart of the World, Scale 1:1,000,000. US Defense Mapping Agency, Fairfax, VA.

US DMA, 1993. Performance Specification: Digital Terrain Elevation Data (DTED). MIL-PRF-89020, US Defense, Mapping Agency, Fairfax, VA.

USGS, 1972. Geologic Atlas of the Moon, Scale 1:1,000,000. US Geological Survey, Washington, DC.

USGS, 1993. Digital Elevation Models: Data Users Guide 5. US Geological Survey, Reston, VA.

USGS, 1996. Global 30 Arc-Second Elevation (GTOPO30). Earth Resources Observation and Science Center, US Geological Survey, Sioux Falls, SD, https://doi.org/10.5066/F7DF6PQS.

USGS, 2000. HYDRO1k Elevation Derivative Database. Earth Resources Observation and Science Center, US Geological Survey, Sioux Falls, SD, https://doi.org/10.5066/F77P8WN0.

USGS, 2011. Global Multi-Resolution Terrain Elevation Data 2010 (GMTED2010). Earth Resources Observation and Science Center, US Geological Survey, Sioux Falls, SD, https://doi.org/10.5066/F7J38R2N.

USGS, 2015. EarthExplorer. Earth Resources Observation and Science Center, US Geological Survey, Sioux Falls, SD. http://earthexplorer.usgs.gov. (Accessed 27 June 2024).

USSR General Headquarters, 1981. Topographic Map, Scale 1:200,000, Sheet 13-36-06 (Lgov). General Headquarters, Moscow, USSR (in Russian).

USSR General Headquarters, 1986. Topographic Map, Scale 1:500,000, Sheet 12-36-4 (Simferopol). General Headquarters, Moscow, USSR (in Russian).

Ustinova, V.N., Ustinov, V.G., 2004. Seismic morphological analysis in mapping of high-capacity collector layers. I. Morphological classification. Manifestation of facies type of a collector layer in paleotopography. Bull. Tomsk Polytech. Univ. 307 (3), 49–53 (in Russian).

Uuemaa, E., Ahi, S., Montibeller, B., Muru, M., Kmoch, A., 2020. Vertical accuracy of freely available global digital elevation models (ASTER, AW3D30, MERIT, TanDEM-X, SRTM, and NASADEM). Remote Sens. (Basel) 12 (21), 3482.

Vajedian, S., Motagh, M., 2019. Extracting sinkhole features from time-series of TerraSAR-X/TanDEM-X data. ISPRS J. Photogrammet. Remote Sens. 150, 274–284.

Vakhtin, B., 1930. On the determination of mathematical characteristics of topography. Geodezist (2–3), 7–16 (in Russian).

Valentine, A.P., Kalnins, L.M., 2016. An introduction to learning algorithms and potential applications in geomorphometry and Earth surface dynamics. Earth Surf. Dyn. 4, 445–460.

Valeriano, M.M., Rossetti, D.F., 2017. Regionalization of local geomorphometric derivations for geological mapping in the sedimentary domain of central Amazônia. Comput. Geosci. 100, 46–56.

Van der Meij, W.M., Meijles, E.W., Marcos, D., Harkema, T.T.L., Candel, J.H.J., Maas, G.J., 2022. Comparing geomorphological maps made manually and by deep learning. Earth Surf. Process. Landf. 47 (4), 1089–1107.

Van der Sluijs, J., Kokelj, S.V., Fraser, R.H., Tunnicliffe, J., Lacelle, D., 2018. Permafrost terrain dynamics and infrastructure impacts revealed by UAV photogrammetry and thermal imaging. Remote Sens. (Basel) 10 (11), 1734.

Van der Veen, C.J., 1999. Crevasses on glaciers. Polar Geogr. 23 (3), 213–245.

Van der Veen, C.J., 2011. Crevasses. In: Singh, V.P., Singh, P., Haritashya, U.K. (Eds.), Encyclopedia of Snow, Ice and Glaciers. Encyclopedia of Earth Sciences Series. Springer, Dordrecht, the Netherlands, pp. 165–168.

Van Driel, J.N., 1989. Three dimensional display of geologic data. In: Raper, J. (Ed.), Three Dimensional Applications in GIS. Taylor and Francis, London, UK, pp. 1–9.

Van Kessel, C., Pennock, D.J., Farrel, R.E., 1993. Seasonal variations in denitrification and nitrous oxide evolution. Soil Sci. Soc. Am. J. 57 (4), 988–995.

Van Kreveld, M., 1997. Digital elevation models and TIN algorithms. In: Van Kreveld, M., Nievergelt, J., Roos, T., Widmayer, P. (Eds.), Algorithmic Foundations of Geographic Information Systems. Lect. Notes Comput. Sci., vol. 1340, pp. 37–78.

Van Niel, K.P., Laffan, S.W., Lees, B.G., 2004. Effect of error in the DEM on environment variables for predictive vegetation modelling. J. Veg. Sci. 15 (6), 747–756.

Van Nieuwenhuizen, N., Lindsay, J.B., DeVries, B., 2021a. Automated mapping of transportation embankments in fine-resolution LiDAR DEMs. Remote Sens. (Basel) 13 (7), 1308.

Van Nieuwenhuizen, N., Lindsay, J.B., DeVries, B., 2021b. Smoothing of digital elevation models and the alteration of overland flow path length distributions. Hydrol. Process. 35 (7), e14271.

Van Rossel, J., 1972. Digital hypsographic map compilation. Photogrammet. Eng. 38 (11), 1106–1116.

Vapnik, V.N., 2000. The Nature of Statistical Learning Theory, second ed. Information Science and Statistics Book Series. Springer, New York, NY.

Vasmut, A.S., 1983. Modelling in Cartography Using Computers. Nedra, Moscow, USSR (in Russian).

Vassilaki, D.I., Stamos, A.A., 2020. TanDEM-X DEM: comparative performance review employing LIDAR data and DSMs. ISPRS J. Photogrammet. Remote Sens. 160, 33–50.

Vasu, N.N., Lee, S.-R., 2016. A hybrid feature selection algorithm integrating an extreme learning machine for landslide susceptibility modeling of Mt. Woomyeon, South Korea. Geomorphology 263, 50–70.

Vaz, D.A., Silvestro, S., Sarmento, P.T.K., Cardinale, M., 2017. Migrating meter-scale bedforms on Martian dark dunes: are terrestrial aeolian ripples good analogs? Aeolian Res. 26, 101–116.

Veitinger, J., Purves, R.S., Sovilla, B., 2016. Potential slab avalanche release area identification from estimated winter terrain: a multi-scale, fuzzy logic approach. Nat. Haz. Earth Syst. Sci. 16 (10), 2211–2225.

Vening Meinesz, F.A., 1947. Shear patterns of the Earth's crust. Trans. Am. Geophys. Union 28 (1), 1–61.

Venteris, E.R., McCarty, G.W., Ritchie, J.C., Gish, T., 2004. Influence of management history and landscape variables on soil organic carbon and soil redistribution. Soil Sci. 169 (11), 787–795.

Verdin, K.L., Greenlee, S.K., 1998. HYDRO1k Documentation. Earth Resources Observation and Science Center. US Geological Survey, Sioux Falls, SD. https://www.usgs.gov/media/files/hydro-1k-readme. (Accessed 27 June 2024).

Veregin, H., 1997. The effects of vertical error in digital elevation models on the determination of flow-path direction. Cartogr. Geogr. Inf. Syst. 24 (2), 67–79.

Verkhovtsev, V.G., 2008. Contemporary Platform Geostructures of the Ukraine and Dynamics of Their Development (D.Sc. thesis). Institute of Geological Sciences, Ukrainian National Academy of Sciences, Kiev, Ukraine (in Russian).

Verleyen, E., Hodgson, D.A., Sabbe, K., Vyverman, W., 2004. Late Quaternary deglaciation and climate history of the Larsemann Hills (East Antarctica). J. Quat. Sci. 19 (4), 361–375.

Vermeulen, D., van Niekerk, A., 2017. Machine learning performance for predicting soil salinity using different combinations of geomorphometric covariates. Geoderma 299, 1–12.

Veronesi, F., Hurni, L., 2014. Changing the light azimuth in shaded relief representation by clustering aspect. Cartogr. J. 51 (4), 291–300.

Vigil, J.F., Pike, R.J., Howell, D.G., 2000a. A Tapestry of Time and Terrain. Geologic Investigations Series Map I-2720, Scale 1:3,500,000. US Geological Survey, Denver, CO.

Vigil, J.F., Pike, R.J., Howell, D.G., 2000b. A Tapestry of Time and Terrain. Pamphlet to accompany Geologic Investigations Series Map I-2720. US Geological Survey, Denver, CO.

Villaça, C.V.N., Crósta, A.P., Grohmann, C.H., 2021. Morphometric analysis of Pluto's impact craters. Remote Sens. (Basel) 13 (3), 377.

Vincent, C., Wagnon, P., Shea, J.M., Immerzeel, W.W., Kraaijenbrink, P., Shrestha, D., et al., 2016. Reduced melt on debris-covered glaciers: investigations from Changri Nup Glacier, Nepal. Cryosphere 10 (4), 1845–1858.

Vincent, S., Lemercier, B., Berthier, L., Walter, C., 2018. Spatial disaggregation of complex soil map units at the regional scale based on soil-landscape relationships. Geoderma 311, 130–142.

Vincenty, T., 1975. Direct and inverse solutions of geodesics on the ellipsoid with application of nested equations. Survey Rev. 23 (176), 88–93.

Vinogradova, A.I., Eremin, V.K. (Eds.), 1971. Aerial Methods for Geological Investigations. Nedra, Leningrad, USSR (in Russian).

Volchanskaya, I.K., 1981. Morphostructural Regularities in Location of Endogenous Mineralization. Nauka, Moscow, USSR (in Russian).

Volkov, N.M., 1950. Principles and Methods of Cartometry. Soviet Academic Press, Moscow, USSR (in Russian).

Volkov, Y.V., 1995. Loxodromy and minerageny (influence of astronomic resonances in the Earth–Moon system on the origin of ore deposits in the Earth's crust). Bull. Moscow Soc. Nat. Geol. Ser. 70 (6), 90–94 (in Russian).

Volkova, N.I., Zhuchkova, V.K., 2000. Critical analysis of the morphoisographic method as applied to soil mapping. Eurasian Soil Sci. 33 (10), 1025–1037.

Vörös, F., Székely, B., 2022. High-resolution DTM-based estimation of geomorphometric parameters of selected putative Martian scoria cones. Icarus 377, 114923.

Vörös, F., van Wyk de Vries, B., Karátson, D., Székely, B., 2021. DTM-based morphometric analysis of scoria cones of the Chaîne des Puys (France)—the classic and a new approach. Remote Sens. (Basel) 13 (10), 1–22.

Vörös, F., van Wyk de Vries, B., Guilbaud, M.N., Görüm, T., Karátson, D., Székely, B., 2022. DTM-based comparative geomorphometric analysis of four scoria cone areas—suggestions for additional approaches. Remote Sens. (Basel) 14 (23), 6152.

Vörösmarty, C.J., Fekete, B.M., Meybeck, M., Lammers, R.B., 2000. Geomorphometric attributes of the global system of rivers at 30-minute spatial resolution. J. Hydrol. 237 (1–2), 17–39.

Vysotsky, G.N., 1906. On oro-climatic bases of soil classification. Pochvovedenie 8 (1–4), 1–18 (in Russian).

Waagen, J., 2019. New technology and archaeological practice. Improving the primary archaeological recording process in excavation by means of UAS photogrammetry. J. Archaeol. Sci. 101, 11–20.

Wabiński, J., Mościcka, A., 2019. Natural heritage reconstruction using full-color 3D printing: a case study of the valley of five Polish Ponds. Sustainability 11 (21), 5907.

Wadoux, A.M.J.-C., 2019. Using deep learning for multivariate mapping of soil with quantified uncertainty. Geoderma 351, 59–70.

Wadoux, A.M.J.-C., Brus, D.J., Heuvelink, G.B.M., 2018. Accounting for non-stationary variance in geostatistical mapping of soil properties. Geoderma 324, 138–147.

Wadoux, A.M.J.-C., Padarian, J., Minasny, B., 2019. Multi-source data integration for soil mapping using deep learning. Soil 5 (1), 107–119.

Wadoux, A.M.J.-C., Minasny, B., McBratney, A.B., 2020. Machine learning for digital soil mapping: applications, challenges and suggested solutions. Earth Sci. Rev. 210, 103359.

Wadoux, A.M.J.-C., Heuvelink, G.B.M., Lark, R.M., Lagacherie, P., Bouma, J., Mulder, V.L., et al., 2021. Ten challenges for the future of pedometrics. Geoderma 401, 115155.

Wadoux, A.M.J.-C., Odeh, I.O.A., McBratney, A.B., 2023. Overview of pedometrics. In: Goss, M.J., Oliver, M. (Eds.), Encyclopedia of Soils in the Environment, second ed., vol. 4: Stockmann, U., Baeumler, R. (Eds.), Pedology and Pedometrics. Academic Press, Oxford, UK, pp. 471–485.

Wahba, G., 1981. Spline interpolation and smoothing on the sphere. SIAM J. Sci. Stat. Comput. 2 (1), 5–16.

Waksman, S.A., Deemer, R.B., 1928. Participants in the official tour of the 1st International Congress of Soil Science through the United States and Canada. In: Waksman, S.A., Deemer, R.B. (Eds.), Proceedings and Papers of the 1st International Congress of Soil Science, Washington, DC, 13–22 June 1927. Transcontinental Excursion and Impressions of the Congress and of America. American Organizing Committee of the 1st International Congress of Soil Science, Washington, DC, pp. 22–39.

Walker, P.H., Hall, G.F., Protz, R., 1968. Relation between landform parameters and soil properties. Soil Sci. Soc. Am. Proc. 32 (1), 101–104.

Walker, S.J., Wilkinson, S.N., van Dijk, A.I.J.M., Hairsine, P.B., 2020. A multi-resolution method to map and identify locations of future gully and channel incision. Geomorphology 358, 107115.

Walter, C., Lagacherie, P., Follain, S., 2007. Integrating pedological knowledge into digital soil mapping. In: Lagacherie, P., McBratney, A.B., Voltz, M. (Eds.), Digital Soil Mapping: An Introductory Perspective. Developments in Soil Science, vol. 31. Elsevier, Amsterdam, the Netherlands, pp. 281–300.

Wang, J., Cheng, W., Zhou, C., Zheng, X., 2017. Automatic mapping of lunar landforms using DEM-derived geomorphometric parameters. J. Geogr. Sci. 27 (11), 1413–1427.

Wang, J., Kreslavsky, M.A., Liu, J., Head, J.W., Zhang, K., Kolenkina, M.M., et al., 2020. Quantitative characterization of impact crater materials on the Moon: changes in topographic roughness and thermophysical properties with age. J. Geophys. Res. Planet. 125 (10), e2019JE006091.

Wang, S., Hu, Q., Wang, F., Ai, M., Zhong, R., 2017. A microtopographic feature analysis-based LiDAR data processing approach for the identification of Chu tombs. Remote Sens. (Basel) 9 (9), 880.

Wang, W., Zhao, Y., Zhang, T., Wang, R., Wei, Z., Sun, Q., et al., 2021. Regional soil thickness mapping based on stratified sampling of optimally selected covariates. Geoderma 400, 115092.

Ward, G., Simmons, M., 1999. The holodeck ray cache: an interactive rendering system for global illumination in nondiffuse environments. ACM Trans. Graph. 18 (4), 361–368.

Warner, W.S., 1995. Mapping a three-dimensional soil surface with hand-held 35 mm photography. Soil Tillage Res. 34 (3), 187–197.

Watson, C.S., Quincey, D.J., Smith, M., Carrivick, J., Rowan, A.V., James, M., 2017. Quantifying ice cliff evolution with multi-temporal point clouds on the debris-covered Khumbu Glacier, Nepal. J. Glaciol. 63 (241), 823–837.

Watson, D., 1992. Contouring: A Guide to the Analysis and Display of Spatial Data. Pergamon, Oxford, UK.

Watts, A.B., Sandwell, D.T., Smith, W.H.F., Wessel, P., 2006. Global gravity, bathymetry, and the distribution of submarine volcanism through space and time. J. Geophys. Res. Solid Earth 111 (B8), B08408.

Watts, A.C., Ambrosia, V.G., Hinkley, E.A., 2012. Unmanned aircraft systems in remote sensing and scientific research: classification and considerations of use. Remote Sens. (Basel) 4 (6), 1671–1692.

Weatherall, P., Marks, K.M., Jakobsson, M., Schmitt, T., Tani, S., Arndt, J.E., et al., 2015. A new digital bathymetric model of the world's oceans. Earth Space Sci. 2 (2), 331–345.

Webster, E.A., Hopkins, D.W., 1996. Contributions from different microbial processes to N_2O emission from soil under different moisture regimes. Biol. Fertil. Soils 22 (4), 331–335.

Webster, R., 1994. The development of pedometrics. Geoderma 62 (1–3), 1–15.

Webster, R., Burrough, P.A., 1974. Multiple discriminant analysis in soil survey. J. Soil Sci. 25 (1), 121–134.

Webster, R., Oliver, M.A., 1992. Sample adequately to estimate variograms of soil properties. J. Soil Sci. 43 (1), 177–192.

Webster, R., Oliver, M.A., 2007. Geostatistics for Environmental Scientists, second ed. Wiley, Chichester, UK.

Wegener, A., 1915. Die Entstehung der Kontinente und Ozeane. Sammlung Vieweg, Braunschweig, Germany.

Wehr, A., Lohr, U., 1999. Airborne laser scanning—an introduction and overview. ISPRS J. Photogrammet. Remote Sens. 54 (2–3), 68–82.

Wei, H., Zhou, G., Fu, S., 2018. Efficient Priority-Flood depression filling in raster digital elevation models. Int. J. Dig. Earth 12 (4), 415–427.

Weibel, R., 1992. Models and experiments for adaptive computer-assisted terrain generalization. Cartogr. Geogr. Inf. Syst. 19 (3), 133–153.

Weibel, R., 1997. Generalization of spatial data: principles and selected algorithms. In: Van Kreveld, M., Nievergelt, J., Roos, T., Widmayer, P. (Eds.), Algorithmic Foundations of Geographic Information Systems. Lect. Notes Comput. Sci., vol. 1340, pp. 99–152.

Weibel, R., Brändli, M., 1995. Adaptive methods for the refinement of digital terrain models for geomorphometric applications. Zeit. Geomorph. Suppl. 101, 13–30.

Weibel, R., Heller, M., 1991. Digital terrain modelling. In: Maguire, D.J., Goodchild, M.F., Rhind, D. (Eds.), Geographical Information Systems: Principles and Applications, vol. 1: Principles. Longman, Harlow, UK, pp. 269–297.

Weijermars, R., 1985/86. The polar spirals of Mars may be due to glacier surges deflected by Coriolis forces. Earth Planet. Sci. Lett. 76 (3–4), 227–240.

Weinberg, B.P., 1934a. Experience of mathematical definition of geomorphological concepts and mathematical characterization of geomorphic peculiarities. In: Proceedings of the 1st Pan-Union Geographical Congress, Leningrad, USSR, 11–18 April 1933, vol. 3. State Geographical Society, Leningrad, USSR, pp. 126–135 (in Russian).

Weinberg, B.P., 1934b. Some peculiarities of the Earth's crust folding. In: Proceedings of the 1st Pan-Union Geographical Congress, Leningrad, USSR, 11–18 April 1933, vol. 3. State Geographical Society, Leningrad, USSR, pp. 123–126 (in Russian).

Weiss, A.D., 2001. Topographic position and landforms analysis. Poster presentation in ESRI User Conference, San Diego, CA. https://www.jennessent.com/arcview/TPI_Weiss_poster.htm. (Accessed 27 June 2024).

Welch, B.C., Pfeffer, W.T., Harper, J.T., Humphrey, N.F., 1998. Mapping subglacial surfaces of temperate valley glaciers by two-pass migration of radio-echo sounding survey data. J. Glaciol. 44 (146), 164–170.

Welch, R., Jordan, T., Lang, H., Murakami, H., 1998. ASTER as a source for topographic data in the late 1990's. IEEE Trans. Geosci. Remote Sens. 36 (4), 1282–1289.

Wertz, J.B., 1974. Intersections linéamentaires et minéralisation dans les Montagnes Rocheuses de l'Amérique du Nord. Geol. Runds. 63 (2), 708–754.

Wessel, B., Huber, M., Wohlfart, C., Marschalk, U., Kosmann, D., Roth, A., 2018. Accuracy assessment of the global TanDEM-X digital elevation model with GPS data. ISPRS J. Photogrammet. Remote Sens. 139, 171–182.

Wessel, B., Huber, M., Wohlfart, C., Bertram, A., Osterkamp, N., Marschalk, U., et al., 2021. TanDEM-X PolarDEM 90 m of Antarctica: generation and error characterization. Cryosphere 15 (11), 5241–5260.

Western, A.W., Grayson, R.B., Blöschl, G., Willgoose, G.R., McMahon, T.A., 1999. Observed spatial organization of soil moisture and its relation to terrain indices. Water Resour. Res. 35 (3), 797–810.

Westoby, M.J., Brasington, J., Glasser, N.F., Hambrey, M.J., Reynolds, J.M., 2012. 'Structure-from-Motion' photogrammetry: a low-cost, effective tool for geoscience applications. Geomorphology 179, 300–314.

Westoby, M.J., Dunning, S.A., Woodward, J., Hein, A.S., Marrero, S.M., Winter, K., et al., 2016. Interannual surface evolution of an Antarctic blue-ice moraine using multi-temporal DEMs. Earth Surf. Dyn. 4 (2), 515–529.

Whelan, M.J., Gandolfi, C., 2002. Modelling of spatial controls on denitrification at the landscape scale. Hydrol. Process. 16 (7), 1437–1450.

White, D., 2000. Global grids from recursive diamond subdivisions of the surface of an octahedron or icosahedron. Environ. Monit. Assess. 64 (1), 93–103.

Whitehead, K., Hugenholtz, C.H., 2014. Remote sensing of the environment with small unmanned aircraft systems (UASs), Part 1: A review of progress and challenges. J. Unmanned Veh. Syst. 2 (3), 69–85.

Whitehead, K., Hugenholtz, C.H., Myshak, S., Brown, O., LeClair, A., Tamminga, A., et al., 2014. Remote sensing of the environment with small unmanned aircraft systems (UASs), Part 2: Scientific and commercial applications. J. Unmanned Veh. Syst. 2 (3), 86–102.

Whitmeyer, S.J., Bailey, J.E., de Paor, D.G., Ornduff, T. (Eds.), 2012. Google Earth and Virtual Visualizations in Geoscience Education and Research. Geological Society of America, Boulder, CO.

Wieczorek, M.A., 2007. Gravity and topography of the terrestrial planets. In: Spohn, T. (Ed.), Treatise on Geophysics, vol. 10. Elsevier, Amsterdam, the Netherlands, pp. 165–206.

Wiekenkamp, I., Huisman, J.A., Bogena, H.R., Lin, H.S., Vereecken, H., 2016. Spatial and temporal occurrence of preferential flow in a forested headwater catchment. J. Hydrol. 534, 139–149.

Wigmore, O., Mark, B., 2017. Monitoring tropical debris-covered glacier dynamics from high-resolution unmanned aerial vehicle photogrammetry, Cordillera Blanca, Peru. Cryosphere 11 (6), 2463–2480.

Willmott, C.J., Rowe, C.M., Philpot, W.D., 1985. Small-scale climate maps: a sensitivity analysis of some common assumptions associated with grid-point interpolation and contouring. Am. Cartograph. 12 (1), 5–16.

Wilson, J.P., 2018. Environmental Applications of Digital Terrain Modeling. Wiley–Blackwell, Chichester, UK.

Wilson, J.P., Gallant, J.C. (Eds.), 2000. Terrain Analysis: Principles and Applications. Wiley, New York, NY.

Wilson, J.P., Spangrud, D.J., Nielsen, G.A., Jacobsen, J.S., Tyler, D.A., 1998. Global positioning system sampling intensity and pattern effects on computed topographic attributes. Soil Sci. Soc. Am. J. 62 (5), 1410–1417.

Wilson, J.P., Aggett, G., Deng, Y., Lam, C.S., 2008. Water in the landscape: a review of contemporary flow routing algorithms. In: Zhou, Q. Lees, B., Tang, G.-A. (Eds.), Advances in Digital Terrain Analysis. Lecture Notes in Geoinformation and Cartography Book Series. Springer, Berlin, Germany, pp. 213–236.

Wilson, J.T., 1941. Structural features in the Northwest Territories. Am. J. Sci. 239 (7), 493–502.

Wing, O.E.J., Bates, P.D., Neal, J.C., Sampson, C.C., Smith, A.M., Quinn, N., et al., 2019. A new automated method for improved flood defense representation in large-scale hydraulic models. Water Resour. Res. 55, 11007–11034.

Wingham, D.J., Francis, C.R., Baker, S., Bouzinac, C., Brockley, D., Cullen, R., et al., 2006. CryoSat: a mission to determine the fluctuations in Earth's land and marine ice fields. Adv. Space Res. 37 (4), 841–871.

Winstral, A., Elder, K., Davis, R.E., 2002. Spatial snow modeling of wind-redistributed snow using terrain-based parameters. J. Hydrometeorol. 3 (5), 524–538.

Wise, D.U., 1969. Regional and sub-continental sized fracture systems detectable by topographic shadow techniques. In: Baer, A.J., Norris, D.K. (Eds.), Research in Tectonics. Geological Survey of Canada Paper 68–52, pp. 175–198.

Wise, D.U., Funiciello, R., Parotto, M., Salvini, F., 1985. Topographic lineament swarms: clues to their origin from domain analysis of Italy. Geol. Soc. Am. Bull. 96 (7), 952–967.

Wise, S., 2000. Assessing the quality for hydrological applications of digital elevation models derived from contours. Hydrol. Process. 14 (11–12), 1909–1929.

Wise, S.M., 2007. Effect of differing DEM creation methods on the results from a hydrological model. Comput. Geosci. 33 (10), 1351–1365.

Wladis, D., 1999. Automatic lineament detection using digital elevation models with second derivative filters. Photogrammet. Eng. Remote Sens. 65 (4), 453–458.

Włodarski, W., Papis, J., Szczuciński, W., 2017. Morphology of the Morasko crater field (western Poland), influences of pre-impact topography, meteoroid impact processes, and post-impact alterations. Geomorphology 295, 586–597.

Wölfl, A.-C., Snaith, H., Amirebrahimi, S., Devey, C.W., Dorschel, B., Ferrini, V., et al., 2019. Seafloor mapping—the challenge of a truly global ocean bathymetry. Front. Mar. Sci. 6, 283.

Wolfman, Y.M., Gintov, O.B., Ostanin, A.M., Kolesnikova, E.Y., Murovskaya, A.V., 2008. On the role of structural kinematic identification of tectonic disruptive faults in formation of the ideas on the structure and geodynamics of the Crimean region. Geophys. J. 30 (1), 49–61 (in Russian).

Wolock, D.M., Price, C.V., 1994. Effects of digital elevation model map scale and data resolution on a topography-based watershed model. Water Resour. Res. 30 (11), 3041–3052.

Wood, E.F., Sivapalan, M., Beven, K.J., Band, L., 1988. Effects of spatial variability and scale with implications to hydrologic modeling. J. Hydrol. 102 (1–4), 29–47.

Wood, E.F., Sivapalan, M., Beven, K.J., 1990. Similarity and scale in catchment storm response. Rev. Geophys. 28 (1), 1–18.

Wood, J.D., 1996. The Geomorphological Characterisation of Digital Elevation Models (Ph.D. thesis). University of Leicester, Leicester, UK.

Wood, J., 2009. Overview of software packages used in geomorphometry. In: Hengl, T., Reuter, H.I. (Eds.), Geomorphometry: Concepts, Software, Applications. Developments in Soil Science, vol. 33. Elsevier, Amsterdam, the Netherlands, pp. 257–268.

Wood, J.D., Fisher, P.F., 1993. Assessing interpolation accuracy in elevation models. IEEE Comput. Graph. Appl. 13 (2), 48–56.

Woodrow, K., Lindsay, J.B., Berg, A.A., 2016. Evaluating DEM conditioning techniques, elevation source data, and grid resolution for field-scale hydrological parameter extraction. J. Hydrol. 540, 1022–1029.

Wu, F., 2003. Scale-dependent representations of relief based on wavelet analysis. GeoSpatial Inf. Sci. 6 (1), 66–69.

Wu, Q., Deng, C., Chen, Z., 2016. Automated delineation of karst sinkholes from LiDAR-derived digital elevation models. Geomorphology 266, 1–10.

Wu, Q., Chen, Y., Wilson, J.P., Liu, X., Li, H., 2019. An effective parallelization algorithm for DEM generalization based on CUDA. Environ. Model. Software 114, 64–74.

Wu, W., Li, A.-D., He, X.-H., Ma, R., Liu, H.-B., Lv, J.-K., 2018. A comparison of support vector machines, artificial neural network and classification tree for identifying soil texture classes in southwest China. Comput. Electron. Agricult. 144, 86–93.

Wu, Y., Zhou, L., Zhong, F., Zhong, K., Yuan, B., Zhou, J., 2017. Fine prediction of reef boundary based on slope attributes: a case study of Permian Changxing Formation in Luodingzhai area, Sichuan Basin, SW China. Petrol. Explor. Dev. 44 (6), 960–971.

Wynn, R.B., Huvenne, V.A.I., le Bas, T.P., Murton, B.J., Connelly, D.P., Bett, B.J., et al., 2014. Autonomous underwater vehicles (AUVs): their past, present and future contributions to the advancement of marine geoscience. Mar. Geol. 352, 451–468.

Wynn, T.J., Stewart, S.A., 2003. The role of spectral curvature mapping in characterizing subsurface strain distributions. In: Ameen, M. (Ed.), Fracture and In-Situ Stress Characterization of Hydrocarbon Reservoirs. Geol. Soc. London Spec. Publ., vol. 209, pp. 127–143.

Xiang, J., Li, S., Xiao, K., Chen, J., Sofia, G., Tarolli, P., 2019. Quantitative analysis of anthropogenic morphologies based on multitemporal high-resolution topography. Remote Sens. (Basel) 11 (12), 1493.

Xiang, T.-Z., Xia, G.-S., Zhang, L., 2019. Mini-unmanned aerial vehicle-based remote sensing: techniques, applications, and prospects. IEEE Geosci. Remote Sens. Mag. 7 (3), 29–63.

Xiong, L.-Y., Tang, G.-A., Strobl, J., Zhu, A.-X., 2016. Paleotopographic controls on loess deposition in the Loess Plateau of China. Earth Surf. Process. Landf. 41 (9), 1155–1168.

Xiong, L.-Y., Tang, G.-A., Zhu, A.-X., Qian, Y.-Q., 2017. A peak-cluster assessment method for the identification of upland planation surfaces. Int. J. Geogr. Inf. Sci. 31 (2), 387–404.

Xiong, L.-Y., Jiang, R.-Q., Lu, Q.-H., Yang, B.-S., Li, F.-Y., Tang, G.-A., 2019. Improved priority-flood method for depression filling by redundant calculation optimization in local micro-relief areas. Trans. GIS 23 (2), 259–274.

Xiong, L., Li, S., Tang, G., Strobl, J., 2022. Geomorphometry and terrain analysis: data, methods, platforms and applications. Earth Sci. Rev. 233, 104191.

Xu, H., van der Steeg, S., Sullivan, J., Shelley, D., Cely, J.E., Viparelli, E., et al., 2020. Intermittent channel systems of a low-relief, low-gradient floodplain: comparison of automatic extraction methods. Water Resour. Res. 56 (9), e2020WR027603.

Xu, T., Moore, I.D., Gallant, J.C., 1993. Fractals, fractal dimensions and landscapes—a review. Geomorphology 8 (4), 245–262.

Xue, Y., Jing, Z., Kang, S., He, X., Li, C., 2021. Combining UAV and Landsat data to assess glacier changes on the central Tibetan Plateau. J. Glaciol. 67 (265), 862–874.

Yabrova, L.A., 1981. Concentrically radial drainage networks as the indicator of ring structures in the Caucasus and other mountain terrains. In: Belyaev, G.K. (Ed.), Nature and Natural Resources of the Central and Eastern Parts of the North Caucasus. Khetagurov North-Ossetian State University, Ordzhonikidze, USSR, pp. 16–25 (in Russian).

Yagodina, L.L., 1972. Modelling of the topographic field by the Chebyshev orthogonal polynomials. Leningrad Univ. Bull. Geogr. Ser. 24 (4), 136–142 (in Russian).

Yamafune, K., Torres, R., Castro, F., 2017. Multi-image photogrammetry to record and reconstruct underwater shipwreck sites. J. Archaeol. Method Theory 24 (3), 703–725.

Yamazaki, D., 2018. MERIT DEM: Multi-Error-Removed Improved-Terrain DEM. Institute of Industrial Sciences, University of Tokyo, Tokyo, Japan. https://hydro.iis.u-tokyo.ac.jp/~yamadai/MERIT_DEM/. (Accessed 27 June 2024).

Yamazaki, D., Ikeshima, D., Tawatari, R., Yamaguchi, T., O'Loughlin, F., Neal, J.C., et al., 2017. A high-accuracy map of global terrain elevations. Geophys. Res. Lett. 44 (11), 5844–5853.

Yang, X., Tang, G., Meng, X., Xiong, L., 2019. Classification of karst Fenglin and Fengcong landform units based on spatial relations of terrain feature points from DEMs. Remote Sens. (Basel) 11 (16), 1950.

Yao, H., Qin, R., Chen, X., 2019. Unmanned aerial vehicle for remote sensing applications—a review. Remote Sens. (Basel) 11 (12), 1443.

Yaroshenko, P.D., 1961. Geobotany. The Main Concepts and Methods. Soviet Academic Press, Moscow, USSR (in Russian).

Yeomans, C.M., Middleton, M., Shail, R.K., Grebby, S., Lusty, P.A.J., 2019. Integrated object-based image analysis for semi-automated geological lineament detection in southwest England. Comput. Geosci. 123, 137–148.

Yilmaz, C.S., Yilmaz, V., Güngör, O., 2018. Investigating the performances of commercial and non-commercial software for ground filtering of UAV-based point clouds. Int. J. Remote Sens. 39 (15–16), 5016–5042.

Yilmaz, V., 2021. Automated ground filtering of LiDAR and UAS point clouds with metaheuristics. Opt. Laser Technol. 138, 106890.

Yoeli, P., 1965. Analytical hill shading. Survey. Mapp. 25 (4), 573–579.

Yoeli, P., 1975. Compilation of data for computer-assisted relief cartography. In: Davis, J.C., McCullagh, M.J. (Eds.), Display and Analysis of Spatial Data. Wiley, London, UK, pp. 352–367.

Yoeli, P., 1983. Digital terrain models and their cartographic and cartometric utilisation. Cartogr. J. 20 (1), 17–22.

Yoeli, P., 1985. The making of intervisibility maps with computer and plotter. Cartographica 22 (3), 88–103.

Yokoyama, R., Shirasawa, M., Pike, R.J., 2002. Visualizing topography by openness: a new application of image processing to digital elevation models. Photogrammet. Eng. Remote Sens. 68 (3), 257–265.

Young, A., 1972. Slopes. Oliver and Boyd, Edinburgh, UK.

Young, M., 1978. Statistical Characterization of Altitude Matrices by Computer. Terrain Analysis: Program Documentation. Report 5 on Grant DA-ERO-591-73-G0040. Department of Geography, University of Durham, Durham, UK.

Young, T., 1805. An essay on the cohesion of fluids. Philos. Trans. R. Soc. Lond. Pt I 95, 55–87.

Yu, A.W., Krainak, M.A., Harding, D.J., Abshire, J.B., Sun, X., Valett, S., et al., 2010. Spaceborne laser instruments for high-resolution mapping. In: Clarkson, W.A., Hodgson, N., Shori, R.K. (Eds.), Solid State Lasers XIX: Technology and Devices. Proc. SPIE 7578, 757802.

Yu, L., Gong, P., 2012. Google Earth as a virtual globe tool for Earth science applications at the global scale: progress and perspectives. Int. J. Remote Sens. 33 (12), 3966–3986.

Yu, Y., Zhang, Z., Shokr, M., Hui, F., Cheng, X., Chi, Z., et al., 2019. Automatically extracted Antarctic coastline using remotely-sensed data: an update. Remote Sens. (Basel) 11 (16), 1844.

Yuan, W., Zhuang, D., Yuan, W., Qiu, D., 2010. Equal arc ratio projection and a new spherical triangle quadtree model. Int. J. Geogr. Inf. Sci. 24 (11), 1703–1723.

Yuan, X., Qiao, G., Li, Y., Li, H., Xu, R., 2020. Modelling of glacier and ice sheet micro-topography based on unmanned aerial vehicle data, Antarctica. In: Paparoditis, N., Mallet, C., Lafarge, F., Jiang, J., Shaker, A., Zhang, H., et al. (Eds.), XXIV ISPRS Congress, Commission III. Int. Arch. Photogramm. Remote Sens. Spatial Inf. Sci. XLIII (B3–2020), pp. 919–923.

Yudin, V.V., Gerasimov, M.E., 2001. Thrusts of the Crimean Mountains. Geophys. J. 23 (2), 121–129 (in Russian).

Yue, L., Shen, H., Zhang, L., Zheng, X., Zhang, F., Yuan, Q., 2017. High-quality seamless DEM generation blending SRTM-1, ASTER GDEM v2 and ICESat/GLAS observations. ISPRS J. Photogrammet. Remote Sens. 123, 20–34.

Yurov, S.I. (Ed.), 1976. Manual on Topographic Surveys in 1:5000, 1:2000, 1:1000, and 1:500 Scales; Elevation Networks. Nedra, Moscow, USSR (in Russian).

Zabihi, M., Mirchooli, F., Motevalli, A., Darvishan, A.K., Pourghasemi, H.R., Zakeri, M.A., et al., 2018. Spatial modelling of gully erosion in Mazandaran Province, northern Iran. Catena 161, 1–13.

Žabota, B., Repe, B., Kobal, M., 2019. Influence of digital elevation model resolution on rockfall modelling. Geomorphology 328, 183–195.

Zadeh, L.A., 1965. Fuzzy sets. Inf. Contr. 8 (3), 338–353.

Zakharchenko, A.V., Zakharchenko, N.V., 2006. Three-dimensional surface morphometry of soil horizons in field studies. Eurasian Soil Sci. 39 (2), 134–140.

Zakharov, S.A., 1911. On the role of micro- and macrotopography in the Podzol region. Pochvovedenie 13 (1), 49–72 (in Russian).

Zakharov, S.A., 1913. On characteristics of soils in mountain regions. Bull. Konstantinovsky Survey. Inst. 4, 1–93 (in Russian).

Zakharov, S.A., 1927. A Course of Soil Science. Gosizdat, Moscow, USSR (in Russian).

Zakharov, S.A., 1940. Importance of slope aspect and gradient for soil and vegetation distribution in the Great Caucasus. J. Botan. URSS 25 (4–5), 378–405 (in Russian).

Zakharov, V.G., 1988. Fluctuations in ice shelves and outlet glaciers in Antarctica. Polar Geogr. Geol. 12 (4), 297–311.

Zakharov, V.G., 1994. Antarctic Glaciers Fluctuations. Akkorinformizdat, Moscow, Russia (in Russian).

Zakharov, V.G., 2002. Dynamics of the Dålk outlet glacier and hydrometeorological conditions in the Prydz Bay (Eastern Antarctica) in 1989–1990 and 1994. Mater. Glaciol. Stud. 93, 169–180 (in Russian).

Zamani, A., Hashemi, N., 2000. A comparison between seismicity, topographic relief, and gravity anomalies of the Iranian Plateau. Tectonophysics 327 (1–2), 25–36.

Zanchi, A., Francesca, S., Stefano, Z., Simone, S., Graziano, G., 2009. 3D reconstruction of complex geological bodies: examples from the Alps. Comput. Geosci. 35 (1), 49–69.

Zaslavsky, D., Rogowski, A.S., 1969. Hydrologic and morphologic implications of anisotropy and infiltration in soil profile development. Soil Sci. Soc. Am. Proc. 33 (4), 594–599.

Zebker, H.A., Goldstein, R.M., 1986. Topographic mapping from interferometric synthetic aperture radar observations. J. Geophys. Res. Solid Earth 91 (B5), 4993–4999.

Zebker, H.A., Farr, T.G., Salazar, R.P., Dixon, T.H., 1994a. Mapping the world's topography using radar interferometry: the TOPSAT mission. Proc. IEEE 82 (12), 1774–1786.

Zebker, H., Werner, C., Rosen, P., Hensley, S., 1994b. Accuracy of topographic maps derived from ERS-1 interferometric radar. IEEE Trans. Geosci. Remote Sens. 32 (4), 823–836.

Zeilik, B.S., Perfil'ev, L.G., Vasilenko, A.N., Seitmuratova, E.Y., 1989. New techniques for geological interpretation of annular structures in semi-covered and covered areas. Sov. J. Remote Sens. 5 (4), 606–621.

Zeraatpisheh, M., Ayoubi, S., Jafari, A., Finke, P., 2017. Comparing the efficiency of digital and conventional soil mapping to predict soil types in a semi-arid region in Iran. Geomorphology 285, 186–204.

Zernitz, E.R., 1931. Drainage patterns and their significance. J. Geol. 40 (6), 498–521.

Zevenbergen, L.W., Thorne, C.R., 1987. Quantitative analysis of land surface topography. Earth Surf. Process. Landf. 12 (1), 47–56.

Zhai, R., Lu, K., Pan, W., Dai, S., 2016. GPU-based real-time terrain rendering: design and implementation. Neurocomputing 171, 1–8.

Zhang, F., Zhou, Q., Li, Q., Wu, G., Liu, J., 2018. An enhanced approach for surface flow routing over drainage-constrained triangulated irregular networks. Trans. GIS 22 (1), 43–57.

Zhang, H., Yao, Z., Yang, Q., Li, S., Baartman, J.E.M., Gai, L., et al., 2017a. An integrated algorithm to evaluate flow direction and flow accumulation in flat regions of hydrologically corrected DEMs. Catena 151, 174–181.

Zhang, H., Zhang, P., Prush, V., Zheng, D., Zheng, W., Wang, W., et al., 2017b. Tectonic geomorphology of the Qilian Shan in the northeastern Tibetan Plateau: insights into the plateau formation processes. Tectonophysics 706–707, 103–115.

Zhang, H.-P., Liu, S.-F., Yang, N., Zhang, Y.-Q., Zhang, G.-W., 2006. Geomorphic characteristics of the Minjiang drainage basin (eastern Tibetan Plateau) and its tectonic implications: new insights from a digital elevation model study. Island Arc 15 (2), 239–250.

Zhang, L., Zhang, L., Du, B., 2016. Deep learning for remote sensing data: a technical tutorial on the state of the art. IEEE Geosci. Remote Sens. Mag. 4 (2), 22–40.

Zhang, Q., Shang, H., 2021. The effect of some upscaling procedures on the slope derived from digital elevation models. Trans. GIS 25 (10, 382–395.

Zhang, S., E, D., Wang, Z., Li, Y., Jin, B., Zhou, C., 2008. Ice velocity from static GPS observations along the transect from Zhongshan station to Dome A, East Antarctica. Ann. Glaciol. 48, 113–118.

Zhang, S., Foerster, S., Medeiros, P., Araújo, J.C., Motagh, M., Waske, B., 2016. Bathymetric survey of water reservoirs in north-eastern Brazil based on TanDEM-X satellite data. Sci. Total Environ. 571, 575–593.

Zhang, S., Liu, G., Chen, S., Rasmussen, C., Liu, B., 2021. Assessing soil thickness in a black soil watershed in northeast China using random forest and field observations. Int. Soil Water Conserv. Res. 9 (1), 49–57.

Zhang, W., Montgomery, D.R., 1994. Digital elevation model grid size, landscape representation, and hydrologic simulation. Water Resour. Res. 30 (4), 1019–1028.

Zhao, B., Liu, Y., Xu, W., Liu, Y., Sun, J., Wang, L., 2019. Morphological characteristics of tidal creeks in the central coastal region of Jiangsu, China, using LiDAR. Remote Sens. (Basel) 11 (20), 2426.

Zhao, D., Wu, Z., Zhou, J., Zhang, K., Luo, X., Wang, M., et al., 2021. From 10 m to 11000 m, automatic processing multi-beam bathymetric data based on PGO method. IEEE Access 9, 14,516–14,527.

Zhao, F., Xiong, L.-Y., Wang, C., Wang, H.-R., Wei, H., Tang, G.-A., 2021. Terraces mapping by using deep learning approach from remote sensing images and digital elevation models. Trans. GIS 25 (5), 2438–2454.

Zhao, W.-F., Xiong, L.-Y., Ding, H., Tang, G.-A., 2017. Automatic recognition of loess landforms using Random Forest method. J. Mountain Sci. 14 (5), 885–897.

Zhao, X., Bai, J., Chen, J., 2008. A seamless and adaptive LOD model of the global terrain based on the QTM. In: Zhou, Q., Lees, B., Tang, G.-A. (Eds.), Advances in Digital Terrain Analysis. Lecture Notes in Geoinformation and Cartography Book Series, Springer, Berlin, Germany, pp. 85–103.

Zhidkov, V.A., Zakharov, V.G., 2009. Results of monitoring of the Fridtjovbreen pulsing glacier (West Spitsbergen Island) for the period from 1996 to 2006. Comp. Res. Spitsbergen Nat. 9, 256–265 (in Russian).

Zhong, C., Liu, Y., Gao, P., Chen, W., Li, H., Hou, Y., et al., 2020. Landslide mapping with remote sensing: challenges and opportunities. Int. J. Remote Sens. 41 (4), 1555–1581.

Zhou, A., Chen, Y., Wilson, J.P., Su, H., Xiong, Z., Cheng, Q., 2021. An enhanced double-filter deep residual neural network for generating super resolution DEMs. Remote Sens. (Basel) 13 (16), 3089.

Zhou, B. Zhang, X.-G., Wang, R.-C., 2004. Automated soil resources mapping based on decision tree and Bayesian predictive modeling. J. Zhejiang Univ. Sci. 5 (7), 782–795.

Zhou, G., Sun, Z., Fu, S., 2016. An efficient variant of the priority-flood algorithm for filling depressions in raster digital elevation models. Comput. Geosci. 90 (A), 87–96.

Zhou, G., Liu, X., Fu, S., Sun, Z., 2017. Parallel identification and filling of depressions in raster digital elevation models. Int. J. Geogr. Inf. Sci. 31 (6), 1061–1078.

Zhou, L., Yuan, B., Song, Y., Wu, Y., Kong, Y., Zhou, J., et al., 2016. A seismic quantitative identification method of slope break belt. Petrol. Explor. Dev. 43 (5), 1025–1034.

Zhou, Q., Chen, Y., 2011. Generalization of DEM for terrain analysis using a compound method. ISPRS J. Photogrammet. Remote Sens. 66 (1), 38–45.

Zhou, Q., Liu, X., 2004. Error analysis on grid-based slope and aspect algorithms. Photogrammet. Eng. Remote Sens. 70 (8), 957–962.

Zhou, X., Li, W., Arundel, S.T., 2019. A spatio-contextual probabilistic model for extracting linear features in hilly terrains from high-resolution DEM data. Int. J. Geogr. Inf. Sci. 33 (4), 666–686.

Zhou, Y., Li, Z., Li, J., 2017. Slight glacier mass loss in the Karakoram Region during the 1970s to 2000 revealed by KH-9 images and SRTM DEM. J. Glaciol. 63 (238), 331–342.

Zhu, A.-X., 2000. Mapping soil landscape as spatial continua: the neural network approach. Water Resour. Res. 36 (3), 663–677.

Zhu, A.-X., Band, L.E., Dutton, B., Nimlos, T.J., 1996. Automated soil inference under fuzzy logic. Ecol. Model. 90 (2), 123–145.

Zhu, A.-X., Hudson, B., Burt, J., Lubich, K., Simonson, D., 2001. Soil mapping using GIS, expert knowledge, and fuzzy logic. Soil Sci. Soc. Am. J. 65 (5), 1463–1472.

Zhu, A.-X., Burt, J.E., Smith, M., Wang, R., Gao, J., 2008. The impact of neighbourhood size on terrain derivatives and digital soil mapping. In: Zhou, Q., Lees, B., Tang, G.-A. (Eds.), Advances in Digital Terrain Analysis. Lecture Notes in Geoinformation and Cartography Book Series. Springer, Berlin, Germany, pp. 333–348.

Zhu, J., Wu, W., Liu, H.-B., 2018. Environmental variables controlling soil organic carbon in top- and sub-soils in karst region of southwestern China. Ecol. Indic. 90, 624–632.

Zhu, J., Song, C., Ke, L., Liu, K., Chen, T., 2021. Remote sensing investigation of the offset effect between reservoir impoundment and glacier meltwater supply in Tibetan Highland catchment. Water 13 (9), 1307.

Zhu, L., Sun, J., Li, C., Zhang, B., 2014. SolidEarth: a new Digital Earth system for the modeling and visualization of the whole Earth space. Front. Earth Sci. 8 (4), 524–539.

Zhu, Q., Lin, H.S., 2010. Comparing ordinary kriging and regression kriging for soil properties in contrasting landscapes. Pedosphere 20 (5), 594–606.

Zhu, Y., Huang, C., 2012. An improved median filtering algorithm for image noise reduction. Phys. Proc. 25, 609–616.

Ziadat, F.M., Taylor, J.C., Brewer, T.R., 2003. Merging Landsat TM imagery with topographic data to aid soil mapping in the Badia region of Jordan. J. Arid Environ. 54 (3), 527–541.

Zibilske, L.M., 1994. Carbon mineralization. In: Weaver, R.W., Angle, S., Bottomley, P., Bezdicek, D., Smith, S., Tabatabai, A., et al. (Eds.), Methods of Soil Analysis, pt. 2: Microbiological and Biochemical Properties. Soil Science Society of America, Madison, WI, pp. 835–863.

Zink, M., Bachmann, M., Brautigam, B., Fritz, T., Hajnsek, I., Moreira, A., et al., 2014. TanDEM-X: the new global DEM takes shape. IEEE Geosci. Remote Sens. Mag. 2 (2), 8–23.

Zlatopolsky, A.A., 1992. Program LESSA (Lineament Extraction and Stripe Statistical Analysis): automated linear image features analysis—experimental results. Comput. Geosci. 18 (9), 1121–1126.

Zlatopolsky, A.A., 2011. New LESSA technology resources and digital terrain map analysis: methodology. Curr. Problems Remote Sens. Earth Space 8 (3), 38–46 (in Russian).

Zuber, M., Smith, D.E., Lemoine, F.G., Neumann, G., 1994. The shape and internal structure of the Moon from the Clementine mission. Science 266 (5192), 1839–1843.

Zuber, M., Smith, D.E., Neumann, G., 1996. Clementine Gravity and Topography Data Archive. NASA Planetary Data System Geosciences Node, Washington University, St. Louis, MO. http://pds-geosciences.wustl.edu/missions/clementine/gravtopo.html. (Accessed 27 June 2024).

Zuber, M.T., Solomon, S.C., Phillips, R.J., Smith, D.E., Tyler, G.L., Aharonson, O., et al., 2000. Internal structure and early thermal evolution of Mars from Mars Global Surveyor topography and gravity. Science 287 (5459), 1788–1793.

Zuchiewicz, W., 1989. Selected geomorphic method applied to neotectonic studies in the Northern Carpathians. Bull. Assoc. Franç. Étude Quat. 26 (3), 131–136.

Zumpano, V., Pisano, L., Parise, M., 2019. An integrated framework to identify and analyze karst sinkholes. Geomorphology 332, 213–225.

Zverev, A.T., Strykov, A.I., 1985. Transformation of geological, geophysical, and topographic maps to reveal pseudo-round anomalies (forms) for automated recognition of ring structures on satellite images. Izvestia Vuzov. Geodesy Aerophotosurveying (5), 46–54 (in Russian).

Zvyagintsev, D.G., 1994. Vertical distribution of microbial communities in soils. In: Ritz, K., Dighton, J., Giller, K.E. (Eds.), Beyond the Biomass: Compositional and Functional Analysis of Soil Microbial Communities. Wiley, Chichester, UK, pp. 29–37.

Zwally, H.J., Bindschadler, R.A., Brenner, A.C., Martin, T.V., Thomas, R.H., 1983. Surface elevation contours of Greenland and Antarctic ice sheets. J. Geophys. Res. Oceans 88 (C3), 1589–1596.

Zwally, H.J., Bindschadler, R.A., Major, J.A., Brenner, A.C., 1987. Ice measurements by Geosat radar altimetry. J. Hopkins APL Tech. Dig. 8 (2), 251–254.

Zwally, H.J., Brenner, A.C., Dimarzio, J., Seiss, T., 1994. Ice sheet topography from retracked ERS-1 altimetry. In: ESA, Proceedings of the 2nd ERS-1 Symposium on Space at the Service of Our Environment, vol. 1. NASA Goddard Space Flight Center, Greenbelt, MD, pp. 159–164.

Zwolak, K., Wigley, R., Bohan, A., Zarayskaya, Y., Bazhenova, E., Dorshow, W., et al., 2020. The autonomous underwater vehicle integrated with the unmanned surface vessel mapping the Southern Ionian Sea. The winning technology solution of the Shell Ocean Discovery XPRIZE. Remote Sens. (Basel) 12 (8), 1344.

Index

Note: Page numbers followed by *f* indicate figures and *t* indicate tables.

A

Accumulation curvature (K_a), 21–24*t*, 26, 34–37*f*, 45–46, 67, 106, 108–109*f*, 209, 268, 281, 286–287*t*, 367
Accumulation, transit, and dissipation zones, 65–67, 65*f*, 263–265, 266*f*, 319–320, 320–321*f*, 321–322*t*, 322–323, 323*f*, 375–376, 377*f*
Accuracy/errors of DEMs produced by
- contour digitizing, 144
- conventional topographic surveys, 139
- data fusion, 145
- echo sounding, 143
- geological and soil core drilling, 144
- interferometric techniques, 142–143, 143–144*f*
- kinematic global navigation satellite system (GNSS) surveys, 139
- lidar surveys, 142, 142*f*
- optical sensing of bathymetry, 143
- photogrammetric techniques, 139–140, 140*f*
- satellite radar altimetry, 143–144
- structure-from-motion (SfM) techniques, 140–141, 141*f*
- three-dimensional seismic survey, 144

Adequate grid spacing, 245–248
Antarctic oases, 365–378, 366*f*, 368–377*f*, 382
Applications of digital terrain analysis
- archeology, 13–14
- avalanche hazard, 12–13
- fire hazard, 13
- forestry, 10–11
- geobotany, 10–11
- geology, 1–2, 9, 295–297, 301–315, 319–320, 322–323, 325–340
- geomorphology, 1–2, 6–8
- glaciology, 2, 9–10, 345–355, 346–347*f*, 350–354*f*, 357–363, 358–362*f*
- hydrology, 2, 8
- landslide science, 12
- metrology, 14
- oceanology, 11
- planetology, 11–12, 325–340, 326–336*f*, 338–341*f*, 340*t*
- soil science, 1–2, 8–9, 241–244, 243*f*
- speleology, 10
- urban studies, 13
- zoogeography, 11

Approximation, 191–193
Archeology, digital terrain analysis applications, 13–14
Artifacts, 140–143, 140–141*f*, 147, 154–156, 155–156*f*, 163–165, 337–338, 367
Artificial neural networks, 259
Aspect. *See* Slope aspect
Autonomous underwater vehicle (AUV), 11, 80
Avalanche hazard, digital terrain analysis applications, 12–13

B

Bathymetry, 3–5, 82–85, 83*f*, 143–144, 210*f*, 221, 221*f*, 225–226*f*
Bayesian analysis of evidence, 258, 260
Break lines, 52, 52*f*, 123, 125

C

Calculation methods
- combined morphometric variables, 129–131, 130–131*f*
- Evans-Young method, 99–100, 100*f*
- landform classifications, 131
- local morphometric variables, plane square grid, 100*f*, 102–106
- local morphometric variables, spheroidal equal angular grid, 100*f*, 112–117
- nonlocal morphometric variables, plane square grid, 118–122, 120*f*
- nonlocal morphometric variables, spheroidal equal angular grid, 120–122, 121*f*
- solar morphometric variables, 125, 125*f*
- structural morphometric variables, 122–125, 123–124*f*
- triaxial ellipsoid, 131–137
- universal spectral analytical method, 191–195
- view morphometric variables, 128–129, 128–129*f*
- wind morphometric variables, 126–128, 126–127*f*

Catchment area (CA), 1, 21–23*t*, 50–51, 50*f*, 121–122*f*, 231–237*f*, 242, 250–252, 268–269, 288, 291, 332–333*f*, 335–336*f*, 337, 353–354, 353–354*f*, 373–374, 374*f*
Chebyshev polynomials, 191–192, 194–195, 380
Classifications, 258
Clementine mission, 80
Clinometry, 81
Cokriging, 259
Combined morphometric variables, 21–24*t*, 61–62
- local-local, 21–24*t*, 62 (*see also* Local-local morphometric variables)
- local-nonlocal, 21–24*t*, 61–62 (*see also* Local-nonlocal morphometric variables)

Concave break line, 52, 52*f*, 123, 125
Contour line, 18–20, 19*f*, 52
Conventional topographic surveys, 73–75, 74*f*
Convex break line, 52, 52*f*, 123, 125
Coordinate systems, 93–94, 94*f*
Course. *See* Thalweg
Crest, 52, 52–53*f*, 123, 123*f*
Crevasse, 349–355, 350–354*f*
Cross section, 213, 214*f*
Curvatures, 31–46, 31*f*
- accumulation (K_a), 21–24*t*, 26, 34–37*f*, 45–46, 67, 106, 108–109*f*, 209, 268, 281, 286–287*t*, 367
- complete system of, 46
- difference (E), 21–24*t*, 26, 34–37*f*, 43, 108–109*f*, 268, 281, 286–287*t*, 288, 367
- Gaussian (K), 21–24*t*, 26, 34–37*f*, 38, 63–64, 65*f*, 108–109*f*, 181*f*, 209, 256, 268, 281, 286–287*t*, 295, 297, 299*f*, 301, 367
- horizontal (k_h), 12, 21–24*t*, 26, 32–33, 33–37*f*, 43, 45, 65*f*, 68, 69–70*f*, 108–109*f*, 119*f*, 151, 153*f*, 154–155, 156*f*, 161–162*f*, 165, 177*f*, 184*f*, 186, 186–187*f*, 188, 198*f*, 201*f*, 206, 209, 231–237*f*, 242, 251–253, 251–252*f*, 252*t*, 268, 271, 274*f*, 277–280*t*, 286–287*t*, 288, 297, 299*f*, 302–305, 306*f*, 307, 311–312, 311*f*, 315, 316*f*, 329–330*f*, 337, 348, 353–354, 353–354*f*, 370, 370*f*
- horizontal excess (k_{he}), 21–24*t*, 26, 34–37*f*, 44, 46, 108–109*f*, 268, 281, 286–287*t*, 288
- maximal (k_{max}), 2, 21–24*t*, 26, 34–37*f*, 38–39, 43, 46, 62, 68, 69–70*f*, 108–109*f*, 119*f*, 181*f*, 231–237*f*, 268, 274*f*, 277–280*t*, 286–287*t*, 291, 296*f*, 297, 301, 353–354, 367, 373–375, 373*f*, 376*f*
- mean (H), 2, 21–24*t*, 26, 33–34, 34–37*f*, 38, 63–64, 65*f*, 108–109*f*, 151, 152*f*, 181*f*, 242, 251–253, 251–252*f*, 252–253*t*, 268–269, 277–280*t*, 286–287*t*, 291, 295, 367
- minimal (k_{min}), 21–24*t*, 26, 34–37*f*, 38–39, 44, 62, 68, 69–70*f*, 108–109*f*, 119*f*, 181*f*, 231–237*f*, 268, 273*f*, 277–280*t*, 286–289*t*, 296*f*, 297, 301, 329*f*, 353–354, 353–354*f*, 367, 372*f*, 373–375, 376*f*
- plan (k_p), 21–24*t*, 26, 31–32, 50, 257, 268*f*
- principal (*see* Maximal curvature *and* Minimal curvature)
- ring (K_r), 21–24*t*, 26, 34–37*f*, 44–45, 108–109*f*, 209, 268, 281, 286–287*t*, 367

Curvatures *(Continued)*
unsphericity *(M)*, 21–24*t*, 26, 34–37*f*, 39–41, 108–109*f*, 268, 281, 286–287*t*, 367
vertical (k_v), 12, 21–24*t*, 26, 33–38, 33–37*f*, 46, 52–53, 63, 65*f*, 67–68, 69–70*f*, 108–109*f*, 119*f*, 154–156, 155*f*, 161–162*f*, 177*f*, 180*f*, 199*f*, 201*f*, 203*f*, 206, 211, 231–237*f*, 242, 251–253, 251–252*f*, 252*t*, 268, 273*f*, 277–280*t*, 286–289*t*, 288, 291, 297, 299*f*, 302–305, 306*f*, 307, 311–312, 311*f*, 315, 316*f*, 329*f*, 331*f*, 337, 353–354, 353–354*f*, 371–372, 371*f*
vertical excess (k_{ve}), 21–24*t*, 26, 34–37*f*, 44, 46, 108–109*f*, 268, 281, 286–287*t*
Curvedness, 21–24*t*, 62, 64*f*
Cutting method, 168, 179–181, 181*f*

D

Datum, 93–94, 94*f*
Decision trees, 259
Decomposition, 167, 172–174, 175*f*, 182–183, 185*f*
Deflections, 21–24*t*, 46–49
horizontal (D_{kh}), 21–24*t*, 46–48, 49*f*, 123–125, 124*f*
vertical (D_{kv}), 21–24*t*, 48–49, 49*f*
Delaunay triangulation, 89, 90*f*
DEM of difference (DoD), 2, 7, 9–10, 145, 146*f*, 345–346, 346*f*
DEMs, global and quasi-global, 4–5, 76–77*t*
ACE, 76–77*t*, 84
ALOS AW3D30, 4–5, 9, 76–77, 76–77*t*, 88, 379
ALOS World 3D, 76–77*t*
ASTER, 2–4, 76–77, 76–77*t*, 78*f*, 88, 140*f*, 146*f*
Copernicus GLO-30 and GLO-90, 5, 76–77*t*, 379
EarthEnv-DEM90, 76–77*t*
ETOPO, 76–77*t*, 84
FABDEM, 5, 76–77*t*
General Bathymetric Chart of the Oceans (GEBCO), 2–4, 11, 76–77*t*
GLOBE, 76–77*t*, 179, 180*f*, 266*f*
GMTED2010, 76–77*t*
GTOPO30, 76–77*t*, 88, 184–185, 184*f*, 263*f*
MERIT, 4, 76–77*t*
NASADEM, 4, 76–77*t*
SRTM1, 2–3, 76–77*t*, 81–82, 87–88, 142–143, 143–144*f*, 146*f*, 175*f*
SRTM3, 4, 76–77*t*, 81–82, 112, 176, 177*f*
SRTM30, 76–77*t*, 81–82, 88, 118
SRTM15_PLUS, 2–4, 76–77*t*, 81–82, 84, 88
SRTM30_PLUS, 2–3, 76–77*t*, 81–82, 84, 88, 195, 196*f*, 228
WorldDEM, 5, 76–77*t*
WorldDEM Neo, 76–77*t*
DEMs, regional, 5
ArcticDEM, 4, 8, 10, 379
BedMachine, 5, 88
BedMachine Antarctica, 5, 88, 382
CryoSat-2 DEM, 5, 84, 88, 382
Gapless-REMA100 DEM, 5, 382
High Mountain Asia DEM, 4
ICESat-2 DEM, 5, 80, 88, 382
International Bathymetric Chart of the Arctic Ocean (IBCAO), 2–3, 5, 88, 221, 221*f*, 226*f*
International Bathymetric Chart of the Southern Ocean (IBCSO), 2–3, 5, 88, 382
TanDEM-X PolarDEM 90m, 5, 382
Denitrification, 271, 289–292
Denoising, 167, 170, 174, 178, 183, 186, 186*f*, 189, 191, 193, 207
Dense point cloud, 92–93, 93*f*, 140–141
Desktop virtual globe, 228–229, 229–232*f*, 235–236*f*
Difference curvature (E), 21–24*t*, 26, 34–37*f*, 43, 108–109*f*, 268, 281, 286–287*t*, 288, 367
Digital elevation model (DEM), 2, 73, 74*f*
Digital soil mapping (DSM), 256, 258–260, 260–261*t*, 265, 267
Digital surface model (DSM), 13, 73, 74*f*
Digital terrain model (DTM), 2, 73
Digitizing of contours, 87
Direct georeferencing, 79, 353
Discrete global grid system (DGGS), 92, 92*f*
Discretization, 160, 161*f*
Discriminant analysis, 258–259
Dispersive area *(DA)*, 21–23*t*, 50–51, 50*f*, 121–122*f*, 268–269, 332*f*, 334*f*, 337, 353–354
Dokuchaev
equation, 255
hypothesis, 255–256
Drilling, 85, 87

E

Earth
global helical structures, 325, 326*f*, 338–340, 338–339*f*, 340*t*
global morphometric maps, 327*f*, 329*f*, 332*f*, 335*f*, 337
morphometric globes, 228, 231–232*f*
Eastness (A_E), 28*f*, 30–31, 269, 286–289*t*
Echo sounding, 82, 83*f*
Efremov-Krcho landform classification, 64–67, 65–66*f*
Eight-node single-flow direction (D8) algorithm, 118, 120–121*f*
Elementary area/volume, 246
Elevation, 17, 94
ellipsoidal, 94
orthometric, 94
Ellipsoid
Krasovsky, 135, 327
Martian, 327
multiaxial, 380–381
triaxial, 131–137, 132*f*, 134*f*, 380
Ensemble modeling, 260
Errors. *See* Accuracy/errors of DEMs produced by *and* Root mean square errors (RMSE)
Evans-Young method, 99–100, 100*f*

F

Fault, 301–302
dip, 307, 307*f*
dip-slip, 303–305, 304*f*, 306*f*, 309, 312, 312*f*
displacement, 307, 308*f*
gaping, 305, 306*f*, 307
intersection, 319–320, 320–321*f*, 321–322*t*, 322–323, 323*f*
oblique-slip, 305, 306*f*, 307, 311–313, 312*f*
overthrust, 305, 306*f*, 307
reverse, 302–305, 309
strike, 307, 307*f*
strike-slip, 302, 304*f*, 305, 306*f*, 307, 309, 311–313, 312*f*
thrust, 304–305, 306*f*, 307
zone, 308–309, 310*f*, 312
Feature-preserving smoothing, 169, 178, 179*f*
Fejér summation, 192–193, 206–207
Field geomorphometric interpretation, 367–368
Filosofov method, 171–172, 171*f*
Filtering, 6, 167–189, 169*f*, 171*f*, 174–175*f*, 177*f*, 179–181*f*, 184–188*f*, 381
Fire hazard, digital terrain analysis applications, 13
Flexure, 305, 315, 317*f*
Flow attributes, 10, 21–24*t*, 26, 64. *See also* Curvatures
Flow line, 19*f*, 21–23*t*, 29, 32, 33*f*, 34, 38, 41, 45, 50
Flow routing algorithms, 118–120, 12[illegible], 132*f*, 380
eight-node single-flow direction (D8), 118, 120–121*f*
multiple-flow direction (MFD), 118, 120, 120*f*
Folds/folding, 295–297, 298–299*f*
Forestry, digital terrain analysis applications, 10–11
Form attributes, 10, 21–24*t*, 26, 63. *See also* Curvatures
Fourier transform. *See* Two-dimensional discrete Fourier transform
Fractals, 18
Fracturing, 296–297, 296*f*
Fusion of data, 87–88
Fuzzy logic, 258–259, 264*f*

G

Gaussian
curvature *(K)*, 2, 21–24*t*, 26, 34–37*f*, 38, 63–64, 65*f*, 108–109*f*, 181*f*, 209, 268, 281, 286–287*t*, 295, 297, 299*f*, 301, 367
landform classification, 63–64, 65–66*f*, 181*f*
smoothing, 168, 176–177, 179*f*
General geomorphometry, 20
Generalization, 168–170, 172, 174–176, 178–179, 181*f*, 185–186, 188, 191–193, 206–207, 380–381
Generalized additive models, 259
Generalized linear models, 259
Generating function. *See* Horizontal deflection
Geobotany, digital terrain analysis applications, 10–11
Geological core drilling, 85
Geology, digital terrain analysis applications, 2, 9, 295–297, 301–315, 319–320, 322–323, 325–340
Geomorphology, digital terrain analysis applications, 2, 6–8
Geomorphometry, 1–2, 20, 379

Geomorphon landform classification, 69–71*f*, 72
Gibbs phenomenon, 157–160, 192–193
Glacier
  catastrophic events, 357–363, 358–362*f*
  crevasses, 349–355, 350–354*f*
  Dålk, 357–363, 358–362*f*
  force balance, 347–348
  mass balance, 9–10, 345–346, 346*f*
  motion, 9–10, 345
Glaciology, digital terrain analysis applications, 2, 9–10, 345–355, 346–347*f*, 350–354*f*, 357–363, 358–362*f*
Global navigation satellite system (GNSS), 75
Gradient, 26–29. *See also* Slope gradient
Gradient boosting, 259
Gravitational acceleration vector, 17–18, 19*f*, 25–26, 25*f*
Gravity, 18, 21–23*t*, 25–26, 25*f*, 33, 33*f*, 38, 43, 50, 54
Grid, 89–94, 90–92*f*
  displacement, 160–163
  irregular, 89, 90*f*
  plane, 89–91, 90*f*
  plane square, 89, 90*f*, 91
  spacing, 94, 381
  spheroidal, 91–92, 91–92*f*
  spheroidal equal angular, 91, 91*f*
Ground control points (GCPs), 3, 18, 79, 345
Ground filtering, 80–81, 93, 93*f*, 140–141, 141*f*
Ground penetrating radars (GPRs), 86
Ground penetrating surveys, 85–86

H

Hachures, 212, 212*f*
Haralick image textural characteristics, 379
Heuristic classification, 258
Heuristic generalization, 169
High-frequency noise, 139, 151, 156, 160, 167, 170, 178, 184–185
Hill shading, 6, 10, 12, 55, 56*f*, 211, 211*f*, 213–214*f*, 228, 229*f*, 301–302, 303*f*
Homosoil method, 263
Horizon angle (σ), 21–23*t*, 54*f*, 57
Horizon angle in the wind direction (σ_w), 21–23*t*, 58, 58*f*
Horizontal curvature (k_h), 12, 21–24*t*, 26, 32–33, 33–37*f*, 43, 45, 65*f*, 68, 69–70*f*, 108–109*f*, 119*f*, 151, 153*f*, 154–155, 156*f*, 161–162*f*, 165, 177*f*, 184*f*, 186, 186–187*f*, 188, 198*f*, 201*f*, 206, 209, 231–237*f*, 242, 251–253, 251–252*f*, 252*t*, 268, 271, 274*f*, 277–280*t*, 286–287*t*, 288, 297, 299*f*, 302–305, 306*f*, 307, 311–312, 311*f*, 315, 316*f*, 329–330*f*, 337, 348, 353–354, 353–354*f*, 370, 370*f*
Horizontal deflection (D_{kh}), 21–24*t*, 46–48, 49*f*, 123–125, 124*f*
Horizontal excess curvature (k_{he}), 21–24*t*, 26, 34–37*f*, 44, 46, 108–109*f*, 268, 281, 286–287*t*, 288
Hybrid geostatistical methods, 259
Hydrology, digital terrain analysis applications, 2, 8
Hypsometric tinting, 209, 210–211*f*

I

Ice flow velocity, 347, 347*f*
Ice penetrating surveys, 85–86, 87*f*
Insolation, 21–24*t*, 55–57, 56*f*
Interferometry, 81
Interferometric SAR (InSAR) techniques, 81, 81*f*
Interpolation, 96, 96*f*
Inverse distance weighted (IDW) interpolation, 96, 96*f*
Inverse geodetic problem, 117, 134–135, 380–381
Irregular grids, 89, 90*f*
Isotropy, 164–165
Iwahashi–Pike landform classification, 72

J

Jenny equation, 255–256

K

Kinematic global navigation satellite system (GNSS) surveys, 75, 75*f*
Kriging with external drift, 259

L

Lambertian model, 23–24*t*, 54, 56*f*, 219, 220*f*
Landform classifications, 62–72
  Efremov–Krcho, 64–67, 65–66*f*
  Gaussian, 63–64, 65–66*f*, 181*f*
  geomorphon, 69–70*f*, 72
  Iwahashi–Pike, 69–70*f*, 72
  MacMillan, 68
  Pennock, 68, 69–70*f*
  Schmidt–Hewitt, 68, 69–71*f*
  Shary, 65–66*f*, 67
  topographic position index-based, 68–72, 69–71*f*
Landslide science, digital terrain analysis applications, 12
Laplacian, 46, 72, 242, 296
Laser altimetry, 80–81, 80*f*
Leveling, 74
Lidar survey, 3, 12, 81, 381
Line of sight (LoS) algorithms, 128, 128*f*
Lineaments, 9, 163*f*, 301–305, 310*f*, 376*f*
Linear artifacts, 163–165
Local-local morphometric variables, 21–24*t*, 62
  curvedness (*C*), 21–24*t*, 62, 64*f*, 131
  shape index (*IS*), 21–24*t*, 62, 64*f*, 131
Local morphometric variables, 21–24*t*, 24–49
  flow attributes, 21–24*t*, 26, 64 (*see also* Curvatures)
  form attributes, 21–24*t*, 26, 63 (*see also* Curvatures)
Local-nonlocal morphometric variables, 21–24*t*, 61–62
  stream power index (*SPI*), 21–24*t*, 62, 63*f*, 129–130, 130–131*f*, 268, 277–280*t*, 284–285*t*, 291, 367
  topographic wetness index (*TWI*), 11–12, 23–24*t*, 62, 63*f*, 129–130, 130–131*f*, 242, 243*f*, 246, 251–252, 256, 262–263*f*, 268, 277–280*t*, 291, 367, 373–374, 375*f*, 378
Lommel-Seeliger model, 55
Lunar Orbiter Laser Altimeter (LOLA), 327

M

Machine learning techniques, 6, 8, 12, 259, 379
MacMillan landform classification, 68
Magellan mission, 327
Mapping, 209–213. *See also* Visualization
Mars
  global helical structures, 339, 340–341*f*
  global morphometric maps, 328*f*, 330–331*f*, 333–334*f*, 336*f*, 337
  morphometric globes, 228, 229*f*, 233–234*f*, 237*f*
Mars Orbiter Laser Altimeter (MOLA), 327
Maximal curvature (k_{max}), 21–24*t*, 26, 34–37*f*, 38–39, 43, 46, 62, 68, 69–70*f*, 108–109*f*, 119*f*, 181*f*, 231–237*f*, 268, 274*f*, 277–280*t*, 286–287*t*, 291, 296*f*, 297, 301, 353–354, 367, 373–375, 373*f*, 376*f*
Mean curvature (*H*), 21–24*t*, 26, 33–34, 34–37*f*, 38, 63–64, 65*f*, 108–109*f*, 151, 152*f*, 181*f*, 242, 251–253, 251–252*f*, 252–253*t*, 268–269, 277–280*t*, 286–287*t*, 291, 295, 367
Mean smoothing, 176, 177*f*, 179*f*
Median smoothing, 176–177, 179*f*
Minimal curvature (k_{min}), 21–24*t*, 26, 34–37*f*, 38–39, 44, 62, 68, 69–70*f*, 108–109*f*, 119*f*, 181*f*, 231–237*f*, 268, 273*f*, 277–280*t*, 286–289*t*, 296*f*, 297, 301, 329*f*, 353–354, 353–354*f*, 367, 372*f*, 373–375, 376*f*
Moon
  global morphometric maps, 328*f*, 330–331*f*, 333–334*f*, 336*f*, 337
  morphometric globes, 228, 229*f*, 233–234*f*, 237*f*
Morphometric variables, 20–23, 21–24*t*
  combined (*see* Combined morphometric variables)
  local (*see* Local morphometric variables)
  nonlocal (*see* Nonlocal morphometric variables)
  solar (*see* Solar morphometric variables)
  structural (*see* Structural morphometric variables)
  two-field-specific (*see* Two-field-specific morphometric variables)
  view (*see* View morphometric variables)
  wind (*see* Wind morphometric variables)
Moving window
  plane square, 99, 100*f*, 102
  spheroidal equal angular, 100*f*, 112, 117–118, 132*f*
Multinomial logistic regression, 259, 263*f*, 265*f*
Multiple-flow direction (MFD) algorithms, 118–120, 120*f*
Multiple-point geostatistics, 259
Multiple regression analysis, 248, 251, 253, 253*t*, 259, 268–269, 276, 281, 288–290*t*, 292
Multiple-source illumination modeling, 216–222, 217*f*, 219*f*, 221*f*, 223–224*f*, 226*f*
Multitemporal DEMs, 2, 7, 10–12, 345–346, 360*f*, 363

N

Nonlocal morphometric variables, 21–24t, 49–52, 50–51f
  catchment area (*CA*), 1, 21–23t, 50–51, 50f, 121–122f, 231–237f, 242, 250–252, 268–269, 288, 291, 332–333f, 335–336f, 337, 353–354, 353–354f, 373–374, 374f
  dispersive area (*DA*), 21–23t, 50–51, 50f, 121–122f, 268–269, 332f, 334f, 337, 353–354
  specific catchment area (*SCA*), 21–24t, 50, 268, 273–274f, 276, 277–280t, 286–290t, 291
  specific dispersive area (*SDA*), 21–23t, 50, 268
  topographic position index (*TPI*), 21–23t, 51f, 52
Normal section, 24, 25f, 26
Northness (A_N), 28f, 30–31, 269, 286–288t

O

Oceanology, digital terrain analysis applications, 11
Optical sensing of bathymetry, 85

P

Paleotopography, 297
Partial derivatives of elevation, 17–18, 26
Pedometrics, 8–9, 260–261t
Pennock landform classification, 68, 69–70f
Photogrammetry, 2, 75–78, 76f, 78f
Planar size limitedness, 18
Plan curvature (k_p), 21–24t, 26, 31–32, 50, 257, 268f
Plane grids, 89–91, 90f
Plane square grid, 89–91, 90f
Plane square moving window, 99–100, 100f, 102
Planetology, digital terrain analysis applications, 11–12, 325–340
Platonic solids, 92
Polarimetry, 81
Post-processed kinematic (PPK), 79, 353
Predictive soil mapping, 255–269, 271, 276, 290f, 292
Principal curvatures. *See* Maximal curvature *and* Minimal curvature
Profile curvature. *See* Vertical curvature

R

Radio-echo sounding (RES), 85–86, 87f
Random forest regression, 259
Reference Elevation Model of Antarctica (REMA), 4, 88, 366–367, 379, 382
Reflectance, 21–24t, 54–55
  Lambertian model, 23–24t, 54, 56f, 219, 220f
  Lommel-Seeliger model, 55
Regression kriging, 258–259, 262–263f
Relative terrain aspect (A_r), 21–23t, 57, 58f
Relief Plasticity method, 257, 268f
Remotely operated vehicle (ROV), 80
Resolution, 94–95
Ridge. *See* Crest
Ring curvature (K_r), 21–24t, 26, 34–37f, 44–45, 108–109f, 269, 272, 367
Rock fracturing, 320–321, 321f, 322t, 323f
Romanova methods, 257
Root mean square error (RMSE), 145
  DEM, 145
  local morphometric variables, 146–151
  mapping, 151, 152–153f
  plane square grid, calculation on, 149–150, 149t
  spheroidal equal angular grid, calculation on, 150–151
Rotor (*rot*), 21–24t, 34–37f, 41–42
Roughness, 379
Row and column elimination, 168, 178, 180f

S

Sampling theorem, 95, 151–156
Satellite laser altimetry, 80–81
Satellite radar altimetry, 82–84, 84f
Scale, 245
Scale dependentness, 18
Schmidt–Hewitt landform classification, 68, 69–71f
SCORPAN equation, 256
Seabed 2030, 11, 82
Selective filtering, 169
Semi-global matching algorithm, 80
Shape index (*IS*), 23–24t, 62, 64f, 131
Shary landform classification, 67
Shuttle Radar Topography Mission (SRTM), 81–82
Single valuedness, 17
Singular viewshed. *See* Terrain visibility
Sky view factor (Ψ_s), 21–23t, 56f, 57
Slope aspect (*A*), 13, 21–24t, 26, 28f, 30, 107f, 212, 212f, 242, 251, 269, 273–274f, 291, 367, 369–370, 369f
Slope gradient (*G*), 1, 13, 21–24t, 26–27, 28f, 107f, 119f, 154–156, 155–156f, 211–212, 212f, 241–242, 251–253, 251–252f, 252–253t, 268–269, 273–274f, 276, 277t, 279t, 286–289t, 288, 291, 301, 348, 367, 368f, 369, 374
Slope line, 19f, 25f, 27–29, 33, 44, 50, 50f, 52–53, 52f
Smoothing, 6, 167, 175–178, 183
  feature-preserving, 169, 178, 179f
  Gaussian, 168, 176–177, 179f
  mean, 176, 177f, 179f
  median, 176–177, 179f
Smoothness, 17–18
Soil core drilling, 85
Soil forming factors, 255–256
Soil properties
  bulk density, 260–261t, 274, 284–285t, 287t, 291
  denitrification rate, 274, 275t, 284–285t, 287t, 291–292
  denitrifier enzyme activity, 274, 275t, 284–285t, 287t, 292
  microbial biomass carbon, 260–261t, 274, 275t, 284–285t, 287t, 290f, 290t, 291–292
  microbial respiration rate, 274, 275t, 287t, 291
  moisture, 241–242, 246, 250–253, 252f, 252–253t, 256–257, 260–261t, 274, 284–285t, 287t, 289, 290f, 291–292
  nitrous oxide (N_2O) flux, 260–261t, 274, 275t, 287t, 291–292
  organic carbon, 260–261t, 263, 265, 266f, 291
  organic matter, 260–261t, 262f, 291
Soil science, digital terrain analysis applications, 1–2, 8–9
Solar morphometric variables, 21–23t, 54–57
  horizon angle (σ), 21–23t, 54f, 57
  insolation (*I*), 21–24t, 55–57, 56f, 125f
  reflectance (*R*), 23–24t, 54–55, 56f
  sky view factor (Ψ_s), 21–23t, 56f, 57
Special points, 18–20, 19f
Specific catchment area (*SCA*), 21–24t, 50, 268, 273–274f, 276, 277–280t, 286–290t, 291
Specific dispersive area (*SDA*), 21–23t, 50, 268
Speleology, digital terrain analysis applications, 10
Spheroidal equal angular grid, 91, 91f
Spheroidal equal angular moving window, 100f, 112, 117–118, 132f
Spheroidal grids, 91–92, 91f
Square grid, 89, 90f, 91
Stratigraphic horizon/surface, 2, 85, 88, 301, 314f
Stream power index (*SPI*), 21–24t, 62, 63f, 129–130, 130–131f, 268, 277–280t, 284–285t, 291, 367
Structural equation modeling, 259
Structural morphometric variables, 52–53
  concave break line, 52–53, 52f, 123, 125
  convex break line, 52–53, 52f, 123, 125
  crest, 52, 52–53f, 123, 123f, 125
  thalweg, 52, 52–53f, 123, 123f, 125
Structural line, 52, 52f
Structure-from-motion (SfM), 2–3, 78–80, 79f, 92, 379
Supervised classification, 258
Support vector machine, 259
Surfaces, 17, 18f
Synthetic aperture radar (SAR), 81–82, 81f

T

Tangential curvature. *See* Horizontal curvature
Taylor polynomial, 99, 102
Terrain visibility (*V*), 21–23t, 59, 60f, 128, 128–129f
Thalweg, 52–53, 52–53f, 123, 123f, 125
Theorem
  of total curvatures, 46
  on accumulation curvature, 45
  on flow deceleration, 37
  on flow divergence, 32
  on Laplacian, 46
  on orthogonality, 29
  on special points, 20
  sampling, 95, 151, 154
Theorema Egregium, 21–23t, 39, 297, 299f
Thin-plate spline interpolation, 96f, 97
Three-dimensional modeling, 214–227
  subsurface, 222–227, 227f
  terrain, 214–222, 215f, 222–223f

Three-dimensional seismic survey, 2–3, 86*f*
TOPMODEL, 2, 8, 62, 257
Topographic openness (*TO*), 21–23*t*, 60, 60*f*, 129, 129*f*
Topographic position index (*TPI*), 21–23*t*, 51*f*, 52
Topographic position index-based landform classification, 68–72, 69–71*f*
Topographic surface, 17
Topographic wetness index (*TWI*), 11–12, 23–24*t*, 62, 63*f*, 129–130, 130–131*f*, 242, 243*f*, 246, 251–252, 256, 262–263*f*, 268, 277–280*t*, 291, 367, 373–374, 375*f*, 378
Topography, 1
Torsion stress/deformation, 325, 326*f*, 338–339
Total viewshed (V_Σ), 21–23*t*, 60, 128
Trend-surface analysis, 168, 170–171, 171*f*
Triangulated irregular network (TIN), 89, 90*f*
Triangulation-based interpolation, 96, 96*f*
Triaxial ellipsoid, 132, 132*f*, 134*f*, 135
Two-dimensional
  discrete Fourier transform, 167, 172–173, 174*f*
  discrete wavelet transform, 167, 172–173, 175*f*
  singular spectrum analysis (2D-SSA), 181–189
Two-field-specific morphometric variables, 21–23*t*, 54–60
  solar, 21–24*t*, 54–57
  view, 21–23*t*, 59–60
  wind, 21–23*t*, 57–59

U

Uniform gravity, 18
Universal spectral analytical method, 191–194
Unmanned aerial system (UAS), 3, 9–10, 78–80, 345, 351, 358
Unmanned aerial vehicle (UAV). *See* Unmanned aerial system (UAS)
Unsphericity curvature (*M*), 21–24*t*, 26, 34–37*f*, 39–41, 108–109*f*, 268, 281, 286–287*t*, 367
Unsupervised classification, 258
Upscaling, 262–267, 266–268*f*
Urban studies, digital terrain analysis applications, 13

V

Variability of soil properties, 271
  depth, 281, 288–289
  temporal, 281
Venus
  global helical structures, 340, 340–341*f*
  global morphometric maps, 328*f*, 330–331*f*, 333–334*f*, 336*f*, 337
  morphometric globes, 229*f*
Vertical curvature (k_v), 12, 21–24*t*, 26, 33–38, 34–37*f*, 46, 52–53, 63, 65*f*, 67–68, 69–70*f*, 108–109*f*, 119*f*, 154–156, 155*f*, 161–162*f*, 177*f*, 180*f*, 199*f*, 201*f*, 203*f*, 206, 211, 231–237*f*, 242, 251–253, 251–252*f*, 252*t*, 268, 273*f*, 277–280*t*, 286–289*t*, 288, 291, 297, 299*f*, 302–305, 306*f*, 307, 311–312, 311*f*, 315, 316*f*, 329*f*, 331*f*, 337, 353–354, 353–354*f*, 371–372, 371*f*
Vertical deflection (D_{kv}), 21–24*t*, 48–49, 49*f*
Vertical excess curvature (k_{ve}), 21–24*t*, 26, 34–37*f*, 44, 46, 108–109*f*, 268, 281, 286–287*t*
View morphometric variables, 21–23*t*, 59–60
  terrain visibility (*V*), 21–23*t*, 59, 60*f*, 128, 128–129*f*
  topographic openness (*TO*), 21–23*t*, 60, 60*f*, 129, 129*f*
  total viewshed (V_Σ), 21–23*t*, 60, 128
Viewshed. *See* Terrain visibility *and* Total viewshed
Virtual globes, 228–234
  desktop, 228–229, 229–234*f*
  web, 229–234, 237–238*f*
Visualization, 6, 209–234

W

Wavelet transform. *See* Two-dimensional (2D) discrete wavelet transform
Web virtual globes, 229–234, 237–238*f*
Wind effect index (*WEf*), 21–23*t*, 58, 126–128, 127*f*
Wind exposition index (*WEx*), 21–23*t*, 58, 126–128, 127*f*, 377, 377*f*
Wind morphometric variables, 21–23*t*, 57–59
  horizon angle in the wind direction (σ_w), 21–23*t*, 58, 58*f*
  relative terrain aspect (A_r), 21–23*t*, 57, 58*f*
  wind effect index (*WEf*), 21–23*t*, 58, 126–128, 127*f*
  wind exposition index (*WEx*), 21–23*t*, 58, 126–128, 127*f*, 377, 377*f*

Z

Zakharov equation, 255
Zoogeography, digital terrain analysis applications, 11

9780443247989